T0180353

Lecture Notes in Artificial Intelligence 10751

Subseries of Lecture Notes in Computer Science

More information about this series at http://www.springer.com/series/1244

Ngoc Thanh Nguyen · Duong Hung Hoang
Tzung-Pei Hong · Hoang Pham
Bogdan Trawiński (Eds.)

Intelligent Information and Database Systems

10th Asian Conference, ACIIDS 2018
Dong Hoi City, Vietnam, March 19–21, 2018
Proceedings, Part I

Editors
Ngoc Thanh Nguyen
Wrocław University of Science and Technology
Wrocław
Poland

Duong Hung Hoang
Quang Binh University
Dong Hoi City
Vietnam

Tzung-Pei Hong
National University of Kaohsiung
Kaohsiung
Taiwan

Hoang Pham
Rutgers University
Piscataway, NJ
USA

Bogdan Trawiński
Wrocław University of Science and Technology
Wrocław
Poland

ISSN 0302-9743 ISSN 1611-3349 (electronic)
Lecture Notes in Artificial Intelligence
ISBN 978-3-319-75416-1 ISBN 978-3-319-75417-8 (eBook)
https://doi.org/10.1007/978-3-319-75417-8

Library of Congress Control Number: 2018931885

LNCS Sublibrary: SL7 – Artificial Intelligence

Printed on acid-free paper

This Springer imprint is published by the registered company Springer International Publishing AG part of Springer Nature
The registered company address is: Gewerbestrasse 11, 6330 Cham, Switzerland

Preface

ACIIDS 2018 was the tenth event in a series of international scientific conferences on research and applications in the field of intelligent information and database systems. The aim of ACIIDS 2018 was to provide an international forum for scientific research in the technology and application of intelligent information and database systems. ACIIDS 2018 was co-organized by Quang Binh University (Vietnam) and Wrocław University of Science and Technology (Poland) in co-operation with the IEEE SMC Technical Committee on Computational Collective Intelligence, European Research Center for Information Systems (ERCIS), University of Newcastle (Australia), Bina Nusantara University (Indonesia), Yeungnam University (South Korea), Leiden University (The Netherlands), Universiti Teknologi Malaysia (Malaysia), Ton Duc Thang University (Vietnam), and Vietnam National University, Hanoi (Vietnam). It took place in Dong Hoi City in Vietnam during March 19–21, 2018.

The conference series ACIIDS is already well established. The first two events, ACIIDS 2009 and ACIIDS 2010, took place in Dong Hoi City and Hue City in Vietnam, respectively. The third event, ACIIDS 2011, took place in Daegu (South Korea), followed by the fourth event, ACIIDS 2012, in Kaohsiung (Taiwan). The fifth event, ACIIDS 2013, was held in Kuala Lumpur in Malaysia while the sixth event, ACIIDS 2014, was held in Bangkok in Thailand. The seventh event, ACIIDS 2015, took place in Bali (Indonesia), followed by the eight event, ACIIDS 2016, in Da Nang (Vietnam). The last event, ACIIDS 2017, was held in Kanazawa (Japan).

We received more than 400 papers from 42 countries all over the world. Each paper was peer reviewed by at least two members of the international Program Committee and international reviewer board. Only 133 papers with the highest quality were selected for an oral presentation and publication in the two volumes of the ACIIDS 2018 proceedings.

Papers included in the proceedings cover the following topics: knowledge engineering and Semantic Web; social networks and recommender systems; text processing and information retrieval; machine learning and data mining; decision support and control systems; computer vision techniques; advanced data mining techniques and applications; multiple model approach to machine learning; intelligent information systems; design thinking-based R&D; development techniques; and project-based learning; modelling, storing, and querying of graph data; computational imaging and vision; computer vision and robotics; data science and computational intelligence; data structures modelling for knowledge representation; intelligent computer vision systems and applications; intelligent and contextual systems; intelligent biomarkers of neurodegenerative processes in the brain; sensor networks and Internet of Things; intelligent applications of Internet of Thing and data analysis technologies; intelligent systems and algorithms in information sciences; intelligent systems and methods in biomedicine; intelligent systems for optimization of logistics and industrial applications; analysis of image, video, and motion data in life sciences.

The accepted and presented papers highlight new trends and challenges facing the intelligent information and database systems community. The presenters showed how new research could lead to novel and innovative applications. We hope you will find these results useful and inspiring for your future research work.

We would like to extend our heartfelt thanks to Jarosław Gowin, Deputy Prime Minister of the Republic of Poland and Minister of Science and Higher Education, for his support and honorary patronage of the conference.

We would like to express our sincere thanks to the honorary chairs, Prof. Cezary Madryas (Rector of Wrocław University of Science and Technology, Poland), Prof. Huu Duc Nguyen (Vice-Rector of National University Hanoi, Vietnam), and Dr. Tien Dung Tran (Vice-President of Quang Binh Province, Vietnam), for their support.

Our special thanks go to the program chairs, special session chairs, organizing chairs, publicity chairs, liaison chairs, and local Organizing Committee for their work for the conference. We sincerely thank all the members of the international Program Committee for their valuable efforts in the review process, which helped us to guarantee the highest quality of the selected papers for the conference. We cordially thank the organizers and chairs of special sessions who contributed to the success of the conference.

We would like to express our thanks to the keynote speakers: Thomas Bäck from University of Leiden, The Netherlands, Lipo Wang from Nanyang Technological University, Singapore, Satoshi Tojo from Japan Advanced Institute of Science and Technology, Japan, and Nguyen Huu Duc from Vietnam National University, Hanoi, Vietnam, for their world-class plenary speeches.

We cordially thank our main sponsors, Quang Binh University (Vietnam), Wrocław University of Science and Technology (Poland), IEEE SMC Technical Committee on Computational Collective Intelligence, European Research Center for Information Systems (ERCIS), University of Newcastle (Australia), Bina Nusantara University (Indonesia), Yeungnam University (South Korea), Leiden University (The Netherlands), Universiti Teknologi Malaysia (Malaysia), Ton Duc Thang University (Vietnam), and Vietnam National University, Hanoi (Vietnam). Our special thanks are due also to Springer for publishing the proceedings and sponsoring awards, and to all the other sponsors for their kind support.

We wish to thank the members of the Organizing Committee for their excellent work and the members of the local Organizing Committee for their considerable effort.

We cordially thank all the authors, for their valuable contributions, and the other participants of this conference. The conference would not have been possible without their support.

Thanks are also due to many experts who contributed to making the event a success.

March 2018

Ngoc Thanh Nguyen
Duong Hung Hoang
Tzung-Pei Hong
Hoang Pham
Bogdan Trawiński

Organization

Honorary Chairs

Cezary Madryas	Rector of Wrocław University of Science and Technology, Poland
Huu Duc Nguyen	Vice-President of Vietnam National University, Hanoi, Vietnam
Tien Dung Tran	Vice-President of Quang Binh Province, Vietnam

General Chairs

Ngoc Thanh Nguyen	Wrocław University of Technology, Poland
Duong Hung Hoang	Quang Binh University, Vietnam

Program Chairs

Tzung-Pei Hong	National University of Kaohsiung, Taiwan
Hoang Pham	Rutgers University, USA
Edward Szczerbicki	University of Newcastle, Australia
Bogdan Trawiński	Wrocław University of Science and Technology, Poland

Special Session Chairs

Manuel Núñez	Universidad Complutense de Madrid, Spain
Andrzej Siemiński	Wrocław University of Science and Technology, Poland
Quang Thuy Ha	Vietnam National University, Hanoi (VNU), Vietnam

Publicity Chairs

Van Dung Hoang	Quang Binh University, Vietnam
Marek Kopel	Wrocław University of Science and Technology, Poland
Marek Krótkiewicz	Wrocław University of Science and Technology, Poland

Liaison Chairs

Quang A. Dang	Vietnam Academy of Science and Technology, Vietnam
Ford Lumban Gaol	Bina Nusantara University, Indonesia
Mong-Fong Horng	National Kaohsiung University of Applied Sciences, Taiwan
Dosam Hwang	Yeungnam University, South Korea
Ali Selamat	Universiti Teknologi Malaysia, Malaysia

Organizing Chairs

Thi Dung Vo	Quang Binh University, Vietnam
Hoai Thu Le Thi	Quang Binh University, Vietnam
Adrianna Kozierkiewicz	Wrocław University of Science and Technology, Poland

Local Organizing Committee

Maciej Huk	Wrocław University of Science and Technology, Poland
Marcin Jodłowiec	Wrocław University of Science and Technology, Poland
Rafał Kern	Wrocław University of Science and Technology, Poland
Marcin Pietranik	Wrocław University of Science and Technology, Poland
Krystian Wojtkiewicz	Wrocław University of Science and Technology, Poland
Xuan Hau Pham	Quang Binh University, Vietnam
Tuan Nha Hoang	Quang Binh University, Vietnam
Van Thanh Phan	Quang Binh University, Vietnam
Van Chung Nguyen	Quang Binh University, Vietnam
Lan Phuong Pham Thi	Quang Binh University, Vietnam
Xuan Hao Nguyen	Quang Binh University, Vietnam

Webmaster

Marek Kopel	Wrocław University of Science and Technology, Poland

Steering Committee

Ngoc Thanh Nguyen (Chair)	Wrocław University of Technology, Poland
Longbing Cao	University of Science and Technology Sydney, Australia
Suphamit Chittayasothorn	King Mongkut's Institute of Technology Ladkrabang, Thailand
Ford Lumban Gaol	Bina Nusantara University, Indonesia
Tu Bao Ho	Japan Advanced Institute of Science and Technology, Japan
Tzung-Pei Hong	National University of Kaohsiung, Taiwan
Dosam Hwang	Yeungnam University, South Korea
Lakhmi C. Jain	University of South Australia, Australia
Geun-Sik Jo	Inha University, South Korea
Hoai An Le-Thi	University of Lorraine, France
Zygmunt Mazur	Wrocław University of Science and Technology, Poland
Toyoaki Nishida	Kyoto University, Japan
Leszek Rutkowski	Częstochowa University of Technology, Poland
Ali Selamat	Universiti Teknologi Malaysia, Malaysia

Keynote Speakers

Thomas Bäck	University of Leiden, The Netherlands
Lipo Wang	Nanyang Technological University, Singapore
Satoshi Tojo	Japan Advanced Institute of Science and Technology, Japan
Nguyen Huu Duc	Vietnam National University, Hanoi, Vietnam

Special Sessions Organizers

1. *Special Session on Multiple Model Approach to Machine Learning (MMAML 2018)*

Tomasz Kajdanowicz	Wrocław University of Science and Technology, Poland
Edwin Lughofer	Johannes Kepler University Linz, Austria
Bogdan Trawinski	Wrocław University of Science and Technology, Poland

2. *Special Session on Analysis of Image, Video, and Motion Data in Life Sciences (IVMLS 2018)*

Konrad Wojciechowski	Polish-Japanese Academy of Information Technology, Poland
Marek Kulbacki	Polish-Japanese Academy of Information Technology, Poland
Jakub Segen	Polish-Japanese Academy of Information Technology, Poland
Andrzej Polanski	Silesian University of Technology, Poland

3. *Special Session on Intelligent and Contextual Systems (ICxS 2018)*

Maciej Huk	Wrocław University of Science and Technology, Poland
Keun Ho Ryu	Chungbuk National University, South Korea
Thai-Nghe Nguyen	Cantho University, Vietnam
Nguyen Hong Vu	Ton Duc Thang University, Vietnam
Goutam Chakraborty	Iwate Prefectural University, Japan

4. *Special Session on Intelligent Systems for Optimization of Logistics and Industrial Applications (ISOLIA 2018)*

Farouk Yalaoui	University of Technology of Troyes, France
Habiba Drias	University of Science and Technology (USTHB), Algeria
Taha Arbaoui	University of Technology of Troyes, France

5. *Special Session on Intelligent Applications of Internet of Thing and Data Analysis Technologies (IoT&DAT 2018)*

Shunzhi Zhu	Xiamen University of Technology, Xiamen, P.R. China
Rung Ching Chen	Chaoyang University of Technology, Taiwan
Yung-Fa Huang	Chaoyang University of Technology, Taiwan

6. *Special Session on Intelligent Systems and Methods in Biomedicine (ISaMiB 2018)*

Jan Kubicek	Technical University of Ostrava, Czech Republic
Marek Penhaker	Technical University of Ostrava, Czech Republic
Ondrej Krejcar	University of Hradec Kralove, Czech Republic
Kamil Kuca	University of Hradec Kralove, Czech Republic

7. *Special Session on Intelligent Systems and Algorithms in Information Sciences (ISAIS 2018)*

Martin Kotyrba	University of Ostrava, Czech Republic
Eva Volna	University of Ostrava, Czech Republic
Ivan Zelinka	VSB - Technical University of Ostrava, Czech Republic

8. *Special Session on Design Thinking-Based R&D, Development Techniques, and Project-Based Learning (2DT-PBL 2018)*

Shinya Kobayashi	Ehime University, Japan
Keiichi Endo	Ehime University, Japan

9. *Special Session on Modelling, Storing, and Querying of Graph Data (MSQGD 2018)*

Jaroslav Pokorny	Charles University, Prague, Czech Republic
Bela Stantic	Griffith University, Australia

10. *Special Session on Data Science and Computational Intelligence (DSCI 2018)*

Veera Boonjing	King Mongkut's Institute of Technology Ladkrabang, Thailand
Ronan Reilly	Maynooth University, Ireland

11. *Special Session on Computer Vision and Robotics (CVR 2018)*

Van-Dung Hoang	Quang Binh University, Vietnam
Chi-Mai Luong	University of Science and Technology of Hanoi, Vietnam

My-Ha Le — Ho Chi Minh City University of Technology and Education, Vietnam
Kang-Hyun Jo — University of Ulsan, South Korea

12. *Special Session on Intelligent Biomarkers of Neurodegenerative Processes in the Brain (InBinBRAIN 2018)*

Andrzej Przybyszewski — Polish-Japanese Academy of Information Technology, Poland

13. *Special Session on Data Structures Modelling for Knowledge Representation (DSMKR 2018)*

Marek Krotkiewicz — Wrocław University of Science and Technology, Poland

14. *Special Session on Computational Imaging and Vision (CIV 2018)*

Jeonghwan Gwak — Seoul National University Hospital, South Korea
Manish Khare — Dhirubhai Ambani Institute of Information and Communication Technology Gandhinagar, India
Jong-In Song — Gwangju School of Electrical Engineering and Computer Science, South Korea

15. *Special Session on Intelligent Computer Vision Systems and Applications (ICVSA 2018)*

Dariusz Frejlichowski — West Pomeranian University of Technology, Szczecin, Poland
Leszek J. Chmielewski — Warsaw University of Life Sciences, Poland
Piotr Czapiewski — West Pomeranian University of Technology, Szczecin, Poland

16. *Special Session on Advanced Data Mining Techniques and Applications (ADMTA 2018)*

Bay Vo — Ho Chi Minh City University of Technology, Vietnam
Tzung-Pei Hong — National University of Kaohsiung, Taiwan
Chun-Hao Chen — Tamkang University, Taiwan

International Program Committee

Salim Abdulazeez	College of Engineering, Trivandrum, India
Muhammad Abulaish	South Asian University, India
Waseem Ahmad	Waiariki Institute of Technology, New Zealand
Toni Anwar	Universiti Teknologi Malaysia, Malaysia
Ahmad Taher Azar	Benha University, Egypt
Amelia Badica	University of Craiova, Romania
Costin Badica	University of Craiova, Romania
Kambiz Badie	ICT Research Institute, Iran
Hassan Badir	École Nationale des Sciences Appliquées de Tanger, Morocco
Emili Balaguer-Ballester	Bournemouth University, UK
Zbigniew Banaszak	Warsaw University of Technology, Poland
Dariusz Barbucha	Gdynia Maritime University, Poland
Ramazan Bayindir	Gazi University, Turkey
Leon Bobrowski	Bialystok University of Technology, Poland
Bülent Bolat	Yildiz Technical University, Turkey
Veera Boonjing	King Mongkut's Institute of Technology Ladkrabang, Thailand
Mariusz Boryczka	University of Silesia in Katowice, Poland
Urszula Boryczka	University of Silesia in Katowice, Poland
Zouhaier Brahmia	University of Sfax, Tunisia
Stephane Bressan	National University of Singapore, Singapore
Peter Brida	University of Zilina, Slovakia
Andrej Brodnik	University of Ljubljana, Slovenia
Piotr Bródka	Wrocław University of Science and Technology, Poland
Grażyna Brzykcy	Poznan University of Technology, Poland
The Duy Bui	University of Engineering and Technology, VNU Hanoi, Vietnam
Robert Burduk	Wrocław University of Science and Technology, Poland
Aleksander Byrski	AGH University of Science and Technology, Poland
David Camacho	Universidad Autonoma de Madrid, Spain
Tru Cao	Ho Chi Minh City University of Technology, Vietnam
Frantisek Capkovic	Institute of Informatics, Slovak Academy of Sciences, Slovakia
Dariusz Ceglarek	Poznan High School of Banking, Poland
Zenon Chaczko	University of Technology, Sydney, Australia
Altangerel Chagnaa	National University of Mongolia, Mongolia
Goutam Chakraborty	Iwate Prefectural University, Japan
Kuo-Ming Chao	Coventry University, UK
Somchai Chatvichienchai	University of Nagasaki, Japan
Rung-Ching Chen	Chaoyang University of Technology, Taiwan
Shyi-Ming Chen	National Taiwan University, Taiwan
Suphamit Chittayasothorn	King Mongkut's Institute of Technology Ladkrabang, Thailand

Sung-Bae Cho	Yonsei University, South Korea
Kazimierz Choroś	Wrocław University of Science and Technology, Poland
Kun-Ta Chuang	National Cheng Kung University, Taiwan
Robert Cierniak	Czestochowa University of Technology, Poland
Dorian Cojocaru	University of Craiova, Romania
Jose Alfredo	Ferreira Costa Federal University of Rio Grande do Norte (UFRN), Brazil
Keeley Crockett	Manchester Metropolitan University, UK
Bogusław Cyganek	AGH University of Science and Technology, Poland
Ireneusz Czarnowski	Gdynia Maritime University, Poland
Piotr Czekalski	Silesian University of Technology, Poland
Quang A Dang	Vietnam Academy of Science and Technology, Vietnam
Paul Davidsson	Malmö University, Sweden
Roberto De Virgilio	Università degli Studi Roma Tre, Italy
Banu Diri	Yildiz Technical University, Turkey
Tien V. Do	Budapest University of Technology and Economics, Hungary
Grzegorz Dobrowolski	AGH University of Science and Technology, Poland
Habiba Drias	University of Science and Technology Houari Boumediene, Algeria
Maciej Drwal	Wrocław University of Science and Technology, Poland
Ewa Dudek-Dyduch	AGH University of Science and Technology, Poland
El-Sayed M. El-Alfy	King Fahd University of Petroleum and Minerals, Saudi Arabia
Vadim Ermolayev	Zaporozhye National University, Ukraine
Nadia Essoussi	University of Carthage, Tunisia
Rim Faiz	University of Carthage, Tunisia
Thomas Fober	University of Marburg, Germany
Simon Fong	University of Macau, Macau, SAR China
Dariusz Frejlichowski	West Pomeranian University of Technology, Szczecin, Poland
Hamido Fujita	Iwate Prefectural University, Japan
Mohamed Gaber	Birmingham City University, UK
Ford Lumban Gaol	Bina Nusantara University, Indonesia
Dariusz Gasior	Wrocław University of Science and Technology, Poland
Janusz Getta	University of Wollongong, Australia
Daniela Gifu	Alexandru loan Cuza University of Iasi, Romania
Dejan Gjorgjevikj	Ss. Cyril and Methodius University in Skopje, Republic of Macedonia
Barbara Gładysz	Wrocław University of Science and Technology, Poland
Daniela Godoy	ISISTAN Research Institute, Argentina
Gergo Gombos	Eotvos Lorand University, Hungary
Antonio Gonzalez-Pardo	Universidad Autónoma de Madrid, Spain
Manuel Grana	University of Basque Country, Spain
Janis Grundspenkis	Riga Technical University, Latvia
Jeonghwan Gwak	Seoul National University, South Korea

Quang-Thuy Ha	VNU University of Engineering and Technology, Vietnam
Sung Ho Ha	Kyungpook National University, South Korea
Dawit Haile	Addis Ababa University, Ethiopia
Pei-Yi Hao	National Kaohsiung University of Applied Sciences, Taiwan
Tutut Herawan	University of Malaya, Malaysia
Marcin Hernes	Wroclaw University of Economics, Poland
Bogumila Hnatkowska	Wrocław University of Science and Technology, Poland
Huu Hanh Hoang	Hue University, Vietnam
Quang Hoang	Hue University of Sciences, Vietnam
Van-Dung Hoang	Quang Binh University, Vietnam
Tzung-Pei Hong	National University of Kaohsiung, Taiwan
Mong-Fong Horng	National Kaohsiung University of Applied Sciences, Taiwan
Eklas Hossain	Oregon University, USA
Jen-Wei Huang	National Cheng Kung University, Taiwan
Maciej Huk	Wrocław University of Science and Technology, Poland
Zbigniew Huzar	Wrocław University of Science and Technology, Poland
Dosam Hwang	Yeungnam University, South Korea
Roliana Ibrahim	Universiti Teknologi Malaysia, Malaysia
Dmitry Ignatov	National Research University Higher School of Economics, Russia
Lazaros Iliadis	Democritus University of Thrace, Greece
Hazra Imran	University of British Columbia, Canada
Agnieszka Indyka-Piasecka	Wrocław University of Science and Technology, Poland
Mirjana Ivanovic	University of Novi Sad, Serbia
Sanjay Jain	National University of Singapore, Singapore
Jaroslaw Jankowski	West Pomeranian University of Technology, Szczecin, Poland
Chuleerat Jaruskulchai	Kasetsart University, Thailand
Khalid Jebari	LCS Rabat, Morocco
Joanna Jedrzejowicz	University of Gdansk, Poland
Piotr Jedrzejowicz	Gdynia Maritime University, Poland
Janusz Jezewski	Institute of Medical Technology and Equipment ITAM, Poland
Geun Sik Jo	Inha University, South Korea
Kang-Hyun Jo	University of Ulsan, South Korea
Janusz Kacprzyk	Systems Research Institute, Polish Academy of Sciences, Poland
Tomasz Kajdanowicz	Wrocław University of Science and Technology, Poland
Nadjet Kamel	University Ferhat Abbes Setif1, Algeria
Mehmed Kantardzic	University of Louisville, USA
Mehmet Karaata	Kuwait University, Kuwait
Nikola Kasabov	Auckland University of Technology, New Zealand
Arkadiusz Kawa	Poznan University of Economics and Business, Poland

Rafal Kern	Wrocław University of Science and Technology, Poland
Manish Khare	Dhirubhai Ambani Institute of Information and Communication Technology, India
Chonggun Kim	Yeungnam University, South Korea
Marek Kisiel-Dorohinicki	AGH University of Science and Technology, Poland
Attila Kiss	Eotvos Lorand University, Hungary
Jerzy Klamka	Silesian University of Technology, Poland
Goran Klepac	Raiffeisen Bank, Croatia
Shinya Kobayashi	Ehime University, Japan
Marek Kopel	Wrocław University of Science and Technology, Poland
Jozef Korbicz	University of Zielona Gora, Poland
Jerzy Korczak	Wroclaw University of Economics, Poland
Raymondus Kosala	Bina Nusantara University, Indonesia
Leszek Koszalka	Wrocław University of Science and Technology, Poland
Leszek Kotulski	AGH University of Science and Technology, Poland
Adrianna Kozierkiewicz	Wrocław University of Science and Technology, Poland
Bartosz Krawczyk	Virginia Commonwealth University, USA
Ondrej Krejcar	University of Hradec Kralove, Czech Republic
Dalia Kriksciuniene	Vilnius University, Lithuania
Dariusz Krol	Wrocław University of Science and Technology, Poland
Marek Krótkiewicz	Wrocław University of Science and Technology, Poland
Marzena Kryszkiewicz	Warsaw University of Technology, Poland
Adam Krzyzak	Concordia University, Canada
Tetsuji Kuboyama	Gakushuin University, Japan
Elżbieta Kukla	Wrocław University of Science and Technology, Poland
Julita Kulbacka	Wroclaw Medical University, Poland
Marek Kulbacki	Polish-Japanese Academy of Information Technology, Poland
Kazuhiro Kuwabara	Ritsumeikan University, Japan
Halina Kwasnicka	Wrocław University of Science and Technology, Poland
Mark Last	Ben-Gurion University of the Negev, Israel
Annabel Latham	Manchester Metropolitan University, UK
Bac Le	University of Science, VNU-HCM, Vietnam
Hoai An Le Thi	University of Lorraine, France
Kun Chang Lee	Sungkyunkwan University, South Korea
Yue-Shi Lee	Ming Chuan University, Taiwan
Chunshien Li	National Central University, Taiwan
Horst Lichter	RWTH Aachen University, Germany
Sebastian Link	University of Auckland, New Zealand
Igor Litvinchev	Nuevo Leon State University, Mexico
Lian Liu	University of Kentucky, USA
Rey-Long Liu	Tzu Chi University, Taiwan
Edwin Lughofer	Johannes Kepler University Linz, Austria
Ngoc Quoc Ly	Ho Chi Minh City University of Science, Vietnam
Lech Madeyski	Wrocław University of Science and Technology, Poland

Bernadetta Maleszka	Wrocław University of Science and Technology, Poland
Marcin Maleszka	Wrocław University of Science and Technology, Poland
Mustafa Mat Deris	Universiti Tun Hussein Onn Malaysia, Malaysia
Takashi Matsuhisa	Karelia Research Centre, Russian Academy of Science, Russia
Tamás Matuszka	Eotvos Lorand University, Hungary
Joao Mendes-Moreira	University of Porto, Portugal
Héctor Menéndez	University College London, UK
Jacek Mercik	WSB University in Wroclaw, Poland
Radosław Michalski	Wrocław University of Science and Technology, Poland
Peter Mikulecky	University of Hradec Kralove, Czech Republic
Marek Milosz	Lublin University of Technology, Poland
Jolanta Mizera-Pietraszko	Opole University, Poland
Leo Mrsic	IN2data Ltd. Data Science Company, Croatia
Agnieszka Mykowiecka	Institute of Computer Science, Polish Academy of Sciences, Poland
Pawel Myszkowski	Wrocław University of Science and Technology, Poland
Grzegorz J. Nalepa	AGH University of Science and Technology, Poland
Mahyuddin K. M. Nasution	Universitas Sumatera Utara, Indonesia
Richi Nayak	School of Electrical Engineering and Computer Science, Australia
Fulufhelo Nelwamondo	Council for Scientific and Industrial Research, South Africa
Huu-Tuan Nguyen	Vietnam Maritime University, Vietnam
Loan T. T. Nguyen	University of Warsaw, Poland
Quang-Vu Nguyen	Korea-Vietnam Friendship Information Technology College, Vietnam
Thai-Nghe Nguyen	Cantho University, Vietnam
Agnieszka Nowak-Brzezinska	University of Silesia in Katowice, Poland
Mariusz Nowostawski	Norwegian University of Science and Technology, Norway
Alberto Núñez	Universidad Complutense de Madrid, Spain
Manuel Núñez	Universidad Complutense de Madrid, Spain
Cheol-Young Ock	University of Ulsan, South Korea
Richard Jayadi Oentaryo	Singapore Management University, Singapore
Kouzou Ohara	Aoyama Gakuin University, Japan
Shingo Otsuka	Kanagawa Institute of Technology, Japan
George Papadopoulos	University of Cyprus, Cyprus
Marcin Paprzycki	Systems Research Institute, Polish Academy of Sciences, Poland
Rafael Parpinelli	Santa Catarina State University (UDESC), Brazil
Jakub Peksinski	West Pomeranian University of Technology, Szczecin, Poland
Danilo Pelusi	University of Teramo, Italy
Hoang Pham	Rutgers University, USA

Xuan Hau Pham	Quang Binh University, Vietnam
Tao Pham Dinh	INSA Rouen, France
Maciej Piasecki	Wrocław University of Science and Technology, Poland
Bartłomiej Pierański	Poznan University of Economics and Business, Poland
Dariusz Pierzchala	Military University of Technology, Poland
Marcin Pietranik	Wrocław University of Science and Technology, Poland
Elias Pimenidis	University of the West of England, UK
Jaroslav Pokorný	Charles University, Prague, Czech Republic
Andrzej Polanski	Silesian University of Technology, Poland
Elvira Popescu	University of Craiova, Romania
Piotr Porwik	University of Silesia in Katowice, Poland
Małgorzata Przybyła-Kasperek	University of Silesia in Katowice, Poland
Andrzej Przybyszewski	Polish-Japanese Academy of Information Technology, Poland
Paulo Quaresma	Universidade de Evora, Portugal
David Ramsey	Wrocław University of Science and Technology, Poland
Mohammad Rashedur Rahman	North South University, Bangladesh
Ewa Ratajczak-Ropel	Gdynia Maritime University, Poland
Manuel Roveri	Politecnico di Milano, Italy
Przemysław Rozewski	West Pomeranian University of Technology, Szczecin, Poland
Leszek Rutkowski	Czestochowa University of Technology, Poland
Henryk Rybiński	Warsaw University of Technology, Poland
Alexander Ryjov	Lomonosov Moscow State University, Russia
Keun Ho Ryu	Chungbuk National University, South Korea
Virgilijus Sakalauskas	Vilnius University, Lithuania
Daniel Sanchez	University of Granada, Spain
Moamar Sayed-Mouchaweh	Ecole des Mines de Douai, France
Rafal Scherer	Czestochowa University of Technology, Poland
Juergen Schmidhuber	Swiss AI Lab IDSIA, Poland
Björn Schuller	University of Passau, Germany
Jakub Segen	Gest3D, USA
Ali Selamat	Universiti Teknologi Malaysia, Malaysia
S. M. N. Arosha Senanayake	Universiti Brunei Darussalam, Brunei Darussalam
Tegjyot Singh Sethi	University of Louisville, USA
Natalya Shakhovska	Lviv Polytechnic National University, Ukraine
Donghwa Shin	Yeungnam University, South Korea
Andrzej Siemiński	Wrocław University of Science and Technology, Poland
Dragan Simic	University of Novi Sad, Serbia
Bharat Singh	Universiti Teknology PETRONAS, Malaysia
Krzysztof Slot	Lodz University of Technology, Poland
Adam Slowik	Koszalin University of Technology, Poland

Vladimir Sobeslav	University of Hradec Kralove, Czech Republic
Kulwadee Somboonviwat	King Mongkut's University of Technology Thonburi, Thailand
Jong-In Song	Gwangju Institute of Science and Technology, South Korea
Zenon A. Sosnowski	Bialystok University of Technology, Poland
Bela Stantic	Griffith University, Australia
Jerzy Stefanowski	Poznan University of Technology, Poland
Serge Stinckwich	University of Caen-Lower Normandy, France
Ja-Hwung Su	Cheng Shiu University, Taiwan
Andrzej Swierniak	Silesian University of Technology, Poland
Edward Szczerbicki	University of Newcastle, Australia
Julian Szymanski	Gdansk University of Technology, Poland
Yasufumi Takama	Tokyo Metropolitan University, Japan
Maryam Tayefeh	Mahmoudi ICT Research Institute, Iran
Zbigniew Telec	Wrocław University of Science and Technology, Poland
Dilhan Thilakarathne	Vrije Universiteit Amsterdam, The Netherlands
Krzysztof Tokarz	Silesian University of Technology, Poland
Bogdan Trawinski	Wrocław University of Science and Technology, Poland
Ualsher Tukeyev	al-Farabi Kazakh National University, Kazakhstan
Aysegul Ucar	Firat University, Turkey
Olgierd Unold	Wrocław University of Science and Technology, Poland
Natalie Van Der Wal	Vrije Universiteit Amsterdam, The Netherlands
Pandian Vasant	Universiti Teknologi PETRONAS, Malaysia
Jorgen Villadsen	Technical University of Denmark, Denmark
Bay Vo	Ho Chi Minh City University of Technology, Vietnam
Gottfried Vossen	ERCIS Münster, Germany
Lipo Wang	Nanyang Technological University, Singapore
Yongkun Wang	University of Tokyo, Japan
Izabela Wierzbowska	Gdynia Maritime University, Poland
Konrad Wojciechowski	Silesian University of Technology, Poland
Michal Wozniak	Wrocław University of Science and Technology, Poland
Krzysztof Wrobel	University of Silesia in Katowice, Poland
Tsu-Yang Wu	Harbin Institute of Technology Shenzhen Graduate School, China
Marian Wysocki	Rzeszow University of Technology, Poland
Farouk Yalaoui	University of Technology of Troyes, France
Xin-She Yang	Middlesex University, UK
Lina Yao	University of New South Wales, Australia
Tulay Yildirim	Yildiz Technical University, Turkey
Slawomir Zadrozny	Systems Research Institute, Polish Academy of Sciences, Poland
Drago Zagar	University of Osijek, Croatia
Danuta Zakrzewska	Lodz University of Technology, Poland

Constantin-Bala Zamfirescu	Lucian Blaga University of Sibiu, Romania
Katerina Zdravkova	Ss. Cyril and Methodius University in Skopje, Republic of Macedonia
Vesna Zeljkovic	Lincoln University, USA
Aleksander Zgrzywa	Wrocław University of Science and Technology, Poland
De-Chuan Zhan	Nanjing University, China
Qiang Zhang	Dalian University, China
Zhongwei Zhang	University of Southern Queensland, Australia
Dongsheng Zhou	Dalian Unviersity, China
Zhandos Zhumanov	al-Farabi Kazakh National University, Guyana
Maciej Zieba	Wrocław University of Science and Technology, Poland
Adam Ziebinski	Silesian University of Technology, Poland
Beata Zielosko	University of Silesia in Katowice, Poland
Marta Zorrilla	University of Cantabria, Spain

Program Committees of Special Sessions

Special Session on Multiple Model Approach to Machine Learning (MMAML 2018)

Emili Balaguer-Ballester	Bournemouth University, UK
Urszula Boryczka	University of Silesia, Poland
Abdelhamid Bouchachia	Bournemouth University, UK
Robert Burduk	Wrocław University of Science and Technology, Poland
Oscar Castillo	Tijuana Institute of Technology, Mexico
Rung-Ching Chen	Chaoyang University of Technology, Taiwan
Suphamit Chittayasothorn	King Mongkut's Institute of Technology Ladkrabang, Thailand
Jose Alfredo F. Costa	Federal University (UFRN), Brazil
Bogustaw Cyganek	AGH University of Science and Technology, Poland
Ireneusz Czarnowski	Gdynia Maritime University, Poland
Patrick Gallinari	Pierre et Marie Curie University, France
Fernando Gomide	State University of Campinas, Brazil
Francisco Herrera	University of Granada, Spain
Tzung-Pei Hong	National University of Kaohsiung, Taiwan
Konrad Jackowski	Wrocław University of Science and Technology, Poland
Piotr Jędrzejowicz	Gdynia Maritime University, Poland
Tomasz Kajdanowicz	Wrocław University of Science and Technology, Poland
Yong Seog Kim	Utah State University, USA
Bartosz Krawczyk	Wrocław University of Science and Technology, Poland
Kun Chang Lee	Sungkyunkwan University, South Korea
Edwin Lughofer	Johannes Kepler University Linz, Austria
Hector Quintian	University of Salamanca, Spain
Andrzej Sieminski	Wrocław University of Science and Technology, Poland
Dragan Simic	University of Novi Sad, Serbia

Adam Slowik	Koszalin University of Technology, Poland
Zbigniew Telec	Wrocław University of Science and Technology, Poland
Bogdan Trawinski	Wrocław University of Science and Technology, Poland
Olgierd Unold	Wrocław University of Science and Technology, Poland
Pandian Vasant	University Technology Petronas, Malaysia
Michal Wozniak	Wrocław University of Science and Technology, Poland
Zhongwei Zhang	University of Southern Queensland, Australia
Zhi-Hua Zhou	Nanjing University, China

Special Session on Analysis of Image, Video, and Motion Data in Life Sciences (IVMLS 2018)

Artur Bąk	Polish-Japanese Academy of Information Technology, Poland
Leszek Chmielewski	Warsaw University of Life Sciences, Poland
Aldona Barbara Drabik	Polish-Japanese Academy of Information Technology, Poland
Marcin Fojcik	Sogn og Fjordane University College, Norway
Adam Gudys	Silesian University of Technology, Poland
Celina Imielinska	Vesalius Technolodgies LLC, USA
Henryk Josinski	Silesian University of Technology, Poland
Ryszard Klempous	Wrocław University of Science and Technology, Poland
Ryszard Kozera	The University of Life Sciences - SGGW, Poland
Julita Kulbacka	Wroclaw Medical University, Poland
Marek Kulbacki	Polish-Japanese Academy of Information Technology, Poland
Aleksander Nawrat	Silesian University of Technology, Poland
Jerzy Pawet Nowacki	Polish-Japanese Academy of Information Technology, Poland
Eric Petajan	LiveClips LLC, USA
Andrzej Polanski	Silesian University of Technology, Poland
Joanna Rossowska	Polish Academy of Sciences, Poland
Jakub Segen	Gest3D LLC, USA
Aleksander Sieron	Medical University of Silesia, Poland
Michat Staniszewski	Polish-Japanese Academy of Information Technology, Poland
Adam Switonski	Silesian University of Technology, Poland
Agnieszka Szczęsna	Silesian University of Technology, Poland
Kamil Wereszczynski	Polish-Japanese Academy of Information Technology, Poland
Konrad Wojciechowski	Polish-Japanese Academy of Information Technology, Poland
Stawomir Wojciechowski	Polish-Japanese Academy of Information Technology, Poland

Special Session on Intelligent and Contextual Systems (ICxS 2018)

Adriana Albu	Politehnica University Timisoara, Romania
Basabi Chakraborty	Iwate Prefectural University, Japan
Goutam Chakraborty	Iwate Prefectural University, Japan
Ha Manh Tran	Ho Chi Minh City International University, Vietnam
Hong Vu Nguyen	Ton Duc Thang University, Vietnam
Hideyuki Takahashi	RIEC, Tohoku University, Japan
Jerzy Świątek	Wrocław University of Science and Technology, Poland
Jozef Korbicz	University of Zielona Gora, Poland
Keun Ho Ryu	Chungbuk National University, South Korea
Kilho Shin	University of Hyogo, Japan
Lina Yang	University of Macau, Macau
Maciej Huk	Wrocław University of Science and Technology, Poland
Masafumi Matsuhara	Iwate Prefectural University, Japan
Michael Spratling	University of London, UK
Nguyen Khang Pham	Can Tho University, Vietnam
Plamen Angelov	Lancaster University, UK
Qiangfu Zhao	University of Aizu, Japan
Quan Thanh Tho	Ho Chi Minh City University of Technology, Vietnam
Rashmi Dutta Baruah	Lancaster University, UK
Takako Hashimoto	Chiba University of Commerce, Japan
Tetsuji Kubojama	Gakushuin University, Japan
Tetsuo Kinoshita	RIEC, Tohoku University, Japan
Thai-Nghe Nguyen	Can Tho University, Vietnam
Yicong Zhou	University of Macau, Macau, SAR China
Yuan Yan Tang	University of Macau, Macau, SAR China
Zhenni Li	University of Aizu, Japan

Special Session on Intelligent Systems for Optimization of Logistics and Industrial Applications (ISOLIA 2018)

Zaki Sari	Abou Bakr Belkaid University of Tlemcen, Algeria
Hicham Chehade	University of Technology of Troyes, France
Yassine Ouazene	University of Technology of Troyes, France
Habiba Drias	University of Science and Technology (USTHB), Algeria
Lionel Amodeo	University of Technology of Troyes, France
Mustapha Nourelfath	University of Laval, Canada
Nathalie Sauer	University of Lorraine, France
Alexandre Dolgui	National Institute of Science and Technology Mines-Telecom, France
Olga Battaia	Institut Superieur de l'Aeronautique et d'Espace, France
Zaki Sari	Abou Bakr Belkaid University of Tlemcen, Algeria
Daoud Ait-Kadi	University of Laval, Canada

Dmitry Ivanov	Hochschule fur Wirtschaft und Recht Berlin, Germany
Lyes Benyoucef	Aix-Marseille University, France
Nidhal Rezg	University of Lorraine, France

Special Session on Intelligent Applications of Internet of Thing and Data Analysis Technologies (IoT&DAT 2018)

Goutam Chakraborty	Iwate Prefectural University, Japan
Bin Dai	University of Technology Xiamen, China
Qiangfu Zhao	University of Aizu, Japan
David C. Chou	Eastern Michigan University, USA
Chin-Feng Lee	Chaoyang University of Technology, Taiwan
Lijuan Liu	University of Technology Xiamen, China
Kien A. Hua	Central Florida University, USA
Long-Sheng Chen	Chaoyang University of Technology, Taiwan
Xin Zhu	University of Aizu, Japan
David Wei	Fordham University, USA
Qun Jin	Waseda University, Japan
Jacek M. Zurada	University of Louisville, USA
Tsung-Chih Hsiao	Huaoiao University, China
Hsien-Wen Tseng	Chaoyang University of Technology, Taiwan
Nitasha Hasteer	Amity University Uttar Pradesh, India
Chuan-Bi Lin	Chaoyang University of Technology, Taiwan
Cliff Zou	Central Florida University, USA

Special Session on Intelligent Systems and Methods in Biomedicine (ISaMiB 2018)

Jan Kubicek	Technical University of Ostrava, Czech Republic
Marek Penhaker	Technical University of Ostrava, Czech Republic
Martin Augustynek	Technical University of Ostrava, Czech Republic
Martin Cerny	Technical University of Ostrava, Czech Republic
Vladimir Kasik	Technical University of Ostrava, Czech Republic
Lukas Peter	Technical University of Ostrava, Czech Republic
Ondrej Krejcar	University of Hradec Kralove, Czech Republic
Kamil Kuca	University of Hradec Kralove, Czech Republic
Petra Maresova	University of Hradec Kralove, Czech Republic
Ali Selamat	Universiti Teknologi Malaysia, Malaysia

Special Session on Intelligent Systems and Algorithms in Information Sciences (ISAIS 2018)

Martin Kotyrba	University of Ostrava, Czech Republic
Eva Volna	University of Ostrava, Czech Republic
Ivan Zelinka	VSB-Technical University of Ostrava, Czech Republic
Hashim Habiballa	Institute for Research and Applications of Fuzzy Modeling, Czech Republic
Alexej Kolcun	Institute of Geonics, AS CR, Czech Republic
Roman Senkerik	Tomas Bata University in Zlin, Czech Republic

Zuzana Kominkova-Oplatkova	Tomas Bata University in Zlin, Czech Republic
Katerina Kostolanyova	University of Ostrava, Czech Republic
Antonin Jancarik	Charles University in Prague, Czech Republic
Igor Kostal	The University of Economics in Bratislava, Slovakia
Eva Kurekova	Slovak University of Technology in Bratislava, Slovakia
Leszek Cedro	Kielce University of Technology, Poland
Dagmar Janacova	Tomas Bata University in Zlin, Czech Republic
Martin Halaj	Slovak University of Technology in Bratislava, Slovakia
Radomil Matousek	Brno University of Technology, Czech Republic
Roman Jasek	Tomas Bata University in Zlin, Czech Republic
Petr Dostal	Brno University of Technology, Czech Republic
Jiri Pospichal	The University of Ss. Cyril and Methodius (UCM), Slovakia
Vladimir Bradac	University of Ostrava, Czech Republic

Special Session on Design Thinking-Based R&D, Development Technique, and Project-Based Learning (2DT-PBL 2018)

Tsuyoshi Arai	Okayama Prefectural University, Japan
Yoshihide Chubachi	Advanced Institute of Industrial Technology, Japan
Keiichi Endo	Ehime University, Japan
Takuya Fujihashi	Ehime University, Japan
Yoshinobu Higami	Ehime University, Japan
Tohru Kawabe	University of Tsukuba, Japan
Shinya Kobayashi	Ehime University, Japan
Hisayasu Kuroda	Ehime University, Japan
Kazuo Misue	University of Tsukuba, Japan
Katsumi Sakakibara	Okayama Prefectural University, Japan
Kiyoshi Sakamori	Advanced Institute of Industrial Technology, Japan
Hironori Takimoto	Okayama Prefectural University, Japan
Toshiyuki Uto	Ehime University, Japan
Chiemi Watanabe	Advanced Institute of Industrial Technology, Japan
Hitoshi Yamauchi	Okayama Prefectural University, Japan

Special Session on Modelling, Storing, and Querying of Graph Data (MSQGD 2018)

Michal Valenta	Czech Technical University, Czech Republic
Martin Svoboda	Charles University, Czech Republic
M. Praveen	CMI, India
Jianxin Li	University of Western Australia, Australia
Konstantinos Semertzidis	University of Ioannina, Greece
Marco Mesiti	DICO - University of Milan, Italy

Vincenzo Moscato	University of Naples, Italy
Cedric Du Mouza	CNAM, France
Jacek Mercik	WSB University in Wroclaw, Poland
Virginie Thion	University of Rennes, France

Special Session on Data Science and Computational Intelligence (DSCI 2018)

Adisak Sukul	Iowa State University, USA
Akadej Udomchaiporn	King Mongkut's Institute of Technology Ladkrabang, Thailand
Jittima Tongurai	Kobe University, Japan
Kulsawasd Jitkajornwanich	King Mongkut's Institute of Technology Ladkrabang, Thailand
Natawut Nupairoj	Chulalongkorn University, Thailand
Peeraphon Sophatsathit	Chulalongkorn University, Thailand
Peerapon Vateekul	Chulalongkorn University, Thailand
Pisit Chanvarasuth	Sirindhorn International Institute of Technology, Thailand
Ronan Reilly	Maynooth University, Ireland
Sanparith Marukatat	National Electronics and Computer Technology Center, Thailand
Sarun Intagosum	King Mongkut's Institute of Technology Ladkrabang, Thailand
Suradej Intagorn	Kasetsart University, Thailand
Turki Talal Salem Turki	King Abdulaziz University, Saudi Arabia
Veera Boonjing	King Mongkut's Institute of Technology Ladkrabang, Thailand

Special Session on Computer Vision and Robotics (CVR 2018)

Van-Dung Hoang	Quang Binh University, Vietnam
Chi- Mai Luong	University of Science and Technology of Hanoi, Vietnam
My-Ha Le	Ho Chi Minh City University of Technology and Education, Vietnam
Kang-Hyun Jo	Ulsan University, South Korea
The- Anh Pham	Hong Duc University, Vietnam
Van-Huy Pham	Ton Duc Thang University, Vietnam
Youngsoo Suh	University of Ulsan, South Korea
Danilo Caceres Hernandez	Universidad Tecnologica de Panama, Panama
Kaushik Deb	Chittagong University of Engineering and Technology, Bangladesh
Ha Nguyen Thi Thu	Electric Power University, Vietnam
Viet-Vu Vu	Thai Nguyen University of Technology, Vietnam
Lan Le Thi	Hanoi University of Science and Technology, Vietnam
Thanh Binh Nguyen	University of Science Ho Chi Minh City, Vietnam
Trung Duc Nguyen	Vietnam Maritime University, Vietnam

The Bao Pham	University of Science Ho Chi Minh City, Vietnam
Yoshinori Kuno	Saitama University, Japan
Heejun Kang	University of Ulsan, South Korea
Wahyono	Universitas Gadjah Mada, Indonesia
Do Van Nguyen	Nagaoka University of Technology, Japan
Truc Thanh Tran	Danang Department of Information and Communications, Vietnam
Long-Thanh Ngo	Le Quy Don Technical University, Vietnam

Special Session on Intelligent Biomarkers of Neurodegenerative Processes in the Brain (InBinBRAIN 2018)

Zbigniew Struzik	RIKEN Brain Science Institute, Japan
Zbigniew Ras	University of North Carolina at Charlotte, USA
Konrad Ciecierski	Warsaw University of Technology, Poland
Piotr Habela	Polish-Japanese Academy of Information Technology, Poland
Peter Novak	Brigham and Women's Hospital, USA
Wieslaw Nowinski	Cardinal Stefan Wyszynski University, Poland
Andrei Barborica	Research and Compliance and Engineering, FHC, Inc., USA
Alicja Wieczorkowska	Polish-Japanese Academy of Information Technology, Poland
Majaz Moonis	UMass Medical School, USA
Krzysztof Marasek	Polish-Japanese Academy of Information Technology, Poland
Mark Kon	Boston University, USA
Lech Polkowski	Polish-Japanese Academy of Information Technology, Poland
Andrzej Skowron	Warsaw University, Poland
Ryszard Gubrynowicz	Polish-Japanese Academy of Information Technology, Warsaw, Poland
Dominik Slezak	Warsaw University, Poland
Radoslaw Nielek	Polish-Japanese Academy of Information Technology, Warsaw, Poland

Special Session on Data Structures Modelling for Knowledge Representation (DSMKR 2018)

Marek Krotkiewicz	Wrocław University of Science and Technology, Poland

Special Session on Computational Imaging and Vision (CIV 2018)

Ishwar Sethi	Oakland University, USA
Moongu Jeon	Gwangju Institute of Science and Technology, South Korea
Jong-In Song	Gwangju Institute of Science and Technology, South Korea

Kiseon Kim	Gwangju Institute of Science and Technology, South Korea
Taek Lyul Song	Hangyang University, South Korea
Ba-Ngu Vo	Curtin University, Australia
Ba-Tuong Vo	Curtin University, Australia
Du Yong Kim	Curtin University, Australia
Benlian Xu	Changshu Institute of Technology, China
Peiyi Zhu	Changshu Institute of Technology, China
Mingli Lu	Changshu Institute of Technology, China
Weifeng Liu	Hangzhou Danzi University, China
Ashish Khare	University of Allahabad, India
Om Prakash	University of Allahabad, India
Moonsoo Kang	Chosun University, South Korea
Goo-Rak Kwon	Chosun University, South Korea
Sang Woong Lee	Gachon University, South Korea
Ekkarat Boonchieng	Chiang Mai University, Thailand
Jeong-Seon Park	Chonnam National University, South Korea
Unsang Park	Sogang University, South Korea
R. Z. Khan	Aligarh Muslim University, India
Sathya Narayanan	NTU, Singapore

Special Session on Intelligent Computer Vision Systems and Applications (ICVSA 2018)

Ferran Reverter Comes	University of Barcelona, Spain
Michael Cree	University of Waikato, New Zealand
Piotr Dziurzanski	University of York, UK
Marcin Iwanowski	Warsaw University of Technology, Poland
Heikki Kalviainen	Lappeenranta University of Technology, Finland
Tomasz Marciniak	UTP University of Science and Technology, Poland
Adam Nowosielski	West Pomeranian University of Technology, Szczecin, Poland
Krzysztof Okarma	West Pomeranian University of Technology, Szczecin, Poland
Arkadiusz Ortowski	Warsaw University of Life Sciences, Poland
Edward Potrolniczak	West Pomeranian University of Technology, Szczecin, Poland
Pilar Rosado Rodrigo	University of Barcelona, Spain
Khalid Saeed	AGH University of Science and Technology Cracow, Poland
Rafael Saracchini	Technological Institute of Castilla y Leon (ITCL), Spain
Samuel Silva	University of Aveiro, Portugal
Gregory Slabaugh	City University London, UK
Egon L. van den Broek	Utrecht University, The Netherlands

Special Session on Advanced Data Mining Techniques and Applications (ADMTA 2018)

Tzung-Pei Hong	National University of Kaohsiung, Taiwan
Tran Minh Quang	Ho Chi Minh City University of Technology, Vietnam
Bac Le	University of Science, VNU-HCM, Vietnam
Bay Vo	Ho Chi Minh City University of Technology, Vietnam
Chun-Hao Chen	Tamkang University, Taiwan
Chun-Wei Lin	Harbin Institute of Technology, China
Wen-Yang Lin	National University of Kaohsiung, Taiwan
Yeong-Chyi Lee	Cheng Shiu University, Taiwan
Le Hoang Son	University of Science, Vietnam
Vo Thi Ngoc Chau	Ho Chi Minh City University of Technology, Vietnam
Van Vo	Ho Chi Minh University of Industry, Vietnam
Ja-Hwung Su	Cheng Shiu University, Taiwan
Ming-Tai Wu	University of Nevada, USA
Kawuu W. Lin	National Kaohsiung University of Applied Sciences, Taiwan
Tho Le	Ho Chi Minh City University of Technology, Vietnam
Dang Nguyen	Deakin University, Australia
Hau Le	Thuyloi University, Vietnam
Thien-Hoang Van	Ho Chi Minh City University of Technology, Vietnam
Tho Quan	Ho Chi Minh City University of Technology, Vietnam
Ham Nguyen	University of People's Security Ho Chi Minh City, Vietnam
Thiet Pham	Ho Chi Minh University of Industry, Vietnam

Contents – Part I

Knowledge Engineering and Semantic Web

Social Networks and Recommender Systems

Decision Support and Control Systems

Computer Vision Techniques

Advanced Data Mining Techniques and Applications

Multiple Model Approach to Machine Learning

Contents – Part II

Data Structures Modelling for Knowledge Representation

Modelling, Storing, and Querying of Graph Data

Data Science and Computational Intelligence

Design Thinking Based R&D, Development Technique, and Project Based Learning

Intelligent and Contextual Systems

Intelligent Systems and Methods in Biomedicine

Intelligent Biomarkers of Neurodegenerative Processes in Brain

Intelligent Systems for Optimization of Logistics and Industrial Applications

Knowledge Engineering and Semantic Web

Solving Inconsistencies in Probabilistic Knowledge Bases via Inconsistency Measures

Van Tham Nguyen[1,2(✉)] and Trong Hieu Tran[2]

[1] Nam Dinh University of Technology Education, Nam Dinh, Vietnam
thamnv.nute@gmail.com
[2] Vietnam National University, Hanoi, Vietnam
{17028002,hieutt}@vnu.edu.vn

Abstract. In most knowledge-based systems, the guarantee of consistency is one of the essential tasks to ensure them to avoid the trivial cases. Because of this reason, a wide range of approaches has been proposed for restoring consistency. However, these approaches often correspond to logical, or probabilistic-logical framework. In this paper, we investigate a model for restoring the consistency of probabilistic knowledge bases by focusing on the method of changing the probabilities in such knowledge bases. To this aim, a process to restore the consistency based on inconsistency measures is introduced, a set of rational and intuitive axioms to characterize the restoring operators is proposed, and several logical properties are investigated and discussed.

Keywords: Probabilistic knowledge bases · Inconsistency measure
Restoring operator

1 Introduction

In the design of expert systems in the artificial intelligence field, one of vital problems is to ensure consistency of the knowledge bases [1]. It is hard to indicate the inconsistency of knowledge as well as to solve this inconsistency. However, the whole management process may become impossible if the inconsistency is not resolved. In literature, some strategies have been developed to address inconsistencies by adjusting the structure of elements in the knowledge base, ruling out parts of the knowledge base [2], or employing inconsistency measures to assess the extent of inconsistency [3–8]. Those strategies often build a family of operators to modify an inconsistent knowledge base into a consistent one. Nowadays, in order to solve inconsistency of knowledge, using inconsistency measures is one of the most common approaches. A class of basic inconsistency measures for probabilistic knowledge bases have been introduced in [9].

The classical approaches of handling inconsistency in propositional knowledge bases through removing formulas were proposed in [10]. Those approaches are performed by building operators which based on discarding some formulas from minimal inconsistent sets [11] but without requiring that the withdrawal be minimal. Minimal inconsistent sets [12] or maximal inconsistent sets [13] have to find before building such operators.

N. T. Nguyen et al. (Eds.): ACIIDS 2018, LNAI 10751, pp. 3–14, 2018.
https://doi.org/10.1007/978-3-319-75417-8_1

In a probabilistic logic environment, the technique of removing formulas has also been applied in [14]. Some algorithms have been developed with linear programming features for calculating the probability of realistic size. The technique of using measures for evaluating the level inconsistency of a probabilistic knowledge base has been presented in [7]. The approach in [5] is similar in nature to that defined in [7]. The idea of this approach is based on the quantification of the minimal adjustment for changing the probabilities in an inconsistent base to obtain a consistent one. In [15], another operator that modifies the original probabilities of the conditionals by employing measures from [6, 7] also developed. The family of operators in [5, 15] satisfies some desirable properties which stem from [16], but it is just applied only to precise the probabilistic knowledge base.

In a probabilistic environment, apart from the removal of probabilistic conditionals, the method of modification of probabilities has been applied in [17]. This work incorporated integrity constraints, applied the family of minimal violation measures from [6], and proposed a family of generalized entailment problems for probabilistic knowledge bases as well as the solutions for such problems. Instead of changing point probability, a class of operators in [3] was proposed by altering probability intervals.

The main contribution of this paper is threefold. First, we make a deeply survey on how to calculate inconsistency measures for the probabilistic framework. Second, we propose two families of operators which are used to deal with the task of resolving the inconsistency. Then, a set of axioms is introduced and the logical properties of the proposed families of operators are investigated and discussed.

This paper is organized as follows. In Sect. 2 we start with some necessary preliminaries about the probability, the probability function, probabilistic constraint, the representation of knowledge bases in probabilistic framework, and some equivalence relations for such knowledge bases. Afterwards, in Sect. 3 we propose a model to change an inconsistent probabilistic knowledge base into a consistent one. This model is based on two inconsistency measures, namely, the minimal violation measure and the unnormalized inconsistency measure. The logical properties for each operator are also introduced and discussed in Sect. 3. Some conclusions and future work are presented in Sect. 4.

2 Probabilistic Knowledge Bases

Let $\mathcal{S}$ be a sample space that includes all possible outcomes of a statistical experiment. Let $\mathbb{E} = \{E_1, \ldots, E_n\}$ be a finite set of events, where each event is a subset of the sample space $\mathcal{S}$. For $F, G \in \mathbb{E}$, the intersection of two events F and G, denoted by FG, is the event containing all elements that are common to Fand G; negation of F, denoted by $\neg F$, is abbreviated by $\overline{F}$. A complete conjunction Θ of $\mathbb{E}$ is an expression of the form $\Theta = \widetilde{E}_1, \widetilde{E}_2, \ldots, \widetilde{E}_n$ with $\widetilde{E}_i = \{E_i, \overline{E}_i\}$. Let $\Gamma(\mathbb{E})$ be the set of all complete conjunctions of $\Gamma(\mathbb{E}) = \{\Theta_1, \ldots, \Theta_{2^n}\}$, therefore $\Gamma(\mathbb{E}) = \{\Theta_1, \ldots, \Theta_{2^n}\}$. A complete conjunction $\Theta \in \Gamma(\mathbb{E})$ satisfies an event F, denoted by $\Theta \models F$, if and only if F positively appears in Θ. Let $SM(H) = \{\Theta \in \Gamma(\mathbb{E}) | \Theta \models H\}$, where H is an event or a set of events. Let $m = |\Gamma(\mathbb{E})|$ be the numbers of complete conjunctions of $\mathbb{E}$. Let $\mathbb{R}_{\geq 0}$ be the set of non-negative real values including +∞. Let $\mathbb{R}_{[0,1]}$ be the set of all real values from 0 to 1. Let : $\Gamma(\mathbb{E}) \rightarrow \mathbb{R}_{[0,1]}$ be a probability function of a complete conjunction. Let $\mathbb{P}(\mathbb{E})$ be the set of all probability functions $\mathcal{P}$, defined by $\mathbb{P}(\mathbb{E}) = \{\mathcal{P}(\Theta_1), \ldots, \mathcal{P}(\Theta_{2^n})\}$. Let $\vec{\omega} = (\omega_1, \ldots, \omega_m)^T$ be a column vector, where an auxiliary variable ω_i corresponding to a probabilisty $\mathcal{P}(\Theta_i)$.

With $F \in \mathbb{E}$,

$$\mathcal{P}(F) = \sum\nolimits_{\Theta \in \Gamma(\mathbb{E}):\Theta \vDash F} \mathcal{P}(\Theta).$$

Definition 1. *Function $\delta(H, \Theta)$ is called an indicator function whenever it is defined as follows:* $\delta : \mathbb{E} \times \Gamma(\mathbb{E}) \rightarrow \mathbb{R}_{[0,1]}$,

$$\delta(H, \Theta) = \begin{cases} 1 & \text{if } \Theta \vDash H \\ 0 & \text{otherwise} \end{cases}$$

Definition 2. *Let $F, G \in \mathbb{E}$ and $\rho \in \mathbb{R}_{[0,1]}$. A probabilistic constraint is an expression of the form $(F|G)[\rho]$.*

If G is tautological, $G \equiv \top$, we abbreviate $(F|\top)[\rho]$ by $(F)[\rho]$. We also denote $\kappa[\beta] = (F|G)[\beta]$ for each constraint $\kappa = (F|G)[\rho]$ and $\beta \in \mathbb{R}_{[0,1]}$.

Definition 3. *Probabilistic knowledge base $\mathcal{K}$ is a finite set of probabilistic constraints defined as: $\mathcal{K} = \langle \kappa_1, \ldots, \kappa_n \rangle$ where $\kappa_i = (F_i|G_i)[\rho_i]$ for all $i = 1, \ldots n$.*

Let $n = |\mathcal{K}|$ be numbers of constraints in $\mathcal{K}$. Let $\mathbb{K} = \{\mathcal{K}_1, \ldots, \mathcal{K}_h\}$ be a set which includes probabilistic knowledge bases. Let $\mathbb{R}_{[0,1]}^{|\mathcal{K}|}$ be a set which includes all real values from 0 to 1 that is the probabilities of the constraints in $\mathcal{K}$. Let $SC(\mathcal{K})$ be a set which includes all probabilistic constraints appearing $\mathcal{K}$, defined by $SC(\mathcal{K}) = \{\kappa_1, \ldots, \kappa_n\}$. Let $\vec{\rho} = (\rho_1, \ldots, \rho_n)^T$ be a column vector, where an auxiliary variable ρ_i corresponding to the probabilistic value of constraint κ_i.

Definition 4 *(Characteristic Function). Let $\mathcal{K} \in \mathbb{K}$. Function $\partial_{\mathcal{K}}\left(\vec{\beta}\right) : \mathbb{R}_{[0,1]}^{|\mathcal{K}|} \rightarrow \mathbb{K}$ is called a characteristic function whenever it is defined as follows: $\partial_{\mathcal{K}}\left(\vec{\beta}\right) = \langle \kappa'_1, \ldots, \kappa'_n \rangle$, where $\kappa'_i = (F_i|G_i)[\rho_i]$ for all $i = 1, \ldots n$ and $\vec{\beta} = (\beta_1, \ldots, \beta_n) \in \mathbb{R}_{[0,1]}^{|\mathcal{K}|}$.*

A probability function $\mathcal{P} \in \mathbb{P}(\mathbb{E})$ satisfies a probabilistic constraint $(F|G)[\rho]$, denoted by $\mathcal{P} \vDash (F|G)[\rho]$, if and only if $\mathcal{P}(FG) = \rho\mathcal{P}(G)$. Let $\vec{x} = (x_1, \ldots, x_n)^T$ be a column vector, where each auxiliary variable x_i corresponds to a probability $\beta_i \in \overrightarrow{\beta}$. A probability function satisfies $\mathcal{K}$, denoted by $\mathcal{P} \vDash \mathcal{K}$, if and only if $\mathcal{P} \vDash \kappa \forall \kappa \in \mathcal{K}$. Let $SM(\mathcal{K}) = \{\mathcal{P} \in \mathbb{P}(\mathbb{E}) | \mathcal{P} \vDash \mathcal{K}\}$. Therefore, if $SM(\mathcal{K}) \neq \varnothing$ then $\mathcal{K}$ is inconsistent, denoted by $\mathcal{K} \vDash \bot$. Otherwise, $\mathcal{K}$ is consistent, denoted by $\mathcal{K} \nvDash \bot$. Let $\kappa_1, \kappa_2 \in \mathcal{K}$, it said that probabilistic constraints are equivalent, denoted by $\kappa_1 \equiv \kappa_2$, if and only if $\text{SM}(\{\kappa_1\}) = \text{SM}(\{\kappa_2\})$. Knowledge bases $\mathcal{K}_1, \mathcal{K}_2$ are extensionally equivalent, denoted by $\mathcal{K}_1 \triangleq \mathcal{K}_2$, if and only if $\text{SM}(\mathcal{K}_1) = \text{SM}(\mathcal{K}_2)$. Knowledge bases $\mathcal{K}_1, \mathcal{K}_2$ are semi-extensionally equivalent, denoted by $\mathcal{K}_1 \mathrel{\overset{\circ}{=}} \mathcal{K}_2$, if $\exists \varphi : \mathcal{K}_1 \to \mathcal{K}_2$ such that $\kappa \equiv \varphi(\kappa)$ for each $\kappa \in \mathcal{K}_1$. Knowledge bases $\mathcal{K}_1, \mathcal{K}_2$ are qualitatively equivalent, denoted by $\mathcal{K}_1 \cong \mathcal{K}_2$, if and only if $|\mathcal{K}_1| = |\mathcal{K}_2|$ and $\exists \vec{\beta} \in \mathbb{R}^{|\mathcal{K}_1|}_{[0,1]} : \mathcal{K}_1 = \partial_{\mathcal{K}_2}\left(\vec{\beta}\right)$.

3 Model for Restoring the Consistency of Probabilistic Knowledge Bases

In this work, consistency restoring problem is defined in the context of priority as follows:

(1) **Input:** An inconsistent probabilistic knowledge base.
(2) **Output:** A consistent probabilistic knowledge base.
(3) **Scope of problem:** The knowledge base is represented in a probabilistic framework.
(4) **Restoring process:**

- Step 1: Computing inconsistency measure
- Step 2: Calculating new probability for each probabilistic constraint

(5) **Result:** A consistent probabilistic knowledge base.

We first study how to compute inconsistency measure:

Definition 5 *(Inconsistency measure). Inconsistency measure* $\mathcal{IM}$ *is a function* $\mathcal{IM} : \mathbb{K} \to \mathbb{R}_{\geq 0}$ *such that* $\mathcal{IM}(\mathcal{K}) = 0$ *if and only if* $\text{SM}(\mathcal{K}) \neq \varnothing$, $\mathcal{K} \in \mathbb{K}$.

3.1 The Minimal Violation Measure

Definition 6 *(Distance Function). Function* $d^p_{\mathcal{K}}$ *is called the distance of a probability function* $\mathcal{P}$ *to a knowledge base* $\mathcal{K}$ *with respect to a p-norm* ($p \geq 1$) *whenever it is defined as follows:*

$$d^p_{\mathcal{K}} = \|(z_1, \ldots, z_n)\|_p = \sqrt[p]{\sum_{i=1}^{n} |z_i|^p}$$

for each $(F_i|G_i)[\rho_i] : \mathcal{P}(F_iG_i) - \rho_i\mathcal{P}(G_i) = z_i$.

Definition 7 *(Characteristic Matrix).* $A_{\mathcal{K}}$ *is called a characteristic matrix of* $\mathcal{K}$ *whenever it is defined as follows:*

$$A_{\mathcal{K}} = (a_{ij}) \in \mathbb{R}^{n \times m}$$

where $a_{ij} = \delta(F_iG_i, \Theta_j)(1 - \rho_i) - \delta(\overline{F}_iG_i, \Theta_j)\rho_i$.

Let $\vec{a}_j$ be the j-th column vector of matrix $A_{\mathcal{K}}$. Let $B_{\mathcal{K}} = A_{\mathcal{K}}^T = (b_{ij}) \in \mathbb{R}^{m \times n}$.

Definition 8 *(Exponential Constraint Vector). Let $\mathcal{K} \in \mathbb{K}$. Function $\vec{\alpha}_{\mathcal{K}} : \mathbb{R}^{n+1} \rightarrow \mathbb{R}^m$ is called an exponential constraint vector of $\mathcal{K}$ whenever it is defined as follows:*

$$\vec{\alpha}_{\mathcal{K}}(\vec{x}, y) = (\alpha_1(\vec{x}, y), \ldots, \alpha_m(\vec{x}, y))^T$$

where

$$\alpha_j(\vec{x}, y) = exp\left(\sum\nolimits_{i=1}^{n} x_i b_{ij} + y - 1\right) \forall j = 1, \ldots, m$$

Definition 9 *(Minimal Violation Inconsistency Measure). $\mathcal{IM}_{\mathcal{K}}^p$ is called minimal violation inconsistency measure of $\mathcal{K}$ with respect to a p-norm ($p \geq 1$) whenever it is defined as follows:*

$$\mathcal{IM}_{\mathcal{K}}^p = min\{d_{\mathcal{K}}^p | A_{\mathcal{K}}\mathcal{P} = \vec{z}\}$$

Let $\vec{\lambda} = (\lambda_1, \ldots, \lambda_n)^T$ be a column vector, where an auxiliary variable λ_i corresponding to a constraint κ_i. Let $\vec{1} \in \mathbb{R}^n$ be a column vector that consists only ones. Let $\vec{0} \in \mathbb{R}^n$ be a column vector that consists only zeros.

Proposition 1. *Minimal Violation Inconsistency Measure $\mathcal{IM}_{\mathcal{K}}^p$ is the solution of the following optimization problem:*

$$min_{\vec{\omega} \in \mathbb{R}^m} \|A_{\mathcal{K}}\vec{\omega}\|_p \tag{1}$$

$$subject\ to: \quad \sum\nolimits_{i-1}^{m} \omega_i = 1; \vec{\omega} \geq \vec{0} \tag{2}$$

Proposition 2. *Minimal Violation Inconsistency Measure $\mathcal{IM}_{\mathcal{K}}^1$ is the solution of the following optimization problem:*

$$\min_{(\vec{\omega}, \vec{\lambda}) \in \mathbb{R}^{m+n}} \sum\nolimits_{i=1}^{n} \lambda_i \tag{3}$$

$$subject\ to: A_{\mathcal{K}}\vec{\omega} - \vec{\lambda} \leq \vec{0}; A_{\mathcal{K}}\vec{\omega} + \vec{\lambda} \geq \vec{0}; \sum\nolimits_{i=1}^{m} \omega_i = 1; \vec{\omega} \geq \vec{0}; \vec{\lambda} \geq \vec{0} \tag{4}$$

Proposition 3. *Minimal Violation Inconsistency Measure $\mathcal{IM}_{\mathcal{K}}^{\infty}$ is the solution of the following optimization problem:*

$$min_{(\vec{\omega}, \lambda) \in \mathbb{R}^{m+1}} \lambda \tag{5}$$

$$subject\ to: A_{\mathcal{K}}\vec{\omega} - \vec{1}\lambda \leq \vec{0}; A_{\mathcal{K}}\vec{\omega} + \vec{1}\lambda \geq \vec{0}; \sum\nolimits_{i=1}^{m} \omega_i = 1; \vec{\omega} \geq \vec{0}; \lambda \geq 0 \tag{6}$$

Proposition 4 *(p-norm Probability Vector of $\mathcal{K}$). p-norm probability vector of $\mathcal{K}$, noted $\vec{\omega}^p_{\mathcal{K}}$, corresponding to $\vec{\omega}^*$ of the solution of the following optimization problem:*

$$arg\ min_{\vec{\omega}\in\mathbb{R}^m} \left\|A_{\mathcal{K}}\vec{\omega}\right\|_p \tag{7}$$

subject to (2).

Proposition 5 *(1-norm Probability Vector of $\mathcal{K}$). 1-norm probability vector of $\mathcal{K}$, noted $\vec{\omega}^1_{\mathcal{K}}$, corresponding to $\vec{\omega}^*$ of the solution of the following optimization problem:*

$$arg\ \min_{(\vec{\omega},\vec{\lambda})\in\mathbb{R}^{m+n}} \sum_{i=1}^{n} \lambda_i \tag{8}$$

subject to (4).

Proposition 6 *(∞-norm Probability Vector of $\mathcal{K}$). ∞-norm probability vector of $\mathcal{K}$, noted $\vec{\omega}^\infty_{\mathcal{K}}$, corresponding to $\vec{\omega}^*$ of the solution of the following optimization problem:*

$$arg\ \min_{(\vec{\omega},\lambda)\in\mathbb{R}^{m+1}} \lambda \tag{9}$$

subject to (6).

Proposition 7 *(Violation Vector). Let $\mathcal{K}$ be a knowledge base. Violation vector of $\mathcal{K}$ with respect to p-norm ($p \geq 1$), noted $\vec{v}^p_{\mathcal{K}}$, defined as:*

$$\vec{v}^p_{\mathcal{K}} = A_{\mathcal{K}}\vec{\omega}^p_{\mathcal{K}}$$

3.2 The Unnormalized Inconsistency Measure

Definition 10. *$\overline{A}_{\mathcal{K}}$ is called a diagonal double matrix of $\mathcal{K}$ whenever it is defined as follows:*

$$\overline{A}_{\mathcal{K}} = \left(\bar{a}_{ij}\right) \in \mathbb{R}^{n\times 2n}$$

where

$$\bar{a}_{ij} = \begin{cases} 1 & \text{if } i = j \text{ and } i,j = 1,\ldots,n \\ -1 & \text{if } i = j \text{ and } i,j = n+1,\ldots,2n \\ 0 & \text{otherwise} \end{cases}$$

Let $\vec{\Delta}\left(\vec{\ell},\vec{\zeta}\right) = (\ell_1,\ldots,\ell_n,\zeta_1,\ldots,\zeta_n)^T$ be a column vector.

Definition 11. *Unnormalized inconsistency measure $\mathcal{IM}^u_{\mathcal{K}}$ is called an unnormalized measure of $\mathcal{K}$ whenever it is defined as follows:*

$$\mathcal{IM}_{\mathcal{K}}^{u} = \gamma$$

where γ is the solution of the following optimization problem:

$$\min_{(\vec{\omega},\vec{\Delta})\in\mathbb{R}^{m\times 2n}}\left(\sum\nolimits_{i=1}^{n}\ell_i + \sum\nolimits_{i=1}^{n}\zeta_i\right) \tag{10}$$

$$subject\ to : \overline{A}_{\mathcal{K}}\vec{\Delta} \leq \vec{1} - \vec{\rho}; \overline{A}_{\mathcal{K}}\vec{\Delta} \geq -\vec{\rho}; \sum\nolimits_{i=1}^{m}\omega_i = 1; \vec{\omega} \geq \vec{0}; \tag{11}$$

$$(\rho_i + \ell_i - \zeta_i)\sum\nolimits_{\Theta\in SM(F_iG_i)}\omega_i - \sum\nolimits_{\Theta\in SM(G_i)}\omega_i = 0, for\ i = 1,\ldots n \tag{12}$$

Proposition 8. *$\mathcal{IM}_{\mathcal{K}}^{u}$ is an inconsistency measure of probabilistic knowledge base $\mathcal{K}$.*

Proposition 9. *Unnormalized probability vector of $\mathcal{K}$, noted $\vec{\omega}_{\mathcal{K}}^{u}$, corresponding to $\vec{\omega}^*$ of the solution of the following optimization problem:*

$$\arg\min_{(\vec{\omega},\vec{\Delta})\in\mathbb{R}^{m\times 2n}}\sum\nolimits_{i=1}^{n}(\ell_i + \zeta_i) \tag{13}$$

subject to (11)–(12).

3.3 The Families of Restoring Operators

We second study the families of operators as well as their logical properties for restoring the inconsistency of a probabilistic knowledge base.

Definition 12 *(Logical Properties of Restoring Operator). Let $\mathcal{K}_1, \mathcal{K}_2, \mathcal{K} \in \mathbb{K}$. Function $\eta : \mathbb{K} \to \mathbb{K}$ is called restoring operator iff the following properties hold:*

(SUC) *$\forall\mathcal{K} \in \mathbb{K} : \eta(\mathcal{K}) \not\models \perp$*
It states that for all probabilistic knowledge bases then after restoring, we should have the result as consistent ones.

(SPR) *$\forall\mathcal{K} \in \mathbb{K} : \eta(\mathcal{K}) = \partial_{\mathcal{K}}\left(\vec{\beta}\right)$ for some $\beta \in \mathbb{R}_{[0,1]}^{|\mathcal{K}|}$*
The property SPR assures that the structure of constraints needs no modification.

(VAC) *If $\mathcal{K} \not\models \perp$ then $\eta(\mathcal{K}) = \mathcal{K}$*
*The property **VAC** requires that result of the restoring not modify a consistent knowledge base.*

(IRS) *If $\mathcal{K}_1 \doteq \mathcal{K}_2$ then $\eta(\mathcal{K}_1) \doteq \eta(\mathcal{K}_2)$*
This property states that if two probabilistic knowledge bases are semi-extensionally equivalent then the result of the two restoring will be semi-extensionally equivalent.

(NOD) *If $\kappa = (F|G)[\rho]$ and $G \not\equiv \top$ then $\exists\mathcal{K} : \kappa \in \mathcal{K}, \kappa \notin \eta(\mathcal{K})$*
The property NOD states that if a probabilistic constraint is non-tautological, it should belong to a probabilistic knowledge base but it does not belong to the result of this ones restoring.

(WIA) *If $SC(\mathcal{K}_1) \cap SC(\mathcal{K}_2) = \varnothing$ then $\eta(\mathcal{K}_1 \cup \mathcal{K}_2) \triangleq \eta(\mathcal{K}_1) \cup \eta(\mathcal{K}_2)$*

Demanding WIA implies that if the restoring of the union of two knowledge bases that does not have a constraint subset should be extensionally equivalent to the restoring of two disjoint knowledge bases.
(MIA) *If* $(\eta(\mathcal{K}_1) \cup \eta(\mathcal{K}_1)) \nvDash \perp$ *then* $\eta(\mathcal{K}_1) \cup \eta(\mathcal{K}_1) \triangleq \eta(\mathcal{K}_1 \cup \mathcal{K}_2)$
This property MIA demands that the restoring of two disjoint knowledge bases that is consistent should be extensionally equivalent to the restoring of of their union.

Definition 13 *(p-norm Restoring Operator). Let* $\mathcal{K} \in \mathbb{K}$*. Function* $\eta^p : \mathbb{K} \to \mathbb{K}$ *is called p-norm restoring operator (*$p \geq 1$*) whenever it is defined as follows:*

$$\eta^p(\mathcal{K}) = \partial_{\mathcal{K}}(\beta_1, \ldots, \beta_n)$$

where

$$\beta_i = \begin{cases} \overline{\mathcal{P}}(F_i|G_i) & \text{if } \overline{\mathcal{P}}(G_i) > 0 \\ \rho_i & \text{otherwise} \end{cases}$$

for i = 1,…,n

Proposition 11. *Vector* $\overrightarrow{\beta}$ *corresponding to* $\overrightarrow{x}^*$ *which is the solution of the following unconstrained optimization problem:*

$$\arg\min_{(\vec{x},y)\in\mathbb{R}^{n+1}} \left(\sum\nolimits_{j=1}^{m} (\vec{\alpha}_{\mathcal{K}}(\vec{x},y))_j - \overrightarrow{x}^T \vec{v}_{\mathcal{K}}^p - y \right) \tag{14}$$

Proposition 12. *The restoring operator* η^p *fulfills SPR, SUC, VAC, WIA, and MIA if* $p \geq 1$ *but only fulfills NOD when* $p > 1$*.*

Definition 14 *(Unnormalized Restoring Operator). Let* $\mathcal{K} \in \mathbb{K}$*. Function* $\eta^u : \mathbb{K} \to \mathbb{K}$ *is called restoring unnormalized operator whenever it is defined as follows:*

$$\eta^u(\mathcal{K}) = \partial_{\mathcal{K}}(\beta_1, \ldots, \beta_n)$$

where $\beta_i = \left| \rho_i + \ell_i^* - \zeta_i^* \right|$ *for* $\ell_i^*, \zeta_i^* \in \vec{\omega}_{\mathcal{K}}^u$ *and i = 1,…,n.*

Proposition 13. *The restoring operator* η^u *fulfills SPR, SUC, VAC, IRS, NOD, MIA, and WIA*

3.4 Example

A hospital makes a survey of drug test. It assigns two independent groups to this survey.

- The first group provides the results in which the probability that people test positively (denoted by T) is P(T) = 0.7; the probability that people are drug user (denoted by D) is P(D) = 0.4; and the probability that a person is a drug user, given that his test is positive, is P(D|T) = 0.8.

- The second group provides the results in which the probability that people test positively (denoted by T) is P(T) = 0.5; the probability that people are drug user is P(D) = 0.5; and the probability that a person test positive when that person who is a drug user is P(T|D) = 1.

According to the above results, we have the probabilistic knowledge bases as follows:

$$\mathcal{K}_1 = \langle (T)[0.7], (D)[0.4], (D|T)[0.8] \rangle; \mathcal{K}_2 = \langle (T)[0.5], (D)[0.5], (T|D)[1.0] \rangle.$$

By Definition 5, because $SM(\mathcal{K}_2) \neq \emptyset$, we have $\mathcal{K}_2 \vDash \top$, and then $\mathcal{IM}^1_{\mathcal{K}_2} = \mathcal{IM}^2_{\mathcal{K}_2} = \mathcal{IM}^{\infty}_{\mathcal{K}_2} = \mathcal{IM}^{u}_{\mathcal{K}_2} = 0$. While $SM(\mathcal{K}_1) = \emptyset$, thus we have $\mathcal{K}_1 \vDash \bot$, and then we will now compute measures, vectors of $\mathcal{K}_1$ as follows:

Step 1: Computing inconsistency measure:

- Finding characteristic matrix $A_{\mathcal{K}_1}$ and diagonal double matrix $\overline{A}_{\mathcal{K}_1}$

$$A_{\mathcal{K}_1} = \begin{pmatrix} 0.3 & 0.3 & -0.7 & -0.7 \\ 0.6 & -0.4 & 0.6 & -0.4 \\ 0.2 & -0.8 & 0 & 0 \end{pmatrix}, \overline{A}_{\mathcal{K}_1} = \begin{pmatrix} 1 & 0 & 0 & -1 & 0 & 0 \\ 0 & 1 & 0 & 0 & -1 & 0 \\ 0 & 0 & 1 & 0 & 0 & -1 \end{pmatrix}$$

- Computing $\mathcal{IM}^1_{\mathcal{K}_1}, \mathcal{IM}^2_{\mathcal{K}_1}, \mathcal{IM}^{\infty}_{\mathcal{K}_1}, \mathcal{IM}^{u}_{\mathcal{K}_1}$
  We have $\vec{\lambda} = (\lambda_1, \lambda_2, \lambda_3)^T, \vec{\omega} = (\omega_1, \omega_2, \omega_3, \omega_4)^T$, $\vec{\Delta} = (\ell_1, \ell_2, \ell_3, \zeta_1, \zeta_2, \zeta_3)$, $\vec{\rho} = (0.7, 0.4, 0.8)$
  - By Proposition 2, $\mathcal{IM}^1_{\mathcal{K}_1}$ is the solution of the following linear problem

$$\min\,(\lambda_1 + \lambda_2 + \lambda_3) \tag{15}$$

$$\text{subject to}: A_{\mathcal{K}_1}.\vec{\omega} - \vec{\lambda} \leq \vec{0}; A_{\mathcal{K}_1}.\vec{\omega} + \vec{\lambda} \geq \vec{0}; \sum\nolimits_{i=1}^{4} \omega_i = 1; \vec{\omega} \geq \vec{0}; \vec{\lambda} \geq \vec{0} \tag{16}$$

  - By Proposition 1, $\mathcal{IM}^2_{\mathcal{K}_1}$ is the solution of the following optimization problem

$$\min \left\| A_{\mathcal{K}}.\vec{\omega}_p \right\| \tag{17}$$

$$\text{subject to}: \sum\nolimits_{i=1}^{4} \omega_i = 1; \vec{\omega} \geq \vec{0} \tag{18}$$

  - By Proposition 3, $\mathcal{IM}^{\infty}_{\mathcal{K}_1}$ is the solution of the following linear problem

$$\min \lambda \tag{19}$$

$$\text{subject to}: A_{\mathcal{K}_1}\vec{\omega} - \vec{1}\lambda \leq \vec{0}; A_{\mathcal{K}_1}\vec{\omega} + \vec{1}\lambda \geq \vec{0}; \sum\nolimits_{i=1}^{4} \omega_i = 1; \vec{\omega} \geq \vec{0}; \lambda \geq 0 \tag{20}$$

  - By Proposition 8, $\mathcal{IM}^{u}_{\mathcal{K}_1}$ is the solution of the following optimization problem

$$\min(\ell_1+\ell_2+\ell_3+\zeta_1+\zeta_2+\zeta_3) \tag{21}$$

$$\text{subject to}: \overline{A}_{\mathcal{K}}\vec{\Delta} \leq \vec{1}-\vec{\rho}; \overline{A}_{\mathcal{K}}\vec{\Delta} \geq -\vec{\rho}; \sum_{i=1}^{4} \omega_i = 1; \vec{\omega} \geq \vec{0} \tag{22}$$

$$0.7+\ell_1-\zeta_1-\omega_1-\omega_3=0; 0.5+\ell_2-\zeta_2-\omega_1-\omega_2=0 \tag{23}$$

$$(0.8+\ell_3-\zeta_3)(\omega_1+\omega_2)=0 \tag{24}$$

Therefore, we have $\mathcal{IM}^1_{\mathcal{K}_1}=0.16, \mathcal{IM}^2_{\mathcal{K}_1}=0.14, \mathcal{IM}^\infty_{\mathcal{K}_1}=0.057$, and $\mathcal{IM}^u_{\mathcal{K}_1}=0.18$.

Step 2: Calculating new probability for each probabilistic constraint

- Finding violation vectors: By Propositions 5, 4, 6, and 9, $\vec{\omega}^1_{\mathcal{K}_1}, \vec{\omega}^2_{\mathcal{K}_1}, \vec{\omega}^\infty_{\mathcal{K}_1}$, and $\vec{\omega}^u_{\mathcal{K}_1}$ are the values of the argument of the solution of the optimization problems (15), (17), (19), and (22), respectively. Therefore, we have $\vec{\omega}^1_{\mathcal{K}_1}$ = (0.56, 0.14, 0, 0.3), $\vec{\omega}^2_{\mathcal{K}_1}$ = (0.46, 0.19, 0, 0.35), $\vec{\omega}^\infty_{\mathcal{K}_1}$ = (0.46, 0.19, 0, 0.36) and $\vec{\omega}^u_{\mathcal{K}_1}=(0,0.18,0,0)$.
  Finding probability vectors: By Proposition 7, $\vec{v}^1_{\mathcal{K}_1}=(0,0.16,0)$, $\vec{v}^2_{\mathcal{K}_1}=(-0.05,0.06,-0.06)$, and $\vec{v}^\infty_{\mathcal{K}_1}=(-0.057,0.057,-0.057)$.
- Finding new probability:
  - By Proposition 11, $\vec{\beta}$ corresponds to $\vec{x}^*$ which is the solution of the unconstrained optimization problems (15), (16), and (17) corresponding to p = 1, 2, and ∞:

$$\arg\min\left(\sum_{i=1}^{4} \alpha_i(\vec{x},y)-0.16x_3-y\right) \tag{24}$$

$$\arg\min\left(\sum_{i=1}^{4} \alpha_i(\vec{x},y)-0.05x_1+0.06x_2-0.06x_3-y\right) \tag{25}$$

$$\arg\min\left(\sum_{i=1}^{4} \alpha_i(\vec{x},y)-0.057x_1+0.057x_2-0.057x_3-y\right) \tag{26}$$

where

$$\alpha_1(\vec{x},y)=\exp(0.3x_1+0.6x_2+0.2x_3+y-1),$$

$$\alpha_2(\vec{x},y)=\exp(0.3x_1-0.4x_2-0.8x_3+y-1),$$

$$\alpha_3(\vec{x},y)=\exp(-0.7x_1+0.6x_2+y-1),$$

$$\alpha_4(\vec{x},y)=\exp(-0.7x_1-0.4x_2+y-1).$$

For p = 1, 2, and ∞, we have: $\vec{\beta}=(0.7,0.6,0.8), \vec{\beta}=(0.65,0.55,0.75)$, and $\vec{\beta}=(0.68,0.56,0.77)$, respectively.

– By Definition 14, we have $\vec{\beta} = (0.7, 0.58, 0.8)$.
We have employed Mathlab Software to compute $\mathcal{IM}^1_{\mathcal{K}_1}, \mathcal{IM}^2_{\mathcal{K}_1}, \mathcal{IM}^\infty_{\mathcal{K}_1}, \mathcal{IM}^u_{\mathcal{K}_1}$, $\vec{\omega}^1_{\mathcal{K}_1}$, $\vec{\omega}^2_{\mathcal{K}_1}$, $\vec{\omega}^\infty_{\mathcal{K}_1}$, and $\vec{\omega}^u_{\mathcal{K}_1}$.

Table 1 shows the new probability of knowledge bases $\mathcal{K}_1$ after using restoring operators $\eta^1, \eta^2, \eta^\infty$, and η^u.

Table 1. New probabilistic of $\mathcal{K}_1$

κ_i	η^1	η^2	η^∞	η^u
(T)[0.7]	0.7	0.65	0.68	0.7
(D)[0.4]	0.6	0.55	0.56	0.58
(D|T)[0.8]	0.8	0.75	0.77	0.8

4 Conclusion

In this paper, we have investigated two inconsistency measures for probabilistic-logical framework and adapted them to the probabilistic framework. We also introduced problems for computing such inconsistency measures and new probability of each probabilistic constraint in original base. In proportion to each inconsistency measure, we proposed a family of restoring operators as well as the assessment of desirable properties for altering probability to obtain a consistent one. However, two such operators are only applicable to a probabilistic knowledge base. Therefore, in the future we will go on investigating to apply these operators to merge a set of inconsistent probabilistic knowledge bases into a consistent one.

Acknowledgment. The authors would like to thank Professor Quang Thuy Ha, Faculty of Information Technology, Hanoi University of Engineering and Technology, Vietnam and Professor Ngoc Thanh Nguyen, Faculty of Computer Science and Management, Wroclaw University of Science and Technology, Poland for their expertise support.

References

1. Nguyen, N.T.: Advanced Methods for Inconsistent Knowledge Management. Springer, London (2008). https://doi.org/10.1007/978-1-84628-889-0
2. Finthammer, M., Kern-Isberner, G., Ritterskamp, M.: Resolving inconsistencies in probabilistic knowledge bases. In: Hertzberg, J., Beetz, M., Englert, R. (eds.) KI 2007. LNCS (LNAI), vol. 4667, pp. 114–128. Springer, Heidelberg (2007). https://doi.org/10.1007/978-3-540-74565-5_11
3. Bona, G.D.: Measuring inconsistency in probabilistic knowledge bases. Ph.D. thesis, University of Sao Paulo (2016)
4. Kern-Isberner, G. (ed.): Conditionals in Nonmonotonic Reasoning and Belief Revision. LNCS (LNAI), vol. 2087. Springer, Heidelberg (2001). https://doi.org/10.1007/3-540-44600-1

5. Muiño, D.: Measuring and repairing inconsistency in probabilistic knowledge bases. Int. J. Approx. Reason. **52**(6), 828–840 (2011)
6. Potyka, N.: Linear programs for measuring inconsistency in probabilistic logics. In: Proceedings of the Fourteenth International Conference on Principles of Knowledge Representation and Reasoning, KR 2014, pp. 568–577. AAAI Press ©2014 (2014)
7. Thimm, M.: Measuring inconsistency in probabilistic knowledge bases. In: Uncertainty in Artificial Intelligence (UAI 2009), pp. 530–537. AUAI Press (2017)
8. Thimm, M.: Inconsistency measures for probabilistic logics. Artif. Intell. **197**, 1–24 (2013)
9. Nguyen, V.T., Tran, T.H.: Inconsistency measures for probabilistic knowledge bases. In: Proceedings of KSE 2017, pp. 156–161. IEEE Xplore (2017). https://doi.org/10.1109/kse.2017.8119450
10. Hansson, S.O.: A Textbook of Belief Dynamics. Springer, Dordrecht (1999). https://doi.org/10.1007/978-94-007-0814-3. pp. 5, 3, 11, 35, 51-55, 69, 107
11. Hunter, A., Konieczny, S.: Measuring inconsistency through minimal inconsistent sets. In: Proceedings of Principles of Knowledge Representation and Reasoning (KR 2008), pp. 358–366. AAAI Press (2008)
12. Marques-Silva, J.: Minimal unsatisfiability: models, algorithms and applications. In: Multiple-Valued Logic (ISMVL), pp. 9–14. IEEE (2010)
13. Liffiton, M.H., Sakallah, K.A.: On finding all minimally unsatisfiable subformulas. In: Bacchus, F., Walsh, T. (eds.) SAT 2005. LNCS, vol. 3569, pp. 173–186. Springer, Heidelberg (2005). https://doi.org/10.1007/11499107_13
14. Klinov, P.: Practical reasoning in probabilistic description logic. Ph.D. thesis, The University of Manchester, Manchester, UK (2011)
15. Potyka, N.: Solving reasoning problems for probabilistic conditional logics with consistent and inconsistent information. Ph.D. thesis, FernUniversitat, Hagen (2016)
16. Potyka, N., Thimm, M.: Consolidation of probabilistic knowledge bases by inconsistency minimization. In: Proceedings of ECAI 2014, pp. 729–734. IOS Press (2014)
17. Potyka, N., Thimm, M.: Probabilistic reasoning with inconsistent beliefs using inconsistency measures. In: International Joint Conference on Artificial Intelligence 2015 (IJCAI 2015), pp. 3156–3163. AAAI Press ©2015 (2015)

The Assessing of Influence of Collective Intelligence on the Final Consensus Quality

Adrianna Kozierkiewicz(✉), Van Du Nguyen, and Marcin Pietranik

Wrocław University of Science and Technology, Wybrzeże Wyspiańskiego 27, 50-370 Wrocław, Poland
{adrianna.kozierkiewicz,van.du.nguyen, marcin.pietranik}@pwr.edu.pl

Abstract. Knowledge integration is a task of providing a unified, single version of knowledge through a process of joining several, independent knowledge bases. It involves not only providing a summary of available information, but also resolving any potential inconsistencies. Our previous research showed that in the context of the increasing amount of heterogeneous data, the approach to the aforementioned integration that decomposes it into smaller subtasks which outcomes are eventually combined in order to achieve the expected result, has been proved useful. This approach is called a multi-level knowledge integration. The biggest problem that we encountered was choosing a criterion on which an initial decomposition should be based on and preliminary we have investigated measuring the inner diversity of created groups. In this paper, we extend our ideas to reflect the intelligence degree of a collective and how it can impact the quality of the final integration. We use our previously developed function that measures such intelligence and perform a series of experiments to determine the impact it can have on the process of multi-level knowledge integration.

Keywords: Intelligent collective · Consensus · Multi-level consensus

1 Introduction

Nowadays, in the context of social media and a high diversity of information sources, we cannot expect that making a decision about some issue will be based solely on an opinion of a one expert. Even a task as simple as checking a weather for the next day, involves analysing several different forecasting systems that can differ in their predictions. The same applies when analysing opinions originating from a group of people e.g. during elections.

Members of such group (hereafter referred to as the collective) can have a plethora of different opinions on a given subject. Every opinion can be equally good or wrong (because we assume that members of a group are actual experts and not random people stating their judgments on an unfamiliar topic) and the final opinion of the collective is a result of the integration of opinions of its members. It can be understood as a method of providing a unified, single version of knowledge through a process of joining several, independent opinions.

N. T. Nguyen et al. (Eds.): ACIIDS 2018, LNAI 10751, pp. 15–24, 2018.
https://doi.org/10.1007/978-3-319-75417-8_2

Performing such integration entails two problems. The first one concerns the necessity of evaluating the obtained consensus. In the literature a wide array of functions that can be indicated the quality of the integrations result can be found. These functions are based on comparing the resulting collective's consensus with opinions of its members. The second major problem may appear as a consequence of a cardinality of the collective. The higher the number of experts the more difficult it may be to perform the integration of their opinions (especially with a complex formal structure used to express such opinions).

A potential remedy can be based on designating a final consensus in stages. In such approach, the collective is initially divided into a several, smaller groups, the integration of opinions of experts from these groups is performed. Eventually, a desired result is the integration of the outcomes of the partial integrations. The biggest difficulty is the initial decomposition of the collective into subgroups.

In our previous research [5], we have proved that grouping the collective of experts into groups that are characterized by high inner diversity gives the best results. In other words – if a group contains as much disagreeing experts as possible the better the outcome and eventually the better the final integration. It is consistent with an intuition – the higher the diversity of every group the better they represent the whole population from which they have been extracted.

In this paper, we show results of a set of experiments that we performed to investigate another grouping criterion. It is based on a function that can measure the intelligence of a collective (that in this context can be interpreted both as a group of experts and any of their subgroups). This function was presented in [10] and is used to accurately reflect the intelligence of a collective. Such intelligence is understood twofold: *(i)* as the degree to which a final consensus of experts' opinions is different to the reality and *(ii)* the degree to which members of the collective were wrong. This approach also takes into a consideration a cardinality of the collective, eventually becoming a very powerful method of measuring the quality of experts in any given collective and any of its sub-collectives.

The paper is organized as follows. Section 2 contains an overview of related works and the research that has been done in the considered field. Section 3 can be treated as an introduction to consensus theory and contains basic notions used throughout the rest of the paper. In Sect. 4 we briefly overview a multi-level approach to consensus determination. Section 5 is a broad description of conducted experiments and the statistical analysis of gathered results. The last section serves as a summary and a brief description of our upcoming research ideas.

2 Related Work

In the age of Big Data and Internet of Thing more and more data gets produced [14]. Using MapReduce, a cloud-based technology, the data processing task is split into smaller tasks that can be run in parallel across several nodes in the cluster. Its effectiveness has been revealed in a wide range research problems such as large-scale machine learning problems, clustering problems, etc. [2]. Recently, many MapReduce architectures have been developed for handling with challenges of Big Data. For instance, Hadoop was inspired by Google's MapReduce for an analysis of large datasets [13] which uses a distributed user-level filesystem to manage the storage of resources across the cluster [7]. Meanwhile, Apache Spark is an open source cluster computing framework. Spark has several advantages compared to other big data and MapReduce technologies like Hadoop and Storm in terms of supporting many data types, processing time [1].

As mentioned earlier, research problems on the multi-level integration have not yet been widely investigated in the literature. Almost all of the consensus-based knowledge integration algorithms are often based on a one-level approach [9]. However, in the age of Big Data, such a naive solution may entail limitations related to a data processing time and a memory usage. In this case, an input is often very large and the multi-level approach can be a useful approach for such consensus determination [8]. The general idea of the multi-level approach is based on dividing a large knowledge profile into smaller ones (clustered profiles). Then the process of consensus determination is applied to each of extracted profiles to determine their representatives. After that, these representatives will be treated as elements of a new profile. These steps are repeated until a predefined number of levels is reached.

The preliminary research problems on multi-level integration have been presented in [3, 4]. In these papers, authors have investigated the problems of a one-level and a two-level consensus determination, which is one of methods of integration. Therefore, these two terms (consensus determination and integration methods) will be used interchangeably. A formal framework for designating the consensus in the one-level and the two-level approaches for the assumed macro- and micro-structures has been developed. The results given by the one-level and the two-level algorithms are worse by 1% and 5% respectively in comparison to the optimal solution. In [11, 12] authors have investigated on the problem of two-level integration. In [11] k-means algorithm is used for clustering a large knowledge profile into smaller ones. Then the process of consensus determination is used to determine the representatives of clustered collectives. In the second level of consensus determination, the representative of each clustered profile has been assigned a weight value depending on the number of members in the corresponding profile. The experimental results have shown that the weighted approach is helpful in reducing the difference between the results of the two-level and the one-level of consensus determinations in comparison with the non-weighted approach. Later, in [12], authors have presented an improvement of the two-level consensus choice by taking into account the susceptibility to consensus. By means of experimental analysis, the proposed method is proved useful.

In [5], the problem of the multi-level integration has been investigated by considering Fleiss' kappa measure. The main concern of the paper is to determine the impact that the internal consistency of the classified profiles on the quality of the final integration. The experimental results have indicated that to achieve a higher quality of the final integration the grouping approach should be based on the lowest value of Fleiss' kappa measure. From this finding, it can be concluded that the diverse collective has a positive impact on the quality of the final integration.

Furthermore, the multi-level idea has also been applied in the ontology integration task [6]. By means of analytical analysis, the results have revealed that there is no difference between the results of the one-level and the multi-level integration processes.

Even the time of processing data in case of using the multi-level approach is decreased by 20% in comparison to the one-level approach. It can be concluded that this finding is especially important in the emerging context of Big Data.

In addition to these research problems, this paper presents the results of applying the intelligence function proposed in [10] used as a grouping criterion in the problem of the multi-level knowledge integration.

3 Basic Notions

Let U be a set of objects representing the potential elements of knowledge referring to a concrete subject in the real world. Let 2^U be the powerset of set U that is the set of all subsets of U. Then the set of all k-element subsets (with repetitions) of set U is $\prod_k(U)$ (for $k \in N$), and let $\prod(U) = \bigcup_{k=1}^{\infty} \prod_k(U)$ be the set of all non-empty finite subsets with repetitions of set U. A set $X \in \prod(U)$ is called a knowledge profile involving the knowledge states given by members on the same subject in the real world. Elements of U have two structures (i.e. macrostructure and microstructure). The microstructure is considered as the representations of elements in the set U such as: linear orders, n-tree, tuples, etc. The macrostructure is understood as relationship between elements and often defined as a distance function with a signature $\delta : U \times U \rightarrow [0, 1]$. In this paper, the function δ only satisfies a part of the metric conditions (without transitive condition), called a *half-metric*. In [9] many consensus-based postulates have been proposed for a consensus determination. However, there exist two most popular ones are "*1-Optimality*" and "*2-Optimality*" (or O_1 and O_2 for short).

Definition 1. *Let* (U, δ) *be a distance space, a consensus of a profile* $X \in \prod(U)$ *is determined based on:*

- *criterion* O_1 *if:* $\delta(x^*, X) = \min_{y \in U} \delta(y, X)$
- *criterion* O_2 *if:* $\delta^2(x^*, X) = \min_{y \in U} \delta^2(y, X)$

where x^* represents the consensus of profile X, $\delta(x^*, X)$ represents the sum of distances from x^* to the knowledge states of collective X. Similarly, $\delta^2(x^*, X)$ is the sum of squared distances from x^* to knowledge states of the collective X.

Definition 2. *Let* (U, δ) *be a distance space, the intelligence degree of a profile X is described by a function Int as follows:*

$$Int : \prod(U) \rightarrow [0,\ 1]$$

As mentioned in [10], the intelligence function should satisfy the following criteria:

1. $\forall X \in \prod(U)$:
   (a) $\forall x_i \in X$: If $(r = x_i) \wedge (r = x^*)$, then $Int(X) = 1$.
   (b) $\forall x_i \in X$: If $(\delta(r, x_i) = 1) \wedge (\delta(r, x^*) = 1)$, then $Int(X) = 0$.
2. $\forall X, Y \in \prod(U)$:

If $(\delta(r,x^*) = \delta(r,y^*)) \wedge \left(\frac{\delta(r,X)}{card(X)} \leq \frac{\delta(r,Y)}{card(Y)}\right)$, then $Int(X) \geq Int(Y)$.
where x^*, y^* represent the consensuses of profiles X and Y respectively, r represents the real state.

Definition 3. The *function Int of a collective X satisfying the above criteria has the following definition:*

$$Int(X) = 1 - \left((1-\alpha) \times \frac{\delta(r,X)}{card(X)} + \alpha \times \delta(r,x^*)\right)$$

where α presents the intelligent coefficient of the group's consensus ($\alpha \in [0,1]$).

Definition 4. *For a given profile $X \in \prod(U)$, the quality 1 of a consensus x^* is defined as follows:*

$$Q_1^i(x,X) = 1 - \frac{\delta^i(x^*,X)}{card(X)}$$

where $i \in \{1,2\}$.

Definition 5. *For a given profile $X \in \prod(U)$, the quality 2 of a consensus x^* is defined as follows:*

$$Q_2^i(x,r) = 1 - d^i(r,x^*)$$

where $i \in \{1,2\}$ and r represents the real state.

The quality 1 measures how the final consensus (final experts' decision) differs from the elements of profile. The quality 2 measures how the final consensus differs from the real value.

4 Multi-level Consensus Determination Method

All systems for a decision support require knowledge databases which are further processing or integrating. In many cases, we need to store and process a big set of data what is a very difficult or even impossible to do in one step. In this work, we assumed that to make a decision regarding some problem we collect experts' opinions from many sources and based on collected data we make a final decision. In this paper, experts' opinions, will be called a knowledge of a collective and the designated decision a consensus. As it was said in the previous Section, these problems are equivalent to the data integration problem and their names can be used interchangeably.

The general idea of the multi-level consensus determination method is based on an initial division of the sequence of n experts' opinions into k classes. For each class, these opinions are integrated using an ordinary integration algorithm (see Algorithm 1). The final consensus is designated as a consensus of these partial consensuses. The procedure can be repeated many times where outputs obtained in the previous stage serve as inputs of subsequent steps. The elements of the multi-level consensus determination approach are described in more details below.

4.1 The Formal Representation of Experts' Opinions

We assume that the profile X contains the experts' opinions stored as binary vectors of the length equal to N. The cardinality of profile X is equal to n. Thus, the microstructure of set U is defined as: $U = \{u_1, u_2, \ldots.\}$ where elements of the universe are binary vectors. By $\delta(x, y) = \sum_{j=1}^{N} |x^j - y^j|$ we denote a macrostructure of set U, where $x = (x^1, x^2, \ldots, x^N), y = (y^1, y^2, \ldots, y^N), x^i, y^i \in \{0, 1\}, i \in \{1, \ldots, N\}$. Therefore, the profile is formally presented as: $X = \{x_1, x_2, \ldots, x_n\} \in \prod(U)$, where $x_i = (x_{i1}, x_{i2}, \ldots, x_{iN}), i \in \{1, \ldots, n\}$.

The assumed distance space (U, δ) can model many objects, situations, tasks taken from any universe of discourse. We can interpret it as the problem considered by the group of workers, experts, etc. where each member of the collective is asked about their opinion concerning N different objects or answering N different questions. Each member of the collective can choose from two possible options(i.e. *yes* or *no)* because the profile consists of binary vectors. The consensus determined based on the collected data is the final opinion of the group of experts.

4.2 Criterion of Classification of Experts' Opinions

In our approach, we assume that the profile X is primarily divided into k classes. The criterion of classification is the *Int*(X) function presented in the previous Section. For simplicity, the classes are created using a brute force method.

4.3 Consensus Determination Method

As it was mentioned in the previous Section, a consensus can satisfy the *1-* or *2-optimality* postulate. For the assumed distance space (U, δ) the consensus determination method is based computing the number of occurrences of *zeros* and *ones* in the experts' opinions. The final decision is constructed from the most popular opinions (most frequently chosen). As easily seen, the method satisfying *1-optimality* consensus is based on a democracy and is very straightforward. Therefore, only this algorithm (presented below) will be considered in this paper.

5 The Results of Experiment

The described approach to the multi-level consensus determination was implemented and tested in a special dedicated environment. The main aim of our research is answer to the question: *how the intelligence of a collective as the classification criterion*

Algorithm 1 Consensus determination method

Input: $X=\{x_1, x_2, \ldots, x_n\}$
Output: consensus x^*

```
for j =1 to N do
    f_j=0;
    for i=1 to n do
        if x_ij=1 then
            f_j++;
        end if
    end for
    if f_j ≥ n/2 then
        x_j*=1;
    else
        x_j*=0;
    end if
end for
```

influences the quality of the final consensus determined using the multi-level method? For this purpose, some experimental assumptions have been done. Firstly, we have assumed that the group of experts could be mistaken maximum 5% to the real state. Additionally, the number of experts $n = 9$ and the number of class $k = 3$ and the number of objects $N = 10$ have been chosen. The odd number of the cardinality of a profile consisting of binary vectors ensure that determined consensus will be reliable [9]. The experiments have been done for a uniform distribution of zeros and ones when the proportion of zeros to ones have been equal 5:5, 6:4, 7:3, 8:2, 9:1, respectively. All calculations have been repeated 100 times.

5.1 The Influence of Grouping Strategy

In the first step the quality of the final consensus has been compared in relations with the distribution of the input experts' opinions and the methods of grouping experts in the first level. The results of conducted experiments have been graphically presented in Fig. 1.

The obtained results have been statistically analysed on the significance level $\alpha = 0.05$. The collected data have been divided into 10 groups composed of three samples each: *Max Collective Intelligence, Min Collective Intelligence, Random*. For each group of data, analysis has been made separately. Before selecting a proper test, we have analysed the distribution of all obtained data by using the Lilliefors' test. At least one sample from each group does not come from the normal distribution (*p-values* have been smaller than 0.00001), therefore for the further analysis the Kruskal-Wallis test has been used. The *p-value* for each ten group is less than 0.00001 then we reject the null hypothesis stating that the criterion of grouping the experts' opinion on the first level has no influence on the quality of the final consensus. The analysis of the means of the ranks allowed to decide that the worst grouping strategy of input experts' opinions is the one based on the minimal value of collective intelligence. The obtained

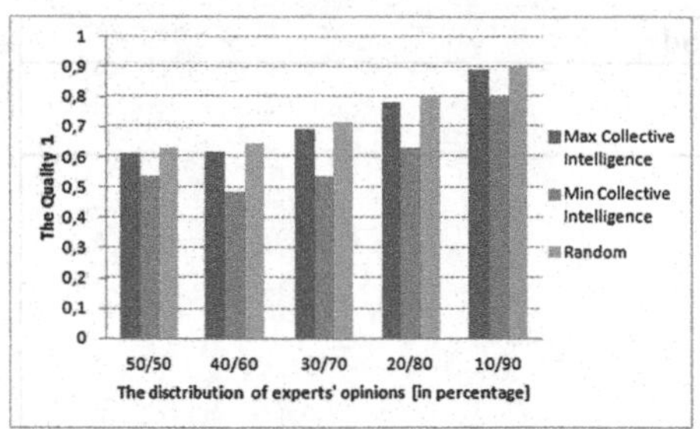

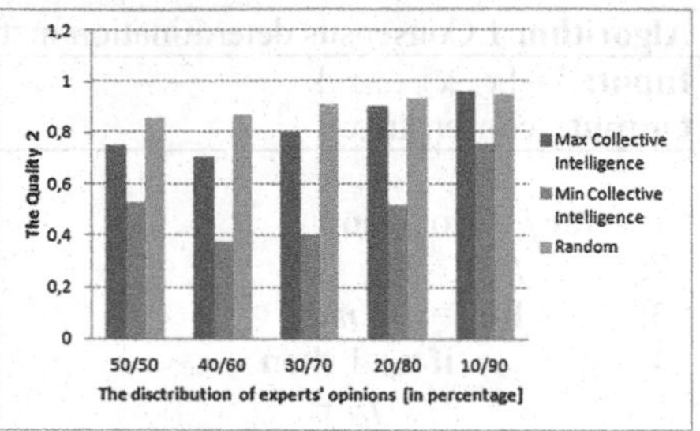

Fig. 1. The influence of grouping strategy

results are presented in Table 1. The post-hoc analysis demonstrated also the statistical difference between *Random* and *Max Collective Intelligence* samples but only in cases where the proportion of zeros to ones are 50%/50%, 60%/40% and 70%/30%. Such results allow to draw a general conclusion that to improve the quality of the final consensus we should consider grouping strategy on the first level based on the collective intelligence function.

5.2 The Influence of Grouping Strategy in Case of Different Group of Experts

In Sect. 5.1 we have assumed that the group of experts could be mistaken by maximum 5% to the real state. In this part of our research we would like to verify how the fallibility of group of experts influences on the final consensus quality. Figure 2 presents obtained results.

Table 1. The mean of the ranks for samples and two quality measures

Distribution	Quality 1			Quality 2		
	Max collective intelligence	Min collective intelligence	Random	Max collective intelligence	Min collective intelligence	Random
50%/50%	171.32	71.86	208.32	161.715	73.09	216.695
60%/40%	178.31	56	217.19	160.77	58.665	232.075
70%/30%	183.15	55.23	213.255	174.52	52.675	224.3
80%/20%	187.14	57.505	206.85	190.29	52.225	208.98
90%/10%	179.69	82.3	189.51	194.74	66.53	190.23

It is obvious, that if experts are more wrong it has no influence on the value of the quality 1. However, the analysis of the graph presented in Fig. 2 (the right sight), allow us to suppose that the probability of experts' mistake in comparison to the real value has a big influence for the value of the quality 2 and for a different grouping strategy. To prove a mentioned hypothesis some analysis has been done. For deeper analysis, we split data into two classes: first containing 6 samples concerning with the *Max Collective Intelligence* and second with 6 samples relative with the *Random* criterion.

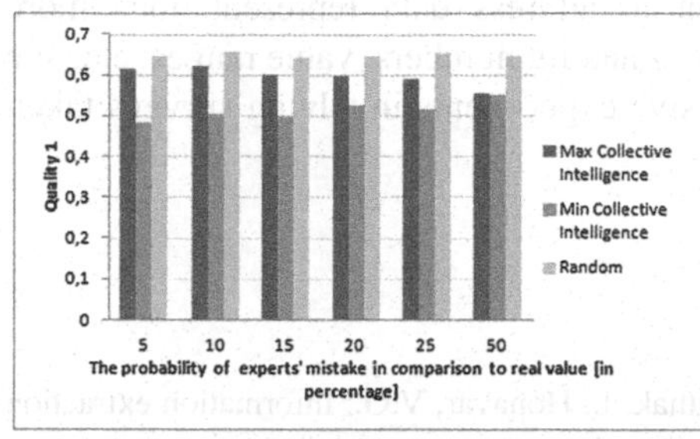

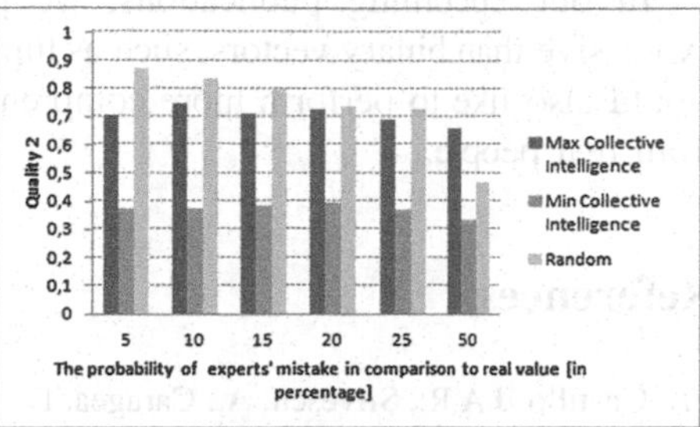

Fig. 2. The influence of grouping strategy in case of different group of experts

In the first step we checked the normal distribution of tested samples. At least one sample in each class does not come from the normal distribution (p-values have been smaller than 0.00001), therefore for the further analysis the Kruskal-Wallis test has been used. We obtain the value of statistical test equal 20.345629, *p-value* 0.0011 for *Max Collective Intelligence* samples and 263.28 and *p-value* less than 0.00001 for *Random* samples, respectively. The results of post-hoc Dunn tests pointed out that for *Max Collective Intelligence* samples only two pair of samples differ statistically: 10% with 50% and 20% with 50%. However, for *Random* samples, all pairs of samples differ significantly. The analysis of the mean of the ranks for *Random* samples presented in Table 2 suggests that if the fallibility of group experts increase then the quality 2 of the final consensus decrease.

Table 2. The mean of the ranks for particular samples and different probability of experts' mistake in comparison to real value

	5%	10%	15%	20%	25%	50%
Max collective	301.235	342.79	313.415	321.81	276.86	246.89
Random	430.72	389.165	348.805	282.63	270.92	80.76

6 Conclusions and Future Works

This paper is devoted to applying a group intelligence function to the multi-level consensus determination method as the criterion of the initial classification of the experts' opinions. The performed experiments and the statistical analysis demonstrated that the choice of grouping strategy on the first level has a statistical influence on the quality of the final consensus. Our research demonstrated that the worst grouping strategy is based on the minimum value of the collective intelligence. Such results are consistent with an intuition stating that experts with the smallest intelligence determine the worst final decisions. The better quality of the final consensus is possible to obtain if on the first level experts' opinions are grouped based on the maximum value of the group intelligence function or simply in a random way, where the intelligence of a collective is diverse. However, this conclusion is valid, only if experts from the group are correct. In a situation, where the group of experts is less reliable, to ensure the highest quality of the final consensus, a classification criterion based on the maximum value of a group intelligence should be used.

In our upcoming publications, we plan to address data representations more expressive than binary vectors, such as tuples of natural numbers, value ranges, etc. We would also like to perform more comprehensive experiments involving surveys taken from real people.

References

1. Castillo, J.A.R., Silvescu, A., Caragea, D., Pathak, J., Honavar, V.G.: Information extraction and integration from heterogeneous, distributed, autonomous information sources-a federated ontology-driven query-centric approach. In: Proceedings of Fifth IEEE Workshop on Mobile Computing Systems and Applications, pp. 183–191 (2003)
2. Dean, J., Ghemawat, S.: MapReduce: simplified data processing on large clusters. Commun. ACM **51**(1), 107–113 (2004)
3. Kozierkiewicz-Hetmańska, A.: Comparison of one-level and two-level consensuses satisfying the 2-optimality criterion. In: Nguyen, N.-T., Hoang, K., Jędrzejowicz, P. (eds.) ICCCI 2012. LNCS (LNAI), vol. 7653, pp. 1–10. Springer, Heidelberg (2012). https://doi.org/10.1007/978-3-642-34630-9_1
4. Kozierkiewicz-Hetmanska, A., Nguyen, N.T.: A comparison analysis of consensus determining using one and two-level methods. In: Proceedings of KES 2012, pp. 159–168 (2012)
5. Kozierkiewicz-Hetmanska, A., Pietranik, M.: Assessing the quality of a consensus determined using a multi-level approach. In: Proceedings of IEEE International Conference on INISTA 2017, pp. 131–136. IEEE (2017)
6. Kozierkiewicz-Hetmanska, A., Pietranik, M.: The knowledge increase estimation framework for ontology integration on the concept level. J. Intell. Fuzzy Syst. **32**, 1161–1172 (2017)
7. Maitrey, S., Jha, C.K.: MapReduce: simplified data analysis of big data. Procedia Comput. Sci. **57**, 563–571 (2015)
8. Maleszka, M., Nguyen, N.T.: Integration computing and collective intelligence. Expert Syst. Appl. **42**, 332–340 (2015)
9. Nguyen, N.T.: Advanced Methods for Inconsistent Knowledge Management. Springer, London (2008). https://doi.org/10.1007/978-1-84628-889-0
10. Nguyen, V.D., Merayo, Mercedes G., Nguyen, N.T.: Intelligent collective: the role of diversity and collective cardinality. In: Nguyen, N.T., Papadopoulos, G.A., Jędrzejowicz, P., Trawiński, B., Vossen, G. (eds.) ICCCI 2017. LNCS (LNAI), vol. 10448, pp. 83–92. Springer, Cham (2017). https://doi.org/10.1007/978-3-319-67074-4_9
11. Nguyen, V.D., Nguyen, N.T.: A two-stage consensus-based approach for determining collective knowledge. In: Le Thi, H.A., Nguyen, N.T., Do, T.V. (eds.) Advanced Computational Methods for Knowledge Engineering. AISC, vol. 358, pp. 301–310. Springer, Cham (2015). https://doi.org/10.1007/978-3-319-17996-4_27
12. Nguyen, V.D., Nguyen, N.T., Hwang, D.: An improvement of the two-stage consensus-based approach for determining the knowledge of a collective. In: Nguyen, N.-T., Manolopoulos, Y., Iliadis, L., Trawiński, B. (eds.) ICCCI 2016. LNCS (LNAI), vol. 9875, pp. 108–118. Springer, Cham (2016). https://doi.org/10.1007/978-3-319-45243-2_10
13. Sugha, P., Gunavathi, R.: A survey paper on map reduce in big data. Int. J. Sci. Res. **5**, 1103–1107 (2016)
14. Vossen, G.: Big data as the new enabler in business and other intelligence. Vietnam J. Comput. Sci. **1**(1), 1–12 (2013)

Design Pattern Ranking Based on the Design Pattern Intent Ontology

Channa Bou, Nasith Laosen, and Ekawit Nantajeewarawat(✉)

School of ICT, Sirindhorn International Institute of Technology, Thammasat University, Pathumthani, Thailand
bou.channa93@gmail.com, nasith@gmail.com, ekawit@siit.tu.ac.th

Abstract. Selecting an appropriate design pattern for a given design problem is often difficult. We propose an automatic approach for recommending and ranking design patterns to facilitate design pattern selection. A similarity score is calculated between an input design problem and the problem types addressed by each design pattern, which are represented in Kampffmeyer's design pattern intent ontology. Design patterns are then ranked according to the obtained similarity scores. Experiments were conducted to evaluate the proposed approach with 24 input problem descriptions. With appropriate parameter settings, the actual answers to 70.83% of the input problems are included by the top five recommended patterns. By incorporation of additional knowledge sources for improving similarity calculation, the actual answers to 91.67% of the input problems can be recommended within the top-five ranks.

Keywords: Design pattern · Design pattern recommendation
Cosine similarity · Design pattern intent ontology

1 Introduction

A design pattern provides a general design solution to a commonly occurring problem in software design [3]. Gang-of-Four (GoF) patterns are the most popular and most widely used design patterns for object-oriented software design. An approach for formalizing GoF design patterns was introduced in [5] and further elaborated in [6], in which design patterns are represented in an ontology, called the design pattern intent ontology (DPIO). The DPIO representation focuses on types of problems that are addressed by design patterns, where a problem type is characterized by pairs of constraints and concepts. A constraint and a concept together describe an action performing some specific task, e.g., 'select algorithm' and 'handle state'.

A design pattern wizard tool was developed in [6] to facilitate design pattern selection using the formalized DPIO. A user is asked to select one or more pairs of constraints and concepts. A query is then constructed to retrieve design patterns that are solution to the problem types characterized by the selected

N. T. Nguyen et al. (Eds.): ACIIDS 2018, LNAI 10751, pp. 25–35, 2018.
https://doi.org/10.1007/978-3-319-75417-8_3

constraint-concept pairs. Selecting appropriate constraints and concepts is however a nontrivial task. There are many constraints and concepts in the DPIO and the user might not understand their meanings clearly. When only a few constraint-concept pairs are selected, too many design patterns may be retrieved since the selected pairs may characterize many design problems, each of which may be solved by many design patterns. When many pairs are selected, there may be no problem type characterized by the conjunction of all the selected pairs and, as a result, no pattern may be retrieved.

To assist a system analyst in designing object-oriented software, we propose an automatic approach for selecting and ranking design patterns based on the DPIO [6]. A vector representing a design pattern is constructed based on the problem types solved by the pattern. When an input design problem from a user is given, an input problem vector is created based on matching key terms extracted from the input problem to constraints and concepts characterizing problem types. Similarity scores between the input problem vector and design pattern vectors are computed. The design patterns are then ranked and recommended based on the computed scores. Experiments were conducted to evaluate the proposed framework. By appropriate parameter settings and incorporation of some additional knowledge sources for enhancement of key term matching, the obtained experimental results are promising. We restrict our attention in this work to the 13 GoF design patterns in the category "performing a domain-specific task other than object creation", which is the largest and most complicated category in the pattern usage hierarchy described in [12].

The paper is organized as follows: Sect. 2 reviews related works on design pattern recommendation. Section 3 describes the proposed framework. Section 4 presents the experiments and results. Section 5 provides conclusions.

2 Related Works

Hasheminejad and Jalili [4] presented a two-phase selection method to assist software developers with design pattern selection. Each design pattern is represented by a vector of words appearing in the descriptions of the pattern taken from design pattern textbooks. In the first phase, a binary classification model is constructed for each category of patterns, e.g., creational, structural, and behavioural GoF patterns. The obtained classifiers are used for predicting the category of a given input design problem, which is encoded as a vector of words occurring in the textual description of the problem. In the second phase, the word vector representing the problem is compared with the word vectors representing design patterns in the predicted category using cosine similarity. Design patterns are recommended based on the resulting similarity scores. An evaluation model was introduced in [4] to select the best learning technique for pattern category classification in the first phase. Evaluation of pattern recommendation in the second phase was however not presented. In contrast to our work, where a design pattern is represented in a conceptual level in terms of the problem types addressed by it, occurrences of words representing a pattern in [4] are low-level features that may not clearly express its true characteristics.

Bouassida et al. [1] proposed an interactive toolset for recommending an appropriate design pattern. Semantic correspondences were determined between element names in an input class diagram and the participant names of design patterns provided by the GoF book [3]. Based on the obtained correspondences, predetermined recommendation rules were used for finding and instantiating a suitable design pattern. A user may interact with the rules to provide additional necessary information for choosing an appropriate pattern. No experimental evaluation was reported in [1].

Palma et al. [11] presented a design pattern recommender using a Goal-Question-Metric (GQM) approach. Knowledge of design patterns, e.g., intents and applicabilities described in the GoF book [3], was transformed into textual conditions and sub-conditions, which were subsequently formulated as questions. A user was asked to answer the formulated questions with three types of answers, i.e., 'yes', 'no', 'do not know', and a total metric weight of the obtained answers for each design pattern is calculated. The design pattern with the highest total weight is recommended. Eight evaluators, i.e., six undergraduate students and two IT professionals, were asked to simulate and evaluate the GQM model with 11 input questions, with the accuracy of 50% being reported.

3 Methodology

Figure 1 shows an overview of the proposed framework. As a preparation process, design pattern vectors (DPVs), representing types of problems solved by design patterns, are constructed based on the design pattern intent ontology (DPIO) provided by [6]. When an input design problem description is presented, an input problem vector (IPV) is created. A cosine similarity is computed between the IPV and the DPV for each pattern. Design patterns are then ranked based on the computed similarity scores. Each process is detailed below.

3.1 Preparation: Design Pattern Vector Construction

A design pattern is a solution to one or more types of design problems. Based on the DPIO [6], Table 1 shows design problem types and Table 2 shows the problem

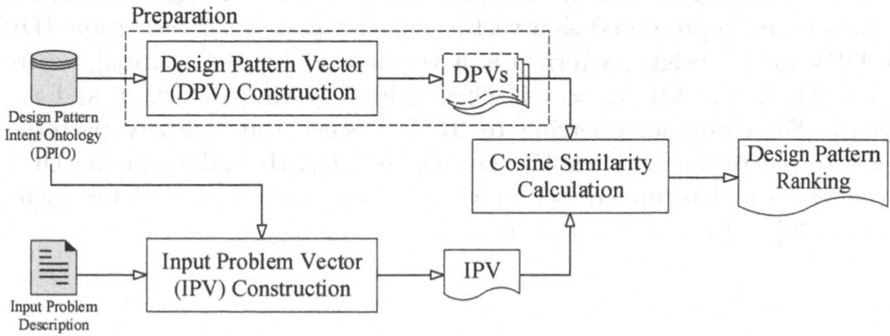

Fig. 1. An overview of the proposed approach

Table 1. Types of design problems

Type	Description	Type	Description
D_1	Abstraction implementation decoupling	D_{16}	Operation decoupling
D_2	Access control	D_{17}	Placeholder decoupling
D_3	Adaption	D_{18}	Protocol variation
D_4	Algorithm decoupling	D_{19}	Request decoupling
D_5	Algorithm selection	D_{20}	Sender/receiver decoupling
D_6	Algorithm variation	D_{21}	State change notification
D_7	Behavioral problem	D_{22}	State control
D_8	Complexity hiding	D_{23}	State dependency
D_9	Control undo	D_{24}	State duplication
D_{10}	Dynamic functionality control	D_{25}	State memorization
D_{11}	Event dependency	D_{26}	State objectification
D_{12}	Event notification problem	D_{27}	Structural problem
D_{13}	Interaction control	D_{28}	Time control
D_{14}	Interface decoupling	D_{29}	Time decoupling
D_{15}	Inversion of control problem	D_{30}	Virtual machine problem

Table 2. Types of design problems solved by each design pattern

Design pattern	Problem types	Design pattern	Problem types
Bridge	D_1, D_{14}, D_{27}	Memento	$D_7, D_{23}, D_{24}, D_{25}, D_{26}, D_{29}$
CoR	D_7, D_{18}, D_{20}	Observer	$D_7, D_{11}, D_{12}, D_{21}, D_{23}$
Command	$D_7, D_9, D_{19}, D_{28}, D_{29}$	Proxy	D_2, D_{17}, D_{27}
Decorator	D_{10}, D_{27}	State	$D_7, D_{22}, D_{23}, D_{26}$
Facade	D_3, D_8, D_{27}	Strategy	D_4, D_5, D_6, D_7
Interpreter	D_{12}, D_{30}	Template method	D_6, D_7, D_{13}, D_{15}
Mediator	D_7, D_8, D_{13}, D_{16}		

types that are solved by each design pattern. The types of design problems solved by a pattern are represented as a vector, called a *design pattern vector* (DPV).

A DPV for a design pattern p is a vector $\boldsymbol{v} = [v_1\, v_2\, v_3\, \ldots\, v_{30}]$, where for each $i \in \{1, 2, \ldots, 30\}$, $v_i = 1$ if D_i is solved by the pattern p, and $v_i = 0$ otherwise. For example, referring to Table 2, since the Strategy pattern is a solution to the problem types D_4, D_5, D_6 and D_7, the values in the DPV for this pattern are determined by: $v_4 = v_5 = v_6 = v_7 = 1$ and for each $j \in \{1, 2, \ldots, 30\} - \{4, 5, 6, 7\}$, $v_j = 0$.

3.2 Input Problem Vector Construction and Cosine Similarity Calculation

An input problem description is represented as an *input problem vector* (IPV), which takes the form $\boldsymbol{u} = [u_1\, u_2\, u_3\, \ldots\, u_{30}]$, where for each $i \in \{1, 2, \ldots, 30\}$, u_i is the score obtained by matching key terms extracted from the input problem description with the problem type D_i. Figure 2 depicts an IPV construction process. We explain key term extraction and key term matching in this process below, followed by cosine similarity calculation between an IPV and DPVs.

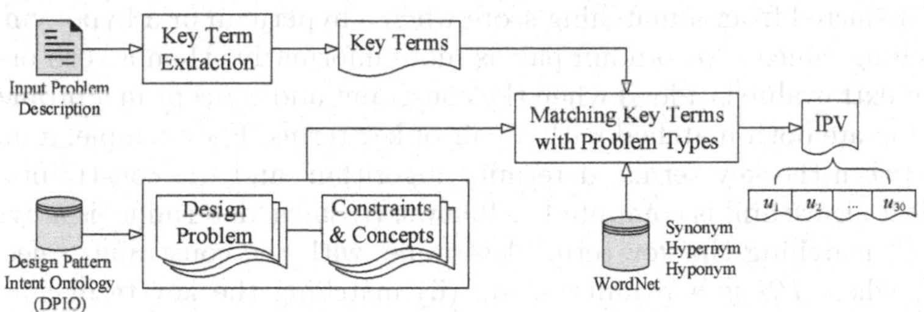

Fig. 2. An IPV construction process

Key Term Extraction. Key terms are extracted from a textual input description by using the Stanford Dependency parser [9] to generate types of dependencies (TDs) representing grammatical relations between pairs of words occurring in each sentence. Among 50 TDs provided by the parser, only four TDs that provide entities and intentions associated with the input problem are considered, i.e., nominal subject (nsubj), nominal subject in passive (nsubjpass), noun modifier (nmod), and direct object (dobj). Word stemming is applied to the extracted key terms to obtain their base forms. Table 3 illustrates grammatical relations and pairs of key terms extracted from an input sentence. The TD *dobj*(selects/VBZ, algorithm/NN), for example, indicates that the noun 'algorithm' acts as the direct object of the verb 'selects'.

Table 3. Key terms extracted from the sentence "*The user selects an algorithm at runtime or defines a configuration in a file*"

TD	Grammatical relation	Extracted key terms
nsubj	*nsubj*(selects/VBZ, user/NN)	'user select'
	nsubj(defines/VB, user/NN)	'user define'
nmod	*nmod:at*(selects/VBZ, runtime/NN)	'select runtime'
	nmod:in(configuration/NN, file/NN)	'configuration file'
dobj	*dobj*(selects/VBZ, algorithm/NN)	'select algorithm'
	dobj(defines/VB, configuration/NN)	'define configuration'

Matching Key Terms with Problem Types. A problem type is described in the DPIO [6] by pairs of constraints and concepts, e.g., the problem type D_5 is described by 'select algorithm', 'control algorithm', and 'control behavior'. A constraint represents an action verb, e.g., 'control', while a concept represents a noun, e.g., 'algorithm'. The pairs of key terms extracted from an input problem are matched with constraints and concepts describing each problem type.

Each of constraint (verb) matching and concept (noun) matching gives a score of 1. To increase the possibility of matching, synonyms, hypernyms (general terms), and hyponyms (specific terms) of constraints and concepts, provided by WordNet [10], which is a lexical database for English, are considered. A penalty value is deducted from a matching score when a hypernym or a hyponym is used for matching. Since a verb-noun pair is more informative than a verb or a noun alone, an extra value is added when the constraint and concept in a problem type description are both matched with a pair of key terms. For example, a matching score between the key terms 'determine algorithm' and the constraint-concept pair 'select algorithm' is computed as follows: (i) since 'determine' is a hypernym of 'select', matching the key term 'determine' with the constraint 'select' gives $1 - PS$, where PS is a penalty score, (ii) matching the key term 'algorithm' with the concept 'algorithm' itself gives 1, and (iii) since both constraint and concept are both matched, an extra value of 1 is added. The resulting score is thus $(1 - PS) + 1 + 1$.

Suppose that m pairs of key terms $pk_1, pk_2, \ldots, pk_m$ are extracted from an input problem description and a problem type D_i, where $i \in \{1, 2, \ldots, 30\}$, is described by n constraint-concept pairs $cc_1, cc_2, \ldots, cc_n$. The value u_i in the IPV $\boldsymbol{u} = [u_1\, u_2\, \ldots\, u_{30}]$ representing the input problem is calculated by

$$\sum_{j=1}^{n} \sum_{k=1}^{m} (match(pk_k, cc_j)), \qquad (1)$$

where for any j and k such that $1 \le j \le n$ and $1 \le k \le m$, $match(pk_k, cc_j)$ is the score obtained by matching pk_k with cc_j.

Cosine Similarity Calculation. The DPV for a design pattern and the IPV representing an input problem are normalized. The cosine similarity score is computed between the normalized DPV for each design pattern and the normalized IPV. Let $\boldsymbol{u} = [u_1\, u_2\, \ldots\, u_{30}]$ be an IPV and $\boldsymbol{v} = [v_1\, v_2\, \ldots\, v_{30}]$ be the DPV for a design pattern p. The normalized IPV and the normalized DPV for p are $\boldsymbol{u'} = [u'_1\, u'_2\, u'_3\, \ldots\, u'_{30}]$ and $\boldsymbol{v'} = [v'_1\, v'_2\, v'_3\, \ldots\, v'_{30}]$, respectively, where for each $i \in \{1, 2, \ldots, 30\}$, $u'_i = u_i/(\sum_{j=1}^{30} u_j)$ and $v'_i = v_i/(\sum_{j=1}^{30} v_j)$. The cosine similarity for the pattern p is then determined by:

$$\frac{\sum_{i=1}^{30}(u'_i \times v'_i)}{\sqrt{\sum_{i=1}^{30}(u'_i)^2} \times \sqrt{\sum_{i=1}^{30}(v'_i)^2}} \qquad (2)$$

> Many different objects within an application may need the ability to log messages. The logging feature may be put into a separate class. A message can be logged to different types of destinations such as a file, console and others by using different algorithms. The algorithm can be selected at runtime. There are two types of the messages; text messages and encrypted messages. These log messages should be displayed on the console or written down to a file, depending on the configuration at runtime.

Fig. 3. An example of an input problem

4 Experiments and Results

4.1 Experiment Settings and Basic Results

An experiment was conducted to evaluate the proposed approach. We collected 24 input problems, referred to as Q_1–Q_{24}, from three software design pattern books [2,7,8]. For each $i \in \{1, 2, \ldots, 24\}$, the pattern that should be applied to the input problem Q_i is called the *actual answer* to Q_i. Figure 3 shows an example of an input problem, to which the Strategy pattern should be applied.

Eight parameter settings, differing on the usage of synonyms, hypernyms, and hyponyms, are considered. They are denoted by P(*syn*, *hyper*, *hypo*), where each of *syn*, *hyper*, and *hypo* is either '0' or '1'. When *syn* (respectively, *hyper* and *hypo*) is '1', synonyms (respectively, hypernyms and hyponyms) are used, and they are not used otherwise. For example, synonyms, hypernyms, and hyponyms are all used in the parameter setting P(1,1,1), while only synonyms and hypernyms are used in the setting P(1,1,0). A penalty score (*PS*) taken from the set {0.2, 0.4, 0.6, 0.8} is used.

Table 4. The results obtained using the parameter settings P(1,1,1) and P(1,1,0)

(a) P(1,1,1)

PS	The percentage of correct answers				
	Top-1 rank	Top-2 ranks	Top-3 ranks	Top-4 ranks	Top-5 ranks
0.2	20.83%	37.50%	54.17%	62.50%	62.50%
0.4	25.00%	45.83%	58.33%	62.50%	62.50%
0.6	33.33%	45.83%	66.67%	66.67%	66.67%
0.8	33.33%	50.00%	66.67%	66.67%	66.67%

(b) P(1,1,0)

PS	The percentage of correct answers				
	Top-1 rank	Top-2 ranks	Top-3 ranks	Top-4 ranks	Top-5 ranks
0.2	29.17%	37.50%	54.17%	66.67%	70.83%
0.4	25.00%	37.50%	58.33%	62.50%	70.83%
0.6	25.00%	33.33%	66.67%	66.67%	70.83%
0.8	25.00%	41.67%	66.67%	66.67%	70.83%

Compared to other parameter settings, P(1,1,1) and P(1,1,0) yield better experimental results. Table 4 shows the results obtained from these two settings with different penalty scores. For each integer n such that $1 \leq n \leq 5$, the column "Top-n Ranks" shows the percentage of problems to which the actual answers are included by the patterns recommended in the top-n ranks. The row in which $PS = 0.6$ in Table 4b, for example, shows that the actual answers to 66.67% (16/24) of the problems are recommended within the top-three ranks, and the actual answers to 70.83% (17/24) of them are recommended within the top-five ranks. The setting P(1,1,0) yields the highest accuracy for the top-five ranks. The penalty score of 0.8 gives better overall performance compared to the other penalty scores.

4.2 IPV Construction with Additional Knowledge Sources

Even with the best setting, i.e., P(1,1,0), the actual answers to seven problems are not recommended within the top-five ranks. To improve the results, we extend IPV construction by incorporation of two additional sources of information.

Additional Constraints and Concepts (ACC). Pattern usage descriptions in the pattern usage hierarchy proposed in our previous work [12] provide an additional knowledge source for determining constraints and concepts. A pattern usage description describes an intention or a characteristic of a problem that is solved by a design pattern. Each pattern usage description is manually and semantically mapped to the problem types in Table 1. Constraints and concepts extracted from a pattern usage description are then associated with its corresponding problem type. Table 5 shows the resulting mapping and the constraints/concepts additionally associated with problem types. For example, the pattern usage description 'Accessing external resources (hard-disk, internet, etc.)' is more specific than and thus mapped to the problem type D_2 (i.e., 'Access control'), and then the constraints/concepts 'access resource', 'hard-disk', and 'internet' extracted from this description are assigned to D_2. Similarly, the pattern usage description 'Working with grammar and text parsing' is mapped to the problem type D_{30} (i.e., 'Virtual machine problem') since the meaning of this description is related to some constraint-concept pair that characterizes D_{30} in the DPIO [6], i.e., 'interpret grammar'.

Additional Term Correspondences by Experts (ATE). From the experiment with the setting P(1,1,0) and $PS = 0.8$ in Sect. 4.1, the actual answers to the input problems Q_6, Q_7, Q_{13}, Q_{16}, Q_{17}, Q_{18} and Q_{24} are not included by the top-five recommended patterns. On closer examination, several of them contain terms that technically correspond to constraints or concepts in the context of object-oriented design, but WordNet does not include their correspondences. For example, the term 'instance' refers to the concept 'object', but they are not related in terms of synonyms, hypernyms, or hyponyms in WordNet. Likewise, the term 'method' technically means an 'algorithm' provided by an object, but this correspondence is not presented in WordNet. To improve key term matching, we provide the additional term correspondences shown in Table 6.

Results. We referred to the original IPV construction described in Sect. 3.2 as 'Core', and refer to its extension with only ACC, with only ATE, and with both ACC and ATE as 'Core+ACC', 'Core+ATE', and 'Core+ACC+ATE', respectively. Table 7 shows the results obtained by the extensions using P(1,1,1) and

Table 5. Mapping pattern usage descriptions in [12] with problem types

No.	Pattern usage description	Problem type	Additional constraints/concepts
1	Accessing external resources (hard-disk, internet, etc.)	D_2	'access resource', 'hard-disk', 'internet'
2	Working with grammar and text parsing	D_{30}	'parse grammar', 'parse text'
3	Reducing retrieval time from external resources/Data caching	D_2 D_{25}	'reduce time', 'resource' 'cache data'
4	Changing an algorithm flow	D_6	'change algorithm', 'change flow'
5	Selecting an algorithm depending on an environment	D_5 D_{23}	'select algorithm' 'depend on environment'
6	Working with many alternative algorithms	D_5	'algorithm'
7	Choosing an appropriate algorithm on object creation	D_5	'choose algorithm'
8	Determining an algorithm at runtime	D_5	'determine algorithm', 'runtime'
9	Changing algorithms based on the current computation state	D_6 D_{23}	'change algorithm' 'based on state'
10	Storing object data	D_{25}	'store data'
11	Storing an object operation	D_{25}	'store operation'
12	Working with an undo functions	D_9	'undo function'
13	Working with communication among objects	D_{13}	'communication'
14	Notification of information change	D_{21}	'notify information', 'notify change'
15	One-to-many object communication	D_{13}	'communication'
16	Centralized object communication	D_{13}	'centralize communication'

Table 6. Additional term correspondences

Constraint/concept	Key term	Constraint/concept	Key term
Algorithm	Method	Interaction	Communication
Control	Handle, management	Method	Operation
Function	Method, operation	Object	Instance
Functionality	Task	Operation	Method

Table 7. The results obtained using the extensions

Parameter setting	Method	The percentage of correct answers				
		Top rank	Top-2 ranks	Top-3 ranks	Top-4 ranks	Top-5 ranks
P(1,1,1)	Core	33.33%	50.00%	66.67%	66.67%	66.67%
	Core+ACC	41.67%	54.17%	70.83%	75.00%	79.17%
	Core+ATE	29.17%	58.33%	83.33%	83.33%	83.33%
	Core+ACC+ATE	29.17%	54.17%	83.33%	87.50%	91.67%
P(1,1,0)	Core	25.00%	41.67%	66.67%	66.67%	70.83%
	Core+ACC	33.33%	50.00%	62.50%	70.83%	70.83%
	Core+ATE	29.17%	58.33%	79.17%	83.33%	83.33%
	Core+ACC+ATE	33.33%	62.50%	79.17%	87.50%	87.50%

P(1,1,0), with the penalty score of 0.8. Compared to Core, each of Core+ACC, Core+ATE, Core+ACC+ATE improves the recommendation performance. The extensions with the setting P(1,1,1) yield better improvement compared to those with P(1,1,0). With P(1,1,1), the results for the top-five ranks are improved from 66.67% (16/24) to 79.17% (19/24), 83.33% (20/24), and 91.67% (22/24) using Core+ACC, Core+ATE, and Core+ACC+ATE, respectively.

5 Conclusions

To assist a system analyst in an object-oriented software design phase, an approach for recommending appropriate GoF patterns is presented. A design pattern is represented as a vector indicating the problem types it addresses. Cosine similarity is calculated between an input problem vector and each design pattern vector. Design patterns are then ranked and recommended based on the similarity scores. Experiments were conducted to evaluate the proposed framework with 24 input problems taken from three design pattern books [2,7,8]. With appropriate parameter settings, 66.67% (16/24) of the actual answers to the input problems are recommended within the top-three ranks and 70.83% (17/24) of them are recommended within the top-five ranks. By using additional knowledge sources to improve key term matching for input problem vector construction, the percentage of recommending correct patterns within the top-three ranks and that within the top-five ranks increase to 83.33% (20/24) and 91.67% (22/24), respectively.

References

1. Bouassida, N., Kouas, A., Ben-Abdallah, H.: A design pattern recommendation approach. In: Proceedings of the 2nd International Conference on Software Engineering and Service Science, pp. 590–593. IEEE (2011)
2. Cooper, J.W.: Java Design Patterns: A Tutorial. Addison-Wesley, Boston (2000)
3. Gamma, E., Helm, R., Johnson, R., Vlissides, J.: Design Patterns: Elements of Reusable Object-Oriented Software. Addison-Wesley, Boston (1994)
4. Hasheminejad, H., Jalili, S.: Design patterns selection: an automatic two-phase method. J. Syst. Softw. **85**, 408–424 (2012)

5. Kampffmeyer, H., Zschaler, S.: Finding the pattern you need: the design pattern intent ontology. In: Engels, G., Opdyke, B., Schmidt, D.C., Weil, F. (eds.) MODELS 2007. LNCS, vol. 4735, pp. 211–225. Springer, Heidelberg (2007). https://doi.org/10.1007/978-3-540-75209-7_15
6. Kampffmeyer, H.: The Design Pattern Intent Ontology. VDM Verlag Dr. Müller e.K., Saarbrücken (2007)
7. Kuchana, P.: Software Architecture Design Patterns in Java. Auerbach Publications, Boca Raton (2004)
8. Lasater, C.G.: Design Patterns. Jones & Bartlett Publishers, Massachusetts (2010)
9. Marneffe, M.C., Maccartney, B., Manning, C.D.: Generating typed dependency parses from phrase structure parses. In: Proceedings of the 5th International Conference on Language Resources and Evaluation, pp. 449–454 (2006)
10. Fellbaum, C.: WordNet: An Electronic Lexical Database. MIT Press, Cambridge (1998)
11. Palma, F., Farzin, H., Guéhéneuc, Y.G., Moha, N.: Recommendation system for design patterns in software development: an DPR overview. In: Proceedings of the 3rd International Workshop on Recommendation Systems for Software Engineering, pp. 1–5. IEEE (2012)
12. Sanyawong, N., Nantajeewarawat, E.: Design pattern recommendation based-on a pattern usage hierarchy. In: Proceedings of the 18th International Computer Science and Engineering Conference, pp. 134–139. IEEE (2014)

Solving Query-Answering Problems with Constraints for Function Variables

Kiyoshi Akama[1] and Ekawit Nantajeewarawat[2(✉)]

[1] Information Initiative Center, Hokkaido University, Hokkaido, Japan
akama@iic.hokudai.ac.jp

[2] Computer Science Program, Sirindhorn International Institute of Technology, Thammasat University, Pathumthani, Thailand
ekawit@siit.tu.ac.th

Abstract. A query-answering problem (QA problem) is concerned with finding all ground instances of a query atomic formula that are logical consequences of a given logical formula describing the background knowledge of the problem. Based on the equivalent transformation (ET) solution method, a general framework for solving QA problems on first-order logic has been proposed, where a first-order formula representing background knowledge is converted by meaning-preserving Skolemization into a set of clauses typically containing global existential quantifications of function variables. The obtained clause set is then transformed successively using ET rules until the answer set of the original problem can be readily derived. In this paper, we extend the space of clauses by introducing constraints and invent three ET rules for simplifying problem descriptions by using interaction between clauses and constraints. This extension provides a general solution for a larger class of QA problems.

Keywords: Query-answering problem · Function variable
Model-intersection problem · Constraint
Equivalent transformation rule

1 Introduction

It is well known that built-in constraint atoms play a crucial role in knowledge representation and are essential for practical applications [5]. In this paper, we consider first-order formulas that possibly include built-in constraint atoms. The set of all such formulas is denoted by $\mathrm{FOL_c}$. A *query-answering problem* (*QA problem*) on $\mathrm{FOL_c}$ is a pair $\langle K, a\rangle$, where K is a formula in $\mathrm{FOL_c}$ and a is a user-defined query atom [2]. The answer to a QA problem $\langle K, a\rangle$ is defined as the set of all ground instances of a that are logical consequences of K. Characteristically, a QA problem is an "all-answers finding" problem, i.e., all ground instances of a given query atom satisfying the requirement above are to be found.

QA problems have been researched extensively in the logic-programming community [7,8] and in the semantic-web community [9,10]. The problem class being addressed has been, however, rather small. For example, Prolog in logic

N. T. Nguyen et al. (Eds.): ACIIDS 2018, LNAI 10751, pp. 36–47, 2018.
https://doi.org/10.1007/978-3-319-75417-8_4

programming deals mainly with only definite clauses, and knowledge representation in semantic web takes rather small expressive power compared with full first-order formulas, although it can also represent some formulas corresponding to non-definite clauses.

The lack of solution for QA problems on full FOL_c in the conventional research stems from the contradiction of the satisfiability-based approach and the conventional Skolemization for FOL_c. The conventional Skolemization does not generally preserve the satisfiability nor the logical meanings of formulas in FOL_c, which is a fatal limitation of the conventional first-order logic and logic programming [1,3]. To overcome the difficulty, we developed a new theoretical framework for computational logic [1,4].

For transformation of QA problems in FOL_c into equivalent clausal forms, meaning-preserving Skolemization has been developed in [1] together with a new extended space, called the $ECLS_F$ space, over the set of all first-order logical formulas. This extended space includes function variables, which are variables ranging over function constants.

When we transformed a QA problem on FOL_c into a new problem on clauses, we discovered *model-intersection problems* (*MI problems*) on $ECLS_F$. A MI problem is a pair $\langle Cs, \varphi \rangle$, where Cs is a set of extended clauses and φ is a mapping, called an *extraction mapping*, used for constructing the output answer from the intersection of all models of Cs. More formally, the answer to a MI problem $\langle Cs, \varphi \rangle$ is $\varphi(\bigcap Models(Cs))$, where $Models(Cs)$ is the set of all models of Cs and $\bigcap Models(Cs)$ is the intersection of all elements of $Models(Cs)$.

The set of all MI problems on extended clauses in $ECLS_F$ constitutes a very large class of problems and is of great importance. The class of MI problems on $ECLS_F$ is the first one that enables structural embedding of the full class of proof problems on FOL_c and the full class of QA problems on FOL_c. As outlined by Fig. 1, all proof problems and all QA problems on FOL_c can be mapped, preserving their answers, into MI problems on $ECLS_F$. By solving MI problems on $ECLS_F$, we can solve proof problems and QA problems on FOL_c.

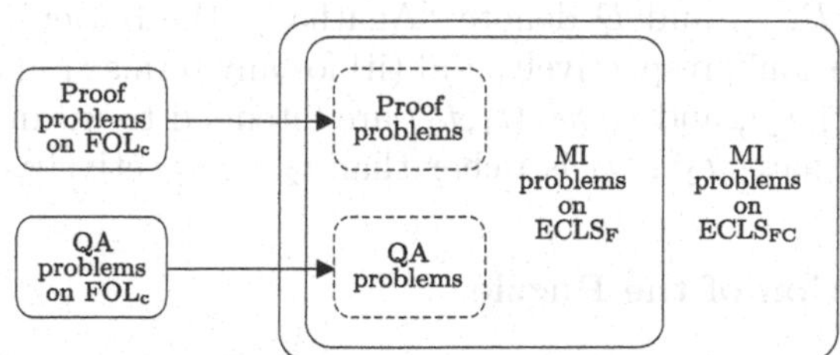

Fig. 1. MI-problem-centered view of logical problems

We have proposed a general schema for solving MI problems on $ECLS_F$ by equivalent transformation (ET), where problems are solved by repeated problem simplification using ET rules [4]. We take a general approach to dealing with MI problems, which is based on the ET solution method, i.e., a given MI problem is transformed equivalently into simpler forms until its answer set can be readily obtained. This will be illustrated by the Agatha puzzle (the "Dreadsbury Mansion Mystery" problem) in Sect. 2 of this paper.

However, there is some MI problem that cannot be solved by equivalent transformation in the space ECLS_F, i.e., ET transformation sequence starting from the MI problem reaches some terminal point in ECLS_F that is not a solution, which will also be shown by the Agatha example. This motivates an extension of the theory by devising a new space and making ET rules in the new space.

This paper extends our theory by (i) introduction of constraints for function variables, resulting in a new clause space, and (ii) invention of new ET rules for dealing with *func*-atoms and constraints. The new clause space is called ECLS_FC. This space and the new rules make it possible to solve a larger class of MI (and thus QA) problems, including the Agatha puzzle.

2 An Introductory Example

2.1 Dreadsbury Mansion Mystery: Formalization

Consider the "Dreadsbury Mansion Mystery" problem, which is described as follows: Someone who lives in Dreadsbury Mansion killed Aunt Agatha. Agatha, the butler, and Charles live in Dreadsbury Mansion, and are the only people who live therein. A killer always hates his victim, and is never richer than his victim. Charles hates no one that Aunt Agatha hates. Agatha hates everyone except the butler. The butler hates everyone not richer than Aunt Agatha. The butler hates everyone Agatha hates. No one hates everyone. The problem is to find who is the killer.

Assume that *eq* and *neq* are predefined binary constraint predicates and for any ground usual terms t_1 and t_2, (i) $eq(t_1, t_2)$ is true iff $t_1 = t_2$, and (ii) $neq(t_1, t_2)$ is true iff $t_1 \neq t_2$. The background knowledge of this mystery is formalized as the conjunction of the first-order formulas in Fig. 2, where (i) the constants A, B, C, and D denote "Agatha", "the butler", "Charles", and "Dreadsbury Mansion", respectively, and (ii) for any terms t_1 and t_2, $live(t_1, t_2)$, $kill(t_1, t_2)$, $hate(t_1, t_2)$, and $richer(t_1, t_2)$ are intended to mean "t_1 lives in t_2", "t_1 killed t_2", "t_1 hates t_2", "t_1 is richer than t_2", respectively.

2.2 Formalization of the Puzzle

We understand the Agatha puzzle ("who killed aunt Agatha?") as finding all persons who killed Agatha. Formalizing this puzzle is to specify the answer (the set of all killers) in terms of the background knowledge of the puzzle.

Let K be the conjunction of the first-order formulas in Fig. 2 and let $q = killer(x)$. The Agatha puzzle is formalized as a QA problem $\langle K, q \rangle$. The answer to this QA problem, denoted by $answer(K, q)$, is the set of all ground instances of q that follows logically from K. K is converted into the set *Cs*

$$\exists x : (live(x, D) \wedge kill(x, A))$$
$$\forall x : (live(x, D) \leftrightarrow (eq(x, A) \vee eq(x, B) \vee eq(x, C)))$$
$$\forall x \forall y : (kill(x, y) \rightarrow hate(x, y))$$
$$\forall x \forall y : (kill(x, y) \rightarrow \neg richer(x, y))$$
$$\neg(\exists x : (hate(C, x) \wedge hate(A, x) \wedge live(x, D)))$$
$$\forall x : ((neq(x, B) \wedge live(x, D)) \rightarrow hate(A, x))$$
$$\forall x : ((\neg richer(x, A) \wedge live(x, D)) \rightarrow hate(B, x))$$
$$\forall x : ((hate(A, x) \wedge live(x, D)) \rightarrow hate(B, x))$$
$$\neg(\exists x : (live(x, D) \wedge (\forall y : (live(y, D) \rightarrow hate(x, y)))))$$
$$\forall x : (kill(x, A) \rightarrow killer(x))$$

Fig. 2. Background knowledge represented by first-order formulas

$$
\begin{array}{ll}
C_1: & live(x, D) \leftarrow func(f_0, x) \\
C_2: & kill(x, A) \leftarrow func(f_0, x) \\
C_3: & \leftarrow live(x, D), neq(x, A), neq(x, B), neq(x, C) \\
C_4: & live(A, D) \leftarrow \\
C_5: & live(B, D) \leftarrow \\
C_6: & live(C, D) \leftarrow \\
C_7: & hate(x, y) \leftarrow kill(x, y) \\
C_8: & \leftarrow kill(x, y), richer(x, y) \\
C_9: & \leftarrow hate(A, x), hate(C, x), live(x, D) \\
C_{10}: & hate(A, x) \leftarrow neq(x, B), live(x, D) \\
C_{11}: & richer(x, A), hate(B, x) \leftarrow live(x, D) \\
C_{12}: & hate(B, x) \leftarrow hate(A, x), live(x, D) \\
C_{13}: & \leftarrow hate(x, y), func(f_1, x, y), live(x, D) \\
C_{14}: & live(y, D) \leftarrow live(x, D), func(f_1, x, y) \\
C_{15}: & killer(x) \leftarrow kill(x, A)
\end{array}
$$

Fig. 3. The initial state with extended clauses

consisting of the fifteen extended clauses C_1–C_{15} in Fig. 3 by applying meaning-preserving Skolemization [1]. Let $\mathcal{G}_u$ be the set of all ground user-defined atoms. The QA problem $\langle K, q \rangle$ is then reformulated as a model-intersection (MI) problem $\langle Cs, \varphi \rangle$, where φ is a mapping from the power set of $\mathcal{G}_u$ to the power set of $\{A, B, C\}$, defined by $\varphi(G) = \{t \mid killer(t) \in G\}$ for any $G \subseteq \mathcal{G}_u$. In other words, we have

$$answer(K, q) = \varphi(\bigcap Models(Cs)).$$

Our plan for solving the Agatha puzzle is to simplify $\varphi(\bigcap Models(Cs))$ mainly by transforming Cs preserving $Models(Cs)$ or $\bigcap Models(Cs)$.

2.3 Computation for Solving the Dreadsbury Mansion Mystery

In order to simplify the set of the clauses in Fig. 3, we use many ET rules, including unfolding, definite-clause removal, and side-change transformation, given in

$f_0 : \{\}$
$f_1 : \{\}$
C_{44}: $\leftarrow neq(x, B), func(f_1, B, x)$
C_{45}: $\leftarrow func(f_1, B, A)$
C_{46}: $\leftarrow func(f_1, B, C)$
C_{47}: $\leftarrow func(f_0, C)$
C_{48}: $killer(x) \leftarrow func(f_0, x)$
C_{49}: $\leftarrow func(f_1, B, x), func(f_0, x)$
C_{50}: $\leftarrow func(f_1, x, A), func(f_0, x)$
C_{51}: $\leftarrow neq(x, A), neq(x, B), neq(x, C), func(f_0, x)$
C_{52}: $\leftarrow neq(x, A), neq(x, B), neq(x, C), func(f_1, C, x)$
C_{53}: $\leftarrow func(f_1, A, A)$
C_{54}: $\leftarrow func(f_1, A, C)$
C_{55}: $\leftarrow neq(x, B), func(f_1, A, x)$

Fig. 4. Clauses obtained by unfolding *live*-atoms after *specAtom*

$f_0 : \{\langle [\,], \{A, B, C\}\rangle\}$
$f_1 : \{\langle [B], \{B\}\rangle, \langle [C], \{A, B, C\}\rangle, \langle [A], \{B\}\rangle\}$
C_{56}: $\leftarrow func(f_1, B, A)$
C_{57}: $\leftarrow func(f_1, B, C)$
C_{58}: $\leftarrow func(f_0, C)$
C_{59}: $killer(x) \leftarrow func(f_0, x)$
C_{60}: $\leftarrow func(f_1, B, x), func(f_0, x)$
C_{61}: $\leftarrow func(f_1, x, A), func(f_0, x)$
C_{62}: $\leftarrow func(f_1, A, A)$
C_{63}: $\leftarrow func(f_1, A, C)$

Fig. 5. Clauses obtained by positive-function-variable restriction

our previous papers [2,4–6]. The transformation process is shown below together with important final steps in Sect. 2.5.

1. By unfolding using the definition of *kill* (C_2), all body atoms with the predicate *kill* are removed. By definite-clause removal, the definition of *kill* (C_2) is removed. By unfolding, three body atoms with patterns $hate(A, x)$ or $hate(C, x)$ are removed.
2. By side-change transformation for *richer*, the *richer*-atoms in some clauses are removed and *not_richer*-atoms are obtained in the other side of these clauses.
3. By unfolding, *not_richer*-atoms and *hate*-atoms in clause bodies are removed. The definitions of *not_richer* and *hate* are removed.
4. Unfolding with respect to *live*-atoms has been suspended since one of the definite clauses defining *live* introduces a new *live*-atom with a pure variable as its first argument. To remedy the situation, the atom $live(x, D)$ in the body

of a clause is specialized into three clauses, which enable further application of unfolding.

5. By unfolding *live*-atoms and definite-clause removal, all *live*-atoms are removed and the clauses in Fig. 4 are obtained.

At this stage, unfolding and definite-clause removal were applied 19 times. Other rules that were also applied and their application counts are as follows: elimination of subsumed clauses 22 times, elimination of true body atoms 10 times, elimination of duplicate atoms 5 times, constraint solving for *neq* 4 times, side-change transformation 1 time, and atom specialization 1 time. Fifteen clauses consisting of 35 atoms are reduced to twelve clauses with 23 atoms.

2.4 The Main Objective of the Paper

Except for the *killer*-atom in C_{48}, the clauses in Fig. 4 contain only two kind of atoms, i.e., *neq*-atoms and *func*-atoms. We cannot apply unfolding to the clauses in Fig. 4 since there is no *killer*-atom in the body of these clauses. We cannot apply the definite-clause-elimination rule to remove the definition of *killer* (C_{48}) since the answer to this QA problem is concerned with *killer*-atoms. We cannot transform these clauses further by the ET rules that have been developed so far.

The main objective of this paper is to extend the theory by

1. introduction of constraints for function variables, and
2. invention of new ET rules for dealing with *func*-atoms and constraints.

This extension makes it possible to solve a larger class of QA problems, including the Agatha puzzle.

2.5 Transformation with Function Variables and Constraints

We will introduce constraints for function variables in Sect. 3 and invent three ET rules, i.e., (i) positive function-variable restriction, (ii) negative function-variable restriction, and (iii) *func*-atom evaluation. After the completion of this theory extension, we can have a complete solution to the Agatha puzzle.

The function variables in Fig. 4 can be freely instantiated without any constraint, which is represented explicitly in the first two lines of the figure. Further transformation is done with the interaction between clauses and constraints as follows:

1. Refer to Fig. 4. By positive function-variable restriction, C_{51} changes f_0 to $f_0 : \{\langle [\,], \{A, B, C\}\rangle\}$, which means $f_0 \in \{A, B, C\}$. By positive function-variable restriction, f_1 is changed sequentially to
   - $f_1 : \{\langle [A], \{B\}\rangle\}$ by C_{55},
   - $f_1 : \{\langle [A], \{B\}\rangle, \langle [B], \{B\}\rangle\}$ by C_{44}, and
   - $f_1 : \{\langle [A], \{B\}\rangle, \langle [B], \{B\}\rangle, \langle [C], \{A, B, C\}\rangle\}$ by C_{52}.

   The constraint on f_1 in the last line above means $f_1(A) \in \{B\}$ and $f_1(B) \in \{B\}$ and $f_1(C) \in \{A, B, C\}$. The clauses C_{44}, C_{51}, C_{52}, and C_{55} are then removed. The clauses in Fig. 5 are obtained.

2. C_{56}, C_{57}, C_{62}, and C_{63} are removed by *func*-atom evaluation.
3. The clause C_{58} changes f_0 into $f_0 : \{\langle[\,], \{A, B\}\rangle\}$ by negative function-variable restriction, and it is removed. At this stage the remaining clauses are C_{59}, C_{60}, and C_{61}.
4. Since $func(f_1, B, B)$ is true, the clause C_{60} is transformed into:

$$C_{64} : \quad \leftarrow func(f_0, B)$$

5. The clause C_{64} changes f_0 further into $f_0 : \{\langle[\,], \{A\}\rangle\}$ by negative function-variable restriction, and the clause is removed.
6. Since $func(f_0, A)$ is true and $func(f_1, A, A)$ is false, C_{61} is removed by *func*-atom evaluation.
7. Since $func(f_0, A)$ is true, C_{59} is transformed into:

$$C_{65} : \quad killer(A) \leftarrow$$

The only remaining clause is C_{65}, from which the answer can be readily obtained, i.e., Agatha is the killer.

3 An Extended Space with Constraints for Function Variables

3.1 User-Defined Atoms, Constraint Atoms, and *func*-Atoms

We consider an extended formula space that contains three kinds of atoms, i.e., user-defined atoms, built-in constraint atoms, and *func*-atoms. A *user-defined atom* takes the form $p(t_1, \ldots, t_n)$, where p is a user-defined predicate and the t_i are usual terms. A *built-in constraint atom*, also simply called a *constraint atom* or a *built-in atom*, takes the form $c(t_1, \ldots, t_n)$, where c is a predefined constraint predicate and the t_i are usual terms. Let $\mathcal{A}_u$ be the set of all user-defined atoms and $\mathcal{A}_c$ the set of all constraint atoms.

A *func-atom* [1] is an expression of the form $func(f, t_1, \ldots, t_n, t_{n+1})$, where f is either an n-ary function constant or an n-ary function variable, and the t_i are usual terms. It is a *ground func*-atom if f is a function constant and the t_i are ground usual terms.

3.2 Constraints for Function Variables

To enrich the expressive power of clauses and obtain more flexible transformation, we introduce constraints for function variables.

Let *FVar* be the set of all function variables and *FCon* the set of all function constants. Let $\mathcal{G}_t$ denote the set of all ground usual terms. Each n-ary function constant is associated with a mapping from $\mathcal{G}_t^n$ to $\mathcal{G}_t$. There are two types of variables: usual variables and function variables. A function variable can be instantiated into any function constant, but not into a usual term.

A *constraint set* is a set of pairs each of which takes the form $\langle seq, G\rangle$, where seq is a sequence in $seq(\mathcal{G}_t)$ and G is a set of terms in $\mathcal{G}_t$. A constraint set S is attached to each occurrence of a function variable $f \in FVar$ and such an occurrence is denoted by the pair $\langle f, S\rangle$. When the attached constraint set S is empty, f is called a *pure* function variable and is often denoted simply by f itself. The set S in $\langle f, S\rangle$ specifies constraints for instantiation that should be satisfied by any substitution σ for function variables that is applied to f, i.e., if $f\sigma$ is a function constant, say f_c, and $\langle [t_1, \ldots, t_n], G\rangle \in S$, then $f_c(t_1, \ldots, t_n) \in G$.

3.3 Extended Clauses

An *extended clause* C on $\mathcal{A}_u \cup \mathcal{A}_c$ is a formula of the form

$$a_1, \ldots, a_m \leftarrow b_1, \ldots, b_n, \mathbf{f}_1, \ldots, \mathbf{f}_p,$$

where each of $a_1, \ldots, a_m, b_1, \ldots, b_n$ is a user-defined atom in $\mathcal{A}_u$ or a constraint atom in $\mathcal{A}_c$, and $\mathbf{f}_1, \ldots, \mathbf{f}_p$ are *func*-atoms. All usual variables occurring in C are implicitly universally quantified and their scope is restricted to the extended clause C itself. The sets $\{a_1, \ldots, a_m\}$ and $\{b_1, \ldots, b_n, \mathbf{f}_1, \ldots, \mathbf{f}_p\}$ are called the *left-hand side* and the *right-hand side*, respectively, of the extended clause C, and are denoted by $lhs(C)$ and $rhs(C)$, respectively. C is said to be *pure* iff each function variable occurring in C is pure.

Let $\mathrm{ECLS_{FC}}$ be the set of all extended clauses, possibly containing function variables with constraints. $\mathrm{ECLS_{FC}}$ is an extension of $\mathrm{ECLS_F}$, which is the space of extended clauses with only pure function variables. Let $Cs \subseteq \mathrm{ECLS_{FC}}$. Cs is said to be *pure* iff all clauses in Cs are pure. Cs is said to be *normal* iff for any function variable $f \in Fvar$, if $\langle f, S_1\rangle$ and $\langle f, S_2\rangle$ are occurrences of f in Cs, then $S_1 = S_2$.

4 ET Rules for Processing Function Variables

4.1 Positive Function-Variable Restriction

Let $\langle Cs, \varphi\rangle$ be a MI problem on $\mathrm{ECLS_{FC}}$. Assume that Cs is normal and contains an extended clause C such that

$$C = (eq(x, t_1), \ldots, eq(x, t_n) \leftarrow func(f, s_1, \ldots, s_m, x)),$$

where x is a usual variable, $t_1, \ldots, t_n, s_1, \ldots, s_m$ are ground terms, and f is a function variable. C gives a positive restriction for f. More precisely, $\langle Cs, \varphi\rangle$ can be equivalently transformed into $\langle Cs', \varphi\rangle$, where Cs' is obtained from Cs as follows:

1. $Cs' = Cs - \{C\}$.
2. For each attached constraint set S of f in Cs', change S as follows:
   (a) If $\langle [s_1, \ldots, s_m], G\rangle \in S$, then change S into

$$(S - \{\langle [s_1, \ldots, s_m], G\rangle\}) \cup \{\langle [s_1, \ldots, s_m], G \cap \{t_1, \ldots, t_n\}\rangle\}.$$

(b) If there is no pair of the form $\langle [s_1, \ldots, s_m], G\rangle$ in S, then change S into

$$S \cup \{\langle [s_1, \ldots, s_m], \{t_1, \ldots, t_n\}\rangle\}.$$

The correctness of this transformation is shown as follows: The constraint imposed by the clause C means that $f(s_1, \ldots, s_m) \in \{t_1, \ldots, t_n\}$. Hence, in Step 2a, elements outside $\{t_1, \ldots, t_n\}$ are removed from G. In Step 2b, nonexistence of $\langle [s_1, \ldots, s_m], G\rangle$ means that $f(s_1, \ldots, s_m)$ is arbitrary. So $\langle [s_1, \ldots, s_m], \{t_1, \ldots, t_n\}\rangle$ should be added to S.

4.2 Negative Function-Variable Restriction

Let $\langle Cs, \varphi\rangle$ be a MI problem on $\mathrm{ECLS_{FC}}$. Assume that Cs is normal and contains an extended clause C such that

$$C = (\leftarrow func(f, s_1, \ldots, s_m, s)),$$

where $s_1, \ldots, s_m$ and s are ground terms and f is a function variable whose attached constraint set contains $\langle [s_1, \ldots, s_m], G\rangle$ for some $G \subseteq \mathcal{G}_t$. C gives a negative restriction for f. More precisely, $\langle Cs, \varphi\rangle$ can be equivalently transformed into $\langle Cs', \varphi\rangle$, where Cs' is obtained from Cs as follows:

1. $Cs' = Cs - \{C\}$.
2. For each attached constraint set S of f in Cs', if $\langle [s_1, \ldots, s_m], G\rangle \in S$, then:
   (a) If $G = \{s\}$, then add the empty clause $(\leftarrow)$ to Cs'.
   (b) If $G \neq \{s\}$, then change S into

$$(S - \{\langle [s_1, \ldots, s_m], G\rangle\}) \cup \{\langle [s_1, \ldots, s_m], G - \{s\}\rangle\}.$$

The correctness of this transformation is shown as follows: Assume that $\langle [s_1, \ldots, s_m], G\rangle \in S$. Since the negative clause C imposes a constraint $f\sigma(s_1, \ldots, s_m) \neq s$ for any substitution σ for function variables such that $f\sigma$ is a function constant, basically G becomes $G - \{s\}$. Since G cannot be the empty set, we have a contradiction when $G = \{s\}$, which is represented by the empty clause $(\leftarrow)$.

4.3 *func*-Atom Evaluation

Assume that f is a function variable and $s_1, \ldots, s_m$ are ground terms in $\mathcal{G}_t$. Suppose that there is a *func*-atom $func(f, s_1, \ldots, s_m, t)$ in the right-hand side of a clause C in a clause set Cs, and that the constraint set S attached to f contains $\langle [s_1, \ldots, s_m], G\rangle$. The set G imposes a constraint on possible values of $f\sigma(s_1, \ldots, s_m)$ for any substitution σ for function variables, i.e., $f\sigma(s_1, \ldots, s_m)$ must belong to G. Since the function variable f is existentially quantified, the condition $G = \emptyset$ means the nonexistence of $f\sigma$, from which we can immediately know the answer to an MI problem concerning Cs, i.e., $\varphi(\bigcap Models(Cs)) = \varphi(\emptyset)$. When G is not empty, we have *func*-atom evaluation rules, which produce the following equivalent transformation:

C'_1: $live(f_0, D) \leftarrow$ C'_{13}: $\leftarrow hate(x, f_1(x)), live(x, D)$
C'_2: $kill(f_0, A) \leftarrow$ C'_{14}: $live(f_1(x), D) \leftarrow live(x, D)$

Fig. 6. Clauses obtained by the conventional Skolemization

1. If there is a ground *func*-atom $func(f, s_1, \ldots, s_m, s)$ in the right-hand side of a clause C in Cs and the constraint set attached to f contains $\langle [s_1, \ldots, s_m], \{s\} \rangle$, then $func(f, s_1, \ldots, s_m, s)$ can be removed from C.
2. If there is a *func*-atom $func(f, s_1, \ldots, s_m, v)$ in the right-hand side of a clause C in Cs and the constraint set attached to f contains $\langle [s_1, \ldots, s_m], \{s\} \rangle$, then C can be specialized by instantiating v into s and the true *func*-atom $func(f, s_1, \ldots, s_m, s)$ in the resulting clause can be removed.
3. If there is a ground *func*-atom $func(f, s_1, \ldots, s_m, s)$ in the right-hand side of a clause C in Cs and the constraint set attached to f contains $\langle [s_1, \ldots, s_m], G \rangle$ such that $s \notin G$, then C can be removed from Cs.

The correctness of this transformation is shown as follows: Assume that the constraint set attached to f contains $\langle [s_1, \ldots, s_m], G \rangle$. The above transformation is correct since $func(f, s_1, \ldots, s_m, s)$ is true if G is a singleton set $\{s\}$ and it is false if $s \notin G$.

5 Experiments

We have constructed an experimental MI-problem solver by directly implementing the theory, and have tried to apply it to solve many QA and proof problems, many of which are not large but impossible to solve with theories and techniques developed in the logic programming and semantic web communities.

The Agatha puzzle, which is formalized as a QA problem (cf. Sect. 2.2), was solved by 102 applications of ET rules, consisting of unfolding and definite-clause removal, constraint solving for *neq*, elimination of subsumed clauses, elimination of duplicate atoms, side-change transformation, elimination of independent satisfiable atoms, finite specialization of atoms, positive function-variable restriction, negative function-variable restriction, and *func*-atom evaluation [2,4–6].

The following three Agatha proof problems were also successfully solved by our MI-solver: (i) "did Agatha kill herself?", (ii) "didn't the butler kill Agatha?", and (iii) "didn't Charles kill Agatha?" These three proof problems are represented as the conjunctions of the first-order formulas in Fig. 2 and $\neg killer(A)$, $killer(B)$, and $killer(C)$, respectively. The empty clause ($\leftarrow$) was produced from each of them, thereby proving that Agatha is the killer, not the other ones.

The Agatha puzzle cannot be truly solved by the conventional methods. When we transform the conjunction of the first-order formulas in Fig. 2 by using the conventional Skolemization, we obtain the clause set consisting of C_3–C_{12} and C_{15} in Fig. 3 and C'_1, C'_2, C'_{13}, and C'_{14} in Fig. 6. Our MI-solver transforms this clause set into the singleton set of the empty clause ($\leftarrow$), which shows that the clause set has a contradiction. This contradiction is due to incorrect transformation by the conventional Skolemization. Moreover, none of the three Agatha proof problems listed above can be solved by the conventional methods.

6 Conclusions

The ET solution method provides a basis for solving a very large class of QA problems on $\mathrm{FOL_c}$. The first step of the solution is to transform QA problems on $\mathrm{FOL_c}$ into a set of extended clauses on $\mathrm{ECLS_F}$ by meaning-preserving Skolemization preserving the logical meanings of formulas. This has led to discovery of a class of MI problems. Each QA problem is transformed into an equivalent MI problem. The conventional Skolemization is a major hindrance to development of a general solution for QA problems on $\mathrm{ECLS_F}$ since it does not generally preserve satisfiability nor logical meanings, and thus does not preserve the answers to QA problems. We can improve solutions for a very large class of QA problems by investigating general solutions for MI problems. Many equivalent transformation rules working on the $\mathrm{ECLS_F}$ space have been invented [2,4–6], such as unfolding, definite-clause removal, resolution, factoring, side-change, elimination by subsumption, which have increased the power of solving a large class of MI (and QA) problems. This paper made clear similar creation steps again, i.e., (i) the space extension from $\mathrm{ECLS_F}$ to $\mathrm{ECLS_{FC}}$ by addition of constraints on function variables, and (ii) rule introduction relating to function variables and constraints. This extension has strengthened the power of solving a large class of MI (and thus QA) problems.

References

1. Akama, K., Nantajeewarawat, E.: Meaning-preserving Skolemization. In: 3rd International Conference on Knowledge Engineering and Ontology Development, Paris, France, pp. 322–327 (2011)
2. Akama, K., Nantajeewarawat, E.: Equivalent transformation in an extended space for solving query-answering problems. In: Nguyen, N.T., Attachoo, B., Trawiński, B., Somboonviwat, K. (eds.) ACIIDS 2014. LNCS (LNAI), vol. 8397, pp. 232–241. Springer, Cham (2014). https://doi.org/10.1007/978-3-319-05476-6_24
3. Akama, K., Nantajeewarawat, E.: Function-variable elimination and its limitations. In: 7th International Joint Conference on Knowledge Discovery, Knowledge Engineering and Knowledge Management, Lisbon, Portugal, vol. 2, pp. 212–222 (2015)
4. Akama, K., Nantajeewarawat, E.: Model-intersection problems with existentially quantified function variables: formalization and a solution schema. In: 8th International Joint Conference on Knowledge Discovery, Knowledge Engineering and Knowledge Management, Porto, Portugal, vol. 2, pp. 52–63 (2016)
5. Akama, K., Nantajeewarawat, E.: Unfolding existentially quantified sets of extended clauses. In: 8th International Joint Conference on Knowledge Discovery, Knowledge Engineering and Knowledge Management, Porto, Portugal, vol. 2, pp. 96–103 (2016)

6. Akama, K., Nantajeewarawat, E.: Side-change transformation. Technical report, Hokkaido University (2017)
7. Lloyd, J.W.: Foundations of Logic Programming, 2nd edn. Springer, Heidelberg (1987). https://doi.org/10.1007/978-3-642-83189-8
8. Robinson, J.A.: A machine-oriented logic based on the resolution principle. J. ACM **12**, 23–41 (1965)
9. Donini, F.M., Lenzerini, M., Nardi, D., Schaerf, A.: $\mathcal{AL}$-log: integrating datalog and description logics. J. Intell. Coop. Inf. Syst. **10**, 227–252 (1998)
10. Motik, B., Sattler, U., Studer, R.: Query answering for OWL-DL with rules. J. Web Semant. **3**, 41–60 (2005)

Heuristic Algorithms for 2-Optimality Consensus Determination

Adrianna Kozierkiewicz(✉) and Mateusz Sitarczyk

Faculty of Computer Science and Management,
Wroclaw University of Science and Technology, Wybrzeze Wyspianskiego 27,
50-370 Wroclaw, Poland
{adrianna.kozierkiewicz,mateusz.sitarczyk}@pwr.edu.pl

Abstract. The data stored in many distributed sources is often used in decision making processes. However, the determination of a one, consistent version of its elements (called in this paper as a consensus) could be a very time- and cost-consuming. Therefore, the balance between these factors and the quality the results is needed. This paper is devoted to designing and analysing some methods of the consensus determination. The main aim of conducted experiments is to find the most efficient algorithm which achieves the best quality of the results in the shortest possible time.

Keywords: Consensus · Data integration · Heuristic
Genetic algorithm

1 Introduction

Nowadays, an importance of the information is very fundamental. Unless the most essential, information is one of the most valuable resources in the World. Information increase speed is enormous high, however the evolution of the technology allows to collect it. The task of processing a large set of data is still a big problem. Information becomes from distributed, inconsistent resources and the form of storage it can be completely different. The determination of a one, consistent and reliable version seems to be an essential problem. It can be solved by using Consensus Theory [17], thus, a result of the aforementioned integration can be called a consensus.

Solving such problem, at first requires establishing a structure which is used to store data. Complex structures allow to reflect the elements and relationships of real world better, however they can also complicate their processing. In this paper, we have assumed that data is represented by binary vectors. Although such structure is quite simple, many processes, that take place in the real world, can be easily and conveniently modelled using them.

The final consensus should represent all of the input data in the best way. In this paper, we would like to obtain the most "fair" consensus. Such result is only possible to obtain if the sum of squared distances between the designated

N. T. Nguyen et al. (Eds.): ACIIDS 2018, LNAI 10751, pp. 48–58, 2018.
https://doi.org/10.1007/978-3-319-75417-8_5

consensus and input data is the smallest. Thus, the mentioned distance is the most uniform. In the literature [18], a consensus which fulfils this condition is called the 2-optimality consensus. Unfortunately, it is the NP-complete problem and only a brute-force method allows to find the optimal solution, which entails that for the large set of input data, the time of execution of such algorithm is not acceptable. Therefore, heuristic algorithms for the determination of the 2-optimality consensus are desirable.

The main aim of this paper is finding the effective heuristic method which allows maintain the balance between the time of performance and the quality of the final consensus. In this work, some algorithms are proposed. The developed methods are compared with heuristics known from the literature [17,18]. All of the obtained results are statistically analysed and general conclusions are drawn.

The article is structured as follows. In the next Section, the short overview about similar researches is described. In Sect. 3 authors present the basic notions of the Consensus Theory. Section 4 contains the description of the proposed heuristics. In Sect. 5 the results of the experimental verification with the statistical analysis are presented. Section 5 concludes the paper.

2 Related Works

The data integration problem is very common and known in the literature for a long time. The main problem of choosing a proper theory can be formulated: *for a given set X being a subset of a universe U the consensus choice concerns on a selection of a subset of X.* The choosing of the subset of X is not random because some criteria must be satisfied.

There are many approaches to data integration. One of them is a widely known the Consensus Theory. Barthelemy and Monjardet [5] proposed two classes of problems connected with this theory:

- problems, in which a certain and a hidden structure is searched
- problems, in which inconsistent data related to the same subject is unified.

In this paper, we focus on problems described in the second class. Considering solving the consensus problem we can follow one of four known approaches: axiomatic, constructive, optimisation and boolean reasoning.

In the axiomatic approach, the set of axioms is given and it specifies all conditions, which must be fulfilled by a consensus choice function. Form and structure of the axioms are related with a problem being solved. In [16,18] 10 postulates for the consensus choice function were proposed: reliability, unanimity, simplification, quasi-unanimity, consistency, Condorcet consistency, general consistency, proportion, *1-optimality*, *2-optimality*.

In the constructive approach, a consensus determination problem is solved by considering a microstructure and a macrostructure of the universe U. A microstructure of U is defined as a structure of its elements and a macrostructure as a relation between elements of U. Many different microstructures have been investigated yet: linear orders [1,15]; semi-lattices [3]; n-tree [2,8]; ordered partitions and coverings [7,18]; non-ordered partitions [4]; weak hierarchies [14], ontologies [10,11] and binary vectors [9,12,13].

The optimisation approach is based on some optimality rules. They are classified into one of the following groups: global optimality rules, Condorcet's optimality rules and maximal similarity rules.

The last approach, the boolean reasoning, look upon the consensus problem as an optimisation problem, which is coded as Boolean formulas. The first part of those determines a solution of the problem [6,19].

Analysis of the *2-optimality* consensus determination problem has been conducted in some articles yet. Due to the fact that determination of the *2-optimality* consensus is NP-complete problem authors proposed heuristic algorithms solving this task [17].

In our previous works [9,12,13], for the assumed micro- and macrostructure, investigations of the mean error for a different number of vectors in the profile and vectors' length was conducted. Researches showed, that the difference between optimal result achieved using the brute-force method and the consensus determined using methods mentioned in [17] is less than 5%. So far, no other methods solving the *2-optimality* consensus problems have not been investigated in details. Therefore, this paper is the extension of the previous works.

3 Basic Notions

The Consensus Theory is widely used in many problems like data and knowledge integration, the determination of a consistent decision based on experts' opinions, the determination the empty value of attributes in databases and many others. Below, some basic definitions which are used in this paper are presented.

By U we denote a finite, nonempty set of a universe of objects. Each object can reflect the potential elements of a knowledge referring to a certain world. The set of all subsets with repetitions of U can be marked as 2^U. Let $\Pi_b(U)$ be the set of all b-element subset (with repetitions) of the set U for $b \in N$. Thus $\Pi(U) = \bigcup_{b \in N} \Pi_b(U)$ is the set of all nonempty subsets with repetitions of the universe U. By a knowledge profile (or shortly, a profile) we call each X which belongs to $\Pi(U)$.

The consensus determination problem requires finding x from the universe of objects U which can be treated as the best representation of a profile X. It can be found only if the micro- and macrostructure of the universe are known.

Definition 1. *The macrostructure of the set U is a distance function $\delta : U \times U \rightarrow [0,1]$ which satisfies the following conditions [17]:*

1. $\forall_{v,u \in U}, \delta(v,u) = 0 \Leftrightarrow v = u$
2. $\forall_{v,u \in U}, \delta(v,u) = \delta(u,v)$

Definition 1 lacks a transitive condition because it is too strong for many practical situations. Therefore, a space (U,d), defined as above, does not need to be a metric space. We call it a distance space.

Definition 2. *For the assumed distance space (U, δ), the consensus choice problem requires establishing the consensus choice function. A consensus choice function in space (U, δ) is defined in the following way:*

$$C : \Pi(U) \to 2^U \tag{1}$$

By $C(X)$ we denote the representation of $X \in \Pi(U)$. By $c \in C(X)$ we call a consensus of a profile X.

In [17,18] authors presented some postulates for a consensus choice functions. In this paper, we have only focus on *2-optimality*. The *2-optimality* criterion can be referred as the most 'fair' consensus. Formally, this postulate is defined in the following way:

Definition 3. *For a profile $X \in \Pi(U)$ a consensus choice function C satisfies the postulate of 2-optimality iff $(x \in C(X) \Rightarrow (\delta^2(x, X) = \min_{y \in U} \delta^2(y, X))$, where $\delta^2(x, X) = \sum_{y \in X} (\delta(x, y))^2$.*

4 Heuristics for the 2-Optimality Consensus Determination

Considering the consensus problem we have to establish a micro- and a macrostructure of universe U. In this paper, we assume that the profile X, belonging to this universe, consists of n binary vectors, which length is equal to N. Formal definition is as follows: universe $U = \{u_1, u_2, \dots\}$, where each element is a binary vector, the profile $X = \{a_1, a_2, ..., a_n\} \in \Pi(U)$, where: $a_i = (a_i^1, a_i^2, ..., a_i^N), i \in \{1, ..., n\}, a_i^j \in \{0, 1\}, j \in \{1, ..., N\}$. The distance function is defined as: $\delta(w, v) = \sum_{j=1}^{N} |w^j - v^j|$ for such $w, v \in U$ that $w = (w^1, w^2, ..., w^N), v = (v^1, v^2, ..., v^N), v^q, w^q \in \{0, 1\}, q \in \{1, ..., N\}$.

For the assumed distance space (U, δ) we use four different algorithms determining the *2-optimality* consensus. The first algorithm [17], denoted in the future as $h1$, is presented as Algorithm 1.

The second heuristic algorithm (denoted as $h2$) considered in this paper, is a modification of the first one. Instead of choosing x in a random way, it computes as the $1-$ *optimality* consensus [17]. Steps of the method are shown as Algorithm 2.

Another method applied to determine the *2-optimality* consensus is a genetic algorithm. The basic idea is presented in Algorithm 3. In this paper we use two different selection methods. In the first example, denoted as $gen1$, we use a roulette-wheel selection and $f(x) = \frac{1}{\delta(x,X)+1}$ as the fitness function. In the second one, denoted as $gen2$, we use a tournament selection with the size of tournament equal to 3 and the fitness function defined as: $f(x) = \delta(x, X)$.

We propose our own heuristic algorithm ($h3$) of determining the *2-optimality* consensus, which is based on the method $h1$. Results are computed using Algorithm 4.

Algorithm 1. Basic heuristic algorithm $h1$

Require: $X = \{a_1, a_2, ..., a_n\}$
Random x
$md = \delta^2(x, X)$
for $i = 1$ to N **do**
    $x[i] = x[i] \oplus 1$
    if $\delta^2(x, X) < md$ **then**
        $md = \delta^2(x, X)$
    else
        $x[i] = x[i] \oplus 1$
    end if
end for

Algorithm 2. Heuristic algorithm $h2$

Require: $X = \{a_1, a_2, ..., a_n\}$
Compute x using 1-optimality algorithm
$md = \delta^2(x, X)$
for $i = 1$ to N **do**
    $x[i] = x[i] \oplus 1$
    if $\delta^2(x, X) < md$ **then**
        $md = \delta^2(x, X)$
    else
        $x[i] = x[i] \oplus 1$
    end if
end for

Algorithm 3. Genetic algorithm determining 2-optimality consensus gen

Require: $X = \{a_1, a_2, ..., a_n\}, pop\_size, iterations, p\_cross, p\_mut$
Fill an initial population P with pop_size random binary vectors
Compute a fitness function for each individuals of population P
for $i = 1$ to $iter\_num$ **do**
    $newP$ = select individuals from P with the best value of fitness function
    Crossover individuals from new population $newP$ in pairs with probability equal to p_cross
    Mutate individuals from new population $newP$ on one position with probability p_mut
    $P = newP$
    Compute a fitness function for each individuals of population P
end for
x = best individual from population P

Algorithm 4. Proposed heuristic $h3$

Require: $X = \{a_1, a_2, ..., a_n\}$

```
sum = 0
for i = 1 to N do
    f[i] = 0
    x[i] = 0
    for all a in X do
        if a[i] = 1 then
            f[i] = f[i] + 1
            sum = sum + 1
        end if
    end for
end for
to_change = round(sum/n)
positions = choose to_change positions from f with the highest value
for all position in positions do
    x[position] = 1
end for
md = δ²(x, X)
for i = 1 to N do
    x[i] = x[i] ⊕ 1
    if δ²(x, X) < md then
        md = δ²(x, X)
    else
        x[i] = x[i] ⊕ 1
    end if
end for
```

5 Experimental Evaluation of Heuristic Algorithms for 2-Optimality Consensus Determination

The experiment was conducted using a special dedicated environment. We performed two experiments: in the first one, we compared genetics algorithms $gen1$ and $gen2$ with the heuristic $h1$ and the brute-force method (bf), which is the only optimal algorithm. In the second, we investigated the performance of methods $h1$, $h2$ and $h3$ in comparison to the brute-force approach. As in the previous works [9,12], we used the length of vectors equal to $N = 12$ and $N = 15$ for the first evaluation. For the second, the parameters where equal to $N = 10$ and $N = 15$. The number of vectors is equal to $n = 3000$ for the first experiment and respectively $n = 15015$, $n = 14175$ for second. The datasets consisted of randomly generated vectors from a uniform distribution. All statistical analysis were made using a significance level $\alpha = 0.01$.

5.1 Statistical Analysis Genetic Algorithms in Comparison to $h1$ and Brute-Force Method

The consideration of the first experiments' results for shorter vectors we start with an analysis of normalised consensus distance from profile, received from algorithms and defined as: $\frac{\sum_{y \in X} \delta^2(x,y)}{n}$. Before selecting a proper test we analysed the distribution of the obtained data by using the Shapiro-Wilk test. The received $p-value$ greater than α suggests, that we cannot reject the null hypothesis and claim that the data comes from a normal distribution. Equality of the variance for each sample allows us to use the one-way ANOVA test for independent samples. The statistic test value $F = 21.557$ and $p - value < 0.000001$ less than α suggests that we should reject the null hypothesis that samples are from the same distribution. In the next step, we analyse the difference between samples' average distance in pairs by using the Tamhane's T2 post-hoc test. As the result, we get that the genetic algorithm $gen1$ is worse than the genetic algorithm $gen2$ by 1.43%, worse then the heuristic $h1$ by 1.34% and 1.43% then the optimal solution. The genetic algorithm $gen2$ returns result equal to brute-force and better than the heuristic $h1$ by 0.1%. The algorithm $h1$ is worse than the brute-force by 0.1%.

Next, we analyse the performance time of this algorithms. The Shapiro-Wilk test for all samples points out that samples do not come from a normal distribution. The statistical value $H = 72.984$ for the Kruskal-Wallis test and $p - value < 0.000001$ suggests, that at least one sample is different. We compare then medians for samples in pairs. Conducted computations show, that the genetic algorithm $gen1$ is faster by 16.98% than the genetic algorithm $gen2$ and by 40.23% than the brute-force, however by 99.47% slower then the heuristic $h1$. The algorithm $gen2$ is slower by 99.56% than the heuristic $h1$ and faster than the brute-force by 28.01%. The heuristic $h1$ is faster by 99.68% than the brute-force.

In a similar way, we conduct tests for longer vectors. We start with an analysis of distributions of samples' distance by using the Shapiro-Wilk test. Results show, that samples come from the normal distribution ($p-value > \alpha$). The one-way ANOVA test for independent samples points out, that samples do not come from the same distribution because $F = 37.457$ and $p - value < 0.000001$. The comparison of averages distance achieved using the Tamhane's T2 post-hoc test shows, that the genetic algorithm $gen1$ is worse than the genetic algorithm $gen2$ by 1.79%, by 1.73% than the heuristic $h1$ and by 1.8% then the optimal result. The genetic algorithm $gen2$ returns result better by 0.07% then the heuristic $h1$ and worse by 0.01% then the brute-force. The algorithm $h1$ is worse than the brute-force by 0.08%.

In the last step of this experiment, we investigate the performance time for algorithms. $P - value > \alpha = 0.01$ achieved for each sample using the Shapiro-Wilk test suggests that samples come from a normal distribution. Results of the multiple-sample test suggest that samples have different variances, therefore for the further analysis we use the Welch and Brown-Forsythe one-way ANOVA test for independent samples. $F = 6523.529$ and $p - value < 0.000001$ allow us to reject the null hypothesis, that all samples come from the same distribution. At the end, we compare samples' average distance in pairs by using the Tamhane's T2 post-hoc test. Results show, that the genetic algorithm $gen1$ is faster than the algorithm $gen2$ by 36.27% and by 90.48% then the brute-force, however slower than the heuristic $h1$ by 99.5%. The genetic algorithm $gen2$ is slower than the heuristic $h1$ by 99.68% and faster then the optimal algorithm by 85.07%. The heuristic $h1$ is faster than the brute-force by 99.95%. Results of the average distance for samples and the average performance time for both vector's length are presented in the Fig. 1.

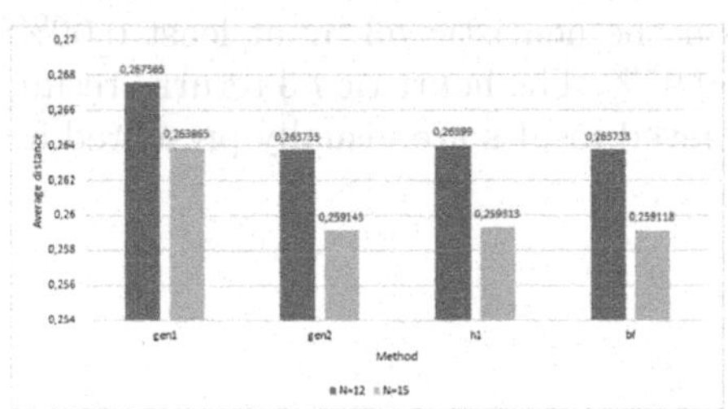

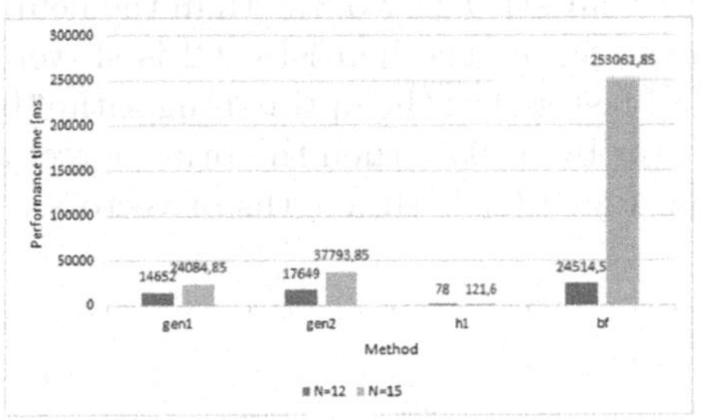

Fig. 1. Average distance between consensus and elements of profile and algorithms performance time for genetic algorithm comparison

5.2 Statistical Analysis the Proposed Heuritic Algorithms

In the second stage of our research, we analyse the distance between consensus and profile's elements achieved using methods $h1$, $h2$, $h3$ and the brute-force, as well as the performance time for these algorithms. We start with a study of the distance for short vectors. The $p-value$ of the Shapiro-Wilk test for all samples is greater than α, therefore we claim that samples come from a normal distribution. The equality of samples' variances allows to use the one-way ANOVA test for independent samples. The statistical value $F = 0.130954$ and $p - value = 0.941$ suggest, that there is no reason to reject the null hypothesis, and we claims, that all samples come from the same distribution. Therefore all methods return results with the similar quality.

Next, we analyse the performance time for all methods. For one sample the Shapiro-Wilk test returns $p - value < \alpha$. This sample does not come from the normal distribution. We use the Kruskal-Wallis test to investigate equality of distributions. $H = 53.037$ and $p - value < 0.000001$ show that samples are different. Then, we compare medians of samples in pairs. Results show, that the heuristic $h1$ is faster by 4.19% then the heuristic $h2$, by 5.92% by the heuristic $h3$ and by 98.92% in comparison to the brute force. The heuristic $h2$ is faster than the heuristic $h3$ by 1.81% and by 98.87% then the brute-force. The heuristic $h3$ is faster than the optimal algorithm by 98.85%.

In the second stage of this experiment, we analyse longer vectors. $p - value$ for each distance sample is greater than α, therefore we claim, that samples come from a normal distribution. All conditions are satisfied to use the one-way ANOVA test for independent samples for further analysis. $F = 0.410$ and $p - value = 0.746$ allow us to claim, that all samples come from the same distribution. There is no significant difference between distances for all methods.

The last investigation concentrates on the performance time. As previous, we start with analysis if samples come from the normal distribution by using the Shapiro-Wilk test. For each method $p-value$ is less than α, therefore we reject a null hypothesis. The Kruskal-Wallis statistical value $H = 57.244$ and $p-value < 0.000001$ suggests, that samples do not come from the same distribution. The comparison of medians in pairs points out, that the heuristic $h1$ is faster then the heuristic $h2$ by 3.8%, than the heuristic $h3$ by 3.71% and by 99.95% than the brute-force. The heuristic $h2$ is slower than the heuristic $h3$ by at least 0.09% and faster then the optimal-algorithm by 99.95%. The heuristic $h3$ returns result faster by 99.95% then the brute-force. Achieved results are visually presented in the Fig. 2 for both lengths of vectors.

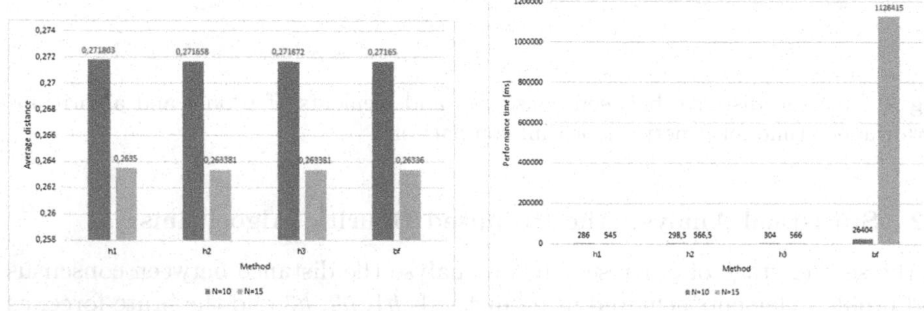

Fig. 2. Average distance between consensus and elements of profile and algorithms performance time for new algorithm comparison

6 Future Works and Summary

In this paper, the comparison of different methods of determining the *2-optimality* consensus has been presented. The analysis has been focused on the distance between the determined consensus and the profile's elements and on the performance time of algorithms.

Our research has shown, that all of the presented heuristics return results with almost the same quality. There are no significant differences between the optimal solution and results of those algorithms. Therefore, a new method, proposed in this paper as Algorithm 4, could be a good alternative. What is more, the genetic algorithms could be successfully used in determining the *2-optimality* consensus, because they are able to give optimal results. However, the time of performance is much higher in comparison to other heuristic methods and selecting good parameters is not a simple problem.

In the future, more research concentrating on the consensus determination is planned. We intend to investigate the influence of a profile modification on the quality of the consensus and propose the best algorithm keeping in balance the time of a performance and the quality of the result. Additionally, the analysis of differences between a one- and a multi-level approach to the consensus determination for different data structures is planned.

References

1. Arrow, K.J.: Social Choice and Individual Values. Wiley, New York (1963)
2. Barthelemy, J.P.: Thresholded consensus for n-trees. J. Classif. **5**, 229–236 (1988)
3. Barthelemy, J.P., Janowitz, M.F.: A formal theory of consensus. SIAM J. Discrete Math. **4**(1991), 305–322 (1991)
4. Barthelemy, J.P., Leclerc, B.: The median procedure for partitions. DIMACS Ser. Discrete Math. Theor. Comput. Sci. **19**, 3–33 (1995)
5. Barthelemy, J.P., Monjardet, B.: The median procedure in cluster analysis and social choice theory. Math. Soc. Sci. **1**, 235–267 (1981)
6. Brown, F.N.: Boolean Reasoning. Kluwer Academic Publisher, Dordrecht (1990)
7. Daniłowicz, C., Nguyen, N.T.: Consensus-based partition in the space of ordered partitions. Pattern Recogn. **21**, 269–273 (1988)
8. Day, W.H.E.: Consensus methods as tools for data analysis. In: Bock, H.H. (ed.) Classification and Related Methods for Data Analysis, pp. 312–324. North-Holland, Amsterdam (1988)
9. Kozierkiewicz-Hetmańska, A.: Comparison of one-level and two-level consensuses satisfying the 2-optimality criterion. In: Nguyen, N.-T., Hoang, K., Jędrzejowicz, P. (eds.) ICCCI 2012. LNCS (LNAI), vol. 7653, pp. 1–10. Springer, Heidelberg (2012). https://doi.org/10.1007/978-3-642-34630-9_1
10. Kozierkiewicz-Hetmańska, A., Pietranik, M.: The knowledge increase estimation framework for ontology integration on the concept level. J. Intell. Fuzzy Syst. **32**(2), 1161–1172 (2017)
11. Kozierkiewicz-Hetmańska, A., Pietranik, M., Hnatkowska, B.: The knowledge increase estimation framework for ontology integration on the instance level. In: Nguyen, N.T., Tojo, S., Nguyen, L.M., Trawiński, B. (eds.) ACIIDS 2017. LNCS (LNAI), vol. 10191, pp. 3–12. Springer, Cham (2017). https://doi.org/10.1007/978-3-319-54472-4_1
12. Kozierkiewicz-Hetmańska, A., Nguyen, N.T.: A comparison analysis of consensus determining using one and two-level methods. In: Advances in Knowledge-Based and Intelligent Information and Engineering Systems, pp. 159–168 (2012)
13. Kozierkiewicz-Hetmańska, A., Sitarczyk, M.: The efficiency analysis of the multi-level consensus determination method. In: Nguyen, N.T., Papadopoulos, G.A., Jędrzejowicz, P., Trawiński, B., Vossen, G. (eds.) ICCCI 2017. LNCS (LNAI), vol. 10448, pp. 103–112. Springer, Cham (2017). https://doi.org/10.1007/978-3-319-67074-4_11
14. McMorris, F.R., Powers, R.C.: The median function on weak hierarchies. DIMACS Ser. Discrete Math. Theor. Comput. Sci. **37**, 265–269 (1997)
15. Mirkin, B.G.: Problems of Group Choice. Nauka, Moscow (1974)
16. Nguyen, N.T.: Using distance functions to solve representations choice problems. Fundamenta Informaticae **48**(4), 295–314 (2001)

17. Nguyen, N.T.: Consensus choice methods and their application to solving conflicts in distributed systems. Wroclaw University of Technology Press (2002). (in Polish)
18. Nguyen, N.T.: Advanced Methods for Inconsistent Knowledge Management. Springer, London (2008). https://doi.org/10.1007/978-1-84628-889-0
19. Pawlak, Z.: Rough Sets-Theoretical Aspects of Reasoning About Data. Kluwer Academic Publisher, Dordrecht (1991)

A Memory-Efficient Algorithm with Level-Order Unary Degree Sequence for Forward Reasoning Engines

Hiromu Hiidome, Yuichi Goto(✉), and Jingde Cheng

Department of Information and Computer Sciences, Saitama University, Saitama 338-8570, Japan
{hiromu,gotoh,cheng}@aise.ics.saitama-u.ac.jp

Abstract. A forward reasoning engine is an indispensable component in many advanced knowledge-based systems with purposes of creation, discovery, or prediction. Time-efficiency and memory-efficiency are crucial issues for any forward reasoning engine. FreeEnCal is a forward reasoning engine for general-purpose, and has been used for several applications, e.g., automated theorem finding. To improve time-efficiency, current implementation of FreeEnCal uses "trie", which is a kind of tree structure, to store all logical formulas that are given or deduced in FreeEnCal. However, the implementation is not so memory-efficient from view point of applications of FreeEnCal. The paper presents a memory-efficient algorithm of FreeEnCal, and shows theoretical evaluation of the algorithm. The algorithm uses level-order unary degree sequence (LOUDS) that is a kind of succinct data structures and is used to represent tree structures concisely. By using LOUDS to construct trie in FreeEnCal, we can improve memory-efficiency of FreeEnCal.

Keywords: Forward reasoning engine · Succinct data structures
Level-order unary degree sequence · Memory efficiency

1 Introduction

A forward reasoning engine is a computer program to automatically draw new conclusions by repeatedly applying inference rules, which are programmed in the reasoning engine or given by users to the reasoning engine as input, to given premises and obtained conclusions until some previously specified conditions are satisfied. It is an indispensable component in many advanced knowledge-based systems with purposes of creation, discovery, or prediction [3]. In any application, the most demanded functionality of a forward reasoning engine is to deduce enough number of conclusions in acceptable time. Therefore, time-efficiency and memory-efficiency are crucial issues for forward reasoning engines.

FreeEnCal [3] is a forward reasoning engine for general-purpose, that provides an easy way to customize reasoning task by providing different axioms,

N. T. Nguyen et al. (Eds.): ACIIDS 2018, LNAI 10751, pp. 59–70, 2018.
https://doi.org/10.1007/978-3-319-75417-8_6

inference rules, and facts and hypotheses as input data of FreeEnCal. FreeEnCal has been used for several applications, e.g., automated theorem finding [1,5], formal analysis with reasoning for cryptographic protocols [11,12], and so on. To improve time-efficiency of FreeEnCal, an algorithm of FreeEnCal with a data structure "trie" that is a kind of tree structures [4] was proposed and its implementation was developed [8,9]. Trie is used to store a set of all logical formulas that are given or deduced in FreeEnCal. However, the implementation is not so memory-efficient from view point of applications of FreeEnCal.

On the other hand, Jacobson proposed succinct data structure that enables to compress data size nearly information-theoretically and support operations without lowering access rate in 1989 [6]. Level-order unary degree sequence (LOUDS) [6] that is a kind of succinct data structure to represent tree structures. Therefore, we can construct a memory-efficient trie by using LOUDS.

This paper presents a memory-efficient algorithm with LOUDS for FreeEnCal. The paper also shows theoretical evaluation of the proposed algorithm. The proposed algorithm is effective not only for FreeEnCal, but also for other forward reasoning engines, because FreeEnCal is a typical forward reasoning engine.

The rest of the paper is organized as follows: Sect. 2 gives brief explanation of LOUDS. Section 3 explains overview of FreeEnCal. Section 4 presents a memory-efficient algorithm with LOUDS. Theoretical evaluation of the algorithm is shown at Sect. 5. Section 6 gives conclusion and future works.

2 LOUDS: Level-Order Unary Degree Sequence

LOUDS represents a tree by using a bit vector that called LOUDS bit-string (LBS). Currently, it is used as space efficient dictionary [10]. Suppose that the number of nodes of a tree is n. LOUDS needs $2n + o(n)$ bits to represent the tree [6]. By contrast, about $128n$ bits is needed if we represent the tree by using pointers, "link-based tree" for short. A link-based tree uses 64 bits per a node and 64 bits per an edge. The number of edges of the tree is $n - 1$. Thus, a link-based tree needs about $128n$ bits. LOUDS is memory-efficient rather than link-based tree.

LBS is constructed as follows. Suppose that a tree has an imaginary super-root node that is connected with its root node like Fig. 1. We start to construct LBS from super-root in breadth first search. In each node, we add "1" into LBS i times if the current node has i child nodes, and then we add "0" into LBS. In this way, "1" corresponds to an edge, and "0" corresponds to a node. Therefore, LBS needs $2n + 1$ bits to store a tree that has n nodes [6].

In this paper, id of an edge is defined as follows. (1) Id of an edge is 1 iff the edge connects the super-root and the root (1st node) in a tree. (2) Id of an edge is j ($1 < j \leq n$) iff the edge connects j-th node and its parent node where the number of nodes in the tree is n. In addition, if ids of a node and an edge are same, the node corresponds to the edge, and vise vasa. Suppose α and β are edges. α is a child/parent/sibling edge of β if a node corresponding to α is a child/parent/sibling node of a node corresponding to β, respectively.

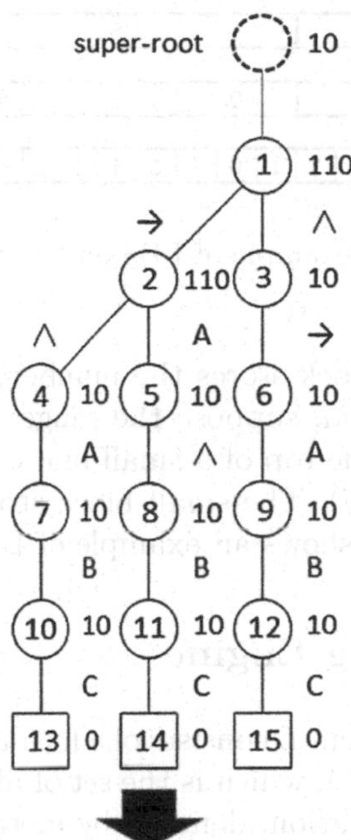

Fig. 1. An example of an original trie and a LOUDS-based trie

On the other hand, trie is used to store a set of logical formulas in FreeEnCal in order to reduce cost of comparison between two logical formulas [8,9]. Trie is a kind of tree structure, and each edge has a label (character). A path from the root to a leaf expresses a logical formula in FreeEnCal. To represent trie by using LOUDS, an array is necessary to store the labels. Figure 1 shows an example of a trie and a LOUDS-based trie storing three logical formulas, i.e., "$\rightarrow \wedge ABC$", "$\rightarrow A \wedge BC$" and "$\wedge \rightarrow ABC$" where those formulas are represented in prefix notation. A label corresponding to i-th edge is stored into i-th element in an array.

Navigation on the LOUDS is performed by *rank* and *select* operations on the LBS [6]. Index data that needs asymptotically $o(n)$ bits is given to execute *rank* and *select* in $O(1)$ times. These operations are executed as follows.

*rank*0(i): return the number of 0 bits to the left of position i in LBS.
*rank*1(i): return the number of 1 bits to the left of position i in LBS.
*select*0(i): return the position of the i-th 0 bit in the LBS.
*select*1(i): return the position of the i-th 1 bit in the LBS.

Index data is created as follows. We divide LBS into some large blocks, and divide large blocks into some small blocks. Suppose the top of a large block is

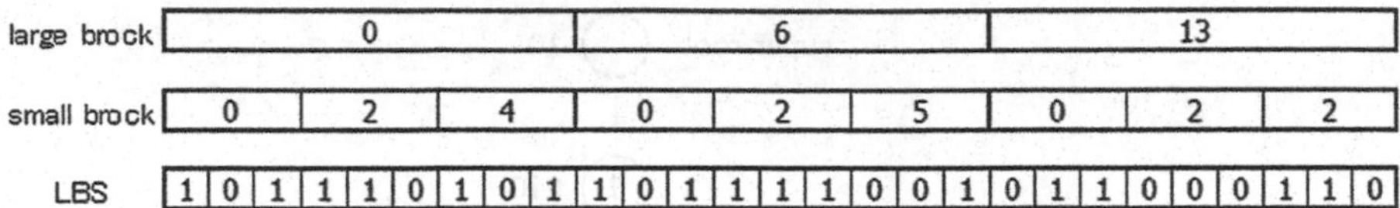

Fig. 2. An example of LBS and its index data

i-th bit of LBS. The large block stores the number of "1" from top of the LBS to i-th bit. On the other hand, suppose the range of a large block is from i-th bit to j-th bit of LBS, and the top of a small block belonging to the large block is k-th bit of LBS ($i < k < j$). The small block stores the number of "1" from i-th bit to k-th bit. Figure 2 shows an example of LBS and its index data.

3 Forward Reasoning Engine

In general, a formal logic system L consists of a formal language, called the object language and denoted by $F(L)$, which is the set of all well-formed formulas of L, and a logical consequence relation, denoted by meta-linguistic symbol $\vdash_L$, such that $P \subseteq F(L)$ and $c \in F(L)$, $P \vdash_L c$ means that within the frame work of L, c is valid conclusion of premises P, i.e., c validly follows from P. For a formal logic system $(F(L), \vdash_L)$, a logical theorem t is a formula of L such that $\phi \vdash_L t$ where ϕ is empty set. Let $Th(L)$ denote the set of all logical theorems of L. $Th(L)$ is completely determined by the logical consequence relation $\vdash_L$. According to the representation of the logical consequence relation of a logic, the logic can be represented as a Hilbert style axiomatic system, a Gentzen natural deduction system, a Gentzen sequent calculus system, or other type of formal systems.

A formal theory with premises P based on L, called a L-theory with premises P and denoted by $T_L(P)$, is defined as $T_L(P) := Th(L) \cup Th^e_L(P)$, and $Th^e_L(P) := \{et \mid P \vdash_L et \text{ and } et \notin Th(L)\}$ where $Th(L)$ and $Th^e_L(P)$ are called the logical part and the empirical part of the formal theory, respectively, and any element of $Th^e_L(P)$ is called an empirical theorem of the formal theory.

According to the above definition, FreeEnCal [3] is a computer program that can automatically obtain a fragment (a subset) of $Th(L)$ and $Th_L(P)$. FreeEnCal uses degrees of nested logical connectives or modal operators in a logical formulas as a restriction to obtain a finite subset of $Th(L)$ or $Th_L(P)$. How to create such subsets is defined in [3].

FreeEnCal mainly consists of following five processes, and performs the processes repeatedly. **Inputting process:** it takes given logical formulas as premises, inference rules, and degrees of nested logical connectives or modal operators as input data, then stores the logical formulas into a formula buffer formed by trie on the memory space. **Derivation process:** it checks whether an inference rule can apply to each of tuples which consist of logical formulas which have not been checked yet in the formula buffer. If the inference rule can apply to a tuple, it deduces logical formulas according to the applied inference rule. Then, deduced logical formulas are stored in the temporal formula buffer. **Duplication checking process:** it finds all of deduced logical formulas which are duplicate of given premises or previously deduced logical formulas. Found duplicate formulas are removed from the temporal formula buffer. **Adding process:** it adds all logical formulas which are not duplicate into the formula buffer. In other words, the temporal formula buffer is merged into the formula buffer. **Outputting process:** it outputs current deduced logical formulas into a file.

The dominant and time-consuming processes of FreeEnCal are unification and pattern matching between two logical formulas. Suppose A and B are logical formulas including meta variables where meta variables are variables that can be replaced by any logical formulas. If A and B are unified then A and B are transformed into a same formula by substituting sub-formulas of A and B into their meta variables recursively. *Unification between* A and B is a process: (1) check whether A and B can be unified or not, (2) if yes, obtain the unified formula of them. If A matches B then A is transformed into B by substituting sub-formulas of B into meta variables in A. *Pattern matching* between A and B is a process to check whether A can match B or not. If A can match B, then A is duplicate of B. Unification is performed frequently in derivation process, and pattern matching is also performed frequently in duplication process. To improving efficiency of FreeEnCal, fast algorithms for unification and pattern matching with trie are proposed and implemented [8,9].

4 A Memory-Efficient Algorithm with LOUDS for Forward Reasoning Engines

As shown in Sect. 2, LOUDS-based trie is more memory-efficient than link-based trie so that we improve the memory-efficiency of FreeEnCal by using LOUDS-based trie.

To realize FreeEnCal with LOUDS-based trie, we should change algorithms of main operations to a set of logical formulas stored as trie. The main operations are creation, unification, pattern matching, deletion, and combination. Creation is to create a trie, e.g., the formula buffer or the temporal formula buffer, from given or deduced logical formulas. Original algorithms of unification and pattern matching on trie are described in [8,9]. Deletion is to delete duplicate logical formulas from the temporal formula buffer after pattern matching on trie in duplication checking process. Combination is to merge the temporal formula buffer with the formula buffer, in other word, to merge two tries.

Primitive methods for operations on LOUDS-based are as follows. Suppose i denotes the position on a LBS. The proposed algorithm traces edges of LOUDS-based trie.

node_id(i, LBS) $= rank1(i)$**:** if the position i points to an edge, the method returns id of the node corresponding to the edge. If not, the return value is not used.

is_edge(i, LBS) $= rank1(i) - rank1(i-1)$**:** if the position i points to an edge, then the method returns true (1). if not, return false (0).

first_child(i, LBS) = $select0(rank1(i)) + 1$: if an edge where the position i points has child edges, then the method returns the position of a child edge whose id is the minimum among ids of the child edges. If not, the method returns the position of a node corresponding to an edge where the position i points.

sibling(i) = $i + 1$: if an edge where the position i points has a right sibling edge, then the method returns the position of the right sibling edge. If not, the return value is not used.

parent(i, LBS) = $select1(rank0(i))$: the method returns the position of a parent edge of an edge where the position i points.

have_sibling(i, LBS) = **is_edge(sibling(i), LBS)**: if an edge where the position i points has a right sibling edge, then the method returns true (1). If not, it returns false (0).

to_edge(i, LBS) = $select1(rank0(i) - 1)$: if the position i points a node, the method returns the position of an edge corresponding to the node.

Creation and Combination algorithms for LOUDS-based trie uses an algorithm to merge two LOUDS-based tries in described in [7]. The combination algorithms is the algorithm of [7] itself. The creation algorithm is as follows: (1) divide a set of given or deduced logical formulas into several subsets whose size is small enough to continue to do forward reasoning; for each divided subset, (2) create a trie represented by some notation except LOUDS; (3) create a LOUDS-based trie from the trie; (4) merge the LOUDS-based trie and the previously created LOUDS-based trie according to the algorithm of [7].

Unification algorithms for LOUDS-based trie are as follows. In following algorithm, "*Idx"denotes a current position in LBS. "FrontIdx" is a constant whose value is 3 because it points to the first edge that has a label in a trie. "*LBS" denotes a variable to store LBS of a trie. "*LabelAry" denotes a array to store labels of a trie. IRLBS and IRLabelAry are used for a trie to represent a logical formula in inference rules. FBLBS and FBLabelAry are used for a trie to represent logical formulas in formula buffer. NewLBS and NewLabelAry are used for a trie to represent deduced logical formulas in derivation process. "BindM" is a sets of tuples: a sentential variable, a sub-formula, and the sub-formula's LBS. For example, suppose BindM is $\{< A, \rightarrow ab, 101010100 >\}$. The tuple means a sub-formula $\rightarrow ab$ is substituted for a sentential variable A and the LBS of the sub-formula is "101010100." **find(c)** is a method of BindM. BindM.find('A') returns $<\rightarrow ab, 101010100 >$. "*Stack" is a stack, and it stores a tuple that consists of a current position in a LBS of a trie, the LBS, and an array to store labels of the trie. A method **get_tail_index(index, sub-formula, LBS)** returns the position of an edge that corresponds to the last letter of sub-formula that starts from index in LBS.

Algorithm 1. *Unify* (IRIdx, &IRLBS, &IRLabelAry, IRStack FBIdx, &FBLBS, &FBLabelAry, FBStack, BindM)

```
if (is_edge(IRIdx, IRLBS) is false) and (IRStack is empty) and
(is_edge(FBIdx, FBLBS) is false) and (FBStack is empty) then
  return BindM
else if is_edge(IRIdx, IRLBS) is false then
  <index, LBS, Ary> ⇐ IRStack.pop
  ChildIRIdx ⇐ first_child(index, LBS)
  Unify(ChildIRIdx, LBS, Ary, IRStack, FBIdx, FBLBS, FBLabelAry,
  FBStack, BindM)
else if is_edge(FBIdx, FBLBS) is false then
  <index, LBS, Ary> ⇐ FBStack.pop
  ChildFBIdx ⇐ first_child(index, LBS)
  Unify(IRIdx, IRLBS, IRLabelAry, IRStack, ChildFBIdx, LBS, Ary,
  FBStack, BindM)
else
  for (i ⇐ FBIdx; is_edge(i, FBLBS) is true; i ⇐ sibling(i)) do
    c ⇐ IRLabelAry[node_id(IRIdx, IRLBS)]
    d ⇐ FBLabelAry[node_id(i, FBLBS)]
    if c or d is variable then
      if c is variable and c is substituted then
        IRStack.push(<IRIdx, IRLBS, IRLabelAry>)
        < SubstiAry, SubstiLBS > ⇐ BindM.find(c)
        Unify(FrontIdx, SubstiLBS, SubstiAry, IRStack, i, FBLBS, FBLa-
        belAry, FBStack, BindM)
      else if c is variable and c is not substituted then
        for each sub-formula f starting from i do
          CpBindM ⇐ BindM
          CpBindM.push(<c, the pointer of f, the pointer of LBS of f>)
          TailIdx ⇐ get_tail_index(i, f, FBLBS)
          ChildIRIdx ⇐ first_child(IRIdx, IRLBS)
          ChildFBIdx ⇐ first_child(TailIdx, FBLBS)
          Unify(ChildIRIdx, IRLBS, IRLabelAry, IRStack, ChildFBIdx,
          FBLBS, FBLabelAry, FBStack, CpBindM)
      else if d is variable and d is substituted then
        FBStack.push(<FBIdx, FBLBS, FBLabelAry>)
        < SubstiAry, SubstiLBS > ⇐ BindM.find(d)
        Unify(IRIdx, IRLBS, LabelAry, IRStack, FrontIdx, SubstiLBS,
        SubstiAry, FBStack, BindM)
      else if d is variable and d is not substituted then
        for each sub-formula f starting from IRIdx do
          CpBindM ⇐ BindM
          CpBindM.push(<c, the pointer of f, the pointer of LBS of f>)
          TailIdx ⇐ get_tail_index(IRIdx, f, IRLBS)
          ChildIRIdx ⇐ first_child(TailIdx, IRLBS)
```

```
            ChildFBIdx ⇐ first_child(i, FBLBS)
            Unify(ChildIRIdx, IRLBS, IRLabelAry, ChildFBIdx, FBLBS,
            FBLabelAry, CpBindM)
    else
        if c == d then
            for (j ⇐ IRIdx; is_edge(j, IRLBS) is true; j ⇐ sibling(j)) do
                ChildIRIdx ⇐ first_child(j, IRLBS)
                ChildFBIdx ⇐ first_child(i, FBLBS)
                Unify(ChildIRIdx, IRLBS, IRLabelAry, IRStack, ChildFBIdx,
                FBLBS, FBLabelAry, FBStack, BindM)
return
```

Pattern matching algorithms for LOUDS-based trie is as follows. In this algorithm, "BindM" is a set of tuples: a sentential variable and a sub-formula. For example, suppose BindM is $\{< A, \rightarrow ab >\}$. BindM.find('$A$') returns $\rightarrow ab$. A method **get_same_subformula** compares a substituted formula in a variable with a sub-formula that starts from a current focused edge. If they are the same, the function returns the position of an edge that corresponds to the last letter of the sub-formula. Otherwise, it returns false (0). If a logical formula is judged that the formula is duplicate by pattern matching, DeleteBitChecker algorithm records bits corresponding to the duplicate formulas into "DeleteCheckBit". DeleteCheckBit is a bit vector as long as a target LBS. A method **brother_node_count** returns the number of sibling edges.

Algorithm 2. *PatternMatching* (FBIdx, &FBLBS, &FBLabelAry, NewIdx, &NewLBS, &NewLabelAry, &DeleteCheckBit, BindM)

```
if is_edge(NewIdx, NewLBS) is false then
    DeleteBitChecker(NewIdx, NewLBS, NewLabelAry, DeleteCheckBit)
    return
else
    for (i ⇐ FBIdx; is_edge(i, FBLBS) is true; i ⇐ sibling(i)) do
        c ⇐ FBLabelAry[node_id(i, FBLBS)]
        if c is variable then
            if c is substituted then
                SubstiFormula ⇐ BindM.find(c)
                ChildNewIdx ⇐ get_same_subformula(SubstiFormula, NewIdx,
                NewLBS, NewLabelAry)
                if ChildNewIdx is not false then
                    ChildFBIdx ⇐ first_child(i, FBLBS)
                    PatternMatching(ChildFBIdx, FBLBS, FBAryLabel, Child-
                    NewIdx, NewLBS, NewAryLabel, DeleteCheckBit, BindM)
            else
                for each sub-formula f starting from NewIdx begin do
                    CpBindM ⇐ BindM
                    CpBindM.push(< c, f >)
```

```
            TailIdx ⇐ get_tail_index(NewIdx, f, NewLBS)
            ChildFBIdx ⇐ first_child(i,FBLBS)
            ChildNewIdx ⇐ first_child(TailIdx, NewLBS)
            PatternMatching(ChildFBIdx, FBLBS, FBLabelary, Child-
            NewIdx, NewLBS, NewLabelAry, DeleteCheckbit, CpBindM)
      else
        for (j ⇐ NewIdx; is_edge(j, NewLBS) is true; j ⇐ sibling(j)) do
          if c == NewLabelAry[node_id(j, NewLBS)] then
            ChildFBIdx ⇐ first_child(i, FBLBS)
            ChildNewIdx ⇐ first_child(j, NewLBS)
            PatternMatching(ChildFBIdx, FBLBS, FBLabelAry, Child-
            NewIdx, NewLBS, NewLabelAry, BindM)
            break
    return
```

Algorithm 3. *DeleteBitChecker*(NewIdx, &NewLBS, &NewLabelAry, &DeleteCheckBit)

```
DeleteCheckBit [NewIdx] ⇐ true
idx ⇐ to_edge(NewIdx, NewLBS)
while do
  DeleteCheckBit[idx] ⇐ true
  NewLabelAry[node_id(idx, NewLBS)] ⇐ empty
  breakFlg ⇐ false
  if have_sibling(idx, NewLBS) is false then
    n ⇐ brother_node_count(idx, NewLBS)
    for (count ⇐ n − 1; 0 ≤ count; count−−) do
      if count == 0 then
        DeleteCheckBit [sibling(idx)] ⇐ true
        idx ⇐ parent(idx, NewLBS)
      else
        if DeleteCheckBit[idx - count] is false then
          breakFlg ⇐ true
          break
    if breakFlg is true then
      break
  else
    break
return
```

After Duplication checking process, duplicate logical formulas are removed from NewLBS by using DeleteCheckBit. Figure 3 shows how to remove bits corresponding to duplicate logical formulas from LBS by using DeleatCheckBit. On DeleatCheckBit of (b), elements that has "1" corresponds to black nodes and their edges in tree of (a). If i-th bit on DeleatCheckBit is "1", then i-th bit on LBS is removed. After duplication checking process, FBLBS and NewLBS,

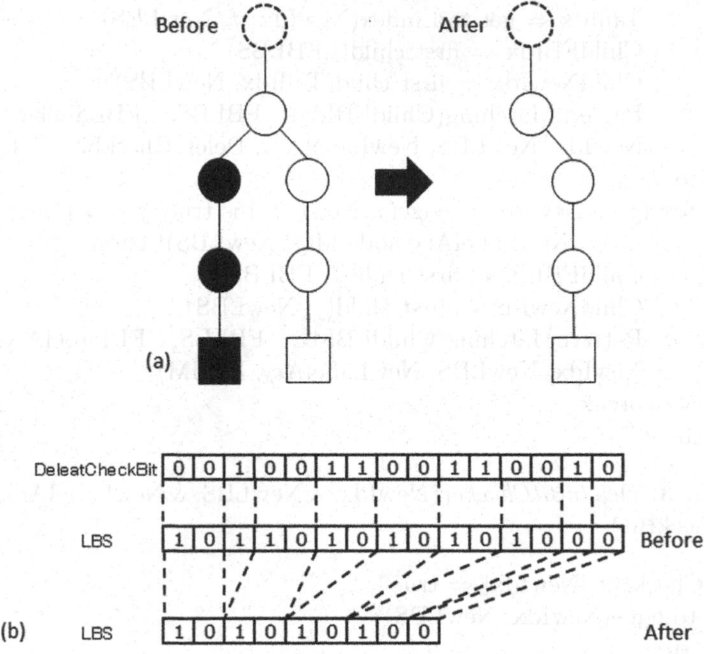

Fig. 3. How to remove bits corresponding to duplicate logical formulas from LBS by using DeleatCheckBit

and FBLabelAry and NewLabelAry are merged respectively according to the algorithm described in [7].

5 Theoretical Evaluation

To confirm effectiveness of LOUDS-based trie from view point of memory-efficiency, we compared amount of memory space of link-based trie and LOUDS-based trie theoretically. The number of nodes of a trie depends on a set of logical formulas given as input data of FreeEnCal. FreeEnCal should deal with various types of logical formulas. Therefore, we counted a number of nodes of a trie that is constructed from a set of logical formulas deduced by FreeEnCal. Then, we calculated amount of memory space of link-based trie and LOUDS-based trie by using the counted number of nodes.

We calculated memory space of link-based trie and LOUDS-based trie as follows. Suppose that the number of nodes of a trie is n. Link-based trie uses 64 bits per a node, 64 bits per an edge, and 32 bits per a label corresponding to an edge. It totally spends $160n$ bits. LOUDS-based trie uses $2n + 1$ bits per a tree, and 32 bits per a label corresponding to an edge. Moreover, LOUDS-based trie needs index data. Although size of a large block and a small block depend on length of LBS, in many implementations, those block size are fixed. Usually, size of a large block is 512 bits and size of a small block is 64 bits. A large block is constructed as a 32-bit array. Because a maximum value stored in a small block is 512, a small block is constructed as a 16-bit array. Thus, LOUDS-based trie uses $2n + 32n + (2n/512 * 32) + (2n/64 * 16) = 34.625$ bits. In addition, duplication checking process uses DeleteCheckBit whose length is as long as LBS, i.e., $2n + 1$ bits. Therefore, the total amount of memory space of LOUDS-based trie is about $36.625n$ bits.

For this theoretical evaluation, we prepared 6 data sets. "Ec", "Ee", and "EcQ" are subsets of logical theorems of strong relevant logic Ec, Ee, and EcQ [2]. "NBG-EcQ" and "NBG-EeQ" are subsets of formal theories of NBG set theory based on strong relevant logic EcQ and EeQ, respectively [5]. "Otwayree" and "Cointoss" are sets of deduced logical formulas in case studies of formal analysis method with reasoning for cryptographic protocols [12]. The first three data sets are sets of logical theorems, and the others are sets of empirical theorems. "NBG-*", "Otwayree", and "Cointoss" use different predicates, functions, and individual constants.

Table 1 shows results of the calculation. "characters" shows a kinds of characters in a data set. "formulas" shows the number of logical formulas in a data set. "nodes" shows the number of nodes of a trie constructed from a set of logical formulas in a data set. "link(KB)" shows amount of memory space of link-based trie. "LOUDS(KB)" shows amount of memory space of LOUDS-based trie. We can reduce 1/4 memory space by using LOUDS-based trie.

Current LOUDS-based trie is not so memory-efficient rather than link-based trie because amount of memory space for storing labels of the trie influences the total amount of memory space of LOUDS-based trie. On the other hand, a kind of characters in each data set is not so many. Suppose the kinds of characters in a set of logical formulas is x, and the number of nodes of a trie is n. x kinds of the characters are stored into a 32-bit array. An index of the array represented as a 8 bit variable is used as a label of the trie. Thus, we can use $8n + 32x$ bits for storing labels/characters of the trie rather than $32n$ in LOUDS-based trie. In Table 1, "LOUDS-2(KB)" shows amount of memory space of the modified LOUDS-based trie. The modified LOUDS-based trie is 10 times memory-efficient than link-based trie.

Table 1. Comparison link-based and LOUDS-based trie

Data set	Characters	Formulas	Nodes	Link(KB)	LOUDS(KB)	LOUDS-2(KB)
Ec	11	2.0×10^3	2.5×10^4	4.9×10^2	1.1×10^2	3.9×10^1
Ee	9	1.7×10^4	1.6×10^5	3.2×10^3	7.3×10^2	2.5×10^2
EcQ	18	1.2×10^6	2.4×10^7	4.6×10^5	1.0×10^5	3.7×10^4
NBG-EcQ	18	3.1×10^2	6.0×10^2	1.1×10^1	2.7×10^0	1.0×10^0
NBG-EeQ	59	9.7×10^2	8.0×10^3	1.6×10^2	3.6×10^1	1.2×10^1
Otwayree	41	8.5×10^3	1.0×10^5	1.9×10^3	4.4×10^2	1.5×10^2
Cointoss	36	1.3×10^3	1.0×10^4	2.1×10^2	4.9×10^1	1.7×10^1

6 Concluding Remarks

We have presented a memory-efficient algorithm with LOUDS for FreeEnCal: a forward reasoning engines for general-purpose. In theoretical evaluation, the proposed algorithm is 10 times memory-efficient rather than traditional algorithm.

This work is just theoretical work. To implement a forward reasoning engine with the proposed algorithm is a future work. After implementation, to confirm effectiveness of the algorithm by measuring amount of memory space and execution time of the implementation is also a future work.

References

1. Cheng, J.: Entailment calculus as the logical basis of automated theorem finding in scientific discovery. In: Raul, V. (ed.) Systematic Methods of Scientific Discovery: Papers from the 1995 Spring Symposium, pp. 105–110. AAI Press (1995)
2. Cheng, J.: A strong relevant logic model of epistemic processes in Scientific discovery. In: Kawaguchi, E., et al. (eds.) Information Modeling and Knowledge Bases XI. Frontiers in Artificial Intelligence and Applications, vol. 61, pp. 136–159. IOS Press, Amsterdam (2000)
3. Cheng, J., Nara, S., Goto, Y.: FreeEnCal: a forward reasoning engine with general-purpose. In: Apolloni, B., Howlett, R.J., Jain, L. (eds.) KES 2007. LNCS (LNAI), vol. 4693, pp. 444–452. Springer, Heidelberg (2007). https://doi.org/10.1007/978-3-540-74827-4_56
4. Fredkin, E.: Trie memory. Commun. ACM **3**(9), 490–499 (1960)
5. Gao, H., Goto, Y., Cheng, J.: A systematic methodology for automated theorem finding. Theoret. Comput. Sci. **554**, 2–21 (2014)
6. Jacobson, G.: Space-efficient static trees and graphs. In: 30th Annual Symposium on FOCS 1989, pp. 549–554 (1989)
7. Koyanagi, T., Yoshida, I., Unno, Y., Shinjo, Y.: Method to build filters for online building of LOUDS trie. In: Computer System Symposium 2011, pp. 33–41 (2011). (in Japanese)
8. Koh, T., Goto, Y., Cheng, J.: A fast duplication checking algorithm for forward reasoning engines. In: Lovrek, I., Howlett, R.J., Jain, L.C. (eds.) KES 2008. LNCS (LNAI), vol. 5178, pp. 499–507. Springer, Heidelberg (2008). https://doi.org/10.1007/978-3-540-85565-1_62
9. Koh, T., Goto, Y., Cheng, J.: A fast algorithm for derivation process in forward reasoning engines. Int. J. Comput. Sci. **4**(3), 219–231 (2010)
10. Kudo, T., Hanaoka, T., Mukai, J., Tabata, Y., Komatsu, H.: Efficient dictionary and language model compression for input method editor. In: Workshop on Advances in Text Input Method, Chiang Mai, Thailand, pp. 19–25 (2011)
11. Yan, J., Wagatsuma, K., Gao, H., Cheng, J.: A formal analysis method with reasoning for cryptographic protocols. In: 12th International Conference on Computational Intelligence and Security, Wuxi, China, pp. 566–570 (2016)
12. Wagatsuma, K., Harada, T., Anze, S., Goto, Y., Cheng, J.: A supporting tool for spiral model of cryptographic protocol design with reasoning-based formal analysis. In: Park, J.J., Chao, H.-C., Arabnia, H., Yen, N.Y. (eds.) Advanced Multimedia and Ubiquitous Engineering. LNEE, vol. 354, pp. 25–32. Springer, Heidelberg (2016). https://doi.org/10.1007/978-3-662-47895-0_4

Methodical Aspects of Knowledge Management in a Contemporary Company

Maciej Malara and Zbigniew Malara(✉)

Faculty of Computer Science and Management,
Wrocław University of Science and Technology, Wrocław, Poland
maciej.malara@pro.wp.pl, zbigniew.malara@pwr.edu.pl

Abstract. The paper presents selected aspects of knowledge management in a company. Components of the knowledge management model were described and the need for their product-based valuation was indicated. Emphasizing the importance of the intellectual capital retention, the principal component analysis method was employed with an intention to use it for the needs of an incentive (rewarding) system.

Keywords: Knowledge management · Intellectual capital
Principal component analysis · Company management

1 Introduction

In the global economy, apart from factors such as raw materials, equipment, machines, finance, energy, land and cheap labour, which determine the competitive capabilities of a company in the traditional economy, additional resources play an important role. They include an intellectual capital, i.e. employees, their knowledge, competences, attitudes and ideas. The creation of this capital requires sophisticated tools and specialists with high competences. Hence, the gathering of knowledge and the ability to use and maintain it seem to be one of the greatest challenges faced by the economy today. In some industries, such as IT, telecommunication, pharmacy, consulting, education, this resource turns out to be so important that managers are looking for methods to determine its value [11, p. 126]. This means that a company has to develop systematic rules and models for knowledge management, which in the modern global economy are the main determinants for the space of freedom for their survival and development as well as an important source of competitive advantage [14]. The previous findings of practitioners and theoreticians of management sciences, which make the knowledge management the main discipline responsible for creating the wealth and employment across all industries, support this thesis [4, p. 140]. Knowledge management, especially in recent years, has been becoming one of the most developed research disciplines, mainly for practical reasons. Knowledge is a goal and a mean to reach the goal. It must be skilfully acquired, explored, exploited and gathered in knowledge bases and employees' minds. Moreover, companies must also demonstrate their capabilities to manage knowledge, including the ability to retain it [2, p. 71].

N. T. Nguyen et al. (Eds.): ACIIDS 2018, LNAI 10751, pp. 71–81, 2018.
https://doi.org/10.1007/978-3-319-75417-8_7

2 Knowledge Management in a Company

Depending on the potential of a company, its current position on the market and strategic intentions, it creates and mobilizes knowledge and competence resources of various degree of significance for the functioning of a company.

The view that the condition for an effective operation of a company is its ability to survive, adapt and develop [6, p. 68] and that the ability to identify the environment and shape the company's activity in accordance with the requirements imposed by it provides a basis for functioning of each company and is the main determinant of its successes [16] has solid grounds in the literature and organizational practice. If we accept this view as correct, it can be concluded that the company should reap benefits from various categories of knowledge. These categories include *basic knowledge* for the needs of rudimental operations, *advanced knowledge* that allows gaining a competitive advantage, and *innovative knowledge* which ensures the position of a leader and is full of solutions that other market participants do not have.

The extent to which companies are able to use their knowledge is determined on the one hand by the abilities to enrich the knowledge possessed and to acquire new knowledge, while on the other hand by the ability to create conditions for the needs of appropriate rationing (distribution) and further maintenance of such knowledge or/and its transfer. This requires both overcoming the barriers in appreciating the importance of knowledge management and creating the conditions for enhancing the ability to bridge the gap between the creation of knowledge resources and their use in the company. Based on a literature query, a list of barriers in knowledge management can be created. These are:

- *incoherence of the management staff*, who in fact thinks only about its position, territory and influences [13], [3, p. 33],
- insufficient use of employees' knowledge [22, p. 33], [15, p. 32],
- information problems and shortage of knowledge resulting from an uneven distribution of knowledge in terms of needs and requirements of all the company's units and employees, which in turn generates problems of an informational nature[1],
- difficulties in acquiring informal knowledge [18, p. 46],
- inadequate organizational culture and a lack of an atmosphere *to share the knowledge openly*[2],
- overestimation or underestimation of the role of technique and technology, especially of the IT infrastructure [16, p. 29],

[1] In order to reduce a deficit of knowledge, the company should, inter alia: acquire knowledge from the environment and adapt it to its needs, e.g. through trade, foreign investments and license agreements, as well as the through the development of own research and the use (development) of experience; *absorb knowledge*, i.e. create conditions for upgrading the qualifications and for intellectual development of employees, for example by training; *transfer knowledge*, that is use new information technologies and provide good access to knowledge resources.

[2] Numerous experiments and research results indicate that this factor plays a fundamental role in the strategic use of skills, information and ideas in employees' minds. A company that wants to achieve successes on the global market should create an organizational culture based on the collective character of the organization, while avoiding solutions based on the individualism.

- useless databases, which should be eliminated basing on the criterion of practical usability [19, p. 29],
- stereotypical thinking that *knowledge means power* [19, p. 30],
- a gap between the concept and the action [21, pp. 11–15],
- rejecting (not taking into account) negative knowledge[3],
- limiting contacts and informal conversations[4],
- limiting to cognitive knowledge[5],
- transferring knowledge through a single medium [21, p. 30].

With regard to the factors conducive to proper knowledge management, apart from acquiring, accumulating, processing and creating knowledge to a possible broad extent, the company must create conditions for its exchange and transfer or retention with the intent to multiply its value (quality) [10].

To meet these requirements, the company's management board should:

- put a considerable emphasis on education and training of employees [9, pp. 99–169],
- continuously develop the intellectual curiosity of employees[6],
- avoid stereotypical thinking[7],

---

[3] Experience is often a result of errors and failures, but in no case should it be omitted in the process of knowledge creation. That's because constructive conclusions can be drawn from negative experiences. This requires creation of an atmosphere of trust that is conducive to the exchange of informative knowledge about errors and failures.

[4] Synergistic effects are a consequence of regular, intense contacts between employees, which are conducive to the emergence of new, often surprising solutions. The fewer limitations and restrictions regarding forms, topics and content, the more opportunities to transfer useful knowledge. According to H. Simon, no more than 50% of the knowledge needed in a company may result from the planned talks and meetings. The remaining half is provided by unplanned conversations which are not aimed at a specific goal (i.e. Asian organizational culture).

[5] In the knowledge management process, especially at the stages of acquisition and transfer of knowledge, an important role is played by the *willingness* and *motivation* as well as the *instruments* that strengthen them, such as *incentives*, *comforting* in the event of failure, *calling* for perseverance and *indicating* obstacles. This means that cognitive aspects should be combined with motivational aspects, otherwise the effects may turn out incomplete.

[6] It is a good practice for companies to transfer the best solutions from their existing plants to new ones, thanks to which the knowledge about the best solutions worked out in previous production processes by own employees and external specialists working at the construction is used when designing a new plant. The employees, knowing that they are building a new division *for themselves*, are self-motivated for good work. During the work they acquire knowledge about industrial process. Immediately after completing construction works, they are prepared to carry out production tasks. The skills acquired during the construction and start-up of a new division are transferred to a set of key competences of the company.

[7] A common way to fight the *syndrome of uneven distribution of knowledge* is exchange of knowledge resources between divisions, which includes a regular provision of information about the results achieved by one division to all other divisions. These results are then analysed at management meetings in divisions in order to share knowledge on the best solutions with other divisions, based on the results of the analysis. Complex and hard-to-grasp knowledge is passed on with the help of employees (specialists) who are delegated to other divisions in order to communicate and consolidate mental patterns together with it.

- consciously create mechanisms for identifying and developing the factors conducive to appearance of systemic effects of knowledge[8],
- create conditions for collective learning[9],
- create conditions for developing the creativity and innovation as it takes place at RANK XEROX [20, pp. 20–22],
- develop knowledge systems (databases and networks) and rules of their use [20, p. 20],
- manage competences in the field of the company's human resource function as it is done in NUCOR STEEL CORPORATION – an American metallurgical company [23, pp. 11–17].

The K'NETIX network in a French Buckman Laboratory is an example that petrifies the findings of this part of the considerations, illustrating a manner of knowledge management. These are perfectly organized knowledge resources, which can be widely used by employees [1, pp. 415–416].

3 Verification of Findings

Knowledge management in a company can be examined in four dimensions determining the space of freedom for this process[10]:

- *degree of knowledge organization* (codification),
- *intensity* of spreading the knowledge and its renewal,
- *ability* to absorb (use) the knowledge,
- *skill* to retain (maintain) it.

As it appears from the above, a useful instrument that effectively supports the use of the intellectual capital in a company is the development and use of a knowledge system, taking into account those dimensions, in which a requirement to build the loyalty is permanently embedded and without which effective knowledge management will not be possible.

[8] Each company has its own, individual knowledge management manner (system). It results, inter alia, from the company's business profile. This means that the knowledge management should be adapted to specific character and needs of a given company. In order to obtain positive effects from the functioning of a knowledge management system, companies pay attention to two main aspects: taking into account their own interests resulting from the strategic direction of their efforts and working on the development of an organizational culture focused on sharing the knowledge and cooperation.

[9] Many various factors are conducive to collective learning. They include mainly cultural ones, such as: creating the right atmosphere for establishing close and direct contacts between employees; getting rid of the culture based on far-reaching individualism; rewarding the employees who willingly share their knowledge with others, and motivating not only individual employees, but also entire teams.

[10] This division was proposed by Zack, who was the first to introduce the concept of the knowledge transfer cycle in a company [17].

A correctly functioning knowledge system supported by a consistent strengthening of loyal behaviours and attitudes of employees requires identification of the premises guiding the employees who decide to stay with the company and work for it for a long period of time. Then, these premises should be used to develop an incentive (reward) system that will allow not only satisfying the expectations of employees and make them stay with the company and work for it with engagement, but will also allow building their loyalty towards the employer and thus preserving the knowledge resources and the intellectual capital. In order to substantiate the thesis formulated in this way, there was presented an experiment carried out by the authors of this study, which allowed identifying the premises that guide the employees when they decide whether to stay with the company.

3.1 The Subject of the Analysis and Presentation of Data

When analysing the literature on the subject, components of the intellectual capital were identified [1] as well as important factors which can significantly affect the development and retention of knowledge in the organization [2]. Based on these factors, the criteria – components of the model of retention and accumulation of knowledge in the company were established. In order to verify the model, data on the American economy from the period of 1995–2015 were collected and used. They are presented in Table 1. The principal component analysis method (PCA) was used in the study, while the calculations were carried out using GRETL statistical software. The criteria were formulated in the following form:

X1 – expenditures on human capital, expressed in average annual remuneration (per capita),
X2 – expenditures on R&D on an annual basis (per capita),
X3 – expenditures on marketing activities (annually, per capita),
X4 – expenditures on PR activities (ratio of annual volume to the total number of business entities registered in the USA),
X5 – annual value of payments of retirement benefits resulting from Social Security (per capita, excluding the 401(k) plan).

All numerical values presented in Table 1 are expressed in thousands of US dollars and in constant prices of 2015.

3.2 Description of Research Tool

Principal component analysis (PCA) is used to determine new variables, a possibly small subset of which will provide as much information as possible about the whole variability in the data set. The new set of variables forms an orthogonal base in the space of features. Variables are selected in such a way that the first variable maps as much variation in the data as possible. After determining the first variable, another one is determined so that it is orthogonal to the first one and explains the remaining variability as much as possible. The next variable is selected in such a way that it is orthogonal to the first two ones etc. The set of vectors obtained in this way forms an

Table 1. Input data of the model

No.	X1	X2	X3	X4	X5	Year
1	44634	24736	42500	5800	8125	1995
2	45524.5	26103	44000	6200	9062.5	1996
3	46415	27599	46700	6900	6562.5	1997
4	48014.5	29129	48000	7300	4062.5	1998
5	49614	31043	50000	7800	7812.5	1999
6	50670	33315	52300	9300	10937.5	2000
7	51726	33868	52000	8700	8125	2001
8	52228.5	33315	55200	10450	4375	2002
9	52731	34293	56400	10800	6562.5	2003
10	53271	34714	59600	11500	8437.5	2004
11	53811	36107	57780	12200	12812.5	2005
12	54795.5	37721	58390	11800	7187.5	2006
13	55780	3955	60410	12100	10312.5	2007
14	55911	41534	60490	13600	18125	2008
15	56042	41137	56020	12700	19375	2009
16	56291	41093	57540	14720	22812.5	2010
17	56540	4211	57770	13200	29062.5	2011
18	56620.5	42049	69530	15700	27187.5	2012
19	56701	43325	70730	14423	28437.5	2013
20	57707.5	44585	73520	16423	31250	2014
21	58714	46277	76440	15370	27500	2015

orthogonal base in the space of features. The purpose of the principal component analysis method is therefore to find a transformation of the coordinate system that will describe the variability between observations in the best way [12].

PCA method maximizes the variance of the first coordinate, then the variance of the second coordinate and the next ones. The coordinate values transformed in such a way are called loadings from generated factors (principal components). In this way, a new observation space is built, in which the initial factors explain the most of the variability. So, the aforementioned operation can be used to understand the structure of the population studied and the nature of the data used. PCA method may be based on a correlation matrix or a covariance matrix created from the input set.

The algorithm in both versions is identical, but the results are different. When a covariance matrix is used, the input set variables with the largest variation in have the greatest impact on the result, which may be advisable, if the variables represent comparable, relatively uniform values. In turn, the use of the correlation matrix corresponds to the initial normalization of the input set, so that each variable has an identical variance at the input (without weights), which may be advisable, if we are unable to ensure the comparability of values of the variables tested.

The PCA algorithm consists of the following steps based on [5]:

- Determination of means for rows

$$u[m] = \frac{1}{N}\sum_{n=1}^{N} X[m,n] \tag{1}$$

- Calculation of the deviation matrix

$$X'[i,j] = -X[i,j] - u[i] \tag{2}$$

- Determination of the covariance/correlation matrix

$$\mathbf{C} = \mathbb{E}[\mathbf{B} \otimes \mathbf{B}] = \mathbb{E}[\mathbf{B} \cdot \mathbf{B}^*] = \frac{1}{N}\mathbf{B} \cdot \mathbf{B}^* \tag{3}$$

- Calculation of eigenvalues of the covariance/correlation matrix

$$\mathbf{V}^{-1}\mathbf{C}\mathbf{V} = \mathbf{D} \tag{4}$$

- Selection of eigenvalues (in order to minimize losses in information, the ones with the highest value are selected)
- Determination of eigenvectors

$$\begin{bmatrix} a_{11} - \lambda & a_{12} & \cdots & a_{1n} \\ a_{21} & a_{22} - \lambda & \cdots & a_{2n} \\ \vdots & \vdots & \ddots & \vdots \\ a_{n1} & a_{n2} & \cdots & a_{nn} - \lambda \end{bmatrix} \cdot \begin{bmatrix} x_1 \\ x_2 \\ \vdots \\ x_n \end{bmatrix} = 0 \tag{5}$$

- Projection onto eigenvectors

$$y = \begin{bmatrix} y_0 \\ y_1 \\ \vdots \\ y_{n-1} \end{bmatrix} = V^T \cdot x = \begin{bmatrix} v_0^T \\ v_1^T \\ \vdots \\ v_{n-1}^T \end{bmatrix} \cdot x \tag{6}$$

where:

V – matrix of eigenvectors,
x – projected vector,
y – vector in the new space,
N – number of eigenvector.

Each principal component is therefore described by:

- eigenvalue,
- eigenvector,

- factor loadings,
- contributions of variables,
- communalities.

3.3 Presentation and Evaluation of Quality of the Results Obtained

The results obtained with the GRETL statistical processor are shown in Tables 2 and 3. Since there is no single universal criterion for selecting the number of principal components, it is justified to use multiple criteria for this purpose [7, pp. 84–89]:

Table 2. Results generated by the GRETL statistical processor (1)

Principal component analysis			
n = 21			
Eigenvalue correlation matrix			
Factor	Eigenvalue	Share	Cumulative share in variance
1	3.6320	0.7264	0.7264
2	1.0166	0.2033	0.9297
3	0.2193	0.0439	0.9736
4	0.0938	0.0188	0.9923
5	0.0384	0.0077	1.0000

Table 3. Results generated by the GRETL statistical processor (2)

Eigenvalue vectors (component loadings)					
	PC1	PC2	PC3	PC4	PC5
X1	0.154	−0.940	0.227	−0.195	0.051
X2	0.503	0.061	0.450	0.482	−0.556
X3	0.488	0.261	0.088	−0.812	−0.162
X4	0.511	0.128	0.184	0.208	0.804
X5	0.473	−0.166	−0.840	0.166	−0.125

Percentage of Explained Variance. The number of principal components, which a researcher should adopt, depends on the extent to which they represent primary variables, i.e. on the variance of primary variables contained in them. All principal components carry 100% of the variance of primary variables. If the sum of the variances for some first components constitutes a significant part of the total variance of primary variables, then these principal components can substitute the primary variables to a satisfactory degree. It is assumed that this variance should be reflected in the principal components in more than 80%.

Kaiser Criterion. The Kaiser criterion says that the principal components, which we want to leave for interpretation, should have at least the same variance as any standardized primary variable. Due to the fact that the variance of each standardized primary variable is 1, according to the Kaiser criterion only the principal components with an eigenvalue that exceeds 1 or is close to it are valid.

Scree Plot. The plot illustrates the rate of decrease of eigenvalues, i.e. the percentage of the variance explained. The point on the plot, at which this process stabilizes and the descending line becomes horizontal, is called the end of the scree (the end of downward trend of the information about the primary variables that is carried by principal components). The components located to the right of the end of the scree represent a negligible variance and mostly present a random noise.

In the light of the above criteria, the decision was made to leave the following variables in the model: **X1** (*'expenditures on human capital'*) and **X2** (*'expenditures on R&D'*) as fully representative and crucial for the explanation of the phenomenon. Since a satisfactory result was obtained, the plans to build a synthetic variable were abandoned.

The variables presented above can therefore be firmly considered as crucial for the retention and accumulation of knowledge in a company operating on the American market.

4 Conclusions and Future Work

Knowledge is subjected to processes of identification, acquisition, exploitation, exploration, development and retention – irrespective of the sources, from which a company draws knowledge, the character and characteristics of knowledge, as well as the conditions, needs, manners and skills that determine the possibilities of managing the knowledge. Each phase of this endless process is a part of a knowledge system that requires the use of appropriate tools to assist its management. This means that modern companies, irrespective of their phase of development, have to go a *long way* and overcome many hindrances to switch from a traditional model of managing to a knowledge-based management model[11].

These phases accurately represent the problem that companies have to solve. The results of the research [8, p. 29] form a view that a vast majority of modern companies is still unaware that knowledge can be managed or has no experience or opportunities to do so.

In fact, the following quotation from a study by Zack refers to this situation: *Although there is much talk about linking the knowledge management with the business strategy, in practice this is widely ignored* [8, p. 30]. Finally, it is worth paying

[11] According to the authors of the studies conducted on a group of 423 companies by KPMG Consulting [8, p. 29], in 1999 nearly 43% of companies were in the phase of chaos, i.e. in the basic phase where there is no correlation between the importance attached to knowledge management and the achievement of its goals. The authors of the studies additionally distinguished four further phases of so-called *knowledge journey*, after going through which a company can achieve the excellence in knowledge management. These are: the phase of *awareness*, the phase of *directing*, the phase of *management*, and the phase of *integrated management*. The authors of the studies classified 32.4% of the companies into the first two phases, 9% of the companies—into remaining two phases, while only 1% of companies qualified to the last phase! [8, p. 29].

attention to two further limitations associated with the philosophy of knowledge management in a company. These limitations are associated with the organizational culture and thus with the system of values and the business model applicable in a given company. It seems that the rigid bureaucratic model with a fixed hierarchy and professional specialization is still prevailing, which hinders its diffusion. An additional, important factor limiting the absorption of the idea of knowledge management is the time pressure associated with quick reactions to the changes occurring in the company's environment. On the one hand, employees, especially managers, do not have time to learn and, on the other hand, both groups do not have time to share their knowledge. If we add to these factors also a lack of willingness to share the knowledge and the connivance for leaks of knowledge outside the company, there should be a call for strong leadership, which should also include the responsibility for the protection of knowledge.

Will it be possible to adopt such a concept in the near future? The companies struggling with the global reality must respond to the problem formulated in such a way. Their response must take into account a number of perspectives, including the responsibility for knowledge, definition of requirements for the knowledge management system, and priorities to be adopted when implementing such a system. In other words, regardless of the type, character and degree of advancement (level) of knowledge, the company must continue efforts to multiply (quantity) and enrich (quality) it, as well as work on methods of its protection (retention). Otherwise, the issue of knowledge management in a company will remain a theoretical problem.

References

1. Brilman, J.: Nowoczesne koncepcje i metody zarządzania. Polskie Towarzystwo Ekonomiczne, Warszawa (2012). (in Polish)
2. Drucker, P.F.: Menedżer skuteczny. Wydawnictwo Akademii Ekonomicznej w Krakowie, Kraków (2004). (in Polish)
3. Friedman, M.: The Social Responsibility of Business into Increase Its Profits. Times Magazine, New York (1973)
4. Herman, A.: Zarządzanie wartością przedsiębiorstwa opartego na wiedzy. In: Przedsiębiorstwo przyszłości. Nowe paradygmaty zarządzania europejskiego. Wydawnictwo ORGMASZ, Warszawa (2003). (in Polish)
5. Krzanowski, W.J.: Principles of Multivariate Analysis: A User's Perspective. Oxford University Press, Oxford (2000)
6. Malara, Z.: Restrukturyzacja organizacyjna przedsiębiorstwa. Oficyna Wydawnicza Politechniki Wrocławskiej, Wrocław (2001). (in Polish)
7. Massart, D.L., Van Der Heyden, Y.: From Tables to Visuals: Principal Component Analysis, Part 2. LC-GC Europe, vol. 18 (2004)
8. Materials of the Conference: Personel XXI wieku, Zarządzanie wiedzą w przedsiębiorstwie. Konferencja Polskiej Fundacji Promocji Kadr oraz Wyższej Szkoły Przedsiębiorczości i Zarządzania im. L. Koźmińskiego, Warszawa (2001). (in Polish)
9. Mikuła, B., Pietruszka-Ortyl, A., Potocki, A.: Zarządzanie przedsiębiorstwem XXI wieku. Wydawnictwo DIFIN, Warszawa (2002). (in Polish)

10. Nonaka, I., Takeuchi, H.: The Knowledge Creating Company. Oxford University Press, Oxford (2005)
11. Obłój, K.: Tworzywo skutecznych strategii. Polskie Wydawnictwo Ekonomiczne, Warszawa (2012). (in Polish)
12. PQStatSoftware Knowledge Repository. Accessed 2 Dec 2017
13. Smith, A.: An Inquiry into Nature and Causes of the Wealth of Nations. Encyclopedia Britannica, Chicago (1952)
14. Strojny, M.: Zarządzanie wiedzą. Wstęp do dyskusji. Personel 6 (2001). (in Polish)
15. Wawrzyniak, B.: Od koncepcji do praktyki zarządzania wiedzą w przedsiębiorstwie. Master of Business Administration 1, (2002). (in Polish)
16. Webber, R.A.: Zasady zarządzania organizacjami. Państwowe Wydawnictwo Ekonomiczne, Warszawa (1994). (in Polish)
17. Zack, M.: Co-evolution. Dynamique Creatrice. Village Mondial, Paris (2007)
18. Zarządzanie na Świecie, no. 2 (2010). (Polish journal)
19. Zarządzanie na Świecie, no. 6 (2010). (Polish journal)
20. Zarządzanie na Świecie, no. 8 (2010). (Polish journal)
21. Zarządzanie na Świecie, no. 12 (2010). (Polish journal)
22. Zarządzanie na Świecie, no. 33 (2010). (Polish journal)
23. Zarządzanie na Świecie, no. 12 (2011). (Polish journal)

Social Networks and Recommender Systems

Fairness in Culturally Dependent Waiting Behavior: Cultural Influences on Social Communication in Simulated Crowds

Sutasinee Thovuttikul(✉), Yoshimasa Ohmoto, and Toyoaki Nishida

Department of Intelligence Science and Technology, Graduate School of Informatics, Kyoto University, Kyoto, Japan
thovutti@ii.ist.i.kyoto-u.ac.jp, {ohmoto,nishida}@i.kyoto-u.ac.jp

Abstract. Difficulties in living in unfamiliar cultures are caused by differences in the patterns of thinking, points of view, and styles of physical action. In this paper, we present our findings on learners' cultural understanding during interaction based on culturally influenced communication in simulated crowds. Participants in the experiment are supposed to live in a shared virtual space. They are asked to obtain multiple tickets available at two service counters in the system. A virtual service person provides a ticket upon request from a customer. Additionally, one or more virtual customers move around in the system to acquire tickets. If a counter is occupied by a customer, the others have to wait. Two types of waiting styles–line and group waiting–and two fairness levels of the service person–fair and unfair service–are configured and evaluated. The counter selection results and reasoning results were analyzed using the ANOVA process. We found that culture in Thailand influences ideas of waiting differently in different first- and third-person point of view (POV) settings. The participants in the first-person POV group show a tendency to focus on the concept of fairness more than the participants in the third-person POV setting, whereas the latter pay more attention to cultural reasoning in their waiting behavior.

Keywords: Cultural learning system · Perception on different culture
Fairness · Waiting behavior · Simulated crowd

1 Introduction

The world seems smaller today than in the past, as traveling to different parts of the world is faster and easier. When a traveler first arrives in a foreign country, such as when an American travels to an Asian country or an Asian student goes to study in a European country, they may encounter many aspects that differ from their home town, such as language, food, the way of thinking, and daily life. Learning to understand people from other cultures is necessary for those who are exposed to different cultures [1]. Dresser [2] provides an example: an American makes the mistake, on first greeting an Asian (such as a Vietnamese), of hugging and kissing in public, thereby insulting

N. T. Nguyen et al. (Eds.): ACIIDS 2018, LNAI 10751, pp. 85–95, 2018.
https://doi.org/10.1007/978-3-319-75417-8_8

her. Even though hugs and kisses are common greeting behaviors in western culture, it is considered rude in Asian culture. Intercultural tourism is superficial [3] because the travelers experience the other culture/country for a short time. Although they might not understand complex situations, they should at least learn appropriate behavior to be able to avoid misunderstandings and miscommunication.

Hofstede et al. [3] describe the intercultural communication learning process that includes three phases: awareness, knowledge, and skill. Awareness enables learners to notice different or strange signals in an environment and recognize the differences to their own cultural background. Knowledge is obtained by establishing the meaning of the behavior in the new culture and updating their own knowledge. By practicing being aware and updating knowledge, they will gain the skill of awareness and the knowledge to understand the situation and behavior in the new cultural place. A simple activity that a tourist cannot avoid is waiting in queues. Waiting is related to the idea of fairness [4]. Although fairness is a simple term, its meaning is deepened by many factors including culture. We cannot consider waiting without taking into account culture. Therefore, this research investigates the details of cultural influence on waiting behavior. Crowded places are suitable for practicing cultural communication because we can easily identify similar behavior of a large number of people. Certain unique behavior may be difficult to observe in a small group of people because too many different behaviors are represented by only a few people. In contrast, in case of a crowd, most people will behave in the same manner so we can more easily observe unique behavior in the crowd. Currently, computer simulation offers a suitable solution for helping people learn to overcome the difficulty in communication in different cultures. We cannot set up a real human crowd but we can simulate the agent and environment in a learning system.

This paper presents the initial state of this research; herein, we focus on the awareness and knowledge processes. We select an international traveler as a case study of misunderstandings in cultural communication. We aim to capture the perception of activities through the cultural filter. We begin the discussion in this paper with Thai culture.

In the next section, we discuss related work on cultural behavior learning and cultural dimension theories. In the Sect. 3, we describe our system and a solution to confirm our hypothesis. In Sect. 4, we present our experimental setting and a concrete result. The experimental results are presented as evidence of our findings. In the Sect. 5, we summarize our findings and propose our future plans based on these results, to achieve the main goal of our research.

2 Related Work

The study of cultural communication has gained popularity over the last decade and many researchers have developed virtual simulation systems to represent different cultural behaviors and communication. These systems are designed to provide learners with an understanding of different cultural behaviors through complex models of a virtual agent's behavior [5–9], useful scenarios [6–9] or powerful interactive tools [6, 9]. The learners observe the situation from a third-person point of view (POV) just as in the real world [5–7] and then they are asked to interact with the agent first-person POV

[7, 9]. By these steps, the participants gain a one-to-one conversation experience with the agents. Awareness and knowledge are necessary for learners to learn the cultural communication. We first hypothesize that fist- and third-person POVs may yield different perceptions based on cultural background. In this study, we aim to explore the different benefits on cultural learning in different POVs.

Hall [10] discusses territoriality and personal space; this is a simple concept but we cannot see it as a physical boundary. There is an invisible bubble around each person that depends on a number of factors: relationships to nearby people, emotion, activities, and culture. For example, people allow friends or family to stay closer than others who are strangers. Culture is an important factor controlling personal space. For example, the bubbles in northern Europe are larger than in southern Europe. Hall [11] guides that culture performs a function of providing a tall screen between a man and his outside world. The screen restricts attention and ignoring at the same time. Thus, we are interested in examining the concrete concepts of attention and ignoring, and applying these concepts to build a system that can produce a suitable process for learners to learn about a different culture. Our goal is to establish the influence of culture on perception. We set two different POVs, first- and third-person, of two participant groups, in order to examine the effect of cultural influence on each POV.

Queue-waiting is a good practical case for studying cultural communication because it involves cultural influence and cultural space handling. The waiting position, waiting queue shape, and waiting process are important factors in the pleasantness of waiting [4]. As mentioned above, the space used is culturally influenced [5]. People have different ways of managing their own space. Thus, waiting is a concrete example of cultural communication in this research. In practice, Thai people form queues in formal places such as banks, hospitals and libraries. However, in informal places that do not have strict rules to control queue waiting, they do not form a queue. Travelers belonging to another culture may have many questions regarding waiting practices in Thailand. How does the service person know who is next? Is the service person fair or not? Such confusion may arise when a traveler with a certain cultural background travels to another culture Hofstede's [3] cultural dimension is used to categorize thinking, belief, and behavior for more than 40 countries. Six dimensions of national culture were defined based on an aspect of each culture when measured relative to other cultures. The dimensions are as follows: power distance, individualism versus collectivism, masculinity versus femininity, uncertainty avoidance, long-term versus short-term orientation, and indulgence versus restraint. We only discuss two dimensions that are relevant to our experiments.

Dimension 2. Individualism vs. collectivism: This dimension represents the difference between people. Collectivism cultures feel and identify others as in-group or out-group, whereas individualism feels that there is no group; everyone is unique (Thai 20).

Dimension 3. Masculinity vs. femininity or achievement-oriented versus cooperation oriented: This dimension describes how gender influences roles. In high-femininity cultures both genders are assumed to be cooperation-oriented (Thai 34).

In this study, we apply these dimensions to create a scenario of waiting and observe the effect of culture on the participant's perception and interpretation. In terms of communication, perception is a cognitive process by which people come to interpret and understand other people, events and objects [12]. People perceive the world differently and we can understand the different behavior of other people by learning how their perception operates. Learning perceptual processes helps us to understand the meaning of behavior and improve communication. We will demonstrate the variety of behavior in a crowd. In general, if a learner sees that the properties of choices are obviously different, they will easily select the better choice by sense of sight. In contrast, if the main properties of these choices are similar, then they more carefully consider the features based on their own priorities. Thus, our second hypothesis is that culture influences how people make a choice or simply ignore a situation.

3 System Architecture

In this research, we introduce an initial scenario of a "simulated crowd" [13–15]. The system provides the agent and environment for practicing cultural communication. In order to develop a learning system, we intend to identify the factors that affect the participant learning process. Culture plays an important role in the communication process, as discussed above. We aim to obtain concrete evidence to confirm that culture is an influence on communication by adopting this setting.

3.1 Virtual Ticket Counter (VTC)

A shared virtual environment was set up on a network for participants to converse as shown in Fig. 1(left). We connected two computers in a client-server network to establish a communication channel to the virtual space. In the VTC, each participant uses a terminal to participate in social activities in a shared virtual space. A simple model was designed to control the agent's behavior: walking, collision avoidance, and waiting (either line or group waiting). A Wizard of Oz (WOZ) system was used to control the avatars in this system [15]. A cultural expert controls the service person avatar based on predefined rules. The service person avatar can thus naturally respond to the participant in real time.

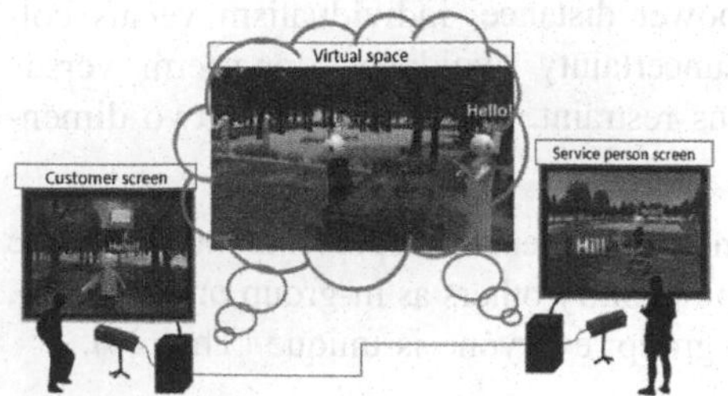

Fig. 1. System setting and a screen shot of the system (Left: system setting, up right: first-person POV, down right: third-person POV)

3.2 Influence of Culture on Perception

In order to learn culture-dependent behavior for communication, the participant should be aware of the differences to be considered in learning a new culture. In previous works, most researchers conducted experiments by providing different behavior for participants from two or more cultures. Following this process, a learner can see the results of different cultural behaviors, but the details of how culture influences perception, interpretation, and behavior remain undiscovered. In this study, we intend to identify the cultural factors that influence perception in the cultural learning process. We develop a system that allows the agent to present examples of nonverbal communication, such as hand gestures and standing distance. The participant can learn and practice behavior with our agent and avatar. Waiting behavior is used as an activity to analyze cultural influence. We select two dimensions that are relevant to waiting behavior for use in our experiments.

Collectivism culture: Waiting behavior is a social activity. Groups are formed based on relationships, such as family, friends or other salient social functions.

Femininity culture: There is not much force or urgency. People relax and enjoy the waiting time.

We use both dimensions to describe the characteristics of waiting in Thailand. Random group waiting represents the relaxed waiting queues and social activities in Thai culture.

3.3 First- and Third-Person POVs

We introduce POV as a key to measure the different perceptions acquired from the target event. The study of the Synthetic Evidential Study (SES) framework [16] examines how people understand different perceptions, feelings and reasoning from the first-person POV and third-person POV. There are two groups of participants: first- and third-person POVs. The same simulation is conducted from the different points of view. The first-person view camera is installed close to the eye of the customer avatar; it tracks the avatar during walking and adjusts to the avatar viewing position. The third-person view camera is installed close to the counter as a static camera which captures all the customer avatars and service persons.

4 Experiment

In this initial stage of the research, we started by identifying a suitable setting that could increase culture-dependent communication awareness and understanding. In this experiment, we aimed to confirm that the participant who attended the event or activity in the virtual space also used their cultural background to communicate and interact with the agent and avatar.

4.1 Task

To design the experiment to find culturally influenced factors of communication, we simulated a simple event in the virtual world: waiting to buy a ticket at a counter. We modeled the agent's behavior to present two different kinds of waiting style depending on culture: line waiting and group waiting. Nonverbal behavior is a useful tool for people to communicate in both the real world and simulated world. Meaningful nonverbal behavior such as standing position, face direction, and hand gestures are the key to communication in this experiment. The communication behavior was represented by the agent and experimenter's avatar. A set of behavior was used for comparing the different perceptions of both participant groups.

A. Customer agents' waiting behavior: Two counters, A and B, represent distinctive styles of waiting. The participant sees both counters at the same time and from the same distance. The participant can understand the waiting style from the positions and walking movements.

Line waiting counter A (Individual): The customer agent stops in front of counter. A behind the previous agent and walks closer to the counter following the line.
Group waiting counter B (Collectivism): The customer agent stops in front of counter B, as close as possible to the counter.

B. Queue-jumping customer: There is a customer agent who is a friend of the service person. He stops in front of the counter directly, faces the service person, and places an order. The participant cannot respond to this behavior but can observe each service person's response.

C. Service person's fairness and morals: Service persons serve all customer agents by the 'first come first serve' rule in all sessions. When the queue-jumping customer arrives at the counter, the service person has two kinds of responses: accept (femininity) or reject (masculinity) the request.

D. Predefined rules for service person and customer behavior: The interaction between the service person and customer was designed based on predefined rules. The service person serves the participant or customer agent a ticket at the counter. The scenario follows these steps: (1) The service person greets and asks for the customer's order. (2) The customer places the order by showing the ticket icon. (3) The service person acknowledges the order, saying "thank you", and moves a hand to prepare a ticket, and after around 5 s of preparing the ticket, passes the ticket to the customer. (4) The customer takes the ticket by holding out a hand to receive the ticket and leaves with the ticket, turning and walking away from the counter.

The participant was asked to imagine that they are visiting an international theme park. People from many countries share the public space together at the theme park. They have to go to the ticket service counter to get a ticket for a waiting time. Sometimes, the staff also goes to the counter to get a ticket and they have to wait same as the customer. Each session comprises three sections: *observation, practice, and interaction.* We started the *observation section* by asking the learner to observe the video of the activities. Then the participant was asked to *practice* getting a ticket at both counters. After the observation and practice sections, the participant was assumed to have enough knowledge to be able to interact with both service persons. In the final section of each session, the *interaction section*, the participant was asked to go to any ticket counter they chose three times to get three tickets. Then, the participant was asked to fill in a questionnaire about their choice of counter to evaluate their interpretations and perceptions.

4.2 Experimental Setting and Participants

During the experiment, the participant sees all the customer agents and service persons at both counter A and B, with line and group waiting, respectively. The queue-jumper who is a friend of the service person comes to both ticket counters in sessions 2, 3, 4, 5. Both service persons serve with the same level of fairness (both serve fairly or both serve unfairly) in sessions 2 and 5. In contrast, they offer a different service in sessions 3 and 4 (one of them fairly and the other unfairly). One participant will join five sessions. The experiment begins with session 1, a basic communication. The other 4 sessions are random so as to remove the effect of order from each session. The participants were 32 Thai students (16 males and 16 females). They were recruited through ads posted in Chiang Mai University, Thailand. Their average age was 22 years (ranging from 19–30, n = 32, SD = 2.94). Eight male and eight female participants joined the experiment in the first-person view setting, whereas eight male and eight female participants joined the experiment in the third-person view setting. The participant's body movements were captured using the Kinect and transferred to control the avatar's movements and posture.

4.3 Experimental Results

The participants had three chances to select a counter, A or B, in the *interaction section* of each session. They could freely select the same or different counter. We asked the participants to select the counter they preferred to wait at. The participant's counter selection and its reasons were used as the key experimental result. We called this selection "counter selection result". After they attended each session, we asked them to write their own reason for each selection. We called this "reasoning result".

A. Counter Selection Results: The counter selection results are shown in Table 1. We applied analysis of variance (ANOVA) to the results to calculate the frequency of selection. However, there was no significant difference in counter selection between first- and third-person POVs in all sessions ($F(1, 30) = 4.006$, $p = 0.0545$).

Table 1. Total selected counter results in each session.

Point of view	Session 1		Session 2		Session 3		Session 4		Session 5	
	Counter A	Counter B	Counter A	Counter B	Counter A	Counter B	Counter A	Counter B	Counter A	Counter B
1st	35(72.9%)	13(27.1%)	32(66.7%)	16(33.3%)	30(62.5%)	18(37.5%)	47(97.9%)	1(2.1%)	32(66.7%)	16(33.3%)
3rd	24(50.0%)	24(50.0%)	29(60.4%)	19(39.6%)	27(56.2%)	21(43.8%)	24(50.0%)	24(50.0%)	22(45.8%)	26(54.2%)

Further analysis of the counter selection result of each participant revealed that the counter selection results were random. We found that the participants could not confirm their perceptions and interpretations because they were not always going to the same counter for the same reason. However, the reasoning results were more clearly defined with respect to the participants' thoughts during selection.

B. Reasoning Results: As mentioned earlier, we controlled two factors in the experiment—customer waiting style and service person fairness behavior—to observe the cultural influence on communication. The agents always wait in line at counter A and in random position at counter B. The service person's fairness behavior in each session is different. We grouped all five sessions into two categories: 'same conditions' of fairness (both service persons fair or unfair in sessions 1, 2, and 5) and 'different conditions' of fairness (one service person is fair and the other is unfair in sessions 3 and 4). The reasoning answers from the questionnaire were evaluated without considering counter selection because we wanted to consider only the stimuli that the participant perceived from our system during counter selection in the experiment. We did not provide any rules or guidance for counter selection; the participant selected the counter freely. We found that selection reasoning could be categorized into the following reasons: (1) queue-jumper/fair or unfair service, (2) waiting style/waiting position, (3) speed (waiting speed, service speed), (4) interaction with service person, and (5) feeling (like, don't like, worried, happy, or angry). We used ANOVA to analyze the reasoning results to determine the participants' perceptions while selecting a counter. All results are shown in Table 2.

Table 2. ANOVA analysis of 'reasoning results'

Reasoning result	First-person view group		Third-person view group	
	Same conditions	Different conditions	Same conditions	Different conditions
Fairness	M = 1.021/SD = 0.812	M = 2.281/SD = 1.060	M = 0.958/SD = 0.735	M = 1.312/SD = 1.130
Waiting style	M = 1.562/SD = 0.926	M = 0.625/SD = 0.927	M = 1.667/SD = 1.087	M = 0.844/SD = 0.630
Speed	M = 0.625/SD = 0.873	M = 0.250/SD = 0.500	M = 0.479/SD = 0.656	M = 0.281/SD = 0.499
Interaction	M = 0.833/SD = 0.825	M = 1.531/SD = 1.352	M = 0.458/SD = 0.686	M = 1.094/SD = 1.093
Feeling	M = 1.625/SD = 0.881	M = 0.750/SD = 1.000	M = 1.354/SD = 0.862	M = 0.562/SD = 0.658

We aim to compare cultural influence on waiting style conditions. The main effect was seen in the significant difference in the fairness and queue-jumper conditions when comparing 'same condition' with 'difference condition' ($F(1, 30) = 3.237, p = 0.0820$). This effect indicates that the tendencies of giving the fairness reasoning in first- and third-person POV groups are different. For further analysis, we plot the ANOVA results of fairness as shown in Fig. 2. When both groups of participants attended the session with the same style of service by the service persons, the participants in both groups did not pay attention to fairness or queue-jumper. In contrast, if only one service person was unfair, most participants in the first-person POV group recognized the unfairness and queue-jumper more than under the same conditions of fairness, while only a small number of the participants in the third-person POV group provided feedback about fairness levels.

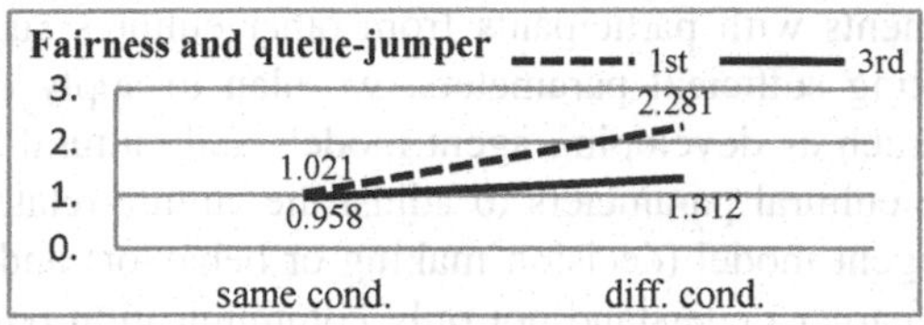

Fig. 2. Fairness reasoning result in 'same conditions' and 'different conditions'

We did not find any significant difference in other reasoning, but we found that most participants in both groups recognize speed more in 'same conditions'. The waiting position was given as an answer by the participants in the first-person POV group, and when one service person becomes unfair, the participants from both groups did not pay attention to the waiting position. As Degens reported in [5], a relationship between the collectivism dimension and proxemics distance was difficult to set and measure in the physical setting in the virtual simulation. We found that some Thai participants in the third-person POV group ignore the queue-jumper and continue to wait at their preferred counter because they think that person may be in a hurry to do their work. Thus, in this case the idea of collectivism may not only affect the physical position but also the role of work status in the activity.

5 Conclusion and Future Work

In this paper, we presented an experiment to find the effect of Thai culture on fairness perception in waiting behavior by using a simulated crowd. The participant in first- and third-person POV groups performed the same activities in the experiment of buying tickets from two counters. Queue waiting and group waiting styles are used in the system. We confirmed with concrete results, that culture has an influence on human behavior. A novel contribution of this paper is that we determined the effect of culture on waiting behavior at different levels of fairness. Fair service should be preferred in general; however, in this experiment we found that Thai participants (15 of 32) accepting the queue-jumper for their own reasons, commiserate that he may be in a hurry (2 in third-person POV), and think that only one queue-jumper in a line is acceptable (7 in first-person POV, 2 in third-person POV) and one queue-jumper is acceptable because they are relaxing in random group waiting (2 in first-person POV, 2 in third-person POV). These behaviors are evidence of the highly feminine characteristics of Thai cultural influence ton the third-person POV group in learning to understand the context and hidden reasons for jumping the queue, whereas the participants in the first-person POV group consider the fairness. In terms of the collectivism dimension, we found that the participants forgive because the queue-jumper is a friend of the service person. We conclude that the collectivism dimension can refer to groups and relationships. Learning communication between people should be based on the message to be transferred. However, during the transfer, there are many processes to produce the message for communication. Discovering cultural factors is one way that can help us understand the ways of other cultures more deeply. In the future, we will conduct experiments with participants from other cultures such as the Japanese culture. After collecting sufficient parameters, we plan to apply our findings to the cultural agent field, such as developing agent models and cultural learning assistance systems, and use the cultural parameters to adjust the culture-related functions in our system such as the agent model (decision making or behavior) and story (scenario or lessons) to help the learner understand not only communication but more importantly culture-specific communication.

References

1. Samovar, L.A., Porter, R.E., McDaniel, E.R., Roy, C.S.: Communication Between Cultures. Cengage Learning, Boston (2016)
2. Dresser, N.: Multicultural Manners: Essential Rules of Etiquette for the 21st Century. Wiley, Hoboken (2011)
3. Hofstede, G., Hofstede, J.G., Minkov, M.: Cultures and Organizations: Software of the Mind, 3rd edn. McGraw Hill Professional, New York (2010)
4. Rafaeli, A., Barron, G., Haber, K.: The effects of queue structure on attitudes. J. Serv. Res. **5** (2), 125–139 (2002). https://doi.org/10.1177/109467002237492
5. Degens, N., Endrass, B., Hofstede, G.J., Beulens, A., André, E.: 'What I see is not what you get': why culture-specific behaviours for virtual characters should be user-tested across cultures. J. AI Soc. **32**, 37–49 (2014). https://doi.org/10.1007/s00146-014-0567-2
6. Kistler, F., Endrass, B., Damian, I., Dang, C.T., André, E.: Natural interaction with culturally adaptive virtual characters. J MUI **6**, 39–47 (2012). https://doi.org/10.1007/s12193-011-0087-z
7. Hall, L., Tazzyman, S., Hume, C., Endrass, B., Lim, M.Y., Hofstede, G., Paiva, A., Andre, E., Kappas, A., Aylett, R.: Learning to overcome cultural conflict through engaging with intelligent agents in synthetic cultures. J. Artif. Intell. Educ. **25**, 291–317 (2015). https://doi.org/10.1007/s40593-014-0031-y
8. Endrass, B., Degens, N., Hofstede, G.J., André, E., Hodgson, J., Mascarenhas, S., Mehlmann, G., Paiva, A., Ritter, C., Swiderska, A.: Integration and evaluation of prototypical culture-related differences. In: Workshop on Culturally Motivated Virtual Characters (CMVC 2011), Conference on IVA 2011, Reykjavík, Iceland (2011)
9. Mascarenhas, S.F., Silva, A., Paiva, A., Aylett, R., Kistler, F., André, E., Degens, N., Hofstede, G.J., Kappas, A.: Traveller: an intercultural training system with intelligent agents. In: Proceedings of 2013 Autonomous Agents and Multi-agent Systems, USA, pp. 1387–1388 (2013)
10. Hall, E.T., Hall, M.R.: Understanding Cultural Differences. Intercultural Press, Yarmouth (1989)
11. Hall, E.T.: Beyond Culture. Anchor, Garden City (1989)
12. Jandt, F.E.: An Introduction to Intercultural Communication: Identities in a Global Community. Sage Publications, Thousand Oaks (2017)
13. Thovuttikul, S., Lala, D., Ohashi, H., Okada, S., Ohmoto, Y., Nishida, T.: Simulated crowd: towards a synthetic culture for engaging a learner in culture-dependent nonverbal interaction. In: Workshop on Eye Gaze in Intelligent Human Machine Interaction, Conference on IUI 2011, California, USA (2011)

14. Thovuttikul, S., Lala, D., Kleef, V.N., Ohmoto, Y., Nishida, T.: Comparing people's preference on culture-dependent queuing behaviors in a simulated crowd. In: Proceedings of ICCI* CC 2012, pp. 153–162. IEEE Press, Japan (2012). https://doi.org/10.1109/ICCI-CC.2012.6311141
15. Lala, D., Thovuttikul, S., Nishida, T.: Towards a virtual environment for capturing behavior in cultural crowds. In: Conference on Digital Information Management, pp. 310–315. IEEE Press, Melbourne (2011). https://doi.org/10.1109/ICDIM.2011.6093362
16. Nishida, T., et al.: Synthetic evidential study as primordial soup of conversation. In: Chu, W., Kikuchi, S., Bhalla, S. (eds.) DNIS 2015. LNCS, vol. 8999, pp. 74–83. Springer, Cham (2015). https://doi.org/10.1007/978-3-319-16313-0_6

Differential Information Diffusion Model in Social Network

Hong T. Tu[1(✉)] and Khu P. Nguyen[2(✉)]

[1] HCMC University of Technology and Education, Thu Duc, Ho Chi Minh City, Vietnam
hongtt@hcmute.edu.vn

[2] University of Information Technology, VNU-HCM, Thu Duc, Ho Chi Minh City, Vietnam
khunp@uit.edu.vn

Abstract. A social network is modeled as a graph of nodes connected through interactions among users, an important medium for the information spreading and influence on users dynamically. Modeling information diffusion is still key problem to predict influences of information on users. In recent years, numerous literatures have proposed models to solve this problem. However, each of models is coming from different points of view. Based on differential equations and with a real mechanism of transferring, exchanging information in network, in this paper, it is proposed a model of temporal-spatial information diffusion, named differential information diffusion or DID model. This model is setup in accordance with topological structure of network, semantic content and interactive activities of users in network. Experimental computations show the feasibility of the proposed model, conformity with network topology and with prospects of scalability for large networks.

Keywords: Social network · Centrality measures
System of linear differential equations · Information diffusion
Spearman rank-order test · TOPSIS

1 Introduction

Social network is a kind of social structure which consists of individuals or organizations called users and a set of relationship between them. Information diffusion is the process that propagates information over time among users in a social network. In this process, users broadcast their information to others by which users influence one another. Such interaction of users has become a new source of information diffusion.

Research into the information diffusion in networks began in the middle of the previous century with the work of Rogers [1] and Granovetter [2]. Currently, there are a variety of diffusion models arising from the communities of sociology, biology, economics, etc. The popular models namely Threshold model and Cascading model are widely used in the social influence problems, [1, 3]. Besides these models, there are many variation models to reflect more complicated realistic situations.

N. T. Nguyen et al. (Eds.): ACIIDS 2018, LNAI 10751, pp. 96–106, 2018.
https://doi.org/10.1007/978-3-319-75417-8_9

A long time in the 19th century, the infectious disease model took up great interest and promoted influenza epidemics to describe the transmission of contagious disease through individuals. In the recent century, some applications of this model have been widely used to model computer virus infections, news and rumor propagations, [4]. The models of this kind were formulated using ordinary differential equations [5]. Then, models based on differential equations of partial derivatives have been appeared [6]. It is only lately that the Hawkes-based framework to model diffusion phenomena has been introduced to consider not only the hidden interactions between users but also the dynamic interactions of the diffusion in networks, [7].

It is dealt with this paper an adapted information diffusion model based on network structure, user interactions and linear differential equations, named differential information diffusion or DID for short. The main contribution of this work is aimed at setting up a dynamic mechanism of information spreading in accordance with network structure, semantic and interactive relation of users using the formulation of linear differential equation system. The proposed model allows spreading information more realistic and more conveniently in processing some related issues.

After presenting some related works in Sect. 2, the proposed model is described in Sect. 3. A case study illustrated in Sect. 4 is to demonstrate the DID model operating and finally to test of significance for matching the proposed model with network structure. The paper ends with a conclusion and future works in Sect. 5.

2 Related Work

Consider a class of information diffusion models based on differential equations. Firstly, in the epidemic models, e.g. SIR, SIS, SIRS, the numbers of Susceptible (S), Infectious (I), and Recovered (R) individuals at certain time are considered to derive ordinary differential equations, [5, 8]. Solutions of these equations are used to predict the numbers of S, I, and R individuals at a time, but do not specify where these objects in network, structures and linkages between individuals are ignored.

In many recent publications, partial differential equations are used to characterize temporal and spatial information diffusion problem over social networks, [9]. Using solution of such equations, the information density of users at a hop-distance in network from a given source at a time is specified. But it is not shown a specific information density of any user in the same hop-distance. Moreover, the solution of these equations depends on constraints of the boundary-valued problem and the assumption that information is a conserved quantity.

Recently, heat diffusion based approaches have been applied in various domains such as classification, dimensionality reduction, ranking algorithm with Diffusion-Rank, [10]. In [11], Ma et al. have modeled diffusion of information as processes of heat diffusion due to the process of people influencing others is near similar to the heat diffusion phenomenon. However, heat transfer follows the energy conservation law while spreading of information is not quite like so. Additionally, some changes in time of information spreading process have not included in the model.

To improve shortcomings in applying the heat transfer process to information diffusion, in [12] Doo has complemented some properties to reflect interactive activities of users in nodes of social network. This makes the model more practical, while the semantic factor, similarity of diffusive information, flexibility in solving problem is still not including.

3 Proposed Model

Let G = (V, E) be a directed graph of network, V is the set of nodes representing users or individuals n = |V|, E the set of directed edges representing relationships among users, m = |E|. Each edge e = (i, j) is an arrow emanating from i, terminating in j. Let $N^{+}(i)$ be out-neighbors of i or the set of terminated nodes after emanating from i, $N^{-}(i)$ in-neighbors or the set of emanated nodes before terminating at i.

3.1 Mechanism of Information Diffusion

In principle, the set of directed edges defines the topological connectivity of nodes in network. But the relationship among users at nodes is not only determined by their connectivity but also the amount of activities done by each node and the interactive between connected nodes. Active nodes are those that are influenced by others active nodes and may influence their inactive neighbors. Inactive nodes are those that are not influenced by their neighbors. If a node is inactive then it has no interaction to others.

Therefore, an active node may only activate to other if the other is also active and both of them are interactive. Hence, the interactive activity between nodes is an important factor, a necessary condition to diffuse information in network.

At a node $i \in V$, an active user can send information to or receive information from others. Due to some reasons, the user at i only sends or receives in part of information with ratios respectively denoted by $\gamma_i, \rho_i \in [0, 1]$. These ratios may be considered as the probabilities of sending or receiving information of i. If γ_i or ρ_i equals zero, then i is inactive. A couple of nodes i and j is interactive active if both of γ_j and ρ_i differ from zero at the same time. In process of diffusion, γ_i and ρ_i are dynamically vary from time to time. In this paper, γ_i and ρ_i are estimated using discrete-numerical valued functions describing actions according to possible events at i.

Let s_{ij} be the ratio of information content the user j sends to $i \in N^{+}(j)$. These ratios s_{ij} are asymmetric for all $i \neq j$, e.g. a famous actor is usually on being informed about his or her activities to admirers but not vice versa. Similarity of the diffusive information content the user j sends to i may be an estimation of s_{ij}. Thus, the ratio of information j sends to $i \in N^{+}(j)$ is $s_{ij}\gamma_j$, and depending on the adoption of i, user at this node only acquires partly of $s_{ij}\rho_i\gamma_j \in [0, 1]$. This last ratio can be considered as the probability that the node i adopts information sent from $j \in N^{-}(i)$.

Let $\delta_j = \sum_{i \in N+(j)} s_{ij}\varepsilon_i\gamma_j$ be the ratio of the residual information at $i \in N^{+}(j)$, $\varepsilon_i = 1 - \rho_i$. In case of $\rho_i = 1$ with $i \in N^{+}(j)$, $\sum_{i \in N+(j)} s_{ij} = 1$. The total amount of un-adopted information w.r.t. ratio δ_j may be stored as an archival part of the node j to keep the balance between sending, receiving information.

In summary, the following entry notated by k_{ij} is used to determine mechanism of sending and receiving information between two given nodes i, j.

$$k_{ij} = \begin{cases} -\gamma_j & \text{if } i = j \\ s_{ij}\rho_i\gamma_j & \text{if } i \in N^+(j) \quad i \neq j, \quad i, j = 1, 2, \ldots, n \\ 0 & \text{otherwise} \end{cases} \tag{1}$$

The square matrix $\mathbf{K}_D = (k_{ij})$ of order n, $k_{ij} \in [0, 1]$, is called kernel of information diffusion mechanism. Each k_{ij} characterizes the probability the node $i \in N^+(j)$ adopts the information received from the node j.

3.2 Differential Equations

Let $f_j(t)$ is the information density of j at t, and $F(t) = (f_1(t), f_2(t), \ldots, f_n(t))^T \in \Re^n$. This node sends a ratio of $\gamma_j f_j(t)$ to $i \in N^+(j)$, then $i \in N^+(j)$ adopts an amount $s_{ij}\rho_i\gamma_j f_j(t)$ from j. In a time interval dt, if the rate of change of density or the information conductivity in network is α, (1) gives the change of information density at the node j as

$$dF = F(t + dt) - F(t) = \alpha \mathbf{K}_D F(t) dt \tag{2}$$

Equation (2) is a system of linear differential equations whose solution w.r.t an initial condition $F(t_0) = (f_1(t_0), f_2(t_0), \ldots, f_n(t_0))^T$ at $t = t_0$ is defined by

$$F(t) = \exp(\alpha t \mathbf{K}_D) F(t_0) \tag{3}$$

In which, the square matrix $\mathbf{R}_\alpha(t) = \exp(\alpha t \mathbf{K}_D)$ is called resolver of (2), [13]. By recurring, Eq. (3) is written in time step δt and at the time $t_h = h\delta t$ as follows,

$$F(t_h) = \mathbf{R}_\alpha(h\delta t) F(t_0) = \mathbf{R}_\alpha(\delta t) F(t_{h-1}) \; h = 1, 2, \ldots \tag{4}$$

So, F(t) can be computed recursively based on Eq. (4) starting from $F(t_0)$. Thus, some changes in the diffusion mechanism $\mathbf{K}_D$ can be updated step-by-step.

Recalculating $\mathbf{R}_\alpha(h\delta t)$ at each step h is very expensive in large-scale datasets. But, Eq. (4) allows to avoid this if $\mathbf{R}_\alpha(\delta t)$ can be approximated near-precisely using the definition of the matrix exponential or the methods of reducing dimension of space.

An algorithm for finding the solution in each time step t_h with Eq. (4) is designed using the following procedures, where i and j designate nodes in network:

i. Initializing the initial condition $F(t_{h-1})$ before doing the step,
ii. Checking the condition of transmitting with the similarities s_{ij} if necessary,
iii. Updating the sending and receiving parameters ρ_i, γ_j, as mentioned in Sect. 3.1,
iv. Estimating the matrix exponential $\mathbf{R}_\alpha(\delta t)$ w.r.t. the parameters above,
v. Computing the matrix multiplication in Eq. (4) to output $F(t_h)$

The complexity of such an algorithm depends on the number n and the method to evaluate $\mathbf{R}_\alpha$. Anyhow, the computing cost will not exceed cubic-polynomial.

3.3 Accordance of the DID Model with Network Topology

Let $C_j = (c_{1j}, c_{2j}, \ldots, c_{nj})^T \in \Re^n$ j = 1, 2, …, k be the jth centrality measures of nodes in G, [14], here c_{ij} may be degree, closeness, or between-ness of each node i, etc. Technique for order preference by similarity to ideal solutions, or TOPSIS [15], is applied to ranking nodes according to the k centrality values. To do so, C_j need be normalized by dividing the sum of entries in each column.

Let w_i be the weight of the jth centrality measure, and

$$b_{ij} = w_j c_{ij} / \sum\nolimits_{v=1\ldots n} c_{vj},\ i = 1,2,\ldots,n, j = 1,2,\ldots,k \quad (5)$$

Then $(b_{ij}) \in \Re^{n\times k}$ is ranking matrix. Let C^+ be the subset of profit centralities, C^- the complement of C^+ in {1, 2, …, k} or the subset of un-profit centralities. Sets of the best, worse values in C^+, C^- denoted $A^+ = \{b_1^+, b_2^+, \ldots, b_k^+\}$ and $A^- = \{b_1^-, b_2^-, \ldots, b_k^-\}$ respectively, are called ideal solutions where

$$\begin{aligned} b_j^+ &= \max\{b_{ij} | i = 1,\ldots,n\}, b_j^- = \min\{b_{ij} | i = 1,\ldots,n\}\ j \in C^+ \\ b_j^+ &= \min\{b_{ij} | i = 1,\ldots,n\}, b_j^- = \max\{b_{ij} | i = 1,\ldots,n\}\ j \in C^- \end{aligned} \quad (6)$$

The distances from the ith node to A^+, A^- are determined by

$$S_i^+ = \left\{\sum\nolimits_{j=1\ldots k} (b_{ij} - b_j^+)^2\right\}^{1/2},\ S_i^- = \left\{\sum\nolimits_{j=1\ldots k} (b_{ij} - b_j^-)^2\right\}^{1/2}\ i = 1,2,\ldots,n \quad (7)$$

The relative closeness R_i^+ or R_i^- of the node $i \in V$ w.r.t A^+, A^- is defined by

$$R_i^+ = S_i^+ / (S_i^- + S_i^+) \text{ and } R_i^- = S_i^- / (S_i^- + S_i^+)\ i = 1,2,\ldots,n \quad (8)$$

These R_i^+ and R_i^- are used to order preference the nodes according to the ideal solutions. The nodes with higher R_i^- are to be un-profit and the nodes with higher R_i^+ are supposed to be played the best and should be higher priority.

To test for accordance of the DID model and a structure of network, Spearman test is applied with a given level significance of a probability.

4 Experiments

To demonstrate clearly and easily seen, the dataset of documents extracted from [16] are assigned to nodes in Fig. 1 according to the similarities s_{ij} between the nodes i, j. These s_{ij} are calculated using the algorithms in hk-LSA model proposed in [17, 18]. The diffusion mechanism is illustrated in the figure with the kernel $\mathbf{K_D}$ described in Fig. 2. It is easily to check that $\sum_{i\in N+(j)} s_{ij} = 1$ for all j as $\rho_i = 1$ with $i \in N^+(j)$.

In this experiment, DID model is setup with $\delta t = 0.5$, $\alpha = 1$ and h = 50 computation steps. To realize possibilities in sending and adopting information at each node

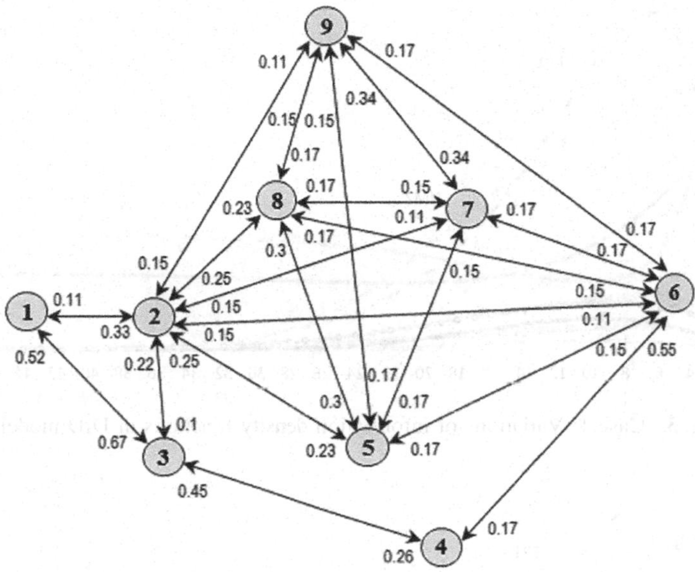

Fig. 1. Graph of network. Numbers close to head of arrow are s_{ij}, e.g. $s_{12} = 0.11$.

$$\begin{pmatrix}
-\gamma_1 & s_{12}\rho_1\gamma_2 & s_{13}\rho_1\gamma_3 & 0 & 0 & 0 & 0 & 0 & 0 \\
s_{21}\rho_2\gamma_1 & -\gamma_2 & s_{23}\rho_2\gamma_3 & 0 & s_{25}\rho_2\gamma_5 & s_{26}\rho_2\gamma_6 & s_{27}\rho_2\gamma_7 & s_{28}\rho_2\gamma_8 & s_{29}\rho_2\gamma_9 \\
s_{31}\rho_3\gamma_1 & s_{32}\rho_3\gamma_2 & -\gamma_3 & s_{34}\rho_3\gamma_4 & 0 & 0 & 0 & 0 & 0 \\
0 & 0 & s_{43}\rho_4\gamma_3 & -\gamma_4 & 0 & s_{46}\rho_4\gamma_6 & 0 & 0 & 0 \\
0 & s_{52}\rho_5\gamma_2 & 0 & 0 & -\gamma_5 & s_{56}\rho_5\gamma_6 & s_{57}\rho_5\gamma_7 & s_{58}\rho_5\gamma_8 & s_{59}\rho_5\gamma_9 \\
0 & s_{62}\rho_6\gamma_2 & 0 & s_{64}\rho_6\gamma_4 & s_{65}\rho_6\gamma_5 & -\gamma_6 & s_{67}\rho_6\gamma_7 & s_{68}\rho_6\gamma_8 & s_{69}\rho_6\gamma_9 \\
0 & s_{72}\rho_7\gamma_2 & 0 & 0 & s_{75}\rho_7\gamma_5 & s_{76}\rho_7\gamma_6 & -\gamma_7 & s_{78}\rho_7\gamma_8 & s_{79}\rho_7\gamma_9 \\
0 & s_{82}\rho_8\gamma_2 & 0 & 0 & s_{85}\rho_8\gamma_5 & s_{86}\rho_8\gamma_6 & s_{87}\rho_8\gamma_7 & -\gamma_8 & s_{89}\rho_8\gamma_9 \\
0 & s_{92}\rho_9\gamma_2 & 0 & 0 & 0 & s_{96}\rho_9\gamma_6 & s_{97}\rho_9\gamma_7 & s_{98}\rho_9\gamma_8 & -\gamma_9
\end{pmatrix}$$

Fig. 2. Kernel matrix $\mathbf{K_D}$ of the DID model.

i, the parameters γ_i and ρ_i are discretized using the values of the exponential power function $EP(a, b, c) \stackrel{def}{=} at^b\exp\{t^c\}$ or the values of the normal distribution function $NM(a, b, c) \stackrel{def}{=} a(1/(\sqrt{2\pi}))\exp\left\{-0.5[(t-b)/c]^2\right\}$, where a, b, c are parameters.

The graphs of EP1 = EP(10,12,4), NM2 = NM(5.6,5,4) are used and discretized with each step $t_h = h\delta t$. In Figs. 3, 4 and 5, they are sketched using dash-dot or dash curves. Their behaviors show changes in ratio of sending or receiving information density to or from a node.

The solid curves in Figs. 3, 4 and 5 represent information densities the results of DID model. There, the abscissae are time steps h = 0(1)50, density scales in the ordinate. Due to space limit, three cases are illustrated to show the efficient of the model.

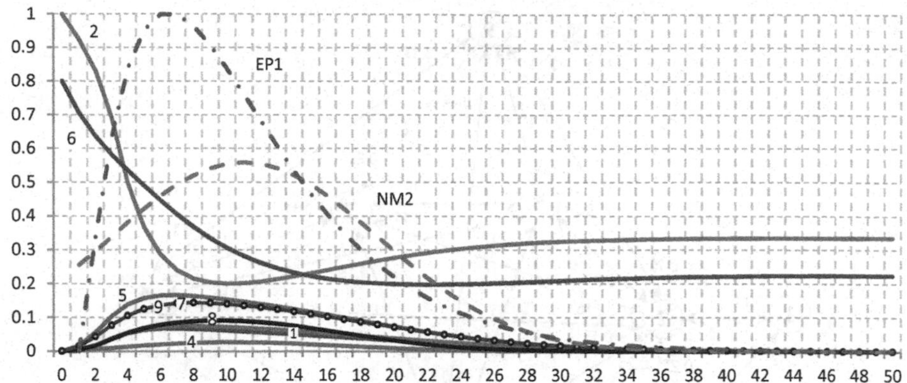

Fig. 3. Case 1: Variations of information density functions in DID model.

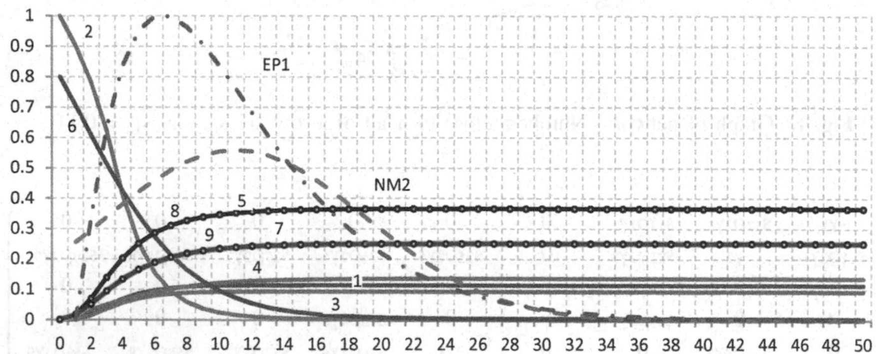

Fig. 4. Case 2: Variations of information density functions in DID model.

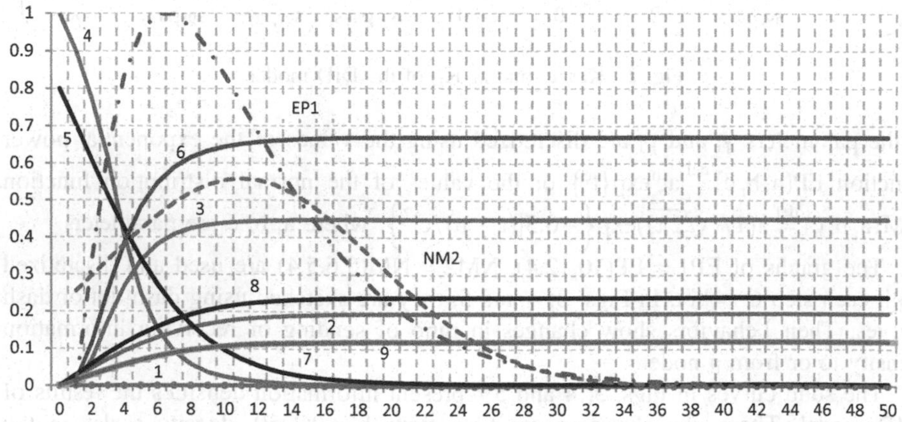

Fig. 5. Case 3: Variations of information density functions in DID model.

4.1 Case 1

In this case, the initial conditions $f_2(0) = 1$, $f_6(0) = 0.8$ with $\gamma_2 = EP_1, \gamma_6 = Nor-02$, others $f_i(0) = 0$, $\gamma_i = 1, i \neq 2, 6$. $\rho_4 = \rho_8 = NM_2$ other $\rho_i = 1$. Figure 3 describes the DID solution. Both nodes 2 and 6 are sending and receiving with nodal degrees greater than 6. After sharing information, in the 25 first time steps the densities of the 2 and 6 become stable. Thus, these nodes activated un-efficiently to the other nodes.

4.2 Case 2

The same setting as the 1st case, but the nodes 2 and 6 are not received, $\rho_2 = \rho_6 = 0$ and $\rho_i = 1, i \neq 2, 6$. These nodes transmit information more efficient as in Fig. 4. The nodes 2, 8 are close neighbor with high similarities, so the latter receives much information from the former. So are the node 2, 9 but with small similarities the node 9 received less information than. Similarly, the remainder nodes are also adopted more information than, while the densities of the nodes 2, 6 are strong decreased.

4.3 Case 3

The initial condition is turning to $f_4(0) = 1$, $f_5(0) = 0.8$ with $\gamma_4 = EP_1, \gamma_5 = NM_2$ and $\rho_4 = \rho_5 = 0$ and $f_i(0) = 0$ $i \neq 4, 5$; otherwise $\gamma_i = \rho_i = 1$. The nodes 4 and 5 only send information, while the others just receive. The node 6 adjacent to 4 and 5 with high similarity, that why it adopts much information than the others. The node 3 is rather far the 5 but very high similarities with the node 4, therefore this node received 45% information. All of the nodes become early stabilization after 12 time-steps.

4.4 Test for Accordance of Model with Network Topology

Spearman rank-order test of significance is applied to test for accordance of the DID model with the network structure. Network topology plays an important role for information diffusion. Therefore, some measures of node centrality such as degree of node, closeness of node, and between-ness centrality, eigenvalue centrality of node need be considered.

Denote Dc, Cc, Bc, and Ec respectively these types of centrality measures, their values are calculated for each node in Fig. 1 and given in the first columns of Table 1.

The TOPSIS allows ranking the nodes w.r.t. their four centralities and listing in the $\mathcal{R}_C$ column of Table 1. The columns f_i−Fk, k = 1, 2, 3 are the stabilized densities beyond 30 time-steps as illustrated in Figs. 3, 4 and 5 and $\mathcal{R}_{Fk}$ the ranking of f_i−Fk, i = 1, 2,…, 9.

Using Spearman's test of rank-correlations with the statistic r is estimated by

$$r_k = 1 - 6\sum\nolimits_{i=1\ldots n} d_{ki}^2/[n(n-1)] \quad (9)$$

Here, n = 9 and d_{ki} is the difference between the ranks of $\mathcal{R}_{Fk}$, $\mathcal{R}_C$ in the ith row of Table 1. Three values of r_k are in the bottom row in the table.

Table 1. Results of calculating and testing the accordance between the centralities and stabilized information diffusion processes

No	Dc	Cc	Bc	Ec	$\mathcal{R}_C$	fi-F1	$\mathcal{R}_{F1}$	fi-F2	$\mathcal{R}_{F2}$	fi-F3	$\mathcal{R}_{F3}$
1.	2	0.571	0.000	0.108	9	0.002	6	0.115	4	0.000	1
2.	7	0.889	0.339	0.429	1	0.334	9	0.001	1	0.196	6
3.	3	0.615	0.054	0.125	7	0.002	7	0.094	3	0.449	8
4.	2	0.571	0.018	0.104	8	0.000	1	0.137	5	0.001	2
5.	5	0.727	0.000	0.391	4.5	0.001	3	0.366	8.5	0.005	3
6.	6	0.800	0.161	0.408	2	0.223	8	0.005	2	0.670	9
7.	5	0.727	0.000	0.391	4.5	0.002	4.5	0.251	6.5	0.120	4.5
8.	5	0.727	0.000	0.391	4.5	0.000	2	0.366	8.5	0.240	7
9.	5	0.727	0.000	0.391	4.5	0.002	4.5	0.251	6.5	0.120	4.5
							r_1 : -0.388		r_2 : 0.250		r_3 : -0.529

Notes: $\mathcal{R}_C$: Ranking of nodes by four centrality measures using TOPSIS technique; fi-F1 in Fig. 1, fi-F2 in Fig. 2, fi-F3: fi Fig. 3; $\mathcal{R}_{F1}$: ranks of fi-F1, $\mathcal{R}_{F2}$: ranks of fi-F2, $\mathcal{R}_{F3}$: ranks of fi-F3.

The critical values w.r.t n = 9 and at levels of significance 1%, 5%, 10% are 0.834, 0.7, 0.6, respectively. It is shown that all $|r_k|$ values in Table 1 are not greater than these critical values. Therefore, the ranking $\mathcal{R}_C$ of the nodes w.r.t. their centralities is statistically in accordance with the ranking of information densities in each case study of the DID model when these diffusion processes stabilized.

Moreover, Table 1 also supports each case study that the nodes with small rank of centrality nearly tended to receive less information than those with greater rank. Therefore, in this case the processes of DID model is in accordance with the network topology.

5 Conclusion and Future Work

It is dealt with this paper a proposed model named DID or differential information diffusion. This model is aimed at estimating information density spreading in a given network, with three unique features:

First, a novel mechanism to setup the model as realistically as possible includes possibilities of sending and adopting, interactive activities, and similarities between nodes in network.

Second, the DID model is introduced to extend the model of differential linear equation system by incorporating the proposed diffusion mechanism. The complexity of solving problems in the model is not exceeding cubic-polynomial, mainly in matrix multiplications and estimating matrix exponential of the resolver.

Also, by splitting-up the DID model in small time intervals and solving the problem in each time step, the diffusion mechanism can be updated according to temporal changes in network.

Third, the compatibleness between the mechanism and the DID model is tested by TOPSIS ranking and Spearman test of significance. At error levels less than 1% or 5%, it is shown that the proposed diffusion mechanism and DID model are in accordance with each other.

Compared to the existing information diffusion model based on network topology and differential equation the DID model with the proposed mechanism is more realistic and flexible in using.

Finally, to check easily and clearly, the dataset used in the experiment is rather small, but commonly. In near future, more experiments will be done to test scalability of the proposed model with some real-larger datasets.

References

1. Rogers, E.M.: Diffusion of Innovations. Free Press, New York (1962)
2. Granovetter, M.S.: Threshold models of collective behavior. Am. J. Sociol. **83**(6), 1420–1443 (1978)
3. Kempe, D., Kleinberg, J., Tardos, E.: Maximizing the spread of influence through a social network. In: 9th ACM SIGKDD, International Conference on Knowledge Discovery and Data Mining, KDD 2003, New York, USA, pp. 137–146 (2003)
4. Guille, A., Hacid, H., Favre, C., Zighed, D.: Information diffusion in online social networks: a survey. SIGMOD Rec. **42**, 17–28 (2013)
5. Kermack, M.: Contributions to the mathematical theory of epidemics. In: Royal Society of Edinburgh, Section A: Mathematics, vol. 115 (1972)
6. Lei, C., Lin, Z., Wang, H.: The free boundary problem describing information diffusion in online social networks. J. Differ. Equ. **254**, 1326–1341 (2013)
7. Pinto, J.C.L.: Information diffusion and opinion dynamics in social networks. Doctor thesis NNT: 2016TELE0001, Pierre and Marie Curie University, Paris, France (2016)
8. Woo, J., Chen, H.: Epidemic model for information diffusion in web forums: experiments in marketing exchange and political dialog. Springer-Open J. Springer-Plus **5**, 66 (2016). https://doi.org/10.1186/s40064-016-1675
9. Wang, H., Wang, F., Xu, K.: Modeling information diffusion in online social networks with partial differential equations. arXiv1310.0505.v1, CS-SI 1 (2013)
10. Yang, H., King, I., Lyu, M.R.: DiffusionRank: possible penicillin for Web spamming. In: Proceedings of the ACM SIGIR Conference, pp. 431–438 (2007)
11. Ma, H., Yang, H., Lyu, M.R., King, I.: Mining social networks using heat diffusion processes for marketing candidate selection. In: CIKM (2008)
12. Doo, M.: Spatial and social diffusion of information and Influence: models and algorithms. Ph.D. thesis. Georgia Institute of Technology, Georgia, USA (2012)
13. Dieudonné, J.: Calcul Infinitésimal. Chapitre XII, Hermann éditeurs des Sciences et des Arts, pp. 381–418, Imprimé en France, Durant Paris (1986)
14. Liua, J., Xiong, Q., Shi, W., Shi, X., Wang, K.: Evaluating the importance of nodes in complex networks. Sci.-Direct Phy. A **452**, 209–219 (2016)
15. Kuo, T.: A modified TOPSIS with a different ranking index. Eur. J. Oper. Res. **260**, 152–160 (2017)
16. Deerwester, S., Dumais, S.T., Furnas, G.W., Landauer, T.K., Harshman, R.: Indexing by latent semantic analysis. J. Am. Soc. Inf. Sci. **41**, 391 (1990)

17. Tu, H.T., Phan, T.T., Nguyen, K.P.: An adaptive latent semantic analysis for text mining. In: The IEEE International Conference on System Science and Engineering – ICSSE 2017. IEEE Xplore, HCM City Vietnam (2017)
18. Tu, H.T., Nguyen, K.P.: Kernel based similarity and discovering documents of similar interests. The 13th IEEE International Conference on: Natural Computation, Fuzzy Systems and Knowledge Discovery - ICNC-FSKD 2017, Guilin, China (2017)

Targeted Misinformation Blocking on Online Social Networks

Canh V. Pham[1,2(✉)], Quat V. Phu[2], and Huan X. Hoang[1]

[1] University of Engineering and Technology, Vietnam National University, Hanoi, Vietnam
maicanhki@gmail.com, huanhx@vnu.edu.vn

[2] Faculty of Information Technology and Security, People's Security Academy, Hanoi, Vietnam
quatphu97mdbg@gmail.com

Abstract. In this paper, we investigate a problem of finding smallest set of nodes to remove from a social network so that influence reduction of misinformation sources at least a given threshold γ, called Targeted Misinformation Blocking (TMB) problem. We prove that TBM is #P-hard under LT model. For any parameter $\epsilon \in (0, \gamma)$, we designed Greedy algorithm which return the solution A with the expected influence reduction greater than $\gamma - \epsilon$, and the size of A is within factor $1 + \ln(\gamma/\epsilon)$ of the optimal size. To speed-up Greedy algorithm, we designed an efficient heuristic algorithm, called STBM algorithm. Experiments were conducted on real-world networks which showed the effectiveness of proposed algorithms in term of both effectiveness and efficiency.

Keywords: Misinformation · Social network
Approximation algorithm

1 Introduction

Besides disseminating official information, Online Social Networks (OSNs) are channels in which also allow spreading misinformation and rumors. In order for social networks as a channel of reliable information for users, many strategies have been proposed to prevent the spread spread of misinformation [3–5,7–11]. Diffusion propagation models are the bases for studying on and identification source of misinformation and restriction the spreading misinformation, in which, there are two most common models, *Linear Threshold (LT)* and *Independent Cascade (IC)* models [13]. Base on that, some authors proposed a mathematical approach to detect misinformation or information sources in the case we known the set of nodes were infected by misinformation [1,2]. Recently, there have been various approaches to decontaminate misinformation by choosing a set nodes to initialize good information and spread it on the same network to convince other users recently [3,4].

In order to block spread of misinformation on OSNs, an effective solution is to remove the important nodes or edges from networks [5,6,12]. Some authors

N. T. Nguyen et al. (Eds.): ACIIDS 2018, LNAI 10751, pp. 107–116, 2018.
https://doi.org/10.1007/978-3-319-75417-8_10

proposed place monitor or immunization vaccines strategies on some nodes to limit the spread of known misinformation/epidemic sources [2,7–10]. Placing monitor or vaccination on a node is equivalent to removing this node from the network during propagation process. Recently, Zhang et al. formulated the problem of placing monitor at a set nodes so that information spreading from known sources of misinformation to protected central nodes no greater than the protection threshold [2]. Zhang and Prakash [8] have developed vaccination strategies for nodes that limit the spread of disease on social networks on the IC model. The similar methods have also been applied for edges and nodes to control propagation at groups under LT model [9]. Later on, Song et al. [10,11] studied the problem of limiting misinformation combining time delay on a various of IC model. They also designed heuristic algorithms that outperform the previous algorithms. However, it is difficult to collect data to establish parameters in their models because they are quite complicated.

Although previous works considered strategies to limit the spreading misinformation, but they do not consider the target for preventing misinformation (i.e., stop it with a given threshold). In reality, to make sure the OSNs are reliable, we need to limit the spread of misinformation so that the number of users not infected by misinformation is greater than a given *threshold*. In other words, this threshold ensures reliability in a social network. Motivated by the phenomenon, in this paper, we investigated the *Targeted Misinformation Blocking* (TMB) problem, in which aim to find the smallest set nodes to remove from the network so that the influence reduction from known misinformation sources at least given threshold γ under LT model. For the complexity, we proved that TMB problem is #P-hard. We proposed a Greedy algorithm which provided a ratio of $1 + \ln(\gamma/\epsilon)$. We further proposed an efficient heuristic algorithm called STMB which is scalable algorithm for TMB on large-scale networks. Experiments were performed on real-world social traces of NetS, AS and NetHEPT datasets show the performance of our proposed algorithms. In each of the network, we observe that STMB is outperform to the other algorithms in terms of minimizing the size of selected nodes while the runtime is faster than Greedy algorithm.

Organiation. The rest of paper is organized as follows. We first introduce propagation model, problem definition in Sect. 2. We prove complexity of TMB in Sect. 3. Section 4 presents our proposed algorithms. The experimental results on several datasets are in Sect. 5. Finally, we give some tasks for future work and conclusion in Sect. 6.

2 Model and Problem Definition

Firstly, we introduce well-know Linear Threshold (LT) diffusion model (see [13]). Based on this, we then formal statement of targeted misinformation blocking problem.

2.1 Diffusion Model

Let $G = (V, E, w)$ is a directed graph represents a social network with a node set V and a directed edge set E, $|V| = n$ and $|E| = m$. Let $N_-(v)$ and $N_+(v)$ are the set of in-neighbors and out-neighbor of node v, respectively. Each directed edge $(u, v) \in E$ is associated with an influence weight $w(u, v) \in [0, 1]$ such that $\sum_{u \in N_-(v)} w(u, v) \leq 1$. Given a subset $S \in V, S = \{s_1, s_2, \ldots, s_k\}$ represents the misinformation sources (as the *seed set* in IM problem [13]). In LT model, each node $v \in V$ has two possible states, *active* and *inactive* and the influence cascades in G as follow. First, every node $v \in V$ uniformly chooses a threshold $\theta_v \in [0, 1]$, which represents the weighted fraction of u's neighbors that must be active to activate u. Next the influence propagation happens in round $t = 1, 2, 3 \ldots$. At round 1, we activate nodes in set S, and set all other nodes inactive. At round $t \geq 1$, an inactive node v is activated if weighted number of its activated neighbors are greater than or equal its threshold, i.e., $\sum_{\texttt{in activated neighbors } u} w(u, v) \geq \theta_v$. Once a node becomes activated, it remains activated in the process of spreading. The influence propagation ends when no more nodes can be activated.

2.2 Problem Definition

Denote $\sigma_S(G)$ is the influence spread of S in G under LT model, i.e., expected number nodes given activated by S. Kempe et al. [13] show that LT model to be equivalent to *live-edge* graph which is constructed by the rules are: (1) for every $v \in V$, select at most one of its incoming edges at random, such that the edge (u, v) is selected with probability $w(u, v)$, (2) and no edge is selected with probability $1 - \sum_{u \in N_-(v)} w(u, v)$. The selected edges are called *live* and all other edges are called *blocked*. By Claim 2.1 in [13], we have: $\sigma_S(G) = \sum_{g \in \mathcal{G}} \Pr[g] R(g, S)$, where $\mathcal{G}$ is set of sample graphs generated from G according *live-edge* model with a probability denoted by $\Pr[g]$ and $R(g, S)$ denotes the set of nodes reachable from S in g (see more detail in [13]). The influence spread from S when remove A is the influence spread of S in induce graph $G[V \setminus A]$, denoted by $\sigma_S(G \setminus A)$. The *influence reduction* of A defined as, $h(A) = \sigma_S(G) - \sigma_S(G \setminus A)$. In this paper, we consider *Targeted Misinformation Blocking* (TMB) which is defined as follows:

Definition 1 (TBM). *Let $G = (V, E, w)$ is a directed graph represents a social network. Given a set of misinformation source $S = \{s_1, s_2, \ldots, s_k\}, S \in V$ and integer number $\gamma \leq |V|$, find a set $A \subset V \setminus S$ of the smallest size nodes to remove form G such that the expected influence reduction, $h(A)$ at least γ.*

3 Complexity

In this section, we show that TMB problem is #P-hard. Note that a #P problem is at least as hard as the corresponding NP problem.

Theorem 1. TMB *problem is #P-hard in LT model.*

Proof. To prove TMB is #P-hard, we reduce from *s*-*t* paths which was proved #P-hard [16], defined as follow:

Definition 2 (*s*-*t* paths problem [16]). *Given a directed graph* $G = (V, E), |V| = n, |E| = m$, *s*-*t* *paths problem ask to compute the number of (directed) paths from node s to node t that visit every node at most once.*

Consider an instance $\mathcal{I}_1$ of *s*-*t* paths problem, where $G = (V, E)$, $s, t \in V$ are given. As Fig. 1 shows, from G, we construct G' as follow: we add a new node u and add two edges $(s, u), (t, u)$ with weights $w(s, u) = w(t, u) = 1/2$, we add more set Q include $2n$ nodes and connect u to them with the same weight is equal to 1. For the others edges, we set the weight is equal to $w = 1/\Delta$, where Δ be the maximum in-degree of any node in G. This assumption still satisfies the LT mode since the total of in-neighbour weight is not greater than 1. Let $\mathsf{P}(G, s)$ is the set of all simple paths from s in graph G, $\mathsf{P}(G, s, t)$ is the set of all simple paths from s to t in graph G. By the equivalence given Claim 2.6 in [13], we have: $\sigma_S(G') = \sum_{x \in \mathsf{P}(G', s)} \prod_{e \in x} w(e)$, and $\sigma_S(G' \setminus \{u\}) = \sum_{x \in \mathsf{P}(G' \setminus \{u\}, s)} \prod_{e \in x} w(e)$. Eliminate the same elements in the two above equations so the remaining paths containing node u. Set these paths divided into two groups: paths have u is the endpoint and the paths have $v \in Q$ is endpoint. Therefore, $h(u) = \sigma_S(G') - \sigma_S(G' \setminus \{u\}) = \sum_{x \in \mathsf{P}(G', s, u)} \prod_{e \in x} w(e) + \sum_{v \in Q} \left(\sum_{x \in \mathsf{P}(G', s, v)} \prod_{e \in x} w(e) \right) = \frac{2n+1}{2} \sum_{i=0}^{n-1} \alpha_i w^i + n$, where $\alpha_i = |\mathsf{P}_i(G, s, t)|$. Let $f(w) = \sum_{i=0}^{n-1} \alpha_i w^i$, on G' we easy see that $0 \le f(w) \le 1$; $n \le h(u) \le 2n + \frac{1}{2}$, and $h(u) = \max_{v \in G'} h(v)$. We first show that if we can determine $f(u) \ge \beta$ for any integer $\beta \in [0, 1]$ in polynomial time, we can solve *s*-*t* paths problem in polynomial time. Since the weigh $w = 1/\Delta$, $f(w)$ is a fraction with a numerator of Δ^{n-1} and the numerator at most Δ^{n-1}. By using binary search from 1 to Δ^{n-1}, we can find value of $f(w)$. This task can be done in $\mathcal{O}(\log(\Delta^{n-1})) = \mathcal{O}((n-1)\log \Delta) = \mathcal{O}(n \log n)$. Hence, we can calculate $f(u)$ in polynomial time. We then the adjust weight w to n distinction values $\frac{1}{\Delta}, \frac{1}{\Delta+1}, \ldots, \frac{1}{\Delta+n-1}$. By using above method, we can find value of $f(w)$ corresponding to each w. Hence, we obtain a set of n linear equations $\sum_{i=0}^{n-1} \alpha_i w^i = f(w), w \in \{\frac{1}{\Delta}, \frac{1}{\Delta+1}, \ldots, \frac{1}{\Delta+n-1}\}$ with $\{\alpha_0, \alpha_2, \ldots, \alpha_{n-1}\}$ as variables. The matrix of this equation is $M_{n \times n} = \{m_{ij}\}$ and $m_{ij} = w^i, i, j = 0, \ldots, n$ so this is Vandermonde matrix and we can easily to compute the unique solution $\{\alpha_0, \alpha_2, \ldots, \alpha_{n-1}\}$ for the linear system of equations. The total of *s*-*t* paths in G is $\sum_{i=0}^{n-1} \alpha_i$. Therefore, if we can determine $f(u) \ge \beta$ for any integer $\beta \in [0, 1]$ in polynomial time, we can solve *s*-*t* paths problem in polynomial time.

We now consider an instance $\mathcal{I}_2$ of TMB where $S = \{s\}$, $\gamma = \beta \frac{2n+1}{2} + n, \beta \in [0, 1]$. Assume that an $\mathcal{A}$ is a polynomial-time algorithm solving TMB problem. Consider two cases: (1) If $\mathcal{A}$ returns the solution set A whose size is equal to 1, we only need to select $A = \{u\}$, infer $f(w) \ge \beta$; (2) If $\mathcal{A}$ returns the solution set A whose size larger than 1. At that besides u, some nodes are chosen into A. We infer $f(w) < \beta$. Therefore, $\mathcal{A}$ can be used to decide $f(w)$ is greater than β,

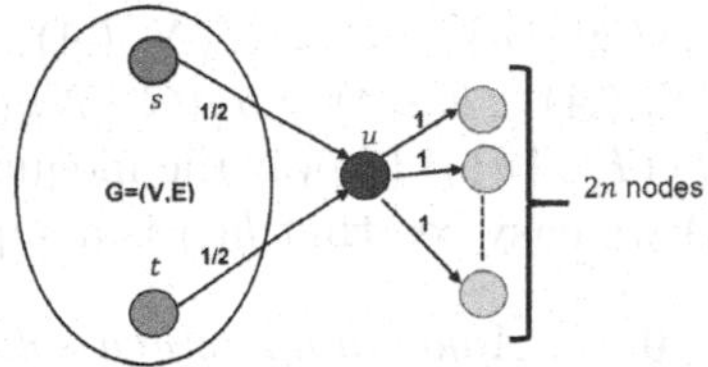

Fig. 1. Reduce from *s*-*t* paths to TMB.

that can also solve the *s*-*t* paths problem. This implying that our TMB problem is at least as hard as *s*-*t* paths problem. □

4 Proposed Algorithms

4.1 Greedy Algorithm

We introduce an approximation algorithm that provide a ratio of $1 + \ln(\gamma/\epsilon)$ base on $h(.)$ is proved *submodular* and *monotone* function, i.e., for $A \subset T, v \notin T$ $h(A + \{v\}) - h(A) \geq h(T + \{v\}) - h(T)$.

Algorithm 1. Greedy Algorithm (GA)

Data: Graph $G = (V, E, w)$, $S = \{s_1, s_2, .., s_k\}$, threshold $< \gamma < |V|$, parameter $\epsilon \in (0, \gamma)$

Result: set of nodes A

$A \leftarrow \emptyset$;

while $h(A) > \gamma - \epsilon$ **do**

  $u = \arg\max_{v \in V \setminus \{A \cup S\}} h(A + \{v\}) - h(A)$; $A \leftarrow A \cup \{u\}$;

end

return A;

Theorem 2. *The function $h(.)$ is submodular and monotone function.*

Proof. Denote $N_E(A)$ is set of edges adjacent with all nodes in A. By Theorem 5 in [12], for $A \subseteq T$ we have $h(T) - h(A) = \sigma_S(N_E(A)) - \sigma_S(N_E(T)) \geq 0$. Therefore $h(.)$ is a monotonically increasing. We then show that the function $\sigma_S(G_i \setminus A)$ is a supermodular function of the set A is the variable, i.e., $\forall A \subseteq T \subset V, \forall v \in T \setminus A$, we have $\sigma_S(G \setminus (A \cup \{v\})) - \sigma_S(G \setminus A) \leq \sigma_S(G \setminus (T \cup \{v\})) - \sigma_S(G \setminus T)$ Let $E_{T,v} = N_E(T + \{v\}) \setminus N_E(T), E_{A,v} = N_E(A + \{v\}) \setminus N_E(A)$ we have $E_{T,v} \subseteq E_{A,v}$ and due to $A \subseteq T$. We obtain $N_E(A) \cup E_{T,v} \subseteq N_E(A + \{v\})$. Let $\sigma_S(G \setminus X)$ is the influence of S for graph G after remove the set edges $X \subset E$, we obtain $\sigma_S(G \setminus A) = \sigma_S(G \setminus N_E(A)$. By Theorem 6 in [12], $\forall X \subseteq Y, e \in Y \setminus X$, we have:

$$\sigma_S(G_i \setminus (X \cup \{e\}) - \sigma_S(G_i \setminus X) \leq \sigma_S(G \setminus (Y \cup \{e\})) - \sigma_S(G_i \setminus Y) \quad (1)$$

Therefore, $\sigma_S(G \setminus A) - \sigma_S(G \setminus (A \cup \{u\})) = \sigma_S(G \setminus N_E(A)) - \sigma_S(G \setminus N_E(A + \{v\})) \geq \sigma_S(G \setminus N_E(A)) - \sigma_S(G \setminus (N_E(A) \cup E_{T,v})) \geq \sigma_S(G \setminus N_E(T)) - \sigma_S(G \setminus (N_E(T) \cup E_{T,v})) = \sigma_S(G \setminus T) - \sigma_S(G \setminus (T \cup \{u\}))$ (Apply the inequality (1)). Combine with $h(A) = \sigma_S(G) - \sigma_S(G \setminus A)$ we easy see that $h(.)$ is a supermodular function. □

Theorem 3. *For any $\epsilon \in (0, \gamma)$, Algorithm 1 return solution A satisfies $h(A) \geq \gamma - \epsilon$, and the size of A is within factor $1 + \ln(\gamma/\epsilon)$ of the optimal size.*

The proof of Theorem 3 straightforward based on [15]. Base on Theorem 2, the greedy algorithm given in Algorithm 1 achieve $1 + \ln(\gamma/\epsilon)$ approximation ratio. The algorithm simply chooses the node that provides maximum largest *incremental influence reduction* in each step, defined as $\delta(A, u) = \min\{\gamma, h(A + \{u\})\} - h(A)$. The main challenge of this algorithm comes from calculate $\sigma_S(.)$ is #P-hard (see [14]). Therefore, we introduce an efficiency algorithm in next subsection.

4.2 Scalable TMB Algorithm

We try to tackle this problem with a speed-up approach proposed by Zhang [9]. This approach use characteristics of the LT model, in which the set nodes that reach from a seed node v in live-edge is a tree root at v. In our proposed algorithm, we first simplify the instance of TMB problem by merging set source $S = \{s_1, s_2, \ldots, s_k\}$ into a *supper source node* I. For each node $v \in N_+(S)$, we assign weight $w(I, v) = \sum_{s \in N_+(v) \cap S} w(s_i, v)$ and remove S after update the new weight set, the result's called *merged graph* G'. Based on the characteristic of LT model, the instances before and after of TMB are equivalence (see more details in [8,9]). Next, we'll generate η sample graphs g from the G'. For each g, we construct an induced tree root at I by removing the edges $(v, I), \forall v \in g$. We obtained set $\mathcal{L}$ which contains η tree (line 3). The influence reduction of a node v on each tree is calculated by using DFS algorithm. We then approximate the marginal influence reduction of node u on G is equal to average influence reduction of node u on all tree $T_I \in \mathcal{L}$ (line 4).

After that, we apply the lazy forward method in [17] to select the solution based on $h(.)$ is submodular function (lines 10–22). Assign $r(u, T_I)$ is the number of all reachable nodes in T_I from node u. The node is selected in each step also removed from each tree $T_I \in \mathcal{L}$ and $r(u, T_I), u \in T_I$ will be updated (line 16) in the way as follows: (1) For children of u, we can remove them because it is not reachable from I, (2) for any ancestor v of u, $r(v, T_I \setminus u) = r(v, T_I) - r(u, T_I)$, which can be done in constant time. The details of algorithm are presented in Algorithm 2.

Complexity. Merge algorithm takes $\mathcal{O}(k + |N_+(S)|)$ (line 3). Generating η sample takes $\mathcal{O}(\eta(m + n))$. Calculating $r(T_I, u), \forall u \in T_I$ can be done in $\mathcal{O}(\eta n)$. For lazy forward phase, the total time needed takes at most $\mathcal{O}(q\eta n)$ where q is the number of iterations of while loop. Therefore, Algorithm 2 runs in $\mathcal{O}(\eta(m + qn))$.

Algorithm 2. Scalable TMB (STMB) Algorithm

Data: Graph $G = (V, E, w)$, $\mathcal{S} = \{s_1, s_2, .., s_q\}$, threshold $\gamma > 0$
Result: set of nodes A

$A \leftarrow \emptyset$; $(G', I) \leftarrow$ `Merge`$(G, \mathcal{S})$.
Remove all node, I can't reach in G.
Generate η live-edge graphs and set η tree $\mathcal{L} = \{T_I^1, T_I^2, \ldots, T_I^{|\eta|}\}$
For each $T_I \in \mathcal{L}$, calculate $r(u, T_I)$ for all $u \in T_I$ (by using DFS algorithm).
for $u \in V$ **do**
    $u.\delta(u) \leftarrow \frac{1}{\eta} \sum_{T_I \in \mathcal{L}} r(u, T_I)$; $u.cur \leftarrow 1$
    Insert element u into Q with $u.\delta(u)$ as the key
end
$h_{max} \leftarrow 0$; $iteration \leftarrow 1$
while $h_{max} < \gamma - \epsilon$ **do**
    $u_{max} \leftarrow dequence\ Q$
    if $u_{max}.cur = iteration$ **then**
        $A \leftarrow A \cup \{u_{max}\}$; $iteration \leftarrow iteration + 1$
        foreach $T_I \in \mathcal{L}_c$ **do**
            If $u_{max} \in T_I$, remove node u_{max} and update $r(v, T_I)$, $\forall v \in T_I$.
        end
        $h_{max} \leftarrow h_{max} + u_{max}.\delta(u_{max})$
    else
        $u_{max}.\delta(u_{max}) \leftarrow \frac{1}{\eta}\left(\sum_{T_I \in \mathcal{L}} r(I, T_I) - \sum_{T_I \in \mathcal{L}} r(I, T_I \setminus u_{max})\right)$
        $u_{max}.cur = iteration$; re-insert u_{max} into Q
    end
end
return A;

5 Experiments

In this section, we show experimental results of proposed algorithms on three real-world datasets to evaluate the performance and compare them with several other baselines algorithm.

5.1 Experiment Setup

Dataset. The three real-world networks we use and their basic statistics are summarized in Table 1. We assign the weights of edges in LT model according to previous studies [12–14]. The weight of the edge (u, v) is $w(u, v) = \frac{1}{|N_-(v)|}$. For the misinformation source, we randomly choose S in 4–6% of the set nodes. The code is written in Python 2.7 using the NetworkX library and all experiments are run on a Linux Server machine with 2.30 GHz Intel® Xeon® CPU E5-2697 and 128 GB of RAM DDR4.

Algorithms Compared. In our experiments, we compare STMB algorithm with other algorithms below: PageRank: Compute a ranking of the nodes in the graph

Table 1. Datasets

Dataset	NetS [18]	AS [19]	NetHEPT [13,14]
Num. of nodes	1.5K	6.4K	15.2K
Num. of edges	5.4K	12.5K	32.2K
Avg. degree	3.8	7.5	4.2
Num. of source nodes	100	300	1000

G based on the structure of the incoming links. It was originally designed as an algorithm to rank web pages. We setup damping parameter for PageRank is 0.85. Because $h(.)$ is monotonic function, we used binary search algorithm to find A set with $|A|$ nodes having highest-ranked. High-Degree: A heuristic based on the notion of degree centrality. We sort all nodes base on degree of each node then making the same to PageRank, we use binary search algorithm to find A. Greedy: The Greedy algorithm (Algorithm 1) with the lazy evaluation optimization in [17]. We run 10,000 simulations to accurately estimate $h(A)$ for every A set obtained for each algorithm.

5.2 Experiment Results

Solution Quality. As demonstrated in the Fig. 2, the number of selected nodes gave by STMB algorithm is the smallest. STMB is up to 39% better than Greedy method, 60%–95% and 57%–87% better than that PageRank and High-Degree respectively. To check A set got from STMB algorithm, we run 10000 times Monte-Carlo simulations to calculate function $h(A)$ and result is shown in Fig. 3. In most cases $h(A)$ is greater than γ.

Table 2. Compare running time between algorithms

Dataset	STMB	Greedy	Page Rank	High-Degree
NetS	**17.57**	14206.80	35.73	30.24
AS	45.70	14074.87	**14.39**	17.85
NetHEPT	**165.12**	582566.74	392.34	374.66

Running Time. The running time of different algorithms on the three networks are given in Fig. 2 and Table 2. On the NetS and NetHEPT dataset, our STMB algorithm is roughly two times faster than the PageRank, High-Degree and 800–3500 times faster than the Greedy. On the AS dataset, STMB algorithm is slower than the PageRank and High Degree but still 300 times faster than Greedy. From the result, we see that STMB algorithm is very competitive in its time efficiency.

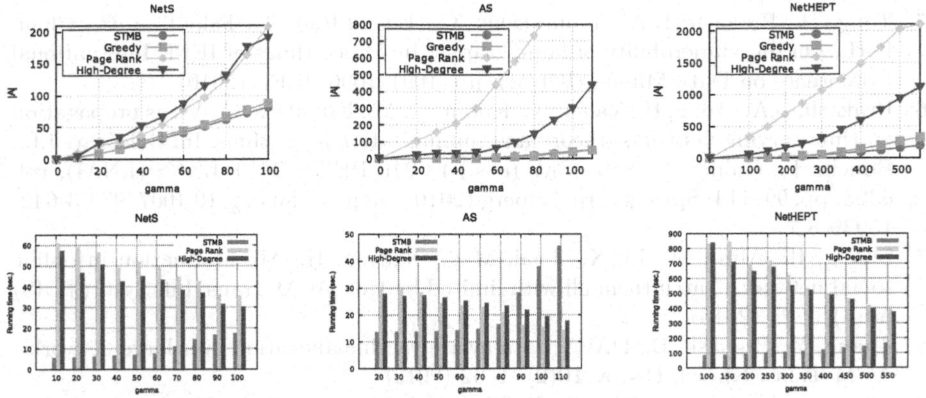

Fig. 2. Comparison of solution quality of algorithms on NetS, AS, NetHEPT networks

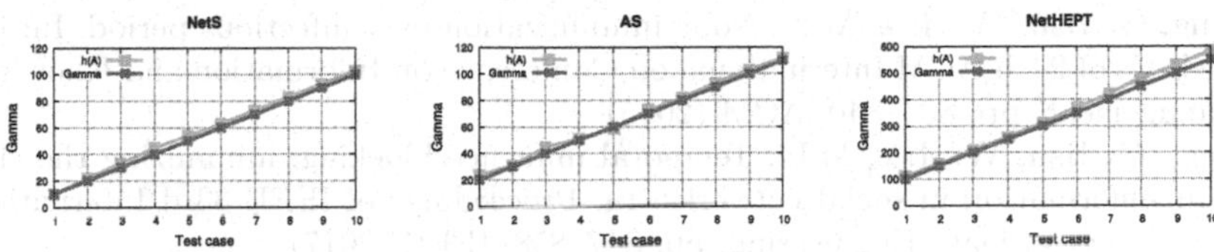

Fig. 3. Check result of STBM on TBM problem.

6 Conclusions

In this paper, we studied the TBM problem, in which aim to finding smallest set nodes to remove from a social network so that the number of influence reduction no less than a given threshold γ. Besides proving the problem is #P-Hard. We proposed two algorithms: Greedy and STMB algorithms. In the future, we will tackle the TBM problem in other diffusion model, especially IC model.

References

1. Nguyen, D.T., Nguyen, N.P., Thai, M.T.: Sources of misinformation in online social networks: who to suspect? In: Proceedings of IEEE Military Communications Conference (MILCOM), pp. 1–6. IEEE (2012)
2. Zhang, H., Kuhnle, A., Zhang, H., Thai, M.T.: Detecting misinformation in online social networks before it is too late. In: Proceedings of 2016 IEEE/ACM International Conference on Advances in Social Networks Analysis and Mining (ASONAM), pp. 541–548. IEEE (2016)
3. Nguyen, N.P., Yan, G., Thai, M.T.: Analysis of misinformation containment in online social networks. Comput. Netw. **57**, 2133–2146 (2013)
4. Budak, C., Agrawal, D., El Abbadi, A.: Limiting the spread of misinformation in social networks. In: Proceedings of 20th International Conference on World Wide Web, pp. 665–674. ACM (2011)

5. Tong, H., Prakash, B.A., Tsourakakis, C., Eliassi-Rad, T., Faloutsos, C., Chau, D.H.: On the vulnerability of large graphs. In: Proceedings of IEEE International Conference on Data Mining (ICDM), pp. 1091–1096. IEEE (2010)
6. Prakash, B.A., Tong, H., Valler, N., Faloutsos, M., Faloutsos, C.: Virus propagation on time-varying networks: theory and immunization algorithms. In: Balcázar, J.L., Bonchi, F., Gionis, A., Sebag, M. (eds.) ECML PKDD 2010. LNCS (LNAI), vol. 6323, pp. 99–114. Springer, Heidelberg (2010). https://doi.org/10.1007/978-3-642-15939-8_7
7. Zhang, H., Alim, M., Li, X., Thai, M.T., Nguyen, H.: Misinformation in online social networks: catch them all with limited budget. ACM Trans. Inf. Syst. (TOIS) **34**(3), 1–24 (2016)
8. Zhang, Y., Prakash, B.: DAVA: data-aware vaccine allocation over large networks. ACM Trans. Knowl. Discov **10**(2), 1–32 (2015)
9. Zhang, Y., Adiga, A., Saha, S., Vullikanti, A., Prakash, B.A.: Near-optimal algorithms for controlling propagation at group scale on networks. IEEE Trans. Knowl. Data Eng. **28**(12), 3339–3352 (2016)
10. Song, C., Hsu, W., Lee, M.L.: Node immunization over infectious period. In: Proceedings of 24th ACM International on Conference on Information and Knowledge Management, pp. 831–840. ACM (2015)
11. Song, C., Hsu, W., Lee, M.L.: Temporal influence blocking: minimizing the effect of misinformation in social networks. In: Proceedings of IEEE 33rd International Conference on Data Engineering, pp. 847–858. IEEE (2017)
12. Khalil, E.B., Dilkina, B., Song, L.: Scalable diffusion-aware optimization of network topology. In: Proceedings of 20th ACM SIGKDD International Conference on Knowledge Discovery and Data Mining, pp. 1226–1235. ACM (2014)
13. Kempe, D., Kleinberg, J., Tardos, E.: Maximizing the spread of influence through a social network. In: Proceedings of 9th ACM SIGKDD International Conference on Knowledge Discovery and Data Mining, pp. 137–146. ACM (2003)
14. Chen, W., Wang, C., Wang, Y.: Scalable influence maximization in social networks under the linear threshold model. In: Proceedings of 2010 IEEE International Conference on Data Mining, pp. 88–97. IEEE (2010)
15. Goyal, A., Bonchi, F., Lakshmanan, L.V.S., Venkatasubramanian, S.: On minimizing budget and time in influence propagation over social networks. Soc. Netw. Anal. Min. **2**(1), 179–192 (2012)
16. Valiant, L.G.: The complexity of enumeration and reliability problems. SIAM J. Comput. **8**(3), 410–421 (1979)
17. Leskovec, J., Krause, A., Guestrin, C., Faloutsos, C., Van-Briesen, J.M., Glance, N.S.: Cost-effective outbreak detection in networks. In: Proceedings of 13th ACM SIGKDD International Conference on Knowledge Discovery and Data Mining, pp. 420–429. ACM (2007)
18. Nguyen, D.T., Zhang, H., Das, S., Thai, M.T., Dinh, T.N.: Least cost influence in multiplex social networks: model representation and analysis. In: Proceedings of IEEE 13th International Conference on Data mining, pp. 567–576. IEEE (2013)
19. Leskovec, J., Kleinberg, J., Faloutsos, C.: Graphs over time: densification laws, shrinking diameters and possible explanations. In: Proceedings of 11th ACM SIGKDD International Conference on Knowledge Discovery in Data Mining, pp. 177–187. ACM (2005)

Conversation Strategy of a Chatbot for Interactive Recommendations

Yuichiro Ikemoto[1(✉)], Varit Asawavetvutt[1], Kazuhiro Kuwabara[2], and Hung-Hsuan Huang[2]

[1] Graduate School of Information Science and Engineering, Ritsumeikan University, Kusatsu, Japan
is0201hh@ed.ritsumei.ac.jp

[2] College of Information Science and Engineering, Ritsumeikan University, Kusatsu, Shiga 525-8577, Japan

Abstract. This paper presents a conversation strategy for interactive recommendations using a chatbot. Chatbots are attracting attention to provide a flexible user interface using natural language for various domains. For a given task, what kind of questions to ask and/or what information should be provided and how to process user responses play a crucial role in developing an effective chatbot. In this paper, we focus on a task of recommending an item that suits a user's preference and propose a conversation strategy where a chatbot combines questions about user's preference and recommendations soliciting user's feedback to them. The balance between questions and recommendations is controlled by changing the parameter values. We target a chatbot that uses a graphical user interface (GUI) and apply approaches proposed in the field of recommendation systems. Preliminary experiment results with a prototype indicate the potential of our proposed approach.

Keywords: Chatbot · Conversation strategy
Interactive recommendation

1 Introduction

This paper presents a conversation strategy for a chatbot by focusing on interactive recommendation tasks. Generally, a chatbot exploits natural language processing, which does not depend on a particular task or domain. Various frameworks have been proposed to facilitate the development of chatbots (for example, [2,12]). For a given task, not only natural language processing but also how to conduct conversations is important. Some chatbots also rely on such a graphical user interface (GUI) as buttons with which a user can send a predefined message by just clicking a button. In such a case, issues include not only what kind of messages are sent to users but also what possible answers are provided to them.

In this paper, we focus on a task that recommends an item that matches a user's preferences and consider a conversation strategy in a chatbot setting.

N. T. Nguyen et al. (Eds.): ACIIDS 2018, LNAI 10751, pp. 117–126, 2018.
https://doi.org/10.1007/978-3-319-75417-8_11

Typical recommender systems use algorithms such as content-based or collaborative filtering [8]. In these systems, the interaction between a system and a user is generally one-shot. On the other hand, in conversational recommender systems, a user repeatedly interacts with a system. For example, critiquing-based recommender systems use user's feedback or *critiques* about recommended items to narrow down suitable items to recommend [3]. Here we apply such methods that have been proposed in recommender systems research to a chatbot setting. We also employ a GUI that is provided on some chatbot platforms to develop a more user-friendly system.

The rest of the paper is organized as follows. The next section describes related work, and Sect. 3 describes a model for interactive recommendations using a chatbot. Section 4 explains the prototype we built using the LINE messaging service, and Sect. 5 presents preliminary evaluation results obtained using our prototype. Finally, Sect. 6 concludes the paper and describes some future work.

2 Related Work

2.1 Recommender Systems

To achieve effective recommendations, user preferences for items must be inferred correctly [9]. A user's preferences are generally represented by a user–item matrix. When a user's previous behaviors are not known beforehand, the so-called *cold start* problem needs to be addressed. To solve this problem, a framework for eliciting user preferences was proposed [4] that identified questions to learn a new user's preferences. Another method was proposed where a series of recommendations is made and the user's preferences are constantly updated to reflect user feedback on recommended items [13].

In a conversational recommender system, obtaining feedback can be categorized into two basic types: *navigation by asking* and *navigation by proposing* [10]. These feedback strategies are analyzed in terms of user efforts and cost. From the viewpoint of user interaction, generalized linear search (GLS) was also proposed that minimizes the number of user interactions to discover items that match a user's interests [6]. Adapting interaction strategies was also proposed in conversational recommender systems [7] that use reinforcement learning techniques. Conversational recommendations are also applied to acquire the functional requirements of products [11]. This proposed framework introduces an ontology structure and explores user preferences through question and answer interactions that resemble those between professional sales people and customers.

In contrast to these studies, we focus on a mechanism that switches between questions and recommendations and apply it to a chatbot setting for natural interaction between a user and a chatbot.

2.2 Word Retrieval Assistant

Systems have been proposed that help aphasia sufferers recall an item's name [1,5]. Such word-finding difficulty is one typical symptom of a person with aphasia and describes the plight of a person with aphasia who has a clear mental image of an item but cannot find its name or language to express it. In an activity that mirrors the traditional game of 20 questions, a human conversation partner usually asks a series of questions to infer the name of the thing an aphasia sufferer wants to say. For example, suppose that an aphasia sufferer wants to say *banana* but cannot recall its name. A conversation partner asks such questions as *Is it food?*, *Is it a fruit?*, *Is it yellow?*. The questions used in this setting are basically multiple choice or yes-no type of queries. If a person with aphasia answers them with *yes*, what he wants to say might eventually be inferred as *banana.* The word retrieval assistance system asks a series of questions instead of a human conversation partner to infer what an aphasia sufferer is struggling to remember.

This word-finding process resembles a method that identifies an item whose characteristics a user knows but not the item itself. In this sense, a word retrieval assistance system shares properties with interactive recommendations.

In word retrieval assistance systems, the order in which questions are posed is critical to efficiently infer what the user has in mind. Typical heuristics calculate the information gain of questions and the question with the biggest information gain is asked next. In a word retrieval assistance system, there is only one correct answer, but in a recommendation system, multiple items might be suggested. Thus, different heuristics are necessary for an interactive recommendation.

In addition, since the target user of a word retrieval assistance system has difficulty answering a question in a free-text form, a graphical user interface (GUI) is deployed, such as buttons. When a user answers a multiple choice question, using GUI is more convenient than a free-text entry. When GUI elements are available on a chatbot platform, they should also be exploited for interactive recommendations.

3 Interactive Recommendations with a Chatbot

3.1 Recommendation Model

We assume n items from which an item(s) is recommended to a user, based on her preferences. Each item is characterized by m *properties.* Item s_i is represented by m-dimensional vector $\boldsymbol{s_i} = (s_{i,1}, \cdots, s_{i,m})$. Similarly, user u's preferences are represented as m-dimensional vector $\boldsymbol{u} = (u_1, \cdots, u_m)$.

The similarity between user u and item s_i, $\text{sim}(u, s_i)$ is calculated based on the Pearson correlation coefficient:

$$\text{sim}(u, s_i) = \frac{\sum_{j=1}^{m}(u_j - \overline{u_j})(s_{i,j} - \overline{s_{i,j}})}{\sqrt{\sum_{j=1}^{m}(u_j - \overline{u_j})^2}\sqrt{\sum_{j=1}^{m}(s_{i,j} - \overline{s_{i,j}})^2}}. \tag{1}$$

The initial value of a user vector is set to $\boldsymbol{u} = (0, \cdots, 0)$, which means that her preference is unknown. Through interactions between a user and a system, the user vector's value is updated.

We consider the following two types of triggers to update the user vector:

- *question*: a user is asked whether she is interested in the property specified.
- *recommendation*: she is presented with a recommendation and is asked whether she likes it.

In the former case, the value of u_j is directly set from the user's answer to the question. Suppose that the possible answers are YES, NO, or NOT SURE. If the user's response is YES, the value is set to 1, and if her answer is NO, the value is set to −1. Otherwise, the value is set to 0.

In the latter case, the value of u_j is updated based on her feedback in response to the recommendation. We categorize the possible feedback from the user as LIKE, DISLIKE, or NOT SURE. If her feedback is LIKE, the property value of the user vector is increased if the same property of the recommended item is positive, and it is decreased if the same property of the recommended item is negative. If her feedback is DISLIKE, the property value of the user vector is decreased if the same property of the recommended item is positive; it is increased if the same property of the recommended item is negative. Otherwise, the user vector is not changed. More specifically, for recommended item s_i, the user vector is updated as follows:

$$u_j \leftarrow \begin{cases} u_j + s_{i,j} \times \beta & \text{(when user's feedback is } \mathsf{LIKE}\text{)} \\ u_j - s_{i,j} \times \beta & \text{(when user's feedback is } \mathsf{DISLIKE}\text{)} \\ u_j & \text{(when user's feedback is } \mathsf{NOT\ SURE}\text{).} \end{cases} \tag{2}$$

Here β determines how much a user's feedback to a recommendation affects her preferences.

In the above model, even though we set the number of answer choices to three to simplify the user interface, increasing the number is easy, as in a typical 5-star rating system, by adjusting the value of β.

The overall processing flow is triggered by a user who inputs a certain text such as Start (Fig. 1). First, the user vector is initialized to $(0, 0, \cdots, 0)$. Then the similarity between an item and a user is calculated by formula (1). When an item's similarity exceeds the *recommendation threshold* (α), the item with the highest similarity is presented as a recommendation. Here the *recommendation threshold* is adjusted as described in the next section. The feedback from the user to the recommended item is reflected on the user vector by formula (2).

3.2 Conversation Strategy

In the above conversation control flow, the *recommendation threshold* (α) determines whether a recommendation is made. A bigger value of α means that a recommendation is only made when an item has been found that strongly matches the user's preference. Thus, more questions will be asked before the first recommendation is made. In this way, we can manipulate the balance between questions and recommendations by properly setting the value of α.

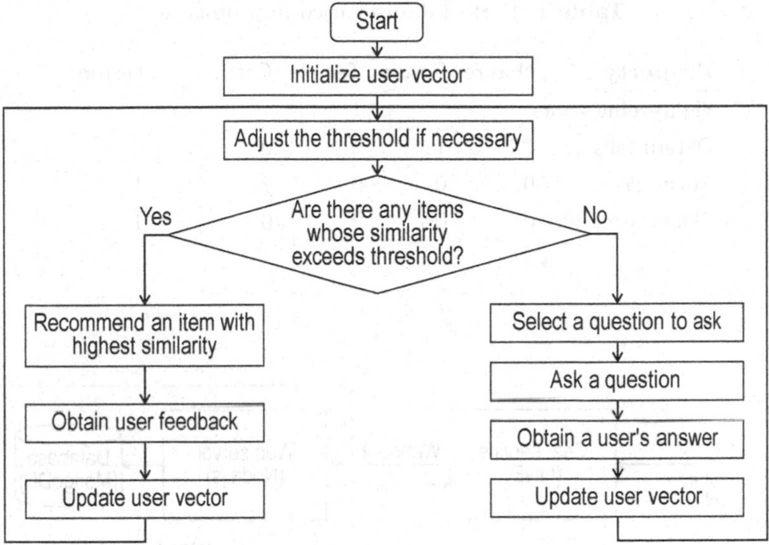

Fig. 1. Overall processing flow

As one heuristic, we decrease the *recommendation threshold* (α) by multiplying by the *threshold adjustment* γ ($\gamma \leq 1$) to increase the chance of successful recommendations when additional recommendations are requested.

The order in which questions are asked is also important. For example, if all the items that might be recommended have value 0 for a certain property, asking a question about that property may be ineffective. Thus, as another heuristic, we calculate a set of items that might be recommended (more specifically, where its similarity value is not negative), and for each property, we count the number of items in that set whose property value is positive. The property with the highest count number is asked next. In this way, we expect that the number of items will be increased that may have higher similarity.

4 Prototype System

4.1 Dataset

We compiled a dataset of sightseeing spots in Kochi prefecture in Japan for our prototype. The number of sightseeing spots in the dataset is 49, and the number of properties is 19. We assigned a value of 2 to the property value of a sightseeing spot if it has a strong tendency about the property, and we assigned a value of 1 if it somewhat has the characteristics of the property. If it lacks such characteristics, the property's value is set to 0.

Table 1 shows part of the data used in the prototype. In addition to the property values of the sightseeing spots, URLs of thumbnail pictures and related web sites are included in the dataset.

Table 1. Part of dataset used in prototype

Property	Nature	Resort	Temple	Castle	...	History
Sightseeing spots						
Ohtaru falls	2	1	0	0	...	0
Kochi castle	0	0	0	2	...	1
Chikurin temple	0	0	2	0	...	1
⋮						

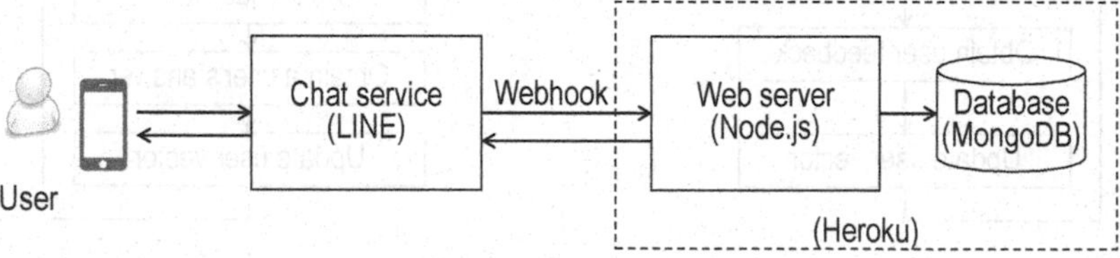

Fig. 2. Overview of prototype system

4.2 System Design

We used the LINE messaging service[1] as a platform to construct a prototype chatbot system. The LINE platform provides a messaging API to facilitate the development of a chatbot. When a message is sent to the chatbot, a registered Webhook is invoked whose return value specifies the message that is returned to a user. In this prototype, we used Node.js to build a web server for Webhook, which is deployed on the Heroku cloud application platform[2] (Fig. 2).

The LINE messaging API allows us to send not only text messages but also to use GUIs like buttons or links to web sites. In the prototype, we used buttons to let a user input her responses. Figure 3(a) shows an example screenshot that asks a question, and Fig. 3(b) shows the recommendation of an item to a user and a request for feedback about it. When a recommendation is made, its thumbnail picture and a link to a relevant web site are also shown to provide information to determine her feedback on a recommended sightseeing spot.

5 Evaluation Experiments

5.1 Recommendation Threshold

To investigate the effects of *recommendation threshold* (α) that determines the balance between questions and recommendations, we conducted a simulation

[1] https://line.me/en/.

[2] https://www.heroku.com/.

(a) Question

(b) Recommendation

Fig. 3. Chatbot user interface

with different user models using the above dataset. With different α values, we counted the number of questions before the first recommendation is made or the system fails to find a suitable recommendation. Since the system has 19 properties, the maximum number of questions is 19.

Figure 4 shows the average, maximum, and minimum counts with eight different user models, which specify the properties that might interest users. Here we presume eight types of typical tourists who visit Kochi prefecture and created a user model by manually defining a user vector for each type. Since a suitable recommendation varies for each user model, the difference between the maximum and minimum counts is rather large. However, the overall result indicates that a higher α generally leads to more questions before the first recommendation is made. Thus, by properly setting the α value, we expect to forge a better balance between questions and recommendations.

5.2 User Study

We also conducted preliminary evaluation experiments with human users. Based on the above simulation results, we set the value of *recommendation threshold* (α) to 0.6. Other parameter values were determined as follows after pretrial experiments: the *threshold adjustment* that determines the amount of decrease of the recommendation threshold value was set to 0.9 ($\gamma = 0.9$) and the feedback coefficient that determines the amount of changes in the user vector after feedback to a recommendation is set to 0.2 ($\beta = 0.2$).

We asked 12 university students in their twenties who also regularly use the LINE service to try our prototype deployed as a LINE bot and to answer questionnaires afterwards. They registered our prototype as a *friend* in their LINE service for their smartphones. The questionnaire consisted of five questions with four choices: strongly agree, agree, disagree, and strongly disagree. The questionnaire also included a free description text format that sought general comments about the prototype system.

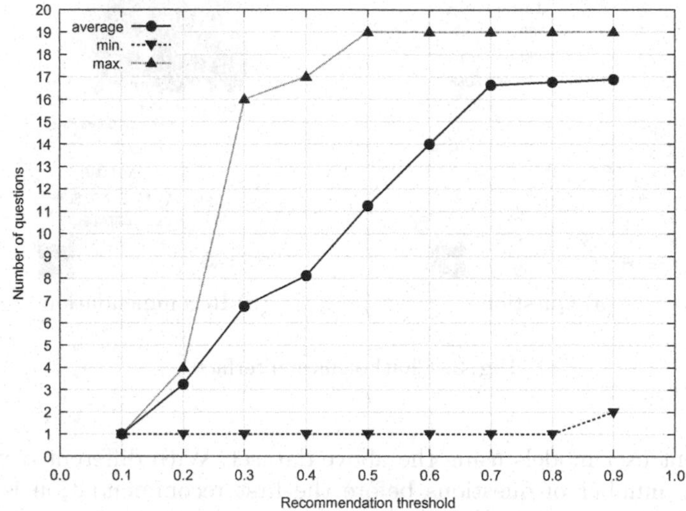

Fig. 4. Number of questions asked before first recommendation

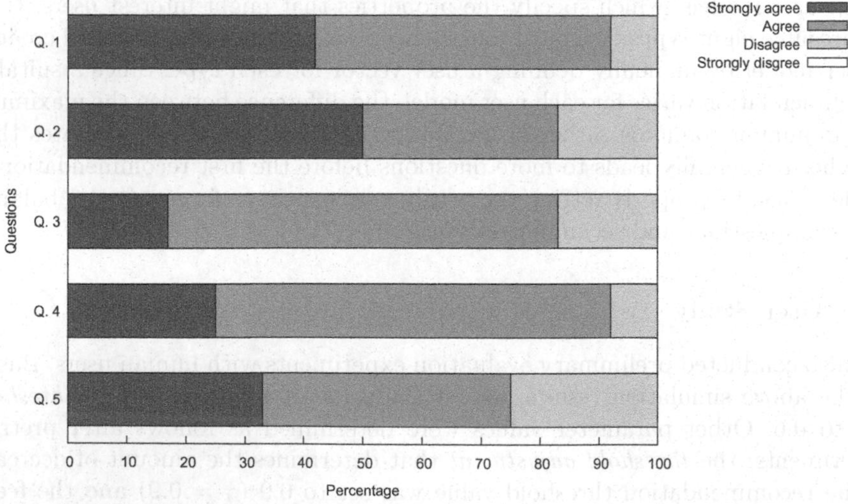

Q. 1: Is the order of questions natural?
Q. 2: Are the questions and recommendations naturally combined?
Q. 3: Is the feedback about recommended items properly considered?
Q. 4: Is the system easy to use?
Q. 5: Do you want to use the system for a different region?

Fig. 5. Questionnaire results

The results of the multiple-choice type questions are shown in Fig. 5. The overall response was positive; combining the questions and recommendations was especially well received by most users (Q. 2). This indicates that the proposed conversation strategy worked well with this domain. Compared with this result, handling feedback to recommendations has room for improvement (Q. 3). The free descriptive question results mostly addressed the viewpoint of a practical system; for example, one comment suggested that not only sightseeing spots but also restaurants in the region should be included in the recommendations.

6 Conclusion and Future Work

This paper described a conversation strategy that intersperses recommendations with questions for user preferences. In addition to directly asking about user preferences, we also considered feedback from users about the recommendations they received. Our simulation results indicate that by changing a parameter value, a balance can be achieved between the number of questions and recommendations. In addition, we implemented a prototype chatbot on the LINE messaging service whose preliminary user evaluation results indicate that it was well received by users.

In the current recommendation model, we did not consider the relationships between properties. In some situations related questions should be asked together. Exploiting the relationships among the properties is one future issue.

In our user evaluation experiments, we defined parameter values based on pretrial results. Since appropriate parameter values depend on the system's dataset, how to determine them is another future issue. When parameter values are determined, subjective impressions about the system must be addressed.

The domain of the current prototype is the recommendations of sightseeing spots in a particular region. Since our proposed approach is applicable to other domains, we plan to expand it to investigate its effectiveness.

Acknowledgment. This work was partially supported by JSPS KAKENHI Grant Number 15K00324.

References

1. Arima, S., Kuroiwa, S., Horiuchi, Y., Furukawa, D.: Question-asking strategy for people with aphasia to remember food names. J. Technol. Persons Disabil. **3**, 10–19 (2015)
2. Augello, A., Scriminaci, M., Gaglio, S., Pilato, G.: A modular framework for versatile conversational agent building. In: 2011 International Conference on Complex, Intelligent and Software Intensive Systems (CISIS), pp. 577–582. IEEE (2011)
3. Chen, L., Pu, P.: Critiquing-based recommenders: survey and emerging trends. User Model. User-Adapt. Interact. **22**(1), 125–150 (2012)
4. Christakopoulou, K., Radlinski, F., Hofmann, K.: Towards conversational recommender systems. In: Proceedings of the 22nd ACM SIGKDD International Conference on Knowledge Discovery and Data Mining, KDD 2016, pp. 815–824. ACM, New York (2016)

5. Kuwabara, K., Iwamae, T., Wada, Y., Huang, H.-H., Takenaka, K.: Toward a conversation partner agent for people with aphasia: assisting word retrieval. In: Czarnowski, I., Caballero, A.M., Howlett, R.J., Jain, L.C. (eds.) Intelligent Decision Technologies 2016. SIST, vol. 56, pp. 203–213. Springer, Cham (2016). https://doi.org/10.1007/978-3-319-39630-9_17
6. Kveton, B., Berkovsky, S.: Minimal interaction search in recommender systems. In: Proceedings of the 20th International Conference on Intelligent User Interfaces (IUI 2015), pp. 236–246. ACM (2015)
7. Mahmood, T., Ricci, F.: Improving recommender systems with adaptive conversational strategies. In: Proceedings of the 20th ACM Conference on Hypertext and Hypermedia, HT 2009, pp. 73–82. ACM, New York (2009)
8. Ricci, F., Rokach, L., Shapira, B. (eds.): Recommender Systems Handbook. Springer, US (2015). https://doi.org/10.1007/978-0-387-85820-3
9. Shi, Y., Larson, M., Hanjalic, A.: Collaborative filtering beyond the user-item matrix: a survey of the state of the art and future challenges. ACM Comput. Surv. **47**(1), 3:1–3:45 (2014)
10. Smyth, B., McGinty, L.: An analysis of feedback strategies in conversational recommenders. In: The 14th Irish Conference on Artificial Intelligence and Cognitive Science (AICS 2003), pp. 211–216 (2003)
11. Widyantoro, D.H., Baizal, Z.: A framework of conversational recommender system based on user functional requirements. In: 2014 2nd International Conference on Information and Communication Technology (ICoICT), pp. 160–165. IEEE (2014)
12. Yan, M., Castro, P., Cheng, P., Ishakian, V.: Building a chatbot with serverless computing. In: Proceedings of the 1st International Workshop on Mashups of Things and APIs, MOTA 2016, pp. 5:1–5:4. ACM, New York (2016)
13. Zhao, X., Zhang, W., Wang, J.: Interactive collaborative filtering. In: Proceedings of the 22nd ACM International Conference on Information and Knowledge Management, CIKM 2013, pp. 1411–1420. ACM, New York (2013)

Competitive Information Diffusion Model in Social Network with Negative Information Propagation

Hong T. Tu[1](✉) and Khu P. Nguyen[2](✉)

[1] HCMC University of Technology and Education, Thu Duc, Ho Chi Minh City, Vietnam
hongtt@hcmute.edu.vn
[2] University of Information Technology, VNU-HCM, Thu Duc, Ho Chi Minh City, Vietnam
khunp@uit.edu.vn

Abstract. Social networks can be used to exchange effectively information among people. But some networks also became the channels for spreading of information competing against to the information being diffused in network. In fact, when people adopted a positively recommend then one may have a high probability to refuse the recommend with a presence of a reasonable negative comment. Such exchanges make modeling information diffusion becomes more difficult. There are no much previous models have studied the spreading of positive and negative information flows simultaneously. In this paper, it is dealt with a proposed model for negative information diffusion in competition against a positive information flow. This problem is referred to the competitive information diffusion model or CID for short. In consideration of mechanism and realization the model, experimental case study shows some feasible contribution of the CID model.

Keywords: Social networks · Competitive information diffusion
Negative opinions · System of linear differential equations
Matrix exponential function

1 Introduction

Social networks are the effective channels for information exchange. However, some networks also became the most effective channel for spreading of information sources that compete against or act in opposition to the information being diffused in network. The information with such property is commonly called negative information.

In Oxford dictionary, the term information is defined as facts provided or learned about something or someone. However, the distinction between facticity and fictitiousness of interested information is sometimes rather fuzzy and so the same with positive and negative information. In literature, the term misinformation is also used to designate false or inaccurate information, especially with deliberately intent to deceive, [1]. While the term negative information appeared naturally and can be understood that it aims simply at non-positive information. In some senses, negative information is

N. T. Nguyen et al. (Eds.): ACIIDS 2018, LNAI 10751, pp. 127–136, 2018.
https://doi.org/10.1007/978-3-319-75417-8_12

near-meaning with misinformation but not only so, [2]. But both of them when diffusing may bring to undesirable consequences.

In real life, when people adopted a positively recommend then one may have a high probability to refuse the recommend if he or she receives a reasonable negative comment. Such exchange of state make modeling information diffusion becomes more difficult with the presence of negative information, [3]. Thus, the understanding of how negative information influenced is still at a complex issue. Consequently, no much previous models have considered such information flows.

In mining social networks for marketing, the authors in [4] suggested the concept of "negative information" and sought to reduce impact of this information source in product marketing. Recently, in Tweeter, noteworthy features have been used to be very relevant for predicting the credibility of tweets are positive or negative sentiments of tweet messages, [1, 5]. These two sentiments may be considered as a couple of features of information in positive or negative tweets.

It is dealt with this paper a proposed model for negative information diffusion, an extension of the DID model [6], in competitive situations with a positive information flow. In the meaning of competition, this proposed model is referred to the competitive information diffusion or CID model. The main contribution of this work is aimed at modifying the DID model appropriately for negative information diffusion, studying mechanism and realization the model in competitive situations.

The rest of the paper is presented as follows. A review related work is in Sect. 2. The proposed model for negative information diffusion is in Sect. 3. In Sect. 4 some results of experimental case are illustrated. Finally, conclusion and future work are given in Sect. 5.

2 Related Works

In 2004, the authors in [7] proposed a propagation model of trust and distrust to answer the question of why people trust and distrust others. However, the work was not addressing the problem of how negative comments diffuse to convert between trust and distrust. Carnes et al. in [8], dealt with the question how to find an initial set of nodes to target for this information flow, given that the initial set of nodes adopting that information flow is known. Some computational experiments showed that their proposed models addressed basically the treatment of such questions but did not showed how spreading these information sources are.

Ma et al. in [4], have modeled diffusion of information as processes of heat transfer and proposed heuristic to simulate product adoptions in the presence of both positive and negative comments. Beside the current positive information flow represented by a positive-valued function, the authors introduced the concept of "negative information diffusion" presented by a negative-valued function and suggested a heuristic to defend against the negative information. But the understanding of how negative information diffused has been still at a crude level and need more analysis.

An extension to the independent cascade model that incorporates the emergence and propagation of negative opinions was proposed in [9]. In [10], the authors incorporated view point of negativity bias in such a way that negative opinions usually is dominated over positive opinions a commonly acknowledged in the social psychology literature. A heuristic algorithm based on cascade model was setup to compute influence spreading maximization. Through simulations, it is shown that their suggested heuristic has matching influence with a standard greedy algorithm.

Not so long ago, many publications have paid attention to game theory in building up models of competitive information diffusion, [11]. The authors in [12] introduced a game-theoretic model of information diffusion and explained how human factors impact competitive information spreading. The process of diffusion was described as the dynamic of a cooperative game and payoff of players was defined by a utility function. However, the game-theoretic models do not address the problem of taking advantage of both the social network and viral marketing when introducing a new technology into a market.

3 Proposed Model

If user adopts some facts of information the node is positively activated. However, due to some defects detected in the received information, user at this node may cancel this positive influence and either adopts the negative information or refuses both of these kinds of information. Therefore, a node in network may be in one of three states positive, negative, or neutral. A user at a node is activated positively or negatively at specific time step if it is neutral in the previous step and becomes positive or negative after that.

When a node is negatively activated, it strictly holds negative state with a probability almost one even if its nearest neighbors are positively influenced or turning to positive, [10]. This situation is similar to the fact that in font of both positive and negative opinions, negative opinions are likely to dominate and be attended. Studies in social psychology it is also shown that "negative events may have more penetrance or contagiousness than positive events", [13]. This manifestation is of negativity bias and dominance phenomenon.

Assume that the positive information source attains a confidence p to users. Such a confidence is considered as a probability user stays in positive state after being activated positively. For example, due to receiving negative information on quality of the product a user may generate negative opinion and refuse his or her confidence in the product. In this case the confidence p reflects the quality of the product or information advertised in the network.

Hence, p is considered as a p-threshold to admit and q = 1−p the q-threshold to deny the received information. These thresholds are also used to specify whether changing state from positive or negative of user at a node.

Given a social network is represented by a directed graph G = (V, E), where V is the set of nodes or users with n = |V| and E the set of directed edges to depict relationship among users, m = |E|. Each edge e is denoted by (i, j) emanating from node i terminating in j, $p_S(e)$ the probability of propagation the state of i to j.

3.1 Probability of Activation

Let A denotes the set of activated nodes the union of the A+, A − which consists of positively, negatively nodes respectively. $\mathcal{P}(\mathrm{a}, \mathrm{i})$ is a shortest directed path in hops that connects a node a ∈ A to i ∈ V with k−1 intermediary nodes i_1, i_1, …, i_{k-1}. This path is denoted by $\mathcal{P}(\mathrm{a}, \mathrm{i}) = \{\mathrm{a} = \mathrm{i}_0, \mathrm{i}_1, \ldots, \mathrm{i}_{k-1}, \mathrm{i}_k = \mathrm{i}\}$, emanates from i_0 and ends at i_k = i. So, the hop-distance or length of $\mathcal{P}(\mathrm{a}, \mathrm{i})$ is $k = |\mathcal{P}(\mathrm{a}, \mathrm{i})| \stackrel{\mathrm{def}}{=} h(\mathrm{a}, \mathrm{i})$.

For a node i, N+(i), N−(i) is defined the set of out-neighbors, in-neighbors of i, respectively. Let $p_{A+}(i)$, $p_{A-}(i)$ be in turn the probability that i is positively, negatively activated by a node a in A+, A−.

Proposition 1. Given a network G = (V, E) of directed edges with a p-threshold of information propagation. Assume $p_S(e) = 1$ for all e ∈ E, the probability a node i ∈ V activated positively by a node a in A+ is determined by

$$\mathrm{p}_{\mathrm{a}}^{+}(\mathrm{i}) = \mathrm{p}^{h(\mathrm{a},\mathrm{i})+1} \tag{1}$$

Indeed, let *h*(a, i) = k, (1) is $\mathrm{p}_{\mathrm{a}}^{+}(\mathrm{i}) = \mathrm{p}^{k+1}$. If k = 0 then i = a ∈ A+ is positively activated with a probability p or $\mathrm{p}_{\mathrm{a}}^{+}(\mathrm{i}) = \mathrm{p}^{0+1}$. By induction, assume every node in N −(i) is positively activated with a probability p^k. Let i ∈ V−A and *h*(a, i) = k > 1. In N −(i) the nodes are at a hop-distance of k − 1 from a ∈ A+. Because $p_S(e) = 1$ for any e ∈ E, so i will clearly be activated positively at a distance k randomly by some nodes in N−(i). As a result, i is activated positively with a probability of p^{k+1} no matter which node in N−(i) activates i. The proposition follows by induction method.

The above result is extended into the general case that the propagation probability of each edge is independently of the former events, as

Proposition 2. Given a network G = (V, E) of directed edges with a p-threshold of information propagation. The probability a node i ∈ V activated positively by a ∈ A+ is

$$\mathrm{p}_{\mathrm{a}}^{+}(\mathrm{i}) = \Pi_{\mathrm{e}\in\mathcal{P}(\mathrm{a},\mathrm{i})}\mathrm{p}_{\mathrm{s}}(\mathrm{e}) \times \mathrm{p}^{h(\mathrm{a},\mathrm{i})+1} \tag{2}$$

In case of a in subset A−, the probability that a node i ∈ V activated negatively is derived from (2) by the following

$$\mathrm{p}_{\mathrm{a}}^{-}(\mathrm{i}) = \Pi_{\mathrm{e}\in\mathcal{P}(\mathrm{a},\mathrm{i})}\mathrm{p}_{\mathrm{s}}(\mathrm{e}) \times \mathrm{p}^{h(\mathrm{a},\mathrm{i})+1} \tag{3}$$

The probability (2) or (3) includes 2*h*(a, i)+1 multiplications of p or q in [0, 1], thus its magnitude is a number multiplied by $10^{-2h(a,i)+1}$. This implies that the longer paths from information source nodes in A to i are, the smaller these probability values. As a matter of fact, when estimating how frequent a node activated positively or negatively using (2) or (3) is possible if the number of hop-distances between i ∈ V and a propagation source node a ∈ A is small enough. Otherwise, it is difficult to estimate the above probabilities because they are getting very smaller and further sources.

It is better to pay attention to the confidence p of the spreading positive information in network. If a user at a given node has a confidence in the positive information with a probability approximate to an upper bound of p, then there is a possibility he or she stays positive state. But how near to determine such a decision. The answer depends how he or she had gone through the mill on both of these information sources.

In experience, there is possible to user at a node staying positive if he or she believes that the confidence of the positive information is still in an upper bound of p. In the other words, user would be in a distinct possibility of leaving his or her current state with a probability less than $p - \varepsilon$ for some $\varepsilon > 0$. However, how much such a possibility is then still all a matter of luck and depends on choosing ε the user has been in front of the diffused information sources.

3.2 Competitive Information Diffusion

The mechanism of diffusion in [6] is applied to the spreading of positive information. For the negative information sources that diffuse simultaneously together with the positive one, the sending ratios γ_j of these negative sources is usually equal to 1 due to the negative bias and dominance. The similarity between the diffusion information content can be measured using the method in [14] or [15]. If s_{ij} is the ratio of information that user at a node i receives from j in N−(i), $p_{ij} = s_{ij}\gamma_j$ can be considered as the propagation probability of information to i from j.

$$k_{ij} = \begin{bmatrix} -\gamma_i & \text{if } j = i \\ s_{ij}\rho_i\gamma_j & \text{if } j \in N-(i) \\ 0 & \text{otherwise} \end{bmatrix} \tag{4}$$

If there is a balance between the information that a node i sends to others and the one that i adopts from nodes j in N−(i) then $\gamma_i = \rho_i \sum_{j \in N-(i)} s_{ij}\gamma_j$. In reality, this is not necessary because that remains to be dependent upon situations of sending or receiving information of a node.

Similarly, when a user at node j sends a ratio of information γ_j to i in N+(j) these nodes i's may adopt in part of it. For convenience, firstly it is assumed temporarily that $\gamma_j = 1$ and $\rho_j^o = 1/\sum_{i \in N+(j)} s_{ij}$. By letting θ_j be $1/\sum_{i \in N+(j)} s_{ij}\rho_j^o$, the sum of $s_{ij}\rho_i\theta_j$ for $i \in N+(j)$ is the total adoption of user at i in the condition of $\gamma_j = 1$. After that, all ratios γ_j can be adjusted - in accordance with possibility of sending information of user at node j, accordingly to the possibility of adoption information of these nodes when receiving information from other nodes in network.

The positive or negative information density vector function at time t is denoted by $F(t) = (f_1(t), f_2(t), \ldots, f_n(t))^T \in \mathfrak{R}^n$. Each non-negative $f_i(t)$ can be a positive or negative information source spreading to others, a component receiving information in network. In a time interval δt, the difference $\delta F = F(t + \delta t) - F(t) = \alpha \mathbf{K}_D F(t)\delta t$ is a system of homogeneous linear differential equations. At the starting time $t = t_0$ with an initialization $F(t_0) = (f_1(t_0), f_2(t_0), \ldots, f_n(t_0))^T$ the solution F(t) is obtained [6], by

$$F(t) = \exp(\alpha t \mathbf{K}_D)F(t_0) \tag{5}$$

This solution $F(t) = (f_1(t), f_2(t), \ldots, f_n(t))^T$ is defined by the matrix exponential $\mathbf{R}_\alpha(t) = (r_{ij}) \stackrel{def}{=} \exp(\alpha t\mathbf{K}_D)$. The dynamic change of the information density vector is depicted thought the discrete time-marks $t_h = h\delta t$ at step h = 0, 1, 2,... by

$$F(t_h) = \mathbf{R}_\alpha(\delta t)F(t_{h-1}) \quad h = 1, 2, \ldots \tag{6}$$

Let $\sigma_i \in \{-1, +1\}$ be the state of information source $f_n(t_h)$ at time step h. As in Sect. 3.1, the proposed procedure with negative shifting named CID-N for changing from positive to negative state is depicted as follows:

i. Letting $\sigma_i \coloneqq 1$ for all node i, and giving a tolerance ε so that $p - \varepsilon > 0$
ii. For i = 1…n do // calculating each component $F(t_h)$
iii. For j = 1…n do if $\sigma_i = -1$: $f_i(t_h) \coloneqq \sum_{v:\sigma v = -1} r_{iv} f_v(t_{h-1})$
iv. else if $p' \coloneqq$ random $(p - \varepsilon, 1) < p$: $f_i(t_h) + = p - p'$ and $\sigma_i \coloneqq -1$
v. else $f_i(t_h) \coloneqq \sum_{v:\sigma v = +1} r_{iv} f_v(t_{h-1})$ // a sum of only positive information sources.

It is necessary to estimate a upper-bound to the value of F (t) in Eq. (6). In doing so, a basic theorem on location of extreme eigenvalues and a weighted log-norm in $\mathfrak{R}^n$ are applied, [16].

Let $\mathbf{A} \in \mathfrak{R}^{n\times n}$ and ξ is the number greater than any eigenvalue of **A**, there exists a positive-defined symmetric matrix $\mathbf{H} \in \mathfrak{R}^{n\times n}$ that satisfies Lyapunov equation [17],

$$\mathfrak{L}(\xi, \mathbf{A}, \mathbf{H}) = 2\mathbf{I} + (\mathbf{A} - \xi\mathbf{I})^T\mathbf{H} + \mathbf{H}(\mathbf{A} - \xi\mathbf{I}) = 0 \tag{7}$$

Where, $\mathbf{I} \in \mathfrak{R}^{n\times n}$ is the identity matrix. The weighted log-norm of **A** based on **H** is defined by $\|\mathbf{A}\|_\mathbf{H} = \xi - 1/\lambda_{max}(\mathbf{H})$, $\lambda_{max}(\mathbf{H})$ is the maximum eigenvalue of **H**.

Let $\xi_0 = \|\mathbf{K}_D\|_1$ is the 1-norm of $\mathbf{K}_D$, $\xi_0 \geq \lambda_{max}(\mathbf{K}_D)$ [16]. Applying the basic theorem in [16], $\mathfrak{L}(\xi_0, \mathbf{K}_D, \mathbf{H}_1) = 0$ gives $\mathbf{H}_1$ and $\xi_1 = \xi_0 - 1/\|\mathbf{H}_1\|_1$. Repeating this process for k = 1, 2, …, it is derived the sequences $\{\xi_k\}$, $\{\mathbf{H}_k\}$ with $\mathfrak{L}(\xi_{k-1}, \mathbf{H}_k) = 0$, where

$$\xi_k = \xi_{k-1} - 1/\|\mathbf{H}_k\|_1 \quad k = 1, 2, \ldots \tag{8}$$

The $\{\xi_k\}$ is monotone decreasing and tends to $\lambda_{max} = \max_{i=1..n}\{\lambda_i(\mathbf{K}_D)\}$ and $\mathbf{H}_k$ is positive defined symmetric, cf. Assertion 1 in Appendix. Since, there are the maximum, minimum eigenvalues $\lambda_{max}(\mathbf{H}_k)$, $\lambda_{min}(\mathbf{H}_k)$, then

$$\eta_k^2 = \lambda_{max}(\mathbf{H}_k)/\lambda_{min}(\mathbf{H}_k) \quad k = 1, 2, \ldots \tag{9}$$

Proposition 3. Let K_D be the diffusion matrix. For each time step δt, with information conductivity α and ξ_k, η_k defined in (8), (9), it is obtained as follows

$$\|\mathbf{R}_\alpha(\delta t)\|_2 \leq \eta_k \exp(\xi_k \alpha\delta t) \quad k = 1, 2, \ldots \tag{10}$$

This statement is derived by using Assertion 2 in Appendix. If k is large enough, letting $\varphi = \eta_k \exp(\alpha\xi_k\delta t)$ and applying (10) repeatedly to (6) give

$$\|F(t_h)\|_2 \leq \varphi\|F(t_{h-1})\|_2 \leq \varphi^2\|F(t_{h-2})\|_2 \leq \cdots \leq \varphi^{h-1}\|F(t_1)\|_2 \leq \varphi^h\|F(t_0)\|_2$$

If the initial condition $F(t_0)$ of (5) or (6) satisfies $\|F(t_0)\|_2 \leq 1$, then at each time step t_h it is obtained $\|F(t_h)\|_2 \leq 1$ with α chosen so that $\varphi^h \leq 1$. In this case, all non-negative components $f_1(t_h), \ldots, f_n(t_h)$ of $F(t_h)$ are upper bounded by 1 and true densities of information diffusion in network.

4 Experimental Study

4.1 Case 1

The graph of spreading mechanism with the diffusion kernel $\mathbf{K}_D$ demonstrated in [6] is used as a typical network. The similarity measures s_{ij} between nodes i and j in the network are calculated using the hk-LSA model in [14] and dataset in [18]. The ratios γ_i and ρ_i that represent the percentages of sending, receiving information of node i are estimated using the heuristic illustrated in the third paragraph of Sub-sect. 3.2. The matrix $\mathbf{K}_D$ of information diffusion and ratios γ_i, ρ_i are presented in Table 1.

Using the 1-norm, $\xi_0 = \|\mathbf{K}_D\|_1 = 3.1423$ and $\lambda_{max}(\mathbf{K}_D) = 6.2939 \times 10^{-11}$. The solution of $\mathcal{L}(\xi_0, \mathbf{H}_1) = 0$ gives a positive-defined symmetric matrix $\mathbf{H}_1$ with $\|\mathbf{H}_1\|_1 = 0.3219$ and its extreme eigenvalues give $\eta_1 = 1.7584$. In this case, α and δt are 1, 0.5, respectively. Let $\varphi_k = \eta_k \exp(\alpha\xi_k\delta t)$, $\varphi_1 = 6.3813$; (8) and (9) run as follows

For $k = 2$: $\xi_2 = 0.00280$, $\eta_2 = 0.7584$, $\varphi_2 = 0.8721$;
For $k = 3$: $\xi_3 = 0.00105$, $\eta_3 = 0.7575$, $\varphi_3 = 0.8704$;
For $k = 4$: $\xi_4 = 4.1 \times 10^{-6}$, $\eta_4 = 0.7575$, $\varphi_4 = 0.8703$.

From (10), this shown $\|R_\alpha(\delta t)\| \leq 0.8703$ which ensures that $\|F(t_h)\|_2 \leq 1$ for all time step $h = 1,2,\ldots$, when the initial condition $F(t_0)$ satisfies $\|F(t_0)\|_2 \leq 1$.

Table 1. Matrix of information diffusion and sending-receiving ratios at nodes in network

i	1	2	3	4	5	6	7	8	9	ρ_i
γ_i	1	1	1.57	1	1	1.16	1	1	1	
1	−1	0.2282	0.8659	0	0	0	0	0	0	0.817
2	0.1779	−1	0.1203	0	0.2173	0.1005	0.1219	0.2173	0.1219	0.278
3	0.8221	0.1465	−1.57	0.6038	0	0	0	0	0	0.642
4	0	0	0.5838	−1	0	0.4877	0	0	0	1.101
5	0	0.1685	0	0	−1	0.1336	0.1620	0.2888	0.1620	0.302
6	0	0.0961	0	0.3962	0.1647	−1.16	0.1847	0.1647	0.1847	0.344
7	0	0.0961	0	0	0.1647	0.1523	−1	0.1647	0.3694	0.344
8	0	0.1685	0	0	0.2888	0.1336	0.1620	−1	0.1620	0.302
9	0	0.0961	0	0	0.1647	0.1523	0.3694	0.1647	−1	0.344

4.2 Case 2

For α = 0.5, the unit densities at positive source nodes 1, 3. Node 5 is negative source. Other nodes are non-sources in Fig. 1. The confidence of the positive information sources is 70%. Nodes 6, 7, 8 changed states to negative from time step 19, 21, 26.

Although the sources are provided high density, but with a low conductivity the densities of non-source nodes are not high and the rate of sending information from source nodes to other nodes looks very slow.

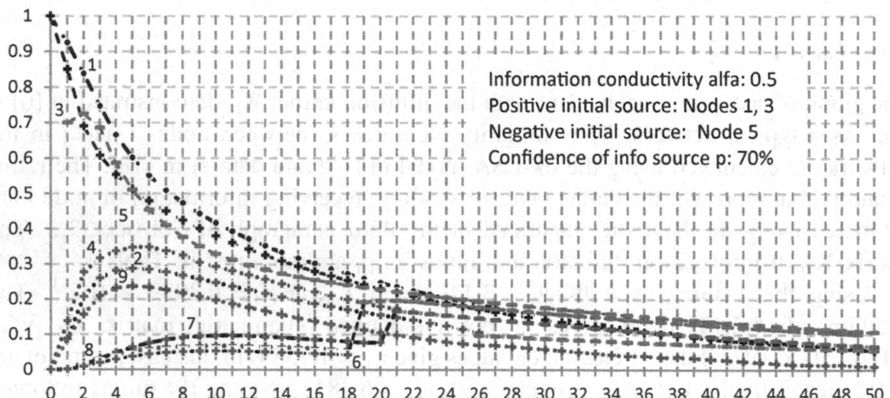

Fig. 1. Case 1: variations of information density and state changes in CID model

4.3 Case 3

By increasing the conductivity to α = 1.0 and the density of diffusion sources is still unit. The positive source nodes are 5, 7 and 9 negative source. The state of nodes in Fig. 2 changed faster with higher density due to the quality of information is too low.

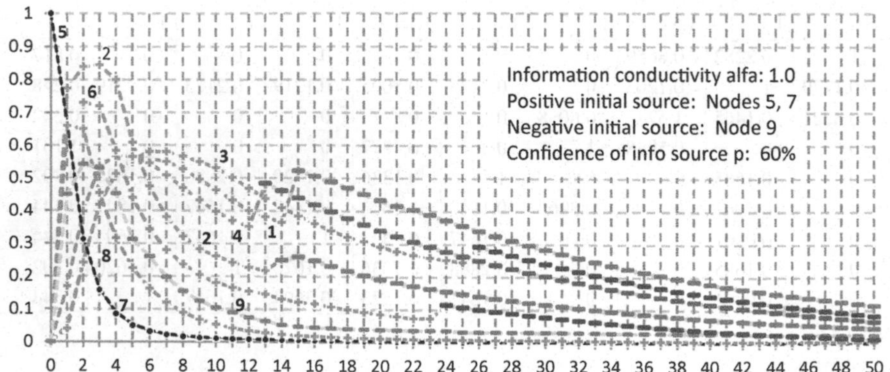

Fig. 2. Case 2: variations of information density and state changes in CID model

5 Conclusion and Future Work

In this paper, a model named CID have been proposed to simulate of competitive information spreading which is a development of the DID in network. Upper bound of information diffusion density is setup with Lyapunov equations. The shifting process of state from positive to negative is dealt with CID-N algorithm. Experiments have shown that the proposed model and remedy for competitive information diffusion are feasible. In next study, the suggested solution will be carried on large-scale datasets and real social networks.

Appendix

Assertion 1. Let $\{\xi_k\}$ is in (8) a monotone decreasing, then this sequence tends to λ_{max} where the square matrix $\mathbf{H}_k$ is positive-defined symmetric, k = 1, 2,…

Indeed, let $\lambda_{max} = \max_{i=1..n}\{\lambda_i(\mathbf{K}_D)\}$, $\xi_0 = \|\mathbf{K}_D\|_1 = \max_{j=1..n}(\sum_{i=1..n}|k_{ij}|) \geq \lambda_{max}$ [16]. Otherwise, by locating of eigenvalues in [16] $\xi_0 > \lambda_{max}$ then there exists ξ_1 so that $\xi_1 = \xi_0 - 1/\|\mathbf{H}_1\|_1 \geq \xi_0 - 1/\lambda_{max}(\mathbf{H}_1) = \|\mathbf{K}_D\|\mathbf{H}_1 \geq \lambda_{max}$ or $\xi_0 \geq \xi_1 \geq \lambda_{max}$. Because of $\xi_1 > \lambda_{max}$, there exists $\mathbf{H}_2$ that $\mathfrak{L}(\xi_1, \mathbf{K}_D, \mathbf{H}_2) = 0(\xi_1, \mathbf{K}_D, \mathbf{H}_2) = 0$, $\xi_2 = \xi_1 - 1/\|\mathbf{H}_2\|_1 \geq \lambda_{max}$ with $\xi_0 \geq \xi_1 \geq \xi_2 \geq \lambda_{max}$ and so on. Repeating this reasoning, it implies

$$\mathfrak{L}(\xi_{k-1}, \mathbf{K}_D, \mathbf{H}_k) = 0,\ \xi_k = \xi_{k-1} - 1/\|\mathbf{H}_k\|_1 \geq \lambda_{max}\ k = 1, 2, ..$$

The sequence $\{\xi_k\}$ is monotone decreasing and approaches to λ_{max}. Next, consider $\mathfrak{L}(\xi_k, \mathbf{H}_{k+1}) - \mathfrak{L}(\xi_{k-1}, \mathbf{H}_k)$ k = 1, 2,…. The difference can be expanded and written in the form $(\mathbf{K}_D - \xi_k\mathbf{I})^T(\mathbf{H}_{k+1} - \mathbf{H}_k) + (\mathbf{H}_{k+1} - \mathbf{H}_k)(\mathbf{K}_D - \xi_k\mathbf{I}) = -2(\xi_{k-1} - \xi_k)\mathbf{H}_k$. This implies to $(\xi_{k-1} - \xi_k)\mathbf{H}_k > 0$. But $\xi_{k-1} > \xi_k$ for all k, then $\mathbf{H}_k$ keeps the partial order relation, and is positive-defined symmetric matrix. This completes the Assertion 1.

Assertion 2. If the matrix $\mathbf{H}_k$ is a positive-defined symmetric, $\lambda_{max}(\mathbf{H}_k)$ and $\lambda_{min}(\mathbf{H}_k)$ are their extreme eigenvalues. Then, $\|\exp(\tau\mathbf{K}_D)\|_2 \leq \eta_k\exp(\xi_k\tau)$ where $\tau \geq 0$, η_k and ξ_k are defined by (8) and (9), respectively.

Due to symmetry and positive-defined of $\mathbf{H}_k$, the eigenvalues $\lambda_{max}(\mathbf{H}_k)$, $\lambda_{min}(\mathbf{H}_k)$ give $\eta_k^2 = \lambda_{max}(\mathbf{H}_k)/\lambda_{min}(\mathbf{H}_k)$. Also, there exists an orthogonal transformation $\mathbf{T}_k$ with $\mathbf{H}_k = \mathbf{T}_k^{-1}\mathbf{T}_k$ and $\mathbf{K}_T = \mathbf{T}_k\mathbf{K}_D\mathbf{T}_k^{-1}$ or $\mathbf{K}_D = \mathbf{T}_k^{-1}\mathbf{K}_T\mathbf{T}_k$. In matrix theory, e.g. Lancaster et al. 1985, it is proved that $\eta_k = \|\mathbf{T}_k^{-1}\|_2\|\mathbf{T}_k\|_2$. So, for a positive number τ

$$\|\exp(\tau\mathbf{K}_D)\|_2 = \|\exp(\tau\mathbf{T}_k^{-1}\mathbf{K}_T\mathbf{T}_k)\|_2 = \|\mathbf{T}_k^{-1}\exp(\tau\mathbf{K}_T)\mathbf{T}_k\|_2 \leq \eta_k\|\exp(\tau\mathbf{K}_T)\|_2$$

Moreover, for u, v $\in \mathfrak{R}^n$ that u = $\mathbf{T}_k$v and the 2-norm with inner product implies

$$\|\exp(\tau\mathbf{K}_T)\|_2^2 = \max_{u\neq 0}\{[\exp(\tau\mathbf{K}_T)u]^T[\exp(\tau\mathbf{K}_T)u]\}/(u^Tu) =$$
$$\max_{v\neq 0}\{[\exp(\tau\mathbf{K}_D)v]^T\mathbf{H}_k[\exp(\tau\mathbf{K}_D)v]\}/(v^T\mathbf{H}_kv) = \|\exp(\tau\mathbf{K}_D)\|\mathbf{H}_k^2 \leq [\exp(\|\mathbf{K}_D\|\mathbf{H}_k\tau)]^2$$

By definition $\|\mathbf{K}_D\|\mathbf{H}_k \leq \xi_k$, so $\|\exp(\tau\mathbf{K}_T)\|_2 \leq \exp(\xi_k\tau)$. This is the Assertion 2.

References

1. Krishna Kumar, K.P., Geethakumari, G.: Detecting misinformation in online social networks using cognitive psychology. Hum.-Cent. Comput. Inf. Sci.: Springer Open J. **4**: 14 (2014). http://www.hcis-journal.com/content/4/1/14
2. Zhang, H., Alim, M.A., Thai, M.T., Nguyen, H.T.: Monitor placement to timely detect misinformation in online social networks. In: 2015 IEEE International Conference on Communications - ICC, pp. 1152–1157 (2015)
3. Amoruso, M., Anello, D., Auletta, V., Ferraioli, D.: Contrasting the spread of misinformation in online social networks. In: Proceedings of 16th International Conference on Autonomous Agents and Multi-agent Systems, São Paulo, Brazil, pp. 1323–1331 (2017)
4. Ma, H., Yang, H., Lyu, M.R., King, I.: Mining social networks using heat diffusion processes for marketing candidate selection. In: CIKM (2008)
5. Castillo, C., Mendoza, M., Poblete, B.: Information credibility on Twitter. In: Proceedings of 20th International Conference on WWW, pp. 675–684. ACM, Hyderabad (2011)
6. Nguyen, K.P., Tu, H.T.: Differential information diffusion in social networks. In: Asian Conference on Intelligent Information and Database Systems – ACIIDS 2018, Dong Hoi, Vietnam. Springer, Heidelberg, 19–21 March 2018 (accepted)
7. Guha, R.V., Kumar, R., Raghavan, P., Tomkins, A.: Propagation of trust and distrust. In: Proceedings of ACM WWW Conference, pp. 403–412 (2004)
8. Carnes, T., Nagarajan, C., Wild, S.M., van Zuylen, A.: Maximizing influence in a competitive social network: a follower's perspective. In: ICEC 2007. ACM 978-1-59593-700-1/07/0008, Minneapolis (2007)
9. Kempe, D., Kleinberg, J., Tardos, E.: Maximizing the spread of influence through a social network. In: 9th ACM SIGKDD, KDD 2003, New York, USA, pp. 137–146 (2003)
10. Chen, W., et al.: Influence maximization in social networks when negative opinions may emerge and propagate, MSR-TR-2010-137. In: Proceedings of SIAM International Conference on SDM (2011)
11. Alon, N., Feldman, M., Procaccia, A.D., Tennenholtz, M.: A note on competitive diffusion through social networks. Inf. Process. Lett. **110**, 221–225 (2010). Elsevier
12. Sun, Q., Yao, Z.: Evolutionary game analysis of competitive information dissemination on social networks: an agent-based computational approach. Math. Probl. Eng. (2015). Article ID 679726, Hindawi Publ. Corporation
13. Rozin, P., Royzman, E.B.: Negativity bias, negativity dominance, and contagion. Pers. Soc. Psychol. Rev. **5**(4), 296–320 (2001)
14. Tu, H.T., Phan, T.T., Nguyen, K.P.: An adaptive latent semantic analysis for text mining. In: IEEE International Conference on: System Science and Engineering - ICSSE 2017. IEEE Xplore, HCM City (2017)
15. Tu, H.T., Nguyen, K.P.: Kernel-based similarity and discovering documents of similar interests. In: 13th IEEE International Conference on: Natural Computation, Fuzzy Systems and Knowledge Discovery - ICNC-FSKD 2017, Guilin, China (2017)
16. Hu, G.-D., Liu, M.: The weighted logarithmic matrix norm and bounds of the matrix exponential. Linear Algebra Appl. **390**, 145–154 (2004). Elsevier
17. Jarlebring, E.: Methods for Lyapunov equations. Lecture Notes on Linear Algebra, KTH Report, Royal Institute of Technology, 10044 Stockholm, Sweden, pp. 03–06 (2017)
18. Deerwester, S., Dumais, S.T., Furnas, G.W., Landauer, T.K., Harshman, R.: Indexing by latent semantic analysis. J. Am. Soc. Inf. Sci. **4** (1990)

Using Differential Evolution with a Simple Hybrid Feature for Personalized Recommendation

Michał Bałchanowski(✉), Urszula Boryczka, and Kamil Dworak

Institute of Computer Science, University of Silesia,
Bedzinska. 39, 41-200 Sosnowiec, Poland
{michal.balchanowski,urszula.boryczka,kamil.dworak}@us.edu.pl

Abstract. This article presents the application of the differential evolution algorithm (DE), the aim of which is to adjust weights to particular features of an active users profile, in order to render his preferences in the best possible way. In our system we applied a popular technique of collaborative filtering, which is used to generate recommendations. The users profile is a vector, which consists of values, which characterize a given user. Using a hybrid feature made it possible to use simple weighted Euclidean distance, which significantly decreased the amount of necessary computations and made it possible to compare particular profiles in a system faster. The results of the conducted experiments were compared with a modified weighted Euclidean distance and Pearsons correlation.

Keywords: Recommendation systems · Collaborative filtering
Differential evolution · Weighted Euclidean distance

1 Introduction

The dynamic development of the Internet made it possible to access materials, products and services that we are interested in. Unfortunately, together with the increase of the accessibility of data the problem of its effective processing appeared. An average user is not capable of analyzing all offers of popular websites. In order to face the increasing expectations posed by the internet services, a need to create a system, which would facilitate the decision making process, appeared. Therefore, Recommendations Systems (RS) were created that based on certain information about a user create personalized recommendations. The features of a good recommendation system is high quality of the generated recommendations, building trust, transparency in the means of generating recommendations and indicating towards new, so far not discovered items [1,2].

Within this article we will present the Euclidean similarity measure using hybrid features of the users profile and will compare it with the one used in our previous article [3]. Apart from the comparison of the generated recommendation, we will also present results of research showing to what degree the change

N. T. Nguyen et al. (Eds.): ACIIDS 2018, LNAI 10751, pp. 137–146, 2018.
https://doi.org/10.1007/978-3-319-75417-8_13

of the Euclidean similarity measure can influence the speed with which an algorithm works. In this article we also applied the differential evolution algorithm, the aim of which is to properly attach the weights to the particular attributes of the users profile.

The successive section of this article is divided in the following way: in the next chapter we present a literature overview together with some elementary information on the topic of Recommendation Systems and the Differential Evolution algorithm. Chapter number 3 is totally dedicated to computing the fitness function, essential for a correct functioning of differential evolution, whereas the architecture of our system is described in chapter number 4. The last two chapters are dedicated to the results of experiments and conclusions.

2 Research Background

At the beginning Recommendation Systems during generating recommendations used only votes given by users to the particular items. However, over time additional information about the user started to be used in order to improve the quality of generated recommendations [4]. Unfortunately, also this approach had some drawbacks, since for particular users a given feature can have a different priority. In [5] a system, the aim of which was to detect these dependencies through attaching weights to particular features and then their adaptation through the Genetic Algorithm (GA), depending on the users preferences, was suggested. The successive article which used this technique were [6] where the Particle Swarm Optimization Algorithm (PSO) was applied to find the proper weights. Then, with a view to rendering the preferences of the user even better, fuzzy sets with new hybrid user model, have been used [7]. Heuristic algorithms were also used in the clustering methods. For instance, to this end in work [8,9] the Genetic Algorithm (GA) was also applied, in order to divide the users into groups of similar interests. Another interesting work is [10], where the Memetic Algorithm, which turned up to be better than the Genetic Algorithm, was used.

2.1 Collaborative Filtering

One of the most popular data filtration techniques in the recommendation systems is the collaborative filtering technique, which is based on a simple observation that it is more probable that people who have similar tastes will rate particular products in a similar way. Formally, in this technique we have some set of users $U = u_1, u_2, \ldots, u_m$ and some set of items $S = s_1, s_2, \ldots, s_n$, every user $u_j, j = 1, 2, \ldots, m$ has rated a subset of items S_j. User u_A rating of an item s_i, $i = 1, 2, \ldots, n$ is denoted as $r_{A,i}$, whereas all available ratings are collected within the matrix which size is $m \times n$. An example of a system using this technique is [11] and it is believed to be the most popular technique used in RS.

2.2 Similarity Measure

Since our system uses the collaborative filtration technique, an extremely significant issue is to define similarities between users of our system. In order to do that certain measures of similarity are introduced, the aim of which is to define similarity between two system users who evaluated the same items. One of the most elementary similarity measures in recommendation systems is the Pearson correlation measure [11]. Additionally, computing the similarity between the users of the system we can also take into consideration users' additional features (attributes) together with appropriate measures, which will define the degree to what a given features influences the similarity measure, according the formula below [5]:

$$Euclidean(A, j) = \sqrt{\sum_{i=1}^{n}\sum_{f=1}^{z} w_f(v_{A,i,f} - v_{j,i,f})^2}, \tag{1}$$

where A is an active user for whom we generate recommendations, j defines the user who has the same items as user A, n defines the number of items both users have rated, z is a total number of features, w_f is the weight of feature f for user A and $w_{(A,i,f)}$ is the value of feature f on the item i for user A.

2.3 Differential Evolution

Differential evolution is an evolution technique, which is used for numerical optimization. It was introduced by Price and Storn in mid 1990s [12]. Individuals are n dimensional vectors of real numbers and both the phenotype and genotype are identical. Every individual is also a potential solution to the optimization problem. At the beginning we initialize certain elementary population, which over the course of time, through the use of operators, is changed in order to improve the quality of the solution. In the basic version of the differential evolution algorithm we distinguish two operators: mutation and crossover. As opposed to the genetic algorithm the superior operator in DE is the mutation operator, which for every individual form the S population creates a new individual and places it in the temporary population V [13]. Creating a new individual can be expressed by the means of the following formula:

$$v_i = \lambda x_{r1} + F(x_{r2} - x_{r3}), \tag{2}$$

where $r1, r2, r3$ are three randomly generated numbers of individuals from the S population, whereas $r1 \neq r2 \neq r3$, and the F parameter is a amplification factor and usually adopts a value from the range of $0 \leq F \leq 1$. Additionally a parameter λ was introduced, which adopts a value from the range of $(0, 1)$.

Then, using the crossover operator a new individual u_i is created, which is built through connecting the genotypes of a parent x_i from population S, and an individual created as a result of applying a v_i mutation operator from population V. Parameter CR is introduced by the user and it defines the probability of crossover. $Rand(j)$ is generated by the means of a random number from the range $(0, 1)$. The crossover process might be expressed by the means of the following formula:

$$u_{i,j} = \begin{cases} v_{i,j} & if(rand(j) \leq CR) \\ x_{i,j} & otherwise \end{cases} \tag{3}$$

3 Fitness Function

The fitness function is a crucial element of our system and owing to it the Differential Evolution algorithm will be able to manage the evolution process. It defines the quality of the generated individual in DE, owing to which the most adapted individuals might get through to the successive generations. In our case an individual consists of a vector of real numbers, which are weights ascribed to the particular features of a users profile. In order to define the quality of the generated individual (vector of weights), it is crucial to start from computing, on the basis of the defined neighborhood, the forecasted votes for the items (movies) from the training set. What was used to this end was a formula from [14], which can be presented as follows:

$$PredictVote(A, i) = \bar{V_a} + k \sum_{j=1}^{n} Euclidean(a, j)(v_{j,i} - \bar{v_j}), \tag{4}$$

where $\bar{V_a}$ is the mean vote for an active user A, $\bar{v_j}$ is the mean vote for an user j, $v_{(j,i)}$ is actual vote for user j on item i, k is a normalizing factor such that the sum of the Euclidean distances is equal to 1 and n is the size of the neighborhood.

Having the votes predicted by the algorithm, we can naturally compare them with the real votes, which the user gave to the particular items. In that way we can compute the fitness function according to the following formula:

$$fitness = \frac{1}{n_r} \sum_{i=0}^{n} \mid r_i - pr_i \mid, \tag{5}$$

where n_r is the cardinality of the training set of votes for an active user, and r_i and pr_i is the real and the predicted rating, respectively.

4 System Overview

The aim of our system is to generate the best possible recommendation for an active user of the system. In the first place it is important to create a profile for every user in the system, which represents his preferences. Then using the similarity measure we can compare the particular profiles and choose the ones which have the highest value of similarity in relation to an active user, which will create the nearest neighborhood. This neighborhood will be then used to predict the votes in a training set for an active user, which will lead the DE algorithm evolution process. The architecture of our system was presented in Fig. 3.

4.1 Creating Users Profile

In order to compare the users the concept of a profile, which represents a given user, is introduced. In the standard approach in the collaborative filtering technique a sparse matrix user-item, which causes in case of this technique a well know problem with scalability, is retained. However, it can be solved by introducing the concept of a hybrid feature, which connects information about votes and genres, owing to which the profile might be represented by a single vector and not like in our previous approach [3], by a set of vectors, representing all movies watched by a user. By applying such solution we can significantly reduce the number of computation and the comparison of profiles will be much faster. Example of such user vector is shown below (Fig. 1):

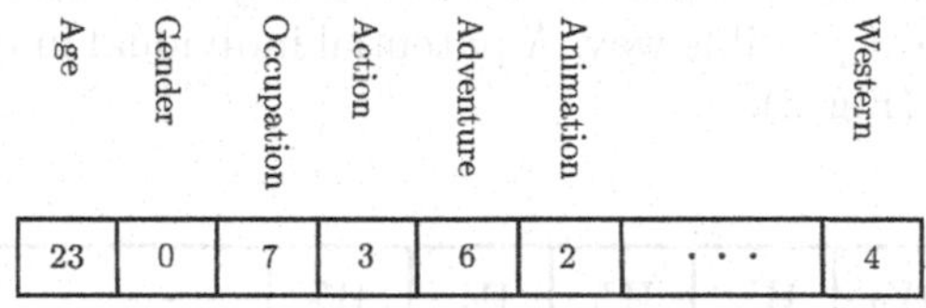

Fig. 1. Example of user profile

The formal definition of a hybrid feature (RGF) was derived from [7] and it represents the relation of highly rated movies of a given genre G_i to all movies rated by a given user u_j and is as follows:

$$RGF(j,i) = \frac{GF(j,i)}{TF(j)}, \tag{6}$$

$$GF(j,i) = \sum_{s \in G_i \subset S_j} \delta(r_{j,s}), k \in 3,4,5 \tag{7}$$

$$\delta(r_{j,s}) = \begin{cases} 1 & k = r_{j,s} \\ 0 & k \neq r_{j,s} \end{cases} \tag{8}$$

where S_j is a set of movies evaluated by u_j, G_i is the genre of a given movie, and $TF(j)$ represents the cardinality of S_j. GF will always define the number of attached movies of a given genre, which the user evaluated by the means of a vote higher or equal to 3. For example, if the user watched 5 horror movies, but he did not like a single one (evaluated them using rating 1 or 2) then for such a user the value of the horror feature will amount to 0. The aim of such an approach is to eliminate or reduce the value of a feature that is of a lower importance for the user.

4.2 Defining the Nearest Neighborhood

Another very important stage of creating recommendation is defining neighborhood, so the most similar user in relation to the active user. Naturally, in order to define similarity between system users we need a similarity measure owing to which we will be able to define the degree of similarity between users. What can be used to this end is the simple Pearson correlation, however, it will not take into consideration the additional information regarding the user. Because of that the modified Euclidean distance was introduced and expressed by the means of Formula 1, which additionally takes into consideration different features of the user, such as age, gender, occupation, etc., as well as weights w_f, which define to what extent a given feature will be taken into consideration. It is not a trivial problem, therefore we have applied the differential evolution algorithm, the aim of which is to adjust the weights w_f in such a way that they will render users preferences in the best possible way. A potential individual in the DE population will look as follows (Fig. 2):

W_1	W_2	W_3	W_4	W_5	W_6	...	W_{21}

Fig. 2. A differential evolution individual, representing weights for the particular features (attributes) of a profile

Additionally, due to the fact that in this article we have introduced to concept of a hybrid feature, the profile of a given user can be represented by a single vector. In order to compute the similarity between the users of the system we can simplify Formula 1 and obtain:

$$EuclideanHybrid(A,j) = \sqrt{\sum_{f=1}^{21} w_f (v_{A,i,f} - v_{j,i,f})^2}, \tag{9}$$

After defining all correlations between the active user A and the remaining system users, it is important to choose N of the nearest neighbors, so users who have the highest similarity measure. Before the application of the above formula, the values of features within the profile of a user were normalized and were in the range between 0 and 1. Also the sum of all weights for 21 features should be equal to one.

4.3 Recommendations

In order to manage evolution process by DE, fitness function has to be computed (Sect. 3). First we calculate similarity between the users of a system, according to Formulas 1 and 9. Next we use we use Formula 4 and compute the predicted rating for all items from the training set, which an active user rated. Then, having the predicted rating and the real rating from the training set, we can compare their values and compute the fitness function (see Formula 5). Owing to that the differential evolution algorithm is capable of defining whether an individual representing the weight properly renders users preferences. In the next step our system choose individual with the highest fitness function, which is then used to generate recommendations on a test set.

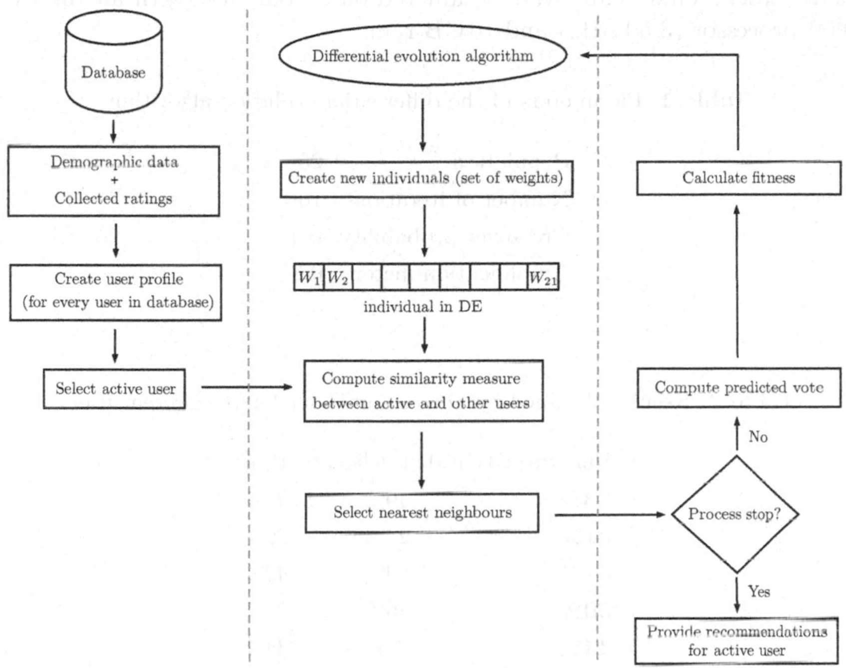

Fig. 3. System architecture

5 Experiments and Results

The experiments conducted in this article were carried out using the MovieLens [15] database. Every user evaluated at least 20 movies and incomplete data was erased. Recommender system uses 22 feature from this data set: movie rating, age, gender, occupation and 18 movie genre frequencies: action, adventure, animation, children, comedy, crime, documentary, drama, fantasy, film-noir, horror, musical, mystery, romance, sci-fi, thriller, war, western. In order to evaluate our system, the set of votes given by an active user was divided into a training set (1/3) and a test set (2/3) (split ratio was derived from [5] and [7]). The training set was used to adapt the weights by the Differential Evolution algorithm and the quality of these weights was tested in a test set. During the research 3 similarity measures were used. The first measure is weighted Euclidean measure *EuclideanHybrid*, which uses profiles with hybrid features. Another is *Euclidean*, where every users profile is defined as a set of vectors [5]. The last similarity measure used in our research is the simple Pearson correlation. Ten system users were randomly chosen for the tests and recommendations were generated for them. The parameters of the Differential Evolution algorithm, which search for the weights for *EuclideanHybrid* and *Euclidean* are presented in Table 1. The size of the nearest neighborhood for all experiments amounted to 50. The experiments below were conducted on a computer with an Intel Core i5-7600 processor (3.50 GHz) and 16 GB ram.

Table 1. Parameters of the differential evolution algorithm

Population	25
Number of iterations	100
Crossover probability	0.9
Amplification factor	0.8

Table 2. Results obtained by DErecommender for 10 different users

Id	EuclideanHybrid	Euclidean	Pearson
1	**53%**	49%	44%
2	**31%**	24%	29%
3	**44%**	39%	42%
4	**50%**	46%	27%
5	**44%**	29%	34%
6	**54%**	51%	43%
7	**47%**	36%	38%
8	31%	**44%**	12%
9	47%	**48%**	41%
10	**26%**	24%	22%

Analyzing the results presented in Table 2 and in the Fig. 4 we can notice that the quality of the generated recommendation using the *EuclideanHybrid* measure is the best for experiments number 1, 2, 3, 4, 5, 6, 7 and 10. A worse value was obtained in experiments number 8 and 9 in relation to the *Euclidean* measure. Even though the difference in the quality of the generated recommendations is quite insignificant, in according to what was presented in Fig. 5 we should notice that when applying the *EuclideanHybrid* the number of computation significantly decreased in relation to the *Euclidean* measure, owing to which the time of browsing for weights by the Differential Evolution algorithm was significantly shorter.

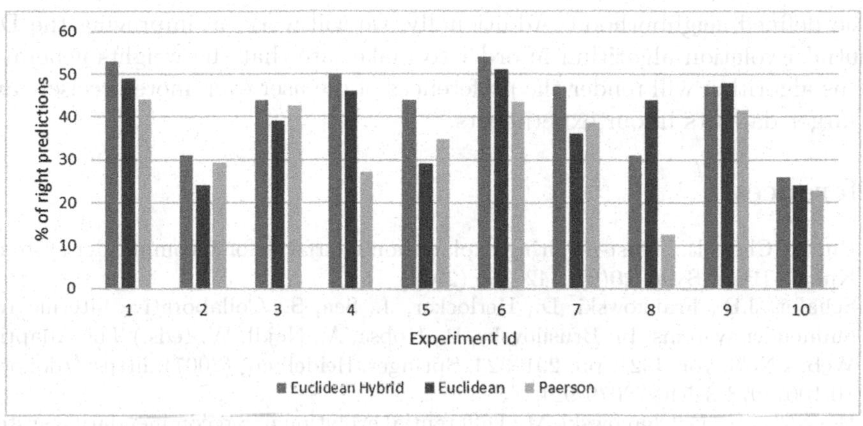

Fig. 4. Comparison of the quality of generated recommendations for 10 users

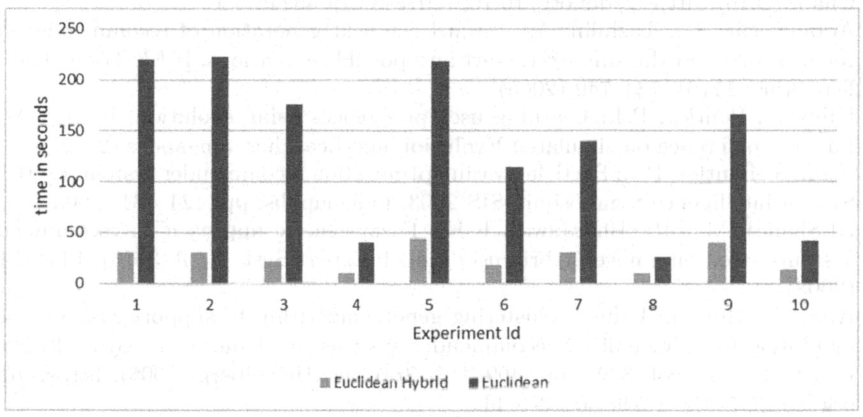

Fig. 5. The speed of the differential algorithm for the particular experiments

6 Conclusion and Future Work

The experiments and their analysis show that it is possible to successfully apply the Differential Evolution algorithm using a simple hybrid feature. Although the improvement of the quality of the generated recommendation is not significant, the speed of computing the similarity between the users improved significantly, which translates into faster defining of the nearest neighborhood and a general improvement of time needed to define the best weights by the Differential Evolution algorithm. Because of that our algorithm requires much less computational overhead, which is usually required in evolutionary approach for real-time recommendation. In the following projects we will concentrate on more complicated hybrid features, which even more precisely will render the preferences of a given user and we will also analyze the influence of particular weights on the quality of the defined neighborhood. Additionally, we will work on improving the Differential Evolution algorithm in order to make sure that the weights generated by this algorithm will render the preferences of the user even more precisely and use larger datasets in our experiments.

References

1. Pu, P., Chen, L.: Trust-inspiring explanation interfaces for recommender systems. Knowl.-Based Syst. **20**(6), 542–556 (2007)
2. Schafer, J.B., Frankowski, D., Herlocker, J., Sen, S.: Collaborative filtering recommender systems. In: Brusilovsky, P., Kobsa, A., Nejdl, W. (eds.) The Adaptive Web. LNCS, vol. 4321, pp. 291–324. Springer, Heidelberg (2007). https://doi.org/10.1007/978-3-540-72079-9_9
3. Boryczka, U., Bałchanowski, M.: Differential evolution in a recommendation system based on collaborative filtering. In: Nguyen, N.-T., Manolopoulos, Y., Iliadis, L., Trawiński, B. (eds.) ICCCI 2016. LNCS (LNAI), vol. 9876, pp. 113–122. Springer, Cham (2016). https://doi.org/10.1007/978-3-319-45246-3_11
4. Adomavicius, G., Tuzhilin, A.: Toward the next generation of recommender systems: a survey of the state-of-the-art and possible extensions. IEEE Trans. Knowl. Data Eng. **17**(6), 734–749 (2005)
5. Ujjin, S., Bentley, P.J.: Learning user preferences using evolution. In: 4th Asia-Pacific Conference on simulated Evolution and Learning, Singapore (2002)
6. Ujjin, S., Bentley, P.J.: Particle swarm optimization recommender system. In: IEEE Swarm Intelligence Symposium, SIS 2003, Indianapolis, pp. 124–131 (2003)
7. Al-Shamri, M.Y.H., Bharadwaj, K.K.: Fuzzy-genetic approach to recommender systems based on a novel hybrid user model. Expert Syst. Appl. **35**(3), 1386–1399 (2008)
8. Kim, K., Ahn, H.: Using a clustering genetic algorithm to support customer segmentation for personalized recommender systems. In: Kim, T.G. (ed.) AIS 2004. LNCS (LNAI), vol. 3397, pp. 409–415. Springer, Heidelberg (2005). https://doi.org/10.1007/978-3-540-30583-5_44
9. Kim, K., Ahn, H.: A recommender system using GA K-means clustering in an online shopping market. Expert Syst. Appl. **34**(2), 1200–1209 (2008)
10. Banati, H., Mehta, S.: A multi-perspective evaluation of MA and GA for collaborative filtering recommender system. Int. J. Comput. Sci. Inf. Technol. (IJCSIT) **2**(5), 103–122 (2010)
11. Resnick, P., Iacovou, N., Suchak, M., Bergstrom, P., Riedl, J.: Grouplens: An open architecture for collaborative filtering of netnews. In: Proceedings of the 1994 ACM Conference on Computer Supported Cooperative Work, CSCW 1994, pp. 175–186 (1994)
12. Storn, R., Price, K.: Differential evolution - a simple and efficient heuristic for global optimization over continuous spaces. J. Global Optim. **11**(4), 341–359 (1997)
13. Boryczka, U., Juszczuk, P., Kłosowicz, L.: A comparative study of various strategies in differential evolution. In: Evolutionary Computing and Global Optimization, KAEiOG 2009, pp. 19–26 (2009)
14. Breese, J.S., Heckerman, D., Kadie, C.: Empirical analysis of predictive algorithms for collaborative filtering. In: Proceedings of the Fourteenth Conference on Uncertainty in Artificial Intelligence, UAI 1998, pp. 43–52 (1998)
15. Harper, F.M., Konstan, J.A.: The movielens datasets: history and context. ACM Trans. Interact. Intell. Syst. **5**(4), 19:1–19:19 (2015)

Text Processing and Information Retrieval

Aspect-Based Sentiment Analysis of Vietnamese Texts with Deep Learning

Long Mai(✉) and Bac Le

University of Science, VNU-HCM, Ho Chi Minh City, Vietnam
mailong225@gmail.com, lhbac@fit.hcmus.edu.vn

Abstract. Aspect-based sentiment analysis (ABSA) is one of the most challenging problems in opinion mining especially for the language with a complex structure like Vietnamese. Many studies tackle this problem by separating it into two subtasks: opinion target extraction and sentiment polarity detection. These subtasks generally addressed by rule-based approaches or conventional machine learning approaches with hand-designed features. By contrast, we propose a sequence-labeling scheme associated with bidirectional recurrent neural networks (BRNN) and conditional random field (CRF) to extract opinion target and detect its sentiment simultaneously. Furthermore, we collect and construct a Vietnamese ABSA dataset specifically for smartphone domain. Experiments on this dataset show that BRNN-CRF architecture achieves a satisfied performance, outperforms CRF with hand-designed features. In addition, our framework requires no feature engineering efforts as well as linguistic resources, allows us to adapt to other languages easily.

Keywords: Aspect-based sentiment analysis · Sequence labeling
Recurrent neural networks · Word embeddings

1 Introduction

With the development of the internet, social media and microblogs these days, people can easily express their opinion towards the products, services, and events they are interested. Therefore, collecting and mining these opinion is essential for companies and organizations which provide these conveniences to fully understand their clients and improve the decision-making process.

The most common problem in opinion mining is document-level sentiment classification in which each document is assigned to the positive, negative, or neutral category. However, this classification does not provide enough information for many cases. For example, a product review can consist of many product aspects, each of which can be expressed by negative or positive sentiment. Therefore, treating the whole review by single sentiment value does not mean that the author has the same sentiment on all aspects. To address this problem, we need to use a deeper analysis which called aspect-based sentiment analysis.

According to International Workshop on Semantic Evaluation 2016, ABSA can be divided into the following subtasks: aspect category detection, opinion target extraction and sentiment polarity detection. This study focuses on opinion target extraction and

N. T. Nguyen et al. (Eds.): ACIIDS 2018, LNAI 10751, pp. 149–158, 2018.
https://doi.org/10.1007/978-3-319-75417-8_14

sentiment polarity detection in which the first subtask aims to extract all aspect terms from a sentence with respect to an opinionated target. An opinionated target can be an entity or an aspect, feature of the entity. The second subtask aims to determine the sentiment polarity (positive/negative) towards each extracted opinion target. For example, given an opinion *I really like iPhone 8, the OLED screen is awesome but the battery is bad*. The system has to extract all the tuples (*iPhone 8,* positive), *(OLED screen,* positive*)* and (*battery*, negative) from this opinion.

Instead of separating the problem into two subtasks, we propose a token-level sequence-labeling scheme to extract opinion target and detect its sentiment simultaneously. BRNN-CRF base on one of the following recurrent units: Simple RNN (SRNN), Long Short-term Memory (LSTM) and Gated Recurrent Unit (GRU) is chosen as the sequence-labeling model. In order to enhance RNN-based models, we gather a large amount of data to train domain-specific word embeddings which were later used for model's weight initialization. To evaluate our approach, we construct a Vietnamese ABSA dataset. The corpus was obtained from Youtube social media, consisting of 2098 review sentences toward smartphone products. According to experiment results, deep BGRU-CRF initialized with domain-specific word embeddings achieves F1-score about 71.79%, outperforms other RNN-based architectures including feature-rich CRF.

The remainder of this paper is organized as follows: Sect. 2 introduces related work on ABSA and our contributions. Section 3 describes the proposed approach. Experiments and discussion are presented in Sect. 4. Finally, we summarize our work and future directions in Sect. 5.

2 Related Works

There are mainly four directions to address ABSA problem: frequency based, syntax based, unsupervised learning and supervised learning.

The first frequency based approach was introduced by Hu and Liu [1], focus on mining product reviews. To address ABSA problem, they defined two subsets: product feature set and opinion word set - a set contains adjectives and theirs sentiment. The product feature set is expanded by adding the frequent noun/noun phrases or that one closest to an opinion word. For opinion word set, they first initialize it with 30 positive and negative adjectives as seeds. After that, Wordnet is used to grow this set by adding its seed words' synonyms. Finally, the sentiment towards each product feature in the sentence is assigned by the dominant sentiment value of opinion words.

The best-known algorithm of syntax-based approaches is Double Propagation, which was proposed by Qiu et al. [2]. Similar to Hu and Liu, they also use two subsets: opinion target set $\boldsymbol{T}$ and opinion word set $\boldsymbol{W}$ which initialized with few opinion word seeds. To expand these two sets, they use bootstrapping strategy with several rules which built on dependency relations between words. For an example rule, *if a noun/noun phrases* $\boldsymbol{n}$ *in the sentence have a dependency relation with an opinion word* $\boldsymbol{o} \in \boldsymbol{W}$ *then* $\boldsymbol{n}$ *is an opinion target*. The bootstrapping process stops when no more opinion word or opinion target is extracted. This approach depends greatly on syntax parsing phase and therefore, generally suffering from non-standard input such as spelling or grammar errors.

Many supervised learning approaches also proposed, normally using CRF to address opinion target extraction which can be regarded as a token-level sequence-labeling problem. Each token in a sentence is labeled using IOB format: B (begin opinion target), I (inside opinion target) and O (outside opinion target). The features for CRF model generally are the word itself, word stylistics, POS tags, the dependency relation between words, etc. Therefore, CRF model depends greatly on these hand-designed features which often requires a lot of feature engineering efforts. By contrast, deep learning approach, with the ability to extract latent features automatically, was emerged and outperformed CRF in many studies [3, 4]. On the other hand, sentiment polarity detection phase can be seen as sentence/phrase level sentiment classification which can be addressed by recursive neural networks for a semantic compositional task [5]. Another approach [6] design rules to detect opinion target boundary then feeding boundary content into a classifier to determine the sentiment.

Unsupervised learning methods were proposed to tackle with domain restriction problem and labeled data requirement of supervised approaches. The most widely used method is LDA (Latent Dirichlet Allocation) which successfully applied in many studies [7, 8]. Lin et al. [9] proposed an extended model of LDA called JST (Joint model of Sentiment and Topic) with the ability to detect topic and its sentiment simultaneously. The drawback of LDA and its variants is that the extracted topic just a words distribution and therefore, it shows no direct relation between the topic and specific opinion target.

Our work is mostly related to the work of Liu et al. [4], using supervised learning approach. However, our contribution differs from Liu in many ways: (1) We perform opinion target extraction and sentiment polarity detection simultaneously with neural CRF model. (2) We use domain-specific embeddings instead of the broad domain like Google and Amazon embeddings. (3) We create a Vietnamese ABSA corpus and conduct several different experiments on this corpus.

3 Methodology

As mentioned before, the objectives of ABSA can be divided into two subtasks: opinion target extraction and sentiment polarity detection. The first subtask can be seen as sequence-labeling problem in which each token in a sentence will be labeled to the following tags: B-T (begin target), I-T (inside target) and O (outside target).

To extract opinion target and detect its sentiment simultaneously, our proposed approach just simply adding sentiment tag (positive/negative) into B-T and I-T tags in opinion target extraction phase. Thus, each token in a sentence is assigned to the following tags: B-PT (begin positive target), I-PT (inside positive target), B-NT (begin negative target), I-NT (inside negative target) and O (outside target). For example from Table 1, there are three opinion targets: *iPhone 8*, *màn_hình OLED* and *pin* attached with positive, positive and negative sentiment respectively.

Table 1. An example of labeled sentence with joint target and sentiment format

Tôi	rất	thích	iPhone	8	,	màn_hình	OLED	rất	tuyệt	nhưng	pin	hơi	yếu
O	O	O	B-PT	I-PT	O	B-PT	I-PT	O	O	O	B-NT	O	O

3.1 BRNN-CRF for Sequence Labeling

To address sequence labeling problem, this study apply BRNN-CRF architecture which is the current state-of-the-art for sequence tagging [10]. Figure 1 demonstrates the architecture.

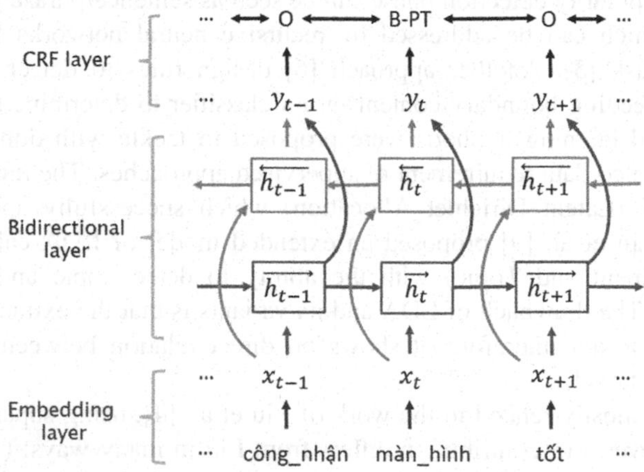

Fig. 1. Sequence labeling with BRNN-CRF architecture

In BRNN-CRF model, we avoid feature engineering efforts by using embedding at the first layer to build word feature representation. Each input word is mapped from word index into D dimensional dense real-valued vectors using look-up table $T \in R^{|V| \times D}$, where V is vocabulary. The weight of this layer (T table) can be initialized randomly then fine-tune during the learning phase or initialized with external word embeddings which were trained from a large corpus (see Sect. 3.3).

The next is bidirectional layer which consists of two sublayers: forward layer (the red line) to summarize information from the past and backward layer (the green line) to summarize information from the future. To calculate hidden state $\overrightarrow{h_t}$ (similar for $\overleftarrow{h_t}$), we use one of the following recurrent units: SRNN, LSTM, and GRU (see Sect. 3.2). After being calculated, these two hidden states then concatenated by single vector $h_t = [\overrightarrow{h_t}, \overleftarrow{h_t}]$ and then passed to the subsequent layer. Figure 1 illustrates the architecture using one bidirectional layer only, instead, we can stack many bidirectional layers on top of each other to get a more complicated model.

Finally, CRF layer is applied to decode the highest probability label sequence $y^* = \{y_1^*, y_2^*, \ldots, y_n^*\}$ for BRNN output sequence $O = \{o_1, o_2, \ldots, o_n\}$. Where $y_t^* \in$

$\{B-PT, I-PT, B-NT, I-NT, O\}$, n is the sentence length. $o_t = Wh_t + b$ in which W is the weight matrix between the hidden layer and CRF layer, b is the bias vector.

The score of given label sequence $y = \{y_1, y_2, \ldots, y_n\}$ can be computed as follow:

$$s(O, y) = \sum_{i=1}^{n} A_{y_i, y_{i+1}} + o_i^{y_i} \tag{1}$$

Where A is the transition matrix. $A_{i,j}$ is the transition score from label i to label j. o_i^j is the score of *j-th* label of *i-th* word in the sentence.

Finally, the probability of label sequence y is computed as formula (2). Where Y_O represents all possible label sequences corresponding to O.

$$P(y|O) = \frac{e^{s(O,y)}}{\sum_{\tilde{y} \in Y_o} e^{s(O,\tilde{y})}} \tag{2}$$

3.2 Recurrent Neural Networks Architectures

Simple RNN. In SRNN architecture [11], hidden state h_t at time step t is directly computed base on current input x_t and previous hidden state h_{t-1}, the result then passed through a non-linear activation function.

$$h_t = f(Ux_t + Wh_{t-1} + b) \tag{3}$$

Where f is a non-linear activation function (e.g. sigmoid). U is weight matrix between the input layer and the hidden layer, W is weight matrix between two hidden states and b is bias vector.

A common problem when training SRNN with BPTT (Backpropagation through time) algorithm is vanishing and exploding gradient. To address this issue, we can simply clip the gradient or utilize truncated BPTT. However, these solutions make SRNN failed to capture long-term dependencies especially for long sequences of data.

Long Short-Term Memory. LSTM was introduced by Hochreiter and Schmidhuber [12] to address the long-term dependencies problem of SRNN. The key idea behind LSTM is the memory cell c to preserves long-range information with only some minor linear interactions. To manipulate this memory cell, LSTM uses the following gates: (1) input gate i to add information into memory cell, (2) output gate o to obtain information from memory cell, (3) forget gate f allows memory cell to flush unnecessary information from the past. The following equations describe how a memory cell is updated at time step t:

$$f_t = \sigma\left(U_f x_t + W_f h_{t-1} + b_f\right) \tag{4}$$

$$i_t = \sigma(U_i x_t + W_i h_{t-1} + b_i) \tag{5}$$

$$c_t = f_t \odot c_{t-1} + i_t \odot tanh(U_c x_t + W_c h_{t-1} + b_c) \quad (6)$$

$$o_t = \sigma(U_o x_t + W_o h_{t-1} + b_o) \quad (7)$$

$$h_t = o_t \odot tanh(c_t) \quad (8)$$

Where U_k and W_k are the weight matrices between the input layer and hidden layer and between two hidden layers, respectively. c_k is memory cell vector, b_k is bias vector, σ is sigmoid activation function, $\odot$ is an element-wise multiplication.

Gated Recurrent Unit. GRU was proposed by Cho et al. [13], is a variant of LSTM. GRU uses two gates: reset gate r and update gate z to control the flow of information at each hidden state. The reset gate r decides what previous memory to keep/ignore while the update gate z defines the combination of current input with the previous hidden state. The detailed calculations for each hidden state are described as follows:

$$z_t = \sigma(U_z x_t + W_z h_{t-1} + b_z) \quad (9)$$

$$r_t = \sigma(U_r x_t + W_r h_{t-1} + b_r) \quad (10)$$

$$\tilde{h}_t = tanh(U_h x_t + W_h(r_t \odot h_{t-1})) \quad (11)$$

$$h_t = (1 - z_t)h_{t-1} + z_t\tilde{h}_t \quad (12)$$

Where U_k, W_k are two weight matrices, b_k, σ, $\odot$ is bias vector, sigmoid function, and element-wise multiplication respectively.

3.3 Pre-trained Word Embeddings

The first pre-trained word embeddings to be examined was proposed by Hong et al. [14]. To create distributed representation of words, the authors use Skip-gram model [15] with the embeddings dimension is set to 300. The training data consisting of 7.3 GB of text from 2 million articles through a Vietnamese news portal. This portal informs about numerous topics and issues, thus, we named this pre-trained word embedding as open domain embeddings for convenience.

On the other hand, we train a domain-specific word embeddings from 400 MB of Vietnamese comments on electronic products. These comments were obtained from Youtube social media (1.081.457 comments) and Vietnamese e-commercial forum (2.044.403 comments). Our embeddings are trained by Skip-gram model with embedding size is set to 100, context window is set to 5. To standardize training input, we preprocessing each comment follows these steps:

- Sentence segmentation and word segmentation using vnTokenizer toolkit [16].
- Eliminating special characters (☺, ♥, etc.) and punctuation except {.,?}.
- Lowercase, replacing consecutive digit characters by single token **NUM**.
- Replacing infrequency token (min count < 5) by single token **UNKNOW**. This token is the representative for the out-of-vocabulary token.

4 Experiments

4.1 Dataset

To construct Vietnamese ABSA dataset, we use Youtube Data API to gather comments from Youtube review videos of 15 well-known smartphone products (e.g. iPhone 7, LG G6, etc.). These comments then standardized through the following preprocessing steps: eliminating non-Vietnamese commentary, performing sentence segmentation and word segmentation, eliminating special characters. After sampling, we obtain a total of 2098 representative sentences. For each of these sentences, opinion targets including theirs sentiment (positive/negative) are manually annotated by an annotator with brat annotation tool [17]. Note that the opinion target is defined as a word or multi-words that represent an entity (e.g. iPhone 7, Samsung, Siri, etc.) or a smartphone aspect, feature (e.g. camera, water resistance, etc.). After annotated, the result is then reviewed and corrected by a second annotator. Uncertain cases were discussed and addressed by both annotators. Finally, we split the result into 1728 and 370 sentences for train and test respectively. Table 2 shows the corpus statistics.

Table 2. Corpus statistics

	Train	Test	Total
Sentences	1728	370	2098
Positive opinion target	1344	322	1666
Negative opinion target	919	184	1103
Total opinion target	2263	506	2769
Distinct opinion target	680	241	774

4.2 Experiment Setting

Evaluation Method. The evaluation metric used in experiments is F1-score: $F1 = 2 * P * R/(P + R)$, where P (Precision) is the percentage of correct opinion target found by the system and R (Recall) is the percentage of opinion target in the corpus was found by the system. A correct prediction occurs when the predicted opinion target span matches exactly with the gold opinion target span annotated by the human. Note that the opinion target in this evaluation metric already includes its sentiment.

RNN Setting. Our RNN-based models are trained by Adam optimizer with 50 iterations. Model weights are initialized as from random uniform distribution and then updated after 64 training sentences. To prevent overfitting, the drop-out technique [18] is applied on embedding layers and all hidden layers before inputting to the others. To tuning other hyperparameters, we do grid search with 5-fold cross-validation on the train set. As a result, we found that the number of bidirectional layers is 2 and dropout rate is 0.5 are the optimal hyperparameters for all RNN-based models. BSRNN-CRF and BGRU-CRF achieve the best result when the hidden node is set to 500, while BLSTM-CRF, with more complicated structure, is 400.

CRF Baseline. In our experiments with the CRF model, each term in the sentence is represented as the following features: item itself, POS tags, word stylistics (letter only, digit only, mixture) of current word and context words (two prefixes and two suffixes). CRF model is trained by L-BFGS algorithm with coefficient is set to 1.0 for L2 regularization.

4.3 Experiment Results

Results and Comparison among RNN-based Models. Figure 2 illustrates the performance of three RNN-based architectures: BSRNN-CRF, BLSTM-CRF, and BGRU-CRF on the test set. The blue, orange and green columns respectively indicate model initialized with random, open domain and domain-specific embeddings. The result shows that two architectures with long-term memory capabilities, BLSTM-CRF and BGRU-CRF, outperform BSRNN-CRF in term of F1-score. BGRU-CRF has less training time but achieves a better result than BLSTM-CRF in most cases, which can be explained by the less complex structure of GRU and therefore, appropriate for small training data in this study.

On the other hand, pre-trained word embeddings have shown its tremendous positive impact on models performance, the highest gained was about 15.22% on BSRNN-CRF follows by 13.36% and 6.20% on BLSTM-CRF, BGRU-CRF respectively. The reason behind this improvement is demonstrated in Table 3. Although trained on a smaller dataset, domain-specific embeddings yield a better result than open-domain embeddings. The maximum gain is about 6.09% on BSRNN-CRF. This demonstrates that the domain of training data is much more important than the amount of training data for this specific task.

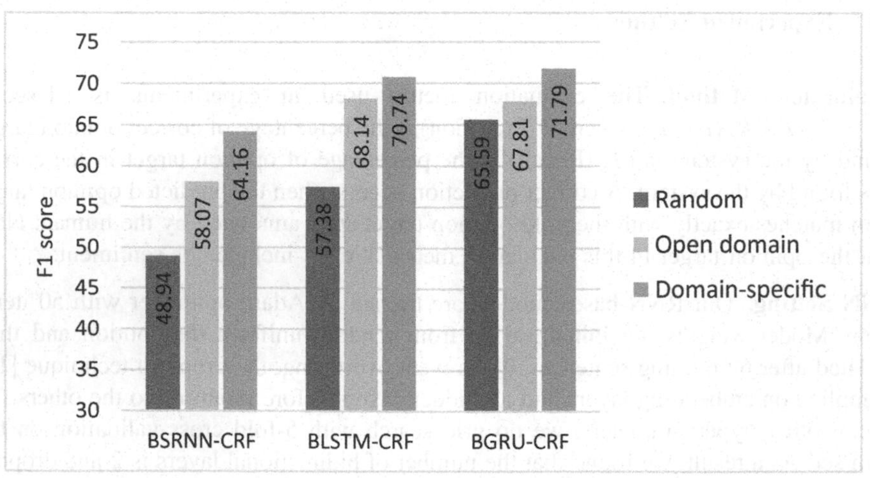

Fig. 2. Comparison of BSRNN-CRF, BLSTM-CRF, and BGRU-CRF

Table 3. Top 5 neighboring words—based on cosine similarity—for random embeddings (column 2), open domain embeddings (column 3) and domain-specific embeddings (column 4) after fine-tuning by BGRU-CRF model. The red color indicates non-relevant word

	Random	Open domain	Domain-specific
tốt (good)	hiện_đại (modern) tiến_bộ (advance) thật_sự (real) rất (extreme) cũng (also)	quan_trọng (important) tuyệt_vời (excellent) hiệu_quả (effective) cải_thiện (improve) ổn (fine)	ngon (good) xuất_sắc (excellent) tuyệt_vời (great) hoàn_hảo (perfect) tuyệt (great)
iphone	hệ_thống (system) vỏ (case) ipad ramNUMgb chính_sách (policy)	ipad apple galaxy smartphone samsung	ip NUMs ipad iphoneNUM NUMgs

BRNN-CRF vs CRF. Table 4 shows the performance of BGRU-CRF and CRF on overall, ability to recognize known and unknown targets. We use term *known targets* to indicate the opinion targets in the test set had appeared in the train set, the opposite for *unknown targets*. According to the experiment result, BGRU-CRF model outperforms CRF in all aspect.

Table 4. Comparison of BGRU-CRF and CRF

Model	Overall			Known targets	Unknown targets
	Precision	Recall	F1	Recall	Recall
BGRU-CRF	68.12	75.87	71.79	87.50	48.97
CRF	65.21	47.43	54.91	52.92	23.95

On overall comparison, BGRU-CRF has a much higher F1-score than the CRF, about 16.88%. Although CRF achieves precision nearly equal BGRU-CRF, its recall is quite low. The reason is that CRF does not perform well in terms of extract unknown target, indicated by the recall is about 23.95%. By contrast, BGRU-CRF works well in this case, its recall is about 48.97% - more than twice CRF. This can be explained by the use of pre-trained word embeddings offers BGRU-CRF a wide-range vocabulary while CRF only observed a small number of words in the train set.

5 Conclusion and Future Direction

We propose a deep learning approach to address ABSA problem without hand-craft rules or feature engineering efforts. Deep BGRU-CRF has shown its strength when achieves the best F1-score, outperforms other RNN-based models including feature-rich CRF. On the other hand, word embeddings, which were trained on a specific domain, have improved performance significantly for all RNN-based architectures. Besides the proposed approach, we also annotated 2098 smartphone review sentences for ABSA task. We expect this corpus to facilitate later studies on ABSA of Vietnamese texts using supervised learning approach.

In the future, we would like to increase corpus size and extend corpus domain. Another direction is to incorporate syntactic features alongside word embeddings and make use of external linguistics resources to boost up model performance.

References

1. Hu, M., Liu, B.: Mining and summarizing customer reviews. In: Proceedings of SIGKDD 2004, pp. 168–177. ACM, New York (2004)
2. Qiu, G., Liu, B., Bu, J., Chen, C.: Opinion word expansion and target extraction through double propagation. Comput. Linguist. **37**(1), 9–27 (2011)
3. Irsoy, O., Cardie, C.: Opinion mining with deep recurrent neural networks. In: Proceedings of EMNLP, pp. 720–728. ACL, Doha (2014)
4. Liu, P., Joty, S., Meng, H.: Fine-grained opinion mining with recurrent neural networks and word embeddings. In: Proceedings of EMNLP, pp. 1433–1443. ACL, Lisbon (2015)
5. Socher, R., Perelygin, A., Wu, J., Chuang, J., Manning, C., Ng, A., Potts, C.: Recursive deep models for semantic compositionality over a sentiment treebank. In: Proceedings of EMNLP, pp. 1631–1642. Citeseer, Seattle (2013)
6. Hamdan, H., Bellot, P., Bechet, F.: Lsislif: CRF and logistic regression for opinion target extraction and sentiment polarity analysis. In: Proceedings of SemEval-2015, pp. 753–758. ACL, Denver (2015)
7. Brody, S., Elhadad, N.: An unsupervised aspect-sentiment model for online reviews. In: Proceedings of NAACL HLT, pp. 804–812. ACL, Los Angeles (2010)
8. Lu, B., Ott, M., Cardie, C., Tsou, B.: Multi-aspect sentiment analysis with topic models. In: Proceedings of ICDM Workshops, pp. 81–88. Washington (2011)
9. Lin, C., He, Y., Everson, R., Ruger, S.: Weakly supervised joint sentiment topic detection from text. IEEE TKDE **24**(6), 1134–1145 (2012)
10. Huang, Z., Xu, W., Yu, K.: Bidirectional LSTM-CRF models for sequencetagging. arXiv arXiv:1508.01991 (2015)
11. Elman, L.: Finding structure in time. Cogn. Sci. **14**(2), 179–211 (1990)
12. Hochreiter, S., Schmidhuber, J.: Long short-term memory. Neural Comput. **9**(8), 1735–1780 (1997)
13. Cho, K., Merrienboer, B., Bahdanau, D., Bengio, Y.: On the properties of neural machine translation: encoder-decoder approaches. arXiv arXiv:1409.1259 (2014)
14. Pham, T., Le, P.: End-to-end recurrent neural network models for vietnamese named entity recognition: word-level vs. character-level. arXiv arXiv:1705.04044 (2017)
15. Mikolov, T., Sutskever, I., Chen, K., Corrado, G., Dean, J.: Distributed representations of words and phrases and their compositionality. In: Proceedings of NIPS, pp. 3111–3119. Nevada (2013)
16. VnTokenizer toolkit. http://mim.hus.vnu.edu.vn/phuonglh/softwares/vnTokenizer
17. Brat rapid annotation tool. http://brat.nlplab.org/
18. Srivastava, N., Hinton, G., Krizhevsky, A., Sutskever, I., Salakhutdinov, R.: Dropout: a simple way to prevent neural networks from overfitting. J. Mach. Learn. Res. **15**(1), 1929–1958 (2014)

Practice of Word Sense Disambiguation

Andrzej Siemiński(✉)

Faculty of Computer Science and Management,
Wrocław University of Science and Technology, Wrocław, Poland
Andrzej.Sieminski@pwr.edu.pl

Abstract. The paper aims at the community of researchers and practitioners that work in the area of natural language processing but do not specialize in the word sense disambiguation (WSD). It contains a brief introduction into WSD and describes the classical approaches to solve the problem. The experimental part reports results of disambiguation that were achieved using a set of methods which are available on a widely acclaimed web site. The data used in the test have been tagged by a professional linguist. The senses were represented by the WordNet 3.1 synsets. The conducted experiment studies the basic and ensemble methods and the effects of sense unification.

Keywords: WSD · Word sense disambiguation · Word sense discrimination
WordNet · Sense annotation · Senseval-2 data

1 Introduction

The task of Word sense disambiguation (WSD) is to decide which sense a word has in a given context. Native language speakers disambiguate words without much effort but automatic disambiguation is very hard. To underline its complexity disambiguation was labeled as an AI-complete problem [1]. An exhaustive ACM Computing Survey of the research work in the WSD area was published by Navigli in [21].

The disambiguation was studied extensively in the last 40 years and the researchers have proposed many diverse methods. They were mainly evaluated in a stand-alone mode. In practice, however, WSD is not used separately but it rather constitutes a part of another application such as machine translation or information retrieval. It could be also argued that a fast, accurate and comprehensive WSD algorithm could finally provide a major breakthrough in the realization of the so-called Semantic Web.

The main aim of this paper is not to present the newest or most efficient algorithms but rather to describe popular approaches to the WSD, their publicly available implementations, and evaluate their efficiency. We hope that it could be useful for researchers that do not specialize in WSD but still need to include it in their projects.

The paper is organized as follows. The Sect. 2 presents briefly the basic problems encountered while disemboguing words. It introduces also the WordNet which plays an essential role in this process. The next 3rd Section gives an overview of some popular disambiguation algorithms. In Sect. 4 we describe the experiment setup. Its results are evaluated in the 5th Section. The paper ends with conclusions.

N. T. Nguyen et al. (Eds.): ACIIDS 2018, LNAI 10751, pp. 159–169, 2018.
https://doi.org/10.1007/978-3-319-75417-8_15

2 Word Disambiguation – An Overview

2.1 Basic Issues

In the disambiguation area there are four issues that have to be settled:

- Disambiguation is not possible without prior definition of the word sense representation. The obvious choice is to select a representation form a finite set of senses. Such a set has to be constantly updated. Alternative approach relays on generating the description by a set of rules [3].
- The senses could be confined to one domain or cover an unrestricted language. Many applications could use successfully the domain restricted WSD, especially when it is combined with the Named Entity Recognition.
- We have to decide whether to use supervised or unsupervised approach. The former requires large sets of manually annotated data. Such data exist it could not match the diversity and scope of natural language texts on the Internet. The popular SemCor corpus contains over 200 000 sense tagged words [4]. The unsupervised approach depends on structured resources such as thesauri [5] machine-readable dictionaries [6] or ontologies [7]. As for now the ontologies could be applied only for domain restricted WSD whereas the first two types of resources cover unrestricted language.
- We have also to specify the granularity level. In theory, the fine-grained approach offers most precise sense identification. In some cases, a very detailed granularity is not desired e.g. while translating a word that that has identical or similar senses in both languages. It could be even harmful. The demarcation of senses as it is proposed by linguists is not understood by average native speakers and could lower the recall of information retrieval.

In what follows WordNet is used as a repository of senses. The disambiguation methods described in Sect. 3 were used to disambiguate an extract of the manually annotated SemCor corpus.

2.2 WordNet

WordNet [8] was chosen because of its wide availability and broad coverage. WordNet like projects exists also for many major languages. It is the de facto standard for the WSD.

In WordNet senses are identified by synsets. A synset is a set of words that have identical or very close meaning. A word usually belongs to many synsets and they are listed in descending popularity. The description of a synset contains among others:

- a gloss: textual definition often with a number of usage examples.
- semantic relations that connect synsets e.g. hypernymy, hyponymy, meronymy.

The disambiguation process uses the elements extensively. WordNet is integrated into many other projects.

Despite all its merits and popularity, WordNet has also disadvantages. Its initial goal was to build a lexical database that is consistent with theories of human semantic memory and not for NLP. The natural language is ever evolving and the newly emerging words are not included. Its latest version 3.1 was released in 2006. Therefore any retrieval system must be augmented by Named Entity Recognition. Many researchers argue that the granularity level is simply too fine. This could be exemplified by the disambiguation of the word sound in the following sentence:

" The sound of bells is a net to draw people into the church," he says.

The experts have provided not just one but three proper meanings described by the following glosses:

- sound (the particular auditory effect produced by a given cause) "the sound of rain on the roof"; "the beautiful sound of music"
- sound, auditory sensation (the subjective sensation of hearing something) "he strained to hear the faint sounds"
- sound (mechanical vibrations transmitted by an elastic medium) "falling trees make a sound in the forest even when no one is there to hear them"

All tested methods were unanimous in providing another one:

- sound (the sudden occurrence of an audible event) "the sound awakened them"

For an average language user, all four of them are acceptable. What matters for him/her is that other synsets like:

- sound (a large ocean inlet or deep bay) "the main body of the sound ran parallel to the coast" or
- sound ((phonetics) an individual sound unit of speech without concern as to whether or not it is a phoneme of some language)

which have clearly different meaning were rejected.

To mitigate the over-specification problem researchers have applied synset grouping and relaxed the rules for result evaluation: a disambiguation is successful if it produces any of the synsets from a proper group.

Despite its complex structure, there are many tools for processing it. A fairly recent paper with a comprehensive overview of 12 Java libraries for accessing WordNet was published by [9]. Five of them provide similarity metrics what makes them attractive for disambiguation tasks.

3 WordNet Based Disambiguation

Although WordNet contains information about nouns, verbs, adjectives, and adverbs in what follows the disambiguation is limited to nous.

For nouns, the most common and useful relation is the is-a or hypernymy relation. It makes up over 70% of total relations of nouns. WordNet based disambiguation starts with the selection of algorithms to compute the similarity of synsets. The algorithms could be divided into path based, info content-based and gloss based approaches. In what follows a typical example from each group is described.

3.1 Path-Based Approach

The is-a relation forms a tree-like structure of synsets. In that approach, we base our estimation of the degree of relatedness of two concepts on the length of the shortest path between them. Unfortunately, the path lengths do not have a consistent interpretation throughout the taxonomy or network. The concepts higher in a hierarchy are more general than those located lower in the hierarchy. For that reason, some corrective measure is required.

The Wu and Palmer [10] algorithm was originally proposed for verbs but now is used mainly for nouns as the WordNet hierarchy of verbs is rather shallow.

Let C_1 and C_2 denote the two synsets and C_{lcs} their lowest common subsumer - the least common synset that surmises both C_1 and C_2. The similarity of C_1 and C_2, $Sim(C_1, C_2)$ is defined by the following formula:

$$Sim(C_1, C_2) = \frac{2Len(C_{lcs}, C_{root})}{Len(C_1, C_x) + Len(C_2, C_x) + 2Len(C_{lcs}, C_{root})} \tag{1}$$

where $Len(C_x, C_y)$ denotes the number of nodes separating the synsets C_x and C_y and the C_{root} is the most upper synset.

3.2 Info Content Based

To estimate the information content of a concept or synset we first count its frequency in a large corpus and then apply the following formula :

$$IC(cpt) = -\log(P(cpt)) \tag{2}$$

Where *P(cpt)* denotes the frequency of the concept *cpt* and all its subordinate concepts. Therefore the most general synset located at the very top of the hierarchy has information content equal to 0.

One of the most popular measures using the information content was proposed by Jiang and Conrath [11]. It defines the similarity of two concepts c_1 and c_2 by the following formula:

$$sim(c_1, c_2) = \frac{1}{IC(c_1) + IC(c_2) - 2IC(lcs(c_1, c_2))} \tag{3}$$

Where $lcs(c_1,c_2)$ denotes the lowest common subsumer of the concepts c_1 and c_2.

It is assumed that the difference between the information content of the individual concepts and that of their lowest common subsumer is related to their similarity.

3.3 Gloss Based

In an apparent contrast to the previous approaches, the gloss based method relies on glosses. The overlaps of glosses are used to measure the semantic relatedness of their respective synsts. The basic version of the Lesk algorithm [12] computes the similarity of two synsets S_1 and S_2 using the formula (4):

$$sim(S_1, S_2) = |gloss(S_1) \cap gloss(S_2)| \quad (4)$$

Where $gloss(S_x)$ is the bag of words in the textual definition of sense S_x. The basic version is very sensitive to the exact wording of definitions. To mitigate the phenomena some extended gloss overlap algorithms were introduced [13]. They expand the glosses to include glosses of concepts that are known to be related through explicit relations in the WordNet, e.g. hypernymy or meronymy.

3.4 Basic Method for Disambiguation

For practical reasons, we do not the disambiguate the whole text in one peace but cut it into text windows. The words in a text window are denoted as $\{w_1, w_2, \ldots, w_n\}$. The possible senses for a word w_x are denoted as $\{s_x^1, s_x^2, \ldots s_x^k\}$. A disambiguation algorithm tries to select only one of the senses from each set of senses in such a manner that it maximizes the total similarity of the chosen senses. In a more formal way the algorithms attempt to maximize the following formula [2005_max]:

$$\sum\nolimits_{j=1, j\neq t}^{n} max_{k=1}^{mj}\, sim\left(s_t^i, s_j^k\right) \quad (5)$$

where n is the size of text window and *mj* is the number of senses of the word w_m and sim is the chosen synset similarity measure.

Popular algorithms for selecting senses are described in [2]. The latter paper uses a novel approach that uses a modification of an ACO [14].

4 Experiment Setup

4.1 Data Acquisition

The processed data came from the Senseval-2 Competition [15]. The purpose of the competition was to evaluate the efficiency of word disambiguation algorithms. Several research groups worked on the same data set.

4.2 Senseval-2 Data Processing

The texts in the Senseval-2 set were annotated by professional linguists. In the process, the words were normalized to a basic form, assigned the POS (Part-of-Speech) tag and the WordNet3.1 synsets IDs. On rare occasions, more than one syset ID was used.

Three files were extracted from the original data: Input text divided into sentences. Each sentence formed a single line and the words were separated by spaces. Nouns extracted from the above sentences. The nouns were in their base form. Nouns with accompanying synset IDs. This file is referred to as ED – expert disambiguation.

4.3 Extracting Method Specific Solutions

The disambiguation was accomplished by algorithms available at [Ped]. Using a web service guarantees that we process the data with the well tested method and there should be no deficiencies due to improper implementation, not proper library versions e.c.t. It enabled us also dodge problems replicating results reported in research papers [16]. The used web service is supervised by Ted Pedersen, one of the key figures in the field of word disambiguation.

The following algorithms were tested:

- Gloss based: LESK (extended gloss overlaps), VECTOR – (gloss vectors)
- Path-based: PATH (path length), WUP (Wu and Palmer)
- Info Content: JCN (Jiang and Cornath), LIN (Lin), RES (Resnik)

They processed original data and only nouns using windows with the size equal to 3 or 10. The algorithms produce a number of output files and one of them contains identified nouns with proposed sysnset ID.

4.4 Mapping Expert and Algorithm Data

In order to facilitate evaluation of algorithms, a mapping was created that assigned to each noun occurrence in ED all disambiguations found by the tested algorithms. The process was only seemingly easy. It turned out that the early stages of text processing were accomplished in different ways by various algorithms. As a result, there was some discrepancy in the number of identified nouns or even sentences. Such differences ware relatively easy to correct manually but the sheer amount data made the process tedious and prone to error. To map the disambiguations:

1. Sentences from the expert file were processed one by one.
2. The Jaccard similarity coefficient was used to select the most similar line in a file produced by an algorithm.
3. The synsets were related to each other using their relative positions. It was necessary because on some occasions the same word appeared more than once in a sentence with different meanings.

5 Evaluation of Algorithms

A test disambiguation run is denoted by its code that consists of: algorithm name, text window size and a type processed data where A stands for original text and N for nouns.

5.1 Global Evaluation

Table 1 shows the coverage of nouns that is the quota of nouns from ED that identified by respective algorithms.

Table 1. Nouns coverage per algorithm

Run code	Cover	No	Run code	No	Run code	Cover
RESNIK10N	0.965	12	VECTOR03N	0.876	LIN03N	0.743
WUP10N	0.965	13	JCN10N	0.838	LIN10A	0.717
CH10N	0.959	14	VECTOR03A	0.835	JCN10A	0.686
RESNIK03N	0.952	15	LIN10N	0.825	RESNIK03A	0.686
WUP03N	0.952	16	RESNIK10A	0.819	WUP03A	0.686
CH03N	0.940	17	WUP10A	0.819	CH03A	0.648
PATH10N	0.927	18	LESK10A	0.816	PATH10A	0.644
LESK03N	0.911	19	LESK03A	0.816	PATH03A	0.610
LESK10N	0.905	20	JCN03N	0.803	JCN03A	0.540
VECTOR10N	0.898	21	VECTOR10A	0.803	LIN03A	0.502
PATH03N	0.892	22	CH10A	0.781		

As you can see some of the algorithms used highly inefficient noun identification methods. In the worst case only just above half of them were found. On the other hand, the quota values exceeding 0.95 exhibited by the top 5 runs are good enough. For all runs, the precision quotas are lower than 0.6 what is unsatisfactory. What is even worse the three most precise runs occupy the 8th, 25th and 31st position in the coverage ranking. The clear winner is the LESK algorithm.

Table 2 contains data on the precision measured by the quota of properly disambiguated DS.

Table 2. Disambiguation precision per algorithm

No	Run code	Prec.	No	Run code	Prec.	No	Run code	Prec.
1	LESK03N	0.585	12	VECTOR03A	0.517	23	PATH03N	0.466
2	JCN10A	0.574	13	LIN10N	0.512	24	CH10N	0.464
3	JCN03A	0.559	14	LESK03A	0.510	25	RESNIK10N	0.461
4	LESK10N	0.558	15	PATH10A	0.507	26	WUP10N	0.461
5	VECTOR10N	0.558	16	CH10A	0.488	27	RESNIK03N	0.430
6	JCN10N	0.549	17	LIN03N	0.483	28	WUP03N	0.430
7	LIN10A	0.544	18	LIN03A	0.481	29	RESNIK03A	0.417
8	VECTOR03N	0.540	19	RESNIK10A	0.473	30	WUP03A	0.417
9	JCN03N	0.534	20	WUP10A	0.473	31	CH03A	0.412
10	LESK10A	0.529	21	CH03N	0.470	32	PATH03A	0.396
11	VECTOR10A	0.526	22	PATH10N	0.466			

5.2 Detailed Disambiguation Analysis

The precision of disambiguation is below expectations. Therefore individual disambiguation cases were analyzed. It was found out that the figures were mainly due to the fine granularity of WordNet synsets. It is demonstrated by the following typical example:

Input sentence: Each variation, or change, can occur only once, the rules state.

The noun in question: change. In WordNet 3.1 the noun has 10 senses. Experts have selected the 4th synset with the following gloss:

change (the result of alteration or modification) "there were marked changes in the lining of the lungs"; "there had been no change in the mountains". The result of alteration or modification.

Unfortunately, none of the algorithms had selected it. They have picked instead synsets with glosses like e.g.:

change, alteration, modification or (an event that occurs when something passes from one state or phase to another) "the change was intended to increase sales"; "this storm is certainly a change for the worse"; "the neighborhood had undergone few modifications since his last visit years ago".

For many English language users, all synsets selected by algorithms are acceptable. The senses with clearly different meaning were rejected. We have encountered a similar situation for many times. In order to provide a precise figure to measure the phenomena one needs a cooperation of professional linguists and it is beyond the scope of work covered in this paper.

5.3 Ensemble Methods

The ensemble methods combine algorithms different characteristics. A composite method takes into account all disambiguation choices and picks the most popular one. If senses have an equal number of occurrences than the most sense popular one is selected. It is not useful to take into consideration all runs. The increased complexity of operation is accompanied by loss of precision.

The ensembles used in the experiment were manually constructed. The rules for selection of algorithms were: diversity of operation modes and reasonably good precision and coverage. The used sets are listed in Table 3.

In order to compare more easily achieved results, we have added two one sets with only one element: Best1proper and Best1Cover. Judging by cover and precision the best performance offers the Best7 set. Its' precision and cover increased by some 7% compared to that of the LESK03N solution, see the Table 4 for results.

Table 3. Ensemble methods used in experiment

Code	Description	Selections
Best3	The first 3 selections with the highest precision	Lesk03N, Jcn10A, Jcn03A
Best5	The first 5 selections with the highest precision	Best3, Lesk10N, Vector10N
Best7	The first 7 selections with the highest precision	Best5, Jcn10N, Lin10A
GroupBest	Best algorithm from each group	Lesk03N, Jcn10A, Path10A
Best3 +Cover	Best3 set and 3 algorithms with the best cover	Best3, Resnik10N, Wup10N, Ch10N
Best1Proper	ACs with the best quota	Lesk03N
Best1Cover	ACs with the best cover	Resnik10N

Table 4. The performance of ensemble methods

	Cover	Precision	Binary prec.	Gain
Best3	0.908	0.563	0.692	0.186
Best5	0.933	0.568	0.684	0.170
Best7	0.971	0.592	0.696	0.149
GroupBest	0.908	0.563	0.692	0.186
Best3+Cover	0.965	0.556	0.661	0.159
All	0.99	0.519	0.628	0.174
Best1Proper	0.908	0.563	0.692	0.186
Best1Cover	0.943	0.448	0.589	0.239

For many applications, the level of granularity of the WordNet is to fine. For that reason, we have studied the consequences of using a binary precision – a simple unification rule. It assumed that the expert and actual AC are the same if the selected the synsets are the same or both of the selected synses do not occupy the first position on the list.

As a result, the synsets are divided into two groups: the first consist of only the first, most probable synset and the second contains all remaining synsets. The rule is more lenient than the original rule but it enables us to specify the quota of not properly identified synsets first synsets. Applying the rule increases considerable precision, the gain could be as much as 24%. The binary precision for just one Best1Proper is the same as for the set Best7, see the Table 4.

6 Conclusions

Despite all the researchers' effort, the task of disambiguation is not fully solved yet. This is especially true when we confine ourselves to the public domain software. The performance of the word disambiguation methods that reported in many research papers is much better. However, the reproduction of their results is far from being straightforward and requires specific linguistic knowledge [17]. Disambiguation forms only a part of larger natural language projects. In such cases, the publicly available software is used. The obtained results are in line with reported in [18].

In the paper, we have analyzed the performance of several disambiguation methods that are available on a widely acclaimed Website. The enables us to draw the following conclusions:

- It is very important to pay a great deal of attention to the early stages of text processing. This includes proper treatment of built-in sentences, word normalization to basic forms and part of speech tagging.
- Using just nouns for disambiguation is to be recommended.
- We should not expect to achieve the quality of disambiguation that approaches the results achieved by research work that concentrates solely on disambiguation.
- The size of text windows plays lower than expected role in the quality of disambiguation.

- Using ensemble methods improves performance but not in a significant manner.
- The granularity of senses of WordNet is too fine for most applications. Using sense unification improves greatly the performance. The resulting decrease of precision may be acceptable to many users.

The first sections of the paper describe several approaches to disambiguation. This paper is primarily addressed to researchers or practitioners working on natural language processing but not specializing on disambiguation. It might give them a clue as to the ways in which the standard algorithms could be modified or for the post-processing of obtained results. Such actions could improve the overall level of performance.

References

1. Mallerey, J.C.: Thinking about foreign policy: finding an appropriate role for artificial intelligence computers. Ph.D. dissertation, MIT, Cambridge (1988)
2. Navigli, R.: Word sense disambiguation: a survey. ACM Comput. Surv. (CSUR) **41**(2), 1–69 (2009). https://doi.org/10.1145/1459352.1459355
3. Yarowsky, D.: Hierarchical decision lists for word sense disambiguation. Comput. Hum. **34** (1–2), 179–186 (2000)
4. Miller, G.A., Leacock, C., Tengi, R., Bunker, R.T.: A semantic concordance. In: Proceedings of the ARPA Workshop on Human Language Technology, pp. 303–308 (1993)
5. Roget, P.M.: Roget's International Thesaurus, 1st edn. Cromwell, New York (1911)
6. Soanes, C., Stevenson, A. (eds.): Oxford Dictionary of English. Oxford University Press, Oxford (2003)
7. Gruber, T.R.: Toward principles for the design of ontologies used for knowledge sharing. In: Proceedings of the International Workshop on Formal Ontology, Padova, Italy (1993). https://doi.org/10.1006/ijhc.1995.1081
8. Fellbaum, C. (ed.): WordNet: An Electronic Database. MIT Press, Cambridge (1998)
9. Finlayson, M.A.: Java libraries for accessing the princeton wordnet: comparison and evaluation. In: Proceedings of the 7th International Global WordNet Conference (GWC 2014), Tartu, Estonia, pp. 78–85 (2014)
10. Wu, Z., Palmer, M.: Proceedings of the 32nd Annual Meeting on Association for Computational Linguistics, ACL 1994, pp. 133–138 (1994)
11. Jiang, J., Conrath, D.W.: Semantic similarity based on corpus statistics and lexical taxonomy. In: Proceedings of the 10th International Conference on Research in Computational Linguistics (1997). arXiv:cmp-lg/9709008
12. Lesk, M.: Automatic sense disambiguation using machine readable dictionaries: how to tell a pine cone from an ice cream cone. In: Proceedings of the 5th SIGDOC, New York, NY, pp. 24–26 (1986)
13. Banerjee, S., Pedersen, T.: Extended gloss overlaps as a measure of semantic relatedness. In: Proceedings of the 18th International Joint Conference on Artificial Intelligence IJCAI, Acapulco, Mexico, pp. 805–810 (2003)
14. Siemiński, A.: WordNet based word sense disambiguation. In: Jędrzejowicz, P., Nguyen, N. T., Hoang, K. (eds.) ICCCI 2011. LNCS (LNAI), vol. 6923, pp. 405–414. Springer, Heidelberg (2011). https://doi.org/10.1007/978-3-642-23938-0_41
15. www.hipposmond.com/senseval2/. Accessed 15 Oct 2017
16. http://maraca.d.umn.edu/allwords/allwords.html. Accessed 15 Oct 2017

17. Fokkens, A., van Erp, M., Postma, M., Pedersen, T., Vossen, P., Freire, N.: Offspring from reproduction problems: what replication failure teaches us. In: Proceedings of the 51st Annual Meeting of the Association for Computational Linguistics, pp. 1691–1701. Association for Computational Linguistics, Sofia (2013)
18. Diab, M., Resnik, P.: An unsupervised method for word sense tagging using parallel corpora. In: Proceedings of the 40th Annual Meeting of the Association for Computational Linguistics (ACL), Philadelphia, pp. 255–262 (2002)

Most Frequent Errors in Digitization of Polish Ancient Manuscripts

Kazimierz Choroś(✉) and Joanna Jarosz

Faculty of Computer Science and Management,
Wrocław University of Science and Technology,
Wybrzeże Wyspiańskiego 27, 50-370 Wrocław, Poland
kazimierz.choros@pwr.edu.pl

Abstract. Ancient manuscripts are extremely important sources of information on our history, past culture, and science. To make these invaluable documents easily accessible in the Web they must be digitized and then stored in electronic collections of archival documents. The automatic indexing and retrieval processes of such ancient digitized documents are more efficient if the contents of manuscripts are converted into editable text forms. Unfortunately, contemporary methods of digitization and character recognition are not sufficient enough. During optical character recognition process, low quality of ancient manuscripts and on the other hand not enough advanced software result in numerous errors in output texts. The paper presents the results of the tests with ancient manuscripts in Polish, Latin, and English languages. These experiments have allowed us to detect the most frequent causes of errors during the digitization of ancient manuscripts and to suggest ways to improve the digitization process of ancient manuscripts.

Keywords: Archival repositories · Digital libraries · Antique books
Historical documents · Ancient manuscripts · OCR quality · OCR errors

1 Introduction

Most old medieval manuscripts were religious books like Bible, prayer books, hagiographies, and other theological works. These medieval books are of great values and are owned by church or monasteries libraries, main national libraries or museums. In the past centuries the sacred and secular themes were usually presented together, so the examination of theological books allows us to gain knowledge about literature, history or science of the ancient ages.

Unfortunately, scholars and researchers meet many restrictions on physical access to ancient manuscripts in libraries. Manuscripts are fragile and irreplaceable, so, they are often not available. They are stored in special rooms or in special secured containers in vaults, delivered for study or exhibition only on the basis of special permits and under controlled conditions for usually short time. The early solution to solve this problem was the preparation of photographic facsimiles or microfiches of ancient manuscripts. However, such solutions were not very useful because special equipment was needed to study microfiches and then there were no possibilities to make any selection or to retrieve desired information.

N. T. Nguyen et al. (Eds.): ACIIDS 2018, LNAI 10751, pp. 170–179, 2018.
https://doi.org/10.1007/978-3-319-75417-8_16

Fortunately, modern computer technologies, digital graphics, information retrieval systems, as well as computer networks offer new opportunities to facilitate the examination of ancient documents. Digital copies can be easily not only stored and distributed, but also visualized, enlarged, or even presented in 3D form. Special formats and standards as well as the viewing software have been developed and applied in the systems providing access to different archives of ancient books.

The traditional or on-line library catalogs inform what the great artifacts are stored in their vaults [1, 2]. However, most of libraries are not willing to make valuable old medieval manuscripts accessible for the regular readers. Nowadays the religious and national institutions as well as international cultural organizations more and more frequently digitize ancient manuscripts to make them available on the Internet. One of the most important is the World Digital Library (https://www.wdl.org/en/) which is an international digital library operated by UNESCO and the United States Library of Congress. The World Digital Library makes available on the Internet, free of charge and in multilingual format, a great number of significant primary documents from the whole world, including manuscripts, maps, rare books, musical scores, recordings, films, prints, photographs, architectural drawings, and other significant cultural materials. In October 2017 there were 16,982 ancient manuscripts from 193 countries between 8000 BC and 2000. Also many local national libraries more and more frequently make digitized ancient manuscripts available on the Internet.

Although the copies in the form of sequences of digital images are very useful for people reading these ancient books, such a technical solution is not practical when we want not only browse and read but also make some searches in these ancient texts. The retrieval methods usually use character strings, words, or language expressions to realize text search operations. Therefore, it is not sufficient to photograph or to scan ancient documents even using digital techniques. We need to convert the image of text to real editable text. The OCR (Optical Character Recognition) software is used for this purpose. The efficiency of OCR software is very high in the case of nowadays documents printed by computer equipment like laser printers, on white sheets of papers, using standard computer fonts. However, the situation occurs much more difficult in the case of ancient texts in old, sometimes destroyed documents when original ancient writing styles and fonts were used. In ancient documents words are often very close one to the other and, moreover, ligatures and abbreviations are frequently used.

In the paper the efficiency of digitization of ancient texts is examined. Some experimental results of the tests with ancient manuscripts in Polish, Latin, and English languages are discussed. These experiments have allowed us to detect the most frequent reasons for errors during the digitization of Polish ancient manuscripts and to suggest ways to improve the digitization process of ancient manuscripts.

The paper is structured as follows. The Sect. 2 describes recent related work on digitization of ancient documents. The tests leading to the recognition of error causes in digitization of Polish ancient manuscripts are described in the Sect. 3. The Sect. 4 discusses the problem of degraded manuscripts and presents the techniques how to process such manuscripts. The final conclusions and suggestions for improving the digitization process are discussed in the last Sect. 5.

2 Related Work

In the early 1990's the techniques of OCR approach have already been used in the projects of historic newspaper digitization. Then in 2008 both the British Library and the National Library of Australia decided to undertake the large scale project of historic newspaper digitization when the OCR technology had reached a sufficiently acceptable level to ensure full text searching. The OCR level was acceptable but it was not good. In the next years much research and many tests have been conducted to improve OCR accuracy. The main projects are Impact (digitization.eu), Medievalists (medievalists.net), and many others.

Although the OCR systems are very popular and widely applied their efficiency for the analysis and indexing of ancient manuscripts is not satisfactory. New solutions are being proposed and the OCR techniques are being improved to ensure better digitization of ancient texts [3–6].

Holley analyzed the factors influencing the OCR accuracy in historic newspapers such as [7]: quality of original source, scanning resolution and file format, bit depth of image, image optimization/binarization process, quality of source (density of microfilm), skewed pages, pages with complex layouts, adequate white space between lines, columns and at edge of page so that text boundaries can be identified, image optimization, quality of source, pattern images in OCR software databases, algorithms in the OCR software, algorithms and inbuilt dictionaries in the OCR software, as well as time available to train the OCR.

Holley has also defined how to measure the OCR accuracy. Although it is very time consuming, the best solution is to do proofreading of ancient document and to manually verify the entire text and compare the output to the OCR result. Three levels of accuracy have been proposed for historical newspapers: good OCR accuracy – from 98 to 99% accurate (12% of OCR incorrect), average OCR accuracy – from 90 to 98% accurate (2–10% of OCR incorrect), and poor OCR accuracy – below 90% accurate (more than 10% of OCR incorrect). In the tests performed with a sample of 45 pages of historical newspapers found to be representative of the libraries digitized newspaper collection 1803–1954, the OCR accuracy varied from 71% to 98.02%. Then some potential methods of improving OCR accuracy have been suggested. One of the best suggested solutions to improve OCR accuracy may not rely on a technical approach done by a computer but on a user involvement in manual correcting the mistakes of the OCR process basing on the idea of Web 2.0 in social networking.

In the other experiments [8] the accuracy of digitization of historical newspapers was measured on the level of characters (all characters excluding spaces), words, significant words (all words excluding stop-listed words), words with capital letter start, and number groups. The accuracy was examined for two databases: the 19th Century British Library Newspapers and Burney Collection. The results were rather not satisfactory. The overall averages for the 19th Century Newspaper Project were: character accuracy: 83.6%, word accuracy: 78%, significant word accuracy: 68.4%, words with capital letter start accuracy: 63.4%, number group accuracy: 64.1%. The overall averages for the Burney Collection were: character accuracy: 75.6%, word accuracy: 65%, significant word accuracy: 48.4%, words with capital letter start accuracy: 47.4%, and number group accuracy: 59.3%. These results clearly signal that the age of document has a great influence on the OCR accuracy. In the case of significant words the OCR accuracy was lower than 50%, it means that most of significant words were not correctly identified.

Other problems occur in the case of degraded documents. These cases have been examined for example in [9]. Diem and Sablatnig analyzed local information to recognize characters that are only partially visible. The local descriptors were initially classified using the support vector machine (SVM) and then identified by a voting scheme of neighboring local descriptors. In the proposed approach degraded handwritten characters were recognized even when characters were partially visible and connected. Then the authors concluded that this approach can be easily adapted to other alphabets and languages, moreover that no dictionary is needed to improve the performance of a recognition process.

Whereas in [10] the problem was discussed of the removal of a text superimposed to a more important one in a document image. Such a situation occurs due to the back-to-front interferences from recto and verso images of archival documents. It also occurs when we are trying to recover the erased text in palimpsests from multispectral images. A palimpsest is a manuscript page, either from a scroll or a book, from which the text has been scraped or washed off. Then this page can be reused for another document. The ancient parchments in Pergamon were made of young lamb. They were very expensive and difficult to get, and therefore for economic reason they were often re-used by scraping the previous writing. Whereas the reuse of papyrus was not frequent because papyrus was quite cheap. To solve the problem of such ancient documents the authors proposed a non-stationary, linear mixing model of the optical densities to describe text overlapping in recto–verso images of archival documents. Based on this model two algorithms were developed to separate the two texts. These algorithms are performed in two phases: first the model parameters are estimated off-line from the data and then the restored images are recovered by inverting the data model in a single step.

Some researchers suggested to restore degraded documents before recognition of the text [11]. Such degradations like cuts, merges, blobs, and erosions can be restored leading to the improvement in OCR performance. Experimental results showed significant improvement in image quality and OCR performance on documents collected from a variety of sources.

The legibility enhancement of ancient and degraded handwritings is also discussed in [12]. Because the handwritings are only partially barely visible under normal white light they have been imaged using special techniques in order to increase their legibility and then to enhance the contrast of the faded-out characters. Two different labeling strategies were proposed. First method was concerned with the enhancement of non-degraded image regions of the text and the other technique was applied on degraded image parts. Then the resulting images were merged into the final enhancement result. This approach was evaluated. The enhancement method ensured the better performance of the OCR process in the case of degraded writings, compared to OCR results gained on unprocessed images.

The method of the identification of words for indexing and retrieval purposes, the method tolerant to errors in character segmentation was proposed in [13] and tested on four copies of the Gutenberg Bibles. Whereas in [14] different methods were examined of processing word images and different pre-processing steps were suggested for keyword spotting and document image retrieval.

Another problem occurs when the OCR techniques are applied for ancient and technical vocabulary. In [15] a method was proposed for analysis of historical medical texts. The method combined rule-based correction of regular errors with a medically-tuned spellchecking strategy. These corrections were based on the information about subject-specific language usage from the publication period of the ancient document to be corrected.

The problem of indexing of vast collections of documents available in image format in information retrieval systems is extensively discussed in [16]. Word spotting is proposed as an alternative solution to optical character recognition (OCR), which authors find as rather inefficient for recognizing text of degraded quality and unknown fonts usually appearing in printed text, or writing style variations in handwritten documents. Also a rich bibliography on the problem of retrieval of ancient documents is presented in this paper.

3 Error Causes in Digitization of Ancient Manuscripts

The main goal of our experiments was to recognize the main causes of errors of OCR software applied to Polish ancient manuscripts in comparison to Latin or English texts (Table 1). The texts used in the tests were of high quality, they were not degraded as it frequently happens in the case of ancient manuscripts. The efficiency was examined on the level of characters but mainly on the level of words.

Table 1. Characteristics of the ancient manuscripts used in the tests.

Language	Polish	Latin	English
Year of publication	1797	1752	1860
Place of issue	Warsaw	Dresden	New York
Number of words in the text	265	255	214

The digitization process was performed using three free OCR software available on the Internet: FineReader 14, Tesseract, and OmniPage Ultimate. The Polish ancient manuscript was a document of a French riding master working on the methods for training horses – François Robichon de La Guérinière (1688–1751) and then translated into Polish language by a doctor of philosophy Jacek Krusiński. The document was issued in 1787 in Warsaw and nowadays is in the library of the Wrocław University of Environmental and Life Science (Poland). Two pages of this document were examined. Although the text is from XVIII century it is written in old Polish language. Many words are not included in contemporary Polish dictionaries. Some characters have different shapes, for example the character ‘s’ looks like the present character ‘f’. The document is of good quality, paper is practically white, so there is a significant contrast difference between text and its background. Text is printed using serif font, normal or italics, both capital and lowercase letters, ligature was not used, frequently the letter thickness changes, and then, although lines of text are straight, characters in words do not always lie on the same level (Fig. 1).

NOWA APTECZKA

poſpolicie nazwiſko *Bolu Głowy*, który
częſtokroć jeſt tylko ſymptomatem
inney choroby: iako to Zapalenia;
którego równie iako i wielu innych
chorób ieſt ſzczególną cechą.

Fig. 1. A small sample of the tested ancient manuscript.

Table 2 presents the list of errors and attempts of explanations why such errors occur. We do not noticed which software generated these errors because the efficiency of OCR systems was not the goal of this research. Is it the question of low efficiency of the software or rather the question of specificity of the printed text?

Table 2. Analysis of errors detected in the digitized Polish ancient manuscript.

Original	Digitized	Comments (cause of an error)
E	K	Character of variable thickness
s	f or l	Character ‘s’ in old centuries was written as ‘f’ or ‘l’ today
B	D	Very weak print of this character
ń	d or J	Another OCR error
ł	L or i	Weak print of the letter ł, i.e. L with stroke – with diacritical sign (a special letter in Polish and some other Slavic languages), mainly weak print of the stroke itself
i	t	Letter ‘i’ is printed in such a way that point is connected to the main part of the character.
i	T	Another OCR error
c	l	Small print stain near the character ‘c’
L	JL	Additional small line at the bottom of the character
r	i	Weak printing of the character ‘r’
ę	g	Another OCR error
,	>	Another OCR error
l	i	Weak printing of the character ‘l’ in its upper part
u	o	Another OCR error
ć	E	Another OCR error
st	Ll, II	Another OCR error
cz	tz	Another OCR error

In the next tests we would like to verify how many words are correctly digitized. Because the problem of the recognition of old letter ‘s’ as the contemporary letter ‘f’ was so common that it can be treated as a rule, so in the Table 3 in the first case such a recognition was counted as an error and in the second these errors were not taken into account. This was the reason that in the third test in the OCR system the learning process was activated but only for the letter ‘s’.

Table 3. Percentage of words correctly digitized in the Polish ancient manuscript.

	Number of words in the text	Percentage of words correctly digitized
‘s’ – ‘f’ error	265	78.68%
‘s’ – ‘f’ non error	265	90.75%
after learning process ‘f’ –> ‘s’	265	96.98%

It should be also noticed that we do not used the dictionaries in the OCR systems. In the ancient texts some words were used according to the old grammatical rules. The activation of dictionaries would result in losing the original spelling. We think that this ancient specific writing should be preserved.

The first observation was a nice surprise. The Polish letters with diacritics were correctly digitized.

The same test have been undertaken for English and Latin ancient manuscripts. The tested Latin manuscript was “Specimen Catalogi Codicum Manuscriptorum Bibliothecae Zaluscianae” of Jan Daniel A. Janocki. This book was issued in Drezden in 1752 and today is stored in the National Ossoliński Institute in Wrocław (Poland). The tested part had 255 words.

The main error occurring in the digitized Latin manuscript was caused by the different shape of the letter ‘s’ in the Latin texts. This is the same cause as in the case of Polish manuscripts (Table 4).

As an English manuscript the book “Life in the desert: or, recollections of travel in Asia and Africa” of Colonel L. Du Couret was chosen. The book was issued in 1860 in

Table 4. Analysis of errors detected in the digitized Latin ancient manuscript.

Original	Digitized	Comments (cause of an error)
e	c	Weak printing of the character ‘e’
s	f	Character ‘s’ in old centuries was written as ‘f’ or ‘l’ today
ɛ	f or C, E, F	Non-standard character
m	rn	Another OCR error
st	ll	Another OCR error
1574	1479	Another OCR error
u	o	Another OCR error

New York. This ancient book is in Library of Congress of the USA. Two pages with 214 words were used in the tests.

The results of the OCR digitization of this English manuscript were much better than for Polish or Latin manuscripts. For Polish and Latin manuscripts the best results were on the level of 90%, whereas for the English manuscript – 99%.

The first conclusion is – what is obvious – that much research is conducted with English texts and most of software is oriented on the English language. The digitization of ancient manuscripts demands software taking into account the specificity of other languages.

4 Digitization of Degraded Manuscripts

The quality of manuscript and in consequence the quality of analyzed document image has a great influence on the results of digitization. In our tests we observed that a good strategy is to convert the document full-color image to grayscale image and then to increase the image brightness and to increase the contrast (Fig. 2). This relatively simple procedure makes the text much better visible and furthermore in many observed cases eliminates or at least reduces the back-to-front interferences from recto and verso images.

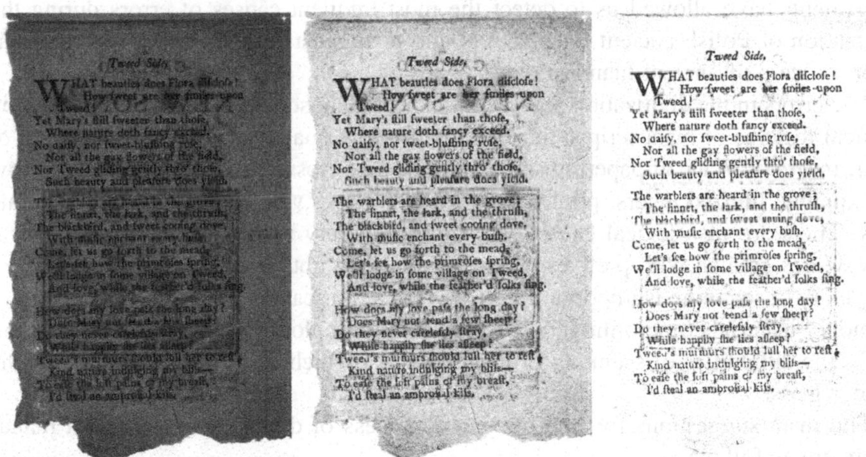

Tweed Side.

WHAT beauties does Flora diſcloſe!
How ſweet are her ſmiles upon Tweed!
Yet Mary's ſtill ſweeter than thoſe,
Where nature doth fancy exceed.
No daiſy, nor ſweet-bluſhing roſe,
Nor all the gay flowers of the field,
Nor Tweed gliding gently thro' thoſe,
Such beauty and pleaſure does yield.

The warblers are heard in the grove;
The linnet, the lark, and the thruſh,
The blackbird, and ſweet cooing dove,
With muſic enchant every buſh.
Come, let us go forth to the mead;
Let's ſee how the primroſes ſpring,
We'll lodge in ſome village on Tweed,
And love, while the feather'd folks ſing.

How does my love paſs the long day?
Does Mary not 'tend a few ſheep?
Do they never careleſsly ſtray,
While happily ſhe lies aſleep?
Tweed's murmurs ſhould lull her to reſt;
Kind nature indulging my bliſs—
To eaſe the ſoft pains of my breaſt,
I'd ſteal an ambroſial kiſs.

Fig. 2. Example of degraded manuscript entitled "A Garland of New Songs" issued in 1805 in Scotland: original full-color image, grayscale image, and bright image with increased contrast.

For the ancient document presented in Fig. 2 the percentage of words correctly digitized increased from 59% to 65% for the grayscale image and to 75% for the bright image with increased contrast. If the recognition of old character 's' as 'f' is not taken into account the percentage of words increased from 67% to 81% and 84% respectively (Table 5).

Table 5. The influence of conversion of the image from full-color image to grayscale image and to image with increased brightness and contrast.

	Percentage of words correctly digitized		
	Full-color image	Grayscale image	Bright image with increased contrast
's' – 'f' error	59%	65%	75%
's' – 'f' non error	67%	81%	84%

5 Conclusions

Ancient manuscripts are extremely important sources of information on our history, past culture, and science. To make these extremely valuable documents easily accessible in the Web they must be digitized and then stored in electronic collections of archival documents. The automatic indexing and retrieval processes of such ancient digitized documents are more efficient if the contents of manuscripts are converted into editable text forms. Unfortunately, contemporary methods of digitization and character recognition are not sufficient enough. During optical character recognition process, low quality of ancient manuscripts and on the other hand not enough advanced software result in numerous errors in output texts. The results of the tests with ancient manuscripts in Polish, Latin, and English languages are presented in the paper. These experiments have allowed us to detect the most frequent causes of errors during the digitization of Polish ancient manuscripts and to suggest ways to improve the digitization process of ancient manuscripts.

The errors in the digitization process of ancient manuscripts are generally caused by physical features of manuscripts: degraded paper, low quality of paper – low weight of paper, uneven surface of paper, in some extreme cases loss of small parts of paper, low print quality: blurred letters, print stains, wavy lines, as well as application of archaic fonts. There are also lexical causes: archaic vocabulary usage, different grammatical rules, words created with prefixes or suffixes that are not currently in use. Moreover, the quality of scanning is important: in many cases the scanning is done on the basis of secondary materials (microfilms), low image resolution, lossy compression format, and finally the difficulties of scanning of ancient books which cannot be straighten on the scanner glass.

The main suggestions for improving the process of digitization of ancient manuscripts are as follows:

- high quality of manuscript images should be ensured by using adequate scanning equipments, high resolution of scanning, lossless compression, manual correction of the document images, manual leveling of the image to make lines of text horizontal, usage of dictionaries with archaic vocabulary;
- manual or automatic correction of the text automatically digitized would be very desirable using special software for such corrections oriented on typical errors, archaic vocabulary, and ancient grammatical rules, as well as respecting the occurrence of diacritic marks in the ancient manuscript written in languages using different diacritic symbols.

References

1. Cashion, D.T.: Cataloging medieval manuscripts, from beasts to bytes. Digit. Philology: J. Medieval Cult. **5**(2), 135–159 (2016)
2. O'Keefe, E.: Medieval manuscripts on the Internet. J. Religious Theological Inf. **3**(2), 9–47 (2000)
3. Springmann, U., Najock, D., Morgenroth, H., Schmid, H., Gotscharek, A., Fink, F.: OCR of historical printings of Latin texts: problems, prospects, progress. In: Proceedings of the First International Conference on Digital Access to Textual Cultural Heritage, pp. 71–75. ACM (2014)
4. Alex, B., Burns, J.: Estimating and rating the quality of optically character recognised text. In: Proceedings of the First International Conference on Digital Access to Textual Cultural Heritage, pp. 97–102. ACM (2014)
5. Järvelin, A., Keskustalo, H., Sormunen, E., Saastamoinen, M., Kettunen, K.: Information retrieval from historical newspaper collections in highly inflectional languages: a query expansion approach. J. Assoc. Inf. Sci. Technol. **67**(12), 2928–2946 (2016)
6. Koistinen, M., Kettunen, K., Pääkkönen, T.: Improving optical character recognition of Finnish historical newspapers with a combination of fraktur and antiqua models and image preprocessing. In: Proceedings of the 21st Nordic Conference on Computational Linguistics, NoDaLiDa. NEALT Proceedings Series, vol. 29, pp. 277–283. Linköping University Electronic Press, Gothenburg, Sweden (2017)
7. Holley, R.: How good can it get? Analysing and improving OCR accuracy in large scale historic newspaper digitisation programs. D-Lib Mag. **15**(3/4) (2009). https://doi.org/10.1045/march2009-holley
8. Tanner, S., Muñoz, T., Ros, P.H.: Measuring mass text digitization quality and usefulness. D-Lib Mag. **15**(7/8) (2009). https://doi.org/10.1045/july2009-munoz
9. Diem, M., Sablatnig, R.: Recognition of degraded handwritten characters using local features. In: Proceedings of the 10th International Conference on Document Analysis and Recognition, ICDAR 2009, pp. 221–225. IEEE (2009)
10. Tonazzini, A., Savino, P., Salerno, E.: A non-stationary density model to separate overlapped texts in degraded documents. Sig. Image Video Process. **9**(1), 155–164 (2015)
11. Kumar, V., Bansal, A., Tulsiyan, G.H., Mishra, A., Namboodiri, A., Jawahar, C.V.: Sparse document image coding for restoration. In: Proceedings of the 12th International Conference on Document Analysis and Recognition (ICDAR), pp. 713–717. IEEE (2013)
12. Hollaus, F., Diem, M., Sablatnig, R.: Improving OCR accuracy by applying enhancement techniques on multispectral images. In: Proceedings of the 22nd International Conference on Pattern Recognition (ICPR), pp. 3080–3085. IEEE (2014)
13. Marinai, S.: Text retrieval from early printed books. Int. J. Doc. Anal. Recogn. (IJDAR) **14**(2), 117–129 (2011)
14. Tan, C.L., Zhang, X., Li, L.: Image based retrieval and keyword spotting in documents. In: Doermann, D., Tombre, K. (eds.) Handbook of Document Image Processing and Recognition, pp. 805–842. Springer, London (2014). https://doi.org/10.1007/978-0-85729-859-1_27
15. Thompson, P., McNaught, J., Ananiadou, S.: Customised OCR correction for historical medical text. Dig. Heritage IEEE **1**, 35–42 (2015)
16. Giotis, A.P., Sfikas, G., Gatos, B., Nikou, C.: A survey of document image word spotting techniques. Pattern Recogn. **68**, 310–332 (2017)

Using the Reddit Corpus for Cyberbully Detection

Tazeek Bin Abdur Rakib and Lay-Ki Soon(✉)

Faculty of Computing and Informatics, Multimedia University, Cyberjaya, Selangor, Malaysia
tazeek.rakib@gmail.com, lksoon@mmu.edu.my

Abstract. With the creation of word embeddings, research areas around natural language processing, such as sentiment analysis and machine translation, have improved. This has been made possible by the limitless amount of text data available on the internet and the usage of a simple, two-layer neural network. However, it remains to be seen if the domain knowledge used to train word embeddings have an impact on the task the embeddings are being used for, based on the domain knowledge of the task itself. In this paper, we extracted and cleaned text data from the Reddit database, followed by training a word embedding model that is based on the word2vec skip-gram model. Then, the features of this model were used to train a random forest classifier for classifying cyberbully comments. Our model was benchmarked with four pre-trained word embeddings, as well as hand-crafted feature extraction methods. The results show that the domain knowledge of word embeddings do play a part in the task it is being used for, as our model has a 2% improvement of precision over the next best score.

Keywords: Cyberbully detection · Data preparation
Textual features · Word embedding

1 Introduction

Cyberbullying takes place when insulting messages or comments are directed towards specific social media users. The abusive users are anonymous most of the time and the contents posted by them are distributed quickly throughout the social media community. These abusive messages are mainly targeted at younger people - mostly children - as most cases of cyberbullying involve the younger person being victimized while the older person is the abuser.

Over the years, methods in classifying cyberbully commments ranges from classic hand-crafted feature extraction to the use of word embedding models. However, most of the methods are usually accompanied by a large static list of profane words, which is updated from time to time. In addition, the words in the dictionary are arbitrary and contribute to no convenient information about the relationship which may exist between them. This extends to data sparsity,

N. T. Nguyen et al. (Eds.): ACIIDS 2018, LNAI 10751, pp. 180–189, 2018.
https://doi.org/10.1007/978-3-319-75417-8_17

which requires more labeled data that is very difficult to acquire and have them labeled. In addition, the word embedding models used in this kind of problem, lack the domain knowledge of social media which consists mainly of slangs and irregular grammar.

In this research, we train a word embedding model from an unexplored corpus of social media data, namely Reddit[1], and benchmark it with existing pre-trained models. To summarize, our main contributions are: (1) Data collection, preparation, and storage for training word vectors; (2) Benchmark and evaluation of our word embedding model with current pre-trained models; (3) Comparison with hand-crafted feature extraction methods in classification of cyberbullying comments.

The flow of the paper is as follows: We will first look at some of the methods used in this research domain, as well as the applications of word embeddings used in other areas of research in Sect. 2. In Sect. 3, the training of our word embedding model pipeline is described, starting from data collection. After training, the model was evaluated with other pre-trained models and benchmarked against existing methods in Sect. 4. Lastly, we give our conclusion and the future works in Sect. 5, that revolves around the research of cyberbully detection.

2 Background Study

Research in classifying cyberbullying comments are mostly based on classical hand-crafted features. One such example is a standard feature extraction method, which consists of strings converted into a feature vector by n-grams, counting the occurrences of words, and the term-frequency inverse-document frequency score. Moreover, the feature vector can be incorporated by additional features, such as capturing second-person pronouns. This is because bullying comments directed towards peers are more likely to be negative. In addition to capturing pronouns, skip-grams can also be used since the features extracted from this method can extract long distance words [2]. [3] developed a model based on textual features that focused on profane words: this involved the total count of profane words and the density of profane words, which was normalized by dividing the total number of profane words by the total number of words in the post. Another feature was also included, which is to measure the overall "badness" of a post by computing the weighted average of profane words. Also, the weight of positive instances were increased by copying the positive training examples several times, in order to overcome the issue of sparsity. [4] proposed the incorporation of users' demographic data, such as gender and age, along with the content of their conversations. In addition to incorporation of gender-specific language, four other features were taken into consideration: Profane words along with their abbreviations and acronyms, second person pronouns for detection of personal involvement, the rest of the pronouns and the term frequency-inverse document frequency value of all the words per post. Also, the profane word count in a post was normalized by division with post length. Later on, word

[1] https://archive.org/download/2015_reddit_comments_corpus/.

embeddings were utilized by [14]. They experimented on a new representation learning method, which is developed by the semantic extension of the deep learning model-stacked denoising autoencoder and it contained two parts: semantic dropout noise and sparsity constraints, where the dropout noise is dependent upon the knowledge domain, as well as the word embeddings. Their methodology is divided into multiple layers: the first layer involved the construction of negative words followed by comparing the word list with their own corpus, with the intersected words being regarded as bullying features; the remaining layers involved feature selection via fisher score to collect "bullying" features.

Dense distributional lexical representations, also known as word embeddings, generally improves performance on a variety of NLP tasks. This is based on the accumulated evidence gathered by [5–7]. Word embeddings are not trained for specific-tasks. Hence, by taking advantage of large amounts of unlabeled training data, word embeddings can be trained quickly and in a scalable manner [8,9]. The word embedding models can then be used on different tasks [10–12]. Some of these groups of word embedding models are word2vec [9]. In word vectors, each unique word from the training corpus is assigned a respective vector in a high-dimensional vector space, which is mostly in the factors of hundreds. These word vectors are arranged such that words having similar background in the corpus are located closer to each other in the vector space.

3 Word Vector Model

3.1 Data Preparation

To train the word embeddings in this research, the reddit corpus is used, which consists of comments from the first five months of 2015. As the scope of the project was focused on english language, non-english sentences were filtered out by checking for ASCII characters. The whole pipeline of the data preparation is based on Fig. 1, concluded by the model being trained:

1. **Comments Extraction:** Metadata was extracted from reddit after web mining was carried out and are in the form of key-value pairs. However, the comments are stored in the 'body' tag.
2. **Segmentation:** This step consisted of two forms of segmentation, each succeeding the other. Firstly, comments were segmented into individual sentences, followed by tokenizing each individual sentence into a list of words, or tokens. The segmentation procedure was done by the NLTK (Natural Language Toolkit) package.
3. **Text Cleaning:** In this step, comments were cleaned and normalized. Firstly, hyperlinks were removed since they did not contribute to any important information. This was followed by standardization of words, such as *Looolll* to *Looll* with the aid of regular expressions, in order to remove as many redundant words as possible. Then, some contractions of words were normalized, such as *you're* expanded to *you are*. Lastly, non-alphabets were removed and the cleaned comment was then lowercased.

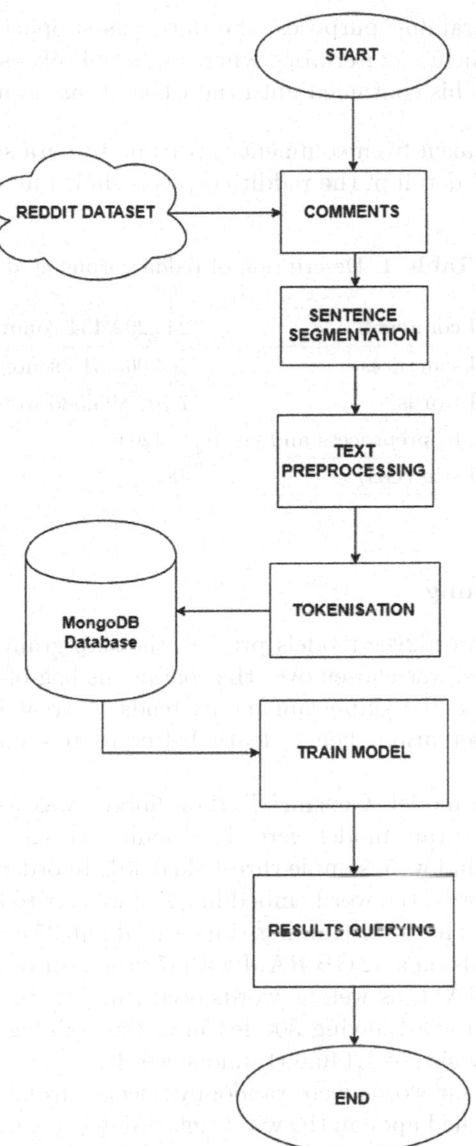

Fig. 1. Data preparation for Word Embeddings (from raw data to word2vec model)

4. **Storage:** Due to the massive amount of comments extracted, they were stored in MongoDB in the form of documents. Each document consisted of two key-value pairs: one for the document number, the other for holding tokenized sentences, which was exactly 5,000 sentences.

5. **Training:** For training purposes, the data was supplied by means of data streaming with the use of iterators, where one set of tokens were processed and then discarded. This continued until the whole training process is completed.

The total time taken from comment extraction to data storage was approximately 120 h. A full detail of the reddit corpus is shown in Table 1.

Table 1. Description of reddit corpus used

Total comments	245,292,434 comments
Total sentences	564,066,513 sentences
Total words	7,702,895,586 words
Time to preprocess and store	$\approx$ 120 h
Total size (GB)	48

3.2 Model Training

From the group of word2vec models present, the skip-gram model architecture was used. This model was chosen over the continuous bag-of-words, another log-linear model because the skip-gram model tends to treat each context-target word as a new observation, hence, learn better representations of infrequent words [1].

For training the model, Gensim[2] Python library was used. The parameter settings for the skip-gram model were the default settings: minimum word frequency 5, context window 5, sample threshold 0.001. In order to capture a better representation of words, the word embedding size was set to 300 and 10 negative samples were used. The total training time was about 25 h after parallelization with 8 worker threads on a 32 GB RAM with i7 processor of 2.2 GHz. Due to the limited size of the RAM, as well as words occurring less than the given threshold of five, and each word having 300 features, the training resulted our model having a vocabulary size of 1,146,604 unique words.

After training, four words were randomly selected from a dictionary of profane words and searched upon in the word embedding vocabulary. Also, a positive sentiment word was used as well. Table 2 shows some of the words that are similar to the five words used. For reproducibility of our research, we have open sourced our source codes[3].

[2] https://radimrehurek.com/gensim/.
[3] https://github.com/tazeek/BullyDetect.

Table 2. Word similarities based on the given words

Word	idiot	dickface	dawg	biatch	wonderful
Similar words	imbecile	fuckface	bruh	bitch	lovely
	moron	fuckstick	homie	biotch	fantastic
	dumbass	dickbag	brah	motherfucker	delightful
	asshat	fuckhead	breh	fucka	wonderfull
	asshole	dickhead	fam	muthafucka	terrific
	dipshit	douchenozzle	bro	mothafucka	great
	fucktard	fucktard	mayne	betch	beautiful
	dolt	shithead	sucka	mutha	marvellous
	ignoramus	jackass	nigga	motha	marvelous
	jackass	douchecanoe	nigguh	beotch	amazing

4 Experiments and Results Analyses

4.1 Experimental Setup

The kaggle dataset[4] was used for experiments. It consisted of 6,594 raw comments, where about 75% of the distribution consists of non-cyberbullying comments, which are true negatives. The data preprocessing for the kaggle dataset is similar to the steps illustrated in Fig. 1, where most of the time was spent on data cleaning. This step consisted of the removal of escape characters, such as tabbed and new-lines, as well as irrelevant UTF-8 (Unicode) characters. During this process, 14 comments were removed as they were either non-english or consisted of only hyperlinks, which contribute to noise.

The dataset was split into two parts: around 4,000 comments for training and the rest for testing. We wanted to check how our model does in detecting cyberbullying comments, which are true positives, which are about 25% of the distribution in the dataset. Hence, precision and the area under curve (AUC) metrics were used in our experiments.

4.2 Baseline Methods

Our newly trained word vector model will be benchmarked against the following pre-trained word embeddings, namely Word2Vec and GloVe (Global Vectors):

- **GoogleNews:** This word2vec model was trained on the Google news dataset, which contains about 100 billion words. The vocabulary size is 3 million words and phrases, each with a feature vector of size 300.
- **Twitter Vectors:** This word2vec model was trained on 400 million twitter posts. The vocabulary size is 3,039,345 words, each with a feature vector of size 400.

[4] https://www.kaggle.com/c/detecting-insults-in-social-commentary/data.

- **Glove.Twitter:** Global vector model trained on 2 billion tweets, which consisted of 27 billion tokens. The vocabulary size is 1.2 million words, with a feature vector of size 200.
- **Glove.Wiki:** Glove vector model trained on Wikipedia-2014 and Gigaword 5 corpus. The corpus consisted of 6 billion tokens, which resulted in the model having a vocabulary size of 400,000 words. The feature vector size for each word was 300.

For fairness of experimenting with word vectors, each comment is represented as the average of the feature vectors from the words in the comment. Then, the average feature vector of a given comment is fed into a random forest classifier with 200 trees. Also, our results will be benchmarked against the hand-crafted feature extraction approaches.

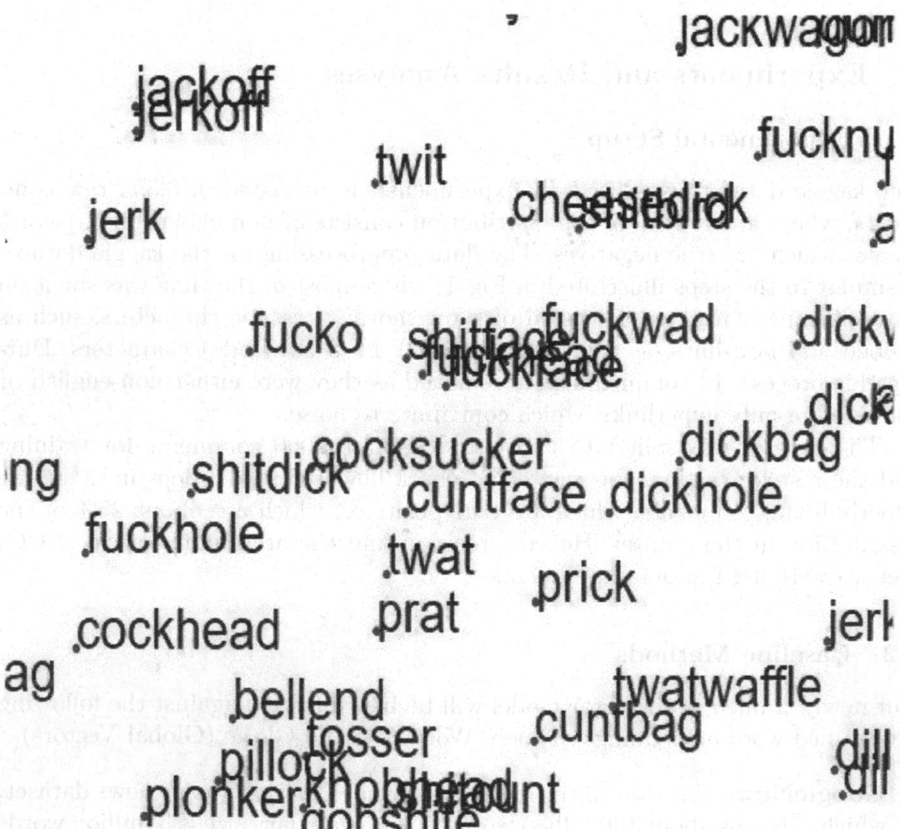

Fig. 2. Visualization of word vectors. Searched word is highlighted in red (Color figure online)

4.3 Results and Findings

From the results shown in Table 3, it can be seen that our model outperforms both the pre-trained word vectors, as well as the hand-crafted methods by 3% more in AUC and 12% more in precision. In comparison to the benchmarked hand-crafted methods, the results show that the features from our word embedding models are more useful, even though the benchmarked methods are aided by a dictionary of profane words. To understand more about the features in our model, we visualize the word vectors using a method called t-distributed stochastic neighbor embedding (t-SNE) by constructing a probability distribution over

Table 3. Results table. The best scores are represented in bold

Method	AUC	Precision
Standard feature extraction [2]	0.83	-
Occurrences pronouns [2]	0.87	-
2,3 Skip Grams [2]	0.87	0.77
GoogleNews [9]	0.88	0.84
Twitter vectors [15]	0.87	0.87
Glove.Twitter [13]	0.89	0.86
Glove.Wiki [13]	0.86	0.85
Our model	**0.90**	**0.89**

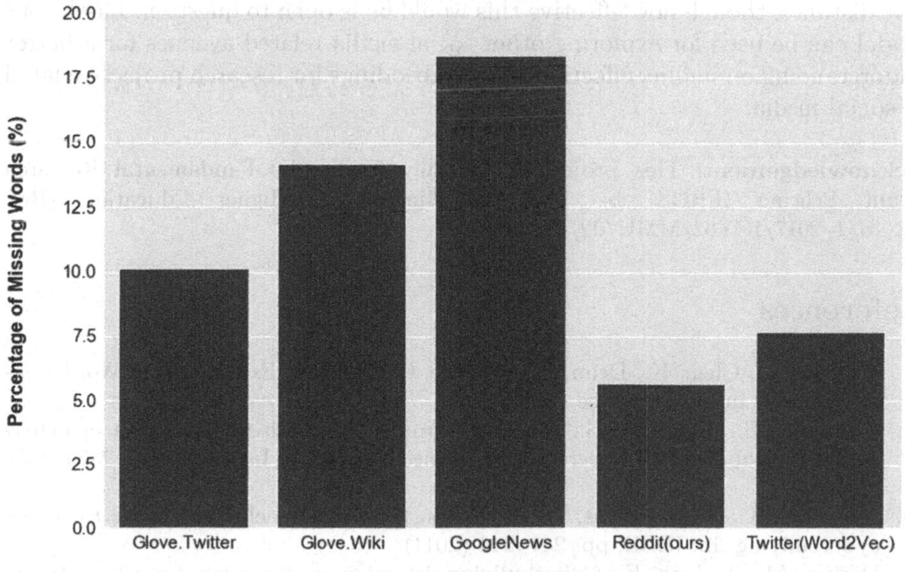

Fig. 3. Percentage of missing words from cyberbully dataset

pairs of words such that similar words with high probabilities will be closer together. Figure 2 shows the results of one such profane word (highlighted in red) and its surrounding words. This shows that the model not only captures profane words that are similar to it by context, but by the grammar as well.

In comparison to word vectors, our model just edges out by 1% in AUC and 2% in precision. In terms of training data, our model did considerably well as only one pre-trained model had less training data compared to ours. Among the five word embeddings, GoogleNews and Glove.Wiki had the lowest scores. This can be down to the differing knowledge, as Fig. 3 would demonstrate; the remaining three models had substantially lower percentage of missing words, with our model having the lowest at about 6%. Hence, the domain in which word embeddings are trained also plays a significant role.

5 Conclusion and Future Work

In this paper, we extracted and cleaned text data from a social media site. This was followed by using the text data to train a word embedding model for classifying cyberbully comments. Our experiments, compared to popular pre-trained models and hand-crafted feature extraction methods, demonstrated the effectiveness of training a model and applying it for classification, where the domains in both training and classification are similar. Also, it would be interesting to see how well our model performs on other cyberbully datasets. Additional features, such as part-of-speech tag and named entities, might help improve the detection task. In addition, one way of capturing missing words would be using edit distance, though how effective this would be is open to question. Lastly, our model can be used for exploring other social media-related avenues for a better understanding on finding effective word embeddings for research projects related to social media.

Acknowledgement. This project is partially funded by Fundamental Research Grant Scheme (FRGS) by Malaysia Ministry of Higher Education (Ref: FRGS/1/2017/ICT02/MMU/02/6).

References

1. Mikolov, T., Chen, K., Dean, J., Corrado, G.: Efficient Estimation of Word Representations in Vector Space (2013)
2. Chavan, V.S., Shylaja, S.S.: Machine learning approach for detection of cyber-aggressive comments by peers on social media network. In: IEEE, pp. 2354–2358 (2015)
3. Reynolds, K., Kontostathis, A., Edwards, L.: Using machine learning to detect cyberbullying. In: IEEE, pp. 241–244 (2011)
4. Dadvar, M., de Jong, F.: Cyberbullying detection: a step toward a safer internet yard. In: WWW 2012 Companion, pp. 121–124 (2012)
5. Bengio, Y., Ducharme, R., Vincent, P., Janvin, C.: A neural probabilistic language model. J. Mach. Learn. Res. 1137–1155 (2003)

6. Turian, J., Ratinov, L.-A., Bengio, Y.: Word representations: a simple and general method for semi-supervised learning. In: ACL, pp. 384–394 (2010)
7. Mikolov, T., Yih, W.-T., Zweig, G.: Linguistic regularities in continuous space word representations. In: NAACL-HLT, pp. 746–751 (2013)
8. Collobert, R., Weston, J.: A unified architecture for natural language processing: deep neural networks with multitask learning. In: Proceedings of the 25th International Conference on Machine Learning, pp. 160–167 (2008)
9. Mikolov, T., Sutskever, I., Chen, K., Corrado, G., Dean, J.: Distributed Representations of Words and Phrases and their Compositionality (2013)
10. Kenter, T., de Rijke, M.: Short text similarity with word embeddings. In: Proceedings of the 24th ACM International on Conference on Information and Knowledge Management, pp. 1191–1200 (2015)
11. Hu, B., Lu, Z., Li, H., Chen, Q.: Convolutional neural network architectures for matching natural language sentences. In: Advances in Neural Information Processing Systems, pp. 2042–2050 (2014)
12. Socher, R., Huval, B., Manning, C.D., Ng, A.Y.: Semantic compositionality through recursive matrix-vector spaces. In: Proceedings of the 2012 Joint Conference on Empirical Methods in Natural Language Processing and Computational Natural Language Learning, pp. 1201–1211 (2012)
13. Pennington, J., Socher, R., Manning, C.D.: GloVe: global vectors for word representation. In: Empirical Methods in Natural Language Processing (EMNLP), pp. 1532–1543 (2014)
14. Zhao, R., Mao, K.: Cyberbullying detection based on semantic-enhanced marginalized denoising auto-encoder. In: IEEE, pp. 1–12 (2015)
15. Godin, F., Vandersmissen, B., De Neve, W., Van de Walle, R.: Multimedia Lab @ ACL W-NUT NER shared task: named entity recognition for Twitter microposts using distributed word representations. In: ACL (2015)

A Positive-Unlabeled Learning Model for Extending a Vietnamese Petroleum Dictionary Based on Vietnamese Wikipedia Data

Ngoc-Trinh Vu[1,2(✉)], Quoc-Dat Nguyen[2], Tien-Dat Nguyen[2], Manh-Cuong Nguyen[2], Van-Vuong Vu[2], and Quang-Thuy Ha[1(✉)]

[1] Knowledge Technology Laboratory, University of Engineering and Technology, Vietnam National University Hanoi, Hanoi, Vietnam
thuyhq@vnu.edu.vn

[2] Vietnam Petroleum Institute, Vietnam National Oil and Gas Group, Hanoi, Vietnam
{trinhvn, datnq, datnt, cuongnm01, vuongvv}@vpi.pvn.vn

Abstract. This study provides a positive-unlabeled learning model for extending a Vietnamese petroleum dictionary based on Vietnamese Wikipedia data. Machine learning algorithms with positive and unlabeled data together with separated and combined between Google similarity distance and Cosine similarity distance, used in this study. The data sources used to integrate are English - Vietnamese oil and gas dictionary and the Vietnamese Wikipedia. In the results, a novelty way for data integration with higher accuracy by using a combination of algorithms. The first Vietnamese oil and gas ontology was built in Vietnam. This ontology is a useful tool for staff in the oil and gas industry in training, research, search daily.

Keywords: Data integration · Ontology · Machine learning
Positive and unlabeled learning · Oil and gas · Wikipedia

1 Introduction

At present, Vietnam does not have oil and gas Ontology in Vietnamese but only English - Vietnamese oil and gas Dictionary by Vietnam Petroleum Institute [1]. The Dictionary includes 11.139 English concepts, these concepts have been translated into Vietnamese and included Vietnamese descriptions, audio, images and video. The number of concepts is also limited by 11.139 concepts, mainly in the early stages of oil and gas industry (Exploration and Production) while there is a lack of concepts in the middle (storage, transport) and the following stages (petrochemical, safety, environmental, economic and petroleum management). In 2010, this dictionary has been transformed to the electronic form dictionary by Petroleum Archive Center (a subsidiary of Vietnam Petroleum Institute) without any changing in the content. However, this is just a desktop-based and unstructured dictionary with separated concepts, and not an ontology. It is also uncomfortable to use and not expose the high value of

N. T. Nguyen et al. (Eds.): ACIIDS 2018, LNAI 10751, pp. 190–199, 2018.
https://doi.org/10.1007/978-3-319-75417-8_18

information inside the dictionary, and not satisfy the requirements of petroleum users in searching, studying, training, data integration, and extension. Therefore, need to build an oil and gas Ontology in Vietnamese, including Vietnamese concepts, has a semantic relationship between concepts and description for these concepts, with the number of concepts expanded beyond the current dictionaries.

Vietnamese Wikipedia, one part of the Wikipedia, is an open data source built in 2003 by the Wikipedia Foundation project [2], and continuously updated by users. In October 2017, there were about 7.155.700 Vietnamese concepts, in which, there are about 1.162.437 concepts with both brief descriptions of concept (abstract) and detailed descriptions (which can be included audio, images, video). There are many petroleum concepts contained in the Vietnamese Wikipedia data, such as "dầu mỏ" (oil), "địa vật lý" (geophysical), "địa chất dầu khí" (petroleum geology), "khai thác dầu khí" (oil and gas production), "khoan" (drill), "hóa dầu" (petrochemical). Wikipedia is a reliable source for references. It can be added, edited by experts, but also by users with incorrect information. However, with the operational principle and a large team of 567.533 members, inaccurate information will be deleted soon. This will be the data source used to integrate with the English – Vietnamese oil and gas dictionary to build the first Vietnamese oil and gas Ontology.

In this study, to integrate data from both sources above, we used machine learning algorithms with positive data as labeled in English – Vietnamese oil and gas dictionary to classify unlabeled data in Vietnamese Wikipedia. The positive dataset is the kernel and extended in the Vietnamese Wikipedia dataset. Machine learning with positive data algorithms combined with Google similarity distance and Cosine similarity distance to improve the accuracy of results. The results of this study, not only offering a new way of data integration with higher accuracy but also built a first Vietnamese oil and gas Ontology, a useful tool for staffs of the oil and gas industry in training, researching, searching, and working daily. Furthermore, the results from this study are the basics for the integration of other valuable specialized data available in Vietnam oil and gas industry.

The rest of this paper organized as follows. The Positive-Unlabeled Learning Model for Extending a Vietnamese Petroleum Dictionary presented in Sect. 2, while the experiments described and evaluated in Sect. 3. In the next section, the related works had been shown. The conclusions and future work presented in the last section.

2 A Positive-Unlabeled Learning Model for Extending a Vietnamese Petroleum Dictionary

2.1 Problem Formulation

Let a Vietnamese Petroleum Dictionary included a set of Vietnamese Petroleum Concepts with their explains. The Dictionary is presented in [1]. Let a Vietnamese Wikipedia on the Petroleum Area[1]. The objective of the problem is to enrich the Dictionary based on some data integration methods on the two datasets.

[1] https://vi.wikipedia.org.

2.2 Proposed Model

Figure 1 presents the proposed model for data integration. The process includes two phases described as follows.

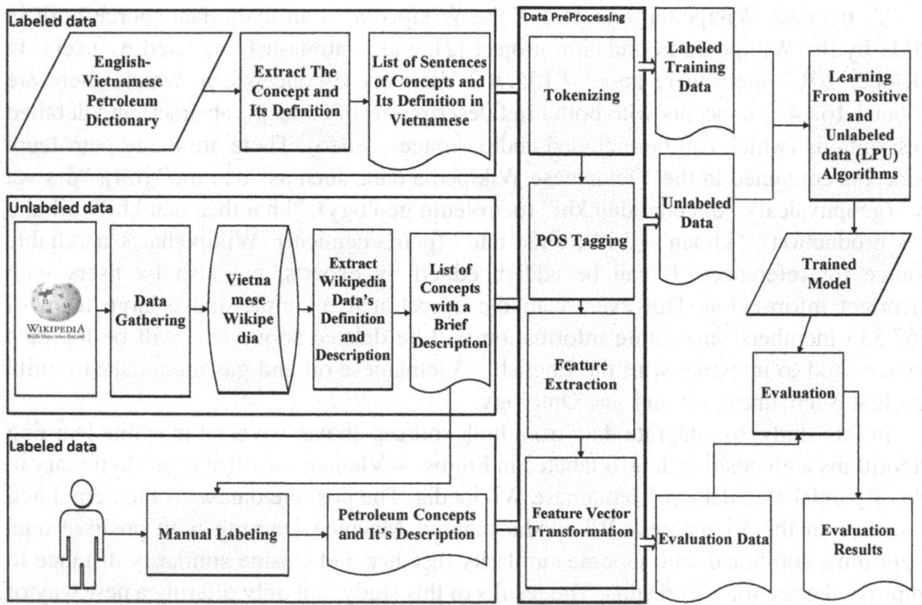

Fig. 1. Proposed data integration model for a Vietnamese oil and gas Ontology.

Phase 1. The integration by directly linguistic matching the concepts of two datasets.

Step 1. Directly linguistic matching each of 11.139 concepts with each of 7.155.700 concepts of Vietnamese Wikipedia to find out the common concepts.

Step 2. Extracting the concepts of the dictionary and Vietnamese Wikipedia, split into words, remove the stop words, meaningless words. Featurization and create the characteristic vector.

Step 3. Directly linguistic matching each of the preprocessed concepts of oil and gas dictionary above with the concepts of Vietnamese Wikipedia to find out the common concepts.

One example for Phase 1 of the model is the completely concept matching of the "Cát kết" (sandstone) concept, was found out in both oil and gas dictionary and Vietnamese Wikipedia. Another example for Phase 1 of the model is the incompletely concept matching of the "bản đồ cấu tạo địa chấn" (seismic structure map) concept of oil and gas dictionary. First, the "bản đồ cấu tạo địa chấn" concept searched in the Vietnamese Wikipedia. There was no concept in Vietnamese Wikipedia have the exactly name "bản đồ cấu tạo địa chấn". After that, JVNTextPro tool[2] used to split "bản đồ cấu tạo địa chấn" concept into three concepts: "bản đồ", "cấu tạo", "địa chấn". Then, each of them searched in Vietnamese Wikipedia data. The result is, the "bản đồ" and "địa chấn" concept is existing in the Vietnamese Wikipedia and the "cấu tạo" concept is not.

Phase 2. The integration based on indirectly linguistic matching the concepts of two data set by using positive – unlabeled learning combination of Normalized Google Distance and Cosine Distance measurement to calculate the similarity between the concept's brief description in oil and gas dictionary and in Vietnamese Wikipedia dataset.

$$SIM_{Total}(C1, C2) = \alpha * SIM_{Cosine}(C1, C2) + (1 - \alpha) * SIM_{NGD}(C1, C2) \quad (1)$$

Here, SIM_{Total} is general measure, C1 and C2 are two concepts need to measure. SIM_{Cosine} is Cosine distance measure. SIM_{NGD} is Normalized Google Distance measure. $\alpha \in [0, 1]$.

Proposed model includes main components as follows.

Wikipedia data processing component. Complete Wikipedia data downloaded from Wikipedia page[3]. The DKPro Java Wikipedia Library[4] used to analyze Wikipedia data. The concept component and its brief introduction extracted as unlabeled data.

Data pre-processing component. Label, unlabeled or evaluation data is passed through pre-processing component and characterization such as sentence separation, word separation, feature extraction, and feature vector creation.

Evaluation dataset construction component. Unlabeled data will be extracted in part to label serves evaluation classification model. Using the concepts of Vietnamese Wikipedia which found out (labeled) in the Phase 1 of the model (a part of 2.500 concepts) as evaluation data set.

Data classification component. Oil and gas labeled and unlabeled data is passed through the classification data component LPU (learning with positive and unlabeled) to create data classification model.

Denoted P is a positive dataset: The set of labeled concepts of oil and gas dictionary. U is an unlabeled dataset: The set of unlabeled concepts of Vietnamese Wikipedia (which may belong to P or not belong to P). The role of the data classification component is building the classifier to classify the concepts belong to U.

The two steps strategy used to solve this problem. In step 1, the "reliable" negative (RN) data set we must be identified. In step 2, a good classifier based on iterative methods will be built and selected. In this study, three algorithms will be implemented, namely PERL, ROC-SVM, and DISTANCE. The LPU[5] tool used to run PERL and ROC-SVM algorithms.

---

[2] http://jvntextpro.sourceforge.net.

[3] https://dumps.wikimedia.org.

[4] https://github.com/dkpro/dkpro-jwpl.

[5] https://www.cs.uic.edu/~liub/LPU/LPU-download.html.

PERL algorithm.

- Step 1. 1-DNF
  - Looking in the feature set for the characteristics which appear frequently in the oil and gas dictionary (P) rather than the unlabeled concept of Vietnamese Wikipedia (U).
  - Extracting the unlabeled concepts in Vietnamese Wikipedia (U) which does not contain the above characteristics and pushing them into the Reliable Negative (RN) set.
- Step 2. Iterative-SVM
  - Use the labeled concepts of oil and gas dictionary (P) and reliable negative data RN to build classifier Q.
  - Use Q to classify the remaining concepts in U, if any concept classified as N then assigned to the RN set.
  - Rebuild the classifier Q from the new set of RNs, the algorithm will stop when no concept is named N in U.

ROC-SVM algorithm.

- Step 1. Use Rocchio method
  - Build two vectors representing two classes P (labeled as +1) and U (labeled as −1).
  - Calculate the similarity between the unlabeled concepts in U and two representative vectors.
  - If the concept nearer the vector represents U than P, then the concept is included in the RN set.
- Step 2: Use Iterative-SVM method

DISTANCE algorithm.

- Step 1: Identify the abnormal element in the depth approach
- Step 2: Use Iterative-SVM
  - Parameters
    - Measure distance:
      - Normalized Google Distance: distance measurement based on the occurrence and co-occurrence statistics of the two concepts
      - Cosine: distance measurement based on the characteristics of two concepts
    - Reliable negative data equal to 15% of the positive data

Evaluation component. Data classification model used for testing on the evaluation data set. The result was evaluated with measures P, R, F on positive data. Here,

$$F = \frac{2 * P * R}{P + R} \tag{2}$$

3 Experiment

3.1 Experiment Data

English – Vietnamese oil and gas dictionary. English-Vietnamese oil and gas dictionary built by Vietnam Petroleum Institute in 1996, included 11.139 English concepts, these concepts have been translated into Vietnamese. Each concept also contained a detail Vietnamese description and audios, images and videos.

Vietnamese Wikipedia. Vietnamese Wikipedia has about 7.155.700 Vietnamese concepts, in which, there are about 1.162.437 concepts with both brief descriptions of concept (abstract) and detailed descriptions (which can be included audio, images, video). There are many oil and gas concepts contained in the Vietnamese Wikipedia data, such as "dầu mỏ" (oil), "địa vật lý" (geophysical), "địa chất dầu khí" petroleum geology, "khai thác dầu khí" (oil and gas production), "khoan" (drill), "hóa dầu" (petrochemicals).

Evaluation data. 1.000 positive data and 1.000 negative data. Labeled data: 11.139 positive data; 60.000 unlabeled data with measure distance Cosine close to positive data. About 140.000 features created. Experimental results on three measures of P, R, F on positive data.

3.2 Experimental Cases

Three experimental cases had been done in this study.

Experimental case 1: Evaluates the integration between the concepts of oil and gas dictionary dataset and Vietnamese Wikipedia dataset by directly concept linguistic matching in Phase 1 of the proposed model.

Experimental case 2: Evaluate different LPU classification methods. Experiment three methods PERL, ROC-SVM, DISTANCE. The measure of ROC-SVM and DISTANCE are Cosine and Normalized Google Distance.

Experimental case 3: Experiment combination measure. Measure combination Cosine and Normalized Google Distance (NGD), α change by the segment.

3.3 Experiment Results and Evaluations

The results of experimental case 1. There are only six concepts are completely similarity in two datasets: "Huệ_biển" (crinoid), "Ổ_đĩa" (disk drive), "Galileo" (galileo), "OPEC" (opec), "Paleogen" (paleogene), "Cát_kết" (sandstone);

When pre-processing the 11.139 Vietnamese concepts: splits the concept into words, remove the stop words, meaningless words, duplicated words, we have 6.000 atom concepts. For each of 6.000 concepts, we compare directly with 7.155.700 concepts in Vietnamese Wikipedia, then we found 2.500 concepts exactly equal. This 2.500 concepts which may include concepts without its brief description (abstract). Vietnamese oil and gas dictionary added the 2.500 concepts (minus 6 same concepts) from Vietnamese

Wikipedia to form Vietnamese ontology with the maximum of 13.633 Vietnamese concepts together with its detailed description.

The results of experimental case 2. The experimental results have been shown in Table 1. The Table 1 presenting that the DISTANCE/ISVM algorithm using the Cosine measurement have the highest P, R, F score and get a pick at 84.17, 80.9, and 82.29 respectively while the ROC/ISVM algorithm using the NGD have the lowest R, P, F score at 67.08, 70.45, and 68.72 respectively. The ROC/ISVM algorithm using the Cosine measurement have the higher R, P, F score than PERL algorithm's R, P, F score.

Table 1. Results of experimental case 2.

Algorithm		P	R	F
PERL		80.24	76.36	78.25
ROC/ISVM	Cosine	82.53	79.21	80.84
	NGD	**67.08**	**70.45**	**68.72**
DISTANCE/ISVM	Cosine	**84.17**	**80.49**	**82.29**
	NGD	73.25	75.61	74.41

The results experimental case 3. The experimental results have been shown in Table 2. According to (1), the value of parameter α belong to [0, 1] with interval is 0.1. If $\alpha = 0$ then algorithm be affected by only NGD whereas if $\alpha = 1$ then algorithm be affected by only Cosine measurement. When the value of α increased from 0 to 1.0 by interval step of 0.1 then the results also changed (Fig. 2).

Table 2. Results of experimental case 3

Method	α										
	0	0.1	0.2	0.3	0.4	0.5	0.6	0.7	0.8	0.9	1
ROC/ISVM (Hybrid)	68.72	72.59	75.67	76.88	78.49	80.36	82.35	**82.41**	80.57	81.29	80.84
DISTANCE (Hybrid)	74.41	79.34	80.46	81.53	82.79	**83.41**	83.17	81.56	82.67	82.19	82.29

3.4 The Result of Building the Vietnamese Oil and Gas Ontology

When applying positive – unlabeled data classification algorithm with the combination of Cosine and Normalized Google Distance measure with $\alpha = 0.50$, to integrate the concept description data in oil and gas dictionary and concept brief description data in the Vietnamese Wikipedia, we find out 5.084 oil and gas concepts. Therefore, the oil and gas dictionary being added this 5.084 concepts and became Vietnamese oil and gas ontology with 16.084 concepts together with its detail description. For examples, 10

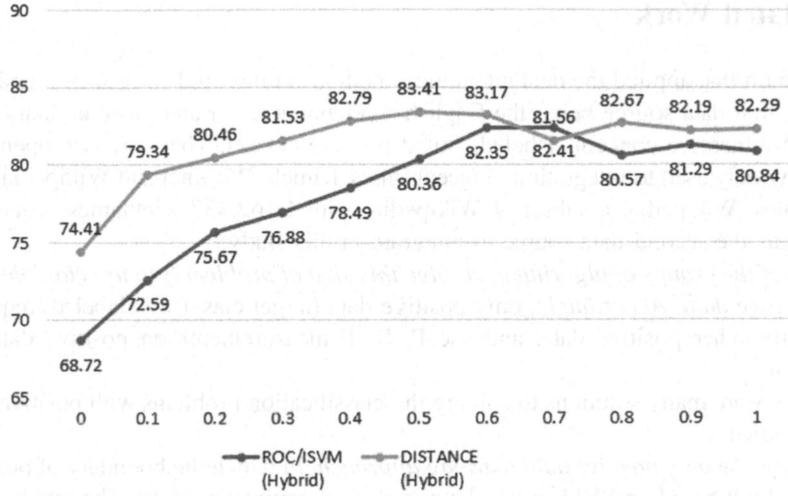

Fig. 2. The depending of F measurement of ROC/ISVM and DISTANCE on α-ratio in the result of experiment 3.

new oil and gas concepts from Vietnamese Wikipedia have been added into the current petroleum dictionary, such as *"Nhiên liệu hoá thạch", "Dầu nhờn", "Độ rỗng", "Nhiên liệu sinh học", "Giếng khoan", "Khí thiên nhiên", "Trầm tích", "Xăng", "Khí đồng hành"*, and *"Hóa dầu"*.

The experts from Vietnam Petroleum Institute checked new concepts and highly appreciate the adding new oil and gas concepts and its description into the current dictionary, these new concepts are correct and reliable.

The results of this study not only enrich the concepts but also make the oil and gas dictionary becomes more complete and useful by a supplement the descriptions from Vietnamese Wikipedia.

3.5 Discussion

The experimental results show that: (i) The distance-based method gives better results than the remaining methods; (ii) The Cosine distance measure is better than NGD due to it based on the features of the two concepts; (iii) The combination of Cosine and Normalized Google distance increased the accurate of results with the mixing parameter $\alpha = 0.5$ for the distance method and 0.7 for the ROC/ISVM method. (iv) data integration using positive – unlabeled data classification algorithm with the combination of Cosine and Normalized Google Distance measure (the second steps of the model) is better than the linguistic matching algorithm (the first step of the model). (v) a completely new Vietnamese oil and gas ontology was built with 16.084 concepts, increased 5.084 concepts compare to old Vietnamese oil and gas dictionary.

4 Related Work

The domain that applied the data integration methods in this study was the oil and gas, with the first data source being the English – Vietnamese oil and gas dictionary with 11.139 Vietnamese concepts labeled, called positive dataset. There are two open data sources widely used to integration in recent times, namely Wordnet and Wikipedia. The Vietnamese Wikipedia, a subset of Wikipedia, with 1.162.437 Vietnamese concepts, selected as the second data source to integrate in this study.

One of the groups of algorithms to solve this kind of problem is to use classification with positive data. Accordingly, only positive data (target class) was labeled, required to identify other positive data, and use P, R, F measurements on positive data for evaluation.

There were many solutions to solving the classification problems with positive data being studied.

Firstly, the only positive data learning approach, in which the boundary of positive data was built based on SVM, using the one class classification [3–9]. The requirement of this algorithm is that the positive dataset must be large.

Secondly, the positive and unlabeled learning approach, in which the "reliable" negative data must be identified. Some typical algorithms for this learning method are Spy-EM, Roc-SVM, LGN [B. Liu, X. Li, W.S. Lee] [10–19].

5 Conclusion

In conclusion, by positive – unlabeled data classification algorithm with combination of Cosine and Normalized Google Distance measure with $\alpha = 0.50$ respectively to calculate the distance, we have successfully integrated the English – Vietnamese oil and gas dictionary with 11.139 concepts with Vietnamese Wikipedia has 1.2 million concepts, to form a completely new Vietnamese oil and gas ontology with 16,000 concepts. However, there are still 11.139 English concepts in the English-Vietnamese oil and gas dictionary that are not yet exploited. In the future, we may consider integrating with other ontologies to extend/enrich this Vietnamese oil and gas ontology and using knowledge from petroleum experts to improve the quality of this ontology.

Acknowledgements. This project has been done by the staffs of Vietnamese Petroleum Institute (VPI), Vietnam National Oil and Gas Group (PetroVietnam).

References

1. Bao, T.C., Bich, P.M., et al.: English – Vietnamese Dictionary of Petroleum. The Science and Technics Publishing House, Ha Noi (1996)
2. Vietnamese Wikipedia page. https://vi.wikipedia.org/wiki/Wikipedia:Giới_thiệu. Accessed 15 Oct 2017

3. Khan, S.S., Madden, M.G.: A survey of recent trends in one class classification. In: Coyle, L., Freyne, J. (eds.) AICS 2009. LNCS (LNAI), vol. 6206, pp. 188–197. Springer, Heidelberg (2010). https://doi.org/10.1007/978-3-642-17080-5_21
4. Khan, S.S., Madden, M.G.: One-class classification: taxonomy of study and review of techniques. Knowl. Eng. Rev. **29**(03), 345–374 (2014)
5. Li, X.-L, Liu, B., Ng, S.-K.: Learning to identify unexpected instances in the test set. In: IJCAI, vol. 7 (2007)
6. Pimentel, M.A., Clifton, D.A., Clifton, L., Tarassenko, L.: A review of novelty detection. Sig. Process. **9999**, 215–249 (2014)
7. Yu, H., Han, J., Chang, K.C.-C.: PEBL web page classification without negative examples. IEEE Trans. Knowl. Data Eng. **16**(1), 70–81 (2004)
8. Fung, G.P.C., Yu, J.X., Lu, H., Yu, P.S.: Text classification without negative examples revisit. IEEE Trans. Knowl. Data Eng. **18**(1), 6–20 (2006)
9. Noto, K., Saier, M.H., Elkan, C.: Learning to find relevant biological articles without negative training examples. In: Wobcke, W., Zhang, M. (eds.) AI 2008. LNCS (LNAI), vol. 5360, pp. 202–213. Springer, Heidelberg (2008). https://doi.org/10.1007/978-3-540-89378-3_20
10. Li, M., Pan, S., Zhang, Y., Cai, X.: Classifying networked text data with positive and unlabeled examples. Pattern Recogn. Lett. **77**, 1–7 (2016)
11. Li, X.-L., Liu, B., Ng, S.-K.: Learning to classify documents with only a small positive training set. In: Kok, J.N., Koronacki, J., Mantaras, R.L., Matwin, S., Mladenič, D., Skowron, A. (eds.) ECML 2007. LNCS (LNAI), vol. 4701, pp. 201–213. Springer, Heidelberg (2007). https://doi.org/10.1007/978-3-540-74958-5_21
12. Li, X.-L, Yu, P.S., Liu, B., Ng, S.-K.: Positive unlabeled learning for data stream classification. In: SDM 2009, pp. 259–270 (2009)
13. Davoudi, H., Li, X.-L., Nhut, N.M., Krishnaswamy, S.P.: Activity recognition using a few label samples. In: Tseng, V.S., Ho, T.B., Zhou, Z.-H., Chen, Arbee L.P., Kao, Hung-Yu. (eds.) PAKDD 2014. LNCS (LNAI), vol. 8443, pp. 521–532. Springer, Cham (2014). https://doi.org/10.1007/978-3-319-06608-0_43
14. Xiao, Y., Liu, B., Yin, J., Cao, L., Zhang, C., Hao, Z.: Similarity-based approach for positive and unlabeled learning. In: IJCAI 2011, pp. 1577–1582 (2011)
15. Sansone, E.: Efficient training for positive unlabeled learning (2016). CoRR abs/1608.06807
16. Kiryo, R., Niu, G., du Plessis, M.C., Sugiyama, M.: Positive-unlabeled learning with non-negative risk estimator (2017). CoRR abs/1703.00593
17. Niu, G., du Plessis, M.C., Sakai, T., Ma, Y., Sugiyama, M.: Theoretical comparisons of positive-unlabeled learning against positive-negative learning. In: NIPS 2016, pp. 1199–1207 (2016)
18. Elkan, C., Noto, K.: Learning classifiers from only positive and unlabeled data. In: KDD 2008, pp. 213–220 (2008)
19. Li, H., Liu, B., Mukherjee, A., Shao, J.: Spotting fake reviews using positive-unlabeled learning. Computación y Sistemas **18**(3), 467–475 (2014)

A New Lifelong Topic Modeling Method and Its Application to Vietnamese Text Multi-label Classification

Quang-Thuy Ha[1], Thi-Ngan Pham[1,2], Van-Quang Nguyen[1,3], Thi-Cham Nguyen[1,4], Thi-Hong Vuong[1], Minh-Tuoi Tran[1], and Tri-Thanh Nguyen[1](✉)

[1] University of Engineering and Technology (UET), Vietnam National University, Hanoi (VNU), Hanoi, Vietnam
{thuyhq,nganpt.di12,quangnv_570,hongvt_57,tuoitm_58, ntthanh}@vnu.edu.vn, nthicham@hpmu.edu.vn

[2] The Vietnamese People's Police Academy, Hanoi, Vietnam

[3] Tohoku University, Sendai, Japan

[4] Hai Phong University of Medicine and Pharmacy, Haiphong, Vietnam

Abstract. Lifelong machine learning is emerging in recent years thanks to its ability to use past knowledge for current problem. Lifelong topic modeling algorithms, such as LTM and AMC, are proposed and they are very useful. However, these algorithms focus on learning bias on the topic level not the domain level. This paper proposes a lifelong topic modeling method, which focuses on learning bias on the domain level based on a proposed domain closeness measure, and an application framework for multi-label classification on Vietnamese texts. Experimental results on three previously solved Vietnamese texts, and five different current Vietnamese text datasets in combination with different topic set sizes showed that our proposed method is better than AMC method for all cases.

Keywords: Close domain · Lifelong topic modeling
Learning bias on the domain level
Similarity measure of weighted word bags · Close topic

1 Introduction

Lifelong Machine Learning (LML) or *Lifelong Learning* (LL) was proposed in 1995 by Thrun and Mitchell [6, 7]. Thrun enunciated that the key scientific concerns that arise in lifelong learning be the acquisition, representation and transfer of domain knowledge and focus on learning bias approaches [7]. In recent years of the fourth industrial revolution, LML becomes an emerging machine learning paradigm thanks to its ability of use knowledge from the past tasks for the current task[1]. According to Chen and Liu [8], LML has three key characteristics, namely, (i) it is a continuous learning process,

[1] http://lifelongml.org/.

N. T. Nguyen et al. (Eds.): ACIIDS 2018, LNAI 10751, pp. 200–210, 2018.
https://doi.org/10.1007/978-3-319-75417-8_19

(ii) it accumulates and retains the learned knowledge and (iii) it has competence to use the past knowledge to help future learning.

Lifelong topic modeling is an kind of *lifelong unsupervised learning* [8–12, 16]. The methods of Lifelong topic modeling are very useful in domains of Text Mining, because of the three reasons described by Chen and Liu [8]. Topics from all the past tasks are stored in a *knowledge base*. When a new task (called the *current task*) represented by a document set (in a new domain) arrives, lifelong topic modeling mines prior knowledge patterns from the knowledge base to help model inference for the current task. Must-links, i.e., word pairs that should belong to the same topics, and cannot-links, i.e., word pairs that should not belong to the same topics, are two typical prior knowledge patterns. When topic modeling on the new domain is completed, the resulted topics are added to the knowledge base for future use [7, 8].

Multi-label learning (MLL) is another emerging supervised framework increasingly demanded by modern applications [2]. Szymanski et al. [1] proposed a solution to partitioning the label space in the task of multi-label classification based on using five data-driven community detection approaches from social networks. The multi-label learning algorithm LIFT (*multi-label learning with Label specIfic FeaTures*), which is proposed by Zhang and Wu [3], builds the specific features of each label by applying clustering analysis on its positive and negative instances, and then carries out training and testing by exploiting the resulted clusters. Developing Zhang et al.'s method in TESC, a single-label text classification algorithm based on using a semi-supervised clustering technique [4], Pham et al. [13, 14] proposed a multi-label classification algorithm named MULTICS (MULTI-label ClaSsification), which could exploit specific features of label/label set based on a semi-supervised clustering technique to extract useful information of both labeled and unlabeled instances together. However, the current ruling machine learning framework learns in isolation. It makes no effort to exploit the learned knowledge in the past to help in future learning. So it requires a large number of training examples.

In this paper, we propose a new lifelong topic modeling method, and its application framework for multi-label classification for Vietnamese texts. This paper has three main contributions: (i) propose a learning bias method on the domain level based on a proposed closeness measure of domains instead of learning bias methods on topic level [8–10], (ii) propose a new closeness measure of two topics based on a cosine measure of two word bags instead of symmetrized Kullback-Leibler (KL) Divergence [9] in topic modeling phase, (iii), an application framework for multi-label classification for Vietnamese texts is also proposed.

The rest of this paper is organized as follows. In the next section, our lifelong topic modeling method is described. A similarity measure of weighted word bags, two concepts of *close topic* and *close domain* are defined. A proposed application framework for multi-label classification for Vietnamese texts also is described. Section 3 shows the experiments with three previous datasets and some cases of the current dataset. Some related work analyzed and compared with this paper is shown in Sect. 4. In the last section, we present conclusions and future work.

2 A New Lifelong Topic Modeling Method and an Application Framework for Multi-label Classification on Vietnamese Texts

2.1 Problem Formulation

Let $T_1, T_2, \ldots, T_N$ be N previously solved topic modeling tasks. Let D_i, V_i, and $Topics_i$ be the dataset, the vocabulary, and the output topics of T_i, correspondingly, for $i = 1, 2, \ldots, N$. Topic models are built by either a hidden topic modeling or a lifelong topic modeling. Let S be the *knowledge base*, which includes all knowledge, information from N previous tasks. S is empty when $N = 0$.

Let T_{N+1} be a new task (called the *current task*), with a dataset D_{N+1}. The problem is to build a lifelong topic model $Topics_{N+1}$ based on the knowledge base S.

This paper proposes a new lifelong topic modeling method based on Chen and Liu's [9, 10], in which, prior knowledge patterns (must-links and cannot-links) are mined by a learning bias method, i.e., only take close domains to the current domain into account, and some closeness measures are defined for domain selection as described in Sect. 2.2.

2.2 Close Topic and Domain

Definition 1. *Similarity measure of weighted word bags*: Given two weighted word bags $A = \{(wa_i, pa_i)\}, B = \{(wb_i, pb_i)\}$, where wa_i and wb_i are words; pa_i and pb_i are their weights. Let C be the vocabulary of words in both A and B, i.e., $C = \{wa_i\} \cup \{wb_i\}$. Let v_A, v_B be weighted vectors (based on C) of A and B, correspondingly, of which a missing word has the weight of 0. The *similarity measure* of A and B, denoted as Similarity (A, B), is defined as:

$$\text{Similarity}(A, B) = \text{cosine}(v_A, v_B) \tag{1}$$

For example,

Given A = {(khách_sạn (*hotel*), 0.0597548), (phòng (*room*), 0.0461742), (nhân_viên (*staff*), 0.0411788), (phục_vụ (*service*), 0.0273162), (sạch (*clean*), 0.0272942)}, and B = {(nhân_viên (*staff*), 0.064638), (resort, 0.058477), (phòng (*room*), 0.050749), (amiana, 0.041064), (phục_vụ (*service*), 0.040162)}.

Then C = {khách_sạn, phòng, nhân_viên, phục_vụ, sạch, resort, amiana}.

Then, v_A = (0.0597548, 0.0461742, 0.0411788, 0.0273162, 0.0272942, 0, 0), v_B = (0, 0.050749, 0.064638, 0.040162, 0, 0.058477, 0.041064), and Similarity(A, B) = cosine (v_A, v_B) = 0.0037588.

The similarity measure of word bags is used to define close topic and domain.

Definition 2. *Close topic*: Given two different topic sets A and B, they are deemed to be close if

$$\text{similarity}(\text{Top}_{\text{M}}(A), \text{Top}_{\text{M}}(B)) \geq \theta, \tag{2}$$

where $\text{Top}_{\text{M}}(A)$ and $\text{Top}_{\text{M}}(B)$ are the sets of top M probability words in A and B, correspondingly; θ is a positive threshold. The *similarity*$(\text{Top}_{\text{M}}(A), \text{Top}_{\text{M}}(B))$ function is called the *close topic measure*.

The close topic measure is used for identifying close domains, i.e., a previous and the current domain.

Definition 3. *Close domain*: Let D_i, D_j be the data sets of tasks T_i, T_j, correspondingly. Let V_i, V_j be the vocabularies of D_i, D_j, correspondingly. Let $Topic_i$, $Topic_j$ be the topic sets discovered from D_i, D_j, correspondingly. D_j is deemed to be close to D_i if they satisfy the following criteria:

(1) *Vocabulary level*:

$$\frac{|V_i \cap V_j|}{|V_i|} + \frac{|V_i \cap V_j|}{|V_j|} \geq \theta_1, \tag{3}$$

where $\frac{|V_i \cap V_j|}{|V_i|}$ indicates the degree of V_j being included in V_i. In terms of information retrieval, this component has the meaning as the precision. Similar interpretation is applied for $\frac{|V_i \cap V_j|}{|V_j|}$, which has the meaning as the recall,

(2) *Top word level*:

$$\text{similarity}\left(\text{Top}_{\text{M}}(V_i), \text{Top}_{\text{M}}(V_j)\right) \geq \theta_2, \tag{4}$$

(3) *Topic level*: T_j is close to T_i if the total of topics in T_j having a similarity topic in T_i over the total number of topics in T_j is greater than or equal to a threshold, i.e.,

$$\frac{\left|\left\{t_2 \in Topics(T_j) \middle| t_1 \in Topics(T_i) \wedge similarity(t_2, t_1) \geq \theta_3\right\}\right|}{\left|Topics(T_j)\right|} \geq \theta_4 \tag{5}$$

where $\theta_1, \theta_2, \theta_3,$ and θ_4 are positive thresholds.

Close domain is used for a bias learning approach, i.e., bias toward the close domain. Concretely, from the previous tasks, only the knowledge from close domains to the current domain is chosen to enrich the current topic learning.

2.3 The Proposed Lifelong Topic Modeling Method

Let $D_1, D_2, \ldots, D_N$ be text datasets of N previously solved topic modeling tasks with N topic sets $Topics_1, Topics_2, \ldots, Topics_N$, correspondingly. Let D_{N+1} be the dataset of a new task T_{N+1} (called the *current task*), and $Topics_{N+1}$ be the topic set built by a hidden topic modeling on the dataset D_{N+1}. $Topics_{N+1}$ is enriched based on using the related knowledge from previous topic sets. Figure 1 describes our lifelong topic modeling method. Important parts in the framework are: (1) an addition of a temporary

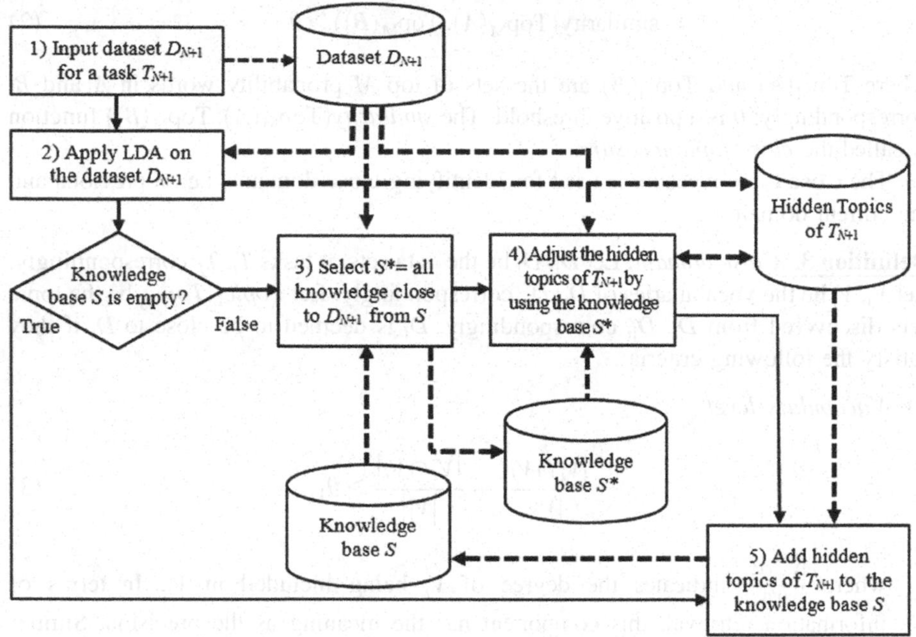

Fig. 1. The proposed lifelong topic modeling framework

knowledge base S* in Step (3) "*Select S* = all knowledge close to* D_{N+1} *from S*", (2) and an investigation of a closeness measure of topics in Step (4) "*Adjust the hidden topics of* T_{N+1} *by using the knowledge base S**".

Step 3 selects all datasets in *N* previous datasets $D_1, D_2, \ldots, D_N$, which are close to D_{N+1}, and copies the related knowledge of the selected datasets to the temporary knowledge base *S** for improving the current task. For each previous dataset *Di*, the selecting process is done in three sub-steps: (i) The first sub-step investigates the closeness of (D_i, D_{N+1}) at the vocabulary level using formula (3); (ii) The second sub-step investigates the closeness of (D_i, D_{N+1}) at the top word level using formula (4); (iii) The third sub-step investigates the closeness of (D_i, D_{N+1}) at the topic level using formula (5). Finally, the dataset D_i is chosen if all the three conditions are satisfied, otherwise D_i is ignored, i.e., one of the above conditions is not satisfied.

All activities of Step (4) are the same as those of AMC [10] except two modifications: (1) the close topic measure is implemented based on formula (2) instead of the symmetrized Kullback-Leibler Divergence [9]; (2) only previous topics (denoted as *p-topics*) in *S** (not *S*) which are close to the current topics are used to mine must-links and cannot-links. By using a multi-generalized Pólya Urn procedure on must-links and cannot-links, current topics are enriched in each iterative Gibbs sampling.

As a result, after the $Topics_{N+1}$ has been built and enriched, it is added to knowledge base *S*, and available as the input for other application.

2.4 An Application Framework for Multi-label Classification on Vietnamese Texts

The enriched current topic set $Topics_{N+1}$ can be used for data representation in the multi-label classification task as depicted in Fig. 2. The set of Hidden Topics from LTM is used to build a new feature set for the texts, i.e., words in a text are replaced by the topic they belong to. To avoid any exception leaks from the future, the testing Dataset D_{test} is not used for enriching the current topic set $Topics_{N+1}$. This solution has an important role in a lifelong machine leaning, because the testing data should be considered as they will be coming in the future. With the help of the knowledge from previous domains, the hidden topics $Topics_{N+1}$ are adjusted and improved to be better than those extracted from LDA [15].

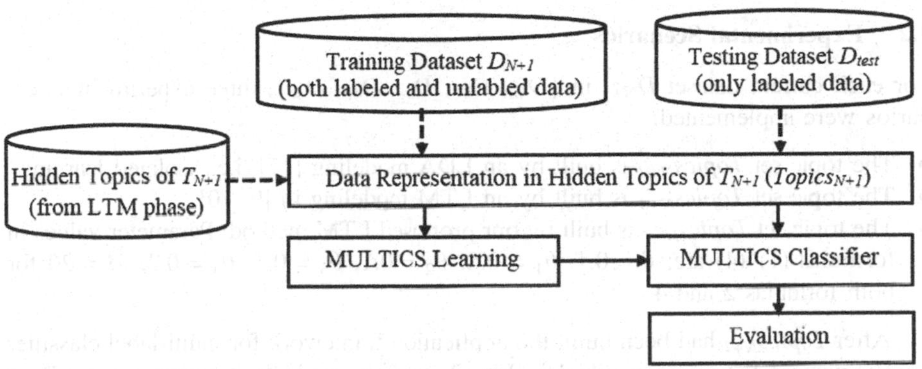

Fig. 2. An application framework for multi-label classifier on Vietnamese texts

Finally, a multi-label classifier based on semi-supervised clustering [14] is built to classify new documents. We use the semi-supervised approach to take full advantages of the unlabeled data available, and to reduce the effort of manually annotating data while still get a better performance.

3 Experimental Results

3.1 The Datasets

We have three *unlabeled* datasets in different domains, i.e., the reviews on (1) tourism and hotels (denoted as D_1), (2) restaurants (denoted as D_2), and (3) mobile phones (denoted as D_3). These three datasets are considered as previous datasets of solved topic modeling tasks. Another dataset of hotel reviews, denoted as D_2, is considered as the current topic modeling task. The reviews in the current domain include both labeled and unlabeled texts, and may have multi-labels on the label set: *Location and price*; *Service*; *Facilities*; *Room standard*; and *Food*. Note that, the labels are used for classification task, not for topic modeling. The details of these datasets are given in Table 1.

Table 1. Datasets from different domains.

Dataset	#reviews	Domain
D_1	26800	Reviews on tourism and hotels
D_2	8093	Reviews on restaurants
D_3	1441	Reviews on mobile phones
D_4	1493	Reviews on hotels

Dataset $D4$ is divided into two parts, i.e., the first includes 300 reviews used as a test dataset $D_{4\text{test}}$ (of the classification); the second includes 1000 remaining reviews used for setting up five current small training datasets, namely $D_{4\text{a}}$, $D_{4\text{b}}$, $D_{4\text{c}}$, $D_{4\text{d}}$, and $D_{4\text{e}}$, with the size of 100, 200, 400, 600, and 1000 reviews, correspondingly.

3.2 Experimental Scenarios

For each current dataset $D_{\text{N+1}}$ in $\{D_{4\text{a}}, D_{4\text{b}}, D_{4\text{c}}, D_{4\text{d}}, D_{4\text{e}}\}$, three experimental scenarios were implemented:

- The topic set $Topics_{N+1}$ is built by an LDA modeling [17], i.e., isolated learning,
- The topic set $Topics_{N+1}$ is built by an LTM modeling in [9, 10],
- The topic set $Topics_{N+1}$ is built by our proposed LTM method. Parameter values in formulas (1)–(4) are: $\theta = 0.1$, $\theta_1 = 0.8$, $\theta_2 = 0.1$, $\theta_3 = 0.1$, $\theta_4 = 0.2$, $M = 20$ for both formulas 2 and 4.

After $Topics_{N+1}$ had been built, the application framework for multi-label classifier on Vietnamese texts, as described in Fig. 2, was run with the training dataset $D_{N\text{+1}}$ along with 3 scenarios for comparison.

Since the average length of the reviews in the dataset D_4 is approximately 8 words, and Pham et al. [13, 14] showed that the performance of MULTICS algorithm got the best results with the number hidden topics of 15 and 25, our experiments also were implemented with the number of topics in {10, 15, 20, 25}.

3.3 Experimental Results and Discussions

Firstly, we identified close domains from previous datasets to current datasets. Table 2 shows the computed results on D_1, D_2, and D_3 to $D_{4\text{a}}$, $D_{4\text{b}}$, $D_{4\text{c}}$, $D_{4\text{d}}$, and $D_{4\text{e}}$. The results show that there are five close domain pairs: (D_1, $D_{4\text{a}}$), (D_1, $D_{4\text{b}}$), (D_1, $D_{4\text{c}}$), (D_1, $D_{4\text{d}}$), and (D_1, $D_{4\text{e}}$). Then, only D_1 was used to adjust hidden topic modeling on each current dataset D_{N+1} in $\{D_{4\text{a}}, D_{4\text{b}}, D_{4\text{c}}, D_{4\text{d}}, D_{4\text{e}}\}$.

The topic set $Topics_{N+1}$ is used for feature representation for multi-label classification. The results of experimental scenarios, i.e., isolated LDA learning, AMC lifelong learning method, and our lifelong learning method are shown in Table 3, where we compare the performance of the systems with the different number of topics added to the feature set of each current dataset.

In most of these scenarios, all the systems often get better results when the training dataset size increases from 100 (D_{4a}) to 1000 (D_{4e}).

Table 2. Close domain checking (**V**: Vocabulary level, **W**: Top word level, **T**: Topic level, **C**: Close, **Y**: Yes, **N**: No)

Dataset	D_1				D_2				D_3			
	V	W	T	C	V	W	T	C	V	W	T	C
D_{4a}	0.95	0.34	0.2	Y	0.86	0	0.05	N	0.41	0.03	0	N
D_{4b}	0.97	0.25	0.45	Y	0.86	0.03	0.1	N	0.46	0	0	N
D_{4c}	0.96	0.38	0.6	Y	0.84	0.03	0.05	N	0.45	0	0	N
D_{4d}	0.95	0.34	0.45	Y	0.82	0.03	0.1	N	0.44	0	0	N
D_{4e}	0.95	0.38	0.5	Y	0.81	0.03	0.1	N	0.44	0	0	N

Table 3. The results of experimental scenarios of the isolated learning LDA, AMC lifelong learning method [10], and our lifelong learning method, where **NoT** is the number of topics, **TDS** is the Training Dataset, **P** is Precision, **R** is Recall, and **F1** is the F-measure.

Methods		Isolated learning LDA			Lifelong learning AMC [10]			Lifelong learning our method		
NoT	TDS	P %	R %	F1 %	P %	R %	F1 %	P %	R %	F1 %
10	D_{4a}	63.26	50.38	56.09	65.19	52.42	58.11	62.94	54.72	58.54
	D_{4b}	70.68	55.24	62.01	74.68	58.52	65.62	75.32	59.03	66.19
	D_{4c}	80.72	67.43	73.48	82.50	67.18	74.05	83.07	67.43	74.44
	D_{4d}	82.42	68.96	75.09	84.38	68.7	75.74	85.05	69.47	76.47
	D_{4e}	82.31	71.50	**76.53**	83.58	71.25	**76.92**	83.28	72.99	**77.80**
15	D_{4a}	62.94	50.13	55.81	63.14	52.72	57.46	62.18	53.64	57.59
	D_{4b}	71.13	54.76	61.88	73.70	57.76	64.76	73.70	58.36	65.14
	D_{4c}	84.01	68.19	75.28	84.01	68.19	75.28	84.01	69.19	75.89
	D_{4d}	84.47	69.21	76.08	84.74	69.21	76.19	84.74	69.21	76.19
	D_{4e}	82.26	72.11	**76.85**	83.58	71.25	**76.92**	84.52	72.26	**77.91**
20	D_{4a}	62.94	50.13	55.81	63.14	51.27	56.59	62.50	52.62	57.14
	D_{4b}	72.70	55.76	63.11	73.70	57.76	64.76	74.03	58.02	65.05
	D_{4c}	84.01	68.19	75.28	84.01	68.19	75.28	84.01	68.19	75.28
	D_{4d}	84.47	69.21	76.08	84.74	69.21	76.19	84.78	69.47	76.36
	D_{4e}	84.23	72.01	**77.64**	83.58	73.68	**78.32**	84.82	73.52	**78.77**
25	D_{4a}	61.18	49.17	54.53	63.14	50.13	55.89	62.50	51.62	56.54
	D_{4b}	72.43	54.70	62.33	73.70	57.76	64.76	73.38	57.51	64.48
	D_{4c}	83.93	67.12	74.59	84.01	68.19	75.28	84.01	68.19	75.28
	D_{4d}	84.35	69.42	76.16	84.74	69.21	76.19	84.78	69.47	76.36
	D_{4e}	83.93	71.52	**77.23**	83.58	71.25	**76.92**	84.82	72.52	**78.19**

In the isolated LDA learning, the best result is 77.64% in the experiment with 20 topics on the training dataset D_{4e} of 1000 reviews.

The performance of the systems is improved in the approach of building topics of current dataset by the AMC modeling in [10] for most cases, and by our proposed approach for all cases. When the size of the current dataset is small (D_{4a}, D_{4b}), the improvement is about 2%. This proves the meaning of the lifelong learning approach.

Our proposed model even gets better results than those of AMC approach in [10] for all groups of experiments. In all cases, the improvement is about 1%. The performances indicate that the size of current dataset affects the performance of classification. In more details, our system outperforms in the smaller number of reviews in training dataset. It means that the features built from our approach give the best support for the classification.

On the other side, our proposed model gets the best result of 78.77% with 20 topics. When the number of topics reaches 25, the system performance in the three approaches (isolated LDA learning, the AMC, and ours) seems to decrease. This can be explained that the larger number of topics affects the relation among topics and the closeness between domain pairs. Therefore, it lowers the bias in learning approach. The reason may also be the characteristics of the datasets, and the above configurations of threshold and 20 topics are suitable for our data.

One-sample test for the hypothesized population mean with the σ unknown in Table 4 shows that the system performance improvement of our method against LDA is about 1.15, and against AMC is about 0.39.

Table 4. One-sample test for the hypothesized population mean with the σ unknown

Parameter	Our method vs. LDA	Our method vs. AMC
One-sample mean $\bar{x}$	1.3875	0.419
One-sample standard deviation s	1.041695133	0.3696065476
The hypothesized population mean μ_o	1.15	0.39
The size of sample n	20	20
One-sample test for mean, σ unknown t	1.019619134	0.3508918972

4 Related Work

Thrun [7] proposed the supervised lifelong machine learning algorithm Explanation-Based Neural Network (EBNN), which does learning bias on both meta-level and base-level. At the meta-level, previous datasets are compiled into the domain knowledge (meta-knowledge), in which training examples are explained and analyzed. At the base-level, the useful (slope) information from the domain knowledge is incorporated into neural network learning. In our method, the knowledge base S plays a role of the domain knowledge, while the knowledge base S^* plays a role of the domain knowledge plus the slope (or bias) information.

The LTM algorithm, proposed in Chen and Liu [9], is an incorporation of LML into topic models for large current datasets. The LMT is the first lifelong modeling algorithm. The knowledge transferred from previous datasets to the current task is only must-links. The set of all previous topics (p-topics), which have symmetrized KL Divergence with a current topic is less then or equal to a threshold π (set to 7.0), is used to mine must-links as 2-length frequent patterns. By using the Pólya Urn procedure, must-links are used as the prior knowledge patterns for enriching current topics in Gibbs sampling.

The AMC algorithm proposed in Chen and Liu [10] uses both must-links and cannot-links. The mining of must-links in AMC is the same as the mining of must-links in LTM [9]. Since a giant set of cannot-links may be mined if using words in previous datasets, then AMC focuses only on pairs of top words in the current dataset. A word pair is called a cannot-link if the number of previous *Topics*, in which the two words are in two different *p-topics*, is bigger than the number of previous *Topics*, in which the two words are in the same *p-topics*. By using the multi-generalized Pólya Urn procedure, must-links and cannot-links are used as the prior knowledge patterns for enriching the current topics in Gibbs sampling.

In comparison with [9, 10], our lifelong modeling method works like AMC except two main differences. Firstly, our method is a learning bias method on the domain level, i.e., only previous datasets close to the current dataset are chosen for enriching current topics. In Definition 3, the close dataset is defined on three levels, in which the close topic is defined based on a cosine measure on word bags. Semantic factor is expressed in cosine measure of close topics. Moreover, a bias learning is base on the close topics to select *p*-topics for mining must-links and cannot-links.

5 Conclusions and Future Work

This paper provides a lifelong topic modeling (LTM) method based on a *learning bias approach on domain level*, and a definition on *close topic* based on a cosine measure on word bags, and an application framework proposed LTM method for building multi-label classifiers on Vietnamese texts. Experimental results on three previous datasets, five cases of current datasets, with four different number of topics showed that the performance of our proposed method is better than that of AMC method for all cases.

More consideration should be studied to upgrade this work. Firstly, in some cases, the similarity measure defined by the cosine measure in formula (1) may be a little strict, thus a looser measure like Precision@K should be chosen. Secondly, the technique of mining cannot-links should be improved, for example, some techniques to mine 2-length negative patterns [5] may be suitable for this task.

References

1. Szymanski, P., Kajdanowicz, T., Kersting K.: How is a data-driven approach better than random choice in label space division for multi-label classification. Entropy **18**(8), 282, 30 p. (2016)
2. Zhou, Z.-H., Zhang, M.-L.: Multi-label learning. In: Sammut, C., Webb, G.I. (eds.) Encyclopedia of Machine Learning and Data Mining, pp. 875–881. Springer, Boston (2017). https://doi.org/10.1007/978-1-4899-7687-1
3. Zhang, M.-L., Wu, L.: LIFT: multi-label learning with label-specific features. IEEE Trans. Pattern Anal. Mach. Intell. **37**(1), 107–120 (2015)
4. Zhang, W., Tang, X., Yoshida, T.: TESC: an approach to text classification using semi-supervised clustering. Knowl.-Based Syst. **75**, 152–160 (2015)

5. Antonie, L., Li, J., Zaiane, O.: Negative association rules. In: Aggarwal, C.C., Han, J. (eds.) Frequent Pattern Mining, pp. 135–145. Springer, Cham (2014). https://doi.org/10.1007/978-3-319-07821-2_6
6. Thrun, S., Mitchell, T.M.: Lifelong robot learning. Robot. Auton. Syst. **15**(1–2), 25–46 (1995)
7. Thrun, S.: Explanation-Based Neural Network Learning: A Lifelong Learning Approach. Springer, US (1996). https://doi.org/10.1007/978-1-4613-1381-6
8. Chen, Z., Liu, B.: Lifelong Machine Learning. Morgan and Claypool Publishers, San Rafael (2016)
9. Chen, Z., Liu, B.: Topic modeling using topics from many domains, lifelong learning and big data. In: ICML 2014, pp. 703–711 (2014)
10. Chen, Z., Liu, B.: Mining topics in documents: standing on the shoulders of big data. In: KDD 2014, pp. 1116–1125 (2014)
11. Wang, S., Chen, Z., Liu, B.: Mining aspect-specific opinion using a holistic lifelong topic model. In: WWW 2016, pp. 167–176 (2016)
12. Chen, Z., Liu, B.: Topic models for NLP applications. In: Sammut, C., Webb, G.I. (eds.) Encyclopedia of Machine Learning and Data Mining, 2nd edn, pp. 1276–1280. Springer, Boston (2017). https://doi.org/10.1007/978-1-4899-7687-1
13. Pham, T.-N., Nguyen, V.-Q., Dinh, D.-T., Nguyen, T.-T., Ha, Q.-T.: MASS: a semi-supervised multi-label classification algorithm with specific features. In: Król, D., Nguyen, N.T., Shirai, K. (eds.) ACIIDS 2017. SCI, vol. 710, pp. 37–47. Springer, Cham (2017). https://doi.org/10.1007/978-3-319-56660-3_4
14. Pham, T.-N., Nguyen, V.-Q., Tran, V.-H., Nguyen, T.-T., Ha, Q.-T.: A semi-supervised multi-label classification framework with feature reduction and enrichment. J. Inf. Telecommun. **1**(2), 141–154 (2017)
15. Blei, D.M.: Probabilistic topic models. Commun. ACM **55**(4), 77–84 (2012)
16. Chen, Z., Ma, N., Liu, B.: Lifelong learning for sentiment classification. In: ACL, pp. 750–756 (2015)
17. Blei, D.M., Ng, A.Y., Jordan, M.I.: Latent Dirichlet allocation. J. Mach. Learn. Res. **3**, 993–1022 (2003)

An English-Vietnamese Translation System Using Artificial Intelligence Approach

Nguyen Van Binh(✉) and Huynh Cong Phap

School of Information and Communication Technology, The University of Danang, Da Nang, Vietnam
{nvbinh,hcphap}@sict.udn.vn

Abstract. The demand of automatic translation is truly emergent, especially in some narrow areas such as health, education, law, etc. However, the quality of machine translation (MT) in general and Vietnamese-related in particular is still very low compared to the need of the practice. There are many factors affecting the quality of machine translation systems, in which the two most important ones are the machine translation model and training corpus. Currently, statistical translation method has been applying for several Vietnamese translation systems, however their results have not been improved significantly. In this paper, we propose a solution of applying new machine translation model using neural network approach in combination with a large and good corpus. Based on the proposed solution, we have implemented and deployed a dedicated English-Vietnamese machine translation system for the legal documents. We have experimented our MT system by using a good corpus of legal domain with 460,000 pairs of Vietnamese-English sentences. Our MT system has produced results in the narrow field of legal documents much better than the two common English-Vietnamese translators using different translation models. Especially, it can translate quite correctly both specialized terms as well as long sentences in this field whereas others can't.

Keywords: Machine translation · Neural machine translation
Automatic translation quality · Vietnamese-related machine translation

1 Introduction

Many translation methods are now being researched and applied in translation systems to translate between hundreds of languages [1]. Common translation models are example-based, rule-based, and statistical. For common languages such as English, French, systems produce acceptable translations in a number of common fields [2], they can be used to refer to the meaning of the target language without translators. However, for under-resourced languages such as Vietnamese, the quality of machine translation (MT) systems is low, even unacceptable in some contexts. Currently, several machine translation systems using statistical or rule-based translation model such as Evtran, Co Viet, Hello Chao, Google translator, etc., supporting Vietnamese-related translation such as Vietnamese-English, Vietnamese-French. However, their translation outputs for Vietnamese-related are quite modest, especially in the specialized collection such as

N. T. Nguyen et al. (Eds.): ACIIDS 2018, LNAI 10751, pp. 211–220, 2018.
https://doi.org/10.1007/978-3-319-75417-8_20

medical, technical, legal documents. Some translation systems can't translate correctly the professional concepts, so the translated text becomes confusing, not valuable. Therefore, it is necessary to study and propose better appropriate translation models in order to improve the quality of the translation systems related to Vietnamese language.

In this article, we propose solutions to improve the quality of Vietnamese-related MT by applying the artificial intelligence method to the translation system, combined with building a large and good quality corpus. We also propose overall solutions to create and improve the quality of the corpus for Vietnamese-related MT. Based on that, we have developed an English-Vietnamese translation system in the field of legal documents and evaluated output results of our MT system with several most common ones to see the potential of the translation model using neural approach to solve the problem of improving the quality of Vietnamese-related MT.

2 State of the Art of Vietnamese-Related Machine Translation

2.1 Current State of Vietnamese-Related Machine Translation

There are many different approaches to improve the quality of statistical translation, including phrase-based, syntax-based statistical translation [3]. The phrase-based statistical translation model translates the source sentence into the target sentence by separating the source sentence into meaningful continuous word chains. Instead of translating on a word-by-word basis, each phrase will be translated into the corresponding phrase in the target sentence, followed by reversing the order of phrases to construct the sentence to be translated [4]. However, this phrase-based MT model may lack information about language such as morphology, part-of-speech tagging, context, etc., therefore the translation system may not translate words that do not appear in the training data set. The new models have proposed a solution that would allow addition of classes of factor information to the training data and such information would be processed through the translation model.

Concerning solutions to build Vietnamese-related MT systems, there are works that have been implemented such as example-based MT, statistical MT, expert-based MT (UNL) [3, 4], etc. However, almost Vietnamese-related MT systems using these solutions associating with small corpora has yielded very modest results, especially in specialized fields.

For automatic translation related to Vietnamese, in addition to Google and Microsoft translation systems, there are also a number of translation systems dedicating to Vietnamese-related such as Co Viet, Evtran, Hello Chao. The Co Viet system provides dictionaries, text translation and online English learning services. The translation system has classified input text to improve the quality of translation, including the fields of computer science, mathematics, accounting, medicine and technology. EVTRAN (some versions known as EV-Shuttle) is a rule-based MT system developed in the 1990s, using hand-built rules to translate text from English to Vietnamese. This is a product that has been commercialized and now the online version is provided by VDict System. However up to now, there is no automatic translation system that dedicatedly supports the good quality of Vietnamese-related translation of documents in the field of legal.

2.2 Evaluation of Vietnamese-Related Machine Translation Quality

The current MT systems still give very modest results: the translation information is incomplete compared to the original, the translated text is not fluent, the grammar is not perfect, the context is not suitable, vocabulary is incomplete. There have been many studies evaluating the quality of current translation systems. Some language pairs result in better translation than other pairs, such as English - French or English – Italian. Studies have also shown that translation of short sentences often produce better results than translation of long sentences [2]. When comparing computer translations and human translations, study at [5] shows that MT systems only produce good translation results when translating individual words or phrases, but poor results for long and complex sentences. Therefore, translation systems can only be used to refer to and collect information about translations. For translation in the specialized fields, study at [6] has evaluated the use of translation systems in medical field, namely communication between the doctor and the patient. The results show that only 57.7% of translated sentences give accurate results, many meaningless sentences or the results are completely wrong against the original content. That shows the limitation in the current automatic translation systems when translating narrow specialized subjects.

In relation to Vietnamese-related automatic translation, we have evaluated the translation quality in a specialized field of the current Vietnamese-related MT systems being used by many users. We use an English-Vietnamese bilingual data for evaluation. These data sets contain 475 pairs of sentences, extracting from legal documents. English sentences are translated into Vietnamese via APIs of the Google and Microsoft systems, using a tool developed by the author team and we collected translated sentences manually for others system. To evaluate the results, we use the Asiya [7] system to measure BLEU, NIST and WER scores between translations and references. The results are shown in Table 1.

Table 1. Evaluation of translation results.

Systems/score	BLEU	NIST	WER
Google	0.40	7.91	0.48
Microsoft	0.36	6.85	0.54
Co Viet	0.27	5.62	0.68
Evtran	0.11	3.32	0.93

Through these figures, it can be seen that the quality of the translations is very low compared to the assessment of the translation quality of common sentences such as English - French and other languages. When looking at results of the translation, we find out many terms are translated incorrectly, so the translation becomes difficult to understand. The translation results can only be used for reference purposes, not be used for important tasks which need precise semantics. Some of the major reasons for these limitations are:

- Inappropriate translation methods: Traditional translation models have many advantages, but when applied to Vietnamese translation, they are still limited, need more evaluation and additional research. Vietnamese differs from several other languages, each word consists of many syllables, while systems work on a single unit, thus reducing the effectiveness of these translation models. Translation methods such as statistical translations only translate well for phrases, short sentences, but less effective when translating long and complex sentences.
- Insufficient corpus: The study at [8] indicates that quality and quantity of the corpus affect the quality of MT systems. However, Vietnamese corpus used for training for automatic translation systems is incomplete, so some words may not be recognized by the systems. Especially in narrow specialized fields such as medical, technical, legal documents, concepts are of great importance, but the systems may still not translate correctly, making translations difficult to understand. There are a number of research projects aiming to improve corpora in quantitative and qualitative terms, unify corpora [9], but not many practical applications.

3 Solution of Improving MT Quality Using Neural Network Machine Translation Model

Selecting a MT model will determine the quality of translation system, thus using the optimal translation model will result in closer translation to human language. In this session, we propose an approach using automatic translation using artificial intelligence (neural network MT) combining a large corpus to improve the quality of Vietnamese-related translation system.

3.1 Solution of Applying Neural Network Machine Translation Model

Neural Network Machine Translation (NMT) has begun to be studied in recent years and is highly appreciated when testing with language pairs such as English - French, English – German [10]. NMT is usually a large-scale neural network that has been trained, stores vector expressing information between words in context, thus being able to translate well long sentences. Unlike the traditional phrase-based translation model, depending on specific phrases for translation and pairing as statistical translation models, NMT will train from input data to create a large-scale neural network which can read source sentences and reproduce target sentences based on the principle of recurrent neural networks. NMT models have an encoder - decoder in which the encoder reads source sentences and builds corresponding information vector, based on which decoder will reproduce the sentence to be translated on the basis of calculating the maximum probability at the output (Fig. 1).

Basically, the general principle of RNN is a neural network capable of handling sequence information (such as one sentence is a sequence of words), in which the state of output at the current time is calculated depending on the outcome of the previous state. Thus, RNN is a memory model and can remember the information of any sequence of any length [11].

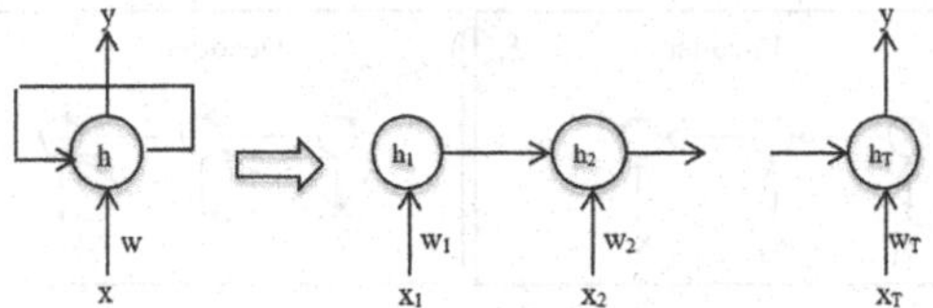

Fig. 1. A recurrent neural network

The RNN model consists of the hidden state h and gives output y when input source sequence $x = (x1, x2, \ldots xT)$. At each time step t, the hidden state $h_{\langle t\rangle}$ of the RNN model is updated according to the formula:

$$h_{\langle t\rangle} = f\left(h_{\langle t-1\rangle}, x_t\right)$$

where f is a non-linear activation function (*tanh, sigmoid*).

From the input training data, the RNN can learn the probability distribution of sequences and predict the next word in a sequence. At time t, the probability of occurrence of the sequence x_t is $p(x_t|x_{t-1} \ldots x_1)$. Considering that in the K word set, the probability of each word appearing is calculated by:

$$p(x_{t,j} = 1|x_{t-1}, \ldots, x_1) = \frac{\exp\left(w_j h_{\langle t\rangle}\right)}{\sum_{j'=1}^{K} \exp\left(w_{j'} h_{\langle t\rangle}\right)}$$

where w_j are the rows of a weight matrix W. Then, the probability that the string x will appear is:

$$p(x) = \prod_{t=1}^{T} p(x_t|x_{t-1}, \ldots, x_1)$$

From the probability distribution model learned, RNN will reproduce step-by-step output sentences through hidden states of the model. Study at [12] builds encoder - decoder to handle any input and output sentence case of any length, encodes sentence and expresses through a fixed length vector by using the extra c matrices to store total context. Then the hidden state at time t will be updated by the formula:

$$h_{\langle t\rangle} = f\left(h_{\langle t-1\rangle}, y_{t-1}, c\right)$$

Similarly, the probability of the condition for appearing the next word y_t will be dependent on the total state c as follows:

$$P(y_t|y_{t-1}, y_{t-2}, \ldots, y_1, c) = g\left(h_{\langle t\rangle}, y_{t-1}, c\right)$$

where f, g are non-linear activation functions.

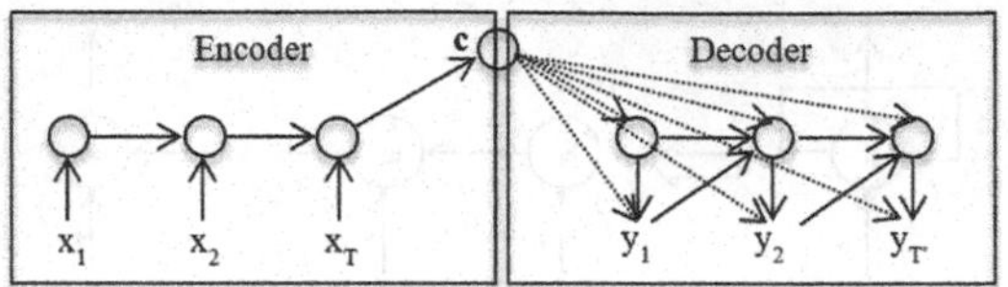

Fig. 2. RNN Encoder–Decoder

Figure 2 depicts the architecture of this RNN Encoder-Decoder model. The encoder and decoder components are trained to find parameters of the model so that the predicted output sequences have the maximum probability:

$$\max_{\theta} \frac{1}{N} \sum_{n=1}^{N} \log p_{\theta}(y_n | x_n)$$

Of which, θ is the parameter of the RNN model and (x_n, y_n) are pairs of training data, in this case bilingual sentence pairs.

In addition, many studies have been undertaken to further improve the RNN model to fit the automatic translation system and to improve the quality of translation systems [11, 13]. While the old algorithm is based on phrases that split sentences into individual words and phrases to translate in a separate way, then the NMT looks at the whole input sentence as a unitary translation. Today, research organizations and major vendors of automatic translation systems have begun to focus on applying neural network modeling to improve the quality of automatic translation for their MT systems. However, this research has only been applied to certain common language pairs.

With advantages of the translation system using neural networks, in this section, we will use this model to build the automatic translation system for pairs of English - Vietnamese sentences in the narrow specialized field for translation of legal documents.

3.2 Solution of Creating and Using a Large Corpus

As discussed above, NMT gives good results and is more likely to replace traditional models. However, a major problem when applying the NMT translation model is how big of a corpus to be used. The quality and volume of used corpus crucially affect the quality of MT systems. In addition, NMT encrypts data through hidden neural layers, using multiple attributes to represent information, thus requiring very large corpus for training to yield satisfactory results. As a result, the application of NMT in the studies on automatic translation for less popular languages such as Vietnamese will have trouble regarding building and standardization of corpus.

In order to build a bilingual corpus of large volume and good quality, in this paper we propose some solutions to build a large corpus based on the synthesis of bilingual data sources and eliminating the data of poor quality.

- Solution of data extraction from trusted sources: At present, there are many sources of bilingual data that have been popularized and can be used as a corpus for MT systems. These data sources include: bilingual websites in a variety of fields such as news, tourism, technology; bilingual language learning materials; translation documents between Vietnamese and other languages; technical translations, film scripts. Using methods of extracting information, we will collect these data sources. It is important to have effective solutions to check similarity between two sentences, separate paragraphs and separate sentences correctly.

- Unification of existing corpora: There are many corpora developed by organizations, individuals, and research groups, but the structure and format of these corpora are different and do not follow a unified standard. It is important to study and propose a common structure of corpus and to build a tool for converting discrete corpora into a unified corpus. The study at [8] suggested some solutions to link data as well as to unify the format and structure of existing corpora. After the unification process, the corpus will be expanded and the size will increase dramatically.
- Solution of corpus expansion by process of post-editing: The process of post-editing translations from the automatic translation system is done by users or experts to help perfection of translation quality. We suggest a solution for building an automatic translation system that integrates translation post-editing functions and encourages users to edit the translations they receive. Post-editing results are considered as new pairs of bilingual sentences and will be updated in the corpus, from which the corpus will gradually be expanded and enhanced the quality by the human direct check and edit.

4 Implementation, Experiment and Evaluation of Results

In the face of globalization and integration trends, Vietnamese enterprises and companies have a need to understand international legal issues when expanding markets and connecting foreign partners. On the contrary, foreign businesses need information about the market as well as regulations in Vietnam before investment. At the same time, legal documents are usually only published in one language, such as English or Vietnamese. At present, some government websites provide translations in English from Vietnamese documents. However, the number of translations is very small, being not able to meet the need to learn administrative or legal information, the law of users as well as businesses. If using human translation services to translate all related documents, it will be extremely costly, time consuming, whereas many contents are overlapped.

Therefore, we propose to build an English-Vietnamese translation system in the field of legal documents using the neural network model combining a large corpus as analyzed above. We also find that the application of neural networks to construct automatic translation systems has been studied in many languages, but so far there have been no concrete experiments in Vietnamese. In this section, after the construction of the translation system, we present the deployment of quality evaluation on the results obtained.

- Machine translation system implementation: As mentioned, we propose the application of computerized neural network translation method to build the automatic translation system for Vietnamese-related to exploit the power of artificial intelligence, bring the best effect to the translation system. In our implementation, we use open-source OpenNMT [14] written by Systran company, a unit in the field of automatic translation, in collaboration with Harvard University's Natural Language Processing Group, for using in automatic translation studies with neural networks. Systran is currently running an OpenNMT automatic translation system, but not yet available for English - Vietnamese language pairs. OpenNMT uses the Torch learning library on the Lua programming language platform, built on traditional NMT modeling enhancements, allowing automatic modeling to observe the entire input sequence to initialize new words in the output. Thus, the translation system can produce better results for long sentences. At the same time, OpenNMT enables memory optimization, accelerates computing when using GPU graphics processors. Our experimental system uses OpenNMT version 0.7 installed on Ubuntu 14.04 operating system on a computer with 1 GPU. For user interface, we built web system using Apache server and ZeroMQ.

- Corpus building: To prepare a corpus for experimenting the implemented MT system, we have collected a corpus of approximately 460,000 pairs of Vietnamese-English sentences from some websites containing bilingual legal documents, such as vbpl.vn, thuvienphapluat.vn. We also use the English-Vietnamese bilingual database OpenSubtitles2016 [15] which is a corpus extracted from the dialogue of the movies which has been translated into Vietnamese.
- Training parameters: we use default parameters of OpenNMT, using 2 RNN layers, word embeddings size 500, vocabulary size 50,000 and train with 13 epochs. The result show perplexity PPL = 5.69 (at epoch 13) (Fig. 3).

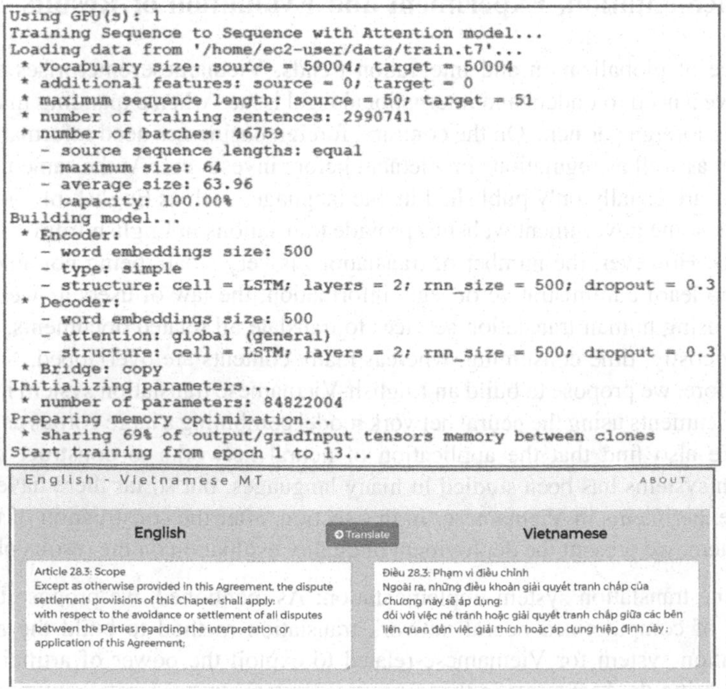

Fig. 3. Training process and user interface

After successfully implementing a legal-dedicated MT system for English-Vietnamese using the neural network model associating with a large corpus, we use the same test sets related to legal documents as described above to experiment and evaluate the translation quality of our system. Translation results are obtained through the OpenNMT "th translate.lua" command, using the model trained in the previous step. We use BLUE, NIST and WER scores to evaluate our system comparing with the two most common English-Vietnamese translators: Co Viet (http://tratu.coviet.vn) and EVTran (https://vdict.com/#translation) (Table 2).

Table 2. Evaluation results

	BLEU	NIST	WER
Our system	0.29	5.78	0.63
Co Viet	0.27	5.62	0.68
Evtran	0.11	3.32	0.93

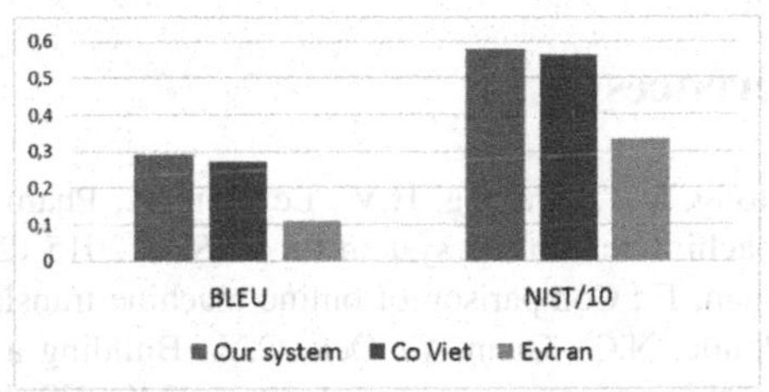

As shown in the evaluation results, our English-Vietnamese MT system based on the neural network model associated with a quite large and good corpus has produced better translation quality than others using different translation models. Especially, with the collected corpus of narrow field like legal documents, our translation system can translate correctly most of the terms, especially long and complex sentences in this field, while others still have very bad quality. However, the translation quality of our MT system is still modest in comparison with other big MT systems or real life demands because the used corpus that we have collected is still small for the need of the neural network machine translation model. Therefore, our MT system probably has good translation quality if the used corpus has larger volume.

5 Conclusion

The quality of Vietnamese-related MT is still very low compared to the need of the practice. It's thus necessary to continue to invest in research to improve two important issues: translation approach and training corpus. Currently, statistical translation method has been applying for several Vietnamese translation systems, however their results have not been improved significantly. In this paper, we have proposed our solutions to improve the quality of MT by applying the neural network translation model in combination with the large and good corpus. We have implemented a Vietnamese-English MT system, using our proposed solutions, dedicating to text translation of legal domain. We have experimented our MT system using the good corpus of legal domain with 460,000 pairs of Vietnamese-English sentences. Our MT system has produced much better results in the narrow field as legal documents comparing to the two most common English-Vietnamese translators using different translation models. Especially, it can translate quite correctly both specialized terms as well as long sentences in this field whereas others can't. However, the translation quality of our MT system is still modest comparing to the user expectation because of the limited volume of used corpus. Therefore, with this experiment we can conclude that our solutions using the neural network translation model in combination with the large and good corpus is the good approach for improving the quality of Vietnamese-related machine translation.

In our perspective, we will continue to expand the collected corpus in terms of quality as well as volume in order to improve the translation quality of our MT system with neural network translation approach.

References

1. Hong, V.T., Thuong, H.V., Le Tien, T., Pham, L.N., Van, V.N.: The English-Vietnamese machine translation system for IWSLT 2015 (2015)
2. Shen, E.: Comparison of online machine translation tools. Tcworld. Czerwiec (2010)
3. Phuoc, N.Q., Quan, Y., Ock, C.Y.: Building a bidirectional English-Vietnamese statistical machine translation system by using MOSES. Int. J. Comput. Electr. Eng. **8**(2), 161 (2016)
4. Nguyen, M.Q., Tran, D.H., Le Pham, T.A.: Using example-based machine translation for English–Vietnamese translation
5. Li, H., Graesser, A.C., Cai, Z.: Comparison of Google translation with human translation. In: FLAIRS Conference, May 2014
6. Patil, S., Davies, P.: Use of Google Translate in medical communication: evaluation of accuracy. BMJ **349**, g7392 (2014)
7. Giménez, J., Màrquez, L.: Asiya: an open toolkit for automatic machine translation (meta-) evaluation. Prague Bull. Math. Linguist. **94**, 77–86 (2010)
8. Pháp, H.C.: Solutions of creating large data resources in natural language processing. In: Król, D., Madeyski, L., Nguyen, N.T. (eds.) Recent Developments in Intelligent Information and Database Systems. SCI, vol. 642, pp. 243–253. Springer, Cham (2016). https://doi.org/10.1007/978-3-319-31277-4_21
9. Pháp, H.C., Thọ, Đ.Đ., Bình, N.V.: Cải tiến chất lượng dịch tự động bằng giải pháp mở rộng kho ngữ liệu. In: Proceeding of Publishing House for Science and Technology (2016)
10. Kalchbrenner, N., Blunsom, P.: Recurrent continuous translation models. In: EMNLP, vol. 3, no. 39, p. 413, October 2013
11. Sutskever, I., Vinyals, O., Le, Q.V.: Sequence to sequence learning with neural networks. In: Advances in Neural Information Processing Systems, pp. 3104–3112 (2014)
12. Cho, K., Van Merriënboer, B., Gulcehre, C., Bahdanau, D., Bougares, F., Schwenk, H., Bengio, Y.: Learning phrase representations using RNN encoder-decoder for statistical machine translation. arXiv preprint arXiv:1406.1078 (2014)
13. Bahdanau, D., Cho, K., Bengio, Y.: Neural machine translation by jointly learning to align and translate. arXiv preprint arXiv:1409.0473 (2014)
14. Klein, G., Kim, Y., Deng, Y., Senellart, J., Rush, A.M.: OpenNMT: open-source toolkit for neural machine translation. arXiv preprint arXiv:1701.02810 (2017)
15. Lison, P., Tiedemann, J.: OpenSubtitles2016: extracting large parallel corpora from movie and TV subtitles. In: LREC, May 2016

Machine Learning and Data Mining

Classification of Bird Sounds Using Codebook Features

Alfonso B. Labao, Mark A. Clutario, and Prospero C. Naval Jr.(✉)

Computer Vision and Machine Intelligence Group, Department of Computer Science, College of Engineering, University of the Philippines Diliman, Quezon City, Philippines
pcnaval@dcs.upd.edu.ph

Abstract. We propose a one-step codebook of frequency based-features to classify bird sounds given Random Forest and Support Vector Machine classifiers. The dataset consists of bird sounds from seventeen (17) bird species, with strong similarities in half of the sounds for a human listener. The codebook acts as a global dictionary that extends extracted sound features in one step from raw audio files and creates clusters to form a high-dimensional feature probability distribution. The one-step codebook approach is compared with other traditional audio features - resulting in six different feature sets. Results indicate that using simple mean frequency and bandwidth or even their multi-modal histogram versions are not accurate enough, performing below 50% if applied to larger 17 class datasets. Accuracies increase if the audio signal's spectral data is transformed to MFCC. The codebook approach on MFCC features with a Random Forest classifier performs best with an accuracy of 93.62%, and with good prediction results for almost all classes.

Keyword: Bird sound classification

1 Introduction

The problem of classifying the species of a bird audio recording has been studied in several papers [3,5,9,10,13,19]. Several recent methods use deep learning [1,18]. But these deep network methods require heavy computation and larger memory requirements. Some recent works [2,16,17] for audio classification still rely on non deep-learning approaches such as support vector machines or random forests for simpler deployment. An early automatic method is [9], which used several segmentation methods to extract the relevant features from a spectrogram transformation of bird audio. Another early method is by [10]. Through preprocessing of features, [9] was able to focus on those segments of a spectrogram that provide the highest discriminatory information for classification. However, the weaknesses of this approach have been identified by [3,19]. Here, having a supervised segmentation in the pre-processing step leads to inaccuracies since it requires knowing apriori the proper way to segment the spectrogram. However, given the wide range of possible spectrograms produced by bird recordings,

N. T. Nguyen et al. (Eds.): ACIIDS 2018, LNAI 10751, pp. 223–233, 2018.
https://doi.org/10.1007/978-3-319-75417-8_21

some relevant features in the spectrogram are left out by incorrect segmentation procedures. To address this problem, [19], used unsupervised feature extraction methods which can learn from data and automatically create new informative features without need for manual segmentation. Their approach uses spherical k-means in which the Mel-frequencies are clustered using the method devised by [6]. After clustering, support vectors of the class-separating sphere are derived, followed by computation of their dot-products with original features. This forms a higher-dimensional feature space which leads to better classification results. But [19] does not use any global dictionaries or codebooks, which could potentially increase accuracies. [3] on the other hand used histograms or a codebook approach in which either Mel-Frequencies Cepstral Coefficients or MFCCs are clustered to form the basis for higher-dimensional probability density functions which becomes the new feature space. MFCC's are further transformations of the audio data's Mel-frequencies, and are computed via log-cosine transformations of mel-spaced filter for dimension reduction. [3] however tested their methods on only six (6) species, while this paper extends the classes on a larger multi-class dataset composed of 17 bird species, along with a smaller training set vis-a-vis a larger test set. In addition, [3] used a two-step codebook clustering approach which first reduces the dimensionality of features into 10 clusters followed by 100 clusters. In this paper, 100 to 500 codebook clusters are directly formed from raw features - hence it is a "one-step" approach. Results indicate that features from the one-step codebook approach provide high classification accuracies even with a larger 17 species dataset. Codebook results are better than point-estimate features or non-codebook raw MFCCs. In summary, the contributions of this paper are:

- A "one-step" feature generation method for bird audio classification that rely on codebook clustering of raw audio data or MFCC information.
- A comparison of the "one-step" feature generation method with other feature sets composed of spectral center and bandwidth [9], histogram approach [3], and summarized MFCC information [19] on a wide 17-species test set given SVM and RF classifiers.

2 Data Sources

Data was derived from an open-source training dataset provided by Kaggle for their 2013 ICML Bird Classification Challenge [8]. The dataset is composed of 30-s audio recording of bird calls and songs produced by 35 bird species. Among the 35, 17 species are taken for the experiment, where 6 bird species, columba palumbus (wood pigeon), corvus corone (carrion crow), cuculus canorus (common cuckoo), garrulus glandarius (eurasian jay), phasianus colchicus (common pheasant), strix aluco (tawny owl) are easy to differentiate from the perspective of human hearing (since they produce very distinct sounds). But the remaining 11 species are harder to differentiate since they sound very similar. To create the testing and training datasets, each 30-s recording is split into 3-s clips. Each 3-s clip is ensured to contain audible bird sounds. Some clips that contained only silence are discarded, but the resulting dataset of split audio-clips still contained around 6 to 8 clips per species. The training dataset is set to be balanced, with each species containing 2 clips. One species, (corvus corone) - has only one (1) available clip, but in total, the training dataset numbers 33 clips. The test set has 94 clips, and is intentionally set to be larger than the training dataset to see if results show good generalization properties (Table 1).

Table 1. Bird species in dataset. Species that are easy to differentiate by humans are marked with asterisk (*).

Aegithalos caudatus (long-tailed tit)	Alauda arvensis (eurasian skylark)
Aarduelis chloris (european greenfinch)	Certhia brachydactyla (short-toed treecreeper)
Columba palumbus (wood pigeon)*	Corvus corone (carrion crow)*
Cuculus canorus (common cuckoo)*	Denrocopos major (great-spotted woodpecker)
Emberiza citrinella (yellowhammer)	Garrulus glandarius (eurasian jay)*
Parus caeruleus (eurasian blue tit)	Phasianus colchicus (common pheasant)*
Picus viridis (european green woodpecker)	Sitta europaea (european nuthatch)
Strix aluco (tawny owl)*	Sturnus vulgaris (common starling)
Turdus viscivorus (mistle thrush)	

3 Methodology

Raw audio data is not suitable as direct features for classification since the dimensionality of raw audio data is of an extremely large scale, and its feature vector centers are very close to each other as not to provide enough discriminatory information [19]. The usual method to address this is to first transform the data into a feature space of reduced dimensions, such as by transforming raw audio data into Short-Fourier Transformed spectrograms that depict the audio data's relative intensities on a frequency scale. Other methods involve Mel scale transformations and its cepstral-transformed cousin, the Mel Frequency Cepstral Coefficients (MFCC's), [21]. However, for some dimensionality reduction methods, the summarization of information may be too drastic to the point that critical information needed for classification is lost. This section describes different feature sets tested in the database. The first feature set, namely spectral center and bandwidths, represent features with drastically reduced dimensionality, to the point that they become summarized point estimates of a frame's position in the frequency range. While these features may be simplistic, [9] referred to these as very useful for bird audio classification. Afterwards, other features of higher dimensionality will then be considered, ending with the codebook approach (Table 2).

Table 2. Feature sets

1. Spectral center and bandwidth	2. Histogram of spectral center and bandwidth
3. Codebook of spectral densities	4. Codebook of Mel frequencies
5. Summarized MFCC coefficients	6. Codebook of MFCC coefficients

3.1 Feature Set 1: Spectral Center and Bandwidth

Feature 1 is a variant of the weighted spectral center and bandwidth proposed by [9], and is the simplest feature type. To derive the spectral center and bandwidth, each training 3-s audio recording is grouped to sample frames. Each sample frame is made up of 512 ms with a Hamming window of 75% overlap. Around 264 frames are produced for each audio recording sample. The spectrogram is constructed using a Discrete Short-Term Fourier Transform (STFT):

$$STFT(x[n])(m,w) \equiv X(m,w) = \sum_{n=-\inf}^{inf} x[n]w[n-m]e^{-jwn}$$

where $x[n]$ denotes the signal and $w[n]$ is the window. The magnitude squared of the STFT yields the spectrogram of the function:

$$spectrogram(x(t))(\tau,\alpha) \equiv |X(\tau,\alpha)|^2$$

The resulting spectrogram will then have a frequency range of 0–20 Khz, with each frame in the recording having their corresponding intensity in the form of relative log-form amplitude (expressed in decibels or dB). The STFT transformation is derived from the R package seewave [20]. The spectral center and bandwidth is derived from the absolute energy of each frame. Only those frames that belong to the upper 40% in terms of energy are retained, where energy is computed as: $Energy = |(ci)|^2$. This simple segmentation follows [3], and the weighted spectral center and weighted bandwidth is computed as:

$$MeanFrequency = \int \frac{freq(x)|c(x)|^2}{\sum |c(x)|^2} dx \; Bandwidth = \int \frac{(freq(x)-M)^2|c(x)|^2}{\sum |c(x)|^2} dx$$

3.2 Feature Set 2: Histogram of Spectral Center and Bandwidth

A simple feature set that transforms the spectral center and bandwidth to higher dimensions is the histogram approach proposed by [3]. It is a multi-modal, where

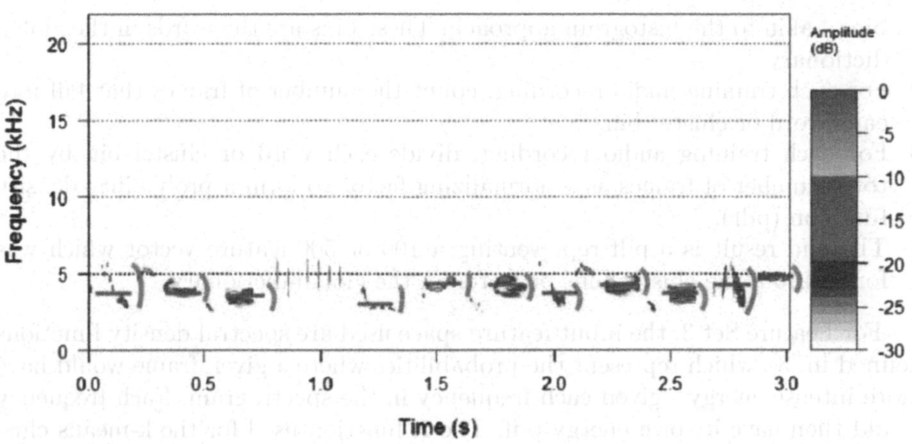

Fig. 1. Spectrogram for carduelis cordis with mean (red) and bandwidth (green) (Color figure online)

features are spread out across the histogram bins and conserves information lost from the point estimate of feature set 1. The histogram is constructed by splitting the range of frequencies into 100 bins, and the bandwidths are split into 50 bins. For each bin, the number of frames that meet the bandwidth and spectral center interval are counted. This creates a 5000-bin 2D histogram which is normalized to a probability-density-function (pdf) to form a higher-dimensional vector (Fig. 1).

3.3 Feature Set 3: Codebook Approach for Spectral Density

The codebook approach makes use of k-means clustering algorithm to group together similar features. K-means clustering involves two main steps, namely, the assignment and update steps:

Assignment Step: $s_i^t = x_p : ||x_p - m_i^t||^2 \leq ||x_p - m_j^t||^2 \quad \forall j, 1 \leq j \leq k$

Update Step: $m_i^{t+1} = \frac{1}{S_i^t} \sum_{x_j \in S_j^t} x_j$

Using the derived clusters, a global-dictionary is formed, where a word in the dictionary is referenced by a cluster, and represents a new dimension in the feature-space. The codebook approach expands an existing input feature space to more dimensions. [3] used a two-step clustering process to create its codebook. But in this experiment, the raw MFCC features are directly grouped into 100 clusters to form the global-dictionary. In summary, the following steps are used in the codebook approach implemented in this paper:

Summary of Codebook Steps:

1. Using an input feature space (e.g. could be any feature drawn from the spectrogram), group the features into clusters (either 100 clusters or 500 clusters) using the k-means clustering algorithm. The clusters would then form

bins - akin to the histogram approach. These bins are the words in the global dictionary.
2. For each training audio recording, count the number of frames that fall into each word or cluster bin.
3. For each training audio recording, divide each word or cluster-bin by the total number of frames as a normalizing factor to form a probability density function (pdf).
4. The end result is a pdf representing a 100 or 500-feature vector which was formedfrom the cluster-bins or words in the global-dictionary.

For Feature Set 3, the input feature space used are spectral density functions defined in [3], which represent the probabilities where a given frame would have more intense energy - given each frequency in the spectrogram. Each frequency would then have its own energy pdf. The R function used for the k-means clustering algorithm [12] is kcca from the R package flexclust [14]. Figure 2 shows a graphical representation of the spectral density probabilities in a spectrogram.

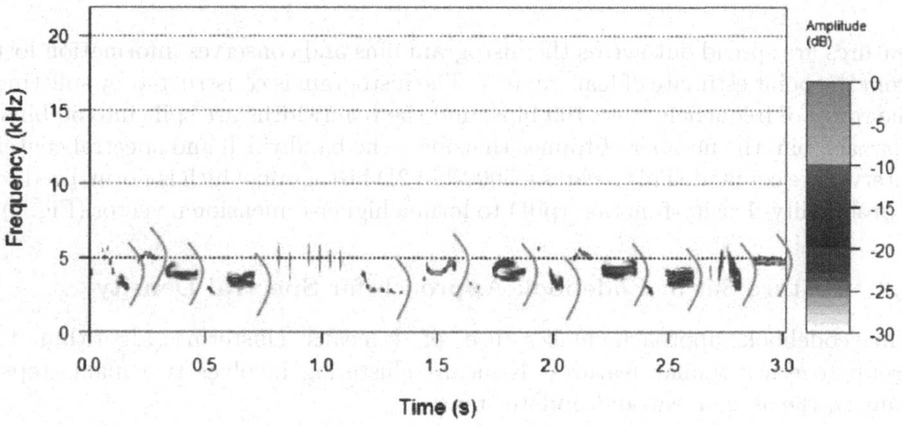

Fig. 2. Spectral-density representation (in the form of pdfs for each frequency)

3.4 Feature Set 4: Codebook for Mel Scale Transformations

The fourth feature approach consists of another codebook approach, but with the inputs being Mel scale transformations of raw audio data. As per [19], the Mel scale represents a higher-dimensional transformation from the usual frequencies expressed in Khz. It more closely represents the sensitivity of the human logarithmic scale of hearing. Computation of the Mel scale is done through the melfcc function, from the R package tuneR.

$$M(f) = 1125ln(1 + (f/700))$$

Where $M(f)$ denotes the Mel scale, and f is the frequency. The resulting feature set is a 40-element vector Mel-transformed auditory spectrum.

3.5 Feature Set 5: Summarized MFCC Information

Feature Set 5 are the derived Mel Frequency Cepstral Coefficients (MFCC) of the audio samples. MFCC's are of lower-dimensions than raw audio data, but contain highly discriminatory information since they are largely uncorrelated (similar to PCA-generated features). MFCC's are further transformations of the audio data's Mel-frequencies, and are computed via log-cosine transformations of mel-spaced filter. Briefly, computation of MFCC's involve the following steps from 1 to 7:

1. Divide the signal into frames. The length of each frame is set at 25 ms.
2. Take the Discrete Fourier Transform of the frame, while filtering out any signals that fall below 500 Hz (to take out noise).
3. Compute the mel-spaced filterbank. The standard is to apply 26 triangular filters to the DFT-transformed signal in step 2.
4. Take the log of filterbank energies to create 26 log filterbank energies
5. The 26 log filterbank energies can be assumed to take the form of a transformed signal, to compute for the MFCC coefficients, take the Discrete-Cosine-Transofrm of the log filterbank energies
6. Keep only the first 13 MFCC coefficients since they contain the most energy (and consequently information).

To form the features for classification, summarized point-estimate information of the MFCC's are taken, which are the mean, standard deviation, maximum, and minimum. [19] also recommended taking the same statistics of the MFCC deltas where the delta's represent the first-differences per frame.

3.6 Feature Set 6: Codebook of MFCC Coefficients

The last feature set implemented in this paper is to expand the MFCC summarized point-estimates to higher dimensions through a codebook approach. This is suggested in [3]. Application of the codebook approach to MFCC coefficients is straightforward, with the inputs being the first 13 MFCC coefficients, and with the number of clusters for the codebook set as 100.

4 Results and Discussion

As mentioned, the easy-to-classify samples from the perspective of human hearing are classes 5, 6, 7, 10, 12, 15. The other samples: 1, 2, 3, 4, 8, 9, 11, 13, 14, 16 and 17 are rather hard to differentiate since they are almost similar. Table 3 shows the accuracy results for each feature set given two strong classifiers, Random Forest (RF) [15] and Support Vector Machines (SVM), [7]. For RF, the R package randomForestSRC [11] is used, while the R package e1071 [] is used for computation of SVM (based on the libsvm package [4]). Accuracy is computed as the number of correctly identified samples divided by the total number of samples (94). Tables 4 and 5 provide the precision and recall measures for each feature set and class. Table 4 uses the Random Forest classifier, while Table 5 uses the Support Vector Machine. SVM hyperparameters within the e1071 R package have sigma set as 1, with a type set as 'C-classification' to denote a classification machine. Other parameters are set as the default of the svm function in e1071, with a radial kernel to transform the input. RF hyperparameters use default settings in the rfsrc function within the R rfsrc package, with the number of trees set at 10,000.

Table 3. Accuracy results for the different feature sets. 4a has 100 codebook clusters while 4b has 500 codebook clusters

Feature set	1	2	3	4a	4b	5	6
RF accuracy	28/94	47/94	55/94	73/94	76/94	84/94	88/94
SVM accuracy	24/94	5/94	57/94	64/94	73/94	85/94	82/94

Table 4. Precision (P) and Recall (R) per Feature Set (S) and Class using RF classifier

	Precision						Recall					
Class	S1	S2	S3	S4	S5	S6	S1	S2	S3	S4	S5	S6
1	33.33	50.00	16.67	50.00	57.14	83.33	16.67	16.67	50.00	66.67	66.67	83.33
2	36.36	57.14	57.14	100.00	100.00	100.00	50.00	100.00	50.00	87.50	87.50	100.00
3	75.00	54.55	66.67	100.00	100.00	85.71	50.00	100.00	33.33	100.00	100.00	100.00
4	75.00	100.00	100.00	71.43	66.67	80.00	60.00	20.00	20.00	100.00	80.00	80.00
5	50.00	100.00	100.00	1000.0	83.33	100.00	60.00	100.00	80.00	100.00	100.00	100.00
6	0.00	100.00	50.00	0.00	100.00	100.00	0.00	50.00	50.00	0.00	50.00	100.00
7	40.00	100.00	100.00	100.00	100.00	100.00	33.33	33.33	100.00	100.00	83.33	83.33
8	20.00	9.09	50.00	33.33	100.00	85.71	33.33	16.67	33.33	66.67	100.00	100.00
9	28.57	50.00	100.00	100.00	100.00	100.00	33.33	50.00	66.67	50.00	83.33	66.67
10	22.22	66.67	57.14	66.67	80.00	100.00	50.00	100.00	100.00	100.00	100.00	100.00
11	16.67	50.00	33.33	75.00	66.67	85.71	16.67	33.33	33.33	50.00	66.67	100.00
12	33.33	42.86	100.00	100.00	100.00	100.00	16.67	50.00	100.00	83.33	100.00	100.00
13	22.22	16.67	42.86	66.67	100.00	85.71	33.33	16.67	100.00	66.67	83.33	100.00
14	33.33	37.50	33.33	75.00	100.00	100.00	16.67	50.00	16.67	100.00	100.00	100.00
15	0.00	100.00	100.00	100.00	100.00	100.00	0.00	100.00	100.00	100.00	100.00	100.00
16	25.00	0.00	0.00	0.00	85.71	100.00	16.67	0.00	0.00	100.00	100.00	83.33
17	0.00	50.00	100.00	71.43	100.00	100.00	0.00	33.33	83.33	83.33	100.00	100.00

Results using only mean and bandwidth in feature set 1 perform poorly. Several of the bird species in the dataset are in the same range of frequencies and bandwidths in the spectrogram, and feature set 1 is too low dimensional and does not provide enough discriminatory information. Results feature set

2's histogram are better for the Random Forest classifier. But accuracies fall short, with an average below 50%. Several of the 0 precision and recall results from feature set 2 are among the harder-to-classify sounds (e.g. 4, 16, and 17). However, among the audibly-distinct sounds, there are also classes that had 0 precision and recall, such as classes 6 and 7. While [3] reported good results using the histogram, his data was limited to six (6) species which had different multi-modal mean frequencies and bandwidths. But in a larger multi-class problem of 17 species, the histogram approach may not provide enough discriminatory information since even the multi-modal range of bandwidths and frequencies remain similar for several birds. The spectral density codebook approach in feature set 3 is more effective than feature sets 1 and 2. Particularly, the audibly distinct sounds (i.e. classes 5, 6, 7, 10, 12, and 15) have very high precision and recall measures ranging from 75 to 100% average. But several of the harder-to-classify samples attained 0 precision and recall (for Random forest, it would be classes 11 and 16, while for SVM it would be classes 6 and 16). For class 11, SVM managed to outperform Random Forest, but at a very bad precision performance of 25% and recall of 17%.

Table 5. Precision (P) and Recall (R) per Feature Set (S) and Class using SVM classifier

	Precision						Recall					
Class	S1	S2	S3	S4	S5	S6	S1	S2	S3	S4	S5	S6
1	0.00	0.00	42.86	40.00	62.50	85.71	0.00	0.00	50.00	33.33	83.33	100.00
2	0.00	0.00	83.33	100.00	100.00	100.00	0.00	0.00	62.50	87.50	87.50	100.00
3	33.33	0.00	66.67	100.00	100.00	100.00	66.67	0.00	66.67	100.00	100.00	100.00
4	60.00	5.32	60.00	80.00	100.00	75.00	60.00	100.00	60.00	80.00	60.00	60.00
5	42.86	0.00	100.00	100.0	100.00	100.00	60.00	0.00	60.00	60.00	100.00	100.00
6	0.00	0.00	0.00	0.00	0.00	0.00	0.00	0.00	0.00	0.00	0.00	0.00
7	66.67	0.00	100.00	100.00	100.00	85.71	33.33	0.00	66.67	100.00	83.33	100.00
8	16.67	0.00	50.00	36.36	85.71	100.00	16.67	0.00	66.67	66.67	100.00	66.67
9	25.00	0.00	100.00	66.67	83.33	80.00	50.00	0.00	16.67	33.33	83.33	66.67
10	25.00	0.00	57.14	100.00	66.67	100.00	50.00	0.00	100.00	100.00	100.00	100.00
11	0.00	0.00	33.33	50.00	83.33	83.33	0.00	0.00	16.67	16.67	83.33	83.33
12	20.00	0.00	100.00	50.00	100.00	100.00	16.67	0.00	100.00	83.33	100.00	100.00
13	33.33	0.00	85.71	42.86	100.00	80.00	16.67	0.00	100.00	100.00	100.00	66.67
14	21.43	0.00	22.73	75.00	100.00	100.00	50.00	0.00	83.33	100.00	100.00	100.00
15	12.50	0.00	100.00	100.00	100.00	66.67	25.00	0.00	100.00	100.00	100.00	100.00
16	0.00	0.00	0.00	0.00	85.71	71.43	0.00	0.00	0.00	0.00	100.00	83.33
17	0.00	0.00	100.00	80.00	100.00	75.00	0.00	0.00	66.67	66.67	100.00	100.00

Feature set 4 uses a similar codebook approach as feature set 3, but with inputs as 40-dimensional Mel Frequencies. Results indicate that the Mel frequency codebook approach out-performed feature sets 1–3. But despite better results, there are still some classes that both classifiers find hard to predict, namely class 6 (which consists of a single training sample), and class 16 (which RF and SVM consistently failed to predict correctly). Feature set 5 consists of point summary estimates of MFCC data. The above results indicate that feature set 5 give out better results for SVM and RF even if they consist of just point estimates. This is a result of MFCC's uncorrelated properties similar to PCA. For MFCC features with a Random Forest, almost all of the easy-to-classify classes churned out very good precision and recall measures. However, even with uncorrelated features, there is still some unsatisfactory performance in class 6 for both Random Forest and SVM. For class 6, Random Forest gave out a low 50% recall, and SVM wasn't able to predict it correctly at all with 0 precision and recall. In feature set 5, the problem encountered before with the hard class 16 has been addressed, with both classifiers being able to predict it correctly above 80%. Feature Set 6 builds upon feature set 5 by applying the code-book approach to MFCC inputs. The best results obtained from Random Forest are in feature set 6, which obtained 88/94 accuracy. In addition, all of the easy-to-classify samples for Random Forest managed to obtain high precision and recall and class 6 managed to be predicted perfectly (with both 100% precision and recall). Several of the harder-to-classify samples are likewise accurately predicted by Random Forest, with classes 1, 2, 3, 8, 10, 11, 13, 14, 16, and 17 providing precision and recall measures that range from 85% to 100%.

5 Conclusion

Our experimental results show that classification performance improves through feature expansion of audio data features to higher dimensional space using a codebook approach. Here, a global-dictionary of words is formed, where each word corresponds to feature cluster. Using a 500-cluster codebook of spectral densities, accuracy metrics are above 70%. The best results among the features involved a codebook global-dictionary approach derived from 100 clusters of the first 13 MFCC's per frame, giving an accuracy of 93.67%, with correct prediction for challenging classes.

References

1. Adavanne, S., Parascandolo, G., Drossos, K., Virtanen, T., et al.: Convolutional recurrent neural networks for bird audio detection. arXiv preprint arXiv:1703.02317 (2017)
2. Bravo, C.J.C., Berríos, R.Á., Aide, T.M.: Species-specific audio detection: a comparison of three template-based classification algorithms using random forests. Technical report, PeerJ Preprints (2017)
3. Briggs, F., Raich, R., Fern, X.Z.: Audio classification of bird species: a statistical manifold approach. In: 2009 Ninth IEEE International Conference on Data Mining, pp. 51–60. IEEE (2009)
4. Chang, C.C., Lin, C.J.: LIBSVM: a library for support vector machines. ACM Trans. Intell. Syst. Technol. (TIST) **2**(3), 27 (2011)

5. Chen, Z., Maher, R.C.: Semi-automatic classification of bird vocalizations using spectral peak tracks. J. Acoust. Soc. Am. **120**(5), 2974–2984 (2006)
6. Coates, A., Ng, A.Y.: Learning feature representations with k-means. In: Montavon, G., Orr, G.B., Müller, K.-R. (eds.) Neural Networks: Tricks of the Trade. LNCS, vol. 7700, pp. 561–580. Springer, Heidelberg (2012). https://doi.org/10.1007/978-3-642-35289-8_30
7. Cortes, C., Vapnik, V.: Support-vector networks. Mach. Learn. **20**(3), 273–297 (1995)
8. Dugan, P., Cukierski, W., Shiu, Y., Rahaman, A., Clark, C.: Kaggle competition. Cornell Univerity, The ICML (2013)
9. Fagerlund, S.: Automatic recognition of bird species by their sounds. Ph.D. thesis, Helsinki University of Technology (2004)
10. Harma, A.: Automatic identification of bird species based on sinusoidal modeling of syllables. In: 2003 IEEE International Conference on Acoustics, Speech, and Signal Processing, (ICASSP 2003), vol. 5, pp. V–545. IEEE (2003)
11. Ishwaran, H., Kogalur, U.B., Blackstone, E.H., Lauer, M.S.: Random survival forests. Ann. Appl. Stat. 841–860 (2008)
12. Jain, A.K.: Data clustering: 50 years beyond k-means. Pattern Recogn. Lett. **31**(8), 651–666 (2010)
13. Lee, C.H., Lee, Y.K., Huang, R.Z.: Automatic recognition of bird songs using cepstral coefficients. J. Inf. Technol. Appl. **1**(1), 17–23 (2006)
14. Leisch, F., Dimitriadou, E.: flexclust: Flexible cluster algorithms. http://cran.r-project.org/package=flexclust, R package version 1.3-1. Cited on p. 194
15. Liaw, A., Wiener, M.: Classification and regression by randomforest. R News **2**(3), 18–22 (2002)
16. Ludeña-Choez, J., Quispe-Soncco, R., Gallardo-Antolín, A.: Bird sound spectrogram decomposition through non-negative matrix factorization for the acoustic classification of bird species. PLoS ONE **12**(6), e0179403 (2017)
17. Qian, K., Zhang, Z., Baird, A., Schuller, B.: Active learning for bird sounds classification. Acta Acust. United Acust. **103**(3), 361–364 (2017)
18. Sevilla, A., Bessonne, L., Glotin, H.: Audio bird classification with inception-v4 extended with time and time-frequency attention mechanisms. In: Working Notes of CLEF 2017 (2017)
19. Stowell, D., Plumbley, M.D.: Automatic large-scale classification of bird sounds is strongly improved by unsupervised feature learning. PeerJ **2**, e488 (2014)
20. Sueur, J., Aubin, T., Simonis, C.: Equipment review: seewave, a free modular tool for sound analysis and synthesis. Bioacoustics **18**(2), 213–226 (2008)
21. Zheng, F., Zhang, G., Song, Z.: Comparison of different implementations of MFCC. J. Comput. Sci. Technol. **16**(6), 582–589 (2001)

A Degenerate Agglomerative Hierarchical Clustering Algorithm for Community Detection

Antonio Maria Fiscarelli[1(✉)], Aleksandr Beliakov[1,2], Stanislav Konchenko[1,2], and Pascal Bouvry[2]

[1] C2DH, CSC-ILIAL, University of Luxembourg,
2, avenue de l'Université, 4365 Esch-sur-Alzette, Luxembourg
antonio.fiscarelli@uni.lu
[2] FSTC-CSC/ILIAS & SnT, University of Luxembourg,
2, avenue de l'Université, 4365 Esch-sur-Alzette, Luxembourg
pascal.bouvry@uni.lu
https://www.c2dh.uni.lu/people/antonio-fiscarelli.0.0,
http://staff.uni.lu/pascal.bouvry

Abstract. Community detection consists of grouping related vertices that usually show high intra-cluster connectivity and low inter-cluster connectivity. This is an important feature that many networks exhibit and detecting such communities can be challenging, especially when they are densely connected. The method we propose is a degenerate agglomerative hierarchical clustering algorithm (DAHCA) that aims at finding a community structure in networks. We tested this method using common classes of graph benchmarks and compared it to some state-of-the-art community detection algorithms.

Keywords: Community detection · Graph clustering · Graph theory

1 Introduction

Many complex systems such as social networks [1], the world wide web [2] and biological networks [3] can be represented using graphs. One of their many properties is the organisation into communities. Often different communities merge and form a hierarchical structure. Also, they differ in size and vertices show different degrees of connectivity. The community structure of a network can give access to relevant information about the dynamics of the network and its characteristics, this is why this has become a very relevant topic in computer science and other disciplines.

The method we propose is a degenerate agglomerative hierarchical clustering algorithm (DAHCA) that aims at finding community structures in networks. We have investigated how effectively our method can detect nested communities and discovered that it can detect both community and subcommunity structure. Next, we have compared our method to some of the state-of-the-art algorithms

N. T. Nguyen et al. (Eds.): ACIIDS 2018, LNAI 10751, pp. 234–242, 2018.
https://doi.org/10.1007/978-3-319-75417-8_22

on the Girvan-Newman benchmark [4] and discovered that it can effectively detect communities and in some cases outperform other algorithms. Finally, our method has been tested on the Zachary karate club network [5], a well known real-world network used as a benchmark for community detection algorithms.

2 Community Detection: Problem Definition and Related Work

This section presents some of the existing community detection algorithms. Community detection consists of grouping related vertices that usually show high intra-cluster connectivity and low inter-cluster connectivity. Many different methods have been proposed over the last years and contributions came from disciplines such as computer science, applied mathematics, physics, biology, economics and so on. However there is no best algorithm. Some algorithms simply perform better or are faster for different types of networks or different applications.

The method proposed by Newman and Girvan in [6] extends the definition of betweeness centrality to edges. Edges connecting communities will have a high edge betweeness and removing them will enhance the community structure of the network (BETW). On the other hand, Clauset et al. [7] use a modularity measure in order to define communities that have many edges within them and few between them (GREEDY). Furthermore, Raghavan et al. [8] use a decentralised technique based on the majority rule to assign vertices to clusters (LAB PROP). The method that Pons and Latapy [9] describe uses random walks in order to define communities. Generally, random walkers tend to stay more in the same community (TRAP). Rosvall and Bergstrom [10] approach the problem using an information theoretic point of view to discover communities by using the probability flow of random walks (INFOMAP). Finally, the method proposed by Newman [11] is based on the eigenspectrum of the modularity matrix in order to maximize the modularity measure (EIGEN).

The most used metric [12,13] to evaluate community detection algorithms is the Normalized Mutual Information (NMI): it measures the agreement between communities and clusters found by a community detection algorithm [14]. NMI = 1 corresponds to perfect assignments, while NMI = 0 corresponds to completely independent assignments. Completeness (COMP) measures how vertices of a community are assigned to the same cluster, while homogeneity (HOMOG) measures how every cluster contains only vertices of the same community. When all vertices are assigned to the same cluster HOMOG = 0 and COMP = 1, whereas if each vertex is assigned to a different cluster HOMOG = 1 and COMP = 0. The Adjusted Random Index (ARI) measures the similarity of the assignments [14]. It ranges from −1 to 1, where ARI = 1 corresponds to perfect assignments, ARI values near 0 correspond to bad assignments and negative values of ARI correspond to independent assignments.

3 The Community Detection Algorithm

DAHCA makes use of the reachability matrix and it contains information about the total number of paths between vertices. This was initially proposed in [15] and it was defined as

$$\mathbf{W} = \sum_{l=0}^{\infty} (\alpha\mathbf{A})^l = [\mathbf{I} - \alpha\mathbf{A}]^{-1} \tag{1}$$

where $\mathbf{A}$ is the adjacency matrix and $\mathbf{I}$ is the identity matrix. The parameter α is tuned so that longer paths contribute less and the sum converges. In our case we defined the reachability matrix as

$$\mathbf{A}^{\mathbf{l}} = \sum_{i=1}^{l} \mathbf{A}^{\mathbf{i}} \tag{2}$$

where every entry $\mathbf{a}^{\mathbf{l}}{}_{\mathbf{ij}}$ represents the exact number of 1-paths, 2-paths,..., l-paths connecting vertex i with vertex j. We decided to use three-length paths because in many networks, communities overlap. Overlapping vertices serve as bridge between them [16], where most vertices can reach others outside their community in just three hops. Numerical tests also confirmed that it performs best for $l = 3$. Every vertex is then characterised by its relative row entry in the reachability matrix: vertices belonging to the same community will be more likely to have common paths. DAHCA starts by assigning a different cluster to each vertex and a value which consists of the sum of its row entry elements. It then selects the vertex having the lowest value, computes the Euclidean distances between it and all its neighborhoods (non-zero entry elements) and assign it the cluster of the most similar vertex. The process iterates until all vertexes have been assigned to a cluster. Next, the algorithm merges vertices belonging to the same cluster in a new vertex, after which the reachability matrix is recomputed as follow:

$$\mathbf{a}^{\mathbf{l}}{}_{ij} = \frac{1}{|\mathbf{c}_i|} \sum_{k \in \mathbf{c}_i} \frac{1}{|\mathbf{c}_j|} \sum_{h \in \mathbf{c}_j} \mathbf{a}_{kh} \tag{3}$$

where $\mathbf{c}_i$ and $\mathbf{c}_j$ are the new clusters obtained and $\mathbf{a}_{kh}$ is an element of the adjacency matrix. Figure 1 shows one iteration of DAHCA. At each step a new cluster assignment will be found. The process iterates until change no longer occurs or until one single vertex remains. This can be seen as a degenerate agglomerative hierarchical clustering: each vertex starts with its own cluster and at each iteration clusters are merged until merging is no longer possible. It is different from a classical agglomerative clustering because more than two vertices can be merged together in one iteration and and it does not always end with a single cluster including all vertices.

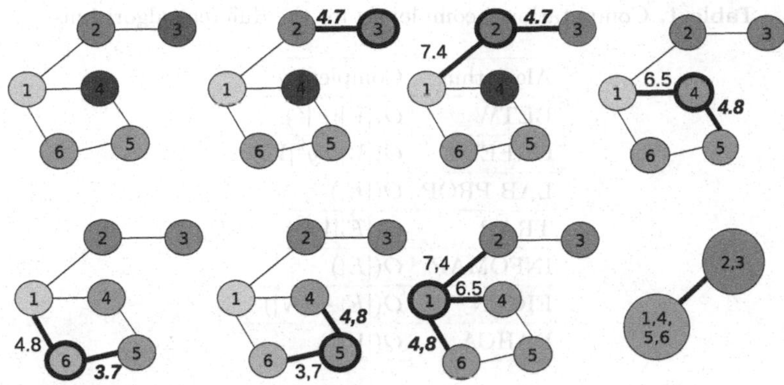

Fig. 1. This figure shows one iteration of DAHCA. A different cluster is initially assigned to each vertex. Vertices are then iteratively selected and they will be assigned the cluster of the closest vertex according to Euclidean distance. Nodes belonging to the same cluster are then merged in a single vertex.

3.1 Complexity

The computational complexity of a community detection algorithm is crucial, especially for large graphs and in cases where the graph is not completely accessible. The complexity analysis of DAHCA can be assessed looking at its phases separately.

- the reachability matrix can be computed in time $O(|V|^l)$ (where $l = 3$ in our case).
- the clustering process can be computed in time $O(|V|)$.
- the merging process can be computed in time $O(|V|^2)$.

Notice that the size of V only corresponds to the actual number of vertices during the first iteration. After that, they are merged according to the cluster they belong to, thus the size of V decreases significantly. The overall complexity of DAHCA is then $O(t(|V|^3 + |V|^2 + |V|)) \simeq O(|V|^3)$ where t is the number of iterations. Table 1 shows the time complexity for all algorithms discussed in this paper.

4 Experiments

We have evaluated how effectively DAHCA can detect both communities and the emergence of nested communities. To do so, we have used a similar benchmark as the one described in [16]. Networks have N number of vertices, are divided in G groups and each group is divided in C communities. Vertex connectivity is set to K, this means that each vertex will be connected to exactly K other vertices in the same community. Benchmark graphs have been generated with $N = 120$, $M = 3$, $C = 2$, $K = 5$. Every edge is then relinked with probability p_r. If so, the

Table 1. Computational complexity for the different algorithms.

Algorithm	Complexity
BETW	$O(\|V\|\|E\|^2)$
GREEDY	$O(\|V\|log^2\|V\|)$
LAB PROP	$O(\|E\|)$
TRAP	$O(\|E\|\|V\|^2)$
INFOMAP	$O(\|E\|)$
EIGEN	$O(\|E\| + \|N\|)$
DAHCA	$O(\|V\|^3)$

vertex is connected to another vertex in the same community with probability p_c, in the same group with probability $(1-p_c)p_g$ or to any vertex in the network with probability $(1-p_c)(1-p_g)$. Edges have been relinked with probability $p_r = 1.0$ and p_g = 0,7, while p_c was dynamically changed to simulate the emergence of communities (notice that this setting is slightly different from the one presented in [16]). For $p_c = 0$ there is no community structure and only the groups are defined, while for $p_c = 1$ the community structure emerges very clearly. Results are shown in Fig. 2. For low values of p_c DAHCA is able to identify the correct number of groups, while for higher values it is able to identify the correct number of groups as well as the correct number of communities.

We also evaluated DAHCA on the Girvan-Newman (GN) benchmark [4] and compared it to some state-of-the-art algorithms used for community detection. Networks have N vertices that are assigned to C equally sized communities. Each vertex has a fixed average degree z. A mixing parameter μ controls the portion of intra-community edges. For $\mu = 0$ communities are completely isolated, for μ = 0.5 vertices will be equally connected to vertices inside and outside their community, while for $\mu = 1$ vertices inside the same communities are not connected at all. Benchmark graphs have been generated with $N = 128$, $C = 4$ and $z = 16$, while μ was dynamically changed. The most used metric for community detection is the Normalized Mutual Information (NMI), but it does not necessarily return zero when the assignment is completely random. In that case it depends on the network size and number of communities [14]. This happens when an algorithm assigns each vertex to a different cluster or all vertexes to the same cluster. Therefore we decided to compute completeness and homogeneity to identify when an algorithm returns these naive assignments. For example, the INFOMAP algorithm scores NMI = 0 for high values of μ (Fig. 3a): one cannot say whether it is due to a very bad assignment, a random assignment or just a naive assignment. Also, using homogeneity and completeness it scores HOMOG = 0 and COMP = 1 (Fig. 3c and d) and clearly assigns every vertex to the same cluster. We also decided to use the ARI because, unlike the NMI, it is always independent of the network size and number of communities.

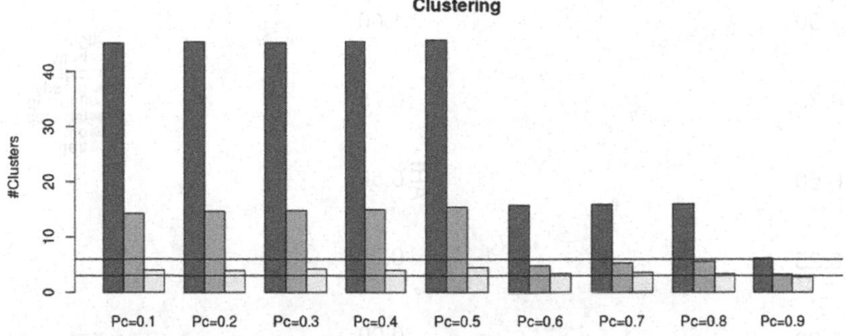

Fig. 2. Clustering over three consecutive iterations. Experiments for $N = 120$, $M = 3$, $C = 2$, $K = 5$, $p_r = 0.2$, $p_g = 0{,}7$. The probability p_c can be found on the x-axes and the number of clusters identified on the y-axes. Each barplot shows the results obtained for a specific p_c value and each single column represents the number of clusters identified at a certain iteration. The two horizontal lines correspond to the total number of groups (3) and the total number of communities (6) in the networks. Experiments have been run 200 times and results averaged.

For low values of μ DAHCA does not perform perfectly, unlike some of the other algorithms. However, for $\mu \in [0.3, 0.6]$ it outperforms GREEDY, INFOMAP, LAB PROP and EIGEN. For higher values of μ it outperforms all other algorithms but BETW. Furthermore, it exhibits an interesting behaviour: for $\mu \in [0.75, 1.0]$ there is an increase in performance. One would assume that performance should decrease for $\mu \geq 0.5$ because communities do not become evident, but as proved in [17] they are actually evident for $\mu \leq 0.75$. Over that range, the number of intercommunity edges becomes much higher than the number of intracommunity edges (an "anti-community" structure), with $\mu = 1.0$ being the point where there are no more edges inside communities. DAHCA is able to detect anti-communities which explains why DAHCA's performance increases.

Finally, we have evaluated DAHCA on a real-world network like the Zachary's karate club network, initially presented in [5] and known to be a vastly used benchmark for community detection algorithms. Every vertex represents members of the club, with 1 and 34 being the administrator and the director (the leaders of the two communities). Edges represent friendship between members. Results for all algorithms are presented in Figs. 4 and 5, while the result obtained using DAHCA is presented in Fig. 6. Nodes belonging to the same communities share the same colour. The algorithms obtain very different results, especially for the number of communities found. Firstly, all algorithms are able to identify vertices 1 and 34 as community leaders and assign them to different communities. Only DAHCA and LAB PROP are able to identify the core sets of vertices $\{1, 2, 3\}$ and $\{33, 44\}$ as nodes with higher connectivity [4] and assign them to different communities. To complete the analysis, Table 2 shows the numerical results of the different algorithms on the Zachary network.

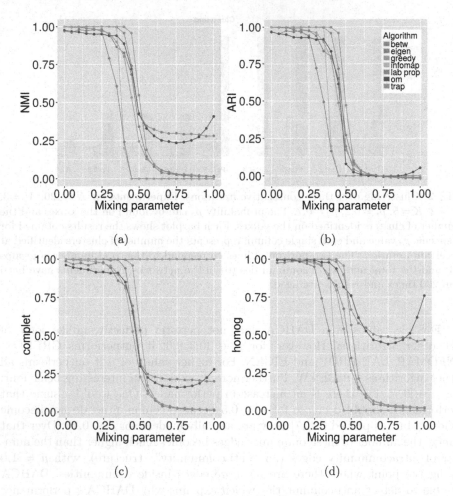

Fig. 3. Experiments on GN benchmark for $N = 128$, $C = 4$, $K = 16$. The Mixing parameter μ can be found on the x-axes and NMI, ARI, completeness and homogeneity on the y-axes. Experiments have been run 50 times and results averaged.

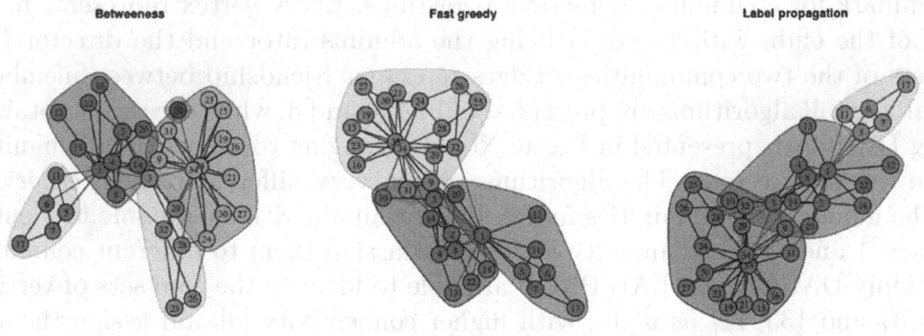

Fig. 4. Communities found by BETW, GREEDY and LAB on the Zachary's karate club network

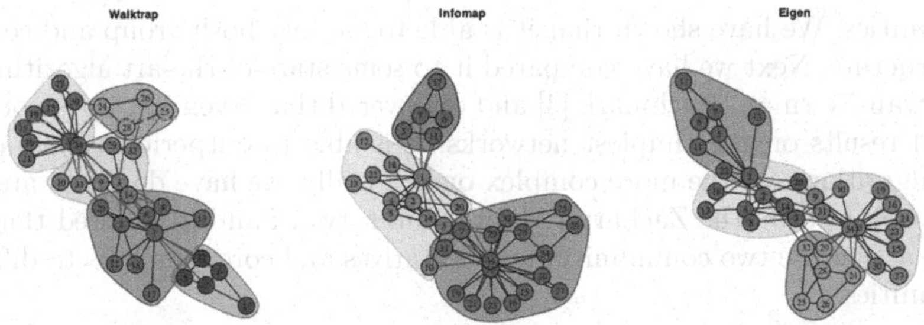

Fig. 5. Communities found by TRAP, INFOMAP and EIGEN on the Zachary's karate club network.

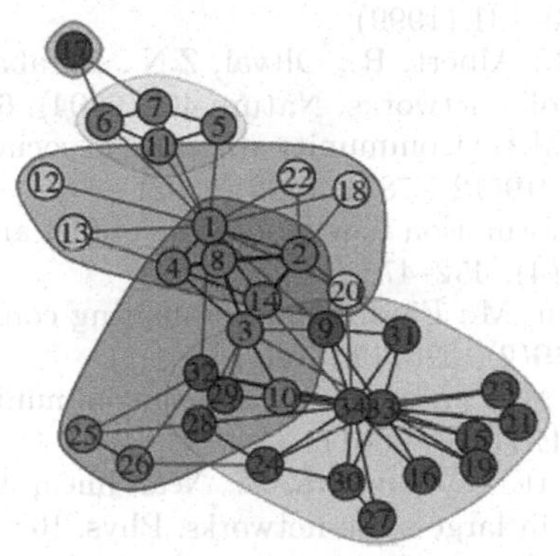

Fig. 6. Communities found by DAHCA on the Zachary's karate club network.

Table 2. Metrics for the different algorithms on the Zachary karate club network

	BETW	GREEDY	LAB PROP	TRAP	INFOMAP	EIGEN	DAHCA
NMI	0.579	0.692	0.548	0.504	0.699	0.677	0.512
ARI	0.469	0.680	0.467	0.333	0.702	0.512	0.452
V measure	0.580	0.692	0.548	0.504	0.699	0.677	0.512
Completeness	0.431	0.577	0.428	0.364	0.593	0.512	0.387
Homogeneity	0.885	0.866	0.763	0.822	0.854	1	0.756

5 Conclusions

In this paper we have proposed a degenerate agglomerative hierarchical clustering algorithm that makes use of the reachability matrix to detect community structures in networks and runs in $O(|V|^3)$. We have tested DAHCA on different benchmarks and settings. First, it has been tested on the benchmark used in [16] where networks are organised in groups and each group is organised in communities. We have shown that it is able to identify both group and community structure. Next we have compared it to some state-of-the-art algorithms on the Girvan-Newman benchmark [4] and discovered that, even if it does not show optimal results on the simplest networks, it is able to outperform most of the other algorithms for the more complex ones. Finally, we have demonstrated the results obtained on the Zachary's karate club network and discovered that it is able to assign the two community representatives and core members to different communities.

References

1. Wasserman, S., Faust, K.: Social Network Analysis: Methods and Applications, vol. 8. Cambridge University Press, Cambridge (1994)
2. Albert, R., Jeong, H., Barabási, A.L.: Internet: diameter of the world-wide web. Nature **401**(6749), 130–131 (1999)
3. Jeong, H., Tombor, B., Albert, R., Oltvai, Z.N., Barabási, A.L.: The large-scale organization of metabolic networks. Nature **407**(6804), 651–654 (2000)
4. Girvan, M., Newman, M.E.: Community structure in social and biological networks. Proc. Natl. Acad. Sci. **99**(12), 7821–7826 (2002)
5. Zachary, W.W.: An information flow model for conflict and fission in small groups. J. Anthropol. Res. **33**(4), 452–473 (1977)
6. Newman, M.E., Girvan, M.: Finding and evaluating community structure in networks. Phys. Rev. E **69**(2), 026113 (2004)
7. Clauset, A., Newman, M.E., Moore, C.: Finding community structure in very large networks. Phys. Rev. E **70**(6), 066111 (2004)
8. Raghavan, U.N., Albert, R., Kumara, S.: Near linear time algorithm to detect community structures in large-scale networks. Phys. Rev. E **76**(3), 036106 (2007)
9. Pons, P., Latapy, M.: Computing communities in large networks using random walks. In: Yolum, I., Güngör, T., Gürgen, F., Özturan, C. (eds.) ISCIS 2005. LNCS, vol. 3733, pp. 284–293. Springer, Heidelberg (2005). https://doi.org/10.1007/11569596_31
10. Rosvall, M., Bergstrom, C.T.: Maps of random walks on complex networks reveal community structure. Proc. Natl. Acad. Sci. **105**(4), 1118–1123 (2008)
11. Newman, M.E.: Finding community structure in networks using the eigenvectors of matrices. Phys. Rev. E **74**(3), 036104 (2006)
12. Danon, L., Diaz-Guilera, A., Duch, J., Arenas, A.: Comparing community structure identification. J. Stat. Mech.: Theory Exp. **2005**(09), P09008 (2005)
13. Yang, Z., Algesheimer, R., Tessone, C.J.: A comparative analysis of community detection algorithms on artificial networks. Sci. Rep. **6**, 30750 (2016)
14. scikit learn: User guide - clustering (2017). http://scikit-learn.org/stable/modules/clustering.html. Accessed 25 Sept 2017
15. Katz, L.: A new status index derived from sociometric analysis. Psychometrika **18**(1), 39–43 (1953)
16. Bagnoli, F., Massaro, E., Guazzini, A.: Community-detection cellular automata with local and long-range connectivity. In: Sirakoulis, G.C., Bandini, S. (eds.) ACRI 2012. LNCS, vol. 7495, pp. 204–213. Springer, Heidelberg (2012). https://doi.org/10.1007/978-3-642-33350-7_21
17. Lancichinetti, A., Fortunato, S.: Erratum: community detection algorithms: a comparative analysis. [Phys. Rev. E 80, 056117 (2009)]. Phys. Rev. E **89**(4), 049902 (2014)

Multiple Function Approximation - A New Approach Using Complex Fuzzy Inference System

Chia-Hao Tu and Chunshien Li(✉)

Laboratory of Intelligent Systems and Applications,
Department of Information Management, National Central University,
Taoyuan, Taiwan
jamesli@mgt.ncu.edu.tw

Abstract. A complex-fuzzy machine learning approach to function approximation for multiple functions is proposed in this paper. The proposed approach involves the utility of complex-valued vector outputs by a novel complex-fuzzy model using complex fuzzy sets and the famous PSO-RLSE hybrid algorithm for machine learning of the model. An experiment was used to test the proposed approach for the ability of approximating four functions simultaneously. With the experimental result, the performance by the proposed model is promising and the proposed approach is compared to other methods. With complex fuzzy sets, the proposed approach has shown the excellent capability of function approximation for multiple functions using one single model with good performance.

Keywords: Multi-target prediction · Fuzzy system · Complex fuzzy set
Function approximation

1 Introduction

In recent years, under the trend of concept of big data, deep learning is widely discussed. The current frameworks of deep learning use neural network technology, due to characteristics of easy to expand and multiple outputs, that increase the flexibility of applications. In contrast, the traditional fuzzy system has the advantages of intuition and easy to interpretation. However, when it is used to produce multiple outputs, the number of parameters increases significantly, limiting its practical scope of application. In this paper, we proposed an approach, which expands the output of complex fuzzy set from a scalar value to vector values and adjusts the model operation and parameter learning algorithm to build a multi-target prediction fuzzy model without increasing the number of parameters. It also means that the model can get more information under the same conditions, which will greatly enhance efficiency of business investment in computing resources. The rest of this paper is organized as follows. Section 2 introduces previous studies in literature on function approximation and complex fuzzy set. Section 3 describes the approach to build multi-target prediction model. Section 4 designs an example to illustrate the performance of the model, and experiment result is compared to other approaches for performance comparison. Finally, the study is discussed and concluded in Sect. 5.

N. T. Nguyen et al. (Eds.): ACIIDS 2018, LNAI 10751, pp. 243–254, 2018.
https://doi.org/10.1007/978-3-319-75417-8_23

2 Literature Review

2.1 Function Approximation

Some real world phenomena can be expressed by functions, such as the speed of free falling objects. The complex phenomena cannot be represented by simple functions, and researchers try to use different techniques to interpret them in different areas. In the field of computer science, one of the most often used methods is to build models based on artificial intelligence. Before researchers apply these models to explain real world phenomena, function approximation capability of them need to be assessed first to determine the quality of them. This is a crucial and fundamental process of solid research. Erdem et al. [1] indicated that soft computing or artificial intelligence methods have been widely used for nonlinear modelling and function approximation. Although the connotation of soft computing continues to expand, fuzzy logic has always been one of the main technologies [2, 3]. Abe and Lan [4], Dickerson and Kosko [5], Chuang et al. [6], Rong et al. [7], Aras et al. [8], have used fuzzy based techniques to implement function approximation at different times and obtained favorable results. The process of function approximation consumes tremendous computing resources. In recent years, high speed computing devices, like GPU [9] and FPGA [10–12], are used to accelerate the process of function approximation, dramatically reduced the time required for function approximation.

2.2 Complex Fuzzy Set

In traditional mathematics, set is known as crisp set. In a crisp set, an element is either a member of the set or not. There is no room for fuzziness. However, the concept of fuzziness exists in the real world. Hence, in 1965, Zadeh [13] first proposed the concept of fuzzy set and its mathematical foundation to deal with the problem of fuzziness. Fuzzy set is a basic component of fuzzy logic system, and it has a membership function to describe the membership degree that elements belong to itself. From crisp set to fuzzy set, it can be seen as extension of integer set $\{0, 1\}$ to real number set $[0, 1]$. However, real number system can continue to extend to complex number system, so many studies try to combine complex number with fuzzy set in different ways. Buckley [14] incorporated the concept of complex number into fuzzy set, creating fuzzy complex number. Ramot et al. [15] proposed complex fuzzy set in 2002. All elements in this set are assigned a complex membership value and the value lies on the unit circle in complex plane. Thus, a complex fuzzy set has greater potential capability for adaptability, if compared to its counterpart in standard type-1 fuzzy set [16]. This view has been confirmed in many studies in recent years [16–19]. Tamir et al. [20], Yazdanbakhsh and Dick [21] all have made a systematic review and introduction to the study of complex fuzzy sets and complex fuzzy logic.

3 Methodology

The proposed approach in this paper is composed of three parts, vector output of complex fuzzy set, the calculation method of model output and corresponding parameter learning algorithm. In the following, we explain each part in detail.

3.1 Vector Output of Complex Fuzzy Set

Gaussian complex fuzzy sets [16, 17] whose membership degrees are more flexible by the utility of phase term to increase their variability in the unit disc of the complex plane, which are beneficial to the membership description. The membership function of Gaussian complex fuzzy set is denoted by cGMF and can be expressed as follow:

$$\mathrm{cGMF}(h, c, \sigma, \lambda) = \mathrm{r}(h, c, \sigma)\exp(j\omega(h, c, \sigma, \lambda)) \tag{1}$$

where h is the feature variable, $\{c, \sigma, \lambda\}$ are the parameters of center, spread and phase frequency factor for the cGMF, $j = \sqrt{-1}$.

$$\mathrm{r}(h, c, \sigma) = \exp\left[-0.5\left(\frac{h-c}{\sigma}\right)^2\right] \tag{2}$$

$$\omega(h, c, \sigma, \lambda) = -\exp\left[-0.5\left(\frac{h-c}{\sigma}\right)^2\right]\left[\left(\frac{h-c}{\sigma^2}\right)\right]\lambda \tag{3}$$

The model must be transformed from 1 output to n outputs to achieve multi-target prediction in single model. By the characteristic that complex number contains real and imaginary parts, we convert the output of membership function, as Eq. (1), from the scalar value to vector, as Eq. (4). The members of the vector include the original complex membership value, with its real part, imaginary part and absolute value.

$$\mathrm{cGMF}(h, m, \sigma, \lambda) = \begin{bmatrix} \mathrm{r}(h, c, \sigma)\exp(j\omega(h, c, \sigma, \lambda)) \\ \mathrm{real}(\mathrm{r}(h, c, \sigma)\exp(j\omega(h, c, \sigma, \lambda))) \\ \mathrm{imag}(\mathrm{r}(h, c, \sigma)\exp(j\omega(h, c, \sigma, \lambda))) \\ |\mathrm{r}(h, c, \sigma)\exp(j\omega(h, c, \sigma, \lambda))| \end{bmatrix} \tag{4}$$

3.2 Computing Model

The fuzzy model in this paper uses T-S type fuzzy rules. Suppose we have a training dataset TD, denoted as follows:

$$\mathrm{TD} = \left\{\left(\vec{h}, \vec{t}\right)^{(i)}, i = 1, 2, \ldots, |\mathrm{TD}|\right\} \tag{5}$$

where $\left(\vec{h},\vec{t}\right)^{(i)}$ is the input and target vector of ith data pair in TD; |TD| is the number of data pairs. The form of T-S type fuzzy rules in our model are as follows:

$$\begin{aligned} &\mathrm{R}^{(j)}\text{: IF } x_1 \text{ is } \mu_1^{(j)}(h_1) \text{ and } x_2 \text{ is } \mu_2^{(j)}(h_2) \text{ and} \ldots \text{and } x_M \text{ is } \mu_M^{(j)}(h_M) \\ &\quad \text{THEN } \vec{y}^{(j)} = a_j^0 + a_j^1 h_1 + \ldots + a_j^M h_M \end{aligned} \tag{6}$$

where $\mathrm{R}^{(j)}$ is the jth rule in the proposed fuzzy model, $j = 1, 2, \ldots, K$, K is the total number of rules; x_l is the lth linguistic variable; h_l is the lth input of input vector; $\mu_l^{(j)}$ is the membership function, as Eq. (4), used in the lth dimension of the jth rule, $l = 1, 2, \ldots, M$; $\vec{y}^{(j)}$ is the model output vector of the jth rule. According to Eq. (4), the firing strength vector $\vec{\beta}_j$ could be found as follows:

$$\vec{\beta}^{(j)} = \left\{\beta_i^{(j)}, i = 1, 2, \ldots, 4\right\} \tag{7}$$

$$\beta_i^{(j)} = \prod_{l=1}^{M} \mu_{l,i}^{(j)}(\mathrm{h}_l) \tag{8}$$

where $\beta_i^{(j)}$ is the ith component of $\vec{\beta}^{(j)}$; $\mu_{l,i}^{(j)}$ is the ith component of $\mu_l^{(j)}$. Once the firing strength vector of a rule is calculated, Eqs. (9) and (10) could be used to calculate the normalized firing strength vector of each rule.

$$\vec{\varphi}^{(j)} = \left\{\varphi_i^{(j)}, i = 1, 2, \ldots, 4\right\} \tag{9}$$

$$\varphi_i^{(j)} = \frac{\beta_i^{(j)}}{\sum_{j=1}^{K} \beta_i^{(j)}} \tag{10}$$

where $\vec{\varphi}^{(j)}$ is the normalized firing strength vector of the jth rule; $\varphi_i^{(j)}$ is the ith component of $\vec{\varphi}^{(j)}$. Then the model output could be calculated as follows:

$$\vec{y} = \{y_i, i = 1, 2, \ldots, 4\} \tag{11}$$

$$y_i = \sum_{j=1}^{K} \varphi_i^{(j)} \left(a_j^0 + a_j^1 h_1 + \ldots + a_j^M h_M\right) \tag{12}$$

The number of elements in $\vec{\beta}$, $\vec{\varphi}$ and $\vec{y}$ is the same as the size in the vector of membership value, with a maximum of four. It means the model has the ability to predict maximum to eight targets if we combine two original real number targets into one complex number target.

3.3 Parameter Learning

The phase of parameter learning utilizes the PSO-RLSE hybrid learning algorithm [18]. It can be understood that this method is very efficient in learning convergence. The original PSO (Particle Swarm Optimization) was first developed by Eberhart and Kennedy [22]. Assume the problem space is with τ dimensions. The PSO can be described as follows:

$$V_i(j+1) = \omega V_i(j) + c_1\rho_1(\text{pbest}_i - L_i(j)) + c_2\rho_2(\text{gbest} - L_i(j)) \tag{13}$$

$$L_i(j+1) = L_i(j) + V_i(j+1) \tag{14}$$

where i is the ith particle and j is the jth iteration for PSO; $V_i(j) = \left[V_{i,1}(j), V_{i,2}(j), \ldots, V_{i,\tau}(j)\right]^T$ is the velocity of the ith particle located in $L_i(j) = \left[L_{i,1}(j), L_{i,2}(j), \ldots, L_{i,\tau}(j)\right]^T$; pbest_i is the best location of the ith particle in history; gbest is the best location of the swarm; $\{\omega, c_1, c_2\}$ are the parameters of PSO; and $\{\rho_1, \rho_2\}$ are random numbers in [0, 1].

The fuzzy model is non-linear in the premise parts. Hence, the PSO is used to optimize the parameters of these parts. Afterwards, it is passed into RLSE (Recursive Least Squares Estimator), to find the optimal solution in the consequent parts via mathematical calculation. For a general least squares estimation problem, the output t of the LSE model can be expressed as below:

$$t = \theta_1 f_1(h) + \theta_2 f_2(h) + \ldots + \theta_\varsigma f_\varsigma(h) + \varepsilon \tag{15}$$

where h is the model's input; $\{f_i(h), i = 1, 2, \ldots, \varsigma\}$ are known function of h; $\{\theta_i, i = 1, 2, \ldots, \varsigma\}$ represent the model parameters to be estimated; and ε is the model error. In this study, $\{\theta_i, i = 1, 2, \ldots, \varsigma\}$ is the parameters of consequent parts of proposed model. Substituting the data pairs of TD, as Eq. (5), into Eq. (15), we can obtain a set of |TD| equations in matrix notation.

$$t = B\theta + \varepsilon \tag{16}$$

where B, θ, ε, and t can be given as follows:

$$\mathrm{B} = \begin{bmatrix} f_1(h^{(1)}) & f_2(h^{(1)}) & \ldots & f_\varsigma(h^{(1)}) \\ f_1(h^{(2)}) & f_2(h^{(2)}) & \ldots & f_\varsigma(h^{(2)}) \\ \vdots & \vdots & \ldots & \vdots \\ f_1(h^{(|\mathrm{TD}|)}) & f_2(h^{(|\mathrm{TD}|)}) & \ldots & f_\varsigma(h^{(|\mathrm{TD}|)}) \end{bmatrix} \tag{17}$$

$$\theta = [\theta_1 \quad \theta_2 \quad \cdots \quad \theta_\varsigma]^{\mathrm{T}} \tag{18}$$

$$\varepsilon = \begin{bmatrix} \varepsilon_1 & \varepsilon_2 & \cdots & \varepsilon_{|TD|} \end{bmatrix}^T \tag{19}$$

$$t = \begin{bmatrix} t_1 & t_2 & \cdots & t_{|TD|} \end{bmatrix}^T \tag{20}$$

The θ can be optimized by the following RLSE equations.

$$P_{j+1} = P_j - \frac{P_j b_{j+1} b_{j+1}^T P_j}{1 + b_{j+1}^T P_j b_{j+1}} \tag{21}$$

$$\theta_{j+1} = \theta_j + P_{j+1} b_{j+1} \left(t_{j+1} - b_{j+1}^T \theta_j \right), \text{for } j = 0, 1, \ldots, (|TD| - 1) \tag{22}$$

where $[b_j^T, t_j]$ is the jth row of [B, t]. To implement RLSE, we initialize θ_0 as a zero matrix, and $P_0 = \alpha I$, where α must be a large positive value and I is the identity matrix. In Eqs. (21) and (22), vector b and vector θ are arranged as follows:

$$b_{j+1} = \begin{bmatrix} bb_1^{(j+1)} & bb_2^{(j+1)} & \cdots & bb_K^{(j+1)} \end{bmatrix} \tag{23}$$

$$bb_i^{(j+1)} = \begin{bmatrix} \vec{\varphi}_i^{(j+1)} & \vec{\varphi}_i^{(j+1)} & h_1^{(j+1)} & \ldots & \vec{\varphi}_i^{(j+1)} & h_M^{(j+1)} \end{bmatrix} \tag{24}$$

$$\theta_j = \begin{bmatrix} \delta_1^{(j)} & \delta_2^{(j)} & \ldots & \delta_K^{(j)} \end{bmatrix} \tag{25}$$

$$\delta_i^{(j)} = \begin{bmatrix} a_{0(i)}^{(j)} & a_{1i}^{(j)} & \ldots & a_{Mi}^{(j)} \end{bmatrix} \tag{26}$$

where i, j is the ith rule in the jth iteration, $i = 1, 2, \ldots, K$, and $j = 0, 1, \ldots, (|TD| - 1)$. The PSO-RLSE hybrid learning method is described as below.

Step 1. Initialize PSO to get the initial parameters of premise part.
Step 2. Calculate the normalized firing strength vectors of fuzzy rules, as (9) and (10).
Step 3. Use RLSE to update the parameters of consequent part, as (21) to (26).
Step 4. Obtain system output by model with training data.
Step 5. Calculate the costs for all particles by repeating step 2–4.
Step 6. Update pbest for every particle, gbest for the swarm and the parameters of premise parts.
Step 7. Go back to step 2 until matching the stop condition.

In addition, the RLSE part needs to adjust matrix size in order to be used in the multi-target prediction model proposed in this study.

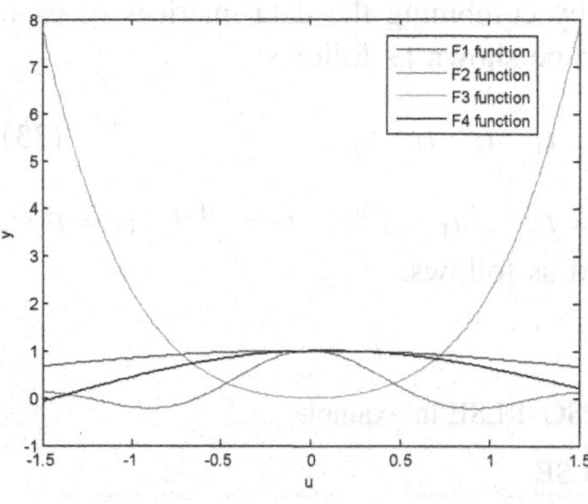

Fig. 1. 4 functions in example

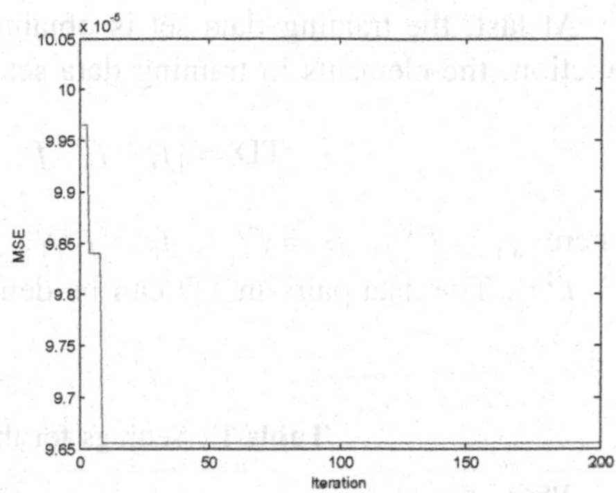

Fig. 2. Learning curve by the proposed method

4 Experimentation

An experiment is designed to verify the capability of the proposed approach to approximate multiple functions simultaneously, and the result is compared to the performance of models from other studies. The non-linear functions selected in the experiment are given as below and shown in Fig. 1.

$$\mathrm{F}_1 : y_1 = \frac{\sin(u)}{u} \tag{27}$$

$$\mathrm{F}_2 : y_2 = \frac{\sin(5u)}{5u} \tag{28}$$

$$\mathrm{F}_3 : y_3 = 1.25u^2 + u^4 \tag{29}$$

$$\mathrm{F}_4 : y_4 = \cos(u) + 0.1u \tag{30}$$

where u is randomly selected from $[-1.5, 1.5]$ with 601 points, the outputs of Eqs. (27) to (30) with these points to form raw data sets are shown as follows:

$$\mathrm{RD}^{(F_i)} = \left\{ x_j^{(F_i)}, j = 1, 2, \ldots, |\mathrm{RD}| \right\}, |\mathrm{RD}| = 601 \tag{31}$$

where $\mathrm{RD}^{(\mathrm{F}_i)}$ is the raw data set of ith function; $x_j^{(\mathrm{F}_i)}$ is the jth output of ith function. Then we calculate the difference based on $x_j^{(\mathrm{F}_i)}$ to create data matrix for each function, each data matrix contains one input and one target variable, denoted as below:

$$\mathrm{DM}^{(\mathrm{F}_i)} = \left[f^{(\mathrm{F}_i)} \quad t^{(\mathrm{F}_i)} \right] \tag{32}$$

where $\mathrm{DM}^{(\mathrm{F}_i)}$ is the data matrix of ith function; $f^{(\mathrm{F}_i)}$ and $t^{(\mathrm{F}_i)}$ are the input and target variable of ith function, $f^{(\mathrm{F}_i)} = x_j^{(\mathrm{F}_i)} - x_{j-1}^{(\mathrm{F}_i)}$ and $t^{(\mathrm{F}_i)} = x_{j+1}^{(\mathrm{F}_i)} - x_j^{(\mathrm{F}_i)}$.

At last, the training data set is obtained by combining the data matrices of each function, the elements in training data set can be shown as follows:

$$\mathrm{TD} = [f_1 \quad f_2 \quad f_3 \quad f_4 \quad t_1 \quad t_2 \quad t_3 \quad t_4] \tag{33}$$

where $f_1 = f^{(\mathrm{F}_1)}$, $f_2 = f^{(\mathrm{F}_2)}$, $f_3 = f^{(\mathrm{F}_3)}$, $f_4 = f^{(\mathrm{F}_4)}$, $t_1 = t^{(\mathrm{F}_1)}$, $t_2 = t^{(\mathrm{F}_2)}$, $t_3 = t^{(\mathrm{F}_3)}$, $t = t^{(\mathrm{F}_4)}$. The data pairs in TD can be denoted as follows:

Table 1. Settings for the PSO–RLSE in example

PSO		RLSE	
# of parameters	24	# of parameters	80
Swarm size	100	P_0	αI
Initial particle positions	Rand in $[0, 1]^{24}$	α	10^9
Initial particle velocities	Rand in $[0, 1]^{24}$	I	80-by-80 identity matrix
$\{w, c1, c2\}$	{0.9, 2, 2}	θ_0	80-by-1 zero vector
Training iterations	200		

$$\mathrm{TD} = \left\{ \left(\vec{h}^{(i)}, \vec{t}^{(i)} \right), i = 1, 2, \ldots, |\mathrm{TD}| \right\}, |\mathrm{TD}| = 600 \tag{34}$$

Table 2. The testing performance of 10 trials in example

F_i	The best MSE	The worst MSE	Mean MSE	Stdv MSE
F_1	3.61×10^{-6}	5.27×10^{-5}	1.63×10^{-5}	2.12×10^{-5}
F_2	1.36×10^{-5}	3.82×10^{-5}	2.7×10^{-5}	2.25×10^{-5}
F_3	5.82×10^{-5}	4.19×10^{-3}	5.35×10^{-4}	1.23×10^{-3}
F_4	2.23×10^{-5}	9.25×10^{-5}	5.82×10^{-5}	7.56×10^{-5}

where $\vec{h}^{(i)}$ and $\vec{t}^{(i)}$ are the ith input and output vector of TD. The first 500 pairs are selected as training data and the others are testing data. The output vectors of membership functions, descripted in Sect. 3.1, contain four membership values; the operation and parameter optimization of the model are the same as the description in Sects. 3.2 to 3.3. The setting for the PSO-RLSE are listed in Table 1.

Table 3. Performance comparison in example

F_i	Method	RMSE	MSE
F_1	ANFIS [23]	7.1×10^{-3}	
	Robust NN [24]	1.1×10^{-3}	
	Proposed approach (training phase)	1.4×10^{-3}	
	Proposed approach (testing phase)	1.9×10^{-3}	
F_2	BP [25] (training phase)		1.53×10^{-4}
	BP [25] (testing phase)		1.47×10^{-4}
	APSOAEFDI-MHLA [25] (training phase)		2.09×10^{-5}
	APSOAEFDI-MHLA [25] (testing phase)		1.9×10^{-5}
	Proposed approach (training phase)		6.45×10^{-5}
	Proposed approach (testing phase)		1.36×10^{-5}
F_3	CNFS with PSO-RLSE [16] (training phase)		1.52×10^{-10}
	CNFS with PSO-RLSE [16] (testing phase)		6.01×10^{-3}
	Proposed approach (training phase)		1.43×10^{-5}
	Proposed approach (testing phase)		5.82×10^{-5}
F_4	Proposed approach (training phase)		1.52×10^{-5}
	Proposed approach (testing phase)		2.23×10^{-5}

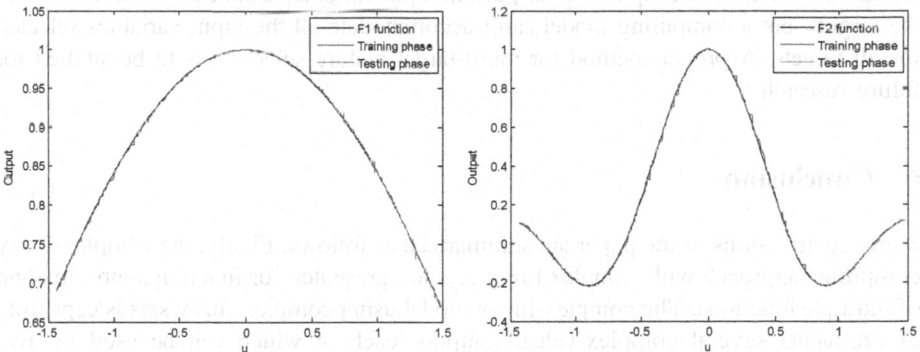

Fig. 3. F1 approximation result **Fig. 4.** F2 approximation result

With 10 repeated trials, the performance of the proposed approach is compared to other single function approximation approaches in Tables 2 and 3. The learning curve and the response by the proposed approach is shown in Figs. 2, 3, 4, 5 and 6.

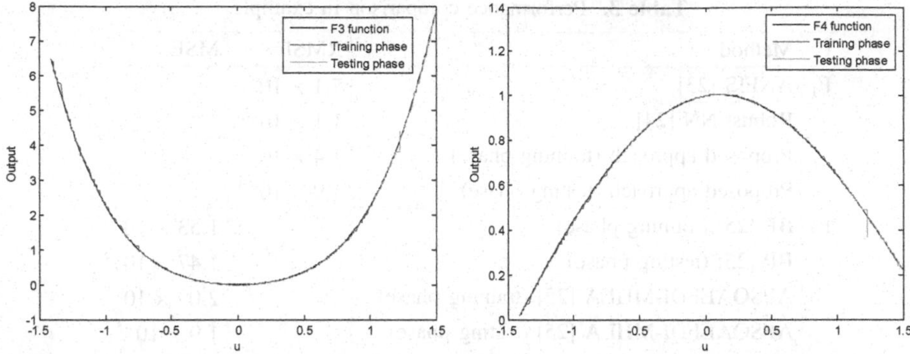

Fig. 5. F3 approximation result

Fig. 6. F4 approximation result

5 Discussion

With the experimental result for function approximation of four functions, the proposed approach has shown good performance, successfully verify the capability of the model for approximating multiple functions simultaneously by one single model. In performance comparison, it is worthy to point out that the proposed model is for four functions simultaneously in one single model while each of the compared methods is for one function only. However, this study also encounters a problem in feature-selection for multiple targets. Certain input variables can be selected for each of the targets, but a computing model can't accommodate all the input variables selected for all targets. A proper method for multi-target feature selection is to be studied for future research.

6 Conclusion

Some contributions of the paper are summarized as follows. Firstly, the complex-fuzzy computing approach with complex fuzzy sets has presented for function approximation of multiple functions. The complex-fuzzy model using complex fuzzy sets is capability of producing several complex-valued outputs, each of which can be used for two real-valued targets for application. This is the so-called dual-output property. Secondly, the famous PSO-RLSE hybrid method has been successfully applied for machine learning of the proposed model for multiple-function approximation. Through the experimentation, it has been shown very efficient, in terms of learning time and performance. Thirdly, complex fuzzy sets used in the proposed approach have been empirically shown useful for function approximation with multiple functions.

Acknowledgments. This work was supported by the Ministry of Science & Technology, Taiwan, under grant no. MOST 105-2221-E-008-091 and grant no. MOST 104-2221-E-008-116.

References

1. Erdem, H., Berkol, A., Sert, M.: Comparative study of universal function approximators (neural network, fuzzy logic, ANFIS) for non-linear systems. Int. J. Sci. Res. Inf. Syst. Eng. (IJSRISE) **1**(2), 45–50 (2015)
2. Konar, A.: Artificial Intelligence and Soft Computing: Behavioral and Cognitive Modeling of the Human Brain. CRC Press, Boca Raton (2000)
3. Tettamanzi, A.G., Tomassini, M.: Soft Computing: Integrating Evolutionary, Neural, and Fuzzy Systems. Springer Science & Business Media, Heidelberg (2013). https://doi.org/10.1007/978-3-662-04335-6
4. Abe, S., Lan, M.S.: Fuzzy rules extraction directly from numerical data for function approximation. IEEE Trans. Syst. Man Cybern. **25**(1), 119–129 (1995)
5. Dickerson, J.A., Kosko, B.: Fuzzy function approximation with ellipsoidal rules. IEEE Trans. Syst. Man Cybern. Part B (Cybern.) **26**(4), 542–560 (1996)
6. Chuang, C.C., Su, S.F., Chen, S.S.: Robust TSK fuzzy modeling for function approximation with outliers. IEEE Trans. Fuzzy Syst. **9**(6), 810–821 (2001)
7. Rong, H.J., Huang, G.B., Sundararajan, N., Saratchandran, P.: Online sequential fuzzy extreme learning machine for function approximation and classification problems. IEEE Trans. Syst. Man Cybern. Part B (Cybern.) **39**(4), 1067–1072 (2009)
8. Aras, A.C., Kaynak, O., Batyrshin, I.: Nonlinear function approximation based on fuzzy algorithms with parameterized conjunctors. In: 2013 IEEE International Conference on Mechatronics (ICM), pp. 81–86, February 2013
9. Berjón, D., Gallego, G., Cuevas, C., Morán, F., García, N.: Optimal piecewise linear function approximation for GPU-based applications. IEEE Trans. Cybern. **46**(11), 2584–2595 (2016)
10. Thomas, D.B.: A general-purpose method for faithfully rounded floating-point function approximation in FPGAs. In: 2015 IEEE 22nd Symposium on Computer Arithmetic (ARITH), pp. 42–49, June 2015
11. Del Campo, I., Echanobe, J., Asua, E., Finker, R.: Controlled-accuracy approximation of nonlinear functions for soft computing applications: a high performance co-proccessor for intelligent embedded systems. In: 2015 IEEE Symposium Series on Computational Intelligence, pp. 609–616, December 2015
12. Karakuzu, C., Karakaya, F., Çavuşlu, M.A.: FPGA implementation of neuro-fuzzy system with improved PSO learning. Neural Netw. **79**, 128–140 (2016)
13. Zadeh, L.A.: Fuzzy sets. Inf. Control **8**(3), 338–353 (1965)
14. Buckley, J.J.: Fuzzy complex numbers. Fuzzy Sets Syst. **33**(3), 333–345 (1989)
15. Ramot, D., Milo, R., Friedman, M., Kandel, A.: Complex fuzzy sets. IEEE Trans. Fuzzy Syst. **10**(2), 171–186 (2002)
16. Li, C., Chiang, T.W.: Function approximation with complex neuro-fuzzy system using complex fuzzy sets–a new approach. New Gener. Comput. **29**(3), 261–276 (2011)
17. Li, C., Chiang, T.-W.: Complex fuzzy computing to time series prediction a multi-swarm PSO learning approach. In: Nguyen, N.T., Kim, C.-G., Janiak, A. (eds.) ACIIDS 2011 Part II. LNCS (LNAI), vol. 6592, pp. 242–251. Springer, Heidelberg (2011). https://doi.org/10.1007/978-3-642-20042-7_25
18. Li, C., Wu, T.: Adaptive fuzzy approach to function approximation with PSO and RLSE. Expert Syst. Appl. **38**(10), 13266–13273 (2011)
19. Li, C., Chiang, T.W.: Complex neurofuzzy ARIMA forecasting—a new approach using complex fuzzy sets. IEEE Trans. Fuzzy Syst. **21**(3), 567–584 (2013)

20. Tamir, D.E., Rishe, N.D., Kandel, A.: Complex fuzzy sets and complex fuzzy logic an overview of theory and applications. In: Tamir, D.E., Rishe, N.D., Kandel, A. (eds.) Fifty Years of Fuzzy Logic and Its Applications. SFSC, vol. 326, pp. 661–681. Springer, Cham (2015). https://doi.org/10.1007/978-3-319-19683-1_31
21. Yazdanbakhsh, O., Dick, S.: A systematic review of complex fuzzy sets and logic. Fuzzy Sets Syst., in Press, Corrected Proof, 23 January 2017
22. Eberhart, R., Kennedy, J.: A new optimizer using particle swarm theory. In: Proceedings of the Sixth International Symposium on Micro Machine and Human Science, MHS 1995, pp. 39–43, October 1995
23. Jang, J.S.R., Sun, C.T., Mizutani, E.: Neuro-Fuzzy and Soft Computing; A Computational Approach to Learning and Machine Intelligence. Prentice-Hall, Upper Saddle River (1997)
24. Foresee, F.D., Hagan, M.T.: Gauss-Newton approximation to Bayesian learning. In: International Conference on Neural Networks, pp. 1930–1935, June 1997
25. Cheng, Y.C., Li, S.T.: Fuzzy time series forecasting with a probabilistic smoothing hidden Markov model. IEEE Trans. Fuzzy Syst. **20**(2), 291–304 (2012)

Decision Tree Using Local Support Vector Regression for Large Datasets

Minh-Thu Tran-Nguyen[1], Le-Diem Bui[1,3], Yong-Gi Kim[3], and Thanh-Nghi Do[1,2](✉)

[1] College of Information Technology, Can Tho University, Can Tho, Vietnam
dtnghi@cit.ctu.edu.vn

[2] UMI UMMISCO 209 (IRD/UPMC), UPMC, Sorbonne University, Pierre and Marie Curie University, Paris 6, France

[3] AI Lab, Computer Science Department, Gyeongsang National University, Jinju, Korea

Abstract. Our proposed decision tree using local support vector regression models (tSVR) is to handle the regression task of large datasets. The learning algorithm tSVR of regression models is done by two main steps. The first one is to construct a decision tree regressor for partitioning the full training dataset into k terminal-nodes (subsets), followed which the second one is to learn the SVR model from each terminal-node to predict the data locally in the parallel way on multi-core computers. The tSVR algorithm is faster than the standard SVR in training the non-linear regression model from large datasets while maintaining the high correctness in the prediction. The numerical test results on datasets from UCI repository showed that the proposed tSVR is efficient compared to the standard SVR.

Keywords: Support vector regression (SVR) · Decision tree
Local support vector regression (local SVR) · Large datasets

1 Introduction

Support vector machines (SVM) proposed by [1] and kernel-based methods have shown the state-of-the-art relevant to many data mining problems, including the classification, the regression and the novelty detection [2]. Nevertheless, the learning problem is accomplished through a quadratic programming (QP), so that the computational cost of a SVM approach is at least square of the number of training datapoints making SVM impractical to handle large datasets. There is a need to scale-up SVM learning algorithms to handle massive datasets.

In this paper, we propose the tSVR (decision tree using local support vector regression) to effectively deal with the non-linear regression task of large datasets. Instead of training a global SVR model, as done by the classical SVR algorithm is very hard to handle large datasets, our tSVR algorithm is to learn in

N. T. Nguyen et al. (Eds.): ACIIDS 2018, LNAI 10751, pp. 255–265, 2018.
https://doi.org/10.1007/978-3-319-75417-8_24

the parallel way an ensemble of local ones that are easily trained by the standard SVR algorithms. The tSVR trains the regression model via two main steps. The first one is to use the decision tree algorithm [3] to partition the large training dataset into k terminal-nodes (subsets). The idea is to reduce the data size for training local non-linear SVR models at the second step. And then, the tSVR learns k non-linear SVR models in the parallel way on multi-core computers in which a SVR model is trained in each terminal-node to predict the data locally. The numerical test results on datasets from UCI repository [4] showed that our proposed tSVR is efficient compared to the standard SVR in terms of training time and prediction correctness. The tSVR algorithm is faster than the standard SVR in the non-linear regression of large datasets while maintaining the high prediction correctness.

The remainder of our paper is organized as follows. In Sect. 2, we briefly present the standard SVR algorithm. In Sect. 3, we illustrate how the tSVR algorithm learns local regression models from large datasets. Section 4 shows the experimental results. Section 5 discusses about related works. We then conclude in Sect. 6.

2 Support Vector Regression

Let us consider a regression task with m datapoints x_i $(i = 1, \ldots, m)$ in the n-dimensional input space R^n, having corresponding targets $y_i \in R$. The support vector regression (SVR) proposed by [1] tries to find the best hyperplane (denoted by the normal vector $w \in R^n$ and the scalar $b \in R$) that has at most ε deviation from the target value y_i. Figure 1 is a simple example of SVR. The training algorithm of SVR pursues this goal with the Lagrangian dual quadratic programming (1) using the Lagrangian multipliers α_i and α_j.

$$\min (1/2) \sum_{i=1}^{m} \sum_{j=1}^{m} (\alpha_i - \alpha_i{}^*)(\alpha_j - \alpha_j{}^*) K\langle x_i, x_j\rangle - \sum_{i=1}^{m} (\alpha_i - \alpha_i{}^*) y_i + \varepsilon \sum_{i=1}^{m} (\alpha_i + \alpha_i{}^*)$$

$$s.t. \begin{cases} \sum_{i=1}^{m} (\alpha_i - \alpha_i{}^*) = 0 \\ 0 \le \alpha_i, \alpha_i{}^* \le C \quad \forall i = 1, 2, ..., m \end{cases} \tag{1}$$

where C is a positive constant used to tune the margin size and the error and a linear kernel function $K\langle x_i, x_j\rangle = \langle x_i \cdot x_j\rangle$ (the dot product of x_i, x_j).

The resolution of the quadratic programming (1) gives $\sharp SV$ support vectors for which $\alpha_i, \alpha_i{}^* > 0$. The predictive hyperplane and the scalar b are determined by these support vectors. And then, the prediction of a new datapoint x is as follows:

$$predict(x) = \sum_{i=1}^{\sharp SV} (\alpha_i - \alpha_i{}^*) K\langle x, x_i\rangle - b \tag{2}$$

Variations on training SVR models use different types of the kernel [5]. It only needs replacing the linear kernel function $K\langle x_i, x_j\rangle = \langle x_i \cdot x_j\rangle$ with other non-linear ones:

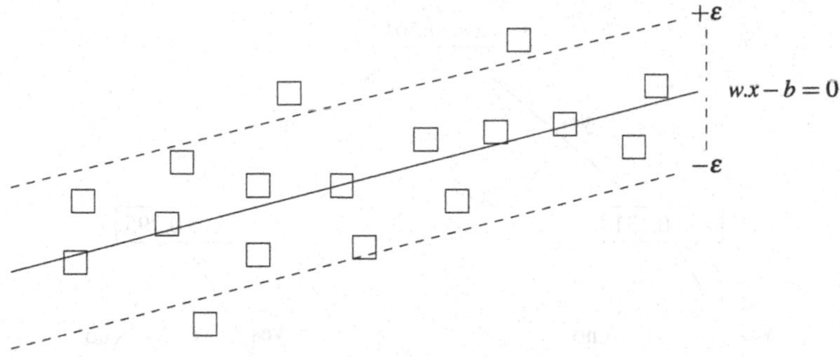

Fig. 1. Linear support vector regression

- a polynomial function of degree d: $K\langle x_i, x_j\rangle = (\langle x_i \cdot x_j\rangle + 1)^d$
- a RBF (Radial Basis Function): $K\langle x_i, x_j\rangle = e^{-\gamma\|x_i - x_j\|^2}$.

The SVR models are most accurate and practical pertinent for many successful applications reported in the classification, the regression, the novelty detection [2].

3 *t*SVR Algorithm for Dealing with Large Datasets

Platt studied the algorithmic complexity of the SVM approach in [6]. His analysis illustrates that the training complexity of the SVM solution in (1) are at least square of the number of training datapoints (i.e. $O(m^2)$), making standard SVM intractable for large datasets. It means that learning the model from the full massive dataset using the SVM algorithm in the usual way is challenge due to the very high computational cost.

tSVR Learning for Local SVR Models
Our proposed *t*SVR algorithm learns an ensemble of local SVR models that are easily trained by the standard SVR algorithm. As illustrated in Fig. 2, the *t*SVR handles the regression task with two main steps. The first one of the algorithm *t*SVR is to learn the decision tree model (using the decision tree training algorithm [3]) to partition the full training dataset into k terminal-nodes (subsets). Thus, the second step of the algorithm *t*SVR is to learn k non-linear SVR models in the parallel way on multi-core computers in which a SVR model is trained in each terminal-node to predict the data locally.

We consider a simple regression task given a target variable y and a predictor (variable) x. Figure 3 shows the comparison between a global SVR model (left part) and 4 local SVR models (right part) for this regression task, using a non-linear RBF kernel function with $\gamma = 10$, a positive constant $C = 10^5$ (i.e. the hyper-parameters $\theta = \{\gamma, C\}$) and a tolerance $\varepsilon = 0.05$.

Since the terminal-node size is smaller than the full training data size, the standard SVR can easily perform the training task from the terminal-node, due

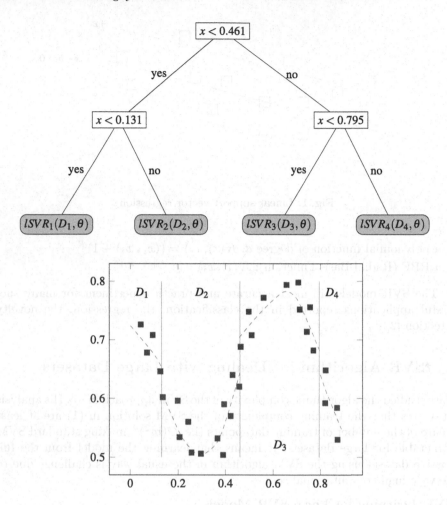

Fig. 2. Decision tree using SV regressors at the terminal-nodes (tSVR)

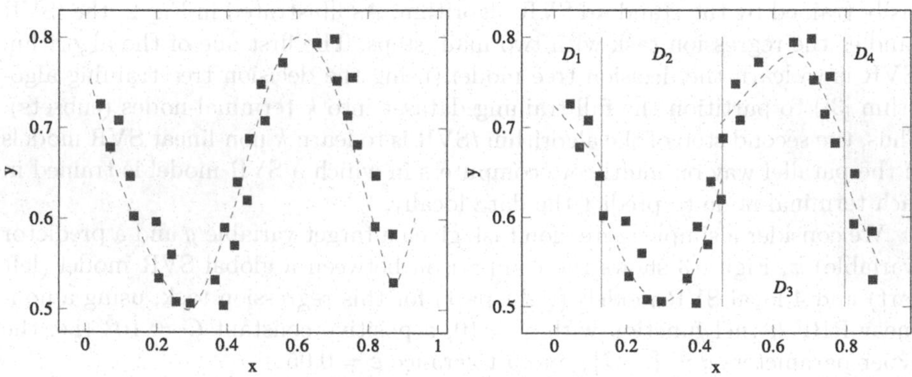

Fig. 3. Global SVR model (left part) versus 4 local SVR models (right part)

to the square reduction of the complexity compared to the learning from the full dataset. In addition, k local models in the training task of tSVR are independently learnt from k terminal-nodes. This is a nice condition to train k local models in the parallel way to take into account the benefits of high performance computing, e.g. multi-core computers or grids. We propose to develop the parallel tSVR algorithm in the simple way based on the shared memory multiprocessing programming model OpenMP [7] for multi-core computers. The parallel training of tSVR is described in Algorithm 1.

Prediction of a New Individual x Using tSVM Model

A new individual x is predicted by the tSVR model as follows. Firstly, it pushes x down the tree t of the tSVR model from the root to the terminal-node D_p. And then the local SVM model $lsvr_p$ learnt from this terminal-node D_p is used to predict the target value of x.

Performance Analysis

The performance analysis starts with the algorithmic complexity of tSVR for building k local SVR models in the parallel way. Suppose that tSVR uses the decision tree to partition the full dataset (with m datapoints) into k balanced terminal-nodes (the terminal-node size is about $\frac{m}{k}$; in other words the minimum number of datapoints for a split to be tried $minobj = \frac{m}{k}$). The training complexity of a SVR for a terminal-node is $O(minobj^2)$ (i.e. $O((\frac{m}{k})^2)$). Therefore, the algorithmic complexity of tSVR for parallel training k local SVR models on a P-core processor is $O(\frac{k}{P}(\frac{m}{k})^2) = O(\frac{m^2}{kP})$. This complexity analysis illustrates that parallel learning k local SVR models in the tSVR algorithm[1] is kP times faster than building a global SVR model (the complexity at least $O(m^2)$).

The performance analysis in terms of the generalization capacity of such tSVR models is illustrated in [8–11]. The parameter $minobj$ is used in the tSVR to give a trade-off between the generalization capacity (the prediction correctness) and the computational cost. This point can be understood as follows:

- If $minobj$ is small then the tSVR algorithm significantly reduces training time. And then, the size of a terminal-node; The locality is extremely with a very low capacity (i.e. the low prediction correctness).
- If $minobj$ is large then the tSVR algorithm reduces insignificant training time. However, the size of a terminal-node is large; It improves the capacity (i.e. the high prediction correctness).

It leads to set $minobj$ large enough (e.g. 200 proposed in [9,10]).

4 Evaluation

We are interested in the performance of the tSVR for large datasets regression. We have implemented the tSVR algorithm in C/C++, OpenMP [7], using the

[1] It remarks that the complexity analysis of the tSVR excepts the tree regressor learnt to split the full dataset. However this training the tree regressor has the very low computational cost compared with the quadratic programming solution required by the SVR learning algorithm.

Algorithm 1. Parallel training tSVR algorithm of local support vector regression

input :
training dataset D
minimum number of individuals *minobj* for a split to be tried
tolerance ε
hyper-parameter γ of RBF kernel function
positive constant C for tuning margin and errors

output:
tree t and k local SVR models

begin
  /*Decision tree algorithm partitions the full dataset D;*/
  learning a tree t for splitting the full dataset D into k terminal-nodes denoted by $D_1, D_2, \ldots, D_k$ (using the early stopped parameter *minobj*)
  #pragma omp parallel for schedule(dynamic)
  for $i \leftarrow 1$ **to** k **do**
    /*learning local support vector regression model from D_i;*/
    $lsvr_i = svr(D_i, \gamma, C, \varepsilon)$
  end
  return tSVR-model = tree t and $\{lsvr_1, lsvr_2, \ldots, lsvr_k\}$
end

standard SVR in the highly efficient library SVM, LibSVM [12]. Our evaluation of the performance is reported in terms of training time and prediction correctness. We are interested in comparing the regression results obtained by our algorithm tSVR with SVR in LibSVM.

All experiments are run on machine Linux Fedora 20, Intel(R) Core i7-4790 CPU, 3.6 GHz, 4 cores and 32 GB main memory.

Datasets

Experiments are conducted with the 5 datasets from UCI repository [4]. Table 1 presents the description of datasets. The evaluation protocols are illustrated in the last column of Table 1. Datasets are already divided in training set (Trn) and test set (Tst). We used the training data to build the SVR models. Then, we predicted the test set using the resulting models.

Tuning Parameters

We propose to use RBF kernel type in tSVR and SVR models because it is general and efficient [13]. The cross-validation protocol (2-fold) is used to tune the regression tolerance ε, the hyper-parameter γ of RBF kernel (RBF kernel of two individuals x_i, x_j, $K[i,j] = exp(-\gamma\|x_i - x_j\|^2)$) and the cost C (a trade-off between the margin size and the errors) to obtain a good correctness. For two largest datasets (Buzz in social media Twitter, YearPredictionMSD), we used a subset randomly sampling about 5% training dataset for tuning hyper-parameters due to the expensive computational cost. Furthermore, our tSVR

Table 1. Description of datasets

ID	Datasets	#datapoints	#dimensions	Target domain	Evaluation protocol
1	Appliances energy prediction	19 735	27	[10.0, 1080.0]	13500 Trn - 6235 Tst
2	Facebook comment volume	40 949	53	[0.0, 1305.0]	27500 Trn - 13449 Tst
3	BlogFeedback	60 021	280	[0.0, 1424.0]	52397 Trn - 7624 Tst
4	Buzz in social media (Twitter)	583250	77	[0.0, 75724.5]	400000 Trn - 183250 Tst
5	YearPredictionMSD	515 345	90	[1922, 2011]	400000 Trn - 115345 Tst

uses the parameter *minobj* (the minimum number of datapoints for a split to be tried by the decision tree regression). We propose to set $minobj = 200$ (according to the advise of [9,10]). Table 2 presents the hyper-parameters of *t*SVR and SVR in the regression.

Table 2. Hyper-parameters of *t*SVR and SVR in LibSVM

ID	Datasets	γ	C	ε	*minobj*
1	Appliances energy prediction	0.02	100000	0.1	200
2	Facebook comment volume	0.001	100000	0.1	200
3	BlogFeedback	0.4	100000	0.05	200
4	Buzz in social media (Twitter)	0.1	100000	0.1	200
5	YearPredictionMSD	0.01	100000	0.1	200

Regression Results

The regression results of SVR of LibSVM and *t*SVR are given in Table 3.

As it was expected, our *t*SVR algorithm outperforms SVR of LibSVM in terms of training time. The average of *t*SVR and SVR of LibSVM training time are 1.19 min and 1551.15 min, respectively. It means that our *t*SVR is 1307.17 times faster than SVR of LibSVM.

For 3 first small datasets, the training speed improvements of *t*SVR versus SVR of LibSVM is 345.80 times. With two large datasets (Buzz in social media - Twitter and YearPredictionMSD), the learning time improvements of *t*SVR against SVR of LibSVM is more significant (about 1348.34 times).

In terms of prediction correctness (mean absolute error - MAE), the error average made by *t*SVR and SVR of LibSVM are 25.16 and 62.01, respectively.

The comparison of prediction correctness, dataset by dataset, shows that *t*SVR is beaten only once (with BlogFeedback dataset) by SVR of LibSVM (4 wins, 1 defeat). It illustrates that our *t*SVR is more accurate than SVR of LibSVM for the prediction.

*t*SVR improves not only the training time, but also the prediction correctness when dealing with large datasets. The regression results allow to believe that our proposed *t*SVR is efficient for handling these data volumes.

Table 3. Results in terms of mean absolute error (MAE) and training time (minutes)

ID	Datasets	Mean absolute error (MAE)		Training time (min)	
		LibSVM	*t*SVR	LibSVM	*t*SVR
1	Appliances energy prediction	47.81	43.56	2.55	0.016
2	Facebook comment volume	8.97	5.31	27.91	0.170
3	BlogFeedback	9.85	23.85	53.78	0.057
4	Buzz in social media (Twitter)	235.25	45.72	5193.59	2.392
5	YearPredictionMSD	8.18	7.38	2477.91	3.297
6	Average	62.01	25.16	1551.15	1.187

5 Discussion on Related Works

Our proposed *t*SVR is in machine learning techniques related to local SVM algorithms. A theorical analysis studied in [9,10] illustrates that there is the trade-off between the generalisation capacity (the predicition correctness) and locality of local learning algorithms. The approaches of local SVMs can be categorized into two kinds of strategies (the hierarchical strategy and the nearest neighbors strategy).

The kind of local algorithms in the hierarchical strategy performs the training task with two main stages (the partition of the training set and learning local models). The first stage of learning task is to split the full training dataset into partitions (subsets) and then the second stage is to train the local supervised models from subsets. The learning algorithm proposed in [14] uses the expectation-maximization (EM) clustering algorithm [15] to split the full training dataset into k joint clusters (the EM clustering algorithm uses the posterior probabilities to make a soft cluster assignment [16]); And then the algorithm learns neural network (NN) models from clusters to locally classify the individuals in clusters. Instead of building local NN models as done by [14], the parallel mixture of SVMs algorithm in [17] proposes to learn local SVM models to predict the individuals in clusters. More recent, the Latent-lSVM [18,19] uses Latent Dirichlet Allocation (LDA [20]) to perform the clustering task for sparse data representation. CSVM [21], kSVR [22], and kSVM [23] propose to uses $k$means algorithm [24] to split the full training dataset into k disjoint clusters; And then kSVR and kSVM learn local non-linear SVM models in the parallel way, instead of training weighted local linear SVMs from clusters as done in CSVM. krSVM [25] is to learn the random ensemble of kSVM models. DTSVM [26,27] and tSVM [11] use decision tree algorithms [3,28] to split the full training dataset into t terminal-nodes (tree leaves); follow which the tSVM algorithm builds local SVM models for classifying impurity terminal-nodes (with mixture of labels) while DTSVM learns local SVM models from all tree leaves. These algorithms are shown to reduce the computational cost for dealing with large datasets while maintaining the prediction correctness.

The kind of local algorithms in the nearest neighbors strategy tries to find k nearest neighbors (kNN) of a new testing individual from the training dataset and then it only learns local supervised model from these kNN to classify the new testing individual. First local learning algorithm proposed in Bottou and Vapnik [9,10] is to train the neural network model from k neighborhoods to predict the label of the testing individual. The investigation of Vincent and Bengio [29] is to train k-local hyperplane and convex distance nearest neighbor (large margin nearest neighbors). Recent local SVM algorithms, including SVM-kNN [30], ALH [31], FaLK-SVM [32] find kNN of the testing individual with different techniques. SVM-kNN tries to use different metrics. ALH algorithm is to combine the weighted distance and features to predict the label of the testing individual. FaLK-SVM uses the cover tree [33] to improve the computational cost for finding the kNN in the feature kernel space.

6 Conclusion and Future Works

We presented the tSVR algorithm of local SVR that achieves high performances for the non-linear regression of large datasets. The training task of tSVR is to partition the full training dataset into k terminal-nodes. This aims at reducing data size in training local SVR. And then it easily learns k non-linear SVR models in the parallel way on multi-core computers in which a SVR model is trained in each terminal-node to predict the data locally. The numerical test results on datasets from UCI repository showed that our proposed tSVR is efficient in terms of training time and prediction correctness compared to the standard SVR in LibSVM. An example of its effectiveness is given with the non-linear regression of YearPredictionMSD dataset (having 400000 datapoints, 90 dimensions) in 3.297 min and 7.38 mean absolute error obtained on the testset.

In the near future, we intend to provide more empirical test on large benchmarks and comparisons with other algorithms. A promising avenue aims at improving the prediction correctness and automatically tuning hyperparameters of tSVR.

References

1. Vapnik, V.: The Nature of Statistical Learning Theory. Springer, Heidelberg (1995). https://doi.org/10.1007/978-1-4757-3264-1
2. Guyon, I.: Web page on SVM applications (1999). http://www.clopinet.com/isabelle/Projects/-SVM/app-list.html
3. Breiman, L., Friedman, J.H., Olshen, R.A., Stone, C.: Classification and Regression Trees. Wadsworth International, Belmont (1984)
4. Lichman, M.: UCI machine learning repository (2013)
5. Cristianini, N., Shawe-Taylor, J.: An Introduction to Support Vector Machines: and Other Kernel-Based Learning Methods. Cambridge University Press, New York (2000)
6. Platt, J.: Fast training of support vector machines using sequential minimal optimization. In: Schölkopf, B., Burges, C., Smola, A. (eds.) Advances in Kernel Methods Support Vector Learning, pp. 185–208 (1999)
7. OpenMP Architecture Review Board: OpenMP application program interface V3.0 (2008)
8. Vapnik, V.: Principles of risk minimization for learning theory. In: Advances in Neural Information Processing Systems 4, NIPS Conference, Denver, Colorado, USA, 2–5 December 1991, pp. 831–838 (1991)
9. Bottou, L., Vapnik, V.: Local learning algorithms. Neural Comput. **4**(6), 888–900 (1992)
10. Vapnik, V., Bottou, L.: Local algorithms for pattern recognition and dependencies estimation. Neural Comput. **5**(6), 893–909 (1993)
11. Do, T., Poulet, F.: Parallel learning of local SVM algorithms for classifying large datasets. T. Large-Scale Data-Knowl.-Cent. Syst. **31**, 67–93 (2016)
12. Chang, C.C., Lin, C.J.: LIBSVM : a library for support vector machines. ACM Trans. Intell. Syst. Technol. **2**(27), 1–27 (2011)
13. Lin, C.: A practical guide to support vector classification (2003)
14. Jacobs, R.A., Jordan, M.I., Nowlan, S.J., Hinton, G.E.: Adaptive mixtures of local experts. Neural Comput. **3**(1), 79–87 (1991)
15. Dempster, A.P., Laird, N.M., Rubin, D.B.: Maximum likelihood from incomplete data via the EM algorithm. J. Roy. Stat. Soc. B **39**(1), 1–38 (1977)
16. Bishop, C.M.: Pattern Recognition and Machine Learning. Springer, New York (2006)
17. Collobert, R., Bengio, S., Bengio, Y.: A parallel mixture of SVMs for very large scale problems. Neural Comput. **14**(5), 1105–1114 (2002)
18. Do, T., Poulet, F.: Classifying very high-dimensional and large-scale multi-class image datasets with latent-LSVM. In: 2016 International IEEE Conferences on Ubiquitous Intelligence & Computing, Advanced and Trusted Computing, Scalable Computing and Communications, Cloud and Big Data Computing, Internet of People, and Smart World Congress (UIC/ATC/ScalCom/CBDCom/IoP/SmartWorld), Toulouse, France, July 18–21, 2016, pp. 714–721 (2016)
19. Do, T., Poulet, F.: Latent-LSVM classification of very high-dimensional and large-scale multi-class datasets. Concurr. Comput.: Pract. Exp. e4224–n/a
20. Blei, D.M., Ng, A.Y., Jordan, M.I.: Latent Dirichlet allocation. J. Mach. Learn. Res. **3**, 993–1022 (2003)

21. Gu, Q., Han, J.: Clustered support vector machines. In: Proceedings of the Sixteenth International Conference on Artificial Intelligence and Statistics, AISTATS 2013, Scottsdale, AZ, USA, 29 April–1 May 2013, vol. 31, pp. 307–315. JMLR Proceedings (2013)
22. Bui, L.-D., Tran-Nguyen, M.-T., Kim, Y.-G., Do, T.-N.: Parallel algorithm of local support vector regression for large datasets. In: Dang, T.K., Wagner, R., Küng, J., Thoai, N., Takizawa, M., Neuhold, E.J. (eds.) FDSE 2017. LNCS, vol. 10646, pp. 139–153. Springer, Cham (2017). https://doi.org/10.1007/978-3-319-70004-5_10
23. Do, T.-N.: Non-linear classification of massive datasets with a parallel algorithm of local support vector machines. In: Le Thi, H.A., Nguyen, N.T., Do, T.V. (eds.) Advanced Computational Methods for Knowledge Engineering. AISC, vol. 358, pp. 231–241. Springer, Cham (2015). https://doi.org/10.1007/978-3-319-17996-4_21
24. MacQueen, J.: Some methods for classification and analysis of multivariate observations. In: Proceedings of 5th Berkeley Symposium on Mathematical Statistics and Probability, vol. 1, pp. 281–297. University of California Press, Berkeley, January 1967
25. Do, T.-N., Poulet, F.: Random local SVMs for classifying large datasets. In: Dang, T.K., Wagner, R., Küng, J., Thoai, N., Takizawa, M., Neuhold, E. (eds.) FDSE 2015. LNCS, vol. 9446, pp. 3–15. Springer, Cham (2015). https://doi.org/10.1007/978-3-319-26135-5_1
26. Chang, F., Guo, C.Y., Lin, X.R., Lu, C.J.: Tree decomposition for large-scale SVM problems. J. Mach. Learn. Res. **11**, 2935–2972 (2010)
27. Chang, F., Liu, C.C.: Decision tree as an accelerator for support vector machines. In: Ding, X., (ed.) Advances in Character Recognition. InTech (2012)
28. Quinlan, J.R.: C4.5: Programs for Machine Learning. Morgan Kaufmann, Burlington (1993)
29. Vincent, P., Bengio, Y.: K-local hyperplane and convex distance nearest neighbor algorithms. In: Advances in Neural Information Processing Systems, pp. 985–992. The MIT Press (2001)
30. Zhang, H., Berg, A., Maire, M., Malik, J.: SVM-KNN: discriminative nearest neighbor classification for visual category recognition. In: 2006 IEEE Computer Society Conference on Computer Vision and Pattern Recognition, Vol. 2, pp. 2126–2136 (2006)
31. Yang, T., Kecman, V.: Adaptive local hyperplane classification. Neurocomputing **71**(1315), 3001–3004 (2008)
32. Segata, N., Blanzieri, E.: Fast and scalable local kernel machines. J. Mach. Learn. Res. **11**, 1883–1926 (2010)
33. Beygelzimer, A., Kakade, S., Langford, J.: Cover trees for nearest neighbor. In: Proceedings of the 23rd International Conference on Machine Learning, pp. 97–104. ACM (2006)

LR-SDiscr: An Efficient Algorithm for Supervised Discretization

Habiba Drias(✉), Hadjer Moulai, and Nourelhouda Rehkab

LRIA Laboratory, Department of Computer Science, USTHB, Algiers, Algeria
{hdrias,hamoulai,nrehkab}@usthb.dz

Abstract. Discretization is the process of transforming continuous attributes into discrete. It has a great importance nowadays, as continuous data are often present in several domains such as health and industry. This paper describes a new supervised discretization method based on a LR (Left to Right) scanning technique called LR-SDiscr (**Left** to **Right** **S**upervised **Discr**etization). Using both merging and partitioning operations, LR-SDiscr discretizes the data in a single pass, which reduces the complexity of the process and ensures scalability. Various discretization measures can be tested and then compared, as the algorithm offers the possibility of introducing any discretization measure as input. The preliminary results of experiments designed for classification purposes are encouraging.

Keywords: Data mining · Data pre-processing
Supervised classification · Supervised discretization
Division and merging framework · Scanner

1 Introduction

In data mining [3], data pre-processing is the first step before launching the knowledge discovery process. It consists of a set of tasks that aim to improve the quality of the data, and hence that of the mining phase. Discretization is a pre-processing task that transforms continuous data into discrete. Its main goal is to reduce the number of values of a continuous attribute by summarizing it to a few intervals and thus ensuring scalability. This process plays a major role in improving the performance of classification algorithms as it makes learning more accurate and faster. In fact, most of the machine learning and data mining algorithms are designed to manipulate discrete data, while in real life, data often comes in a continuous form. Moreover, discrete data are easier to use and understand comparing to continuous data. A typical discretization process consists of four steps; (1) sorting the continuous values, (2) looking for the best cut-point in case of division, or the best adjacent intervals for merging, (3) partitioning or merging the data according to some criterion, (4) halting the process when a stop criterion is met. Cut-points are determined using discretization measures

N. T. Nguyen et al. (Eds.): ACIIDS 2018, LNAI 10751, pp. 266–275, 2018.
https://doi.org/10.1007/978-3-319-75417-8_25

such as binning, entropy, χ^2...etc. These measures differ on whether they use the class information or not.

Discretization methods are roughly categorized in two classes: supervised and unsupervised. Supervised methods takes account of the class information, while unsupervised methods ignore it. In the early days of discretization, unsupervised binning methods like equal-width and equal-frequency were widely used. Binning is a technique that aims to fairly distribute the sorted values into several bins or partitions, according to a specified width or frequency. In equal-width, all the bins have the same width or size. In equal-frequency, all the bins include the same number of values. Equal-width and equal-frequency did not give satisfactory results. So, supervised methods were then introduced with the purpose of achieving better results through the use of class information.

On the other hand, there are two approaches the process of discretization may undertake: a top-down strategy or a bottom-up design. The methods that use the top-down strategy are based on division operations. The process starts with one partition containing all continuous values, then looks for the best cut-point to divide it into two sub-intervals. This process is reiterated on the resulting intervals until a certain stop criterion is met. The bottom-up discretization methods are based on merging actions. They consider each distinct value as a potential cut-point, then the adjacent values are merged according to their similarity degree. The process is reiterated until no merging is possible or some stop criterion is satisfied.

In this paper, we present a new supervised discretization method called LR-SDiscr, a truly original discretization algorithm of type LR (Left to Right) that reads the data sequence from left to right and merge two adjacent values in one interval if they satisfy the merging measure, otherwise, a cut-point is determined. LR-SDiscr was implemented and extensive experiments for classification purposes were undertaken on large data sets. The results show the dominance of the proposed algorithm on the well-known ChiMerge method and recent state-of-the-art algorithms.

The remaining of this paper is organized as follows. The next section reviews related work. Section 3 provides an overview of the ChiMerge algorithm, as our proposal is challenging this technique in particular. In Sect. 4, the suggested algorithm is presented. Section 5 reports the outcomes of the extensive performed experiments. The achieved results are then compared with those of ChiMerge and recent state-of-the-art algorithms. In Sect. 6 we conclude this work and discuss about future researches.

2 Related Work

In this section, a non-exhaustive chronological survey of the most relevant works in discretization research is presented.

ChiMerge [6], is the first method to adopt the merging approach for discretization. It is a supervised bottom-up technique that uses the χ^2 measure to evaluate the relationship between two adjacent intervals based on their dependency on the class. The number of intervals is limited by the threshold but another parameter, maximum number of intervals, can be used to specify the number of constructed intervals.

Chi2 [9] is an improved version of ChiMerge where the threshold varies automatically to yield better results. It uses the χ^2 measure for discretization and the data inconsistency level as a stop criterion. Also, the number of inconsistencies is used to eliminate noise.

An expansive synthesis on discretization techniques is presented in [10]. A detailed overview of existing methods is done and a comparative study is performed using different discretization methods. The authors concluded that despite the important work that has been achieved, there still are unsolved issues and new methods are needed.

In [7], the authors presented a supervised discretization method, called CAIM (Class-Attribute Interdependence Maximization). They describe an optimal discretization as maximization of the class-attribute interdependence and minimization of the number of intervals. The experiments showed superiority of CAIM compared to state-of-the-art algorithms at that time. In [11] the authors proposed another version of the Chi2 method, that considers the effect of variance in merging two intervals, and determines the predefined inconsistency rate based on the least upper bound of data misclassification error. It uses the inconsistency level of the data as a stop criterion.

In [8], the authors proposed an improved version of the CAIM algorithm, called Class-Attribute Contingency Coefficient (CACC). It takes in consideration the data distribution in order to yield better classification results. Experimental results showed that CACC yields a better performance for classification purposes than CAIM.

In [2], the author designed a hybrid algorithm called ChiD, based on ChiMerge and Chi2. It uses the logworth criterion to determine the best set of cut-points.

In [4], the authors proposed four discretization techniques based on the entropy measure. Experimental results showed that the multiple scanning technique is the best.

In [5], the authors undertook a comparative study of four discretization methods based on the entropy measure: the C4.5 approach, the equal interval width method, the equal frequency interval method and the multiple scanning method. The methods are evaluated using two parameters: an error rate determined by 10-fold cross validation, and the size of the tree generated by C4.5. The experimental results showed that the multiple scanning method is the best.

In this work, we present a new original method based on a left to right scanning framework and ChiMerge algorithm. The next section gives an overview of the most important features of the latter.

3 A Brief Overview of ChiMerge

ChiMerge is a supervised and bottom-up discretization method. It is the first discretization technique that used the merging approach and the χ^2 measure to decide whether two adjacent intervals are to be merged or not.

The author defines an accurate discretization, as that where intra-interval uniformity and inter-interval difference are ensured. In other words, an interval must have a fairly consistent relative class frequency, if not it should be split, and two adjacent intervals should have different relative class frequencies, otherwise they should be merged.

ChiMerge performs discretization by starting to sort continuous values in an ascending order and considers each value in an interval. Then it performs the χ^2 test for all the adjacent intervals, and merge those with the smallest test value. The process stops when there are no more intervals to merge. Note that the sorting step is very important, as the choice of the sorting algorithm can influence the overall time complexity of ChiMerge. The χ^2 test is performed using Eq. (1).

$$\chi^2 = \sum_{i=1}^{2} \sum_{j=1}^{p} \frac{(A_{ij} - E_{ij})^2}{E_{ij}} \tag{1}$$

where:

p is the number of classes,
A_{ij} is the number of distinct values in the interval i of class j,
R_i is the number of examples in the interval i, which is equal to $\sum_{j=1}^{p} A_{ij}$
C_j is the number of examples in class j, which is equal to $\sum_{i=1}^{m} A_{ij}$
n is the total number of examples, which is equal to $\sum_{j=1}^{p} C_j$
E_{ij} is the expected frequency of A_{ij} computed as: $(R_i * C_j)/n$.

If two adjacent intervals have different relative frequencies, χ^2 determines whether the relationship between the continuous attribute and the class is real or the result of luck.

The process of discretization stops when all adjacent intervals have a χ^2 value greater than a certain χ^2 threshold. The latter is set according to a significance level. However, selecting a significance level can be challenging. In fact, too high, it makes the process of discretization last longer and produces few intervals, resulting in an over-discretization. And too small, it causes an under-discretization and produces a big number of intervals. In order to address this issue, two other parameters can be used: min-intervals (the minimal number of intervals) and max-intervals (the maximum number of intervals).

ChiMerge is an easy to use robust algorithm, which uses a supervised discretization measure to construct intervals and produces a concise summarization of continuous attributes to help the experts understand the relationship between numerical attributes and the class in a simpler way. In addition, it is facile to find a suitable setup. Indeed, tuning the threshold at .90 or .99 significance level, and fixing the maximum number of intervals between 10 and 15 generally produces good discretization.

One of the biggest disadvantages of ChiMerge is its worst case time complexity. In fact, the latter is of $O(n^2)$ which represents the number of calls to the χ^2 test function, where n is the number of instances. With serious optimizations, it can though be reduced to $O(nlogn)$. For example, it is not necessary to repeat the χ^2 test on pairs that have not been merged at the previous iteration. Also, taking in account the merging of non-adjacent intervals can yield a better performance. Another disadvantage of ChiMerge is its noise tolerance.

Chi2 [9], is an improved version of ChiMerge that automatically changes the χ^2 significance level and takes account of the inconsistency rate in order to yield better results and avoid noise. However, the user has to provide the maximum inconsistency rate to stop the discretization process. In addition, Chi2 ignores the independence of the intervals, which has a significant impact on the discretization scheme. Modified Chi2 is a fully automated discretization method, which considers the independence and replaces the inconsistency test by the level of consistency coined from Rough Sets Theory, resulting in a better predictive accuracy than Chi2. Extended Chi2 is another improved version of Chi2, which takes into account the effect of the variance when merging two intervals and determines the predefined inconsistency rate of the data, allowing it to cope with uncertain data. Extended Chi2 is known to outperform other bottom-up algorithms.

4 Our Proposal

The originality of our work is the integration of a lexical analyser in the discritization process with the aim of reducing time and hence gainning in scalability. The second originality is that our method performs merging and division operations in the same process, a technique that was never undertaken before. Indeed, the main drawback of ChiMerge is its repetitive calculation of the χ^2 test at each iteration, resulting in a quadratic time complexity. In order to palliate this issue, we propose a new supervised discretization method based on a scanning technique, that discretizes the data in one pass only by browsing it in a LR (Left to Right) manner. Chimerge executes the discritization algorithm in a number of passes equal to the number of generated intervals. The advantage of our contribution is therefore very significant with respect to the previous efforts.

The main principle of LR-SDiscr is as follows; after setting the empirical parameters (threshold, maximum number of intervals), the attribute values are sorted using the heap-sort algorithm, that we extended to count the frequency of each value. Then, each value is considered within a distinct range of equal limits (i.e. v_i in the interval $[v_i]$) and written, in order, to the input file with its corresponding frequency. After that, the discretization program is launched. The process stops when it reaches the end of file or if the maximum number of intervals is reached.

4.1 Discretization Using χ^2

The first variant of LR-SDiscr uses the χ^2 statistic as a division/merging measure. The discretization program reads the input file and discretizes the data as follows:

1. Read the first two values v_0, v_1 and their respective frequencies f_0, f_1.
2. Compute the χ^2 test for v_0 and v_1.
3. If the χ^2 test is bigger than the χ^2 threshold then
   (a) v_1 is recognized as a cut-point.
   (b) read the next value v_2 and its frequency f_2.
   (c) Compute the χ^2 test for v_1 and v_2.
   ...
4. If the χ^2 test is lower than the χ^2 threshold then
   (a) merge v_0 and v_1 in one interval $[v_0, v_1]$.
   (b) read the next value v_2 and its frequency f_2.
   (c) Compute the χ^2 test for $[v_0, v_1]$ and v_2.
   ....

The process goes on until there are no more values in the input file, or a stop criterion is satisfied.

4.2 Discretization Using the Entropy

The entropy is one of the most used measure in supervised discretization methods. Entropy of a variable X is defined by Eq. (2):

$$H(X) = -\sum_x p_x \log p_x \tag{2}$$

where: x represents a value of X and p_x the estimated probability of occurrence of x. This is the average amount of information per event, where the information is defined by Eq. (3).

$$I(x) = -\log p_x \tag{3}$$

Information is high for unlikely events and low otherwise. Consequently, the entropy H is high when all events are equally likely ($p_{x_i} = p_{x_j}$), and low when the probability of an event is equal to 1 ($p_x = 1$) and that of the others to 0.

The discretization using the entropy is done in the following way:

1. Read the first two values v_0, v_1.
2. Compute the entropy of $[v_0]$ and $[v_1]$.
3. If the entropy of $[v_0]$ and $[v_1]$ is bigger than the entropy threshold then
   (a) v_1 is recognized as a cut-point.
   (b) read the next value v_2.
   (c) Compute the entropy of $[v_1]$ and $[v_2]$.
   ...
4. If entropy of $[v_0]$ and $[v_1]$ is lower than the entropy threshold then
   (a) merge $[v_0]$ and $[v_1]$ in one interval $[v_0, v_1]$.
   (b) read the next value v_2.
   (c) Compute the entropy of $[v_0, v_1]$ and v_2.
   ....

The process is repeated until it reaches the end of the file or the maximum number of intervals.

5 Experimental Results

In order to validate our approach, extensive experiments were undertaken. The obtained results are exposed, commented and thereafter compared with ChiMerge and recent works in terms of accuracy, speed and scalability. The method was implemented using C language on a machine of Processor Intel Core i5-3317U CPU @ 1.70 GHz x 4 and a RAM of 4.00 GB under the Ubuntu 14.04 LTS 64-bit operating system.

5.1 Evaluation Set up

Large public data sets [1], illustrated in Table 1 have been used to demonstrate the performance of our method.

Table 1. Large data sets

Data set	#instances	#attributes
Ionos	351	34
German	1000	20
Yeast	1484	8
Hypothyroid	3772	29
Abalone	4177	8
satimage	6435	36
Handwritten digits	10992	16
Letter recognition	20000	16
Shuttle	57999	9

For the evaluation of our discretization scheme, we used the 10-fold cross validation technique in combination with C4.5, as this algorithm is mostly used in machine learning. It consists of splitting the data into 10 subsets, and repeating the following process 10 times: select one subset as a validation set and consider the remaining 9 subsets as a training set, then calculate the error rate using C4.5. The prediction error is estimated by the average of the 10 errors. In addition to the computed prediction error, the execution time of the discretization process is saved.

5.2 The Comparison of LR-SDiscr with Other Methods

LR-SDiscr is compared with ChiMerge and two recent supervised discretization methods: ChiD [2] and Multiple scanning [4].

Table 2 presents a comparison of the error rates for LR-SDiscr using χ^2, LR-SDiscr using the entropy, Multiple Scanning, ChiD and ChiMerge. It is clear

Table 2. Comparing C4.5 error rates for LR-SDiscr using χ^2, LR-SDiscr using the entropy, multiple scanning, ChiD and ChiMerge

Data set	LR-SDiscr χ^2	LR-SDisrc entropy	Multiple scanning	ChiD	ChiMerge
Ionos	**8,92**	10,226	10,137	17,558	11,683
German	28,389	**27,42**	28,22	30	29,41
Yeast	**59,896**	64,458	66,309	68,38	68,809
Hypothyroid	**0,844**	7,719	-[1]	3,01	1,3
Abalone	76,097	81,275	76,945	-	**74,974**
Satimage	**14,85**	63,239	-	-	19,949
Handwritten digits	**6,761**	29,49	-	-	12,539
Letter	**20,253**	20,969	-	-	22,139
Shuttle	**0,063**	4,51	-	-	0,13

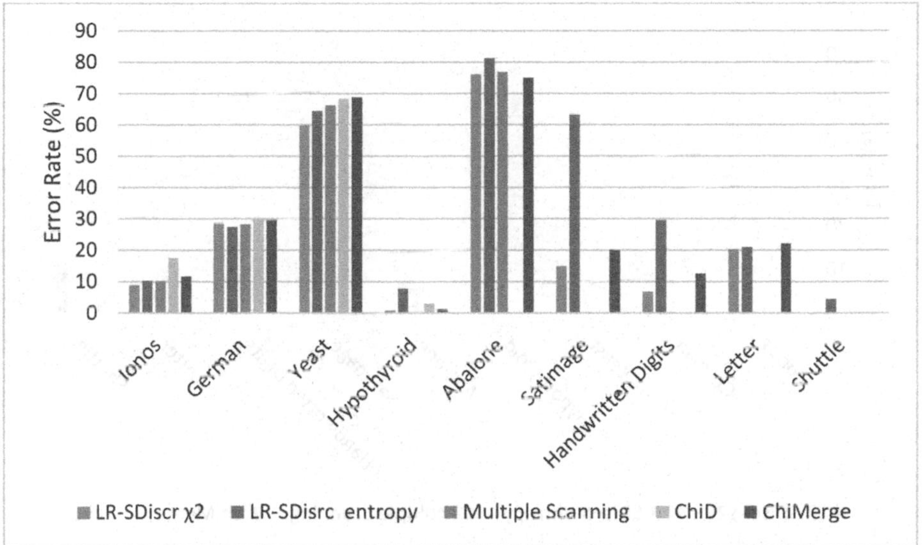

Fig. 1. C4.5 error rates for LR-SDiscr using χ^2, LR-SDiscr using the entropy, multiple scanning, ChiD and ChiMerge

from Table 2 that LR-SDiscr using χ^2 achieves the best outcomes as it registers the lowest error rate in 7 cases over 9, whereas ChiMerge is the lowest in one data set only. On the other hand, we observe that Multiple Scanning and ChiD ran out of memory when the number of instances or/and attributes becomes too big. Figure 1 exhibits a graphical view of the error rates, which confirms that LR-SDiscr using χ^2 is the most effective overall. We also observe that when the size of the data sets increases, LR-SDiscr using the entropy alongside ChiMerge

Table 3. Comparing execution time for LR-SDiscr using χ^2, LR-SDiscr using the entropy, multiple scanning, ChiD and ChiMerge

Data set	LR-SDiscr χ^2	LR-SDisrc entropy	Multiple scanning	ChiD	ChiMerge
Ionos	0,056	**0,027**	1,341	51,903	0.792
German	0,017	**0,005**	0,234	10,518	0.157
Yeast	0,058	**0,027**	0,637	8,561	0.152
Hypothyroid	0,381	**0,028**	-	57,413	1.050
Abalone	0,906	**0,068**	6,354	-	7.341
Satimage	1,671	**0,441**	-	-	3.213
Handwritten digits	2,363	**0,274**	-	-	3.095
Letter	**0,487**	0,608	-	-	1.028
Shuttle	3,384	**0,558**	-	-	11.959

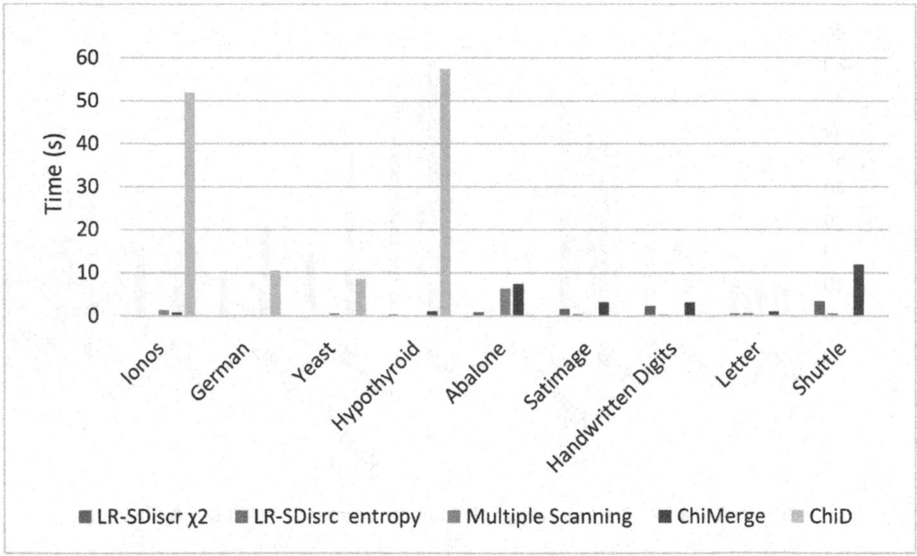

Fig. 2. Execution time for LR-SDiscr using χ^2, LR-SDiscr using the entropy, multiple scanning, ChiMerge and ChiD.

are the less effective. A comparison of the amount of time taken for discretization by each of the five algorithms is shown in Table 3.

From this table we notice that LR-SDiscr, using both measures, is faster than ChiMerge, Multiple Scanning and ChiD. LR-SDiscr using the entropy is the fastest in 8 data sets over 9, but the gap between it and the other variant of LR-SDiscr is small. Also, we observe that ChiD and Multiple Scanning ran out of memory when a data set have a large number of instances, while LR-SDiscr still perform discretization in a reasonable amount of time. Figure 2 gives an overall view of the execution time for the five algorithms. We observe that ChiD takes the most time for discretization, and when the latter ran out of memory, ChiMerge is the slowest comparing to the other methods.

6 Conclusion and Future Work

In this paper we presented a new supervised, bottom-up discretization method integrating a lexical analyser while keeping the discretization idea of Chimerge. The objective was the reduction of time complexity of ChiMerge, which is of $O(n^2)$, regardless to the data sorting of the initial step. For this purpose, we designed a new discretization framework that uses both division and fusion operations. The data is discretized in a single pass reducing the time complexity to $O(n)$, without considering the first step of data sorting. In addition, the discretization measure can easily be modified.

The measures of entropy and χ^2 were exploited, and experiments showed that both achieved promising results comparing to ChiMerge, ChiD and Multiple Scanning. And from the comparative results, it was concluded that the χ^2 measure is the best.

As future researches, we propose to investigate other discretization measures in order to compare their performance and see if the χ^2 measure remains the best.

References

1. Aha, D., et al.: UCI repository of machine learning databases (2017). http://www.ics.uci.edu/mlearn/MLRepository.html
2. Bettinger, R.: A chi^2-based discretization algorithm, a modern analytics. In: Proceedings of WUSS (2011)
3. Han, J., Kamber, M.: Data Mining Concepts and Techniques. Morgan Kaufmann Publishers, Burlington (2011)
4. Grzymala-Busse, J.W.: Discretization based on entropy and multiple scanning. Entropy **15**, 1486–1502 (2013)
5. Grzymala-Busse, J.W., Mroczek, T.: A comparison of four approaches to discretization based on entropy. Entropy **18**, 69 (2016)
6. Kerber, R.: ChiMerge discretization of numeric attributes. In: AAAI Proceedings, pp. 123–128 (1992)
7. Kurgan, L., Cios, K.J.: CAIM discretization algorithm. IEEE Trans. Knowl. Data Eng. **16**, 145–153 (2004)
8. Lee, C.I., Tsai, C.J., Yang, Y.R., Yang, W.P.: A top-down and greedy method for discretization of continuous attributes. In: Proceedings of ICFSKD, pp. 145–153 (2007)
9. Liu, H., Setiono, R.: Feature selection and discretization. IEEE Trans. Knowl. Data Eng. **9**, 1–4 (1997)
10. Liu, H., Hussain, F., Tan, C.L., et al.: Discretization: an enabling technique. Data Min. Knowl. Discov. **6**, 393–423 (2002)
11. Su, C.T., Hsu, J.H.: An extended Chi2 algorithm for discretization of real value attributes. IEEE TKDE **17**, 437–441 (2005)

Path Histogram Distance for Rooted Labeled Caterpillars

Taiga Kawaguchi[1], Takuya Yoshino[1], and Kouichi Hirata[2(✉)]

[1] Graduate School of Computer Science and Systems Engineering, Kyushu Institute of Technology, Kawazu 680-4, Iizuka 820-8502, Japan
{kawaguchi,yoshino}@dumbo.ai.kyutech.ac.jp

[2] Department of Artificial Intelligence, Kyushu Institute of Technology, Kawazu 680-4, Iizuka 820-8502, Japan
hirata@dumbo.ai.kyutech.ac.jp

Abstract. In this paper, we focus on a *caterpillar* as a rooted labeled unordered tree (a tree, for short) transformed to a path after removing all the leaves in it. Also we introduce a *path histogram distance* between trees as an L_1-distance between the histograms of paths from the root to every leaf. Whereas the path histogram distance is not a metric for trees, we show that, for caterpillars, it is always a metric, simply linear-time computable and incomparable with the edit distance. Furthermore, we give experimental results for caterpillars in real data of comparing the path histogram distance with the isolated-subtree distance as the most general tractable variation of the edit distance.

1 Introduction

Comparing tree-structured data such as HTML and XML data for web mining or DNA and glycan data for bioinformatics is one of the important tasks for data mining. The most famous distance measure between *rooted labeled unordered trees* (*trees*, for short) is the *edit distance* [7]. The edit distance is formulated as the minimum cost of *edit operations*, consisting of a *substitution*, a *deletion* and an *insertion*, applied to transform from a tree to another tree.

Unfortunately, it is known that the problem of computing the edit distance between trees is MAX SNP-hard [11]. Furthermore, this statement also holds even if trees are binary [3].

Many variations of the edit distance have developed as more structurally sensitive distances (*cf.*, [5,9]). Almost variations are metrics and the problem of computing them is tractable (cubic-time computable) [8–10]. In particular, the *isolated-subtree distance* (or *constrained distance*) [10] is the most general tractable variation of the edit distance [9].

This work is partially supported by Grant-in-Aid for Scientific Research 17H00762, 16H02870, 16H01743 and 15K12102 from the Ministry of Education, Culture, Sports, Science and Technology, Japan.

N. T. Nguyen et al. (Eds.): ACIIDS 2018, LNAI 10751, pp. 276–286, 2018.
https://doi.org/10.1007/978-3-319-75417-8_26

On the other hand, the histogram distances based on local information [1, 4,6] are also known as the constant-factor lower bounding distances of the edit distance. Whereas we can compute them more efficiently (linear-time computable in almost cases) than the variations of the edit distance, none of them is a metric.

In order to compare trees with preserving both the efficiency and the metricity, in this paper, we introduce a *path histogram distance* as one of the histogram distances and restrict trees to *caterpillars*.

A *path histogram* of a tree consists of pairs of a path from the root to a leaf and its frequency in the tree. Then, a *path histogram distance* for two trees is an L_1-distance between the path histograms for the two trees. Note that the path histogram distance is not a metric for trees in general.

A *caterpillar* (*cf.* [2]) is a tree transformed to a path after removing all the leaves in it. Whereas the caterpillars are very restricted and simple, there are many caterpillars in real dataset, see Table 1 in Sect. 4. Then, it is meaningful to formulate a metric between caterpillars.

Hence, we investigate the path histogram distance for caterpillars. First, for caterpillars, we show that the path histogram distance is always a metric, we can compute it in linear time and it is incomparable with the edit distance and its variations. Furthermore, we give experimental results for caterpillars in real data of comparing the path histogram distance with the isolated-subtree distance.

2 Preliminaries

A *tree* T is a connected graph (V, E) without cycles, where V is the set of vertices and E is the set of edges. We denote V and E by $V(T)$ and $E(T)$. The *size* of T is $|V|$ and denoted by $|T|$. We sometime denote $v \in V(T)$ by $v \in T$. We denote an empty tree $(\emptyset, \emptyset)$ by $\emptyset$. A *rooted tree* is a tree with one node r chosen as its *root*. We denote the root of a rooted tree T by $r(T)$.

Let T be a rooted tree such that $r = r(T)$ and $u, v, w \in T$. We denote the unique path from r to v, that is, the tree (V', E') such that $V' = \{v_1, \ldots, v_k\}$, $v_1 = r$, $v_k = v$ and $(v_i, v_{i+1}) \in E'$ for every i $(1 \leq i \leq k-1)$, by $UP_r(v)$. The *parent* of $v(\neq r)$ is its adjacent node on $UP_r(v)$ and the *ancestors* of $v(\neq r)$ are the nodes on $UP_r(v) - \{v\}$. We say that u is a *child* of v if v is the parent of u and u is a *descendant* of v if v is an ancestor of u. We use the ancestor orders $<$ and $\leq$, that is, $u < v$ if v is an ancestor of u and $u \leq v$ if $u < v$ or $u = v$. We say that w is the *least common ancestor* of u and v, denoted by $u \sqcup v$, if $u \leq w$, $v \leq w$ and there exists no node $w' \in T$ such that $w' \leq w$, $u \leq w'$ and $v \leq w'$. The *degree* of v is the number of children of v and the *degree* of T, denoted by $d(T)$, is the maximum number of degrees for every node in T. We call a node with no children a *leaf* and denote the set of all the leaves in T by $lv(T)$.

We say that u is *to the left of* v in T if $pre(u) \leq pre(v)$ for the preorder number pre in T and $post(u) \leq post(v)$ for the postorder number $post$ in T. We say that a rooted tree is *ordered* if a left-to-right order among siblings is given; *unordered* otherwise. We say that a rooted tree is *labeled* if each node is assigned a symbol from a fixed finite alphabet Σ. For a node v, we denote the label of

v by $l(v)$, and sometimes identify v with $l(v)$. In this paper, we call a rooted labeled unordered tree a *tree* simply.

Next, we introduce an *edit distance* and a *Tai mapping*.

Definition 1 (Edit operations [7]**).** The *edit operations* of a tree T are defined as follows, see Fig. 1.

1. *Substitution*: Change the label of the node v in T.
2. *Deletion*: Delete a node v in T with parent v', making the children of v become the children of v'. The children are inserted in the place of v as a subset of the children of v'. In particular, if v is the root in T, then the result applying the deletion is a forest consisting of the children of the root.
3. *Insertion*: The complement of deletion. Insert a node v as a child of v' in T making v the parent of a subset of the children of v'.

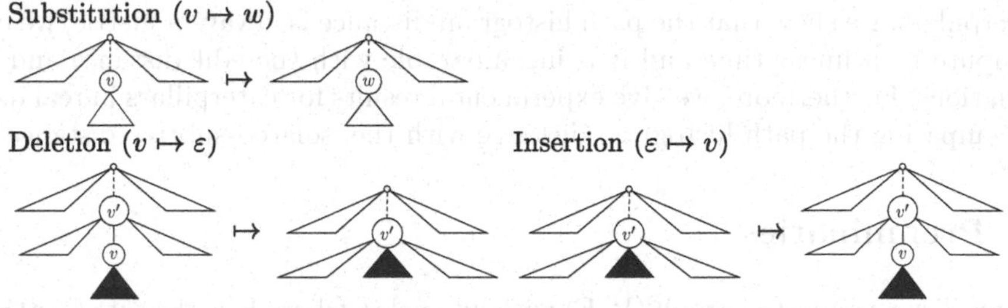

Fig. 1. Edit operations for trees.

Let $\varepsilon \notin \Sigma$ denote a special *blank* symbol and define $\Sigma_\varepsilon = \Sigma \cup \{\varepsilon\}$. Then, we represent each edit operation by $(l_1 \mapsto l_2)$, where $(l_1, l_2) \in (\Sigma_\varepsilon \times \Sigma_\varepsilon - \{(\varepsilon, \varepsilon)\})$. The operation is a substitution if $l_1 \neq \varepsilon$ and $l_2 \neq \varepsilon$, a deletion if $l_2 = \varepsilon$, and an insertion if $l_1 = \varepsilon$. For nodes v and w, we also denote $(l(v) \mapsto l(w))$ by $(v \mapsto w)$. We define a *cost function* $\gamma : (\Sigma_\varepsilon \times \Sigma_\varepsilon \setminus \{(\varepsilon, \varepsilon)\}) \mapsto \mathbf{R}^+$ on pairs of labels. We often constrain a cost function γ to be a *metric*, that is, $\gamma(l_1, l_2) \geq 0$, $\gamma(l_1, l_2) = 0$ iff $l_1 = l_2$, $\gamma(l_1, l_2) = \gamma(l_2, l_1)$ and $\gamma(l_1, l_3) \leq \gamma(l_1, l_2) + \gamma(l_2, l_3)$. In particular, we call the cost function that $\gamma(l_1, l_2) = 1$ if $l_1 \neq l_2$ a *unit cost function*.

Definition 2 (Edit distance [7]**).** For a cost function γ, the *cost* of an edit operation $e = l_1 \mapsto l_2$ is given by $\gamma(e) = \gamma(l_1, l_2)$. The *cost* of a sequence $E = e_1, \ldots, e_k$ of edit operations is given by $\gamma(E) = \sum_{i=1}^{k} \gamma(e_i)$. Then, an *edit distance* $\tau_{\text{TAI}}(T_1, T_2)$ between trees T_1 and T_2 is defined as follows:

$$\tau_{\text{TAI}}(T_1, T_2) = \min \left\{ \gamma(E) \;\middle|\; \begin{array}{l} E \text{ is a sequence of edit operations} \\ \text{transforming } T_1 \text{ to } T_2 \end{array} \right\}.$$

Definition 3 (Tai mapping [7]**).** Let T_1 and T_2 be trees. We say that a triple (M, T_1, T_2) is a *Tai mapping* (a *mapping*, for short) from T_1 to T_2 if $M \subseteq V(T_1) \times V(T_2)$ and every pair (v_1, w_1) and (v_2, w_2) in M satisfies that (1) $v_1 = v_2$ iff $w_1 = w_2$ (one-to-one condition) and (2) $v_1 \leq v_2$ iff $w_1 \leq w_2$ (ancestor condition). We will use M instead of (M, T_1, T_2) when there is no confusion denote it by $M \in \mathcal{M}_{\text{TAI}}(T_1, T_2)$.

Let M be a mapping from T_1 to T_2. Let I_M and J_M be the sets of nodes in T_1 and T_2 but not in M, that is, $I_M = \{v \in T_1 \mid (v, w) \notin M\}$ and $J_M = \{w \in T_2 \mid (v, w) \notin M\}$. Then, the *cost* $\gamma(M)$ of M is given as follows.

$$\gamma(M) = \sum_{(v,w)\in M} \gamma(v, w) + \sum_{v\in I_M} \gamma(v, \varepsilon) + \sum_{w\in J_M} \gamma(\varepsilon, w).$$

Trees T_1 and T_2 are *isomorphic*, denoted by $T_1 \equiv T_2$, if there exists a mapping $M \in \mathcal{M}_{\text{TAI}}(T_1, T_2)$ such that $I_M = J_M = \emptyset$ and $\gamma(M) = 0$.

Theorem 1 (Tai [7]**).** $\tau_{\text{TAI}}(T_1, T_2) = \min\{\gamma(M) \mid M \in \mathcal{M}_{\text{TAI}}(T_1, T_2)\}$.

Unfortunately, the following theorem holds for computing τ_{TAI}.

Theorem 2 (*cf.*, [3,11]). *Let T_1 and T_2 be trees. Then, the problem of computing $\tau_{\text{TAI}}(T_1, T_2)$ is* MAX SNP*-hard, even if both T_1 and T_2 are binary.*

Finally, we introduce an *isolated-subtree mapping* and an *isolated-subtree distance* as the variations of the Tai mapping and the edit distance.

Definition 4 (Isolated-subtree mapping and distance [10]**).** Let T_1 and T_2 be trees and $M \in \mathcal{M}_{\text{TAI}}(T_1, T_2)$. We say that M is an *isolated-subtree mapping*, denoted by $M \in \mathcal{M}_{\text{ILST}}(T_1, T_2)$, if M satisfies the following condition.

$$\forall (v_1, w_1)(v_2, w_2)(v_3, w_3) \in M \Big(v_3 < v_1 \sqcup v_2 \iff w_3 < w_1 \sqcup w_2 \Big).$$

Furthermore, we define an *isolated-subtree distance* $\tau_{\text{ILST}}(T_1, T_2)$ as follow.

$$\tau_{\text{ILST}}(T_1, T_2) = \min\{\gamma(M) \mid M \in \mathcal{M}_{\text{ILST}}(T_1, T_2)\}.$$

It is obvious that $\mathcal{M}_{\text{ILST}}(T_1, T_2) \subseteq \mathcal{M}_{\text{TAI}}(T_1, T_2)$ and then $\tau_{\text{TAI}}(T_1, T_2) \leq \tau_{\text{ILST}}(T_1, T_2)$. In contrast to Theorem 2, the following theorem also holds.

Theorem 3 (*cf.*, [8]). Let T_1 and T_2 be trees. Then, we can compute $\tau_{\text{ILST}}(T_1, T_2)$ in $O(n^2 d)$ time, where $n = \max\{|T_1|, |T_2|\}$ and $d = \min\{d(T_1), d(T_2)\}$.

It is known that τ_{ILST} is the most general tractable variation of τ_{TAI} [9].

3 Path Histogram Distance for Caterpillars

We say that a tree C such that $r = r(C)$ is a *caterpillar* (*cf.* [2]) if C is transformed to a path $UP_r(v)$ for some $v \in C$ after removing $lv(C)$. For a caterpillar C, we call the remained path $UP_r(v)$ a *backbone* of C and denote it by $bb(C)$. It is obvious that $V(C) = V(bb(C)) \cup lv(C)$. We can determine whether or not a tree T is a caterpillar in $O(|T|)$ time.

Let T be a tree such that $r = r(T)$. Then, for $v \in lv(T)$, we regard the path $P = UP_r(v)$ such that $V(P) = \{v_1, \ldots, v_k\}$, $v_1 = r$, $v_k = v$ and $(v_i, v_{i+1}) \in E(P)$ for every i ($1 \leq i \leq k-1$) as a string $l(v_1) \cdots l(v_k)$ on Σ and denote it by $s(r, v)$. Also we say that a string $s \in \Sigma^*$ *occurs* in T if there exists a leaf $v \in lv(T)$ such that $s = s(r, v)$ and denote the number of occurrences of s in T by $f(s, T)$. Furthermore, we define $S(T)$ as $\{s(r, v) \mid r = r(T), v \in lv(T)\}$.

Definition 5 (Path histogram and path histogram distance). For a tree T, a *path histogram* $H(T)$ of T consists of pairs $\langle s, f(s, T)\rangle$ for every $s \in S(T)$.

For trees T_1 and T_2, a *path histogram distance* $\delta_{\mathrm{PH}}(T_1, T_2)$ between T_1 and T_2 is defined as an L_1-distance between $H(T_1)$ and $H(T_2)$, that is:

$$\delta_{\mathrm{PH}}(T_1, T_2) = \sum_{s \in S(T_1) \cup S(T_2)} |f(s, T_1) - f(s, T_2)|.$$

Example 1. For trees, δ_{PH} is not a metric in general. For trees T_1 and T_2 in Fig. 2, it holds that $\delta_{\mathrm{PH}}(T_1, T_2) = 0$ (because $H(T_i) = \{\langle aba, 2\rangle\}$) but $T_1 \not\equiv T_2$.

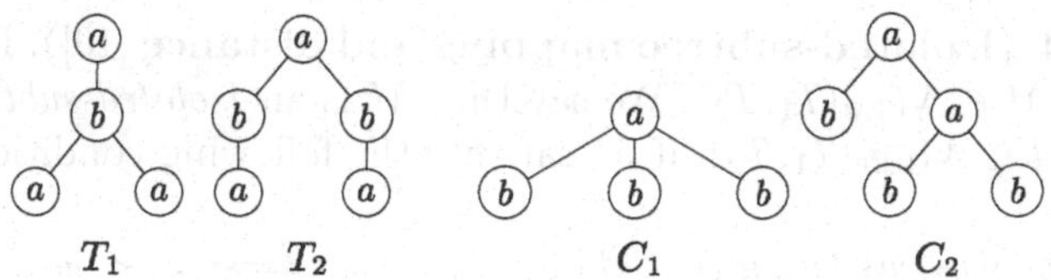

Fig. 2. Trees T_1 and T_2 and caterpillars C_1 and C_2.

In Fig. 2, T_1 is a caterpillar but T_2 is not. If both trees are caterpillars, then the following theorem holds.

Theorem 4. *For caterpillars, δ_{PH} is always a metric.*

Proof. By the definition, it is sufficient to show that $\delta_{\mathrm{PH}}(C_1, C_2) = 0$ iff $C_1 \equiv C_2$ for caterpillars C_1 and C_2. In other words, it is sufficient to show that we can transform C from $H(C)$ uniquely.

Let C be a caterpillar such that $r = r(C)$. Suppose that the backbone $bb(C)$ of C consists of nodes $v_1, \ldots, v_k$ such that $v_1 = r$. We denote a string $l(v_1) \cdots l(v_k)$ representing $bb(C)$ by $s(C)$. Then, for every leaf $v \in lv(C)$, $s(r, v)$ is of the form $l(v_1) \cdots l(v_k)$ such that $v = v_k$ and $l(v_1) \cdots l(v_{k-1})$ is a prefix of $s(C)$. Hence, consider the following procedure:

First, select the longest string $s = l(v_1)\cdots l(v_n)$ in $H(C)$ and set $v_1, \ldots, v_{n-1}$ to a backbone. Next, for every $\langle s, f(s,C)\rangle \in H(C)$ such that $s = l(v_1)\cdots l(v_k)$ $(1 \leq k \leq n)$, set v_k to $f(s,C)$ children of v_{k-1} in $bb(C)$.

Since every caterpillar has just one backbone, the above procedure constructs a caterpillar C from $H(C)$ uniquely. □

Theorem 5. *We can compute $\delta_{\rm PH}(C_1, C_2)$ in $O(|C_1| + |C_2|)$ time.*

Proof. Let C be a caterpillar such that $r = r(C)$ and suppose that $bb(C)$ consists of nodes $r = v_1, \ldots, v_n$. Then, repeat the following procedure from $i = 1$ to n:

For every leaf v which is a child of v_i, store $s = l(v_1)\cdots l(v_i)l(v)$ and the number of leaves with the label $l(v)$ as the children of v_i as $f(s, C)$.

Since $V(C) = V(bb(C)) \cup lv(C)$ and $V(bb(C)) \cap lv(C) = \emptyset$, the repetition traverses every node in C just once, so we can compute $H(C)$ in $O(|C|)$ time. □

Theorem 6. *There exist caterpillars C_1 and C_2 satisfying the following statements. These statements also hold even if $\tau_{\rm TAI}$ is replaced with $\tau_{\rm ILST}$.*

1. *$\tau_{\rm TAI}(C_1, C_2) = 1$ but $\delta_{\rm PH}(C_1, C_2) = O(\lambda)$, where $\lambda = \max\{|lv(C_1)|, |lv(C_2)|\}$.*
2. *$\delta_{\rm PH}(C_1, C_2) = 2$ but $\tau_{\rm TAI}(C_1, C_2) = O(n)$, where $n = \max\{|C_1|, |C_2|\}$.*

Proof. The caterpillars C_1 and C_2 that are isomorphic but just labels of the roots are different satisfy the statement 1. On the other hand, the paths C_1 and C_2 with the same length such that all the labels of nodes in C_1 is a and all the labels of nodes in C_2 are b satisfy the statement 2. □

Hence, $\tau_{\rm TAI}$ or $\tau_{\rm ILST}$ and $\delta_{\rm PH}$ are incomparable metrics between caterpillars.

Note that $\tau_{\rm TAI} \leq \tau_{\rm ILST}$ for caterpillars in general. For caterpillars C_1 and C_2 in Fig. 2, it holds that $1 = \tau_{\rm TAI}(C_1, C_2) < \tau_{\rm ILST}(C_1, C_2) = 3$. Also, since $H(C_1) = \{\langle ab, 3\rangle\}$ and $H(C_2) = \{\langle ab, 1\rangle, \langle aab, 2\rangle\}$, it holds that $\delta_{\rm PH}(C_1, C_2) = 4$.

4 Experimental Results

In this section, we give experimental results of comparing $\delta_{\rm PH}$ with $\tau_{\rm ILST}$ for caterpillars. We use N-glycans and all of the glycans (we refer to all-glycans) from KEGG[1], CSLOGS[2] and dblp[3] datasets. Table 1 illustrates the number (#cat) of caterpillars in the datasets whose number of data is denoted by #data.

We deal with caterpillars for N-glycans, all-glycans, CSLOGS and the selected 50,000 caterpillars in dblp (we refer to dblp$^-$). Table 2 illustrates the information of such caterpillars. Here, $([a, b]; c)$ means that a, b and c are the minimum, the maximum and the average number.

Table 3 illustrates the running time to compute $\delta_{\rm PH}$ and $\tau_{\rm ILST}$ for all the pairs in caterpillars in Table 2. Here, we assume that a cost function in $\tau_{\rm ILST}$ is a unit cost function. The computer environment is that CPU is Intel Xeon E51650 v3 (3.50 GHz), RAM is 1 GB and OS is Ubuntsu Linux 14.04 (64 bit).

[1] Kyoto Encyclopedia of Genes and Genomes, http://www.kegg.jp/.
[2] http://www.cs.rpi.edu/~zaki/www-new/pmwiki.php/Software/Software.
[3] http://dblp.uni-trier.de/.

Table 1. The number of caterpillars in N-glycans and all-glycans from KEGG, CSLOGS and dblp datasets.

Dataset	#cat (c)	#data	%
N-glycans	514	2,142	23.996
all-glycans	8,005	10,704	74.785
CSLOGS	41,592	59,691	69.679
dblp	5,154,295	5,154,530	99.995

Table 2. The information of caterpillars in N-glycans, all-glycans, CSLOGS and dblp$^-$.

Dataset	#node (n)	Degree (d)	Height (h)	#leaves (λ)	#labels (β)
N-glycans	([6,15];6.40)	([1,3];1.84)	([1,9];4.22)	([1,7];2.18)	([2,8];4.50)
all-glycans	([1,24];4.74)	([0,5];1.49)	([0,15];3.02)	([1,14];1.72)	([1,9];2.84)
CSLOGS	([2,404];5.84)	([1,403];3.05)	([1,70];2.20)	([1,403];3.64)	([2,168];5.18)
dblp$^-$	([7,244];11.96)	([6,243];10.94)	([1,3];1.02)	([6,243];10.94)	([7,13];9.86)

Table 3 shows that the running time of δ_{PH} is smaller than that of τ_{ILST}. By comparing the ratios of information in N-glycans to that in all-glycans, c is more influent to compute the distances than other information and more influent to compute δ_{PH} than to compute τ_{ILST} for N-glycans and all-glycans. By comparing the ratios of information in CSLOGS to that in dblp$^-$, the average values of d and λ are more influent to compute the distances for CSLOGS and dblp$^-$.

Figure 3 illustrates the distribution of δ_{PH} (dashed line) and τ_{ILST} (solid line) for pairs of caterpillars in N-glycans, all-glycans, CSLOGS and dblp$^-$. Here, the x-axis is the value of the distance and the y-axis is the percentage of pairs with the distance pointed by the x-axis. We divide the distributions of CSLOGS and dblp$^-$ into detailed three graphs for distances from 0 to 20, from 20 to 100 and from 100 to 500 for CSLOGS and distances from 0 to 10, from 10 to 100 and from 100 to 300 for dblp$^-$, because almost pairs have small distances.

Table 3. The running time (msec.) to compute δ_{PH} and τ_{ILST} for the caterpillars in N-glycans, all-glycans, CSLOGS and the selected 50,000 caterpillars from dblp.

Dataset	#cat (c)	δ_{PH}	τ_{ILST}	$\tau_{\mathrm{ILST}}/\delta_{\mathrm{PH}}$
N-glycans	514	300	6,035	20.12
all-glycans	8,005	53,170	718,807	13.53
CSLOGS	41,592	3,674,530	25,511,072	6.94
dblp$^-$	50,000	11,496,570	144,162,902	12.54

Table 4 illustrates the maximum and the peak values of δ_{PH} and τ_{ILST} in Fig. 3 and the percentage of the number of pairs with the peak values for N-glycans, all-glycans, CSLOGS and dblp$^-$. All the minimum values are 0.

Figure 3 and Table 4 show that both distributions for N-glycans and all-glycans are not similar but near to normal distributions. On the other hand, both distributions for CSLOGS and dblp$^-$ are similar and almost pairs of caterpillars have small distances, but there exist pairs having a large distance.

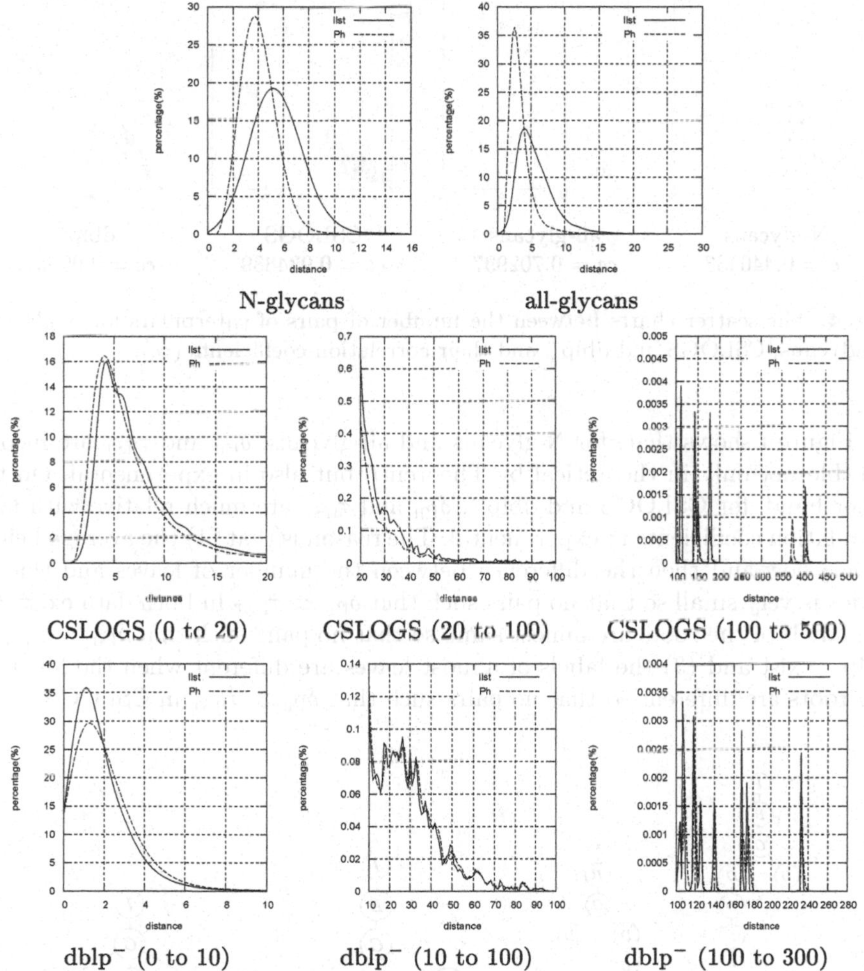

Fig. 3. The distribution of δ_{PH} and τ_{ILST} in N-glycans, all-glycans, CSLOGS and dblp$^-$.

Figure 4 illustrates the scatter charts between the number of pairs of caterpillars with τ_{ILST} pointed at the x-axis and that with δ_{PH} pointed at the y-axis for N-glycans, all-glycans, CSLOGS and dblp$^-$ and their correlation coefficients.

Table 4. The maximum and the peak values in Fig. 3 and the percentage of the number of pairs with the peak value of δ_{PH} and τ_{ILST}.

	N-glycans			all-glycans			CSLOGS			dblp$^-$		
	Max.	Peak	%	Max.	Peak	%	Max.	Peak	%	Max.	Peak	%
δ_{PH}	12	4	28.19	28	3	33.37	775	4	16.46	274	1	28.88
τ_{ILST}	16	5	19.20	29	4	18.58	486	4	15.77	240	1	35.62

N-glycans $cc = 0.440433$ all-glycans $cc = 0.702997$ CSLOGS $cc = 0.934889$ dblp$^-$ $cc = 0.996572$

Fig. 4. The scatter charts between the number of pairs of caterpillars for N-glycans, all-glycans, CSLOGS and dblp$^-$ and their correlation coefficients (cc).

Figure 4 shows that, for N-glycans and all-glycans, δ_{PH} and τ_{ILST} are incomparable not only in theoretical by Theorem 6 but also in experimental. On the other hand, for CSLOGS and dblp$^-$, δ_{PH} and τ_{ILST} are much relative with high correlation coefficients in experimental. The reason is that (1) the average height is very low and then the difference between the number of leaves and that of nodes is very small so that no pairs such that $\delta_{\text{PH}} \ll \tau_{\text{ILST}}$ in both data exist, (2) the labels of the roots are almost same so that no pairs such that $\delta_{\text{PH}} \gg \tau_{\text{ILST}}$ in dblp$^-$ exist and (3) the labels of almost leaves are different when the labels of the roots are different so that no pairs such that $\delta_{\text{PH}} \gg \tau_{\text{ILST}}$ in CSLOGS exist.

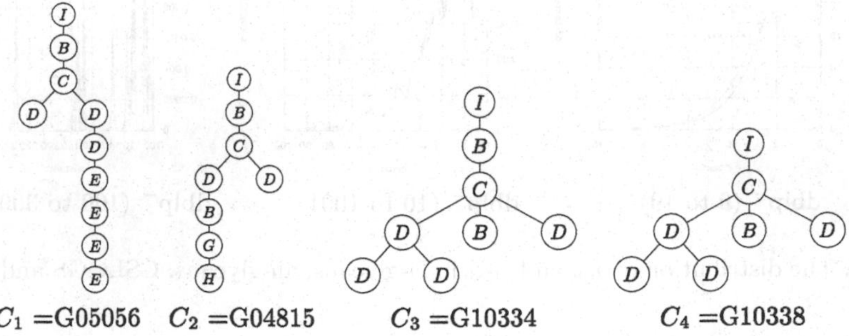

Fig. 5. Caterpillars such that $\delta_{\text{PH}} \ll \tau_{\text{ILST}}$ and $\delta_{\text{PH}} \gg \tau_{\text{ILST}}$.

Figure 5 illustrates caterpillars in N-glycans such that $\delta_{\rm PH} \ll \tau_{\rm ILST}$ and $\delta_{\rm PH} \gg \tau_{\rm ILST}$. Here, it holds that $\delta_{\rm PH}(C_1, C_2) = 2$ but $\tau_{\rm ILST}(C_1, C_2) = \tau_{\rm TAI}(C_1, C_2) = 5$, and $\tau_{\rm ILST}(C_3, C_4) = \tau_{\rm TAI}(C_3, C_4) = 1$ but $\delta_{\rm PH}(C_3, C_4) = 8$. Concerned with Theorem 6, the pairs of caterpillars such that $\delta_{\rm PH} \ll \tau_{\rm ILST}$ tend to have the small number of leaves in contrast to the number of nodes. Also the pairs of caterpillars such that $\delta_{\rm PH} \gg \tau_{\rm ILST}$ tend to be almost isomorphic and either the labels near to the root are different or a node in a caterpillar is inserted to another caterpillar.

5 Conclusion and Future Works

In this paper, we have introduced the path histogram distance and shown that it is a metric, linear-time computable and an incomparable metric with the edit distance and its variations. We have given experimental results of comparing the path histogram distance with the isolated-subtree distance for caterpillars.

Since it is possible to compute the edit distance for caterpillars efficiently, it is a future work to design the algorithm of computing it. Also it is an important future work to extend the idea of this paper to non-caterpillar trees and introduce an incomparable metric to the edit distance between trees if possible.

Acknowledgment. The authors would like to thank anonymous referees of ACIIDS '18 for valuable comments to revise the submitted version of this paper.

References

1. Aratsu, T., Hirata, K., Kuboyama, T.: Sibling distance for rooted labeled trees. In: Chawla, S., Washio, T., Minato, S., Tsumoto, S., Onoda, T., Yamada, S., Inokuchi, A. (eds.) PAKDD 2008. LNCS (LNAI), vol. 5433, pp. 99–110. Springer, Heidelberg (2009). https://doi.org/10.1007/978-3-642-00399-8_9
2. Gallian, J.A.: A dynamic survey of graph labeling, Electorn. J. Combin. **14**, DS6 (2007)
3. Hirata, K., Yamamoto, Y., Kuboyama, T.: Improved MAX SNP-hard results for finding an edit distance between unordered trees. In: Giancarlo, R., Manzini, G. (eds.) CPM 2011. LNCS, vol. 6661, pp. 402–415. Springer, Heidelberg (2011). https://doi.org/10.1007/978-3-642-21458-5_34
4. Kailing, K., Kriegel, H.-P., Schönauer, S., Seidl, T.: Efficient similarity search for hierarchical data in large databases. In: Bertino, E., Christodoulakis, S., Plexousakis, D., Christophides, V., Koubarakis, M., Böhm, K., Ferrari, E. (eds.) EDBT 2004. LNCS, vol. 2992, pp. 676–693. Springer, Heidelberg (2004). https://doi.org/10.1007/978-3-540-24741-8_39
5. Kuboyama, T.: Matching and learning in trees, Ph.D. thesis, University of Tokyo (2007)
6. Li, F., Wang, H., Li, J., Gao, H.: A survey on tree edit distance lower bound estimation techniques for similarity join on XML data. SIGMOD Rec. **43**, 29–39 (2013)
7. Tai, K.-C.: The tree-to-tree correction problem. J. ACM **26**, 422–433 (1979)
8. Yamamoto, Y., Hirata, K., Kuboyama, T.: Tractable and intractable variations of unordered tree edit distance. Int. J. Found. Comput. Sci. **25**, 307–330 (2014)

9. Yoshino, T., Hirata, K.: Tai mapping hierarchy for rooted labeled trees through common subforest. Theory Comput. Syst. **60**, 759–783 (2017)
10. Zhang, K.: A constrained edit distance between unordered labeled trees. Algorithmica **15**, 205–222 (1996)
11. Zhang, K., Jiang, T.: Some MAX SNP-hard results concerning unordered labeled trees. Inf. Process. Lett. **49**, 249–254 (1994)

Deep 3D Convolutional Neural Network Architectures for Alzheimer's Disease Diagnosis

Hiroki Karasawa[1(✉)], Chien-Liang Liu[2], and Hayato Ohwada[1]

[1] Department of Industrial Administration, Tokyo University of Science, Tokyo, Japan
7416615@ed.tus.ac.jp, ohwada@rs.tus.ac.jp

[2] Department of Industrial Engineering and Management, National Ciao Tung University, Hsinchu, Taiwan
clliu@mail.nctu.edu.tw

Abstract. Dementia has become a social problem in the aging society of advanced countries. Currently, 46.8 million people have dementia worldwide, and that figure is predicted to increase threefold to 130 million people by 2050. Alzheimer's disease (AD) is the most common form of dementia. The cost of care for AD patients in 2015 was 818 billion US dollars and is expected to increase dramatically in the future, due to the increasing number of patients as a result of the aging society. However, it is still very difficult to cure AD; thus, the detection of AD is crucial. This study proposes the use of machine learning to detect AD using brain image data, with the goal of reducing the cost of diagnosing and caring for AD patients. Most machine learning algorithms rely on good feature representations, which are commonly obtained manually and require domain experts to provide guidance. Feature extraction is a time-consuming and labor-intensive task. In contrast, the 3D Convolutional Neural Network (3DCNN) automatically learns feature representation from images and is not greatly affected by image processing. However, the performance of CNN depends on its layer architecture. This study proposes a novel 3DCNN architecture for MRI image diagnosis of AD.

Keywords: Alzheimer's disease diagnosis · 3D Convolutional Neural Network Deep residual network · Image processing · Machine learning

1 Introduction

1.1 Research Background

Dementia, a disease involving cognitive impairment and memory disorder, has become a social problem within the aging society of advanced countries. Currently, 46.8 million people suffer from dementia worldwide, and that figure is predicted to increase threefold to 130 million people by 2050 [14]. Alzheimer's disease (AD) is a progressive brain disorder and the most common case of dementia in late life. Seventy percent of dementia cases are AD. AD leads to the death of nerve cells and tissue loss

N. T. Nguyen et al. (Eds.): ACIIDS 2018, LNAI 10751, pp. 287–296, 2018.
https://doi.org/10.1007/978-3-319-75417-8_27

throughout the brain, and thus dramatically reduces the brain volume with time and affects most of its functions [14]. Although we understand AD better than ever before, means of preventing or curing AD are still unavailable. Alzheimer's medications currently being developed suppress progression of the disease. Therefore, early detection is crucial. However, the current method of diagnosis is burdensome to medical institutions and patients, since it involves three steps: interview, examination, and testing. The cost of care for AD patients in 2015 was 818 billion US dollars and is expected to increase dramatically in the future, due to the increasing number of patients as a result of the aging society.

1.2 Purpose

To reduce the cost of care for AD patients, the introduction of computer-supported systems is important. AD causes loss of nerve cells due to the accumulation of β-amyloid peptide in the brain; symptoms usually begin with mild memory impairment and turn into severe brain injury within several years. The AD neurodegenerative process gradually affects different brain functions; thus, imaging such as Single Emission Computerized Tomography (SPECT) and Positron Emission Tomography (PET) are widely used in computer-aided diagnosis (CAD) systems. However, machine learning, learning from data, gives computers the ability to learn without being explicitly programmed. Given sufficient training samples, machine learning builds a predictive model from data, and the learned model can make predictions for subsegment unseen data. Consequently, this study proposes the use of machine learning along with brain image data to construct a detection model for AD in an effort to reduce the burden of medical institutions and patients, and to reduce the cost of caring for AD patients.

In machine learning, each data sample should be represented as a feature vector; therefore, many studies have proposed the extraction of various features from 3D images of magnetic resonance imaging (MRI) and then classification of obtained vectors. However, the qualities of the extracted feature vectors depend on image preprocessing due to registration errors and noise; hence, domain knowledge is required to extract discriminative features. Hosseini-Asl et al. [12] applied the 3D Convolutional Neural Network (3DCNN) with nine layers for AD diagnosis in an effort to tackle this problem. However, the performance of CNN is highly dependent on its layer architecture. It is expected that higher classification accuracy can be obtained using a more suitable layer structure for MRI images. In this study, we propose a novel 3DCNN architecture for MRI image diagnosis of AD.

2 Related Work

2.1 Convolutional Neural Network

The convolutional neural network (CNN) is a feed-forward artificial neural network in which individual neurons are arranged to correspond with the visual cortex. CNN is influenced by biological processing, and a multilayer perceptron was designed to perform with comparatively less preprocessing. The network learns a filter that extracts the feature from the input data used by the last classification layer. The advantage of CNN is that it considers feature learning and classification model simultaneously. The performance of CNN depends greatly on the architecture of the layer and the settings of the filters; thus, many researchers have focused on devising novel architectures to improve performance.

LeCun et al. proposed the first successful application of convolutional networks, known as LeNet5, for handwritten digits and character recognition [2]. LeNet5 is comprised of seven layers, and the input is a 32 × 32 pixel image. LeNet5 consists of two convolutional layers, two sub-sampling layers, and two fully connected (FC) layers. Krizhevsky et al. proposed a CNN architecture known as AlexNet [3], which was the first work to popularize convolutional networks in computer vision. AlexNet was submitted to the ImageNet Large-Scale Visual Recognition Challenge (ILSVRC) 2012 and significantly outperformed the second runner-up. The input image is 224 × 224 with three color channels. Its structure is deeper and larger than that of LeNet 5, and the convolutional layer and the max-pooling layer are alternately followed. Zeiler et al. proposed an improved architecture of AlexNet by tweaking the architecture hyperparameters, expanding the size of the middle convolutional layers, and decreasing the stride and filter on the first layer [4]. Simonyan and Zisserman demonstrated that the depth of the network is a critical component for good performance; they proposed a best network with 16 CONV/FC layers and featured an extremely homogeneous architecture that performs only 3 × 3 convolutions and 2 × 2 pooling from the beginning to the end [7]. Szegedy et al. proposed a deep CNN architecture known as Inception, which was responsible for setting the new state of the art for classification and detection in the ILSVRC 2014 [8]. The Inception module reduces the number of parameters in the network. Additionally, they used an average pooling layer instead of FC layers at the top of the ConvNet, eliminating many parameters that do not seem to matter much. He et al. proposed a residual learning framework (ResNet) to ease the training of networks that are substantially deeper than those used previously [9]. ResNet makes it easier to optimize by learning the residual function with reference to layer input instead of learning unreferenced functions.

2.2 Alzheimer's Disease Diagonois by Machine Learning

It is necessary to extract feature vectors from 3D images of MRI in machine learning for AD diagnosis. Since the result of machine learning depends on the extracted feature vectors, various approaches for feature extraction have been proposed over the last decade.

Liu et al. used multiple selected templates to extract multiview feature representations for subjects, and clustered subjects within a specific class into several subclasses in each view space [10]. They used ensemble learning based on a support vector machine (SVM). Suk et al. proposed a deep-learning-based feature representation with a stacked auto-encoder that uses multi-task and multi-kernel SVM learning as the learning method [11]. However, qualities of the extracted features depend on image preprocessing due to registration errors and noise; hence, domain knowledge is crucial in extracting discriminative features. Obtaining hand-crafted features is labor-intensive and time-consuming; more importantly, hand-crafted features normally do not generalize well. Therefore, this study proposes the use of deep learning to obtain features from data.

Hosseini-Asl et al. applied 3DCNN with nine layers for MRI diagnosis of AD. However, as mentioned in Sect. 2.1, the performance of CNN depends greatly on its layer structure [12]. Since a 3D image typically has more features than a 2D image, a simple 3DCNN architecture of nine layers is insufficient for 3D images. Therefore, we propose a novel 3DCNN architecture for AD diagnosis.

3 Method

The proposed method first preprocesses the MRI image of the data set, then learns the data set using 3DCNNResNet, and finally classifies the test data using the last learned model.

This study proposes a novel 3DCNN architecture and applies the proposed architecture to MRI images. The training images have different sizes and different numbers of slices; therefore, we preprocess the images to rescale them to the same size. Once the training process is completed, we can use the learned model to make predictions based on test data.

3.1 Problem Specification

Let $\mathbf{I} = \{i^{(1)}, \ldots, i^{|\mathbf{I}|}\}$ be a set of MRI brain images. Given images $i \in \mathbf{I}$, the goal is to diagnose the patient's class. The number of outcomes could be 2 to 4, depending on the diagnosis result. We can transform the problem into a machine-learning problem. In CNN, there is no preprocessing to extract the feature vectors from the images; the goal is to learn a scoring function $\mathrm{f} : \mathbf{I} \rightarrow y$, where y is the patient's class.

3.2 Preprocessing

In the preprocessing phase, we resize the images so that they are the same size. Samples of MRI images are presented in Fig. 1. The image size is changed from $256 \times 256 \times 166$ to $96 \times 96 \times 64$ after image resizing.

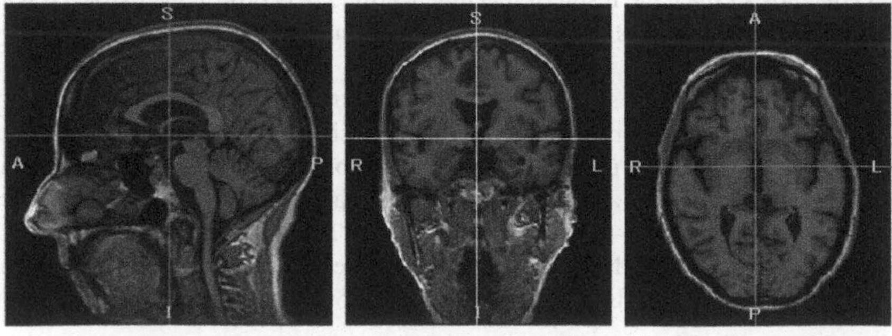

Fig. 1. 3-way orthogonal images of a 3D neuro image of MRI

3.3 3D Convolutional Neural Network

In 2D CNNs, convolutions are applied on the 2D feature maps to compute features from the spatial dimensions only. However, since the MRI image is a 3D image and there is a spatial relationship in the image, we propose convolution using a cubic filter from each layer of the 3D image. Convolution of 3DCNN is illustrated in Fig. 2.

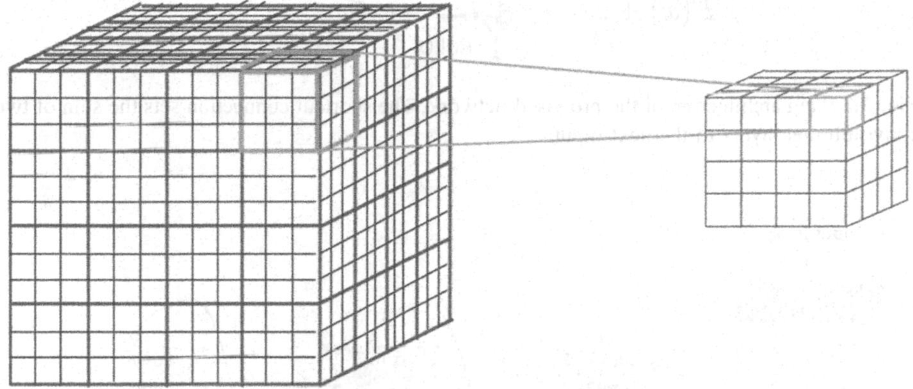

Fig. 2. Simplified figure of 3D convolutional layer

3.4 Our Proposed Architecture

Our proposed architecture is based on ResNet, which learns the residual function with reference to the layer input instead of learning a non-referenced function to make optimization easier. If we define the objective function of learning as H(x), then the difference from the input is F(x) = H(x) − x, and the objective function is redefined as H(x) = F(x) + x. That is, the sum of the current convolutional layer's output and the previous convolutional layer's output is the input of the next convolutional layer.

The proposed architecture uses plain architecture for the residual learning model. Plain architecture uses two 3 × 3 × 3 kernel convolutional layers in one residual block (Fig. 3). The proposed network is comprised of 36 convolutional layers, 1 dropout layer, 1 average-pooling layer, and 1 FC layer. We perform down-sampling directly by convolutional layers that have a stride of 2. We then perform regularization by the dropout layer [5], which prevents over-fitting, by removing some of the nodes in the layer. Our proposed network removes 50% of the nodes. Finally, the network ends with a global average pooling layer and an FC layer with softmax. The architecture of the entire proposed network is presented in Fig. 4, and details of each layer are listed in Table 1.

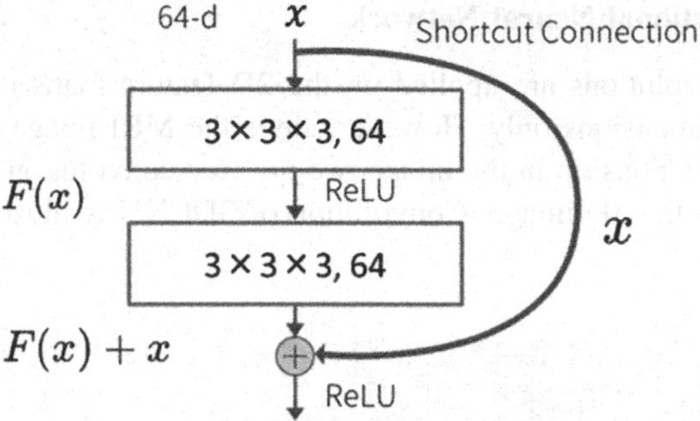

Fig. 3. Plain architecture of the proposed network. The shortcut connection sets the sum of two convolutional layers to the next input.

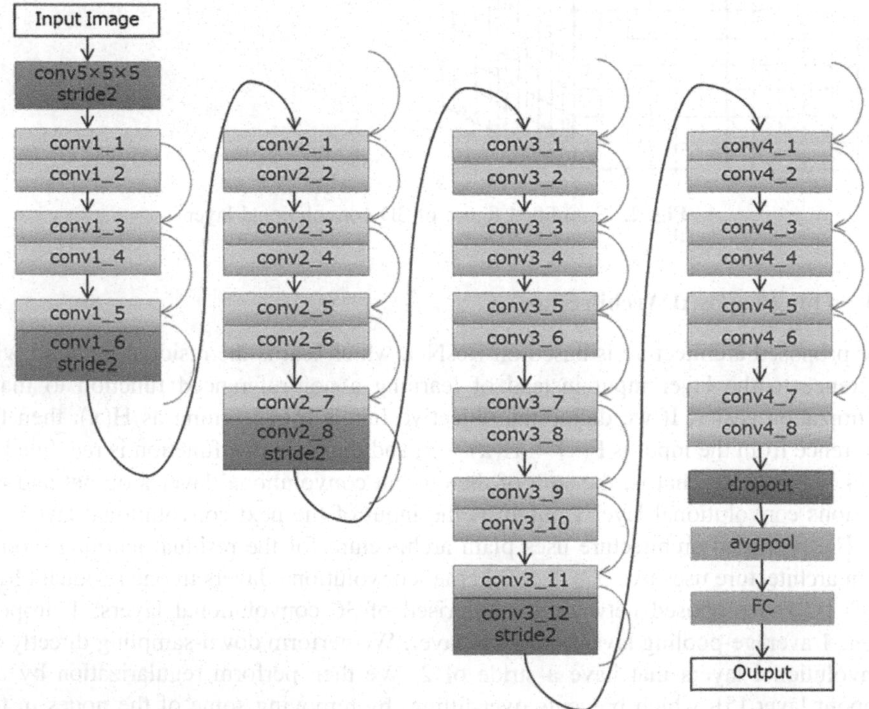

Fig. 4. Our proposed network architecture. The network is comprised of 36 convolutional layers, 1 dropout layer, 1 average pooling layer, and 1 FC layer.

Table 1. Parameters of each layer of our proposed architecture

Layer name	Output size	Our proposed architecture
conv_1	48 × 48 × 32	$5 \times 5 \times 5, 16$ stride2
conv_2	24 × 24 × 16	$\begin{pmatrix} 3 \times 3 \times 3, 16 \\ 3 \times 3 \times 3, 16 \end{pmatrix} \times 3$
conv_3	12 × 12 × 8	$\begin{pmatrix} 3 \times 3 \times 3, 32 \\ 3 \times 3 \times 3, 32 \end{pmatrix} \times 4$
conv_4	6 × 6 × 4	$\begin{pmatrix} 3 \times 3 \times 3, 64 \\ 3 \times 3 \times 3, 64 \end{pmatrix} \times 6$
conv_5	3 × 3×2	$\begin{pmatrix} 3 \times 3 \times 3, 128 \\ 3 \times 3 \times 3, 128 \end{pmatrix} \times 4$
Dropout layer	3 × 3 × 2	Remove 70% nodes
Average pooling layer	1 × 1 × 1	Average pooling $3 \times 3 \times 2$
Fully connected layer		128 → class

4 Experiments

We obtained an experiment data set for AD diagnosis from the Alzheimer's Disease Neuroimaging Initiative (ADNI) database. The data set consists of 574 normal cognitive (NC), 450 mild cognitive intensive (MCI), 358 late MCI, and 346 AD, for a total of 1728 MRI images. Details of the data set are presented in Table 2. We compared five learning models in four classification tasks: two-class classification (NC/AD, NC/MCI), three-class classification (NC/MCI/AD), and four-class classification (NC/MCI/LMCI/AD). Evaluation experiments were conducted by dividing 90% of the data set into training data and 10% into test data. The results of the experiment are presented in Table 3 and Fig. 5.

Table 2. Dataset summary

	Normal (NC)	Mild cognitive impairment (MCI)	Late MCI (LMCI)	AD
Number of subjects	574	450	358	346
Male/female	291/283	272/178	258/100	181/165
Patient age	77.02	76.23	75.07	75.94

5 Discussion

5.1 Discussion of Learning Models

Results of the experiment indicated that our proposed model had the highest classification accuracy among learning models. Comparison of the learning models of 3DCNN and SVM indicated that the accuracy of SVM is high in two-class

Table 3. Experiment results of each learning model for four classification tasks

	2class (NC/AD)	2class (NC/MCI)	3class (NC/MCI/AD)	4class	Learning model	Number of layers
Liu et al.	0.91	0.77	0.54	NA	SVM	
Hosseini-Asl et al. architecture	0.76	0.74	0.57	0.43	3D Convolutional Neural Network	9
VGGNet	0.63	0.59	0.61	0.52		16
ResNet-50	0.81	0.78	0.67	0.61		51
Our propose architecture	0.94	0.90	0.87	0.66		39

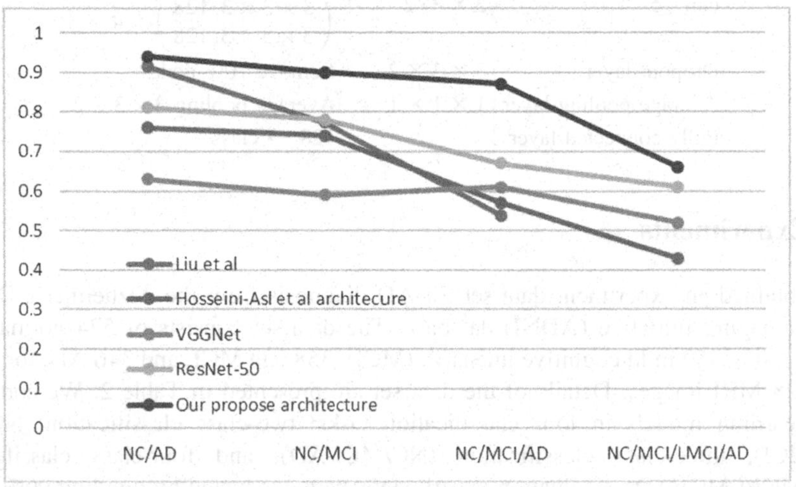

Fig. 5. Experiment results of each learning models for four classification tasks

classification and that the classification accuracy of 3DCNN is high in three-class classification. This is because SVM is suitable for two-class classification but not for multi-class classification.

5.2 Discussion of 3DCNN Architecture

Comparison of the 3DCNN architectures indicates that our proposed architecture has the highest accuracy in all classification tasks. 3DCNN accuracy differs greatly depending on the architecture. Although the deeper architecture tends to have higher accuracy, our proposed architecture with 39 layers has higher accuracy than ResNet-51 with 52 layers. It is assumed that this is because as the layer becomes deeper, the backward parameter of each convolutional layer is multiplied and the extracted feature amount of the previous layer disappears. Residual connection prevents disappearance of feature amount by adding the feature amount extracted in the previous layer to the feature amount extracted in the next layer. However, when the depth exceeds a certain value, the disappearance of the feature goes beyond the capacity of the residual connection, and accuracy begins to decrease.

6 Conclusion and Future Work

In this study, we proposed a novel 3D CNN architecture for AD diagnosis. In related research, only a simple 3DCNN architecture was used, whereas our proposed architecture used residual connection to prevent feature loss and make deeper architecture applicable to the MRI image. Experiments conducted to verify the accuracy of our proposed architecture confirmed that the performance of 3DCNN depends greatly on architecture, and our proposed architecture had the highest accuracy. However, in this study, due to the capacity of GPU memory, it was necessary to make the image smaller by resizing. It is assumed that resizing results in loss of information. Our proposed architecture with 39 layers has higher accuracy than ResNet-51 with 52 layers. It is assumed that as the layer becomes deeper, the backward parameter of each convolutional layer is multiplied and the extracted feature amount of the previous layer disappears. In the future, we will conduct experiments to investigate the effect of image resizing in the preprocessing stage, and experiment to make deeper architecture applicable using an architecture with denser residual connection.

References

1. LeCun, Y., Boser, B., Denker, J.S., Henderson, D., Howard, R.E., Hubbard, W., Jackel, L. D.: Backpropagation applied to handwritten zip code recognition. Neural Comput. **1**(4), 541–551 (1989)
2. LeCun, Y., Bottou, L., Bengio, Y., Haffner, P.: Gradient-based learning applied to document recognition. Proc. IEEE **86**(11), 2278–2324 (1998)
3. Krizhevsky, A., et al.: Imagenet classification with deep convolutional neural networks. In: Neural Information Processing Systems, vol. 25, pp. 1097–1105 (2012)
4. Zeiler, M.D., Fergus, R.: Visualizing and understanding convolutional networks. In: Fleet, D., Pajdla, T., Schiele, B., Tuytelaars, T. (eds.) ECCV 2014. LNCS, vol. 8689, pp. 818–833. Springer, Cham (2014). https://doi.org/10.1007/978-3-319-10590-1_53
5. Srivastava, N., et al.: Dropout: a simple way to prevent neural network from overfitting. J. Mach. Learn. Res. **15**, 1929–1958 (2014)
6. Goodfellow, I.J., et al.: Maxout networks. In: Proceedings on the 30th International Conference on Machine Learning, JMLR: W&CP, vol. 28 (2013)
7. Simonyan, K., Zisserman, A.: Very deep convolutional networks for large-scale image recognition (2014). CoRR, abs/1409.1556
8. Szegedy, C., Liu, W., Jia, Y., Sermanet, P., Reed, S., Anguelov, D., Erhan, D., Vanhoucke, V., Rabinovich, A.: Going deeper with convolutions. (2014). CoRR, abs/1409.4842
9. He, K., et al.: Deep residual learning for image recognition (2015). arXiv preprint arXiv: 1512.03385
10. Liu, M., et al.: Inherent structure-based multiview learning with multitemplate feature representation for Alzheimer's disease diagnosis. IEEE Trans. Biomed. Eng. **63**(7), 1473–1482 (2016)
11. Ji, S., et al.: 3D convolutional neural networks for human action recognition. IEEE Trans. Pattern Anal. Mach. Intell. **35**(1), 221–231 (2013)

12. Hosseini-Asl, E., et al.: Alzheimer's disease diagnostics by adaptation of 3D convolutional network. https://doi.org/10.1109/ICIP.2016.7532332
13. Phillips, O.R., et al.: Major superficial white matter abnormalities in Huntington disease. Front. Neurosci. **10**(0197) (2016). https://doi.org/10.3389/fnins.2016.00197
14. Alzheimer's Disease International: The International Federation of Alzheimer's Disease and Related Disorders Societies

Kernel and Acquisition Function Setup for Bayesian Optimization of Gradient Boosting Hyperparameters

Andrzej Szwabe(✉)

Institute of Control, Robotics and Information Engineering,
Poznan University of Technology, Piotrowo 3a, 60-965 Poznan, Poland
andrzej.szwabe@put.poznan.pl

Abstract. The application scenario investigated in the paper is the bank credit scoring based on a Gradient Boosting classifier. It is shown how one may exploit hyperparameter optimization based on the Bayesian Optimization paradigm. All the evaluated methods are based on the Gaussian Process model, but differ in terms of the kernel and the acquisition function. The main purpose of the research presented herein is to confirm experimentally that it is reasonable to tune both the kernel function and the acquisition function in order to optimize Bayesian Gradient Boosting hyperparameters. Moreover, the paper provides results indicating that, at least in the investigated application scenario, the superiority of some of the evaluated Bayesian Optimization methods over others strongly depends on the amount of the optimization budget.

Keywords: Binary classification · Gradient Boosting
Hyperparameters · Bayesian Optimization · Gaussian Process
Kernel function · Acquisition function · Bank credit scoring

1 Introduction

Until quite recently, automatic hyperparameter optimization was not widely regarded as a mainstream area of the machine learning research [1]. Careful tuning of hyperparameters of machine learning models was frequently considered as unavoidable black art – requiring expert experience, rules of thumb, or brute-force search [2,3]. However, in the recent years, a significant progress has been achieved in the field of the research on automatic hyperparameter optimization methods (sometimes referred to as automatic machine learning [4]) – especially on methods based on the Bayesian Optimization (BO) paradigm and the Gaussian Process modelling principle [4].

Despite this progress, grid search and randomized search are still the most popular methods for non-manual tuning of machine learning algorithms [3,4]. Therefore, one may infer that even the state-of-the-art hyperparameter optimization methods need deeper, empirical investigation – in particular as far as their applicability in domain-specific scenarios are concerned.

N. T. Nguyen et al. (Eds.): ACIIDS 2018, LNAI 10751, pp. 297–306, 2018.
https://doi.org/10.1007/978-3-319-75417-8_28

2 Related Work

Very recently, a paper has been published that focuses on Bayesian hyperparameter optimization of a Gradient Boosting (GB) classifier for credit scoring [3]. Authors of [3] presented results that has served as the main starting point for the research presented herein; the author of this paper acknowledges a strong correspondence between his research and the research presented in [3]. On the other hand, the author of this paper, suggests that a stronger evidence should be provided in order to justify the application of a Tree Parzen Estimator (TPE) as means for optimization of GB hyperparameters. As indicated in [4] and in [3], although TPE addresses the need for so-called conditional parameter spaces (for the case of which GPs are not immediately suitable), exploiting this property requires availability of a useful definition of such conditional spaces – one that is truly suitable for solving the given optimization problem.

Nearly the whole literature on successful applications of GB refers to the case corresponding to $A = 1$ (i.e. to the case of using the regression tree as the base learner for the ensemble model of GB) [3,5]. For this reason the "mainstream" approach to BO that does not require an explicit definition of any conditional relation between any hyperparameter subspaces [2,4] has been followed in the research presented in this paper.

3 Bayesian Optimization Problem

As presented in many papers, e.g. in [4], mathematically, the Bayesian Optimization problem may be expressed as finding a global maximizer of an unknown objective function f:

$$\mathbf{x}^\star = \underset{\mathbf{x}\in\mathcal{X}}{\arg\max}\, f(\mathbf{x}) \tag{1}$$

where X is some design space of interest.

Applying Gaussian Process (GP) regression is widely regarded as the state-of-the-art method for iterative update of an unknown objective function f model under the assumption that functions $f_{1:n}$ are jointly Gaussian and that the observations $y_{1:n}$ are normally distributed given f, resulting in the model:

$$\mathbf{f} \mid \mathbf{X} \sim \mathcal{N}(\mathbf{m}, \mathbf{K}) \tag{2}$$

$$\mathbf{y} \mid \mathbf{f}, \sigma^2 \sim \mathcal{N}(\mathbf{f}, \sigma^2\mathbf{I}) \tag{3}$$

where the elements of the mean vector are:

$$m_i = \mu_0(x_i) \tag{4}$$

and the elements of the covariance matrix are:

$$K_{i,j} = k(x_i, x_j). \tag{5}$$

4 Research Motivation

The literature relevant to the main research problem investigated in this paper, indicates that the leading BO methods, although basically being applied in order to "take the human out of the loop", have their own hyperparameters [2–4, 6]. Therefore, an important question arises: is the practical advantage from using the modern BO methods for automatic hyperparameter optimization significantly dependent on configuration of these methods?

The key elements of the configuration of any of the leading BO methods are: the kernel function type, the kernel function parameters (such as the "width" coefficient or coefficients), the acquisition function type and the value of the acquisition function coefficient that determines the fundamental "exploration vs exploitation" trade-off [2,4]. The research presented in this paper focuses on experimental evaluation of the impact that these elements of the BO algorithm configuration have on the quality of fully automatic bank credit scoring based on Gradient Boosting.

Despite the popularity of GB as a "universal" classification algorithm, alternative approaches have been proposed as means for more specialized binary classification tasks, such as those based on the tensor data processing paradigm [7]. Such approaches seem to form an especially interesting new area of research on domain-specific applications of the Bayesian hyper-parameter optimization methods.

5 Scope of the Experimentation

The experiments presented herein involved application of the kernel and acquisition functions that have been already recognized as useful in many BO application scenarios [2,4].

5.1 Kernel Functions Under Evaluation

As stressed by authors of many papers on BO and GP, including [2,4,8], in GP regression, the kernel function "dictates the structure of the response functions we can fit" [4]. Therefore, it seems to be important to include different types of kernel function in experimental evaluation of the practical value of BO methods.

The RBF kernel, known also as the squared exponential kernel, is one of the most popular choices for GP regression within the BO literature [4,9]. It has the form of the function:

$$k_{\text{sq-exp}}(\mathbf{x}, \mathbf{x}') = \theta_0^2 \exp(-1/2r^2), \tag{6}$$

where

$$r^2 = (x - x')^T \Lambda (x - x') \tag{7}$$

and Λ is a diagonal matrix of d squared length scale parameters ${\theta_i}^2$. The first of these parameters, ${\theta_1}^2$, determines the amount by which the locality influences the covariance, while the following parameters $\theta_{1\ldots d}{}^2$ determine the so-called smoothness of the kernel with respect to each dimension of the input. Often they are set to the same value (e.g. to 1), resulting in a isotropic (spherical) kernel [9], as in the case of the experiments presented in this paper.

As the RBF kernel is infinitely differentiable, its smootheness (more precisely: the smootheness of sample functions of this covariance function) is sometimes regarded as too high for BO applications [2,4]. For this reason, in the BO literature, another type of kernel functions is presented as more suitable for solving practical optimization problems – the family of the Matérn kernel functions:

$$\begin{aligned} k_{\text{Matérn1}}(\mathbf{x}, \mathbf{x}') &= \theta_0^2 \exp(-r) \\ k_{\text{Matérn3}}(\mathbf{x}, \mathbf{x}') &= \theta_0^2 \exp(-\sqrt{3}r)(1+\sqrt{3}r) \\ k_{\text{Matérn5}}(\mathbf{x}, \mathbf{x}') &= \theta_0^2 \exp(-\sqrt{5}r)\left(1+\sqrt{5}r+\frac{5}{3}r^2\right). \end{aligned} \tag{8}$$

The Matérn kernel is a generalization of the RBF kernel. It has an additional parameter ν that controls the smoothness of the resulting function. In particular, the Matérn 3/2 kernel (i.e. the Matérn kernel of $\nu = 3/2$) is once differentiable while the Matérn 5/2 ($\nu = 5/2$) is twice differentiable. As far as differentiability is concerned, the Matérn 5/2 may be regarded as the most similar to the RBF kernel. The use of the Matérn 5/2 as the default kernel function for BO, originally proposed in [2], is now widespread in the BO literature [4,6].

It should be noted that using values of ν outside the $(0.5, 1.5, 2.5, inf)$ set leads to a considerably higher (approximately ten times higher) computational cost, since it requires evaluation of a modified Bessel function. For this reason, only the selected values of ν are considered useful in practical BO application scenarios – including the one investigated herein.

5.2 Acquisition Functions Under Evaluation

Apart form the kernel function selection, the choice of the acquisition function is widely regarded as the most important factor influencing the practical value of BO [2,4]. Any acquisition function is conditioned on the set of observations denoted by $D_n = \{(x_i, y_i)\}_{i=1}^n$ and is used for evaluation of an arbitrary test point x.

Probability of Improvement. As presented in [4], one of the "classic" proposals of a BO acquisition function, probability of improvement (PI) [10], measures the probability that the evaluation of the point x will lead to an improvement upon τ:

$$\alpha_{\mathrm{PI}}(\mathbf{x}; \mathcal{D}_n) := \mathbb{P}[v > \tau] = \Phi\left(\frac{\mu_n(\mathbf{x}) - \tau}{\sigma_n(\mathbf{x})}\right), \tag{9}$$

where Φ is the standard normal cumulative distribution function. In general, the PI heuristics is regarded as strongly favouring exploitation over exploration

[2,4]. To the knowledge of the author, there are two main ways of defining the PI function. The more popular one may be found e.g. in [2,4] and follows the assumption that $\tau = f(argmax_{x_n} f(x_n))$. The more general one introduces an additional parameter ξ [9] that controls the trade-off between preferring higher posterior mean and preferring higher posterior variance:

$$\tau = f(argmax_{x_n} f(x_n)) + \xi. \tag{10}$$

However, as the majority of the literature on BO for hyperparameter optimization follows the simpler way of defining the PI function (i.e. the first one), in the experiments presented herein, the value of ξ has been set to zero.

Expected Improvement. Let Φ denote the standard normal cumulative distribution function. As proposed in [11], a BO acquisition function may be defined as the expectation of the improvement:

$$\begin{aligned} \alpha_{\mathrm{EI}}(\mathbf{x}; \mathcal{D}_n) &:= \mathbb{E}[max\{0, f(x) - f(argmax_{x_n} f(x_n)) - \xi\}] = \\ &= (\mu_n(\mathbf{x}) - f(argmax_{x_n} f(x_n)) - \xi)\,\Phi\left(\frac{\mu_n(\mathbf{x}) - f(argmax_{x_n} f(x_n)) - \xi}{\sigma_n(\mathbf{x})}\right) \\ &\quad + \sigma_n(\mathbf{x})\phi\left(\frac{\mu_n(\mathbf{x}) - f(argmax_{x_n} f(x_n)) - \xi}{\sigma_n(\mathbf{x})}\right). \end{aligned} \tag{11}$$

Analogically to the above-specified definition of PI function, the value of the ξ variable of the EI function, has been set to zero.

Upper Confidence Bound. Another popular way of "negotiating" exploration and exploitation of BO is the upper confidence bound (UCB) criterion [12] which is referred to as being "optimistic in the face of uncertainty" [4]. The "optimism" of (UCB) may indeed be seen in the way the variance component (i.e. σ_n) is used to linearly trade exploration for exploitation:

$$\alpha_{\mathrm{UCB}}(\mathbf{x}; \mathcal{D}_n) := \mu_n(\mathbf{x}) + \beta_n \sigma_n(\mathbf{x}). \tag{12}$$

Obviously, tuning the hyperparameter β_n may lead to a performance boost. Therefore in the experiments presented herein, two values of different orders of magnitude have been used: 1 an 10.

6 Evaluation Methodology

For the purpose of the comparability with the very recent results presented in [3], the results of the classification quality evaluation presented in this paper are expressed using the accuracy metric [1,3]. However, in order to "simulate" a real-world application scenario – in which typically a single data set is used – the research presented herein involves the use of a single data set – the German data set, which is the most widely referenced bank credit scoring data set [13–17].

For the sake of the results comparability, in the experiments presented in this paper, the same hyperparameter space was used as this used by the authors of [3], although some practitioners may find it as too narrow. This space, specified in accordance to the XGBoost parameters names was: 'min_child_weight': (0, 4), 'colsample_bytree': (0.9, 1), 'max_depth': (1, 12), 'subsample': (0.9, 1), 'gamma': (0, 0.01), 'max_delta_step': (0,1). As in the case of any of the widely-referenced BO method based on GP modeling, all the hyperparameters including integer ones – were treated as continuous variables. Also for the sake of the comparability of the results with those presented in [3], the same prediction quality evaluation measure (i.e. accuracy) has been used.

In accordance with the common practice followed by the authors of [3,18], the value of the training ratio (determining the training/testing set partitioning) has been set to 0.8. The experiments have been repeated 20 times to provide appropriately reliable results (although obtaining them using a desktop computer required more than 200 computation hours). Finally, the results of the unit experiments have been averaged to alleviate the impact of the data set partitioning randomization. As in the typical case of a BO experiment, the training set and the testing set of the supervised learning system (being iteratively optimized by the BO system) remains the same (during the sequence of function evaluations constituting the unit experiment), and the function f is calculated using both the training set and the testing set of the supervised learning system. While the author is aware of the availability of other (newer and larger) data sets, the experiments presented herein were performed with the use of the German data set only. This assumption was made because the German data set is definitely the most widely referenced bank credit scoring data set and because the total time of all the repeated experiments was very long, even despite using a single data set.

7 Results of the Experiments

To the author's opinion, in many real-world application cases, the ability of obtaining a configuration of hyperparameters resulting in a 'close-to-optimal' quality in a small number of function evaluations may be as important as the ability of finding the optimal configuration in the (specified a priori) total number of iterations. Clearly, it may be the case when an optimized classifier is being used before the optimization process is completed. From this perspective, the assumption of presenting only the final results of the BO procedure (results obtained after 100 iterations) that was made by the authors of [3] seems to be only partially justified.

Figure 1 presents all the per-iteration results of the random search algorithm application together with the corresponding results of the two selected BO algorithms: 'BO-Matérn(2.5)-UCB(1.0)' and 'BO-Matérn(2.5)-UCB(10.0)'. These two algorithms are based on the combination of the kernel function (Matérn

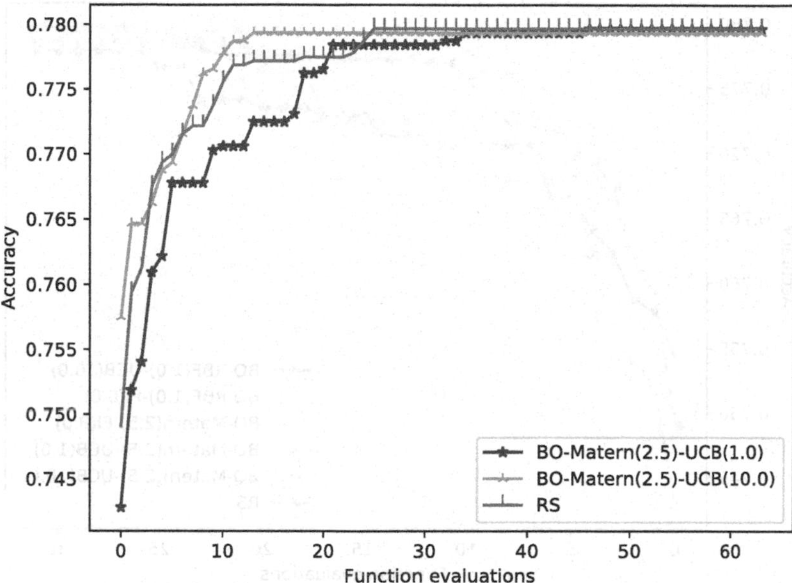

Fig. 1. The advantage of an effectively tuned Bayesian Optimization algorithm over the random search algorithm.

5/2) and the acquisition function type (UCB) that is widely regarded as highly effective [2,4]. It may be seen that applying BO algorithm that is appropriately strongly "oriented on the exploration" (in the case of the UCB – having the β_n parameter set to an appropriately high value) enabled achieving an improvement of the classification accuracy at a lower number of function evaluations (i.e. BO algorithm iterations) than the number necessary for obtaining a similar result while using the random search algorithm. It is worth pointing out that the maximum accuracy values presented in the Fig. 1 are even higher than those reported in [3]. From this perspective, the very small differences between the accuracy levels of the three algorithms observed after 45 (or even after 32 iterations) may be considered neglectable. In consequence, the ostensible superiority of the random search may also be considered neglectable. Moreover, it should be taken into consideration that the Fig. 1 presents maxima observed for each algorithm and for given number of already performed function evaluations, rather than the outcome of the given function evaluation. It is worth mentioning that it is likely that the performance of the evaluated BO algorithms would increase if some form of Markov Chain Monte Carlo had been used as means for the so-called marginalization of the hyperparameters of the kernels (not to be confused with GB hyperparameters) [2]. Such an extension of the "classical" approach (based solely on the likelihood maximization) could probably prevent the BO algorithm from behaving in a way similar to this of an optimizer being stuck in a local optimum.

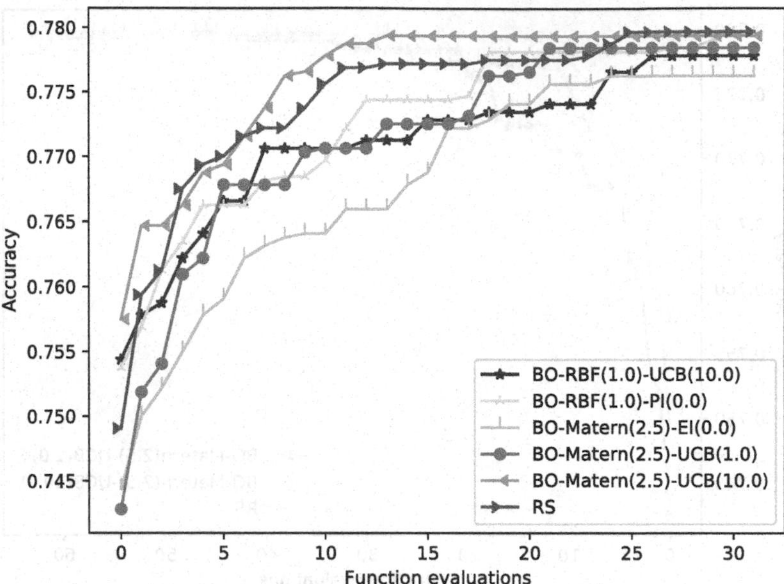

Fig. 2. The impact of the kernel function selection on the performance of the selected Bayesian Optimization algorithms.

Figure 2 presents the best algorithm configurations for each of the evaluated kernel functions – across the first 32 function evaluations (the most important from the practical point of view). The content of the Fig. 2 confirms the superiority of the Matérn 5/2 kernel (i.e. this of $\nu = 2.5$), over the RBF kernel and the high potential of the UCB acquisition function – both observations being reported in [2,4]. However, the surprisingly low effectiveness of both the 'BO-Matérn(2.5)-UCB(1.0)' and the 'BO-RBF(1.0)-UCB(10.0)' algorithms indicates that the selection of the kernel function should be integrated with the acquisition function tuning.

It should be taken into consideration that in contrast to the results presented in [3] the results presented herein have been obtained as a result of 20 repetitions performed in the experimental scenario involving the use of a single data set only (due to the lengthy experiments repetitions and the paper's page limit).

8 Conclusions

As shown in the paper, an appropriate application of the state-of-the-art BO methods may reduce the number of the hyperparameter optimization algorithm iterations required to achieve the accuracy of classification (based on the GB algorithm) that is close to the optimum. The results presented in the paper indicate also that, in order to obtain a high hyperparameter optimization performance, one should consider the use of the state-of-the-art kernel function –

the Matérn kernel – together with the "optimistic" UCB acquisition function. Another practically useful observation from the results reported in this paper is the general importance of integrated tuning of both the kernel function and the acquisition function.

To the author's best knowledge, the paper provides results of the first experimental evaluation of the applicability of the state-of-the-art hyperparameter optimization methods to a bank credit scoring system configuration. These results may be especially helpful for anyone working on tuning hyperparameters of a bank credit scoring system based on the GB algorithm - an algorithm that has been popularized by numerous winning solutions of Kaggle competitions – known as frequently based on the XGBoost implementation of the GB algorithm [5].

Acknowledgments. This work was supported by the Polish National Science Centre, grant DEC-2011/01/D/ST6/06788, and by Poznan University of Technology under grant 04/45/DSPB/0163.

References

1. Flach, P.: Machine Learning: The Art and Science of Algorithms That Make Sense of Data. Cambridge University Press, New York (2012)
2. Snoek, J., Larochelle, H., Adams, R.P.: Practical Bayesian optimization of machine learning algorithms. In: Proceedings of the 25th International Conference on Neural Information Processing Systems, NIPS 2012, USA, vol. 2, pp. 2951–2959. Curran Associates Inc. (2012)
3. Xia, Y., Liu, C., Li, Y., Liu, N.: A boosted decision tree approach using Bayesian hyper-parameter optimization for credit scoring. Expert Syst. Appl. **78**(Suppl. C), 225–241 (2017)
4. Shahriari, B., Swersky, K., Wang, Z., Adams, R.P., de Freitas, N.: Taking the human out of the loop: a review of Bayesian optimization. Proc. IEEE **104**(1), 148–175 (2016)
5. Chen, T., Guestrin, C.: XGboost: a scalable tree boosting system. In: Proceedings of the 22nd ACM SIGKDD International Conference on Knowledge Discovery and Data Mining, KDD 2016, pp. 785–794. ACM, New York (2016)
6. Brochu, E., Cora, V.M., de Freitas, N.: A tutorial on Bayesian optimization of expensive cost functions, with application to active user modeling and hierarchical reinforcement learning, December 2010. arXiv:1012.2599
7. Szwabe, A., Misiorek, P., Walkowiak, P.: Reflective relational learning for ontology alignment. In: 9th International Conference on Distributed Computing and Artificial Intelligence, DCAI 2012, Salamanca, Spain, 28–30th March 2012, pp. 519–526 (2012)
8. Bergstra, J., Bengio, Y.: Random search for hyper-parameter optimization. J. Mach. Learn. Res. **13**, 281–305 (2012)
9. Lizotte, D.J., Greiner, R., Schuurmans, D.: An experimental methodology for response surface optimization methods. J. Glob. Optim. **53**(4), 699–736 (2012)
10. Kushner, H.J.: A new method of locating the maximum point of an arbitrary multipeak curve in the presence of noise. J. Basic Eng. **86**(1), 97–106 (1964)

11. Močkus, J.: On Bayesian methods for seeking the extremum. In: Marchuk, G.I. (ed.) Optimization Techniques 1974. LNCS, vol. 27, pp. 400–404. Springer, Heidelberg (1975). https://doi.org/10.1007/3-540-07165-2_55
12. Srinivas, N., Krause, A., Kakade, S., Seeger, M.: Gaussian process optimization in the bandit setting: no regret and experimental design. In: Proceedings of the 27th International Conference on International Conference on Machine Learning, ICML 2010, USA, pp. 1015–1022. Omnipress (2010)
13. University of California, Irvine (UCI), Machine Learning Repository (MRI): German Credit dataset (2017). https://archive.ics.uci.edu/ml/datasets/Statlog+(German+Credit+Data)
14. Lessmann, S., Baesens, B., Seow, H.V., Thomas, L.C.: Benchmarking state-of-the-art classification algorithms for credit scoring: an update of research. Eur. J. Oper. Res. **247**(1), 124–136 (2015)
15. Brown, I., Mues, C.: An experimental comparison of classification algorithms for imbalanced credit scoring data sets. Expert Syst. Appl. **39**(3), 3446–3453 (2012)
16. Harris, T.: Credit scoring using the clustered support vector machine. Expert Syst. Appl. **42**(2), 741–750 (2015)
17. Huang, C.L., Chen, M.C., Wang, C.J.: Credit scoring with a data mining approach based on support vector machines. Expert Syst. Appl. **33**(4), 847–856 (2007)
18. Finlay, S.: Multiple classifier architectures and their application to credit risk assessment. Eur. J. Oper. Res. **210**(2), 368–378 (2011)

Low-Feature Extraction for Multi-label Patterns Analyzing in Complex Time Series Mining

Ngoc Anh Thi Nguyen[1], Trung Hung Vo[2], Sun-Hee Kim[3], Tran Quoc Vinh Nguyen[1], A-Ran Oh[4], and Hyung-Jeong Yang[4(✉)]

[1] Faculty of Information Technology, The University of Danang – University of Education, Danang City, Vietnam
{ngocanhnt,ntquocvinh}@ued.udn.vn

[2] The University of Danang, 41 Leduan Street, Danang City, Vietnam
vthung@dut.udn.vn

[3] Department of Brain and Cognitive Engineering, Korea University, Seoul 136-713, South Korea
sunheekim@korea.ac.kr

[4] Department of Computer Science, Chonnam National University, Gwangju 500-757, South Korea
dhdkfks@chonnam.ac.kr, hjyang@jnu.ac.kr

Abstract. With an objective of discovering compact features of unlabeled time series data and assessing the performance of grouping homogeneous patterns, this paper thus propose a mathematical modeling that could be capable of the general problem of multi-label time series clustering with high accuracy and reliability. The proposed method is an extension of original Kalman Filter algorithm that offers its main advantages by handling the complex characteristic of time series including: noise, temporal dynamics, correlation, time-lags, and harmonics. Consequently, the vital low-rank features extracted from the proposed approach have benefit properties including comprehensible, interpretable and visualizable that boost clustering quality. The experimental result on real-world dataset is presented to verify the contributions of the proposed method via improving significant performance compared with well-known competitors in both terms of its effectiveness and scalability.

Keywords: Time series · Clustering · Kalman Filter
Expectation-maximization · Feature extraction · Data mining · K-mean

1 Introduction

Research on time series clustering has been considered as challenges issue in data mining due to its effectiveness in providing useful information in many high-impact applications, ranging from biology recognition such as microarray time series clustering, gene expression data learning for functional group or functionally related genes identifications, climate data, environment, finance, medicine, speech/voice/gesture recognition, art/entertainment or biomedical signals and so on [1–6]. Clustering is a

N. T. Nguyen et al. (Eds.): ACIIDS 2018, LNAI 10751, pp. 307–317, 2018.
https://doi.org/10.1007/978-3-319-75417-8_29

solution of unsupervised learning techniques for grouping similar patterns together without any early knowledge [4]. In particular, a set of unlabeled time series is grouped into the same cluster if those patterns share the same strongly similarity structures each other within the group, and otherwise. Such ubiquitous applications, it's worth exploring on time series data to model, process and discover useful knowledge for clustering problem.

With the evolving of data storage increasingly and processors as well, real-world phenomena in time series data maintains the complex characteristics due to itself recording mechanism, such as large amount of noise, temporal dynamics, high correlation among multiple time series, time-lags, and harmonics. In particular, noise in time series is frequently occurring because of quality of measurement recording or transferring [9]. Temporal dynamics, or known as smoothness characteristic, referrers to characterizing the trend of time series exhibited via the movement of each time points. In other words, the evolving behaviors among the neighbors' time ticks points out particular trend of time sequence that derived in different temporal dynamics among the time sequences [9–11]. Temporal components therefore play an important role in clustering of time series whereby sharing the same dynamics, an illustration of temporal dynamics and correlations is shown in Fig. 1(a). Another property in time series often considered is time-lags. Figure 2(a) shows an intuitive example: given a time series 2 deriving from sin function with period of 150, a time series 1 is time-lag of 1, time series 3 is nearby frequency with signal 2. Time-shift is late phase in time between two time sequences. With different lags of signal 1, nearby frequency of signal 3 compared with signal 2, but they are expected to be clustered into the same group as the Fig. 2 presented. The last characteristic we mentioned as above is harmonics which combines two frequencies with different periods is presented. Those signals are also expected to detect and groups into a cluster together which shared the similarity characteristic of periodicities and frequencies. For instance from Fig. 2(b), time series 4 and 5 are expected to fall into the same cluster.

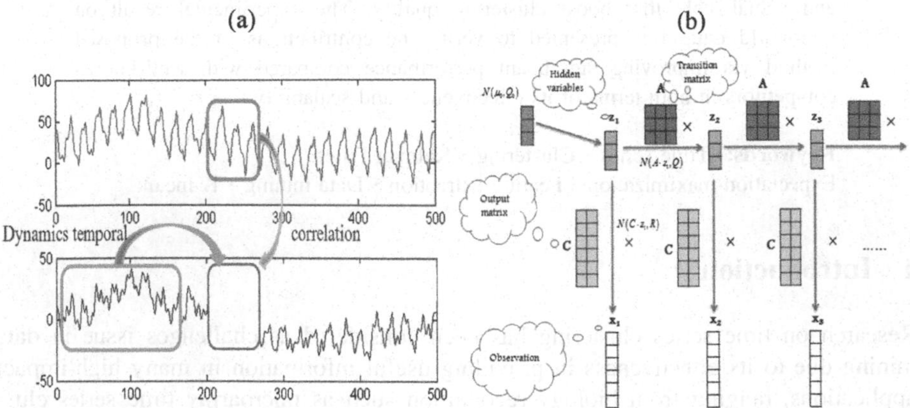

Fig. 1. (a) An illustration about temporal dynamics and correlations characteristics (b) Mathematical modeling of the proposed method

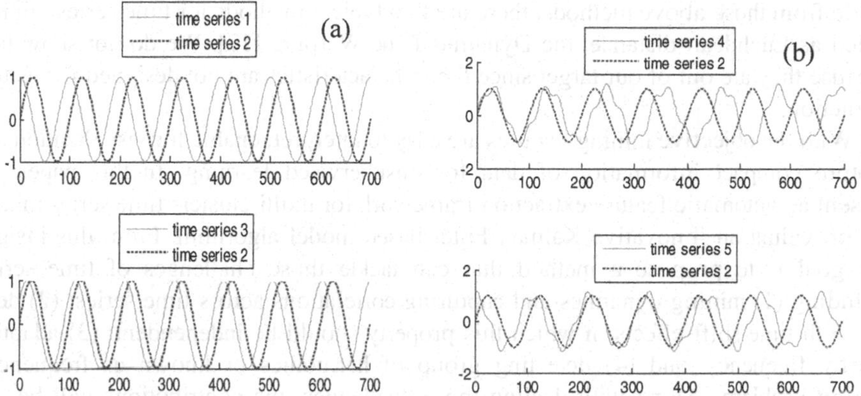

Fig. 2. (a) An illustration time-shift and frequency in close proximity between three time series 1, 2, and 3; (b) An illustration of harmonics grouping of time series 4 and time series 5

In order to mining such time series clustering efficiently, variety of approaches has been developed to identify structure of similar/different patterns which is classified into one of three major categories, namely raw dataset- based clustering directly, feature-based approaches and model-based algorithms [1]. In this article, we investigate on features-based approach for mining compact features that handling all main properties above of time series to extract useful knowledge data for clustering purpose whereby minimization of the within-group samples similarity and maximization of the between-group samples dissimilarity.

Revisiting literature methods for feature extraction in unsupervised learning, a very powerful is known as Principal Component Analysis (PCA) that is valuable in capturing linear correlations among time series [12]. This method derives latent features successfully in a low-rank representation of the original dataset and then could be enabling for decision support in clustering mining sequences. However, PCA ignores temporal evolving trends of patterns due to itself characteristics is not designed to monitoring the ordering of matrix dimensionality. Features extracted from PCA then could not be easy to interpret and lead to poor quality in clustering performance. Alternative feature-based approach is Discrete Fourier Transform (DFT) is also the popular and widely applied algorithm of clustering category. The mechanism working of DFT is first to transforms the time series into the frequency domain [13]. Consequently, the more discriminative features extracted of DFT in new space can capture frequencies in a single time series. Unfortunately, it lacks dynamics discovering. Another method, named is Linear Predictive Coding Cepstrum (LPCC), also is applied to discover distinguished features via extracting a few coefficients. Again, this method meets a drawback for the clustering performance due to its cepstrum features are hard to interpret. Kalman Filter, also known as Linear Dynamical System has been developed for time series recognition, especially for forecasting objective [14]. The advantage of the Kalman Filter is to capture both correlations and temporal smoothness of the time series. However, as we mentioned, its strongpoint is for prediction. Then the hidden variables are extracted from Kalman Filter is not suitable for clustering purpose.

Aside from those above methods, there are two typical methods for time series mining, called as Euclidean distance, the Dynamic Time Warping [17]. We do not show here because they are out of our target since their characteristics are not designed for feature extraction.

With the objective mining features are easy to interpret, enable for visualization and capture compact information of data for unsupervised learning. In this paper, we present an automatic feature extraction framework for multi-clusters time series mining via providing an innovative Kalman Filter-based model algorithm. From this insight, our goal is to propose a method that can tackle those challenges of time series, including (1) mining dynamics and capturing correlations across time series; (2) dealing with time-shift effects in which this property should be independent; (3) adapting nearby frequency and (4) detecting group of harmonics or known as frequencies mixing problem. Along with dealing those challenges, the contributions will be distributed into two axes: (1) refining learning mechanism toward transitions and projection matrix and careful implementing relying on designing model for extract deeper features helpful in clustering; (2) applying to multiple clusters that can provide an unsupervised tool in generally that can be easily and broadly adopted by different applications.

The rest of paper is structured as follows: In Sect. 2, two steps of proposed systems setup for time series clustering is presented, including feature extraction step and clustering algorithm. Experimental results and discussion of the proposed system compared with competitors is given in Sect. 4. The last section presents the summary of the article and finds out our research works in future.

2 Proposed Method: Low-Rank Feature Extraction

Given an unlabeled dataset of *m* multidimensional time series ***X*** that is formulated as an ordered sequence of observations with share the same length ***T***, formed as the matrix of $\mathbf{X}_{m\times T}$, our prime objective is to find a new feature subspace $F' = \{f_1, f_2, \ldots, f_{T'}\}$ representing a low-dimensional reduction of original dataset $F = \{f_1, f_2, \ldots, f_T\}$ whereby $T' \leq T$.

Original Kalman Filter assumes that all observations of time sequences x_n are generated from the series of latent variables z_n via projection matrix **C** or also known as output matrix, and the hidden variables z_n are connected over time ticks throughout the transition matrix **A** [9, 10]. The mathematical equations are used to model Kalman Filter are given as following:

$$z_1 = \mu_0 + \omega_0; \quad z_{n+1} = A \cdot z_n + \omega_n; \quad \mathrm{x_n} = \mathrm{C} \cdot \mathrm{z_n} + \varepsilon_\mathrm{n} \tag{1}$$

Where x_n, $\mathrm{z_n}$ denoted for vector of observations and hidden variables at time tick n, respectively. Transition matrix **A** captures dynamics of time sequences and will predict the hidden variables for the next time points with the initialization state of vector μ_0. Output matrix **C** represents the projection matrix that maps the observation to hidden states at each time tick. In our implementation, matrix **C** denotes for the features extracted from the Kalman Filter where each rows represents a time series and column implies the number of features. Due to the page limitation, we skip the detail of the Kalman Filter model, its tutorial can be found in [11, 18]. Our target here is to motivate to improve the drawback of matrix **C** since it cannot clearly discovery the characteristics of the particular series for clustering purpose. Our graphical model is shown in Fig. 1(b).

At the first step, we motivate to handle the frequency mixing form of latent variables, which are derived from the Kalman Filter. As mentioned above, each row of output projection matrix **C** does not clearly define characteristics of the corresponding time series. Therefore, each row must be defined in form of more prominent and unique representation. In our implementation, we work as a step toward identifying stronger latent variables for each time sequences of Kalman Filter by refining the straight forward transition matrix **A** and output projection matrix **C**. The hidden variables rely on the eigenvalues of the transition matrix **A** as shown the Eq. (1). Then, normalization of the transition matrix **A** will directly discover the frequency proximity and amplitude since eigenvalues present the underlying characteristics of time series including amplitude and frequency mixing. Consequently, in order to discover frequency combinations and its mixing weight, we improve Kalman Filter by applying eigen decomposition on **A**, shown in Eq. (2).

$$\mathbf{A} = \mathbf{V}\boldsymbol{\Lambda}\mathbf{V}^* \tag{2}$$

Where V denotes for eigenvector and $\boldsymbol{\Lambda}$ is the diagonal matrix.

In the case of the projection matrix **C** of Kalman Filter, we need to compensate to conduct the matrix of frequency combination from the eigen-dynamics $\boldsymbol{\Lambda}$, given as:

$$\mathbf{C}_h = \mathbf{C} \cdot \mathbf{V} \tag{3}$$

The frequency proximity is derived as following:

$$\begin{aligned} z_0^{new} &= \mathbf{V}^* \cdot \mu_0 \\ z_n^{new} &= \mathbf{V}^* \cdot z_n \end{aligned} \tag{4}$$

Herein, V maintains conjugate pairs of eigenvalues in Λ. Therefore, we derived Eq. (4) as given:

$$\begin{aligned} z_n^{new} &= \mathbf{V}^* \cdot z_n = \mathbf{V}^* \cdot \mathbf{A} \cdot z_{n-1} + \text{noise} \\ &= \mathbf{V}^* \cdot \mathbf{V} \cdot \boldsymbol{\Lambda} \cdot \mathbf{V}^* \cdot z_{n-1} + \text{noise} \\ &= \boldsymbol{\Lambda}^{n-1} \cdot \mu_0^{new} + \text{noise} \end{aligned} \tag{5}$$

$$x_n = \mathbf{C}_\mathrm{h} \cdot \boldsymbol{\Lambda}^{n-1} \cdot \mu_0^{new} + \text{noise} \tag{6}$$

As can be seen from Eqs. (5) and (6), the characteristics of time series related to frequencies and amplitudes are completed defined via the eigenvalues of transition matrix **A**. From here, the contribution regarding to frequency combination is found.

In the next step, we focus on how to eliminate time shift effects. With the achievement of the frequencies mixing from the matrix above of $\mathbf{C}_h$, each row of this matrix can be applied directly for clustering performance. Unfortunately, this matrix fails to identify similar time series due to its time-shift effect. To this end, we need to take the magnitude of $\mathbf{C}_h$, the same column of the conjugate column of $\mathbf{C}_h$ can be explored. Then, a step of omitting these duplicated columns, the frequency magnitude matrix is now calculated which shows the strong features and time shift now is handed.

In summary, the proposed method consists of following steps: (a) hidden variables are first obtained via original Kalman Filter; (b) the eigen-decomposition then is calculated directly from transition matrix $\mathbf{A}$ so as to figure out the frequency and amplitude of hidden variables of step a; (c) the magnitude of the frequency matrix is formulated in order to eliminate time-shift effect.

With the optimal features extracted above, the clustering phase then is demonstrated. At this step, a popular method is k-means algorithm is applied for real-world application time series recognition.

3 Experimental Results

In this paper, a real dataset taken from UCI repository[1], known as the three-cluster of Cylinder (C)- Bell (B)- Funnel (F) dataset, is carried out in order to evaluate the efficiency and effectiveness of the proposed system. The CBF dataset is a kind of artificial dataset that belongs to temporal domain [20]. In this paper, grouping in three different CBF cluster are expected. A pictorial representation of each sample of Cylinder, Bell and Funnel is shown in Fig. 3.

In order to assess the quality evaluation of clusters, we used the cluster similarity measurement [21]. The first contribution of the proposed method is presented in its clustering quality. Specifically, the proposed method demonstrates a significant performance with 92.84% clustering accuracy, compared with 86.84% of original Kalman Filter, 69.77% of PCA, 90.142% of LPCC and 78.976% of DFT. The reason leads to the highest performance of the proposed method with 4 features is because it can explore deeper hidden variables and handling all characteristics of time series. Specifically, PCA, in our implementation, just keeping the top two hidden variables, or known as principal components, that shows the effectiveness of dimensionality reductions for CBF. Those score features however applied for clustering get poor clustering due to it does not care about the temporal smoothness cross over time, only preserve the Euclidean distance as much as possible. Similarity with original Kalman Filter, the top four hidden variables is chosen for the best clustering. The derived output

[1] https://archive.ics.uci.edu/ml/datasets.html.

matrix of Kalman Filter then is directly used as features input in k-mean algorithm for clustering. However, time-shift effects cannot be handed. Therefore, this issue leads to those features are not easy to be interpretable and meets drawback of clustering. Similarity with LPCC and DFT, we calculate the LPC cepstrum features and Fourier coefficients, respectively. Then apply k-means on those features and get the lower quality than the proposed method.

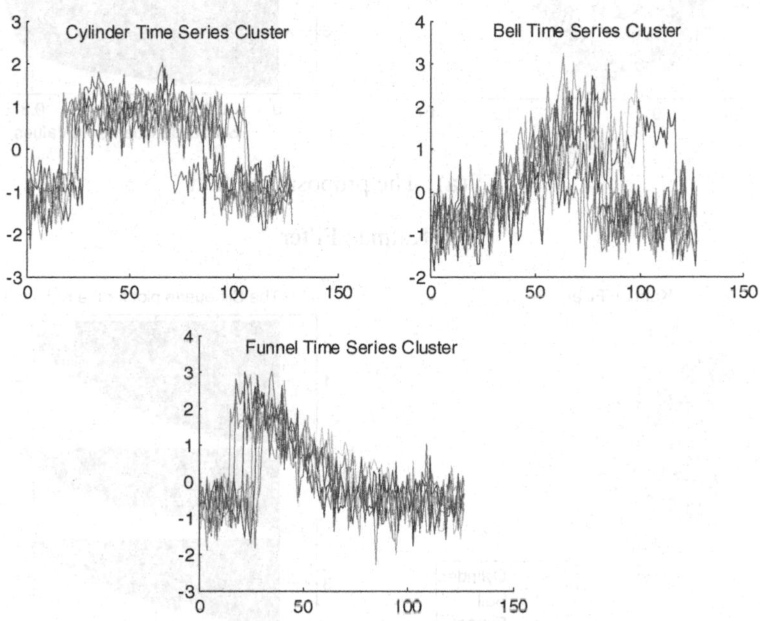

Fig. 3. Samples for each cluster of Cylinder, Bell and Funnel, respectively

The second contribution lies on the visualization aspect. With the interpret features, the proposed method could be clearly provide a good clustering performance via scatter plot and the silhouette coefficient values, presented in Fig. 4. It is noted that, the silhouette coefficients have a values ranging from -1 to 1, where 1 corresponds to the case of better clustering wherein the pattern is clearly identical with neighboring group, a value of 0 indicates the case of patterns is very close to the decision boundary or overlap each other, and negative values denotes those patterns have been assigned into the wrong group.

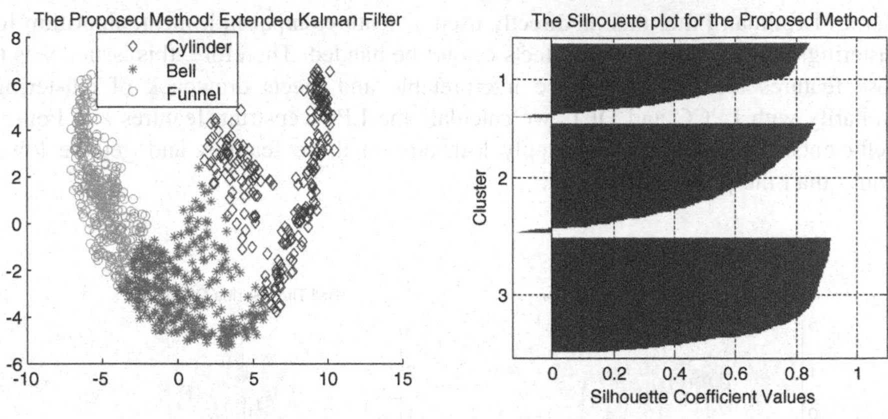

(a) The proposal method

(b) Kalman Filter

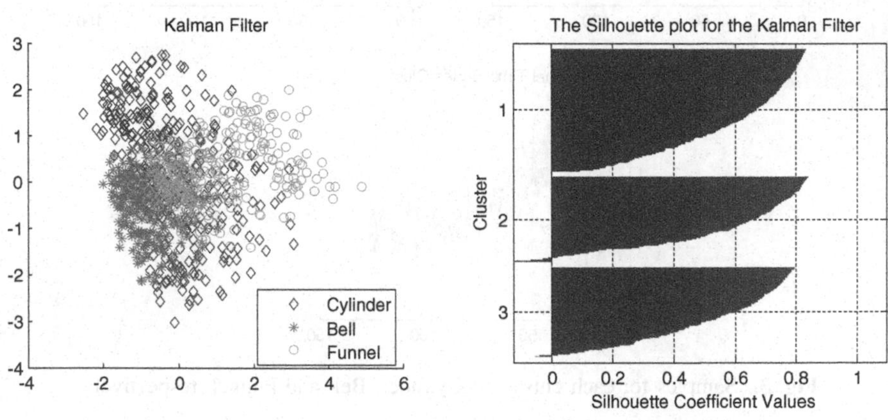

(c) Linear Predictive Coding Cepstrum

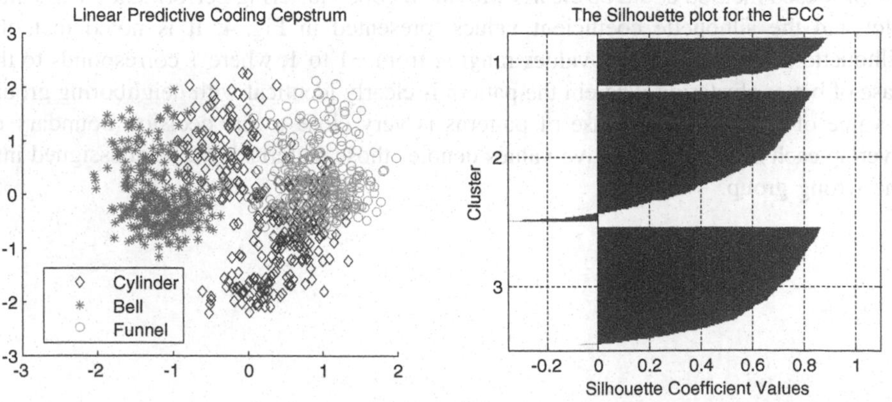

Fig. 4. Visualization for clustering performance based on the first-two features for all approaches: (a) the proposal method, (b) original Kalman Filter, (c) LPCC, (d) PCA, and (e) DFT, respectively.

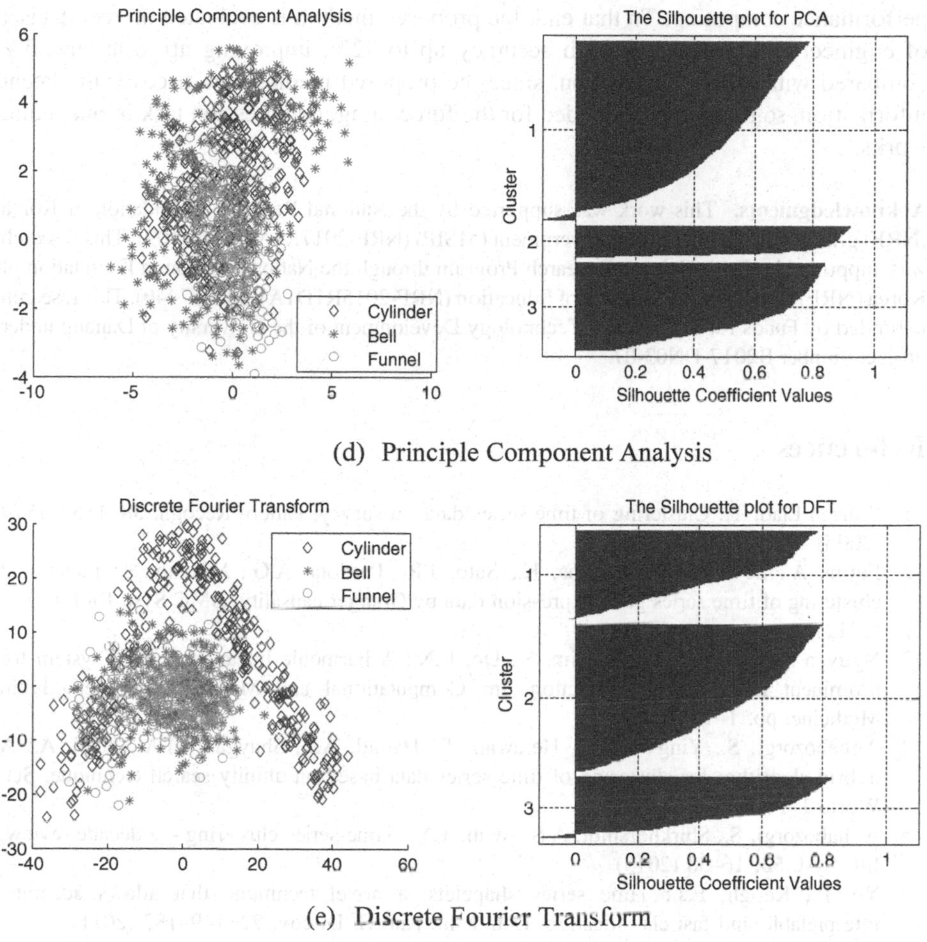

(d) Principle Component Analysis

(e) Discrete Fourier Transform

Fig. 4. (*continued*)

4 Conclusion

This paper is interested with respect to clustering modeling of the time series, understanding ongoing characteristics to obtain better hidden structure of multiple-label time sequences to propose a new approach for feature extraction. In particular, the proposed feature extraction method is based on original Kalman Filter with extension via refining parameter learning mechanism and careful implementing relying on designing model to provide a generality solution for multi-label objects clustering. The contributions of feature-based framework for clustering time course dataset are distributed along main objectives is dimensionality reduction, improving clustering quality and providing high performance computing. To that end, the proposed method is evaluated on real dataset of engineering domain with high accuracy up to 92%, improving up to at least 3% compared with others. In addition, since the proposed method can discover the latent information, so it could be extended for the forecasting, compression task in our future works.

Acknowledgments. This work was supported by the National Research Foundation of Korea (NRF) grant funded by the Korea government (MSIP) (NRF-2017R1A2B4011409). This research was supported by Basic Science Research Program through the National Research Foundation of Korea (NRF) funded by the Ministry of Education (NRF-2015R1D1A1A01057440). This research is funded by Funds for Science and Technology Development of the University of Danang under project number B2017-ĐN03-07.

References

1. Warren Liao, T.: Clustering of time series data—a survey. Pattern Recogn. **38**, 1857–1874 (2005)
2. Fujita, A., Severino, P., Kojima, K., Sato, J.R., Patriota, A.G., Miyano, S.: Functional clustering of time series gene expression data by Granger causality. BMC Syst. Biol. **6**, 137 (2012)
3. Nguyen N.A.T., Yang, H.J., Kim, S., Do, L.N.: A harmonic linear dynamical system for prominent ECG feature extraction. In: Computational and Mathematical Methods in Medicine, pp. 1–10 (2014)
4. Aghabozorgi, S., YingWah, T., Herawan, T., Hamid, A.J., Shaygan, M.A., Jalali, A.: A hybrid algorithm for clustering of time series data based on affinity search technique. Sci. World J. 1–12 (2014)
5. Aghabozorgi, S., Shirkhorshidi, A.S., Wah, T.Y.: Time-series clustering – a decade review. Inf. Syst. **53**, 16–38 (2015)
6. Ye, L., Keogh, E.J.: Time series shapelets: a novel technique that allows accurate, interpretable and fast classification. Data Min. Knowl. Discov. **22**, 149–182 (2011)
7. Khaleghi, A., Ryabko, D., Mary, J., Preux, P.: Consistent algorithms for clustering time series. J. Mach. Learn. Res. **17**, 1–32 (2016)
8. Nguyen, N.A.T., Yang, H.J., Kim, S.: HOKF: high order Kalman filter for epilepsy forecasting modeling. BioSystems **158**, 57–67 (2017)
9. Li, L., Prakash, B.A.: Time series clustering: complex is simpler! In: Proceedings of the 28th International Conference on Machine Learning (2011)
10. Zhou, J., Jia, L., Hu, G., Menenti, M.: Evaluation of Harmonic Analysis of Time Series (HANTS): impact of gaps on time series reconstruction. In: Earth Observation and Remote Sensing Applications (2012)
11. Liu, Z., Hauskrecht, M.: A regularized linear dynamical system framework for multivariate time series analysis. In: Proceedings of Conference on AAAI Artificial Intelligence, pp. 1798–1804 (2015)
12. Jollife, I.T.: Principal Component Analysis. Springer, New York (2002). https://doi.org/10.1007/b98835
13. Jahangiri, M., Sacharidis, D., Shahabi, C.: Shif-split: I/O efficient maintenance of wavelet-transformed multidimensional data. In: Proceedings of the ACM SIGMOD, pp. 275–286 (2005)

14. Navnath, S.N., Raghunath, S.H.: DWT and LPC based feature extraction methods for isolated word recognition. EURASIP J. Audio, Speech, Music Process. (2012)
15. Ralaivola, L., E-Buc, F.D.: Time series filtering, smoothing and learning using the kernel Kalman flter. In: Proceedings of the International Joint Conference on Neural Networks, pp. 1449–1454 (2005)
16. Gunopulos, D., Das, G.: Time series similarity measures and time series indexing. In: Proceedings of ACM SIGMOD (2001)
17. Zhang, X., Liu, J., Du, Y.: A novel clustering method on time series data. Expert Syst. App. **38**, 11891–11900 (2011)
18. Sumway, R.H., Stoffer, D.S.: An approach to time series smoothing and forecasting using the EM algorithm. J. Time Ser. Anal. 253–264 (1982)

Analysis on Hybrid Dominance-Based Rough Set Parameterization Using Private Financial Initiative Unitary Charges Data

Masurah Mohamad[1,3] and Ali Selamat[1,2(✉)]

[1] Faculty of Computing, Universiti Teknologi Malaysia, 81310 Skudai, Malaysia
aselamat@utm.my

[2] University of Hradec Kralove, Rokitanskeho 62, 500 03 Hradec Kralove, Czech Republic

[3] Faculty of Computer and Mathematical Sciences, Universiti Teknologi MARA, Perak Branch, Tapah Campus, 35400 Tapah, Perak, Malaysia
masur480@perak.uitm.edu.my

Abstract. This paper evaluates the capability of the hybrid parameter reduction approach in handling private financial initiative (PFI) unitary charges data to increase the classification performance. The objective of this study is to analyse the performance of the proposed hybrid parameter reduction approach in assisting the neural network classifier to classify complex data sets that might contain uncertain and inconsistent problems. The proposed hybrid parameter reduction approach consists of several methods that will be executed during the data analysis process. Slicing technique and dominance-based rough set approach (DRSA) are the two techniques that play important roles in the proposed parameter reduction process. In order, to analyse the performance of the proposed work, the PFI data that covers all regions in Malaysia is applied in the experimental works. Besides, several standard data sets have also been used to validate the obtained results. The results reveal that the hybrid approach has successfully assisted the classifier in the classification process.

Keywords: DRSA · Parameterization · Hybrid · Uncertainty Inconsistency

1 Introduction

In recent years, many researchers had focused on dealing with uncertainty and inconsistency problems. These problematic data need to be processed or analysed before a good decision can be made. Without selecting the appropriate methods, inaccurate or absolutely incorrect solution might be generated during the decision-making process. Uncertainty issues occur when there are several values of the

N. T. Nguyen et al. (Eds.): ACIIDS 2018, LNAI 10751, pp. 318–328, 2018.
https://doi.org/10.1007/978-3-319-75417-8_30

parameters or attributes which are parts of the information table are unknown or recognised. Uncertainty can be categorised into two classes: external and internal. External uncertainty is related to the attribute which is known as an action that needs to be identified by using incoming events. Meanwhile, internal uncertainty is related to the preferences or factors of the problems that are being decided by the decision-makers. Inaccurate perception, ambiguous information and contrary preference approaches are also categorised as internal uncertainties. Internal uncertainty is always being solved by many research works compared to the external uncertainty [1]. The negligence of considering the whole information during the decision-making process will also cause the inconsistency problem. Inconsistent of data problem might occur when there is a situation that has been misinterpreted by the decision-maker. The object is supposed to have a better decision from a list of condition. However, the object has been assigned as bad because of the wrong judgmentally process done by the decision-maker [2]. The incorrect information will create an inaccurate and inappropriate solution especially when dealing with high dimensional data. Thus, it is important to ensure an efficient approach with a good pre-processor method and with the help of the best classifier to be implemented in order to achieve the best performance in the decision-making process.

Recently, there are many research works related to rough set theory which considers the uncertainty and inconsistency issues in the decision-making process. Some of the research works highlighted the frameworks and the characteristics of the data [3–5], some of the works proposed new and extended methods such as dominance-based rough set (DRSA) [6] and variable precision dominance-based rough set (VP-DRSA) [7] and trapezoidal fuzzy soft set [8]. Moreover, most of the researchers had integrated several rough sets with other methods to overcome the uncertainty and inconsistency issues in different approaches. All the novel, extended and integration works were done in order to increase the capability of the selected methods and approaches in dealing with specified decision analysis problems.

DRSA is known as one of the best rough set approaches in dealing with different types of data problems especially in uncertainty and inconsistent data problems. It has been proven by many research works and also in the author previous work [9]. Inspired by the previous work, this study has extended the capability of DRSA by integrating it with a slicing method as an hybrid parameter reduction method and neural network as a classifier in dealing with problematic data. The private financial initiative unitary charges data was used in the analysis work in order to test the performance of the proposed method in the classification process.

This paper is structured into 5 sections where Sect. 1 introduces the highlighted issues which need to be investigated. Section 2 provides the brief concept on dominance-based rough set approach, parameter reduction method and neural network technique. Meanwhile, Sect. 3 presents the implemented methodology in order to execute the analysis process. In Sect. 4, the conducted experimental work is demonstrated and supported by the results discussion. As a conclusion, Sect. 5 finalises the overall proposed work.

2 Background Knowledge

This section discusses several basic concepts of related topics to provide the understanding on parameter reduction method, dominance-based rough set approach and neural network technique.

2.1 Parameter Reduction Method

Parameter reduction method is a method that is used to remove unwanted parameters or attributes or features in a data set. Besides, parameter reduction is also used to identify the important parameters or any beneficial patterns in a data set during the decision-making process [10]. It is also defined as an attribute reduction or a feature extraction in some application area. Implementing an inefficient and inappropriate parameter reduction method will affect the decision-making process especially when dealing with complex data [11]. There are many kinds of parameter reduction methods that could solved many different issues. For example, rough set theory, soft set theory and fuzzy set theory. These theories are formerly known and usually being implemented to solve uncertainty, inconsistency and vague issues [12,13].

2.2 Dominance-Based Rough Set Approach (DRSA)

In 2010, Slowinski had proposed the dominance-based rough set approach (DRSA) which is an extended version of classical rough set approach (CRSA) that was initiated by Pawlak [14]. The classical rough set approach can only deal with nominal data instead of ordinal data. Moreover, CRSA can support classification task only. DRSA was proposed in order to manage with ordinal data that normally occurred in any of application area. DRSA is also able to handle the data that needs a different kind of criteria to be analysed and considered. This different kind of criteria is also called as multi-attribute criteria. DRSA uses "*if..else*" representation form in the data analysis process and provides two types of models to be considered before the analysis begins. The analysis process is either for the classification task or for optimisation and ranking tasks. DRSA defines the condition attributes as a criteria meanwhile decision classes are defined as a preference ordered. It represents the knowledge as a collection of upward and downward unions of classes, and sets of objects or also known as granules of knowledge are defined as dominance relation. The data is presented by the decision table which can have more than one reduction set. The intersection of the table is known as core. The process of approximation for the dominance cones is the beginning process of initiating the decision rules.

3 Methodology

The purpose of this study is to develop an hybrid approach in dealing with uncertain and inconsistent data. It combines three methods as the main processor, where DRSA and slicing method are implemented in the parameter reduction process and a neural network acts as a classifier. The aim of this proposed approach is to provide an effective approach in handling the highlighted issues which were mentioned in the previous section. The neural network classifier is included in the proposed approach which could assist the parameter reduction process in order to return the high accuracy rate during the classification process. Figure 1 demonstrates the framework of the proposed hybrid DRSA parameterization approach. It comprises of five main phases starting from the data pre-processing phase until the result is obtained from the classification task. The details of all phases will be discussed in the following subsections.

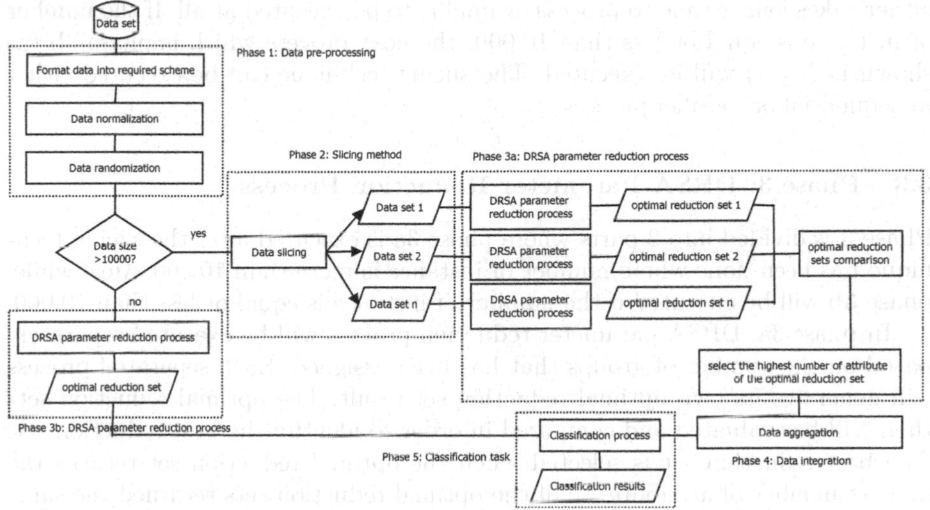

Fig. 1. Proposed framework of the hybrid parameter reduction approach

3.1 Phase 1: Data Pre-processing

The collected data will go through several processes in order to be prepared for the classification task. The processes are data formatting, data normalisation and data randomisation. The raw data is formatted into a required scheme according to the methods or software which are used during the classification task. Basically, the data is presented by using $m \times n$ matrix including the decision class at the end of the data column. The formatted data then is normalised in order to standardise the value of each column and to increase the computer processing performance and also to decrease the memory usage. In addition, the normalised data will be randomised to avoid any unbiased issues and also to increase the accuracy rate of the classification task.

3.2 Phase 2: Slicing Technique

This technique is applied after the data has gone through the pre-processing task and also when the size of the data or instance is more than 10,000. The data will be fragmented into a number of groups by dividing the total number of instance with 10,000. If the calculation contains remainder, the number of group will be added another 1.

The slicing technique will be executed only when the number of instance is more than 10,000. 10,000 is a constant value. The reason of defining the constant value as 10,000 is that, most of the parameter reduction method which are executed by the normal processor (not super computer) can only afford this value or less. If more than 10,000 of instances are being analysed, the processor either takes longer time to process or unable to be executed at all. If the number of instance is equal or less than 10,000, the next process which is phase 3b (as shown in Fig. 1) will be executed. The slicing technique can be executed either in sequential or parallel process.

3.3 Phase 3: DRSA Parameter Reduction Process

Phase 3 is divided into 2 parts where phase 3a is executed after the slicing technique has been done where number of instance is more than 10,000. Meanwhile, phase 3b will be executed if the number of instance is equal or less than 10,000.

In phase 3a, DRSA parameter reduction process will be executed separately based on the number of groups that has been assigned. Each separated process will generates its own optimal reduction set result. The optimal reduction sets then will be evaluated and compared in order to identify the best reduction set. The best reduction set is selected when the optimal reduction set returns the highest number of attribute. If all the optimal reduction sets returned the same number of the highest attribute, the first set of the optimal reduction sets will be selected. This concept has been implemented in the previous works [15].

In phase 3b, DRSA parameter reduction process will be executed directly to the data set. The same concept as implemented in phase 3a has been applied where-as the highest number of attribute will be selected from the generated optimal reduction sets. The first set of the highest attribute will be chosen if the optimal reduction set returns the same number of the highest attribute.

3.4 Phase 4: Data Integration

Data integration phase will only be executed when the data set has gone through the phase 3a. The data set which has been sliced during phase 2 will be integrated for the classification purpose. The data set is organised according to the required format and applied only the selected reduction set attribute for the classification purpose. Data integration phase is done in order to increase the classification performance where large data will provide more information to the classifier when compared to small data. This can be proven at the result section below.

3.5 Phase 5: Classification Task

Classification task is the last phase that needs to be executed. Cleaned and simplified data set was ready to be used as an input to the task. Any classifier can be applied to execute the classification task. The results obtained from this phase will be evaluated using several standard evaluation measures.

4 Experimental Work and Results Discussion

Experimental works had been conducted according to the proposed framework. A few software were used to execute the experimental works, jMAF was used for the DRSA parameter reduction process, Matlab R2014a was implemented to accomplish the classification task and Ms. Excel 2016 was applied to prepare the data in the pre-processing phase and to execute the slicing work. Software such as WEKA was used to execute the other parameter reduction method named as correlation-based feature subset selection (CFS). CFS is applied to validate the obtained results of the proposed work. CFS identifies the optimal attributes of the data set by considering the predictive capability of each of the attribute and also the degree of redundancy among the attributes. Meanwhile, neural network which was executed in Matlab R2014a was being selected as the classifier due to its capability in classifying any kinds of data and had been successfully proven by many research works.

4.1 Data Description

Private financial initiative unitary charges data (PFI) was selected to test the performance of the proposed framework. It had also been used in many research works such as in [16]. The data, however, had been tested only in the forecasting area but not in the classification process. The data was about the construction material price of six regions in Malaysia and was collected from year 1980 until 2012. In addition, another standard five data sets that had been downloaded from UCI Machine Learning Data Repository were also used in this experimental works for the result validation purpose. Table 1 lists the characteristics of all data sets.

Table 1. Data sets characteristics

Data sets	Number of instances	Number of attributes	Data type	Missing values
PFI	6,528	9	Integer, Real	Yes
Dota	102,944	116	Integer	No
CNAE	1,080	857	Integer	No
Semeion	1,593	257	Integer	No
Buzz	140,707	77	Integer, Real	No
HAPT	7,767	562	Real	No

4.2 Results Discussion

The performance of the proposed approach in the classification process is compared by using different data sets and different parameter reduction methods. The proposed approach which is a combination of DRSA, slicing technique and neural network is also being compared with other two sets of combination: (i) without any parameter reduction method, slicing technique and neural network and, (ii) CFS, slicing technique and neural network. The proposed approach is labelled by A1, approach number two which is combination without any parameter reduction method, slicing technique and neural network is labelled by A2 and third approach which is a combination of CFS, slicing technique and neural network is labelled by A3.

Results on Number of Parameter After Going Through the Parameter Reduction Process. As being mentioned in the previous section, three different approaches have been implemented in the experimental works including the proposed approach. In order to evaluate the performance of the proposed approach in the parameter reduction process, the number of optimised attributes are being monitored and compared with the other two benchmark approaches. The proposed approach (A1) which implements DRSA as the parameter reduction process has reduced more than 70% of the original attribute to obtained the optimised attribute set. This contradicts with the A3 approach which is considered more than 3% of the attributes to be selected. The number of attribute in the data set really helps the parameter reduction method to analyse the pattern of the data. However, too large or too small of the data size will affects the whole process of the decision-making.

Performance of the Proposed Approach on Each Data Set. The performance of the proposed approach based on the six standard evaluation measures are presented at the following tables. The values are presented starting from accuracy value (ACC) that will highlight the performance of the classification process in correctly classifying the data set. Then, followed by sensitivity (SENS) value or also known as recall to evaluate the ability of the classifier to identify the positive value of the data set. Meanwhile, specificity (SPEC) will evaluate the opposite value which is negative value of the data set. PPV or positive predictive value is used to measure the percentage of the positive value data set over all value of data set that are classified as positive. It is also known as precision. Negative predictive value (NPV) returns the percentage of the data set over all values of data set that are classified as negative. Moreover, F-measure score is used to recognise any shortcoming problems and overall performance of the proposed approach in the classification process. It is also used to measure whether the PPV and SENS are evenly weighted [17,18].

Referring to Table 2, approach A2 has performed well in classifying all data set compared to A1 and A3. Even though approach A2 performs well in the classification task, it still requires large memory usage and processing time. Surprisingly, the proposed approach has returned quite promising results for all data sets except for Dota data set. All approaches also do not perform well for Dota data set where the average score for the classification accuracy is only 58.25%. However, F-measure score for approach A1 does not return a significant result compared to the accuracy rate for all data sets. These insignificant values have indicated that some issues had been raised up either the data sets used in the experimental works contain imbalance data division or the selected parameter reduction method itself is unable to analyse the problematic and complex data set.

The opponent approach (A3) has shown a quite significant results for all data sets. The obtained values of accuracy rate and the F-measure score for this approach are almost equal for all data sets except for PFI and HAPT data sets.

Table 2. Classification results on all data set

Approaches	ACC	SENS	SPEC	PPV	NPV	F-measure
PFI						
A1	88.2	3.01	93.72	5.30	93.68	3.13
A2	96.03	68.48	97.91	68.61	97.87	67.67
A3	91.53	33.27	95.48	30.52	95.52	28.77
Semeion						
A1	86.71	32.56	92.70	32.12	92.57	32.50
A2	98.16	90.16	99.02	91.03	98.94	90.44
A3	98.08	90.57	98.93	90.44	98.93	90.30
CNAE						
A1	83.81	46.28	90.92	25.49	90.77	24.13
A2	98.63	94.70	99.23	93.36	99.22	93.83
A3	97.23	90.63	98.44	87.72	98.43	88.43
Dota						
A1 (10k instances)	57.33	53.82	53.82	52.87	52.87	50.43
A1 (90,625 instances)	54.28	53.82	53.82	52.87	52.87	50.43
A2	60.93	60.49	60.49	60.07	60.07	60.01
A3 (10k instances)	59.73	59.26	59.26	58.86	58.86	58.76
A3 (90,625 instances)	58.91	57.88	57.88	57.62	57.62	57.48
Buzz						
A1	96.00	95.65	95.65	92.15	92.15	93.77
A2	95.8	94.67	94.67	92.72	92.72	93.65
A3 (460 instances)	95.53	93.88	93.88	91.69	91.69	92.73
A3 (140,707 instances)	95.27	94.74	94.74	91.58	91.58	93.04
HAPT						
A1	93.64	58.41	96.25	49.78	96.19	51.69
A2	98.99	83.21	99.41	80.61	99.38	81.62
A3	97.14	68.56	98.32	68.35	98.27	68.40

The significant scores for the SENS and PPV values indicate that CFS is a good parameter reduction method that can manage any kinds of data sets. Unfortunately, this approach is also unsuccessful in managing PFI data for the similar reasons as approach A1. It can be proven by the SENS and PPV scores which are 33.27% and 30.52% obtained during the classification task.

Overall Results. Three approaches which comprise of three different types of parameter reductions (without any parameter reduction method, CFS and DRSA) are compared. It can be concluded that all selected data sets have performed more than 80% except for Dota in the classification task. The most significant result generated by all approaches is obtained from Buzz data set. Both accuracy rates and F-measure scores for all approaches reach more than 95%. This result indicates that the Buzz data itself is equally divided and not too complex to be analysed. The overall performances of these three approaches on classifying all selected data sets are shown by the average scores of the accuracy rate where A1 obtains 84.28%, A2 89.83% and A3 91.43%. The results had shown that the proposed approach has done quite a good job in handling uncertain and inconsistent data.

5 Conclusion

This paper has presented an analysis on hybrid parameterization approach between dominance-based rough set approach (DRSA) as a parameter reduction method, slicing technique as a data separator and neural network as a classifier. The performance of this hybrid approach has been tested in the classification process by using private financial initiative (PFI) unitary charges data. The obtained results have shown that the proposed approach has successfully managed the PFI data by returning a significant result even though the accuracy rate is lower than the other two benchmark approaches. The obtained results also show that the PFI data which is usually applied in the forecasting and prediction process is able to be used in the classification process. Moreover, the results have proven that the proposed approach is suitable to be implemented for any data that faces uncertainty and inconsistency problems. The flexibility of the data analysis framework offered by this approach makes it capable to deal with different sizes of data sets. It is more beneficial if the proposed approach can be implemented via high performance processor in order to decrease the processing cycle time and also to decrease the size of the computational memory. As a conclusion, the implementation of an efficient parameterization approach will facilitate the classifier to generate a good and reliable decision. Moreover, the types of data such as nominal, cardinal or ordinal might also affect the result generation in the decision-making process. A proper selection on method or approach also needs to be considered in order to obtain the best result.

Acknowledgments. The authors wish to thank Universiti Teknologi Malaysia (UTM) under Research University Grant Vot-02G31, and the Ministry of Higher Education

Malaysia (MOHE) under the Fundamental Research Grant Scheme (FRGS Vot-4F551) for the completion of the research.

References

1. Durbach, I.N., Stewart, T.J.: Modeling uncertainty in multi-criteria decision analysis. Eur. J. Oper. Res. **223**(1), 1–14 (2012)
2. Greco, S., Matarazzo, B., Słowiński, R.: Multicriteria classification by dominance-based rough set approach. In: Handbook of Data Mining and Knowledge Discovery, pp. 1–14 (2002)
3. Borgonovo, E., Marinacci, M.: Decision analysis under ambiguity. Eur. J. Oper. Res. **244**(3), 823–836 (2015)
4. Karami, J., Alimohammadi, A., Seifouri, T.: Water quality analysis using a variable consistency dominance-based rough set approach. Comput. Environ. Urban Syst. **43**, 25–33 (2014)
5. Li, P., Wu, J., Qian, H.: Ground water quality assessment based on rough sets attribute reduction and TOPSIS method in a semi-arid area. China Environ. Monit. Assess. **184**(8), 4841–4854 (2012)
6. Greco, S., Matarazzo, B., Słowiński, R.: Dominance-based rough set approach to decision under uncertainty and time preference. Ann. Oper. Res. **176**(1), 41–75 (2010)
7. Inuiguchi, M., Yoshioka, Y., Kusunoki, Y.: Variable-precision dominance-based rough set approach and attribute reduction. Int. J. Approx. Reason. **50**(8), 1199–1214 (2009)
8. Xiao, Z., Xia, S., Gong, K., Li, D.: The trapezoidal fuzzy soft set and its application in MCDM. Appl. Math. Modell. **36**(12), 5844–5855 (2012)
9. Mohamad, M., Selamat, A.: An analysis of rough set-based application tools in the decision-making process. In: Saeed, F., Gazem, N., Patnaik, S., Saed Balaid, A.S., Mohammed, F. (eds.) IRICT 2017. LNDECT, vol. 5, pp. 467–474. Springer, Cham (2018). https://doi.org/10.1007/978-3-319-59427-9_49
10. Jing, Y., Li, T., Huang, J., Zhang, Y.: An incremental attribute reduction approach based on knowledge granularity under the attribute generalization. Int. J. Approx. Reason. **76**, 80–95 (2016)
11. Meng, Z., Shi, Z.: On quick attribute reduction in decision-theoretic rough set models. Inf. Sci. **330**, 226–244 (2016)
12. Anisseh, M., Piri, F., Shahraki, M.R., Agamohamadi, F.: Fuzzy extension of TOPSIS model for group decision making under multiple criteria. Artif. Intell. Rev. **38**, 325–338 (2012)
13. Feng, F., Li, C., Davvaz, B., Ali, M.I.: Soft sets combined with fuzzy sets and rough sets: a tentative approach. Soft. Comput. **14**(9), 899–911 (2010)
14. Słowiński, R.: New applications and theoretical foundations of the dominance-based rough set approach. In: Szczuka, M., Kryszkiewicz, M., Ramanna, S., Jensen, R., Hu, Q. (eds.) RSCTC 2010. LNCS (LNAI), vol. 6086, pp. 2–3. Springer, Heidelberg (2010). https://doi.org/10.1007/978-3-642-13529-3_2
15. Mohamad, M., Selamat, A.: A new hybrid rough set and soft set parameter reduction method for spam e-mail classification task. In: Ohwada, H., Yoshida, K. (eds.) PKAW 2016. LNCS (LNAI), vol. 9806, pp. 18–30. Springer, Cham (2016). https://doi.org/10.1007/978-3-319-42706-5_2

16. Kamaruddin, S.B.A., Ghani, N.A.M., Ramli, N.M.: Best forecasting models for private financial initiative unitary charges data of east coast and southern regions in Peninsular Malaysia. Int. J. Econ. Stat. **2**, 119–127 (2014)
17. Idris, I., Selamat, A., Omatu, S.: Hybrid email spam detection model with negative selection algorithm and differential evolution. Eng. App. Artif. Intell. **28**, 97–110 (2014)
18. Masetic, Z., Subasi, A.: Congestive heart failure detection using random forest classifier. Comput. Methods Progr. Biomed. **130**, 54–64 (2016)

Decision Support and Control Systems

Smart Lighting Control Architecture and Benefits

Igor Wojnicki and Sebastian Ernst(✉)

Department of Applied Computer Science, AGH University of Science and Technology, Al. Mickiewicza 30, 30-059 Krakow, Poland
{wojnicki,ernst}@agh.edu.pl

Abstract. The paper discusses results of a pilot project aimed at providing dynamic control for 4,000 LED luminaires in Kraków, Poland. The main research goal is to provide a flexible and highly scalable system architecture. Thus, the work regards both theoretical model improvements, comparative technology research, and architectural flexibility assessment based on available third-party systems that must be interacted with. The structure of the system is presented in detail, along with relation to previously-published theoretical work, technologies used and important implementation details. The underlying theoretical concepts, based on the dual graph grammars, are also presented. A critical discussion presents the benefits of the proposed solution in the light of various practical problems encountered during implementation of the pilot project.

Keywords: Smart city · Smart lighting · Dynamic lighting control Distributed system · Dual graph grammars · System integration

1 Introduction and Motivation

The benefits of modern, adaptable lighting systems are hard to question. Such systems provide substantial energy savings, resulting both from technological shift and intelligent control [1]. Saved energy can be directly translated into avoided carbon dioxide emissions. Furthermore, keeping the architecture open enables integration with other smart city systems. The additional value provided by fusing multiple systems operating in a common area are virtually limitless.

Street lighting control systems are often convoluted and complex, especially with regard to their scalability, capability to integrate with other smart city systems and flexibility of control rule definition [2,3]. Research on an architecture that is clear, can be easily integrated with external systems, such as traffic and ambient light sensors, and an ability to control the actual luminaires in real time has been conducted over the span of the several preceding years in numerous projects, including a practical deployment in the city of Krakow.

A theoretical multi-agent architecture for flexible outdoor lighting control system was introduced in [4]. This paper presents the results of further research,

N. T. Nguyen et al. (Eds.): ACIIDS 2018, LNAI 10751, pp. 331–340, 2018.
https://doi.org/10.1007/978-3-319-75417-8_31

focusing on the transformation of the formal model in to a real-life system, as well as the consequences of architectural and technological decisions made in the process.

Using an agent-based background as the basis of the system allows it to remain reliable and adaptable. Reliability is guaranteed by well-tested supervision mechanisms, also used in critical telecommunications systems. The proposed approach binds heterogeneous components, enabling them to flawlessly cooperate with each other. Adaptability is provided through well-defined interfaces and components which support translation of different protocols. Heterogeneity is a result of pragmatic use of the best suited languages and technologies to do particular job.

The structure of the paper is as follows. Section 2 presents the research problems which drove the work presented in this paper. The formal model, being an evolution of previously presented concepts, as well as the primary means of improving performance, is discussed in Sect. 3, along with complexity estimations. Section 4 presents the architectural and technological details, also discussing the feasibility of various tested solutions for implementation of large-scale, graph-based control systems. Lessons learned during the pilot project in Krakw are presented in Sect. 5, followed by conclusions and future work plans.

2 Challenges

Development of an outdoor lighting control system which fulfills the requirements described in Sect. 1 poses several challenges. First of all, there is the problem of scale. A city with a population of 1 million is expected to have 80,000 light points, powered by 1,600 circuits. There will be thousands of ambient light and traffic intensity detectors, which requires the control system to be efficient, or at least scalable. Furthermore, deployment or modernization of lighting infrastructure is an ongoing process, which consists in replacing several thousand light points every few years. Luminaire, sensor and communications equipment models change often, and there never a guarantee a new batch of replaced lamps will use the same solutions as the preceding ones. Thus, the control system has to be modifiable and flexible, to be capable of integration with diverse products.

Because of that, an extensible formal model with performance-wise processing capabilities is needed. A graph-based model serves a formal background to support such scalability and flexibility. As the actual control is provided by graph transformations, a dual graph grammar concept was introduced to boost their performance. Details of the model are presented in Sect. 3.

Furthermore, to create a control system, actual software technologies supporting graph transformations have to be verified and benchmarked. The findings, together with the proposed flexible architecture, are presented in Sect. 4.

To summarise, the main research challenges are threefold:

- to increase the performance of the model,
- to find the most suitable technologies that can be used to implement the control agent, as the decision-making core of the system,

– to propose an architecture of the system, designed with modularity and heterogeneity in mind.

3 Formal Model

The GEM-CA graph represents both the physical environment (General Environment Model) and the aspects related directly to luminaire control (Control Availability). During run-time, only the CA graph (CAG for short) is actively modified.

An example of a CAG is given in Fig. 1. It is a undirected, labeled and attributed graph. A certain formal notation is used. Graph labels are given at the vertices or edges e.g.: *dvt*, or *l*. Optionally, there are also indices preceded by a slash, e.g.: *dvt1*, *l3*. In particular, the indices are arbitrary symbols, not only integers. The presented example represents a single *lighting segment*, which is defined as part of a street with uniform structure: the same number of lanes, width, speed limit, etc.[1] A segment is indicated by a vertex labeled with *s*, in this case it is *s1*. There are two sensors detecting certain environmental parameters for this segment, hence the edges between *s1* and *dvt1* and *dva1*. The *dvt1* is labeled with *dvt*, which indicates that it is a virtual traffic intensity sensor. Since it is virtual, its value is subject to be calculated based on values coming from real sensors.

The concept of a virtual sensor was introduced in order to compensate for a situation of a complex sensor infrastructure, e.g. multiple traffic intensity sensors on the same lane and multiple lanes on the same street. Similarly, *dva1* is labeled with *dva*, which indicates a virtual ambient light sensor.

Furthermore, there are vertices labeled *c*, which represent lighting configurations. For example, *c1* defines a lighting configuration which provides proper lighting to segment *s1*, which complies with the lighting norm *me3*, hence the label between *c1* and *s1*. The configuration sets lumiaires *l1*, *l2* and *l3* to 100% of their nominal power. The luminaires are labeled with *l*, while the power of a particular luminaire is indicated by the attribute *p* (e.g. set to *1.0*, denoted $p(1.0)$).

The sensor data processing view is presented in Fig. 2; it is a directed, labeled and attributed graph. Formally, it is a separate graph, with a distinctive grammar. However, both graphs, namely the control view and the sensor data processing view, are synchronised with use of the dual graph grammar. That enables graph processing, namely the application of graph transformations, to be independent on both graphs, and only require synchronisation when needed. This decreases the computing power necessary to provide the actual transformations and improves system scalability. The synchronization points are the virtual sensor vertices, labeled *dvt1* and *dva1*.

[1] At any given time, the entire lighting segment can be assigned only a single lighting class. Thus, even an uniform part of a street can be split into several lighting segments in order to allow them to be controlled independently.

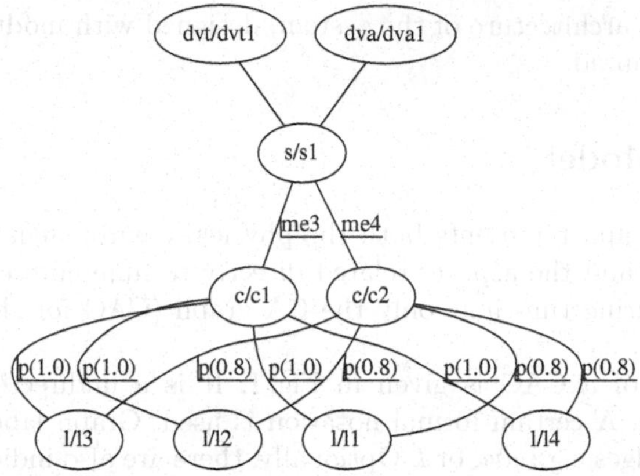

Fig. 1. An example control availability graph, control view.

The presented example graph represents an arithmetical expression which has to be performed over the actual values coming from real sensors. The sensors are labeled *dt* or *da*, representing the traffic intensity or ambient light, respectively. There are also labels representing mathematical operators such as +, *min*, *max*, etc. For example, *dvt1* can be reduced to an expression: $dvt1 = max(dt5 + dt4, dt2) + max(dt3, dt1)$. This reduction can be performed before run-time which reduces the graph size and increases its operational processing performance.

Introducing the concept of virtual sensors significantly decreases the effort needed to process the input data. It reduces the problem size for run-time, in terms of the graph size, measured as the number of its edges. Since the computational complexity of the transformations which provide actual control is $O(n^3)$, the described modification results in the computation time being reduced by a factor of 2.8.

The control process modifies the CAG. It propagates the sensor data and activates appropriate configurations. Having a configuration active triggers the dimming commands being sent to individual luminaires.

4 Architecture and Implementation

This section presents the architecture of the system, providing details on the structure of modules, as well as measures taken to guarantee appropriate reliability.

The implementation described in the paper was developed within a pilot project in Krakw, Poland. It involved replacing almost 4,000 soda-based with state-of-the-art LED lamps and developing an innovative dynamic control system covering the project area. The control decisions are based on the following sensor inputs:

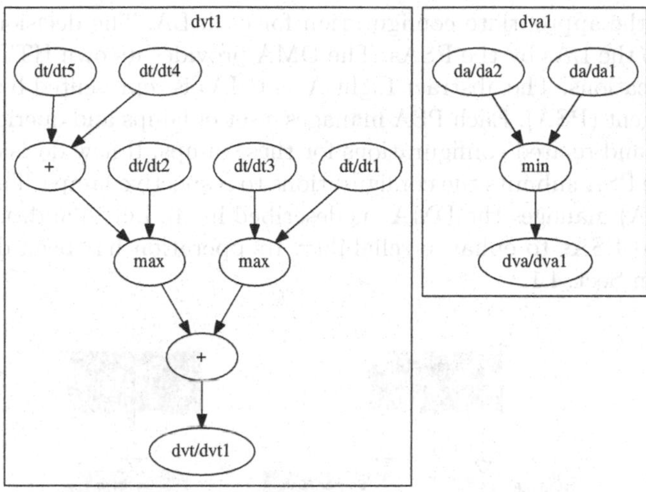

Fig. 2. An example control availability graph, sensor data processing view.

- traffic intensity, measured by induction loops installed in part of the roads in the project area,
- ambient light level, measured by photocells installed in circa 10% of the replaced luminaires.

Sensor data updates trigger decision-making processes, which determine the optimal configuration for each luminaire using pre-calculated photometric data. Luminaire configurations are transferred to the lamps using an API offered by the basic controlled system, supplied together with the lamps.

4.1 System Structure

The overall structure of the solution reflects the abstract one presented in [4] (Fig. 2, page 4) with some modifications determined by the technologies used and the architectures of third-party components. The relation between technologies used and the system structure has been presented in Fig. 3.

The role of the individual modules is as follows. The Sensor Push Agent (SPA) reflects the abstract Sensor Agent (SA) and is responsible for collecting sensor data. As there are many types of sensors, in practical implementations there will be more than one type of SPA. In the Krakw project, the traffic sensors and light intensity sensors are handled by completely separate IT systems, and hence have separate SPAs. They feature communication methods specific to given third-party APIs and make HTTP calls to the DMA. The Lamp Agents (LA) take the form of the control system supplied with the lamps. They accept requests using a HTTP interface, transmit them to the driver built into each luminaire and transform them into control signals for the power supply. The Decision Making Agent (DMA) receives sensor value updates from the SPAs and chooses the appropriate configuration for each LA. The decisions are communicated to the LAs by the PSAs. The DMA provides its own HTTP interface for communications. The abstract Light Agent (IA) is represented by the Profile Switching Agent (PSA). Each PSA manages a set of lamps and queries the DMA for decisions and related configurations for these lamps. If new decisions become available, the PSA submits the configurations to respective lamps. The Supervising Agent (SA) manages the DMA, as described in [4], but in addition manages the SPAs and PSAs to enhance reliability. Its operation has been described in more detail in Sect. 4.1.

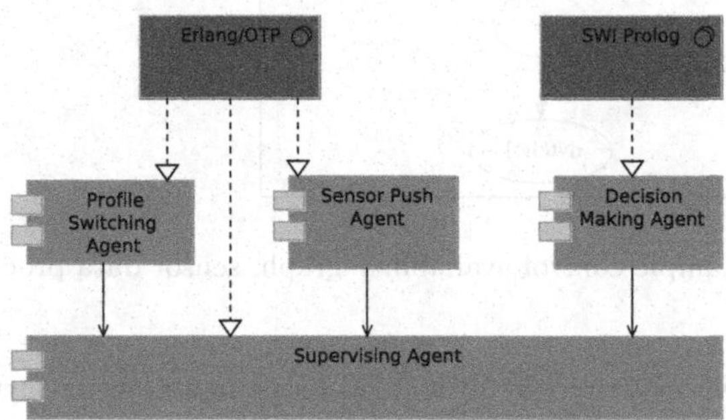

Fig. 3. Deployment diagram, including technologies used

The entire control system follows the principle of Erlang OTP supervision trees [5]. The concept is based on a hierarchical structure, consisting of *supervisors* and *workers*. The role of supervisors is to monitor the execution of workers and restarting them according to the specified strategy. Workers perform actions and, as a rule, cannot be trusted, i.e. they may fail or succeed. This follows Erlang's "let it crash" philosophy by allowing processes to be unreliable (due to any reason), rather than building a complex system of guard mechanisms to prevent errors. Because of this, the system is more lightweight, flexible and easier to maintain.

4.2 Technological Considerations

With the exception of the DMA, all modules are implemented using Erlang/OTP. Because of its suitability for implementation of rule-based systems (RBSs), Prolog [6] has been chosen as the implementation method of the DMA. Inter-module communications can use native Erlang process messaging. Other communication standards are necessary only for the following components:

- the DMA, which provides a HTTP-based REST interface used to submit sensor readings and retrieve lighting configurations,
- the external (third-party) lighting management system, which provides a HTTP-based REST interface,

- the external (third-party) ambient light sensor system, which provides a HTTP-based REST interface,
- the external (third-party) traffic management system, which communicates its reading by making HTTP calls to a pre-configured server address.

All SPAs uses a cache of previously-recorded values to avoid unnecessary calls to the DMA.

The PSA is responsible for making HTTP calls to the DMA and forwarding configurations via the lighting management system's HTTP API. Since the DMA's REST API returns decisions on any sensor value update query, it may be assumed that there is no need to make calls to the DMA on any other occasion. Therefore, a PSA is created upon every update by a SPA. However, an additional ("standalone") PSA may be used in certain situations, as outlined in Sect. 4.1.

The DMA is implemented in Prolog. Its main task is to deliver information about proper setting of each of the luminaires under control. The inference engine updates the GEM-CA graph, which is the formal model of the control system, as described in Sect. 3. To implement the inference engine and the DMA REST interfaces, SWI Prolog was used. There were two reasons to support this choice: performance and language expressive power. The performance was verified against other systems and languages including graph databases. The benchmark took into consideration a complex lighting infrastructure consisting of 1,000 luminaires and 1,400 detectors, running 10 control scenarios, for which execution time was measured. The SWI Prolog based inference engine clearly outperformed other competitors, see Table 1 for details.

5 Practical Experiences and Discussion

An idealistic view on Smart City systems, where all modules are deployed in harmony, with careful selection of technologies used and unified protocols to guarantee interoperability, rarely finds materialisation in real life. When dealing with real-life projects – and often, public tenders – one needs to deal with solutions readily available in the market.

In the aforementioned 4,000-luminaire pilot project (see Sect. 1) alone, the University team also assisted the city officials in integrating the existing systems and specifying requirements for ones to be retrofitted. The goal was to guarantee integration and interoperability while maintaining a reasonable price to assure an appropriate ROI (return on investment). In that project alone, the following problems have been encountered (as outlined in Sect. 4.2):

- One of the existing traffic monitoring systems did not have an API (Application Programming Interface) or other means of integrating with third-party IT systems. The vendor was able to provide this functionality within a separate order, but the system needed to make queries (HTTP) to another machine within the same local network. Thanks to selection of Erlang as the implementation technology, it was possible to provide an appropriate node

Table 1. Performance benchmark, results comparison, less is better.

Technology	Time [s]
SWI Prolog	0.10
Erlang (optimized)	0.30
TinkerGraph graph database	0.99
Erlang	1.50
BitsyGraph graph database	1.53
Neo4j graph database (supercomplex)	1.73
PostgreSQL relational database	3.28
OrientGraph graph database	5.93
MySQL	7.37
InfiniteGraph graph database	7.96
DEX graph database	8.11
Java (HashMap)	9.00
Neo4j graph database	11.92
FoundationDB database	47.00
OrientDB graph database	299.20

(machine) to receive the requests within the institution's local infrastructure and use built-in distributed application integration mechanisms.

- Some lighting infrastructure managements systems in the market required a direct connection to each lighting cabinet controller. However, network restrictions did not allow external connections to all cabinets from the Internet. Because of modularity, it would still be possible to integrate them, providing a simple embedded computer (e.g. a Raspberry Pi[2]) to function as a local Erlang node.
- Finally, the provided lighting management system turned out to provide a cloud-based REST API, albeit with a proprietary structure of JSON data. This facilitated deployment regarding connection structure, and the generation and parsing of JSON data was handled by a separately-implemented, lightweight Erlang module.

As shown in these examples, the modular structure of the proposed solutions made it possible to adapt to various restrictions and conventions imposed by existing systems, either ones already possessed by the city or bought within a public tender.

The concept is inherently heterogeneous with regard to technologies used, and the distributed architecture was developed with traversing various network obstacles, such as firewalls or NATs, in mind. Moreover, the dynamic lighting control system is intended to be extended to other parts of the city as well.

[2] https://www.raspberrypi.org.

This means the system will need to "learn" to communicate with other sensor systems, and possibly with lighting control systems from other vendors.

Experiences from the pilot project have shown that in a real-life scenario, this approach is the only economically viable way of implementing Smart City projects if all systems are not designed from scratch at the same time. Modularity and heterogeneity helps to avoid vendor lockdown and maximise reuse of systems already deployed in the city.

Field testing within the pilot project confirmed the applicability of the dual graph grammar theory for large-scale control systems and demonstrated the performance benefits of systems based on this theoretical approach. While the decision-making core utilises Prolog for its flexibility and expressive power, the other modules were implemented in Erlang/OTP and use the supervision scheme for reliability and easy adaptation to distributed deployment environments.

It must also be noted that the developed system, thanks to a multi-variant photometric design and utilisation of real-time sensor data, is unique due to being fully compliant with the lighting norms [7] and providing dynamic control at the same time.

However, there still is room for improvement in various aspects of Smart City system integration. For instance, the external systems (lighting management, sensors) need to provide some means of communication with third-party systems. Also, the problem of vendor lockdowns still persists – in such case, the city may be forced to purchase additional functionality or licences from a given vendor. This, however, can be mitigated by following appropriate openness guidelines.

6 Conclusions and Future Work

The paper presents the architecture of a lighting control system, developed within a pilot project targeted at providing dynamic control for 4,000 LED luminaires in Krakw, Poland. The main goals for design of the architecture was modularity and heterogeneity, in order to satisfy the performance, scalability, distributability and reliability requirements. All modules of the system were developed using Erlang/OTP, apart from the core decision module, developed in Prolog. This decision proved beneficial not only with regard to expressive power, but also performance, as shown in Sect. 4.2.

Dual graph grammars constituted the theoretical basis for the AI-based decision-making module. The use of this formalism decreased the processing time by a factor of 2.8 (see Sect. 3). The described project proved its feasibility for implementation of high-performance, large-scale control systems.

Having rule-based control on top of a mathematical graph-oriented model driven by graph transformations supports both scalability and reliability. The scalability benefits are threefold:

1. There is no limit to the scale of a system that can be built with the use of the complimentary graphs approach [8].

2. The choice of rules as the method to define control strategy guarantees reusability. The graph transformations providing actual control define user requirements which determine the system behaviour. This remains unchanged regardless of the deployment details. Having subsequent deployments requires only model adjustments, since it reflects the physical infrastructure [9].
3. The formal and technological decisions make the run-time components highly efficient. Their performance is boosted thanks to the dual graph grammar, which separation of the control model from calculations on the sensor data streams.

The technologies used to implement the system were carefully selected and tested, with results presented in Table 1 (page 7).

Finally, real-life experiences from the pilot project have been discussed in Sect. 5. These are valid not just for implementation of lighting control systems, but for integration of various systems in general.

Future work primarily involves further extension and scaling of the project. With regard to the sensor layer, work is performed to maximise the controlled area in situations with limited traffic sensor coverage by means of traffic prediction and interpolation. The control layer is being extended with new modules to adapt to lighting management systems from other third-party vendors.

References

1. Wojnicki, I., Ernst, S., Kotulski, L.: Economic impact of intelligent dynamic control in urban outdoor lighting. Energies **9**(5), 314 (2016)
2. Guo, L., Eloholma, M., Halonen, L.: Intelligent road lighting control systems. Technical report, Helsinki University of Technology, Department of Electronics, Lighting Unit (2008)
3. Highways England: motorway road lighting control system (MoRLiCS) - project details (2012). http://www.highways.gov.uk/knowledge/projects/morlics
4. Wojnicki, I., Kotulski, L., Ernst, S.: On scalable, event-oriented control for lighting systems. In: Barbucha, D., Le, M.T., Howlett, R.J., Jain, L.C. (eds.) KES-AMSTA. Frontiers in Artificial Intelligence and Applications, vol. 252, pp. 40–49. IOS Press, Amsterdam (2013)
5. Armstrong, J.: Programming Erlang: Software for a Concurrent World. Pragmatic Bookshelf, Raleigh (2013)
6. Nillson, U., Maluszynski, J.: Logic, Programming and Prolog, 2nd edn. Linköpings Universitet, Linköpings (2000)
7. CEN: CEN/TR 13201–1:2004, Road lighting. Selection of lighting classes. Technical report, European Commitee for Standardization, Brussels (2004)
8. Kotulski, L.: On the control complementary graph replication. In: Jacek Mazurkiewicz, E.A. (ed.) Models and Methodology of System Dependability. Monographs of System Dependability, vol. 1. Oficyna Wydawnicza Politechniki Wrocławskiej, Wrocław
9. Wojnicki, I., Kotulski, L., Sdziwy, A., Ernst, S.: Application of distributed graph transformations to automated generation of control patterns for intelligent lighting systems. J. Comput. Sci. **23**, 20–30 (2017)

The Adaptation of the Harmony Search Algorithm to the ATSP

Urszula Boryczka and Krzysztof Szwarc(✉)

Institute of Computer Science, University of Silesia,
Bedzinska 39, 41-205 Sosnowiec, Poland
{urszula.boryczka,krzysztof.szwarc}@us.edu.pl
http://ii.us.edu.pl/

Abstract. The paper presents a modification of the Harmony Search algorithm, adapted to the effective resolution of the asymmetric case of the traveling salesman problem. The efficacy of the proposed approach was measured with benchmarking tests and in a comparative study based on the results obtained with the Nearest Neighbor Algorithm, Greedy Local Search and Hill Climbing. The publication also includes the comparison of the results with solutions proposed in literature, which were developed with different metaheuristic techniques.

Keywords: Harmony Search
Asymmetric traveling salesman problem · Metaheuristics

1 Introduction

The Harmony Search (HS) algorithm is a promising metaheuristic used to solve a variety of optimization problems (it has been successfully used in the design of steel frames [3] and the optimization of container storage in a harbor area [2]). Its results are usually characterized with the favorable value of the objective function, while at the same time they are achieved in a relatively short time. The nature of the algorithm, which determines the preferential use of the method for continuous optimization problems, requires the application of sophisticated approaches that allow for its adaptation to other ways of representing a number of important issues in business practice.

The traveling salesman problem (TSP) is a classical combinatorial optimization problem that involves finding the shortest Hamiltonian cycle in a complete weighted graph. Its popularity stems from the fact that it belongs to the class of $\mathcal{NP}$-hard problems and that it can model a variety of utilitarian issues – its asymmetric variant (characterized by the possibility of varying weights of edges connecting the same nodes), representing line infrastructure located in urban areas, has become the basis for many logistical problems (it models, for example, the process of planning the mobile collection of waste electrical and electronic equipment [7] and the process of transport activities related to the acquisition of municipal waste [11]).

N. T. Nguyen et al. (Eds.): ACIIDS 2018, LNAI 10751, pp. 341–351, 2018.
https://doi.org/10.1007/978-3-319-75417-8_32

Taking into account the relatively good research results concerning the use of the Harmony Search algorithm to solve many practical problems, a number of controversies over HS (as the method lacking innovativeness and being only a special case of Evolution Strategies [14]), and the utilitarian importance of the asymmetric variant of the traveling salesman problem, we chose to conduct a study aimed at adapting the cited metaheuristics to the combinatorial optimization problem. The proposed topic was also discussed in paper [1], but the results presented there show the ineffectiveness of the method adapted to TSP, implying the need for an innovative approach to the design of the algorithm facilitating the process of planning the traveling salesman's route.

The paper consists of the following parts: the first part introducing the topic, the second part presenting the Harmony Search algorithm, the third part formulating the asymmetric problem of the traveling salesman, the fourth part proposing the approach to adapt the metaheuristic, the fifth part describing the methodology of the study, the sixth part discussing the results, and the seventh part concentrating on the conclusions and planned work.

2 Harmony Search Algorithm

The Harmony Search technique, proposed in paper [4], is based on the similarity of the jazz improvisation process to the search for a global optimum by algorithmic methods. It assumes the existence of a HM structure (referred to as harmony memory), which stores HMS harmonies (usually from 4 to 10 [9]), consisting of a specified number of pitches (representing the values of the decision variables of a given result). Each HM element is interpreted as a complete solution to a problem whose objective function is determined based on its components.

The initial harmony memory content is generated randomly and later aligned according to the appropriate objective function values (so that the first result is the most favorable). These steps initiate the iterative creation of successive solutions.

The procedure for creating a new solution uses the knowledge accumulated in HM and is based on the analogy to the improvisation of harmony in music. The development of a solution involves an iterative selection of the next pitch, according to two parameters – $HMCR$ (a harmony memory considering rate; its value is usually within the range of 0.7 to 0.99 [2]) and PAR (a pitch adjustment rate; often from 0.1 to 0.5 [2]). Based on the probability of $HMCR$, the pitch i is selected and, using the values in the i position in harmonies belonging to HM (otherwise the value is generated randomly). Creating a solution based on the HM component, the pitch can be modified with a defined probability of PAR (the change in value is based on the bw parameter, the value of which depends on the representation of a problem).

When the next solution is generated, a comparison of its objective function value with the relevant parameter describing the HM component in the last position is made. When a more favorable result is identified, it replaces the worst result in harmony memory and HM elements are rearranged.

The procedure for generating a new solution is performed by IT iterations, and then the best result (in the first position in harmony memory) is returned. The pseudocode of the method is shown in Algorithm 1.

Algorithm 1. The Harmony Search pseudocode based on [4, 5]

```
function HS(HMS, HMCR, PAR, IT, bw)
    iterations = 0
    for i = 0; i < HMS; i++ do
        HM[i]=stochastically generate feasible solution
    end for
    Sort HM
    while iterations < IT do
        H = ∅
        for i = 0; i < n; i++ do                          ▷ n - number of pitches
            Choose random r ∈ (0, 1)
            if r < HMCR then
                H[i]=choose randomly available pitch on position i in HM
                Choose random k ∈ (0, 1)
                if k < PAR then
                    α = bw· random ∈ (−1, 1)              ▷ bw - range of changes
                    H[i]=H[i] + α
                end if
            else
                H[i]=choose randomly available pitch
            end if
        end for
        if f(H) is better than f(HM[HMS − 1]) then
            HM[HMS − 1] = H
            Sort HM
        end if
        iterations++
    end while
    return HM[0]
end function
```

3 The Formulation of the Asymmetric Traveling Salesman Problem

Based on paper [10], the following TSP formulation was adapted: for a directed graph $D = (N, A)$, with weighted arcs represented as c_{ij} (where $i, j \in \{1, 2, \ldots, n\}$), a route (a directed cycle comprising all n cities) of minimal length is sought. The asymmetric variant of the Traveling Salesman Problem (ATSP) is characterized with the possibility of the occurrence of inequality $c_{ij} \neq c_{ji}$.

A decisive variable x_{ij} representing the edge between vertices i and j in the solution found adapts the following values:

$$x_{ij} = \begin{cases} 1 \text{ when the edge } (i,j) \text{ is part of the route constructed,} \\ 0 \text{ otherwise.} \end{cases} \tag{1}$$

The objective function was formulated in the following way:

$$\sum_{i,j} c_{ij} x_{ij} \rightarrow min. \tag{2}$$

The constraints ensuring exactly one visit of the travelling salesman to every city were presented in the following way:

$$\sum_{i} x_{ij} = 1 \quad \forall_{j \in \mathbb{N}}, \quad \sum_{j} x_{ij} = 1 \quad \forall_{i \in \mathbb{N}}. \tag{3}$$

Such a formulation of the Traveling Salesman Problem could result in the occurrence of solutions representing separate cycles instead of 1 cycle, so it is necessary to introduce additional constraints (MTZ):

$$u_1 = 1, \quad 2 \leq u_i \leq n, \quad u_i - u_j + 1 \leq n(1 - x_{ij}), \quad \forall_i \neq 1, \quad \forall_j \neq 1. \tag{4}$$

4 The Proposal of the Approach to HS Design

According to the proposed approach to the design of the HS method (tailored to solve the asymmetric traveling salesman problem), each pitch is represented by integers corresponding to the number of the individual cities visited by the sales agent. The order of their occurrence - in harmony representing the complete route - indicates the sequence of a journey.

Taking into account the nature of the optimization problem under study, the position occupied by a given city in harmony is deemed irrelevant. It is necessary, however, to consider the sequence of vertices, by selecting the next pitch value based on the generated list of available nodes, occurring in the solutions recorded directly after the last city that belongs to the constructed result. Based on the structure created, the city is selected according to the roulette wheel method (the probability of acceptance of a given item is dependent on the value of the objective function of the solution represented by the length of the route, similarly to the approach discussed in paper [6]), or any unvisited node is drawn (when the list of vertices is empty). As a modification of the pitch – related to the *PAR* parameter – we opted for the choice (made within the nodes available) of the city nearest to the last visited site in the solution being created (the results of empirical studies assuming the alignment of the representation problem to continuous space and the use of the *bw* parameter showed the ineffectiveness of this approach in the case of ATSP).

In order to avoid premature convergence, we introduced the option of resetting the *HM* elements at the time of execution of a specified number of *R* iterations from the last replacement of the result in *HM*. This mechanism assumes that the best result is kept and the remaining solutions are drawn. The proposed approach to the HS design is presented in Algorithm 2.

Algorithm 2. The Harmony Search pseudocode for ATSP

```
function HS(HMS, HMCR, PAR, IT, R, first city)
    iterations = 0
    iterationsFromTheLastReplacement = 0
    for i = 0; i < HMS; i++ do
        HM[i]=stochastically generate feasible solution
    end for
    Sort HM
    while iterations < IT do
        H = ∅
        H[0]=first city
        for i = 1; i < n; i++ do                                    ▷ n - number of cities
            Choose random r ∈ (0, 1)
            if r < HMCR then
                list=generate list containing vertices occurring after H[i−1] in HM
                if list.length > 0 then
                    H[i]=choose element ∈ list according to the roulette wheel method
                else
                    H[i]=choose randomly available city ∉ H
                end if
                Choose random k ∈ (0, 1)
                if k < PAR then
                    H[i]=find nearest and available city from H[i − 1]
                end if
            else
                H[i]=choose randomly available city ∉ H
            end if
        end for
        if f(H) is better than f(HM[HMS − 1]) then
            HM[HMS − 1] = H
            Sort HM
            iterationsFromTheLastReplacement = 0
        else
            iterationsFromTheLastReplacement++
        end if
        if iterationsFromTheLastReplacement = R then
            for i = 1; i < HMS; i++ do
                HM[i]=stochastically generate feasible solution
            end for
            Sort HM
            iterationsFromTheLastReplacement = 0
        end if
        iterations++
    end while
    return HM[0]
end function
```

5 Empirical Research Methodology

The nineteen tasks representing the asymmetric traveling salesman problem were selected as the test bed (their characteristics are available in paper [8]). It is assumed that statistically significant results – for non-deterministic algorithms – can be achieved by repeating calculations at least ten times for each instance of the problem [12], so each test was solved 30 times.

The values of the HS parameters were as follows: $R = 1000$, $HMS = 5$, $HMCR = 0.98$, $PAR = 0.25$ (empirical studies were carried out for their designation, the fragmented results of which are shown in Table 1). It was assumed that the algorithm would be executed for 10 min, during which the value of the objective function of the best solution found after 2, 6, and 10 min would be determined (the same time interval between measurements allows for the observation of changes in the dynamics of the improvement of the solution). Average error was calculated as follows: $((result - optimum)/optimum) \cdot 100\%$.

Table 1. The impact of parameter values on the average error obtained after 1 min.

Test name	Average error [%]																			
	R						*HMS*				*HMCR*					*PAR*				
	lack	250	500	750	1000	1250	3	5	7	10	0.95	0.97	0.98	0.99	1	0.2	0.24	0.25	0.26	0.3
p43	0.07	0.04	0.05	0.05	0.05	0.05	0.05	0.05	0.05	0.05	0.05	0.06	0.05	0.05	0.05	0.05	0.05	0.05	0.05	0.05
ry48p	3.28	1.74	1.66	2.17	1.63	1.64	1.74	1.63	1.96	2.29	1.63	1.88	1.35	1.8	3.99	2.14	1.68	1.35	1.89	1.36
ft70	5.69	6.03	5.76	5.45	5.49	5.58	5.73	5.49	5.65	5.59	5.49	5.07	4.75	4.7	4.41	4.74	4.89	4.75	4.75	5.09
Total	3.01	2.6	2.49	2.56	**2.39**	2.42	2.51	**2.39**	2.55	2.65	2.39	2.34	**2.05**	2.18	2.82	2.31	2.21	**2.05**	2.23	2.17

Algorithms were implemented in *C#* and the study was conducted on a Lenovo Y50-70 laptop, the parameters of which are presented in Table 2.

Table 2. The parameters of the laptop used to conduct the study

No	Parameter	Value
1	Processor	Intel Core i7-4720HQ (4 cores, from 2.60 GHz to 3.60 GHz, 6 MB cache)
2	RAM	16 GB (SO-DIMM DDR3, 1600 MHz)
3	Hard drive	1000 GB SATA 5400 rev. Express Cache 8 GB
4	Operating system	Windows 7 Professional N Service Pack 1 64-bit

For comparative purposes – to provide background for the results generated by HS - the solutions obtained by the Nearest Neighbor Algorithm (NNA), Greedy Local Search (GLS) and Hill Climbing (HC) were selected. The first method always started constructing a route from city number one.

The Greedy Local Search variant is characterized with the acceptance of the first designated neighborhood solution, which is described with the more favorable value of the objective function, while Hill Climbing grants acceptance when the entire neighborhood is reviewed. The local search methods used in the study were based on the result found by the Nearest Neighbor Algorithm, and the neighborhood of the current solution was defined as a set of routes differing from it by two cities only, while the initial vertex remains unchanged.

The obtained results are also compared with the solutions presented in literature [8]. Due to the different formulation of the algorithm stop condition, they should not be treated as the basis for a direct comparison of the efficiency of metaheuristics, but should only be used to estimate the quality of the proposed solutions.

6 Results

The percentage surplus of the objective function (compared to the optimum) determined by the methods studied is shown in Table 3. The table also includes

Table 3. Compilation of the percentage surplus of the objective function

Test name	Average error [%]							
	NNA	GLS	HC	AMCPA [8]	GA [8]	HS		
						2 min	6 min	10 min
br17	135.9	7.69	7.69	0.26	1.54	**0**	**0**	**0**
ftv33	30.87	23.64	23.64	7.77	7.79	2.47	**2.2**	**2.2**
ftv35	21.59	21.38	21.38	6.52	6.2	1.21	1.01	**0.94**
ftv38	16.21	10	10	5.33	6.92	1.39	0.94	**0.64**
p43	2.63	0.53	0.96	0.15	0.27	0.05	0.05	**0.03**
ftv44	24.86	24.18	24.18	10.49	8.38	2.05	1.63	**1.45**
ftv47	33.67	28.89	28.89	7.21	4.86	1.91	1.57	**1.44**
ry48p	16.19	15.21	13.81	2.79	4.67	1.11	0.8	**0.69**
ft53	37.78	30.93	30.14	12.64	11.98	10.39	7.65	**6.7**
ftv55	25.12	23.2	23.2	11.41	14.25	3.41	2.5	**1.98**
ftv64	43.50	36.54	36.22	13.18	14.86	3.57	2.4	**1.7**
ft70	11.67	8.15	8.98	4.4	5.45	4.49	4.04	**3.81**
ftv70	31.85	23.85	23.28	13.06	9.97	5.54	4.85	**4.23**
kro124	31.12	26.82	24.77	**7.67**	10.58	11.07	8.94	8.12
ftv170	42.40	38.73	36.88	46.02	43.28	35.95	23.56	**21**
rbg323	30.77	**12.37**	12.67	43.40	60.11	60.61	57.7	56.71
rbg358	55.80	**17.02**	20.55	64.05	74.89	86.84	83.31	82
rbg403	43.41	6.98	**4.87**	17.52	20.83	32.67	31.67	31.05
rbg443	44.19	**7.21**	7.57	25.56	24.29	34.11	32.87	32.61
Total	35.77	19.12	18.93	15.76	17.43	15.73	14.09	**13.54**

the reference results obtained with the Genetic Algorithm (GA) and the Adaptive Multi-Crossover Population Algorithm (AMCPA; results were processed based on [8]).

Based on the results obtained, it was found that the proposed approach to the design of the HS algorithm is characterized with high efficacy. For most instances of the problem – the number of cities below 100 – the average percentage surplus of the objective function value, in relation to the optimum, was decidedly less than 5% (regardless of the time needed to apply the method). HS also rated the best results (among the analyzed algorithms; in terms of the objective function value) for 74% of the benchmarking tests.

For each of the analyzed time intervals in which the measurement was performed, HS obtained the most favorable results in terms of the total percentage surplus of the objective function values (within the methods under study), ranging from 15.73% to 13.54% (respectively for 2 and 10 min).

It is of particular interest that the efficiency of GLS and HC for tasks described by a minimum of 323 vertices is relatively high. Due to the multitude of permissible solutions, the proposed method, together with AMCPA and GA, yielded results with significantly higher values of the objective function.

Table 4 presents the values of the objective function determined by the tested solution methods. Accordingly, it was found that NNA, GLS and HC did not provide the optimal solution in any of the analyzed benchmarking tests. The proposed approach to the HS design made it possible to obtain the most favorable result for seven tasks, every time finding a given result in the first two minutes of running the algorithm. In time, one can observe a decrease in the dynamics of the improvement of the solution by HS (the visualization of the dependence is shown in Fig. 1), but the process still produces the expected effects while avoiding stagnation – it would probably be possible to determine solutions optimal for each test by performing the method longer. In addition, the best results for the ftv38, p43, ry48p, ftv64, and ftv70 tasks are characterized with the objective function value diverging from the optimum only to a small extent, which might demonstrate the need to improve the mechanism of exploiting the method (for example, by hybridizing it with other techniques).

Based on the above compilation, we argue that the algorithm is characterized by a significant standard deviation from the objective function value of the obtained solutions, thus implying the observable non-determinism of the method leading to significantly different results after the same period of time (in consequence, preventing their prediction).

Table 4. Detailed compilation of the results (objective function value)

Test name	NNA	GLS	HC	HS											
				2 min				6 min				10 min			
				Avg	Min	Max	S. dev	Avg	Min	Max	S. dev	Avg	Min	Max	S. dev
br17	92	42	42	39	39	39	0	39	39	39	0	39	39	39	0
ftv33	1683	1590	1590	1317.73	1286	1355	27.12	1314.27	1286	1339	26.89	1314.27	1286	1339	26.89
ftv35	1791	1788	1788	1490.87	1473	1499	8.23	1487.87	1473	1499	10.01	1486.87	1473	1499	10.38
ftv38	1778	1683	1683	1551.23	1532	1580	11.69	1544.4	1532	1566	7.83	1539.77	1532	1549	7.34
p43	5768	5650	5674	5623.07	5620	5627	1.57	5622.53	5620	5627	1.78	5621.97	5620	5627	1.75
ftv44	2014	2003	2003	1646.13	1613	1728	25.75	1639.27	1613	1683	19.39	1636.4	1613	1683	18.27
ftv47	2374	2289	2289	1809.93	1776	1845	11.77	1803.9	1776	1814	11.28	1801.63	1776	1814	11.43
ry48p	16757	16615	16413	14582.27	14481	14921	111.41	14537.43	14459	14867	92.79	14522.07	14459	14707	51.86
ft53	9514	9041	8986	7622.1	7320	7989	193.33	7433.47	7177	7768	157.91	7367.6	7101	7666	137.33
ftv55	2012	1981	1981	1662.83	1608	1712	28.07	1648.13	1608	1699	27.65	1639.9	1608	1692	21.78
ftv64	2639	2511	2505	1904.67	1856	1963	29.56	1883.13	1850	1951	24.55	1870.30	1850	1901	17.03
ft70	43186	41824	42144	40411.1	39958	41042	223.50	40233.5	39567	40730	238.49	40146.03	39427	40586	250.41
ftv70	2571	2415	2404	2058	1973	2096	30.74	2044.57	1973	2096	35.5	2032.57	1973	2096	31.15
kro124	47506	45947	45205	40239.1	39071	41305	588.51	39470.5	38240	40603	566.2	39171.53	38000	40330	579.56
ftv170	3923	3822	3771	3745.5	3270	4613	348.61	3404.17	3103	4038	192.18	3333.6	3101	4038	186.75
rbg323	1734	1490	1494	2129.7	2065	2208	39.99	2091.13	2007	2155	37.91	2077.93	2007	2146	33.51
rbg358	1812	1361	1402	2172.93	2068	2258	44.51	2131.9	2061	2215	40.5	2116.63	2044	2198	38.96
rbg403	3535	2637	2585	3270.43	3190	3323	29.63	3245.6	3185	3295	35.68	3230.37	3158	3292	30.57
rbg443	3922	2916	2926	3647.87	3576	3709	38.18	3613.93	3562	3664	25.92	3607.03	3562	3660	24

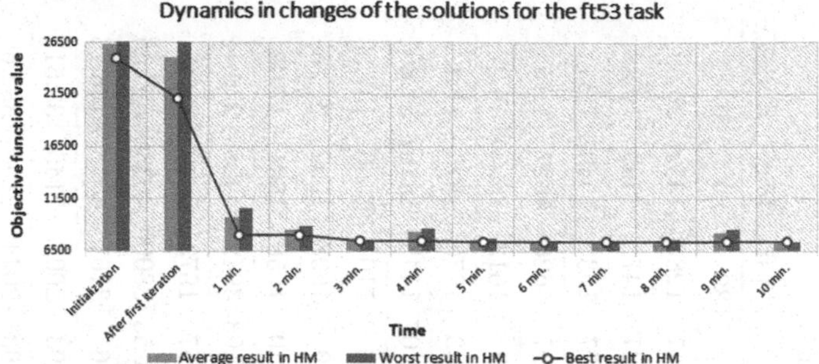

Fig. 1. Dynamics in changes of the solutions for the ft53 task

7 Conclusions and Future Work

The obtained results indicate the relatively good effectiveness of the proposed approach (within the methods under study, it obtained the best – in terms of the total percentage surplus of the objective function – results, characterized with the deviation from the optimum ranging from 13.54% to 15.73%, depending on the running time of the algorithm), encouraging further work on its refinement and the adaption of the method to solving other combinatorial problems.

Accounting for a variety of the obtained results close to the optimum and the observations discussed in paper [13], it is assumed that HS has a weak mechanism of exploitation (its inefficiency was revealed in particular for the tasks where the number of vertexes was above 300, for which the method had far worse results than the results obtained by simple local search methods that were based on the NNA), implying the need to conduct research on the hybridization of this method with other techniques to eliminate the imperfection.

References

1. Antosiewicz, M., Koloch, G., Kamiński, B.: Choice of best possible metaheuristic algorithm for the travelling salesman problem with limited computational time: quality, uncertainty and speed. JTACS **7**(1), 46–55 (2013)
2. Ayachi, I., Kammarti, R., Ksouri, M., Borne, P.: Harmony search algorithm for the container storage problem. In: MOSIM 2010, Hammamet, Tunisia (2010)
3. Degertekin, S.: Optimum design of steel frames using harmony search algorithm. Struct. Multidiscip. Optim. **36**(4), 393–401 (2008)
4. Geem, Z., Kim, J., Loganathan, G.: A new heuristic optimization algorithm: harmony search. Simulation **76**(2), 60–68 (2001)
5. Hetmaniok, E., Jama, D., Słota, D., Zielonka, A.: Application of the harmony search algorithm in solving the inverse heat conduction problem. Zeszyty Naukowe. Matematyka Stosowana, Wydawnictwo Politechniki Slaskiej, Zeszyt, vol. 1, pp. 99–108 (2011)
6. Komaki, M., Sheikh, S., Teymourian, E.: A hybrid harmony search algorithm to minimize total weighted tardiness in the permutation flow shop. In: CIPLS, pp. 1–8 (2014)

7. Mrówczyńska, B., Nowakowski, P.: Optymalizacja tras przejazdu przy zbiorce zuzytego sprzetu elektrycznego i elektronicznego dla zadanych lokalizacji punktow zbiorki. Czasopismo Logistyka **2**, 593–604 (2015)
8. Osaba, E., Diaz, F., Onieva, E., Carballedo, R., Perallos, A.: A population metaheuristic with adaptive crossover probability and multi-crossover mechanism for solving combinatorial optimization problems. IJAI **12**, 1–23 (2014)
9. Panchal, A.: Harmony search in therapeutic medical physics. In: Geem, Z.W. (ed.) Music-Inspired Harmony Search Algorithm. Studies in Computational Intelligence, vol. 191, pp. 189–203. Springer, Berlin (2009). https://doi.org/10.1007/978-3-642-00185-7_12
10. Pataki, G.: The bad and the good-and-ugly: formulations for the traveling salesman problem. Technical report CORC 2000–1 (2000)
11. Płaczek, E., Szołtysek, J.: Wybrane metody optymalizacji systemu transportu odpadow komunalnych w Katowicach. LogForum **4**(2), 1–10 (2008)
12. Talbi, E.: Metaheuristics: From Design to Implementation. Wiley Publishing, Hoboken (2009)
13. Wang, X., Gao, X., Zenger, K.: An Introduction to Harmony Search Optimization Method. Springer, Heidelberg (2015). https://doi.org/10.1007/978-3-319-08356-8
14. Weyland, D.: A rigorous analysis of the harmony search algorithm: how the research community can be misled by a "novel" methodology. IJAMC **1**(2), 50–60 (2010)

Algorithms for Solving the Vehicle Routing Problem with Drones

Daniel Schermer, Mahdi Moeini(✉), and Oliver Wendt

BISOR, University of Kaiserslautern, Postfach 3049,
Erwin-Schrödinger-Str., 67653 Kaiserslautern, Germany
{daniel.schermer,mahdi.moeini,wendt}@wiwi.uni-kl.de

Abstract. The *Vehicle Routing Problem (VRP)* and its variants are well-studied problems in Operations Research. They are related to many real-world applications. Recently, several companies like Amazon, UPS, and Deutsche Post AG showed interest in the integration of autonomous drones in delivery of parcels. This motivates researchers to extend the classical VRP to the *Vehicle Routing Problem with Drones (VRPD)*, where a drone works in tandem with a vehicle to reduce delivery times. In this paper, we focus on solving the VRPD. In particular, we introduce two heuristic algorithms for solving this problem and, through numerical experiments on large-scale instances, we evaluate the performance of the heuristics and show the potential benefit that can be expected when using drones.

Keywords: Last-mile Logistics · Vehicle Routing Problem
Drone Delivery · Large-scale Instances · Heuristic

1 Introduction

The *Unmanned Aerial Vehicles (UAVs)*, i.e., drones, have started to play an increasing role in delivery applications from the point of view of both, researchers and large companies, such as Amazon Inc., Deutsche Post AG, UPS, and Google (see e.g., [7] and references therein).

The UAVs are not considered as an alternative to the conventional delivery vehicles such as truck, but rather as complementary delivery tools. Table 1 highlights the complementary features of trucks and drones for delivery applications. Since drones are not restricted to the road network or congestion, they can generally move, between two locations, faster than trucks. Furthermore, drones and their payload are far more lightweight than trucks', which causes drones to consume much less energy for the movement between two points. However, a drone's carrying capacity is typically limited to one or few parcels. Furthermore, since drones rely on comparatively small batteries for powering their flight, their range is somewhat limited compared to a truck using a fossil fuel.

The *Vehicle Routing Problem (VRP)* is one of the most well-studied problems in operations research. As a generalized case of the well known *Traveling*

N. T. Nguyen et al. (Eds.): ACIIDS 2018, LNAI 10751, pp. 352–361, 2018.
https://doi.org/10.1007/978-3-319-75417-8_33

Table 1. Qualitative differences between trucks and drones.

	Speed	Weight	Capacity	Range	Energy consumption
Truck	Low	Heavy	Many	Long	High
Drone	High	Light	One	Short	Low

Salesman Problem (TSP), for a given fleet of vehicles and a set of customers, the VRP seeks for the optimal set of routes in order to deliver goods to the customers and satisfy their demand. In [9], Toth and Vigo provide an extensive overview of the problem and its variants.

In the academic literature, the interest in integrating drones in delivery applications has surged, following Murray and Chu, who introduced two new NP-hard problems related to deliveries with a vehicle working in tandem with a drone [6]. In their *flying sidekick traveling salesman problem (FSTSP)*, a drone travels along with a vehicle, both of which start from a common *distribution center (DC)* and must return to the same DC at the end of the tour. At any customer, the drone might be launched from the vehicle, starting a sortie. In this case, it will begin delivery to a customer and will then rendezvous with the vehicle at a later customer. In the *parallel drone scheduling TSP (PDSTSP)*, it is assumed that the DC is in close proximity to most of the customers. In this case, it can be beneficial to let the drone make its deliveries independently from the vehicle. In both problems, the objective consists in minimizing the time required to serve all customers and the drone and vehicles must return to the DC. Murray and Chu proposed Mixed Integer Linear Programs (MILP) as well as two simple heuristic methods for solving the FSTSP and PDSTSP, respectively. They note, that only small size instances can be solved up to optimality due to the NP-hard nature of the proposed problems. Later, Ha et al. [2] employed two heuristics to solve the FSTSP: *route first, cluster second* and *cluster first, route second*. The latter heuristic performed better according to their numerical results.

Mathew et al. [4] introduced the *Heterogeneous Delivery Problem (HDP)* which shares most features of the previously introduced FSTSP. They propose a solution by reducing the HDP to the *Generalized Traveling Salesman Problem (GTSP)*. They suggest splitting the HDP into two traceable sub-problems: first a TSP is used to generate the optimal tour, then convex optimization is applied to compute the specific deployment points for the drone and vehicle.

Mourelo et al. [5] worked on a variant of the FSTSP, where they allow multiple drones per vehicle. They introduced a heuristic based on *K-means* for clustering purposes and a *Genetic Algorithm* for routing. They conducted their computational experiments taking into account various drone speeds and they noticed that the drones should be at least twice as fast as the vehicles in order to allow for *significant* improvements.

Wang et al. [12] introduced the *Vehicle Routing Problem with Drones (VRPD)*, where a fleet of trucks, each truck equipped with a given number of drones, delivers packages to customers. According to the classification of Toth and Vigo [9], the VRPD can be classified as a variant of the *Distance-Constrained Capacitated Vehicle Routing Problem (DCVRP)* with a set of heterogeneous vehicles. The objective consists in minimizing the completion time, i.e., the time required to serve all customers and return the fleet to the depot. Wang et al. introduced several lower bounds on the amount of time that can be saved by employing drones. The lower bounds depend on the relative speed of the drones compared to the vehicles and the number of drones per vehicle.

Ulmer and Barrett proposed the *same-day delivery routing problems with heterogeneous fleets (SDDPHF)*, where drones and vehicles serve customers without a need for synchronization [10]. The customers are not known a priori; hence, dynamic routing of the heterogeneous fleet based on the incoming requests is required. Therefore, the previously introduced PDSTSP, where all customers are known a priori, can be viewed as a relaxed version of the SDDPHF. Using computational experiments, Ulmer et al. can show that the combination of vehicles and drones can reduce the delivery costs significantly.

Jiang et al. [3] applied drone logistics to the *Vehicle Routing Problem with Time Windows (VRPTW)* and derived an MILP model. Their problem consists of assigning a swarm of drones to serve customers within predefined time frames. They used an adapted version of *Particle Swarm Optimization* for drone routing. They conclude that their implemented heuristic is well suited for solving their version of the VRPTW.

In this article, we focus on the VRPD introduced by Wang et al. [12]. Our contributions are twofold: first, we introduce two heuristics for solving VRPD. Second, in order to assess the performance of the implemented heuristics, we carried out numerical experiments on *large-scale* TSP instances that we adjusted for using in the context of the VRPD. On the one hand, our numerical results highlight usefulness of incorporating drones in last-mile logistics. On the other hand, based on our experiments, we can derive implications for the way future heuristics should be designed when solving the large-scale VRPD instances.

This paper is organized as follows. Section 2 introduces the notation and describes the VRPD. In Sect. 3, we present our heuristics for solving the VRPD. Section 4 is dedicated to the computational experiments and the numerical results. Finally, we draw some concluding remarks and derive some future research questions in Sect. 5.

2 The Vehicle Routing Problem with Drones

Suppose that a set of n customers and a fleet of m homogeneous trucks, each carrying k homogeneous drones, are given. We might assume that the capacity of each truck is C. The *Vehicle Routing Problem with Drones (VRPD)* looks for minimizing the maximum completion time, i.e., the time required to serve all customers using the trucks and the drones such that, by the end of mission all trucks (carrying drones as well) must be at the depot [7,12]. By t^T_{n+1}, we denote the objective value of the VRPD solution indicating the time that the set of trucks and drones needs for serving all customers and then, returning to the depot. This time corresponds to the mission time of the latest truck or drone that arrives to the depot, as soon as all customers have been served by the set of trucks and drones (other trucks and/or drones arrive ahead of this time). In t_{n+1}^{T}, T stands for the *total* time and $n+1$ indicates that the solution includes the set of n customers and the depot.

Furthermore, we make the following assumptions about the drone's behavior in a risk-free environment [6,7,12]:

- A drone can carry exactly one parcel when airborne.
- A drone has a limited battery life of $\mathcal{E}$ time units. After returning to the truck, the battery life of the drone is recharged instantaneously with no service delay.
- The trucks and drones follow the same distance metric.
- The service time to launch and reunite with a drone is assumed to be negligible.
- The service time required to serve a customer is assumed to be negligible.
- Without loss of generality, the speed of each truck is set to 1 and the speed of each drone is assumed to be α times the speed of the truck.
- A drone must always return to the truck from which it was launched.
- Drones may only be dispatched and picked up from vertices, i.e., at the depot or any other customer location. Furthermore, a drone may not be picked up from the same vertex, where it has been launched from.
- Since the drones are assumed to be in constant flight and can not conserve battery while in flight; consequently, if a truck arrives earlier to a pick-up node, then the truck has to wait for its corresponding drone.
- We assume that when a drone is launched, then its delivery will be successful.

3 Algorithms for Solving the VRPD

In this section, we present two heuristic algorithms that we introduce for solving the VRPD. The heuristics are composed of several components; in particular, they have two main stages: an initialization step and an improving (optimization) phase.

Indeed, as a natural approach, we need an initial solution in order to start the optimization phase. For this purpose, we choose a *route-first cluster-second (RFCS)* heuristic. The RFCS heuristics has different components. First, we create a single tour using the *nearest neighbor heuristic (NHH)* [8]. Afterwards, having created a single tour that contains all vertices, the tour is split equally in m segments, where m is the number of available trucks.

The improvement phase of the heuristics is inspired from the VRP literature. In fact, improvement heuristics for the VRP can be categorized into *single-route improvement* and *multi-route improvement* heuristics [9]. Single-Route improvements focus on improving a single tour at a time. Multi-Route improvements, on the other hand, try to improve the objective function by considering multiple distinct tours simultaneously. In [11], van Breedam has classified the multi-route improvement operations into *String Cross (SC)*, *String Relocation (SR)*, *String Exchange (SE)*, and *String Mix* (SM), where *string* stands for a tour. While SM is a combination of SE and SR, SC is a generalization of the k-opt operator [1].

Since the VRPD imposes additional constraints, that do not allow for a direct application of the introduced single- and multi-route improvement operators, we introduce a variation of the 2-opt and SM operators with the aim of handling the additional constraints imposed by the VRPD. More precisely, the launch and rendezvous vertices must be updated when inverting a sub-sequence of the tour in order to maintain synchronization of the truck and the drone. Additionally, endurance constraints must not be violated when applying the 2-opt or any other operator.

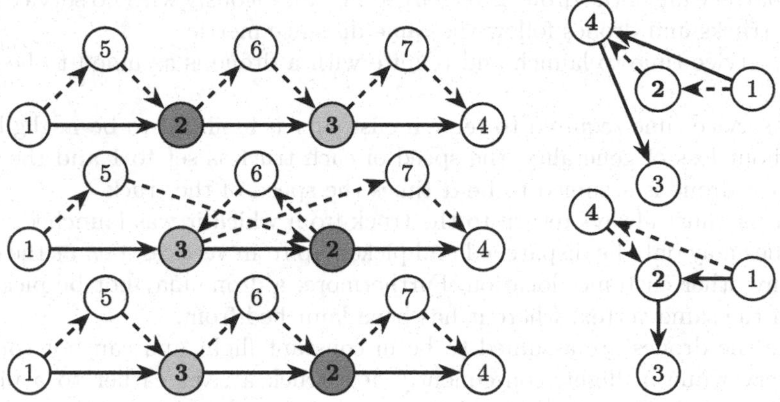

Fig. 1. A visual example of the implemented single-route improvement heuristics: 2-opt (left) and delivery exchange (right). Truck routes are indicated by solid lines and drone routes indicated by dashed lines. In the case of 2-opt, inverting a sub-sequence (top to middle) requires adjusting delivery and rendezvous vertices (bottom) to maintain feasible routing. In all cases, any change must not violate the endurance constraints of the drone.

Furthermore, in order to explore the solution space of the problem, we define two operators that focus mainly on drones: *Drone Insertion (DI)* operator and *Delivery Exchange (DE)* operator. The DI operator is based on the concept of *first-improvement*. More precisely, starting from the first vertex in a tour, whenever the first feasible sortie is possible and reduces the time required to complete the tour, the drone is inserted. The DE operator works as follows: Assume that both, the drone and the truck, serve exactly one customer after a launch and before meeting at a rendezvous vertex, then the DE operator will attempt to change the method of delivery. If the exchange is feasible and the time required to complete the tour decreases, then the change is kept; otherwise, the move is reverted.

Figure 1 shows a visual example of the single-route improvement operators: 2-opt and DE. Figure 2 visualizes the *String Exchange (SE)* and the *String Relocation (SR)* operators. The SE operator exchanges two random customers between two distinct tours. The SR moves a vertex from one tour to another one. The affected customers can be served by drone or truck before the move and will always be assigned to a truck after the move.

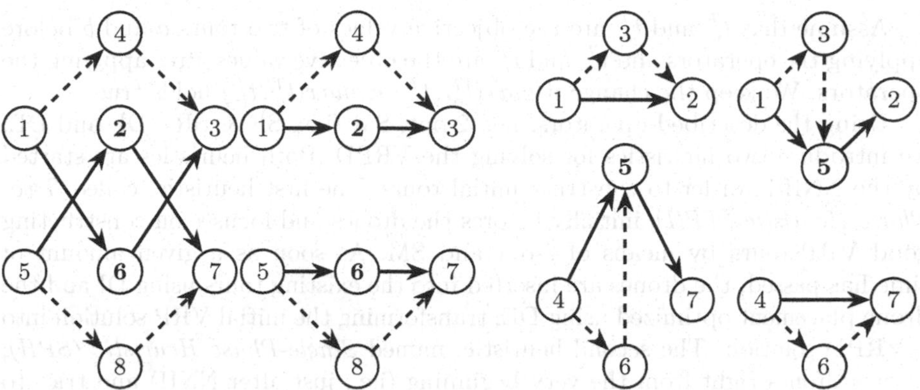

Fig. 2. A visual example of the implemented multi-route improvement operators: string exchange (left) and string relocation (right). In the case of string exchange, the vertices labeled 2 and 6 are exchanged between two tours. In the case of string relocation, the vertex labeled 5 is moved from one tour to another tour.

Algorithm 1. Two-Phase Heuristic (TPH)

```
procedure OPTIMIZE(m, α, β, Instance)
    solution ← routeFirstClusterSecond(Instance,m);
    while !stoppingCriterion do
        if phaseOne then
            solution.twoOpt();
            solution.stringMix();
        else
            solution.insertDrones();
            solution.deliveryExchange();
        end if
    end while
    return solution;
end procedure
```

Algorithm 2. Single-Phase Heuristic (SPH)

```
procedure OPTIMIZE(m, α, β, Instance)
    solution ← routeFirstClusterSecond(Instance,m);
    solution.insertDrones();
    while !stoppingCriterion do
        solution.twoOpt();
        solution.stringMix();
        solution.insertDrones();
        solution.deliveryExchange();
    end while
    return solution;
end procedure
```

Assume that t_a^T and t_b^T are the objective values of two tours a and b before applying the operators and $t_{a'}^T$ and $t_{b'}^T$ are the objective values after applying the operators. We keep the change if $max(t_{a'}^T, t_{b'}^T) < max(t_a^T, t_b^T)$ holds true.

Using the described operators, i.e., 2-opt, SM (i.e., SE & SR), DI, and DE, we introduce two heuristics for solving the VRPD. Both heuristics are started by the NNH in order to construct initial tours. The first heuristic, called *Two-Phase Heuristic (TPH)*, initially ignores the drones and focuses on constructing good VRP tours by means of 2-opt and SM. As soon as a given amount of time has passed, the drones are inserted into the existing tours using DI and the drone placement optimized using DE, transforming the initial VRP solution into a VRPD solution. The second heuristic, named *Single-Phase Heuristic (SPH)*, inserts drones right from the very beginning (i.e., just after NNH) and tries to create good VRPD solutions by using a combination of 2-opt, SM, DI, and DE. Since some sorties might be removed after using 2-opt or SM, due to violation of the endurance constraint, it is necessary to continuously utilize DI. Algorithms 1 and 2 illustrate the pseudo-code of the introduced heuristics (where, α and β are instance and test parameters. See also Sect. 4).

4 Computational Experiments

This section is dedicated to the presentation of our computational experiments and their numerical results. In particular, we have carried out a total of 750 experiments. For this purpose, we rely on instances from the TSPLIB[1] in order to generate VRPD instances. For this purpose, we start by introducing the following metric for setting comparable endurance constraints among all instances. Since each instance can be described by a graph $G(V, E)$, we first consider the adjacency matrix $\mathcal{A}(G)$. Then, through introducing a new parameter β, we set the endurance to $\mathcal{E} = \beta \cdot max(\mathcal{A})$ and use the following parameter values and TSP instances in our experiments:

- Six TSP instances: $rd400$, $att532$, $u574$, $gr666$, $rat783$, $dsj1000$.
- Three different values for $\alpha \in \{2, 3, 4\}$.
- Five different values for $\beta \in \{0.0, 0.25, 0.5, 0.75, 1.0\}$.
- Five runs per problem instance and each pair of parameters of α and β. The case $\beta = 0.0$ corresponds to the classical VRP (i.e., there is no drone). Hence, for the case of $\beta = 0.0$, we only test for one value of α, as no drones will be used, regardless of the value of α.
- We test with $m = 3$ trucks and $k = 1$ drone per truck.
- There is no limit on the capacity C of the trucks but the capacity of each drone is limited to 1.

We implemented the algorithms in Java SE 8. We limit the run time of each algorithm to 5 min and record the best value achieved after passing the time-limit. The algorithms were run on an Intel 5200U CPU limited to the base clock

[1] https://www.iwr.uni-heidelberg.de/groups/comopt/software/TSPLIB95/.

speed of 2.2 GHz with a maximum of 8 GB of RAM available. In the case of TPH, we ran the tour phase for 4 min and used the remaining time to insert drones and applied the delivery exchange heuristic afterwards. For the sake of comparison, we computed reference values, generated using TPH with a tour phase of 5 min and $\beta = 0$ (no drones are inserted). For all instances, we set the first vertex as the depot. Figure 3 shows sample output for an instance with 51 vertices and limited drone endurance. Table 2 shows the numerical results of the experiments. For each instance, the objective value of the VRP solution ($\beta = 0$) is taken as the reference (base) value and the remaining values in the same row are the average values (of five runs) scaled relative to the VRP solution.

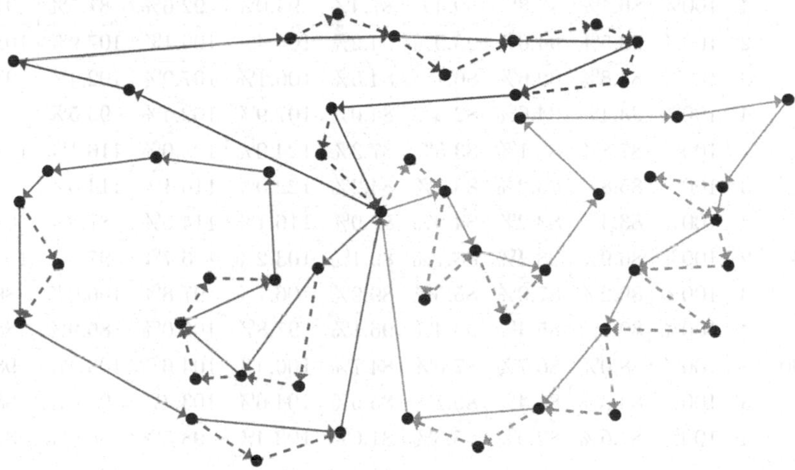

Fig. 3. Sample output for a problem with 51 vertices and limited drone endurance. The truck's and the drone's paths are shown using solid lines and dashed lines, respectively.

According to the numerical results, we can make the following observations:

- The TPH can typically achieve good results among a wide range of α and β values. Additionally, the TPH always produces better results than the base value ($\beta = 0$).
- The SPH only achieves good results for large values of α and β. Additionally, the SPH often produces worse results than the base value ($\beta = 0$).
- Our experiments indicate, that the TPH creates better solutions than the SPH in most cases - in particular on large-scale instances (i.e., u574, gr666, dsj1000). Nevertheless, in some instances (e.g., some cases of rd400, att532, rat783), the SPH can provide better or competitive results in comparison to the TPH. A possible explanation might be due to the presence of the additional constraints imposed by the VRPD. In fact, these constraints limit the search space and make it more difficult to find valid and high-quality VRPD solutions by means of a single-stage heuristic.

Table 2. Summary of the numerical results of the TPH and the SPH algorithms across different instances.

Instance	α	Two-Phase Heuristic (TPH)					Single-Phase Heuristic (SPH)			
		β					β			
		0.00	0.25	0.50	0.75	1.00	0.25	0.50	0.75	1.00
rd400	2	100%	84.5%	85.1%	81.6%	84.6%	108.0%	99.7%	99.3%	99.8%
	3	100%	81.2%	82.1%	84.4%	81.4%	99.7%	99.3%	99.8%	84.5%
	4	100%	82.6%	83.2%	83.1%	83.5%	99.3%	99.8%	84.5%	85.1%
att532	2	100%	87.5%	91.9%	85.9%	87.7%	92.6%	92.6%	94.0%	92.6%
	3	100%	87.6%	87.7%	87.8%	85.6%	92.6%	94.0%	92.6%	87.5%
	4	100%	86.4%	88.8%	89.4%	87.4%	94.0%	92.6%	87.5%	91.9%
u574	2	100%	93.5%	94.6%	93.9%	94.2%	104.4%	106.1%	107.9%	102.4%
	3	100%	84.8%	82.6%	86.1%	84.5%	106.1%	107.9%	102.4%	93.5%
	4	100%	83.4%	84.6%	82.5%	84.0%	107.9%	102.4%	93.5%	94.6%
gr666	2	100%	87.8%	88.1%	83.5%	87.3%	124.0%	122.0%	116.1%	114.5%
	3	100%	85.8%	83.2%	83.1%	84.2%	122.0%	116.1%	114.5%	87.8%
	4	100%	83.4%	83.2%	84.9%	82.9%	116.1%	114.5%	87.8%	88.1%
rat783	2	100%	86.9%	88.4%	88.7%	89.1%	103.2%	106.7%	97.8%	106.0%
	3	100%	89.2%	87.9%	85.5%	89.9%	106.7%	97.8%	106.0%	86.9%
	4	100%	89.5%	85.4%	93.4%	96.1%	97.8%	106.0%	86.9%	88.4%
dsj1000	2	100%	88.0%	86.7%	87.0%	84.7%	106.4%	104.6%	103.4%	98.5%
	3	100%	83.5%	81.4%	86.1%	85.6%	104.6%	103.4%	98.5%	88.0%
	4	100%	84.6%	82.9%	85.3%	81.6%	103.4%	98.5%	88.0%	86.7%

- Although the TPH has a higher-average solution quality, we notice that the TPH fails to provide better solutions in most of the instances that correspond to larger values of α and β. However, the SPH shows a clear progression towards higher objective values with an increase in α and β.

5 Conclusion

In this paper, we introduced two new heuristics, *Two-Phase Heuristic (TPH)* and *Single-Phase Heuristic (SPH)*, for solving the VRPD. To the best of our knowledge, it is the first time that numerical results for large-scale VRPD instances are presented. While the TPH, as a two-stage approach, is based on first creating good VRP tours and then creating a VRPD solutions through insertion of drones, the SPH focuses on creating good VRPD solution right from the scratch. Our observations, based on preliminary numerical results, confirm that the TPH provides better results than the SPH in most cases, and there are a few cases where the SPH produces better or competitive results. Hence, it seems reasonable to claim that good VRPD solutions can be constructed by using a two-stage heuristic, starting from good VRP solutions.

Future research might focus on heuristics that allow for a good exploration of the search space by means of more effective methods for handling additional constraints of the VRPD. Metaheuristics, such as Simulated Annealing, seem to be a reasonable choice for this task. Finally, design of algorithms for solving VRPD in presence of more realistic conditions, e.g., limited truck capacity or multiple drones per truck, can also be considered as interesting future research plans. The research in these directions are in progress and the results will be published in future.

Acknowledgements. The authors acknowledge the chair of Business Information Systems and Operations Research (BISOR) at the TU-Kaiserslautern (Germany) for the financial support, through the research program "CoVaCo".

References

1. Gendreau, M., Potvin, J.-Y.: Handbook of Metaheuristics. Springer, Heidelberg (2010). https://doi.org/10.1007/978-1-4419-1665-5
2. Ha, Q.M., Deville, Y., Pham, Q.D., Ha, M.H.: Heuristic methods for the traveling salesman problem with drone, pp. 1–13. Technical report, CTEAM/INGI/EPL (2015)
3. Jiang, X., Zhou, Q., Ye, Y.: Method of task assignment for UAV based on particle swarm optimization in logistics. In: Proceedings of the 2017 International Conference on Intelligent Systems, Metaheuristics & Swarm Intelligence - ISMSI 2017, pp. 113–117. ACM Press (2017)
4. Mathew, N., Smith, S.L., Waslander, S.L.: Planning paths for package delivery in heterogeneous multirobot teams. IEEE Trans. Autom. Sci. Eng. **12**(4), 1298–1308 (2015)
5. Mourelo, F.S., Harbison, T., Weber, T., Sturges, R., Rich, R.: Optimization of a truck-drone in tandem delivery network using k-means and genetic algorithm. J. Ind. Eng. Manag. **9**(2), 374–388 (2016)
6. Murray, C.C., Chu, A.G.: The flying sidekick traveling salesman problem: optimization of drone-assisted parcel delivery. Transp. Res. Part C: Emerg. Technol. **54**, 86–109 (2015)
7. Poikonen, S., Wang, X., Golden, B.: The vehicle routing problem with drones: extended models and connections. Networks **70**(1), 34–43 (2017)
8. Solomon, M.M.: Algorithms for the vehicle routing and scheduling problems with time window constraints. Oper. Res. **35**(2), 254–265 (1987)
9. Toth, P., Vigo, D.: The Vehicle Routing Problem. Society for Industrial and Applied Mathematics, Philadelphia (2002)
10. Ulmer, M., Barrett, W.T.: Same-day delivery with a heterogeneous fleet of drones and vehicles. Technical report, pp. 1–30 (2017)
11. van Breedam, A.: An analysis of the behavior of heuristics for the vehicle routing problem for a selection of problems with vehicle-related, customer-related, and time-related constraints contents. Ph.D. dissertation, University of Antwerp (1994)
12. Wang, X., Poikonen, S., Golden, B.: The vehicle routing problem with drones: several worst-case results. Optim. Lett. **11**(4), 679–697 (2016)

A Local Search Heuristic for Solving the Maximum Dispersion Problem

Mahdi Moeini([✉]), David Goerzen, and Oliver Wendt

BISOR, University of Kaiserslautern, Postfach 3049,
Erwin-Schrödinger-Str., 67653 Kaiserslautern, Germany
{mahdi.moeini,wendt}@wiwi.uni-kl.de, goerzen.david@gmail.com

Abstract. In this paper, we are interested in studying the *Maximum Dispersion Problem (MaxDP)*. In this problem, a set of objects are given such that each object has a non-negative weight. The objective of the MaxDP consists in partitioning the given set of objects into a predefined number of classes. The partitioning is subject to some conditions. First, the overall dispersion of objects, assigned to each class, must be maximized. Second, there is a predefined target weight assigned to each class and the total weight of each class must belong to an interval surrounding its target weight. It has been proven that the MaxDP is NP-hard and, consequently, difficult to solve by classical exact methods. In this paper, we provide a Variable Neighborhood Search (VNS) algorithm for solving the MaxDP. In order to evaluate the efficiency of the introduced VNS, we carried out numerical experiments on randomly generated instances. Then, we compared the results of our VNS algorithm with those provided by the standard solver Gurobi. According to our results, our VNS algorithm provides high-quality solutions within a short computation time and dominates the solver Gurobi.

Keywords: Maximum Dispersion Problem · Heuristic
Variable Neighborhood Search · Local Search

1 Introduction

As an important branch of mathematical optimization, combinatorial optimization contains many interesting optimization problems with academic or practical background. It is well known that most of these problems are NP-hard; consequently, it is difficult to solve their large instances within a reasonable computation time. An important category of combinatorial optimization problems deals with classification of objects. For example, we might be interested in partitioning a set of objects into disjoint classes with the aim of minimizing or maximizing an objective function [1,4,8,11,13–15]. The description of *objective function* depends on the context or application; more precisely, the objective might be minimization or maximization of a function that defines the pairwise distance of objects or their dissimilarity in a given set (see, e.g., [6,8,11,13], and

N. T. Nguyen et al. (Eds.): ACIIDS 2018, LNAI 10751, pp. 362–371, 2018.
https://doi.org/10.1007/978-3-319-75417-8_34

references therein). While minimizing the distance or dissimilarity, we are interested in identifying similar objects in order to put them in a same class. However, if we seek to maximize the distance, then our focus is on dispersed classes. In the scientific literature, there are several examples of such problems. One of the recently introduced dispersion problems is called *Maximum Dispersion Problem (MaxDP)*. In this problem, the objective consists in creating a partition on a given set of objects with the aim of maximizing the dispersion defined as the pairwise distance between objects that belong to a same class [6].

There are several practical situations in which the MaxDP appears. For example, consider a class of students that must be partitioned into a given number of learning groups [1,6]. We might be interested in putting students with different abilities in the same group in order to create heterogeneous groups including students with different academic or cultural backgrounds. This diversity will help the members of the group in gaining more from their peers. The MaxDP is closely related to the *Maximum Diversity Problem* in which the goal is to select a maximally diverse subset of objects and the diversity is measured by the *sum* of distances between chosen objects [8,11].

Since MaxDP is NP-hard, there is no polynomial-time algorithm for solving this problem. Hence, it is difficult to solve large instances of the MaxDP by means of classical exact methods [6,13]. Consequently, we need to design efficient algorithms in order to address this issue. Efficient heuristics are natural choice in this kind of situations with the aim of providing high-quality solutions in short computation time. In this context, Variable Neighborhood Search (VNS) provides a framework for designing efficient heuristics. The VNS has been used, with success, for solving a large variety of combinatorial optimization problems [3,10]. Therefore, in this paper, we introduce a VNS algorithm for solving the MaxDP. Our contribution consists in introducing new neighborhood structures that fit to the special structure of the MaxDP and are obtained after analyzing the characteristics of the problem. In order to evaluate the efficiency of the introduced VNS, we conducted computational experiments using randomly generated medium-sized and large-scale instances. Then, we compared the results of the algorithm with those provided by the standard solver Gurobi. According to the numerical results, the introduced algorithm provides high-quality solutions and outperforms the solver Gurobi.

This paper is organized as follows. Section 2 presents the MaxDP problem and its mathematical programming models. In Sect. 3, we introduce our VNS algorithm. Section 4 is devoted to the numerical results of our computational experiments. Finally, some concluding remarks are drawn in Sect. 5.

2 The Maximum Dispersion Problem

In this section, we present the notation that we are going to use in this paper and provide mathematical formulations for the MaxDP. We assume that a set $V = \{1, ..., n\}$ of n objects is given such that each object i has a non-negative weight $a_i \geq 0$, where $i \in V$ and we set $A = \sum_{i \in V} a_i$. We want to partition this set of objects into m classes (groups) $c \in C = \{1, ..., m\}$. The partition must meet some conditions: the sum of weights of the objects in any class c must belong to the interval $[(1-\alpha)M_c, (1+\alpha)M_c]$, where $M_c \geq 0$ is the *target weight* of class $c \in C$. We assume that there is no single object having a weight which allows it to make a singleton class on its own. Finally, the pairwise distance between objects i and j, denoted by d_{ij}, is symmetric.

In [6], Fernández et al. present three mathematical models for the MaxDP. For each $i \in V, c \in C$, we define the binary decision variables x_{ic} as follows:

$$x_{ic} = \begin{cases} 1 & : \text{ if object } i \text{ is assigned to class } c, \\ 0 & : \text{ otherwise.} \end{cases} \tag{1}$$

Using the presented notation, we have the following *nonlinear* mathematical programming model for the maximum dispersion problem, denoted by $MaxDP_{NLP}$:

$$\text{Max Min}_{i,j \in V, c \in C} \quad \frac{d_{ij}}{x_{ic}x_{jc}} \tag{2}$$

$$\text{s.t.} \quad \sum_{c \in C} x_{ic} = 1, \quad \forall i \in V, \tag{3}$$

$$\sum_{i \in V} a_i x_{ic} \geq (1-\alpha)M_c, \quad \forall c \in C, \tag{4}$$

$$\sum_{i \in V} a_i x_{ic} \leq (1+\alpha)M_c, \quad \forall c \in C, \tag{5}$$

$$x_{ic} \in \{0,1\}, \quad \forall i \in V, c \in C. \tag{6}$$

In this mathematical model, the objective function (2) is nonlinear and maximizes the minimal pairwise distance between objects that belong to a same class $c \in C$. According to the constraints (3), each object can only be assigned to a single class. The constraints (4) and (5) are called the *balancing constraints* and restrict each class to meet the weight limits around its target weight.

We can linearize $MaxDP_{NLP}$ and use any standard MILP solver such as Gurobi or IBM Cplex for solving the problem. In fact, $MaxDP_{NLP}$ can be linearized in different ways (see [2,6,13]). One approach consists in reformulating the $MaxDP_{NLP}$ as a Mixed-Integer Linear Program (MILP). This approach is based on the idea of replacing the quadratic term $x_{ic}x_{jc}$ by an additional variable z_{ijc} i.e., $z_{ijc} := x_{ic}x_{jc}$. In this case:

$$z_{ijc} = \begin{cases} 1 & : \text{ if objects } i \text{ and } j \text{ are both assigned to class } c, \\ 0 & : \text{ otherwise.} \end{cases} \tag{7}$$

We observe that z_{ijc} variables are symmetric with respect to i and j, that is $z_{ijc} = z_{jic}$; hence, we can assume that $i < j$. Furthermore, we need to add the following additional constraints into the model:

$$x_{ic} + x_{jc} \leq 1 + z_{ijc} \quad \forall i, j \in V, \quad i < j, \quad c \in C.$$

Indeed, when objects i and j belong to a same class $c \in C$, these constraints force the variable z_{ijc} to be 1 [6,7]. Finally, we need to add a continuous variable

u that represents the minimum distance between each pair of objects in the same class [6]. To sum up, the $MaxDP_{NLP}$ is reformulated as the following MILP, that we denote it by $MaxDP_L$:

$$\text{Max } u \tag{8}$$

$$\text{s.t.} \quad (3) - (5), \tag{9}$$

$$u \leq d_{ij}\, z_{ijc} + D(1 - z_{ijc}), \quad \forall i,j \in V, \;\; i < j, \;\; c \in C, \tag{10}$$

$$x_{ic} + x_{jc} \leq 1 + z_{ijc}, \quad \forall i,j \in V, \;\; i < j, \;\; c \in C, \tag{11}$$

$$x_{ic}, z_{ijc} \in \{0,1\}, \quad \forall i,j \in V, \; c \in C, \tag{12}$$

$$u \geq 0, \tag{13}$$

where $D = max_{i,j}\, d_{ij}$. The aim of the $MaxDP_L$ is to maximize the minimum pairwise distance between members of each class. For this purpose, for each class c, the objective function of the $MaxDP_L$ along with the constraints (10) provide the smallest pairwise distance between the members of the class. Indeed, a constraint (10) becomes binding only if $z_{ijc} = 1$, i.e., for any class c, the variable u is bounded by d_{ij} for all objects i and j in the class c. According to the definition of the variables z_{ijc} i.e., $z_{ijc} := x_{ic}x_{jc}$, maximization of u yields, for an optimal solution, a value for u i.e., u^* that equals the minimum of d_{ij}.

3 A Heuristic Solution Method

Due to NP-hardness of the MaxDP, solving its large-scale instances is a big challenge [5,13]. In order to address this issue, we need to design efficient methods able to provide high-quality solutions within short computation time. In this context, the focus of this paper is to use the framework of Variable Neighborhood Search (VNS) in order to introduce an efficient algorithm. The VNS was first introduced by Hansen and Mladenović [12] and it has been used for solving a large variety of (combinatorial) optimization problems for which this approach often provides high-quality solutions (see e.g., [3,9,10], and references therein).

The general structure of a VNS algorithm is composed of an initialization phase and an optimization stage. In particular, the VNS uses a sequence of *neighborhood structures* in order to explore the solution space of the problem. The general frame of VNS includes several neighborhood structures; however, in most cases, the local search heuristics might use a single neighborhood structure [10]. Another feature of VNS is the *shake* operation that consists of random moves inside the feasible region of the problem with the aim of escaping from local optima and preventing premature convergence of the algorithm.

By taking into account the structure of the MaxDP, we define *swap of two elements* between two disjoint classes as the main neighborhood structure that permits to move from one solution of MaxDP to another one. The simplicity of swap operation permits an effective exploration of the solution space. The feasibility of a swap is restricted by the balancing constraints i.e., swapping two elements should not hurt the bounds on their corresponding classes. In order to present the VNS algorithm that we introduce for solving the MaxDP, we need to define some additional notation, where $k \in \{1, ..., m\}$.

- $minDist_{c_k}$: Denotes the minimal distance among all objects of class $c_k \in C$.
- $absMinDist$: Denotes the current absolute minimal distance within all classes i.e., the current objective value.
- MDE_{c_k}: Denotes the set of elements of class k which are involved in the minimal distance of this class.
- $C_{minDist}$: Denotes the set of all classes c_k for which $minDist_{c_k}$ is equal to $absMinDist$.
- κ_{max}: Indicates the limit on the number of shake procedures.

The VNS algorithm that we introduce is composed of three main parts. The first part concerns the initialization stage in which we generate a solution by random distribution of objects among the given classes. Then, we check the feasibility of the random solution through verifying the balancing constraints. If the solution is not feasible, then we simply move the objects among different classes in order to attain the feasibility. As soon as a feasible solution is obtained, we compute the sets MDE_{c_k}, for all $c_k \in C$, and store in $C_{minDist}$ all classes having the minimal distance $absMinDist$.

The second part is a shake procedure, where we choose randomly a class c_k out of $C_{minDist}$ as well as a class c_l out of all remaining classes in C. Each class has its own minimum distance, MDE_{c_k} and MDE_{c_l}, respectively, in which some objects are involved. We select two of such objects e.g., $e_{k,i}$ and $e_{l,j}$. First, we check the feasibility of swapping these objects as well as possibility of obtaining an improvement, either in the current best-known minimum distance or only in the minimum distance of c_k or c_l. In the latter case, by accepting solutions that deteriorate the current best solution, we privilege diversification in order to explore the solution space in a more effective way. If swapping conditions are met, then we perform the swap and update the state of each class. We repeat this procedure κ_{max} times, where κ_{max} indicates the limit on the number of possible shakes. Algorithm 1 shows the pseudo-code of the shake procedure.

Algorithm 1. Shake Procedure

```
procedure SHAKE(κ_max, x)
    κ ← 1;
    while κ ≤ κ_max do
        Select randomly classes c_k and c_l from C_minDist;
        Select randomly e_{k,i} and e_{l,j} from MDE_{c_k} and MDE_{c_l}, respectively;
        if it is feasible to swap e_{k,i} & e_{l,j}, and the swap improves the objective
        value or the minimum distance of one of the classes k and l then
            PerformSwap(e_{k,i}, e_{j,l});
            Update the information of classes and/or the best minimum distance;
            κ ← κ + 1;
        end if
    end while
    return x.
end procedure
```

Algorithm 2. Swap Procedure

```
1: procedure SWAP(x)
2:     improvement ← true;
3:     while improvement == true do
4:         for c_k ∈ C_minDist do
5:             for c_l ∈ C such that l ≠ k do
6:                 for e_{k,i} ∈ MDE_{c_k} do
7:                     for e_{l,i} ∈ MDE_{c_l} do
8:                         if it is feasible to swap e_{k,i} & e_{l,j} and the swap improves
9:                          the objective value or the swap improves the minimum
10:                         distance of one of the classes k and l then
11:                            Swap e_{k,i} and e_{j,l};
12:                            Update the information of classes and/or
13:                            the best minimum distance;
14:                            improvement ← true;
15:                        else
16:                            improvement ← false;
17:                        end if
18:                    end for
19:                end for
20:            end for
21:        end for
22:    end while
23:    return x.
24: end procedure
```

The third part of the algorithm is a deterministic search within the solution space in order to improve the current solution. This task is performed through the *swap* neighborhood structure. In fact, we examine the possibility of improving the best current solution by perturbing the composition of the classes involved in the best solution found so far. To this end, we iterate over all classes $c_k \in C_{minDist}$ and $c_l \in C$ (for all $l \neq k$) and pick up objects $c_{k,j}, e_{l,j}$ that are involved in the minimum distance of their corresponding classes i.e., $e_{k,i} \in MDE_{c_k}$ and $e_{l,j} \in MDE_{c_l}$. If swapping $e_{k,i}$ and $e_{l,j}$ is feasible, then we check whether both $minDist_{c_k}$ and $minDist_{c_l}$ obtained by the swap of these two objects improves the current $absMinDist$. If this condition holds, we perform the swap (line 11 of Algorithm 2), update all necessary information, and start a new search for the classes of $C_{minDist}$. We iterate this procedure until it can no more find any improvement. Algorithm 2 presents the pseudo-code of the swap procedure.

Finally, Algorithm 3 sums up the introduced VNS algorithm for solving the MaxDP. The algorithm iterates for a predefined maximum number of iterations. Within the algorithm, the starting solution x is passed to the shake step in order to obtain x'. This one is passed to the swap operation in order to get x''. If x'' is an improving assignment of objects, we store it, update the information of the solution, and set $\kappa = 1$; otherwise, we increase κ by 1. If κ reaches κ_{max}, then we set $\kappa = 1$.

Algorithm 3. VNS for MaxDP

```
procedure VNS
    x ← randomInitialSolution(V, C);
    Create feasible partition (x);
    Compute C_minDist and MDE_{c_k} : c_k ∈ C;
    iteration ← 0;
    κ ← 1;
    while iteration ≤ iteration_Max do
        absMinDistStart ← absMinDist;
        x' ← shake(κ_max, x);
        x'' ← swap(x');
        iteration ← iteration + 1;
        if absMinDist > absMinDistStart then
            x ← x'';
            κ ← 1;
        else
            κ ← κ + 1;
            if κ > κ_max then
                κ ← 1;
            end if
        end if
    end while
    return Solution.
end procedure
```

4 Numerical Experiments

In order to evaluate the performance of the introduced VNS algorithm in solving the MaxDP, we conducted numerical experiments on randomly generated instances. In this section, we present the test setting, the numerical results, and comments on our observations.

We generated random instances[1] in a similar way as Fernández et al. [5]. More precisely, each object has a coordinate (x, y) randomly distributed in the square $Q = ((0,0),(10,10))$. The associated weight of each object is a real-valued number uniformly distributed over the interval $[1000, 4000]$. The target weights of the classes are drawn uniformly from the interval $[0.75(A/m), 1.25(A/m)]$, and then normalized in order to obtain an overall sum of A. The instances are generated with the size of $n \in \{100, 200, 300, 400, 500\}$ and we partition them into $m \in \{3, 5\}$ classes. For each pair of (set of objects, number of classes), i.e., (n, m), we have generated 3; hence, in total, there are 30 instances. We solved the MaxDP for each instance by setting $\alpha \in \{0.01, 0.05\}$, that means 60 experiments overall.

Our VNS algorithm has some parameters that need to be set. We define the maximal number of shakes to be $\kappa_{max} = 3$. The stopping condition of the algo-

[1] The instances are publicly available on: https://sites.google.com/site/mahdimoeini2013/test-instances.

rithm is based on the maximal permitted number of iterations and is determined by means of the number of objects i.e., $iteration_{Max} = n/3$.

In order to evaluate the quality of the results provided by our VNS algorithm, we compared them by those obtained by the Gurobi Optimizer Version 7.5.1. For this purpose, we implemented the mixed-integer linear programming (MILP) formulation of MaxDP, i.e., the model (8)–(13), and solved it by Gurobi on the same instances as we used for the VNS. As time limit for Gurobi, we chose $t_{max} = 1200$ s. If the solver could not provide an optimum within the determined time limit, we noted the best solution (if there is any) found so far.

We implemented the algorithm as well the MILP model in Java. In order to have a fair comparison, all experiments were done under same conditions and on an AMD Phenom II X4 945 Processor (3.0 GHz) with 4 GB of RAM under Windows 10 operating system.

Table 1 shows the results of the experiments. In the first three columns of this table, instance information such as number of objects n and number of classes m, as well as tolerance value α for balancing constraints are presented. In the remaining part of the table, for each values of m, n, and α, the results of Gurobi as well as the VNS algorithm are reported. In fact, for each of these cases, we tested the methods on three instances, and the average of best objectives value provided by each method i.e., *obj.val.*, the average computation time of each method in seconds i.e., *t(s)* are shows in the table. Furthermore, for each case of m, n, and α, the average number of iterations that the VNS algorithm needed to find the solution is given in the last column of Table 1. The column *Opt?* tells if Gurobi could provide the optimum. Unfortunately, in most of the cases, Gurobi fails to provide the optimal solution within the time-limit. In many cases, the solver aborts its search even within the time-limit (denoted by *O.M.*); however, the solver always provides solutions computed by its internal heuristics.

Comments on the Results

According to the results, we make the following observations:

- Our VNS algorithm provides high-quality solutions or optima for all instances in very short computation time (less than 10 s). In fact our algorithm dominates for all instances, with a large gap, the standard commercial solver Gurobi. Apart from $n = 100$, Gurobi fails to find even good-quality solutions.
- For some instances with $n = 200$, $n = 300$, and $n = 400$, Gurobi can find some solutions (even not optimal) but we had to stop it due to the time limit. For $n = 200$ with $m = 5$ and $n = 300$ with $m = 5, \alpha = 0.05$, the solver stops its solution search at some point within the time limit with a message of *gurobi-out-of-memory exception*. That is why, in these cases, the average computation time of Gurobi is less than the time limit.
- When $n = 500$ or $n = 400$ with $m = 5$, Gurobi provides only heuristic solutions found by its internal heuristic approaches and the solver itself fails in finding any solution. In fact, the solver starts with the *presolve process* and, after a while, stops with the *gurobi-out-of-memory exception* message.

Table 1. Computational results.

Instance			Gurobi			VNS algorithm		
n	m	α	obj.val.	t(s)	Opt?	obj.val.	t(s)	Iter.
100	3	0.01	0.52	122.73	Yes	0.52	0.02	6
	3	0.05	0.52	83.23	Yes	0.52	0.01	3
	5	0.01	1.07	247.98	Yes	1.02	4.03	28
	5	0.05	1.07	253.06	Yes	1.04	1.23	17
200	3	0.01	0.28	1200.09	No	0.32	1.00	25
	3	0.05	0.31	1200.10	No	0.34	0.07	19
	5	0.01	0.33	730.33	No	0.67	1.62	42
	5	0.05	0.15	609.33	No	0.68	0.05	14
300	3	0.01	0.12	1200.09	No	0.21	0.03	5
	3	0.05	0.12	1200.09	No	0.23	0.03	5
	5	0.01	0.12	1200.40	No	0.50	0.49	57
	5	0.05	0.09	1141.83	No	0.52	0.42	26
400	3	0.01	0.07	1200.43	No	0.22	0.27	29
	3	0.05	0.07	1200.28	No	0.19	0.26	28
	5	0.01	0.03	O.M.	No	0.37	0.31	41
	5	0.05	0.02	O.M	No	0.39	0.94	39
500	3	0.01	0.02	O.M.	No	0.17	0.65	67
	3	0.05	0.01	O.M.	No	0.16	0.29	38
	5	0.01	0.04	O.M.	No	0.32	1.70	110
	5	0.05	0.02	O.M.	No	0.36	1.13	89

In these cases, no actual computation time can be provided. In Table 1, these cases are denoted by *O.M.*

5 Conclusion

In this paper, we introduced an effective VNS algorithm for solving the Maximum Dispersion Problem (MaxDP). According to the numerical results of the computational experiments, we observe that our heuristic is able to provide high-quality solutions in short computation time. The proposed algorithm highly dominates the commercial solver Gurobi.

The MaxDP is a recently introduced problem and the avenue of research for solving the MaxDP is quite wide. Design of alternative heuristics or introducing valid inequalities and facets for the MaxDP can be interesting research perspectives. The research in these directions is in progress and we hope to publish the results in a close future.

Acknowledgements. The authors acknowledge the chair of Business Information Systems and Operations Research (BISOR) at the TU-Kaiserslautern (Germany) for the financial support, through the research program "CoVaCo".

References

1. Baker, K.R., Powell, S.G.: Methods for assigning students to groups: a study of alternative objective functions. J. Oper. Res. Soc. **53**(4), 397–404 (2002)
2. Balas, E., Mazzola, J.: Nonlinear 0–1 programming: I. Linearization techniques. Math. Program. **30**(1), 1–21 (1984)
3. Brimberg, J., Mladenović, N., Urošević, D.: Solving the maximally diverse grouping problem by skewed general variable neighborhood search. Inf. Sci. **295**, 650–675 (2015)
4. Erkut, E.: The discrete p-dispersion problem. Eur. J. Oper. Res. **46**(1), 48–60 (1990)
5. Fernández, E., Kalcsics, J., Nickel, S., Ríos-Mercado, R.Z.: A novel maximum dispersion territory design model arising in the implementation of the WEEE-directive. J. Oper. Res. Soc. **61**(3), 503–514 (2010)
6. Fernández, E., Kalcsics, J., Nickel, S.: The maximum dispersion problem. Omega **41**(4), 721–730 (2013)
7. Glover, F., Wolsey, R.: Converting the 0–1 polynomial programming problem to a 0–1 linear program. Oper. Res. **22**(1), 455–60 (1974)
8. Glover, F., Ching-Chung, K., Dhir, K.: A discrete optimization model for preserving biological diversity. Appl. Math. Model. **19**(11), 696–701 (2010)
9. Goeke, D., Moeini, M., Poganiuch, D.: A variable neighborhood search heuristic for the maximum ratio clique problem. Comput. Oper. Res. **87**, 283–291 (2017)
10. Hansen, P., Mladenović, N., Brimberg, J., Moreno Prez, J.A.: Variable neighborhood search. In: Gendreau, M., Potvin, J.Y. (eds.) Handbook of Metaheuristics, pp. 61–86. Springer, Heidelberg (2010). https://doi.org/10.1007/978-1-4614-6940-7_12
11. Martí, R., Gallego, M., Duarte, A.: A branch and bound algorithm for the maximum diversity problem. Eur. J. Oper. Res. **200**(1), 36–44 (2010)
12. Mladenović, N., Hansen, P.: Variable neighborhood search. Comput. Oper. Res. **24**(11), 1097–1100 (1997)
13. Moeini, M., Wendt, O.: A heuristic for solving the maximum dispersion problem. In: Fink, A., Fügenschuh, A., Geiger, M.J. (eds.) Operations Research Proceedings 2016. ORP, pp. 405–410. Springer, Cham (2018). https://doi.org/10.1007/978-3-319-55702-1_54
14. Palubeckis, G., Karčiauskas, E., Riškus, A.: Comperative performance of three metaheuristic approaches for the maximally diverse grouping problem. Inf. Technol. Control **40**(4), 277–285 (2011)
15. Prokopyev, O., Kong, N., Martinez-Torres, D.: The equitable dispersion problem. Eur. J. Oper. Res. **197**(1), 59–67 (2009)

Using a Multi-agent System for Overcoming Flickering Effect in Distributed Large-Scale Customized Lighting Design

Adam Sędziwy(✉) and Leszek Kotulski

AGH University of Science and Technology,
al.Mickiewicza 30, 30-059 Kraków, Poland
{sedziwy,kotulski}@agh.edu.pl

Abstract. Designing of the large scale roadway lighting installations is a complex task for the sake of its computational complexity. Additionally, each project has to meet the mandatory lighting standard requirements and business constraints such as investment costs of an installation, its energy efficiency and so on. The important issue one has to face to comply with lighting standards in the large-scale projects is resolving all conflicts arising in areas where two or more streets with different lighting requirements meet. The common problem for such shared areas is that the installation's adjustments depend on for which street they were calculated. This behavior may cause the flickering effect (Note that the term *flickering* refers in this article to a non convergent behavior of an optimization process and not to a light flickering).The agent-based method of lighting design proposed in the previous works assumed optimization with agents traversing a global structure representing an urban area and adjusting subsequent installations. Now we propose yet another method relying on a multi-agent system approach, assuming that agents operate on the local portions of global data. We also discuss the flickering effect being a result of conflicts among agents and propose methods of their resolution. It has to be stressed that overcoming this problem is necessary for making the proposed approach to be applicable in practical solutions.

Keywords: Multi-agent system · Custom lighting design · LED
Flickering effect

1 Introduction

In this work we consider a scenario of a roadway lighting design process performed by agents, which does not converge (a halting condition of a process is not satisfied) yielding to the flickering effect. We propose the methods of resolving such issues.

The correct preparation of an optimal project of outdoor lighting installation, fulfilling the energy efficiency criteria and being consistent with the mandatory

N. T. Nguyen et al. (Eds.): ACIIDS 2018, LNAI 10751, pp. 372–381, 2018.
https://doi.org/10.1007/978-3-319-75417-8_35

lighting performance requirements [1,2] is a very complex task. In this work we limit our considerations to roadway lighting although there exist related, nontrivial problems like road tunnels lighting [3,4].

Thanks to the graph transformation mechanisms which support parallel computations [5,6] an artificial intelligence system is capable of making it much faster and more efficient way than a human designer. In the implemented system we design a lighting installation separately for each street but there still remain some influences on lighting conditions of some regions caused by installations standing alongside the neighboring streets (e.g., in cross section areas).

Actually we are able [7] to prepare a lighting project in a situation when a street is influenced by any arbitrary set of lighting points. The problem considered in this work concerns a situation when an infinite cycle (in terms of steps required to find a solution for all cycled lighting installations) of such influences appears. It is a common phenomenon in our context because streets form many cycles. An agent system performing parallel computations has to be able to handle such a scenario in a reasonable time and with acceptable energy costs.

2 Related Works and Basic Notions

In this section we present the basics of lighting design related problems which are a background for the further considerations.

2.1 Optimization Process – Main Facts

The objective of an optimization in a roadway lighting design process is determining such settings of particular luminaires constituting a roadway lighting installation, that some objective function for a problem is *optimized*, i.e., minimized or maximized, dependently on the context. For example the objective may be minimizing an annual power usage of a city streetlight installation. Then this function may return a power usage (see for example [8] for energy optimization), investment costs but also take into account another, more complex criteria either business or technical ones [9]. In more advanced (real-life) cases one deals with a multi-criteria optimization with an objective function compromising several factors (e.g., the power usage and investment costs). The mentioned luminaire's settings may cover a range of parameters such as pole height, arm length, fixture model, fixture inclination angle or luminous flux dimming. The area of an optimization may be either a single street (then the problem gets trivial) or an entire city containing tens of thousands of lighting points.

An important constraint imposed on an optimization process is that the obtained solutions have to fulfill mandatory lighting standard requirements (e.g., for standard EN-13201:2 in Europe [1]). The optimal adjustments of luminaires in a large-scale design are searched for all installations located within a city, district or other area. The crucial fact here is that each luminaire is regarded separately, i.e., luminaire's adjustments are found luminaire by luminaire rather than globally, for an entire roadway. This optimization method was described in depth in [7] so it will not be discussed here as being out of the scope. The important technical consequence of such an approach is that this optimization method implies a significant computational overhead. The order of magnitude of a computational complexity can be assessed as follows. Suppose that one optimizes N luminaires in such a way that for each of them 5 parameters can be varied (a number of variants is given in parenthesis): pole height (7), arm length (8), fixture model (2000), fixture installation angle (7), luminous flux level (50). Thus for each single luminaire one has 3.92×10^7 possible variants. To achieve even the partial reduction of this flood of possible states to be checked, the appropriate heuristics have to be applied (not to be discussed here). Unfortunately, it is still not sufficient for making the method to be practically applicable. As it was already shown in other works [6,10] the fundamental workaround is parallelization of computations using for example a multi-agent system. Application of intelligent agents seems to be a suitable approach here, as such a complex optimization problem requires decision making based not only on a built-in agent's knowledge but also a knowledge acquired dynamically from a surrounding environment.

2.2 Lighting Classes

The basic criterion of a considered lighting design optimization is the compliance with lighting standards established for public spaces including highways, roadways, residential areas, road junctions and another conflict areas, bike lanes, walkways and so on. Public lighting standardization is made in two steps. First, it defines lighting classes as such and specifies which types of road situations can be assigned with them. Those assignments depend on numerous different factors such as dominant roadway users, traffic speed and intensity, number of intersections and so on (see Fig. 1). In the second step the performance requirements (e.g., required illuminance level for an area) are set for each class (see Tables 1 and 2).

Table 1. M-lighting classes (for traffic routes, dry surface condition) according to the EN 13201-2:2016 standard. L_{avg}—min. average luminance maintained, U_o—min. overall uniformity, U_l—min. longitudinal uniformity, f_{TI}—max. disability glare, R_{EI}—min. lighting of surroundings.

Class	L_{avg} $[cd/m^2]$ (min. *)	U_o (min.)	U_l (min.)	f_{TI} [%] (max. **)	R_{EI} (min.)
M1	2.00	0.40	0.7	10	0.35
M2	1.50	0.40	0.7	10	0.35
M3	1.00	0.40	0.6	15	0.30
M4	0.75	0.40	0.6	15	0.30
M5	0.50	0.35	0.4	15	0.30
M6	0.30	0.35	0.4	20	0.30

* min.: minimum allowed value; ** max.: maximum allowed value

Parameter	Options	Description [a]		Weighting Value V_W [a]
Design speed or speed limit	Very high	v ≥ 100 km/h		2
	High	70 < v < 100 km/h		1
	Moderate	40 < v ≤ 70 km/h		-1
	Low	v ≤ 40 km/h		-2
Traffic volume		Motorways, multilane routes	Two lane routes	
	High	> 65 % of maximum capacity	> 45 % of maximum capacity	1
	Moderate	35 % - 65 % of maximum capacity	15 % - 45 % of maximum capacity	0
	Low	< 35 % of maximum capacity	< 15 % of maximum capacity	-1

Fig. 1. The scrap of the table for evaluating a lighting class on the basis of a roadway properties (EN 13201-1:2016)

Table 2. C-lighting classes on conflict areas (e.g., shopping streets, road intersections, queuing areas etc.) according to the EN 13201-2:2016 standard

Class	Horizontal illuminance	
	E_{avg} $[lx]$	U_o
C0	50.0	0.40
C1	30.0	0.40
C2	20.0	0.40
C3	15.0	0.40
C4	10.0	0.40
C5	7.50	0.40

2.3 Lighting Class Assignment

In the sequel we assume for simplicity and without the loss of generality that each roadway has some lighting class ascribed (usually Mx, for $x = 1, 2, \ldots, 6$). Similarly, each road junction area is ascribed with some lighting class (usually Cx, for $x = 0, 1, \ldots, 5$). This assumption is illustrated by Fig. 2. It should be remarked that such an approach is static. For the case of a dynamic lighting control it is admitted be the standard that a roadway lighting class can be dynamically changed due to a changeable traffic flow for instance. In those circumstances we obtain multiple performance patterns for a single roadway, dependently on an actual environment state. The detailed discussion concerning a dynamic lighting control can be found in [11].

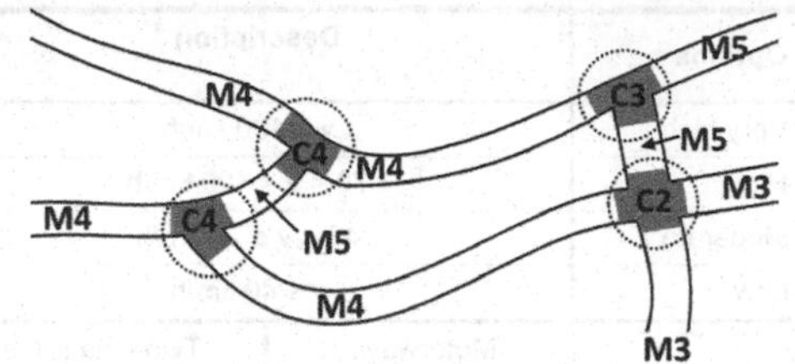

Fig. 2. The sample lighting class assignment. Dotted circles enclose junction areas

3 Large-Scale Optimization - Computation Scheme

In this section we briefly sketch how the customized photometric calculations underlying an optimization process performed by agents, are carried out.

Let $L(R_i) = \{L_1^i, L_2^i, \ldots L_n^i\}$ be a set of luminaires located along a roadway R_i, H be an average mounting height in the set $L(R_i)$ and S be an average spacing in $L(R_i)$. We assume that luminaires L_j^i located along a roadway R_i are indexed (with j) in the order of their natural sequence perceived by an observer walking down the street. The photometric computations for those luminaires are made on so called *calculation fields*. For each of them the appropriate quantities are computed (as those specified in Tables 1 and 2). A calculation field F is a roadway section, sometimes taken together with adjacent walkways, located between two subsequent luminaires. A **relevant area** for a computation filed F is an area covering all points located not further than $12H$ from F in the direction of increasing luminaire indices and $5H$ in side and backward directions (see Fig. 3). **Relevant luminaires** for a computation filed F are all luminaires contained in a relevant area for F.

According to the EN-13201:3 standard we define a number of **relevant forward luminaires** as

$$N_{fw} = \left\lceil \frac{12H}{S} \right\rceil$$

and a number of **relevant side/backward luminaires** as

$$N_{sb} = \left\lceil \frac{5H}{S} \right\rceil.$$

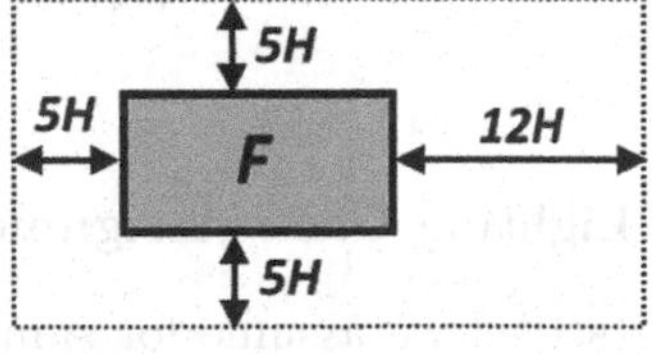

Fig. 3. The relevant area for a calculation field F

If R_i terminates (on any side) and either (i) it does not join any other illuminated road, walkway or any other area (Fig. 4), or (ii) it does and a lighting class changes ($R_1, \ldots, R_4$ in Fig. 5) then the final calculation fields of R_i will be referred to as *terminating* ones.

Remark. The above definition does not clarify practically when can we regard F as a *terminating field.* This decision is taken arbitrary on the basis of the actual installation's layout data. Usually, luminaires located alongside R_i have a "regular" layout in the sense that they are spaced with some $s \pm \Delta s$ and their distance form s roadway is $a \pm \Delta a$. If one takes the very last calculation field F then some luminaires can be "missed" (a number of front luminaires selected for calculation is less than N_{fw}, see Fig. 4) and thus standard requirements cannot be met. The same can apply to the first calculation field for which a number of backward luminaires is less than N_{sb}.

Having the above remark made the following problem has to be addressed prior to launching a customized lighting design: how to proceed with the *terminating fields* of R_i? This problem will be solved in the following way. First, a more operational definition of a terminating field will be introduced. F will be regarded as such if (i) a number of remaining forward luminaires if less than N_{fw} or (ii) lighting performance requirements cannot be met for F. The proceeding on terminating fields, in turn, will be accomplished by changing a lighting class for which photometric computations are made. For example, if a lighting class of R_i is M4 then for terminating fields it is changed to C4 (or the another, appropriate Cx class).

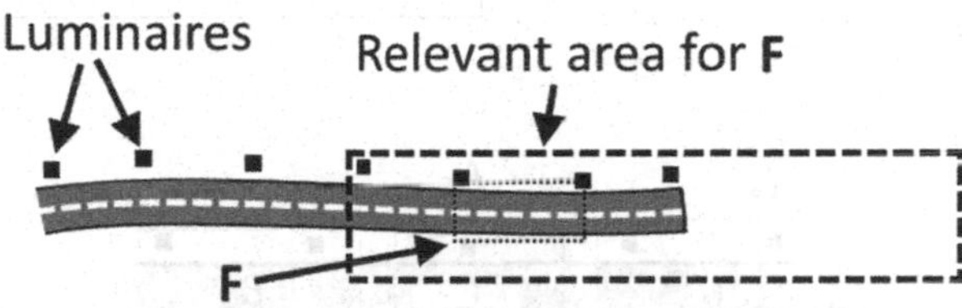

Fig. 4. A street with the sample terminating field F (the inner dotted rectangle) and the relevant area of F (the outer dashed rectangle)

4 Multi-agent system

The scheme of a multi-agent system performing customized optimization is following. There are two types of agents only: a registry agent (RA) and the optimizing agents (OA). A registry agent maintains a database of roadways and luminaires being processed. We do not specify in depth the structure of this database which can reflect the graph structure of the street layout [6,12]. The excerpts important for the future considerations are presented in Table 3 which illustrates dependencies for the sample scene shown in Fig. 5.

Each area, R, is allotted with a list of luminaires affecting it, $L(R)$. The ellipses present in rows 1–4 of Table 3a indicate some remaining luminaires belonging to the $L(R_i)$ set but being out of the area of interest for this example. An RA creates an optimizing agent for each area and updates the corresponding database record with an agent's name (the third column in Table 3a). Thus, in the moment of its creation an OA is assigned with the corresponding area, R_i, and a list of the relevant luminaires $L(R_i)$. Note that for each luminaire $L^i_j \in L(R_i)$ the set $A(L^i_j)$ of agents hosting L^i_j can be fetched on the fly from a database by sending a query to an RA. For the case of sample data shown in Table 3a the agent A_3 is assigned with the following data set: $A_3 \leftarrow \{R_3; [L_1, L_2, L_3, L_4]\}$ and the agents hosting the luminaire L_1 are $A(L_1) = [A_3, A_4, A_5]$.

Table 3. The sample database tables for the scene shown in Fig. 5

(a) Scene components assignment

Area	Relevant luminaires	Agent
R_1	$L_7, L_8, L_9, \ldots$	A_1
R_2	$L_2, L_5, L_6, L_7, \ldots$	A_2
R_3	$L_1, L_2, L_3, L_4, \ldots$	A_3
R_4	$L_1, L_{11}, L_{12}, \ldots$	A_4
C	L_1, L_2, L_7, L_{10}	A_5

(b) Luminaire settings changelog

S/N	Luminaire	Settings (luminous flux)	OA
⋮	⋮	⋮	⋮
198	L_7	0.60	A_1
199	L_2	0.85	A_3
200	L_7	0.70	A_5
201	L_7	0.65	A_2
202	L_1	0.20	A_4
203	L_7	0.60	A_1
⋮	⋮	⋮	⋮

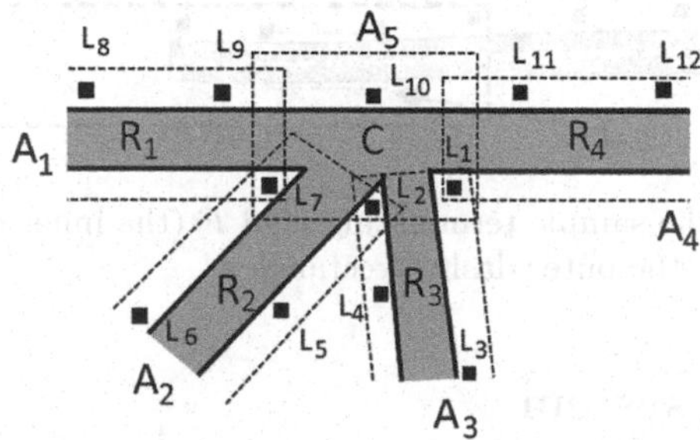

Fig. 5. The sample scene consisting of terminating sections and the junction area

An OA in turn, say A_i, optimizes an installation's adjustments by performing the suitable photometric computations. Once an optimization process is completed, resultant adjustments for each luminaire are registered in a database managed by an RA (see Table 3b).

The important task of a registry agent is discovering and resolving all conflicts in the areas shared by two ore more agents. Such a conflict arises when two (or more) agents register a luminaire(s) settings which cannot be agreed among areas due to the lighting performance standard violation. The records with serial numbers (S/N) 200, 201 and 203 respectively (Table 3b) are the example of such a conflict. The luminous flux value 0.60 (S/N = 203) adjusted for L_7 by A_1 in R_1 area is too low for complying with a standard requirements defined for the C and R_2 areas, managed by A_5 and A_2 respectively which adjusted previously the luminous flux levels to 0.70 (S/N = 200) and 0.65 (S/N = 201). Analogously, the record with the S/N = 201 clashes with the entry 200.

5 MAS Life-Cycle. Conflict Resolution

Figure 6 depicts a multi-agent system life-cycle. Dotted lines delimit states of the particular agent types; dashed arrows with the envelope symbol represent the interactions among agents accomplished by sending the request/update/information messages.

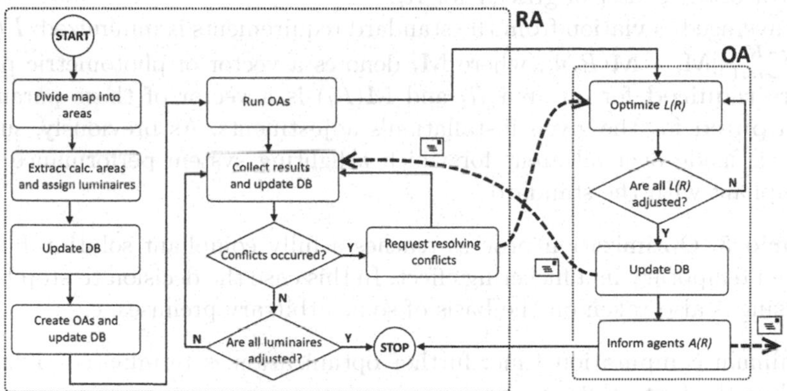

Fig. 6. The life-cycle of a multi-agent system

When a multi-agent system performs a parallel optimization processes the problem of stopping can arise. For example, if all luminaires located along R_k are already adjusted by some agent A_k and an agent A_i responsible for R_i processing reaches its terminating fields located by the junction area then it can modify settings for luminaires already adjusted by A_k. This modification, in turn, requires verification and re-optimizing R_k. Note that this simplified scenario can occur for more complex layouts containing multiple roadways arranged in the circular way. For such complex inputs a multi-agent system may evolve in the following ways.

Scenario 1. Optimization process reaches a fully compliant solution in the result of a convergent optimization process.

Scenario 2. Optimization problem cannot be resolved at all for this set of parameters because there does not exist a set of adjustments in a browsed space of configurations, satisfying the performance requirements. In this situation processing has to be terminated arbitrary if some of below conditions is satisfied. We denote the termination testing function as F and a vector of road situations as $\mathbf{R}$.

1. A number of conflicts (not standard-compliant fields) is less than some number n: $F(\mathbf{R}) = |\{R_i :$ performance requirements are not satisfied for $R_i\}|$.
2. All non-compliant fields posses lighting classes belonging to some set, say {M5, M6}. This approach to conflict resolution has the practical motivation

as it aims at "moving" non-compliant fields from most important roadways to the minor ones.

3. All non-compliant fields are grouped together in terms of mutual geodetic proximity. If all such installations are enclosed in relatively small area then an optimization process can be relaunched with new set of parameters: $F(\mathbf{R}) = \sum_{i<j} |\mathbf{r}(R_i) - \mathbf{r}(R_j)|$, where the summation is made over all areas for which a lighting system performance is not compliant with the standard; $\mathbf{r}(R_i)$ is a vector of the center of gravity for R_i.
4. An averaged deviation from the standard requirements is minimized: $F(\mathbf{R}) = \frac{1}{K}\sum_{i=1}^{K} ||\mathbf{M}_i - \mathbf{M}(R_i)||$, where $\mathbf{M}_i$ denotes a vector of photometric parameters **required** for an area R_i and $\mathbf{M}(R_i)$ is a vector of those parameters **computed** for the given installation's adjustments. As previously, summation is made over all areas for which a lighting system performance is not compliant with the standard.

Scenario 3. Optimization process reaches a fully compliant solution but it is obtained temporary in a flickering effect. In this case the decision to stop further processing is also taken on the basis of some arbitrary premises, e.g.:

1. minimum computation time: further optimization is terminated right after finding the first solution,
2. energy efficiency of a solution,
3. total costs of all resultant installations.

In Fig. 6 it is encapsulated in the decision node inside an RA activity area. Selecting an appropriate resolution strategy depends on a context, i.e., a particular design problem, existing constraints and assumptions. It can be achieved by following one of the approaches presented in Scenarios 2 and 3. Note that we admit for some cases that the `Conflicts occurred?` node can be left with **Y** decision even when some settings are non complying with the performance requirements. It occurs for all cases of Scenario 2.

6 Discussion and Conclusions

Preparation of optimized large-scale street lighting projects is a complex task not only due to the computational complexity but also because of the flickering effect which makes a computation process to be not convergent.

It has to be remarked that considered computations are made for the *static* scenario, i.e., once a lamp is switched on its luminous flux level does not change. The problem gets more complicated when one considers lighting control [11,13]. Then luminous fluxes may change following the variable traffic flows and other variable environment conditions (ambient light, rain etc.). Then the setups of particular installations have to be recalculated accordingly.

To handle such complex computations one can either try to extend the search space but with no guarantee of success or to apply some heuristic approaches.

The first of them is a conditional acceptance for non standard-compliant solutions. The second method is grouping together all non-compliant regions in a single area which can be proceed separately for another set of settings.

The above methods allow to complete computations in a reasonable time and make the agent-based approach applicable in practical use. In particular, the software tool relying on presented concepts for preparing the large-scale projects, is developed at the AGH University. Once the development phase is completed the results of tests will be published in the next works.

References

1. CEN 13201–2, E.C.F.S.: Road lighting - part 2: Performance requirements. (2003) Ref. No. EN 13201–2:2003 E
2. Commission Internationale de l'Eclairage: Lighting of roads for motor and pedestrian traffic, CIE 115:2010. CIE, Vienna (2010)
3. Peña-García, A., Gil-Martín, L., Hernández-Montes, E.: Use of sunlight in road tunnels: an approach to the improvement of light-pipes' efficacy through heliostats. Tunn. Undergr. Space Technol. **60**, 135–140 (2016)
4. Fan, S., Yang, C., Wang, Z.: Automatic control system for highway tunnel lighting. In: Li, D., Liu, Y., Chen, Y. (eds.) CCTA 2010. IAICT, vol. 347, pp. 116–123. Springer, Heidelberg (2011). https://doi.org/10.1007/978-3-642-18369-0_14
5. Sędziwy, A., Kotulski, L.: Graph-based optimization of energy efficiency of street lighting. In: Rutkowski, L., Korytkowski, M., Scherer, R., Tadeusiewicz, R., Zadeh, L.A., Zurada, J.M. (eds.) ICAISC 2015. LNCS (LNAI), vol. 9120, pp. 515–526. Springer, Cham (2015). https://doi.org/10.1007/978-3-319-19369-4_46
6. Sędziwy, A.: Sustainable street lighting design supported by hypergraph-based computational model. Sustainability **8**(1), 13 (2016)
7. Sędziwy, A.: A new approach to street lighting design. LEUKOS **12**(3), 151–162 (2016)
8. Mahoor, M., Salmasi, F.R., Najafabadi, T.A.: A hierarchical smart street lighting system with brute-force energy optimization. IEEE Sens. J. **17**(9), 2871–2879 (2017)
9. Feng, X., Murray, A.T.: Spatial analytics for enhancing street light coverage of public spaces. LEUKOS. https://doi.org/10.1080/15502724.2017.1321486 (2017)
10. Sędziwy, A., Kotulski, L.: Multi-agent system supporting automated gis-based photometric computations. Procedia Comput. Sci. **80**, 824–833 (2016)
11. Wojnicki, I., Ernst, S., Kotulski, L., Sędziwy, A.: Advanced street lighting control. Expert Syst. Appl. **41**(4, Part 1), 999–1005 (2014)
12. Angles, R., Gutierrez, C.: Survey of graph database models. ACM Comput. Surv. **40**(1), 1:1–1:39 (2008)
13. Shahzad, G., Yang, H., Ahmad, A.W., Lee, C.: Energy-efficient intelligent street lighting system using traffic-adaptive control. IEEE Sens. J. **16**(13), 5397–5405 (2016)

The Optimal Control Problem with Fixed-End Trajectories for a Three-Sector Economic Model of a Cluster

Zainelkhriet Murzabekov[1], Marek Miłosz[2], and Kamshat Tussupova[1(✉)]

[1] Al-Farabi Kazakh National University, Ave.al-Farabi 71, 050040 Almaty, Kazakhstan
{murzabekov-zein,kamshat-0707}@mail.ru
[2] Institute of Computer Science, Lublin University of Technology, Nadbystrzycka 36B, 20-618 Lublin, Poland
m.milosz@pollub.pl

Abstract. For the mathematical model of a three-sector economic cluster, the problem of optimal control with fixed ends of trajectories is considered. An algorithm for solving the optimal control problem for a system with a quadratic functional is proposed. Control is defined on the basis of the principle of feedback. The problem is solved using the Lagrange multipliers of a special form, which makes it possible to find a synthesising control.

Keywords: Optimal control problem · Three-sector economic cluster
Lagrange multiplier method · Dynamical system · Quadratic functional

1 Introduction

The problem of optimal control for dynamical systems can be formulated as the problem of finding program control or constructing a synthesising control that depends on the state of the system and the current time. In the first case, the problem can be solved using the Pontryagin maximum principle [1, 2]. In the general case, the Pontryagin maximum principle gives the necessary conditions for optimality and allows one to obtain program control, depending on the current time. In the second case, Bellman's dynamic programming method [3] or Krotov sufficient optimality conditions [4] can be used.

In practice, there is a large number of optimal control problems for economic systems that are nonlinear systems with coefficients that depend on the state of the control object. In economic systems, it is required to achieve a certain level of economic development on a given planning horizon.

A three-sector (i.e. materials, labour resources, and production assets) economic model and the necessary conditions for an optimal balanced growth of the economy are given in Kolemayev [5]. Various aspects of the analysis of economic growth with the development of deterministic and stochastic three-sector dynamical models of open and

N. T. Nguyen et al. (Eds.): ACIIDS 2018, LNAI 10751, pp. 382–391, 2018.
https://doi.org/10.1007/978-3-319-75417-8_36

closed types are presented in the works of Dzhusupov et al. [6], De [7], Dobrescu et al. [8], Zhang [9], Zhou and Xue [10], Sen [11].

The fundamental work of Aseev et al. [12] provides the foundations of the mathematical theory of optimal control of dynamical systems on an infinite interval using the Pontryagin maximum principle. As an example, a two-sector model of optimal economic growth with a random price jump is considered.

The article of Shnurkov and Zasypko [13] studies the optimal control problem (OCP) for a dynamic three-sector model of the economy on the basis of the maximum principle. The OCP considered by them is a problem with free right ends of trajectories with scalar control, representing specific investments in the fund-creating sector of the economy. Note that in this article a frequent case is considered, when the investment changes only in the fund-creating sector.

In contrast to the above works, we consider the optimal control problem (OCP) with fixed ends of trajectories, in a finite time interval. In this paper, we propose to use an approach based on sufficient optimality conditions using Lagrange multipliers of a special type, which allows us to represent the desired control in the form of synthesising control, depending on the state of the system and the current time. In addition, this method makes it possible to take into account the existing restrictions on the values of controls. It should also be emphasised that this study considers the setting of the OCP for a three-sector economic model of a cluster in which the shares of labour and investment resources for all three sectors of the economy can simultaneously change.

Note that the peculiarity of the OCP considered in this paper is that the trajectories of the system must pass through given points at the initial and final instants of time (i.e. the left and right ends of the trajectories are fixed). The problem is considered on a finite time interval, there are restrictions on the values of controls, the task of constructing a synthesising control is posed. To solve this problem, the method of Lagrange multipliers is used [14], with the use of multipliers of a special form, which allows to obtain optimal control in the form of a control sum with feedback and program control.

The proposed approach is used to solve the problem of optimal distribution of investment and labour resources in a three-sector economic model of a cluster.

2 Statement of the OCP for a Three-Sector Economic Model of a Cluster

Consider a three-sector economic model of a cluster, described by a system of six differential and algebraic equations [5]:

$$\begin{aligned} \dot{k}_i &= -\lambda_i k_i + (s_i/\theta_i)x_1, \quad k_i(0) = k_i^0, \quad \lambda_i > 0, \quad (i = 0, 1, 2), \\ x_i &= \theta_i A_i k_i^{\alpha_i}, \quad A_i > 0, \quad 0 < \alpha_i < 1, \quad (i = 0, 1, 2), \end{aligned} \tag{1}$$

as well as three balance conditions:

$$\begin{aligned} &s_0+s_1+s_2=1, \quad s_0\geq 0, \quad s_1\geq 0, \quad s_2\geq 0,\\ &\theta_0+\theta_1+\theta_2=1, \quad \theta_0\geq 0, \quad \theta_1\geq 0, \quad \theta_2\geq 0,\\ &(1-\beta_0)x_0=\beta_1x_1+\beta_2x_2, \quad \beta_0\geq 0, \quad \beta_1\geq 0, \quad \beta_2\geq 0. \end{aligned} \tag{2}$$

Here the state of the system is described by the vector (k_0, k_1, k_2), and $(s_0, s_1, s_2, \theta_0, \theta_1, \theta_2)$ is the vector of control. The initial state of the system is (k_0^0, k_1^0, k_2^0), where $k_i^0 = k_i(0)$ – the capital-labour ratios of i-th sector ($i = 0$ – material, $i = 1$ – creation of funds, and $i = 2$ – production) at $t = 0$. We will consider the problem of transferring the system to the state (k_0^*, k_1^*, k_2^*) for the interval $[0, T]$. As the desired final state (k_0^*, k_1^*, k_2^*), we choose the equilibrium state of the system, which is determined by equating the right-hand sides of the differential equations (1) to zero, i.e.:

$$k_1^* = \left(\frac{s_1A_1}{\lambda_1}\right)^{\frac{1}{1-\alpha_1}}, \quad k_0^* = \frac{s_0\theta_1A_1(k_1^*)^{\alpha_1}}{\lambda_0\theta_0}, \quad k_2^* = \frac{s_2\theta_1A_1(k_1^*)^{\alpha_1}}{\lambda_2\theta_2}. \tag{3}$$

The values of capital-labour ratios $k_i^*(i = 0, 1, 2)$ in the steady state (3) depend on the controls $(s_0, s_1, s_2, \theta_0, \theta_1, \theta_2)$, for which the authors of [15] determined the values of $(s_0^*, s_1^*, s_2^*, \theta_0^*, \theta_1^*, \theta_2^*)$, solving the nonlinear programming in problem to maximise the specific consumption: $x_2 \to \max$.

In the state (k_0^*, k_1^*, k_2^*) (3), the right-hand sides of the differential equations (1) vanish, which means the constant in time of the values of capital-labour ratios $k_i^*(i = 0, 1, 2)$ are constant in the equilibrium state.

Using three balance relations (2), a problem with six controls $(s_0, s_1, s_2, \theta_0, \theta_1, \theta_2)$ can be reduced to a task with three controls, denoted later through (s_1, v_2, θ_1), using the controls:

$$s_0 = v_2(1 - s_1), \; s_2 = (1 - v_2)(1 - s_1).$$

We write the system of differential equations (1) in deviations with respect to the equilibrium state of the system using the following notation:

$$\begin{aligned} &y_1 = k_0 - k_0^*, y_2 = k_1 - k_1^*, y_3 = k_2 - k_2^*,\\ &u_1 = s_1 - s_1^*, u_2 = v_2 - v_2^*, u_3 = \theta_1 - \theta_1^* :\\ &\dot{y}_i = f_i(y, u), \quad y_i(0) = y_i^0, \quad (i = 1, 2, 3), \end{aligned} \tag{4}$$

Here, $y = (y_1, y_2, y_3)'$ denotes the state vector of the object, $u = (u_1, u_2, u_3)'$ denotes the control vector. Linearising the system (4), we obtain a vector differential equation of the form:

$$\dot{y}(t) = Ay(t) + Bu(t), \quad t \in [0, T], \tag{5}$$

where the elements of the matrices $A = \|a_{ij}\|_{3\times 3}$ and $B = \|b_{ij}\|_{3\times 3}$ are determined by the formulas:

$$a_{ij} = \frac{\partial f_i(y,u)}{\partial y_j}, \quad b_{ij} = \frac{\partial f_i(y,u)}{\partial u_j}, \quad (i,j = 1,2,3) \tag{6}$$

with $y = (0,0,0)'$ and $u = (0,0,0)'$.

It should be noted that the controllability criteria for nonlinear systems of the form (4) were obtained in the works of Klamka [16], and for discrete systems in [17]. The system (5) is controllable, i.e. matrices A and B satisfy the controllability criterion defined in [16].

The initial and final states of the system are given as:

$$y(0) = y_0, \quad y(T) = 0. \tag{7}$$

Note that the desired final state of the system $y(T) = 0$ is an equilibrium state in which the specific consumption is maximised and a balanced growth of the sectors of the economy is ensured.

The control vector components $u = (u_1, u_2, u_3)'$ satisfy two-way constraints of the following type:

$$-s_1^* \le u_1 \le 1 - s_1^*, \quad -v_2^* \le u_2 \le 1 - v_2^*, \quad -\theta_1^* \le u_3 \le 1 - \theta_1^*, \tag{8}$$

which are derived from the source constraints $0 \le s_1 \le 1,\ 0 \le v_2 \le 1,\ 0 \le \theta_1 \le 1$.

We consider the OCP: it is required to find the control $u(t)$, which takes the system (5) from the given initial state $y(0) = y^0$ to the equilibrium state $y(T) = 0$ for the interval $[0,\ T]$, while minimising the target functional:

$$J(u) = \frac{1}{2}\int_0^T [y'(t)Qy(t) + u'(t)Ru(t)]dt, \tag{9}$$

where Q and R are positive semidefinite and positive definite (3×3)- matrices, respectively.

Thus, we obtain the so-called LQ-problem (linear-quadratic OCP) (5)–(9), in which the ends of the trajectories of the system are fixed: $y(0) = y^0$, $y(T) = 0$, i.e. it is required to ensure the optimal path through specified start and end points.

3 Solution of the LQ OCP in the Presence of Constraints on the Values of Controls

Consider a control system described by a differential equation of the form:

$$\dot{x}(t) = A(t)x(t) + B(t)u(t) + f(t), \quad (t_0 \le t \le T), \tag{10}$$

with a given initial $x(t_0) = x_0$ and final $x(T) = 0$ states, with constraints on the values of the control:

$$u(t) \in U(t) = \{u|\alpha(t) \le u(t) \le \beta(t)\}, \quad (t_0 \le t \le T). \tag{11}$$

Here, $x(t)$ is the n-vector of the state of the object; $u(t)$ is the m-vector of piecewise-continuous control actions; $A(t)$, $B(t)$ - are matrices of dimensions $(n \times n)$, $(n \times m)$, respectively (the elements of these matrices are continuous functions); $f(t)$ is the n-vector of continuous functions; $\alpha(t)$, $\beta(t)$ are m-vectors whose components are piecewise-continuous functions; t_0 and T are predetermined initial and final moments of time. It is assumed that system (10) is completely controllable at time t_0.

The quality of management is described by the target functional:

$$J(u) = \int_{t_0}^{T} [0.5x'(t)Q(t)x(t) + x'(t)P(t)u(t) + 0.5u'(t)R(t)u(t) + x'(t)p(t) + u'(t)r(t) + \sigma(t)]dt, \tag{12}$$

where $Q(t)$, $P(t)$, $R(t)$ - are matrices of dimensions $(n \times n)$, $(n \times m)$, $(m \times m)$, respectively, $p(t)$, $r(t)$ - are vectors of dimensions $(n \times 1)$, $(m \times 1)$, respectively; $\sigma(t)$- is a scalar function; and a stroke means the transpose operation.

Statement of the problem: it is required to find a synthesising control $u = u(x, t)$ that satisfies the constraint (11) and transforms the system (10) from the given initial state x_0 to the final state x_T for a fixed interval $[t_0, T]$, while minimising the target functional (12).

In the case where the final condition $x(T) = 0$ is given in the form $x(T) = x_T$, we can make the substitution $\tilde{x}(t) = x(t) - x_T$, and again we obtain a problem of the form (10)–(12). In many practical problems, we have $P(\cdot) \equiv 0, p(\cdot) \equiv 0, r(\cdot) \equiv 0, \sigma(\cdot) \equiv 0$, but here we consider a functional of the form (12), since such a formulation of the problem may be required for the realization of some numerical solution algorithms of solving the OCP of a more general form.

We shall consider a symmetric $(n \times n)$-matrix $K(t)$ that satisfies the Riccati differential equation:

$$\dot{K}(t) = -\tilde{A}'(t)K(t) - K(t)\tilde{A}(t) + K(t)S(t)K(t) - \tilde{Q}(t), \quad K(T) = K_T, \tag{13}$$

where $\tilde{A}(t) = A(t) - B(t)R^{-1}(t)P'(t), S(t) = B(t)R^{-1}(t)B'(t), \tilde{Q}(t) = Q(t) - P(t)R^{-1}(t)P'(t)$.

Let us denote by $W(t, T)$ the symmetric $(n \times n)$-matrix of the form:

$$W(t, T) = \int_{t}^{T} \Phi(t, \tau)S(\tau)\Phi'(t, \tau)d\tau$$

Here $\Phi(t, \tau) = \Theta(t)\Theta^{-1}(\tau)$ is the matrix of dimension $(n \times n)$; $\Theta(t)$ - is the fundamental matrix of solutions of a differential equation of the form of a $\dot{y}(t) = \hat{A}(t)y(t)$, where $\hat{A}(t) = \tilde{A}(t) - S(t)K(t)$.

We shall consider the n-vector-valued function $q(t)$, which satisfies the differential equation:

$$\dot{q}(t) = -[\tilde{A}(t) - S(t)K(t)]'q(t) + W^{-1}(t,T)B(t)\varphi(x(t),t) - K(t)\tilde{f}(t) - \tilde{p}(t),\\ q(t_0) = q_0, \tag{14}$$

where

$$\lambda_1(x,t) = R(t)\max\{0; \alpha(t) - \omega(x,t)\},\quad \lambda_2(x,\ t) = R(t)\max\{0;\ \omega(x,t) - \beta(t)\},\\ \omega(x,t) = -R^{-1}(t)\{[K(t)B(t) + P(t)]'x + B'(t)q(t) + r(t)\},\quad \varphi(x,t) = R^{-1}(t)[\lambda_1(x,t) - \lambda_2(x,t)],\\ \tilde{f}(t) = f(t) - B(t)R^{-1}(t)r(t),\quad \tilde{p}(t) = p(t) - P(t)R^{-1}(t)r(t). \tag{15}$$

The solution of the considered OCP (10)–(12) using the above notation can be formulated as the following theorem.

Theorem 1. Let the matrix $R(t)$ with nonnegative elements be positive definite in the interval $t_0 \le t \le T$ and the matrix $\tilde{Q}(t)$ be nonnegative definite. In addition, assume that $W_0 = W(t_0, T) > 0$ and system (10) is completely controllable at time t_0. Then

1. the optimal motion trajectory of the system $x^*(t)$, $(t_0 \le t \le T)$ in problem (10)–(12) is determined from the differential equation:

$$\dot{x}(t) = [\tilde{A}(t) - S(t)K(t)]x(t) + B(t)\varphi(x(t),t) - S(t)q(t) + \tilde{f}(t),\quad x(t_0) = x_0, \tag{16}$$

   where matrix $K(t)$ and vector $q(t)$ satisfy the differential Eqs. (13) and (14), respectively;
2. the optimal control is:

$$u^*(x(t),t) = \omega(x^*(t),t) + \varphi(x^*(t),t),\quad (t_0 \le t \le T), \tag{17}$$

where the values of vector functions $\omega(x^*(t),t)$ and $\varphi(x^*(t),t)$ are calculated using formula (15).

Proof of Theorem 1 is obtained on the basis of sufficient optimality conditions for dynamical systems with fixed-end of trajectories, similar to those obtained in [14] using Lagrange multipliers of a special form.

For an arbitrary continuous function $x(\cdot)$ and an arbitrary piecewise continuous function $u(\cdot)$, defined in the interval $[t_0, T]$, the Lagrange functional is compiled:

$$L(x(\cdot),u(\cdot)) = \int_{t_0}^{T} \{0.5\,x'(t)Q(t)x(t) + x'(t)P(t)u(t) + 0.5\,u'(t)R(t)u(t)\\ + x'(t)p(t) + u'(t)r(t) + \sigma(t) + \lambda_0'(x(t),t)\,[A(t)x(t) + B(t)u(t) + f(t) - \dot{x}(t)]\\ + \lambda_1'(x(t),\ t)\,[\alpha(t) - u(t)] + \lambda_2'(x(t),\ t)\,[u(t) - \beta(t)]\}\,dt, \tag{18}$$

where λ_0, λ_1, λ_2 – multipliers of Lagrange, and $\lambda_1(x(t),\ t) \ge 0$, $\lambda_2(x(t),\ t) \ge 0$, $(t_0 \le t \le T)$; $\lambda_0(x(t),\ t)$ is given in the form $\lambda_0(x(t),\ t) = K(t)x(t) + q(t)$, $(t_0 \le t < T)$;

the $K(t)$ matrix and the $q(t)$ vector satisfy the differential Eqs. (13) and (14), respectively.

Choose the Lagrange multipliers $\lambda_1(x, t)$, $\lambda_2(x, t)$ in such a way that the so-called complementary nonrigidity conditions are satisfied:

$$\lambda_1'(x, t)\,[\alpha(t) - u] = 0, \quad \lambda_2'(x, t)\,[u - \beta(t)] = 0, \quad (\forall\, x \in E^n, \;\; t \in [t_0, T]). \quad (19)$$

For this, we choose $\lambda_1(x, t)$ and $\lambda_2(x, t)$ in the form (15).

4 Solution of the Problem for a Three-Sector Economic Model of a Cluster

As an example, let us consider a mathematical model of a three-sector cluster economy model described by differential equations of the form (4) in the interval $[t_0, T]$. The calculation of the elements of the matrices A and B by formulas (6) to reduce the system to the form (5) was performed on a computer using the possibilities of symbolic computations in the MatLab medium.

Note that getting the above cumbersome formulas without using the possibilities of symbolic computing on a computer would be very difficult.

The numerical calculations were performed on a computer with the different values of the parameters (see Table 1):

Table 1. Parameter values for a three-sector cluster economic model

i	α_i	β_i	λ_i	A_i	s_i^*	θ_i^*	k_i^*
0	0.46	0.39	0.05	6.19	0.2763	0.3944	966.4430
1	0.68	0.29	0.05	1.35	0.4476	0.2562	2410.1455
2	0.49	0.52	0.05	71	0.2761	0.3494	1090.1238

In the last line of Table 1, values of capital-labour ratios k_i^* $(i = 0, 1, 2)$ in the equilibrium state are calculated by the formula (3), in which the values of the controls $(s_0^*, s_1^*, s_2^*, \theta_0^*, \theta_1^*, \theta_2^*)$ are chosen from the condition of maximising the specific consumption. With these values of the model parameters, the following matrices were obtained:

$$A = \begin{pmatrix} -0.0345 & 0.0084 & -0.0063 \\ 0.0 & -0.0160 & 0.0 \\ -0.0197 & 0.0220 & -0.0420 \end{pmatrix}, \quad B = \begin{pmatrix} -87.4858 & 96.6194 & 156.5572 \\ 269.2298 & 0.0 & 0.0 \\ -98.6600 & -109.039 & 409.5059 \end{pmatrix}.$$

The matrices $K(t)$ and $R(t)$ are chosen in the form of matrices stationary in the $[t_0, T]$ interval: $K = 50E_3, \quad R = 10^7 \cdot E_3$.

Since the $K(t)$ matrix is stationary, its derivative with respect to time will be zero. Then it follows from the differential equation (13) that in $Q(t)$ it will also be a stationary matrix in the $[t_0,\ T]$ interval equal to $Q = -A'K - KA + KSK$, where $S = BR^{-1}B'$. Hence:

$$Q = \begin{pmatrix} 13.8274 & -6.3093 & 16.8514 \\ -6.3093 & 19.7212 & -7.7414 \\ 16.8514 & -7.7414 & 51.5278 \end{pmatrix}.$$

the initial state of the system is set to: $y(t_0) = y^0 = (-80, -560, -70)'$. It is required to transfer the system (12) to the equilibrium state: $y(T) = (0, 0, 0)'$ over the interval $[t_0,\ T] = [0,\ 10]$, while minimising the target functional (9).

The numerical calculations were carried out on a computer using the above algorithm for solving OCP (10)–(12), in which, with the help of Lagrange multipliers, it is possible to take into account constraints on the values of controls. In addition, the proposed method of solving the OCP allows us to represent the desired optimal control in the form of synthesising control. As an example, a three-sector economic model of a cluster (5), (7), (9) was considered, with additional constraints (8) on the values of control actions. For the example under consideration, these restrictions have the form:

$$-0.4476 \le u_1 \le 0.5524, \quad -0.5002 \le u_2 \le 0.4998, \quad -0.2562 \le u_3 \le 0.7438. \quad (20)$$

The obtained optimal trajectories and optimal controls are presented in Figs. 1 and 2. As can be seen from Fig. 1, the optimal controls found ensure that the trajectories of the system (5) are brought to the equilibrium state: $y(T) \approx (0,\ 0,\ 0)'$. The optimal controls found (see Fig. 2) do not go beyond the region U, defined by the constraints (20), where the control vector component $u_1(t)$ lies on the boundary of the U region in the interval $[0, t_1]$, then at $t \in (t_1, T]$ enters the interior of the region U. Control switching occurs at time $t_1 \approx 0.673$.

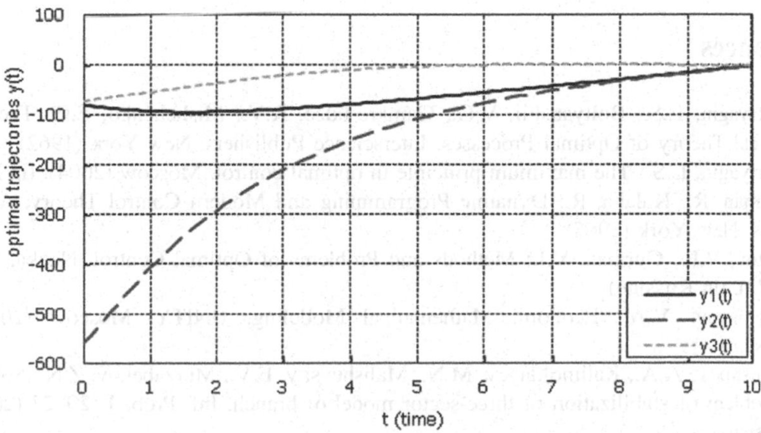

Fig. 1. Graphs of optimal trajectories using Lagrange multipliers

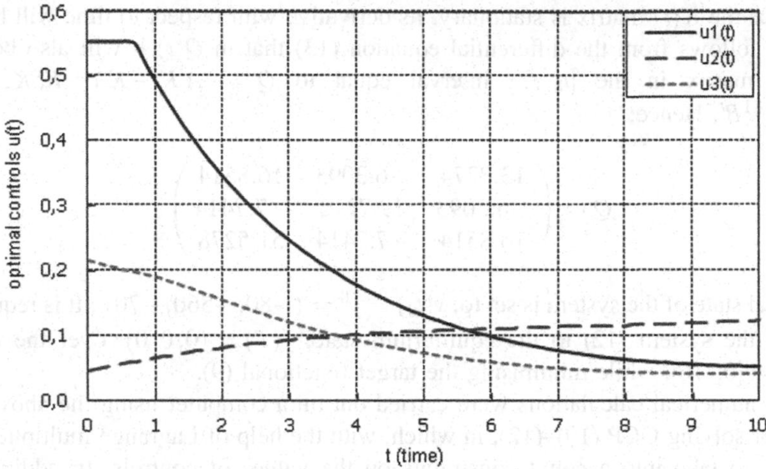

Fig. 2. Graphs of optimal controls using Lagrange multipliers (subject to management restrictions)

5 Conclusion

An algorithm for solving the problem posed was developed and a control was found on the basis of the feedback principle. The problem was solved using Lagrange multipliers of a special type. Numerical calculations were made using the above algorithm, in which Lagrange multipliers managed to take into account constraints on the values of controls. In addition, the proposed method of solving the OCP allows us to represent the desired optimal control in the form of synthesising control. As an example, a three-sector economic model of a cluster with additional restrictions on the values of control actions was considered. The control parameters were selected in such a way as to satisfy the algebraic conditions for the controls and the states of the system.

References

1. Pontryagin, L.S., Boltyanskii, V.G., Gamkre-lidze, R.V., Mishchenko, E.F.: The Mathematical Theory of Optimal Processes. Interscience Publishers, New York (1962)
2. Pontryagin, L.S.: The maximum principle in optimal control, Moscow (2004). (in Russian)
3. Bellman, R., Kalaba, R.: Dynamic Programming and Modern Control Theory. Academic Press, New York (1965)
4. Krotov, V.F., Gurman, V.I.: Methods and Problems of Optimal Control. Nauka, Moscow (1973). (in Russian)
5. Kolemayev, V.A.: Economic-Mathematical Modeling. UNITY, Moscow (2005). (in Russian)
6. Dzhusupov, A.A., Kalimoldayev, M.N., Malishevsky, E.V., Murzabekov, Z.N.: Solution of a problem of stabilization of three-sector model of branch. Inf. Prob. **1**, 20–27 (2011). (in Russian)

7. De, S.: Intangible capital and growth in the 'new economy': implications of a multi-sector endogenous growth model. Struct. Change Econ. Dynam. **28**, 25–42 (2014)
8. Dobrescu, L., Neamtu, M., Opris, D.: Deterministic and stochastic three-sector dynamic growth model with endogenous labour supply. Econ. Rec. **89**(284), 99–111 (2013)
9. Zhang, J.: The analytical solution of balanced growth of non-linear dynamic multi-sector economic model. Econ. Model. **28**(1), 410–421 (2011)
10. Zhou, S., Xue, M.: A model of optimal allocations of physical capital and human capital in three sectors. Wuhan Univ. J. Nat. Sci. **12**(6), 997–1002 (2007)
11. Sen, P.: Capital accumulation and convergence in a small open economy. Rev. Int. Econ. **21** (4), 690–704 (2013)
12. Aseev, S.M., Besov, K.O., Kryazhimskii, A.V.: Infinite-horizon optimal control problems in economics. Russ. Math. Surv. **67**(2), 195–253 (2012)
13. Shnurkov, P.V., Zasypko, V.V.: Analytical study of the problem of optimal investment management in a closed dynamic model of a three-sector economy. Bull. MSTU **2**, 101–115 (2014). (in Russian)
14. Aipanov, S., Murzabekov, Z.: Analytical solution of a linear quadratic optimal control problem with control value constraints. Comput. Syst. Sci. Int. **53**(1), 84–91 (2014)
15. Murzabekov, Z.N., Milosz, M., Tussupova, K.B.: Solution of steady state search problem in three-sector economic model of a cluster. Actual Probl. Econ. **3**(165), 443–452 (2015)
16. Klamka, J.: Constrained controllability of dynamics systems. Int. J. Appl. Math. Comput. Sci. **9**(2), 231–244 (1999)
17. Klamka, J.: Controllability of nonlinear discrete systems. Int. J. Appl. Math. Comput. Sci. **12** (2), 173–180 (2002)

Knowledge Representation of Cognitive Agents Processing the Economy Events

Marcin Hernes(✉) and Andrzej Bytniewski

Wrocław University of Economics,
ul. Komandorska 118/120, 53-345 Wrocław, Poland
{marcin.hernes, andrzej.bytniewski}@ue.wroc.pl

Abstract. Nowadays, companies use accounting IT systems in which the registration of economic events is most often performed with the use of relational or object databases. More often, it becomes necessary not only to register the value of economic events attributes but also to automatically analyze their meaning. These functions can be realized by using, among others, the cognitive agents. Their knowledge related to economic events shall be represented using semantic methods. Therefore the aim of this paper is to develop a semantic method for knowledge representation of cognitive agents processing economic events. This method will facilitate registration of attributes of economic events and the analysis of their meaning.

Keywords: Accounting IT systems · Economic events · Cognitive agents Semantic knowledge representation

1 Introduction

IT systems facilitate efficient functioning of accounting mainly by registering economic events (every event - fact, phenomenon- related to the business conducted by business organizations) and processing information in near-real time, which consequently has a positive effect on the effectiveness and efficiency of decision taking. However, it needs to be stressed that nowadays, companies use accounting IT systems in which the registration of economic events is most often performed with the use of relational or object databases. Analysis of meaning of these events by decision-makers is very time-consuming process. Nowadays, however, in turbulent economic environment business decisions should be taken near real time in order to achieve high efficiency of business organizations' functioning. Therefore, it becomes necessary not only to register by IT systems the values of economic events attributes but also to automatically analyze their meaning. It is important to interpret economic events in the context of supporting decisions, and realizing unexpected information management needs [1]. These functions can be realized by using, among others, the cognitive agents [2, 3]. More often the knowledge of such agents is represented by using semantic methods (e.g. ontologies, semantic nets). Therefore, economic events shall also be represented using semantic methods (enabling representation of semantically complex data) and saved in NoSQL database.

N. T. Nguyen et al. (Eds.): ACIIDS 2018, LNAI 10751, pp. 392–401, 2018.
https://doi.org/10.1007/978-3-319-75417-8_37

The aim of this paper is to develop a semantic method for knowledge representation of cognitive agents processing economic events. This method will facilitate registration of attributes of economic events and the analysis of their meaning.

The first part of the article contains a review of literature on methods for representing agents' knowledge and basic notions. The second part presents the developed method for representing such events. The last part of the article contains a case study which verifies the applicability of the developed method in practical solutions.

2 Related Works and Basic Notions

The literature of subject presents many different methods for agents' knowledge representation used in the mentioned groups. The main of them include first-order predicate logic, production systems, artificial neural networks, frame representation, ontologies such as semantic web, semantic networks and topic maps, multi-attributes and multi-values structures, multi valued logic includes a three valued logic and a fuzzy logic [e.g. 4, 5, 6]. However in relation to economic event representation semantic methods are mainly used in order to detection of these events [e.g. 7] but in small degree in order to automatically analysis of meaning of these events. The use of traditional, relational databases implies the method of representation of attributes of economic events in the form of atomic values. Atomization involves registering an economic event in an indivisible form. An atom consists of three attributes [7]: symbol of a phenomenon, value of a phenomenon, date of phenomenon.

An advantage of such an approach is the ease of processing and searching for data, however it allows for saving in a computer system only simple (atomic) values of phenomena, which is a simplification in modeling the real world in which attributes of economic phenomena are often complex. Additionally, such a representation of economic phenomena does not allow for the analysis of their meaning.

The application of object database enables representation of economic phenomena attributes with the use of complex values. An object is treated as pair [8]: Identification (unique in the whole system) and value.

Both data, as well as their processing procedures, are treated as the value of an object, which is an advantage of such a model. Another advantage of such a model is the possibility of processing incomplete data [9]. However, the analysis of economic phenomena in such a model is also limited.

The formal definition of an economic event, referring to relational and object-oriented databases, is presented, for example, in the works [7, 9–11]. It is presented as follows:

Definition 1.
An economic event is called a set:

$$ZG = \{z, c, p, s, l, d, t, w^q, w^v\} \tag{1}$$

where:

z – symbol of the event ($z \in Z$ – set of events),
c – symbol of the purpose (e.g. cost carrier, product, order; $c \in C$ – set of purposes),
p – symbol of the business entity (e.g. liability center, employee, contractor; $p \in P$ – set of entities),
s – type feature (catalog of types - materials, fixed assets, products; $s \in S$ – set of types),
l – individual catalog numbers (list of individual items: Inventory numbers of fixed assets, commodity-file catalog numbers; $l \in L$ – set of individual catalog numbers),
t – date of event ($t \in T$ – set of dates),
d – accounting document's number ($d \in D$ – set of document's numbers),
w^q – measurement value in natural units,
w^v – measurement value in value units.

The elements of the set representing the economic event are called the dimension of the event. Event representation as defined in Definition 1 may also be extended to include other dimensions depending on the level of detail of data necessary to realization a specific business process.

When an event is recorded in an accounting computer system, it is automatically categorized using the knowledge base and automatically recorded in the accounting books [7]. The definition of an economic event is thus extended to include information on accounting accounts as follows:

Definition 2.

$$ZG = \{ki, kj, z, c, p, s, l, d, t, w^q, w^v\} \quad (2)$$

where:

$ki, kj \in K$ – accounts DT, CT included in the plan of accounts.

3 Method for Knowledge Representation

Cognitive agent's knowledge representation related to economic event is based on semantic network and defined as follows:

Definition 3.

An economic event is called a quadruple:

$$ZG = \langle N, I, R, Z \rangle \quad (3)$$

where:

N – set of dimensions,
I – set of dimensions' instances,

R – set of relationships defined on the set N,
Z – set of axioms i.e. the set of relations between N and I.

The next part of the paper will outline the specific elements of an economic event.

It is assumed that the real world is represented by a pair <O, V> where O is a finite set of objects (e.g., contractor-entity, generic-commodity), and V is the domain of set O (e.g. Alpha-contractor), and

$$V = \cup_{o \in O} V_o \tag{4}$$

where: V_o is the domain of object $o \in O$.

It is assumed that an economic event refers to the real world <O, V>. A detailed definition of an economic event is considered at four levels: dimensions, instances, relationships and axioms.

3.1 Definition on the Dimensions Level

Definition 4.
The dimension of the economic event defined in the world <O, V> is defined as triple:

$$n = \langle idn, O^n, V^n \rangle \tag{5}$$

where *idn* is a unique dimension name (e.g. commodity, date), $O^n \in O$ is the object represented by the dimension (e.g. contractor-entity, generic-commodity), and $V^n \in V$ is the domain of object (e.g. Alfa, Beer):

$$V^n = \cup_{\langle o,v \rangle \in O^n} V_o. \tag{6}$$

Tuple $\langle O^n, V^n \rangle$ is called a structure of dimension n. All dimensions belonging to the same event differ from one another.

Each economic event should contain at least the dimensions specified in Definition 1 or 2, and may additionally include other dimensions (e.g. in the case of products - dimension "packaging").

3.2 Definition on Instance Level

Definition 5.
The n-dimension instance is described by objects from a set O^n and values from the set V^n and defined as a pair:

$$\mathrm{i} = \langle idi, v \rangle \tag{7}$$

where *idi* is a unique identifier of an instance in a real word <O, V>, instead v is the value of the object instance (np. Beer-Heineken) O^n determined as function: $v : O^n \rightarrow V^n$, such, as $v(o) \in V_o$ for each $o \in O^n$.

Value v is also called a description of the instance of the object. Dimension can be interpreted as a set of all instances described in its structure.

In order to describe the fact that i is an instance of dimension n, following notation can be use: $i \in n$. All instances of the same dimension within a given business event should have a different identifier. Two or more instances may have the same values. The given instance can belong to different dimensions.

3.3 Definition on Relationships Level

In an economic event between two dimensions you can define one or more relationships. Relationships describe the relationship between dimensions. For example, a "whole-part" relationship can be defined between two dimensions. Relationship is defined as follows:

Definition 6.
Let set of dimensions N is given. The relationship is called the following relation:

$$R = \langle n, n' \rangle \tag{8}$$

where $n, n' \in N$.

3.4 Definition on Axioms Level

The set Z can be interpreted as a condition of data integrity or relationships between instances and dimensions that cannot be expressed in a set of relations R.

Definition 7.
Let set of dimensions N and set of instances I are given. The axiom is called the following relations:

$$Z = \langle N, I \rangle \tag{9}$$

in dimension: $N \times I$
and

$$Z = \left\langle I, I' \right\rangle \tag{10}$$

in dimension: $I \times I$.

3.5 Graphical Representation of the Economic Event

The developed definition of economic event can be presented graphically. An example of such representation is illustrated in Fig. 1. Oval denotes dimensions, rectangles denote instances, relationships are drawn with a solid line, and dashed lines indicate axioms. The presented economic event contains three dimensions. The interpretation is as follows: Dimension1 is related to Dimension2 and Dimension3. Dimension1 is connected to Instance1 via an axiom. Dimension1 is also related to Dimension3 by axiom, while Dimension3 is also related to Instance3.

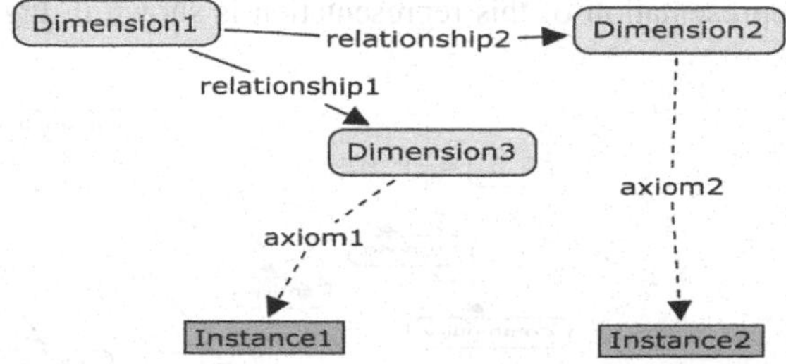

Fig. 1. An example of a graphic representation of a semantic representation of an economic event

4 Verification of Developed Method – Case Study

In order to verify the method developed, the following assumptions were made:

1. The accounting information system is made up of cognitive agents in the LIDA architecture [13]. The knowledge of these agents is represented by a semantic network and they have built-in mechanisms for processing such knowledge.
2. User experience tests will be performed (3 expert persons and 3 non-expert persons will be employed in tests, they task will be to analysis of meaning of 5 different economic events representing by traditional methods and developed method – the measure of tests will be a time of performing analysis).
3. In order to illustrate analyzing process the following exemplary economic event can be considered: "On 30-10-2017, shipping note (Sn) No. 1/2016 for the Alpha-contractor was issued, containing the following items: Wine1 20 pcs, price 3.20 (box: box) and Wine2 6 pcs., price 3,30 (packaging: film)".

Representation of the exemplary event in semantic form is as follows:

1. Dimensions: Issuing_an_document, Date, Purpose, Commodity, Package, Wine, Entity, Sn, Number, Value, Natural Units, Pcs, Value Units, USD.
2. Instances: 30-10-2017, 1/2017, Alpha-Contractor, Wine1, Wine2, Film, Box, 20; 6; 3,20; 3,30; 730; 330.
3. Relationships: Issuing_an_document→Date, Issuing_an_document→Purpose, Issuing_an_document→Entity, Issuing_an_document→Cn, Issuing_an_document→Value, Purpose→Commodity, Commodity →Wine, Commodity →Package, Cn→Number, Value→Natural_units, Value→Value_units, Natural_units→Pcs, Value_units→USD,
4. Axioms: Date→30-10-2017, Package→Film, Package→Box, Wine→Wine1, Wine→Wine2, Entity→Alpha_contractor, Number→1/2017, Pcs.→20, Pcs.→6, USD→3,20, USD→3,30, Wine2→Film, Wine2→6, Wine2→3,30, Wine1→Box, Wine1→20, Wine1→3,20.

The graphic representation of this representation is shown in Fig. 2.

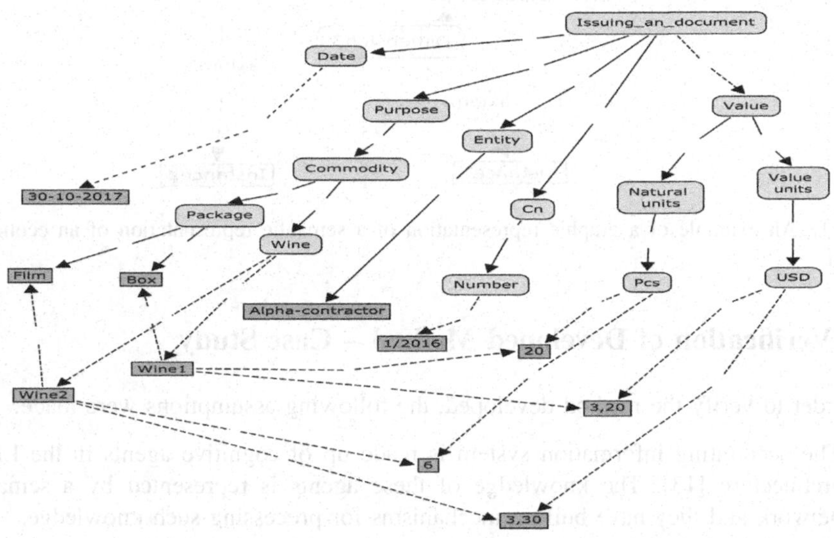

Fig. 2. Graphical form of the considered economic event's semantic representation.

When an event is logged on the system, cognitive agent programs automatically classify the economic event using a knowledge base that contains the event accounting patterns that are also stored as semantic networks. An example of such pattern (graphical form) with reference to the Cn document is shown in Fig. 3.

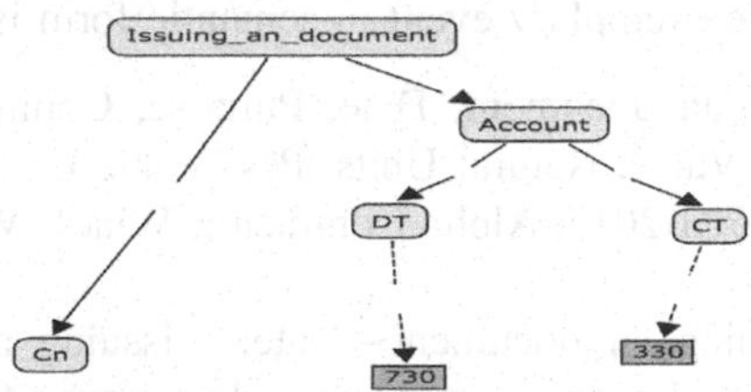

Fig. 3. A Cn document accounting pattern in the knowledge base.

Classification is made using comparative semantic networks, which are described in greater detail in the work [12].

As a result of the classification, the economic event is extended by the following dimensions: Account, DT, CT, instances: 730, 330, relationships: Issuing_an_document→Account, Account→DT, Account→CT, as well as axioms: DT→730, CT→330. Ultimately, the representation (graphical form) of the considered economic event is presented on Fig. 4. In classic IT accounting systems, booking a Cn document would be made on an analytical account (330-100-200, where that segment 100 means

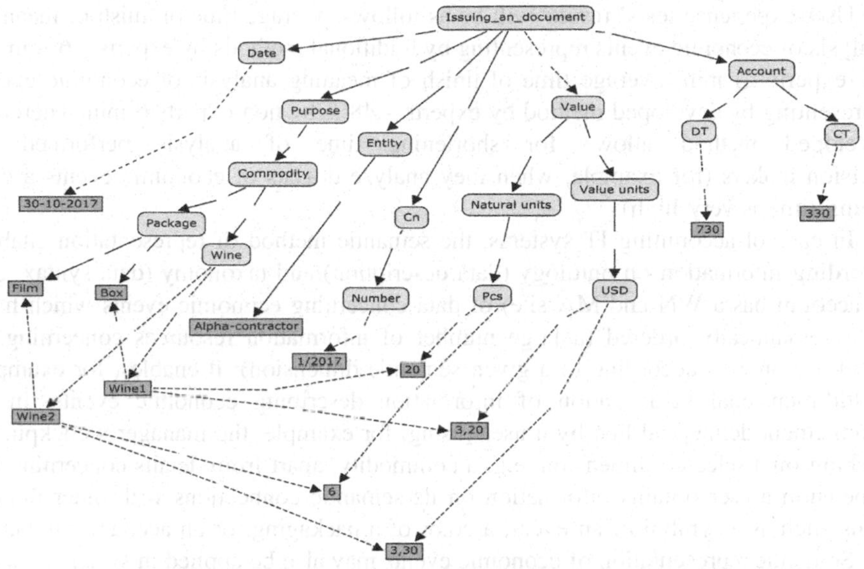

Fig. 4. Representation of the economic events after the process of its classification

a commodity, and 200 - Wine). Instead, by using the semantic representation, the analytical part of the account is expressed directly by the dimensions of the Commodities and Wine (330-Commodity-Wine).

As this form of representation is close to the representation of reality in a human mind, it is possible to determine the meaning, context of a given economic event in a natural way. There is no need to perform any complicated analyses using time and resource consuming methods of data exploration. Events represented with the use of the semantic method can be directly visualized, e.g. in the form of a manager's cockpit. The semantic method is characterized by total openness. The process of determining dimensions, instances, relationships and axioms is not limited by the structure of a database. For example, if there was a need to record the color of a packaging, it would require just adding a new dimension: Color (and relationship it with the dimension: Packaging), and instances: Red, Green (and relationship it using Color dimension axiom with the instances, and join using the axiom e.g. instance: Box with instance: Green).

The developed method enables saving economic events in NoSql databases, mainly graph ones (data is saved in the form of directed graphs, making it possible to model a semantic network). The advantages of the semantic method of representation include:

- flexibility of data model,
- possibility of modifying the structure of data while a system is being started (in runtime) or used by workers,
- effective processing of queries on semantically complex data,
- speed of finding data paths,
- possibility of representing rich data semantics in a graphic form.

User experience tests' results will be as follows: average time of finish of meaning analysis of economic events representing by traditional methods by experts – 6 min, by non-experts 21 min; average time of finish of meaning analysis of economic events representing by developed method by experts – 48 s, by non-experts 6 min. Therefore developed method allows for shortening time of analysis performed by decision-makers (for example, when they analyze dozens of economic events a day, saving time is very high).

In case of accounting IT systems, the semantic method of representation enables recording information on ontology (data description) and taxonomy (data syntax, e.g. an account has a WN and MA site) of data concerning economic events which have been semantically ordered (a large number of information resources concerning an event are ordered according to a given semantic dimension). It enables, for example, multidimensional visualization of information describing economic events, in an arrangement defined ad hoc by a user (using, for example, the manager's cockpit, by clicking on a selected dimension, e.g. a commodity, apart from details concerning the dimension a user obtains information on its semantic connections with other dimensions such as a symbol of an event, a color of a packaging, or an account number).

Semantic representation of economic events may also be applied in systems related to other areas of company operation, e.g. in a controlling system in realization of the function of planning and monitoring performance of a plan. Since planning results are determined with a certain degree of probability, in a controlling subsystem it is possible to additionally use the semantic network with levels of activation of dimensions (e.g. Issuing_an_document, Commodity, Beer) and relationships.

The main limitation of the method is the complexity of a data model, which is why an application of such a method requires using hardware with parameters which guarantee better efficiency than in case of applying the classic method.

5 Conclusions

Representation of knowledge of cognitive agents performing economic events based on the semantic network enables not only processing information connected with these events, but also automatically analysis of their meaning. The method described in the paper may be applied e.g. to represent knowledge of agents in a financial and accounting subsystem of a cognitive integrated management information system [2]. Events represented in such a manner may be saved in a NoSQL database. It may leads, in consequence, to taking faster and more effective decision by decisions-makers.

Further research is going to focus on developing methods for analysis, by cognitive agents, the meaning of economic events, and on developing a prototype of the financial and accounting subsystem using semantic method.

References

1. Owoc, M.L., Weichbroth, P., Zuralski, K.: Towards better understanding of context-aware knowledge transformation. In: Proceedings of FedCSIS 2017, pp. 1123–1126 (2017)
2. Hernes, M.: A cognitive integrated management support system for enterprises. In: Hwang, D., Jung, J.J., Nguyen, N.-T. (eds.) ICCCI 2014. LNCS (LNAI), vol. 8733, pp. 252–261. Springer, Cham (2014). https://doi.org/10.1007/978-3-319-11289-3_26
3. Bytniewski, A., Chojnacka-Komorowska, A., Hernes, M., Matouk, K.: The implementation of the perceptual memory of cognitive agents in integrated management information system. In: Barbucha, D., Nguyen, N.T., Batubara, J. (eds.) New Trends in Intelligent Information and Database Systems. SCI, vol. 598, pp. 281–290. Springer, Cham (2015). https://doi.org/10.1007/978-3-319-16211-9_29
4. Kadhim, M.A., Alam, A., Harleen, M.K.: A multi-intelligent agent architecture for knowledge extraction: novel approaches for automatic production rules extraction. Int. J. Multimedia Ubiquitous Eng. **9**(2), 95–114 (2014)
5. Badawy, O., Almotwaly, A.: Combining neural network knowledge in a mobile collaborating multi-agent system. In: International Conference on Electrical, Electronic and Computer Engineering, ICEEC 2004, pp. 325–328 (2004)
6. Nguyen, N.T.: Processing inconsistency of knowledge in determining knowledge of collective. Cybern. Syst.: Int. J. **40**(8), 670–688 (2009)
7. Hogenboom, A., Hogenboom, F., Frasincar, F., Schouten, K., Meer, O.: Semantics-based information extraction for detecting economic events. Multimedia Tools Appl. **64**(1), 27–52 (2013)
8. Bytniewski, A.: Automation of the accounting system as a way to support management accounting and controlling. In: Nowak, E., Nieplowicz, M. (eds.) Research Papers of Wrocław University of Economics, Accounting and controlling, Wrocław, no. 251, pp. 81–95 (2012). (in Polish)
9. Du, H., Wang, T.J.: A mechanism for converting a relational database into an object-oriented model: an AIS application. Rev. Bus. Inf. Syst. **9**(1), 55–67 (2005)
10. Shawhan, Y.: Data modelling and accounting information system. Rev. Bus. Inf. Syst. **10**(3), 23–28 (2006)
11. Król-Stępień, M.: Accounting information system as a tool to assist enterprise management - statutory requirements and their practical application. Finanse, Rynki Finansowe, Ubezpieczenia, Uniwersytet Szczeciński **58**, 75–81 (2013)
12. Hernes, M.: Performance evaluation of the customer relationship management agent's in a cognitive integrated management support system. In: Nguyen, N.T. (ed.) Transactions on Computational Collective Intelligence XVIII. LNCS, vol. 9240, pp. 86–104. Springer, Heidelberg (2015). https://doi.org/10.1007/978-3-662-48145-5_5
13. Franklin, S., Patterson, F.G.: The LIDA architecture: adding new modes of learning to an intelligent, autonomous, software agent. In: Proceedings of the International Conference on Integrated Design and Process Technology. Society for Design and Process Science, San Diego, CA (2006)

Collinearity Models in the Eigenvalue Problem

Leon Bobrowski(✉)

Faculty of Computer Science, Białystok University of Technology, Białystok, Poland
l.bobrowski@pb.edu.pl

Abstract. Solution of the eigenvalue problems can be based on inverting matrices built from regularized vectors. The regularization parameters are equal to the eigenvalues of the given matrix after fitting in accordance to the collinearity models. In this approach the eigenvectors are equal to the columns of the inverted matrix.

A new procedure of matrix inversion with the basis exchange algorithm has been recently proposed. In accordance with this procedure the inverse matrix is computed in an iterative manner by gradual replacement of unit vectors by successive regularized vectors. Such replacement is impossible if a new regularized vector depends linearly on the vectors which are already in the basis.

Keywords: Eigenvalue problem · Iterative inversion of matrices
Fitting eigenvalues · Collinearity models

1 Introduction

Computational techniques of the eigenvalue problem solution are used for exploring large data sets [1]. One of the most successful and widely used methods of data exploration is the principal component analysis (PCA) [2]. The PCA procedure involves computation of the eigenvalues and eigenvectors of the covariance matrix. The covariance matrix is symmetrical and can be efficiently computed on the basis of large data sets. The eigenvalue computations are also used in other methods of data exploration [3].

A new method of the eigenvalue problem solution has been recently proposed [4]. In accordance with this method eigenvalues are selected through inducing linear dependency between regularized vectors. This method uses the iterative inversion of matrices composed from regularized vectors [5]. The inverse matrix is obtained from a unit matrix in an iterative manner through successive replacement of unit vectors by regularized vectors. As a result, the transformed column of the inverted matrix becomes gradually equal to the desired eigenvector.

The replacement of a successive unit vector by the regularized vector becomes impossible if the new vector depends linearly on regularized vectors which are already in the basis. Collinearity models formulated in the paper allow to specify conditions when the above replacement cannot be executed. A procedure based on collinearity models aimed at finding eigenvalues is proposed in the paper.

N. T. Nguyen et al. (Eds.): ACIIDS 2018, LNAI 10751, pp. 402–409, 2018.
https://doi.org/10.1007/978-3-319-75417-8_38

2 Principal Component Analysis and Eigenvalue Problem

Let us consider a data set $\boldsymbol{C}$ of m feature vectors $\mathbf{x}_j = [x_{j,1}, \ldots, x_{j,n}]^T$ [3]:

$$C = \{\mathbf{x}_1, \ldots, \mathbf{x}_m\} \tag{1}$$

where the feature vectors $\mathbf{x}_j$ ($j = 1, \ldots., m$) can be considered as points in the n-dimensional feature space $F[n]$ ($\mathbf{x}_j \in F[n]$).

We assume that m objects O_j have been represented in a standardazied manner by the n - dimensional feature vectors $\mathbf{x}_j$ ($j = 1, \ldots., m$). Components $x_{j,i}$ of the feature vector $\mathbf{x}_j$ may be numerical results of measurements of n different features x_i ($i = 1, \ldots, n$) of the j - th object O_j ($x_{j,i} \in \{0,1\}$ or $x_{j,i} \in R^1$).

The data set $\boldsymbol{C}$ (1) can be described by the mean vector $\mathbf{m}$ and the covariance matrix:

$$\mathrm{m} = \sum\nolimits_j \mathrm{x_j}/m \tag{2}$$

and

$$S = \sum\nolimits_j (\mathrm{x_j} - \mathrm{m})(\mathrm{x_j} - \mathrm{m})^T/(m-1) \tag{3}$$

The *PCA* method can reduce the dimensionality of multivariate data $\boldsymbol{C}$ (1) whilst preserving as much of the variance as possible. The basic idea of the *PCA* method is to describe the variation of a set of multivariate data in terms of a set of new, uncorrelated variables $z_{i'}$ each of which is a particular linear combination of the n features x_i. The linear transformation of the feature vectors $\mathbf{x}_j$ (1) used in the *PCA* gives such new variables $z_{i'}$ which have the greatest variability [2].

The *PCA* method is based on the solution of the eigenvalue problem with the symmetric covariance matrix $\boldsymbol{S}$ (3) of the dimension n x n [1]:

$$S\,\mathrm{k_i} = \lambda_i \mathrm{k_i} \tag{4}$$

where $\mathbf{k}_i = [k_{i,1}, \ldots, k_{i,n}]^T$ is the i-th eigenvector ($i = 1, \ldots, n$) and λ_i is the i-th eigenvalue ($\lambda_i \geq 0$).

The eigenvectors $\mathbf{k}_i$ should have the unit length:

$$(\forall i \in \{1, \ldots, n\}) \quad \mathrm{k_i^T k_i} = 1 \tag{5}$$

Two eigenvectors $\mathbf{k}_i$ and $\mathbf{k}_{i'}$ (4) of the symmetrical matrix $\boldsymbol{S}$ (3) are perpendicular if their eigenvalues λ_i and $\lambda_{i'}$ are different ($\lambda_i \neq \lambda_{i'}$) [2]:

$$(\forall i \neq i') \quad \mathit{if}\, \lambda_i \neq \lambda_{i'},\ \mathit{then}\, k_i^T \mathrm{k_{i'}} = 0 \tag{6}$$

The eigenvectors $\mathbf{k}_i$ are typically set in descending order of the eigenvalues λ_i:

$$\mathbf{k}_1, \mathbf{k}_2, \ldots, \mathbf{k}_n$$
$$\lambda_1 \geq \lambda_2 \geq \ldots \geq \lambda_n \geq 0 \tag{7}$$

The eigenvector $\mathbf{k}_i$ allows to define the *i*-th principal component z_i:

$$(\forall i \in \{1, \ldots, n\})\, z_i = \mathbf{k}_i^T \mathbf{x} = k_{i,1}x_1 + \ldots + k_{i,n}x_n \tag{8}$$

The eigenvalues λ_i and the eigenvectors $\mathbf{k}_i$ are computed on the basis of the Eq. (4). The Eq. (4) can be provided in the below form:

$$(S - \lambda_i I)\mathbf{k}_i = 0 \tag{9}$$

The determinant $| \boldsymbol{S} - \lambda_i \boldsymbol{I} |$ equal to zero is the necessary condition for a non-trivial solution $\mathbf{k}_i \neq \mathbf{0}$ of the Eq. (9) [1]:

$$|S - \lambda_i I| = 0 \tag{10}$$

The *characteristic* Eq. (10) is solved in order to find the matrix's $\boldsymbol{S}$ eigenvalues λ_i. The eigenvectors $\mathbf{k}_i$ (7) can be directly computed based on the eigenvalues λ_i [1].

The Singular Value Decomposition (*SVD*) is currently the primary method of the eigenvalue problem solution used for the Principal Components Analysis [3].

3 Iterative Matrix Inversion

Let us name the columns of the matrix $S = [s_1, \ldots, s_n]$ (3) as vectors $s_j = [s_{j,1}, \ldots, s_{j,n}]^T$ ($s_j \in R^n$) regularized in the below manner:

$$(\forall j \in \{1, \ldots, n\}) \quad z_j = s_j - \lambda e_j \tag{11}$$

where $\lambda \in R^1$ and e_j is the *j*-th unit vector.

The below matrix S_λ can be defined for an arbitrary value of the *regularizing parameter* λ ($\lambda \in R^1$):

$$S_\lambda = [s_1 - \lambda e_1, s_2 - \lambda e_2, \ldots, s_n - \lambda e_n] = [z_1, z_2, \ldots, z_n] \tag{12}$$

Let us assume that the matrix S_λ (12) was created in an iterative manner from the unit matrix $I = [e_1, \ldots, e_n]$ [5]. During the first step ($k = 1$) the matrix $S_\lambda(1) = [z_1, e_2, \ldots, e_n]^T$ was created by changing the vector e_1 to z_1. Similarly, in the step k, the matrix $S_\lambda(k)$ was created by changing the vector e_k to $z_k = s_k - \lambda\, e_k$ (11):

$$(\forall k = 1, \ldots, n - 1) \quad S_\lambda(k) = [z_1, \ldots, z_k, e_{k+1}, \ldots, e_n]^T \tag{13}$$

We can assume without additional restrictions that during the first $n - 1$ steps k the matrices $S_\lambda(k)$ (12) are non-singular and the inverse matrices $S_\lambda(k)^{-1}$ can be computed for a given value of the parameter λ (13):

$$(\forall k = 1,\ldots,n-1)\ \ S_\lambda(k)^{-1} = [\mathrm{r}_1(k),\ldots,\mathrm{r}_n(k)] \tag{14}$$

The vectors $\mathbf{z}_j$ (11) and the columns $\mathbf{r}_j(k)$ of the inverse matrix $S_\lambda^{-1}(k)$ (13) fulfil the below equations:

$$\begin{aligned}(\forall j \in \{1,\ldots,k\}) \quad & \mathbf{r}_j(k)^T\mathbf{z}_j = \mathbf{r}_j(k)^T(\mathbf{s}_j - \lambda\mathbf{e}_j) = 1, \textit{and} \\ (\forall j' \in \{1,\ldots,j-1\}) \quad & \mathbf{r}_j(k)^T\mathbf{z}_{j'} = \mathbf{r}_j(k)^T(\mathbf{s}_{j'} - \lambda\mathbf{e}_{j'}) = 0\end{aligned} \tag{15}$$

The eigenvalue Eq. (9) can be represented similarly to (15) by using the regularized vectors $z_j = s_j - \lambda\ e_j$ (11):

$$\begin{gathered}(\forall j \subset \{1,\ldots,n\}),\ \textit{and}(\forall j' \in \{1,\ldots,n\}) \\ \mathrm{r}_j(k)^T z_{j'} = \mathrm{r}_j(k)^T(\mathrm{s}_{j'} - \lambda_j \mathrm{e}_{j'}) = 0\end{gathered} \tag{16}$$

We can see that the major difference between Eqs. (15) and (16) is that all products $r_j(k)^T(s_{j'} - \lambda\ e_{j'})$ are equal to zero in the Eq. (16).

If during the successive step $k + 1$ the vector e_{k+1} is replaced by z_{k+1} (11) then the columns $r_j(k + 1)$ of the inverse matrix $S_\lambda(k + 1)^{-1}$ (14) can be computed based on the columns $r_j(k)$ in accordance with the Gauss - Jordan transformation [8]:

$$\mathbf{r}_{k+1}(k+1) = (1/\mathbf{r}_k(k)^T\mathbf{z}_{k+1})\,\mathbf{r}_k(k) \tag{17}$$

and

$$\begin{aligned}(\forall j \neq (k+1)) \quad \mathbf{r}_i(k+1) &= \mathbf{r}_j(k) - (\mathbf{r}_j(k)^T\mathbf{z}_{k+1})\,\mathbf{r}_{k+1}(k+1) = \\ &= r_j(k) - (\mathbf{r}_j(k)^T\mathbf{z}_{k+1}/\mathbf{r}_k(k)^T\mathbf{z}_{k+1})\,\mathbf{r}_k(k)\end{aligned} \tag{18}$$

The basis exchange algorithms were based on the Gauss-Jordan transformation (17), (18) [8]. The inverse matrix $S_\lambda(k)^{-1}$ can be computed efficiently in successive steps k by using the basis exchange algorithm even in case of high-dimensional vectors.

4 Eigenvectors Obtained from Inverted Matrices

Let us suppose that during the n-th step we try to replace the unit vector $\mathbf{e}_n$ in the non-singular matrix $S_\lambda(n - 1) = [\mathbf{z}_1, \ldots, \mathbf{z}_{n-1}, \mathbf{e}_n]^T$ (13) by the regularized vector $\mathbf{z}_n = \mathbf{s}_n - \lambda\ \mathbf{e}_n$ (11). The matrix $S_\lambda(n) = [\mathbf{z}_1, \ldots, \mathbf{z}_{n-1}, \mathbf{z}_n]^T$ will become singular if the n-th regularized vector $\mathbf{z}_n = \mathbf{s}_n - \lambda\ \mathbf{e}_n$ (13) is linear combination of the remaining regularized vectors $\mathbf{z}_j = \mathbf{s}_j - \lambda\ \mathbf{e}_j$ ($j = 1, \ldots, n - 1$) [6]:

$$\mathrm{s}_n - \lambda\,\mathbf{e}_n = \alpha_1(\mathrm{s}_1 - \lambda\,\mathbf{e}_1) + \ldots + \alpha_{n-1}(\mathrm{s}_{n-1} - \lambda\,\mathbf{e}_{n-1}) \tag{19}$$

where $(\forall j \in \{1, \ldots, n - 1\})\ \alpha_j \in R^1$.

Lemma 1: If the matrix $S_\lambda(n-1)^{-1}$ (16) is non-singular and the regularized vector $\mathbf{z}_n = \mathbf{s}_n - \lambda\,\mathbf{e}_n$ (13) is a linear combination (19) of the remaining regularized vectors $\mathbf{z}_j = \mathbf{s}_j - \lambda\,\mathbf{e}_j$ ($j \neq n$), then

$$\mathbf{r}_n(n-1)^T(\mathbf{s}_n - \lambda\,\mathbf{e}_n) = 0 \tag{20}$$

where $\mathbf{r}_n(n-1)$ is the n-th column of the inverse matrix $S_\lambda(n-1)^{-1}$ (16).

Proof: The n-th column $\mathbf{r}_n(n-1)$ of the inverse matrix $S_\lambda(n-1)^{-1}$ (14) fulfills the Eq. (15). Therefore:

$$(\forall j \in \{1, \ldots, n-1\})\;\; \mathbf{r}_n(n-1)^T(\mathbf{s}_j - \lambda\,\mathbf{e}_j) = 0 \tag{21}$$

Based on this and the assumption (19) we can infer that the Eq. (20) holds. □

The Eqs. (20) and (21) can be aggregated in the below manner:

$$(\forall j \in \{1, \ldots, n\})\;\; \mathbf{r}_n(n-1)^T(\mathbf{s}_j - \lambda\mathbf{e}_j) = 0 \tag{22}$$

The Eq. (22) means that $\mathbf{k}_n = \mathbf{r}_n(n-1)\,/\|\,\mathbf{r}_n(n-1)\,\|$ (5) is the eigenvector of the matrix $S = [\mathbf{s}_1, \ldots, \mathbf{s}_n]$ with the eigenvalue λ.

The *Lemma* 1 can be generalized to a case when the l-th vector $\mathbf{z}_l = \mathbf{s}_l - \lambda_l\,\mathbf{e}_l$ (11) with the *regularizing parameter* λ_l is a linear combination (19) of the remaining vectors $\mathbf{z}_j = \mathbf{s}_j - \lambda_l\,\mathbf{e}_j$ ($j \neq l$):

$$\begin{aligned}&(\forall l \in \{1, \ldots, n\})\\ &\mathbf{s}_l - \lambda_l\mathbf{e}_l = \alpha_{l,1}(\mathbf{s}_1 - \lambda_l\mathbf{e}_1) + \ldots + \alpha_{l,n}\alpha_n(\mathbf{s}_n - \lambda_l\mathbf{e}_n)\end{aligned} \tag{23}$$

where $(\forall j \in \{1, \ldots, n\})\; \alpha_{l,j} \in R^1$, and $\alpha_{l,l} = 0$.

The Eq. (23) can be formulated in the below manner:

$$\begin{aligned}&(\forall l \in \{1, \ldots, n\})\\ &\mathbf{s}_l - \alpha_{l,1}\mathbf{s}_1 - \ldots. - \alpha_{l,n}\mathbf{s}_n = \lambda_l(\mathbf{e}_l - \alpha_{l,1}\mathbf{e}_1 - \ldots - \alpha_{l,n}\mathbf{e}_n)\end{aligned} \tag{24}$$

where $(\forall j \in \{1, \ldots, n\})\; \alpha_{l,j} \in R^1$, and $\alpha_{l,l} = 0$.

If the Eq. (24) holds for the l-th regularized vector $\mathbf{z}_j = \mathbf{s}_j - \lambda_l\,\mathbf{e}_j$ (11), then the matrix $S_\lambda = [\mathbf{s}_1 - \lambda_l\mathbf{e}_1, \mathbf{s}_2 - \lambda_l\mathbf{e}_2, \ldots, \mathbf{s}_n - \lambda_l\mathbf{e}_n]$ (14) is singular. In this case, the equation similar to (22) has the below form:

$$\mathbf{r}_l^T(\mathbf{s}_l - \lambda_l\mathbf{e}_l) = 0 \tag{25}$$

where $\mathbf{r}_l$ is the l-th column of the inverse matrix $S_\lambda(l)^{-1} = [\mathbf{r}_1, \ldots, \mathbf{r}_n]$ (16), where:

$$S_\lambda(l) = [\mathbf{s}_1 - \lambda_l\mathbf{e}_1, \ldots, \mathbf{s}_{l-1} - \lambda_l\mathbf{e}_{l-1}, \mathbf{e}_l, \mathbf{s}_{l+1} - \lambda_l\mathbf{e}_{l+1}, \ldots, \mathbf{s}_n - \lambda_l\mathbf{e}_n] \tag{26}$$

Lemma 2: If the l-th column $\mathbf{r}_l$ of the inverse matrix $S_\lambda(l)^{-1}$ (26) fulfills the Eq. (25), then $\mathbf{k}_l = \mathbf{r}_l/\| \mathbf{r}_l \|$ (5) is the eigenvector of the matrix $S = [\mathbf{s}_1, \ldots, \mathbf{s}_n]$ linked to the eigenvalue λ_l (25) [6].

5 Induced Collinearity of Regularized Vectors

The linear dependence (23) of the regularized vector $\mathbf{z}_l = \mathbf{s}_l - \lambda_l\, \mathbf{e}_l$ (11) on the remaining $n - 1$ vectors $\mathbf{s}_j - \lambda_l\, \mathbf{e}_j$ is ($j \neq l$) can be *induced* by an adequate choice of the *regularizing parameter* λ_l. The l-th collinearity model (24) can be formulated as a set of n non-linear equations [6]:

$$\lambda_l = s_{l,l} - \alpha_{l,1}s_{l,i} - \ldots - \alpha_{l,n}s_{l,n} \tag{27}$$

And

$$\begin{aligned}
&s_{l,1} - \alpha_{l,1}s_{1,1} - \ldots - \alpha_{l,n}s_{n,1} = -\lambda_l\alpha_{l,1}\\
&s_{l,2} - \alpha_{l,1}s_{1,2} - \ldots - \alpha_{l,n}s_{n,2} = -\lambda_l\alpha_{l,2}\\
&\ldots\\
&s_{l,n-1} - \alpha_{l,1}s_{1,n-1} - \ldots - \alpha_{l,n}s_{n,n-1} = -\lambda_l\alpha_{l,n-1}\\
&s_{l,n} - \alpha_{l,1}s_{1,n} - \ldots - \alpha_{l,n}s_{n,n} = -\lambda_l\alpha_{l,n}
\end{aligned} \tag{28}$$

where $\mathbf{s}_l = [s_{l,1}, \ldots, s_{l,n}]^T$ ($\mathbf{s}_l \in R^n$).

The l-th set of n Eqs. (27) and (28) contains n unknown variables λ_l and $\alpha_{l,j}$ (without the coefficient $\alpha_{l,l}$ which is equal to zero ($\alpha_{l,l} = 0$)).

6 Fitting Eigenvalues with Collinearity Models

In accordance with the *Lemma* 2, the l-th eigenvalue λ_l of the matrix $S = [\mathbf{s}_1, \ldots, \mathbf{s}_n]$ (11) is equal to the solution λ_l of the collinearity Eqs. (13) and (14). In this case the l-th eigenvector $\mathbf{k}_l$ of the matrix S is defined by the column $\mathbf{r}_l$ of the inverse matrix $S_\lambda(l)^{-1}$ (14) [6]:

$$\mathbf{k}_l = \mathbf{r}_l/\|\mathbf{r}_l\| \tag{29}$$

The solution λ_l of the non-linear Eqs. (27) and (28) can be used in an experiment with many values λ' of the regularizing parameter λ (11). The l-th collinearity model is specified by the Eq. (26). In accordance with the previous remarks, an adequate value λ_l of the regularizing parameter λ (11) should assure both the nonsingularity of the matrix $S_\lambda(l)$ (13) and fulfilling the condition (14).

An iterative procedure of matrix inversion has been described in the recent paper [7]. In accordance with this procedure the non-singular matrices $S_\lambda(k)$ (13) are designed gradually from the unit matrix $I = [\mathbf{e}_1, \ldots, \mathbf{e}_n]$ through successive replacement of the unit vectors $\mathbf{e}_k$ by the regularized vectors $\mathbf{z}_k = \mathbf{s}_k - \lambda'\, \mathbf{e}_k$ (11) with a priori selected value

λ' of the regularizing parameter λ. After the replacement of the vector $\mathbf{e}_k$ by the regularized vector $\mathbf{z}_k = \mathbf{s}_k - \lambda' \mathbf{e}_k$ (11) the columns $\mathbf{r}_j(k)$ of the inverse matrix $S_\lambda(k)^{-1}$ (16) are modified in accordance with the Gauss - Jordan transformation (17), (18).

The transformation (17) can not be performed if the condition (24) is met. The condition (25) means division by zero in the Eq. (17). In accordance with the previous considerations, the condition (25) also means that the l-th vector $\mathbf{z}_l = \mathbf{s}_l - \lambda' \mathbf{e}_l$ (13) is a linear combination (23) of the earlier vectors $\mathbf{z}_j = \mathbf{s}_j - \lambda' \mathbf{e}_j$ $(j < l)$. Therefore, the regularizing parameter λ' which fulfils the condition (25) can be the l-th eigenvalue λ_l of the matrix $S = [\mathbf{s}_1, \ldots, \mathbf{s}_n]$ (11).

The proposed experimental evaluation of particular values λ' of the regularizing parameter λ (11) involves iterative inversion of matrices $S_\lambda(l)$ (26) with the columns $\mathbf{s}_j - \lambda' \mathbf{e}_j$. Two values λ' and λ'' of the regularizing parameter λ (13) can be compared on the basis of the below rule:

$$\textit{if}\ \mathrm{r}_l^{\mathrm{T}}(\mathrm{s}_l - \lambda'\mathrm{e}_l) > \mathrm{r}_l^{\mathrm{T}}(\mathrm{s}_l - \lambda''\mathrm{e}_l), \textit{then}\ \lambda'' \textit{is better than}\ \lambda' \tag{30}$$

The evaluation rule (30) allows to find experimentally the optimal value λ^* of the regularizing parameter λ (13) which assures the condition (25). Therefore, the l-th eigenvalue λ_l of the matrix $S = [\mathbf{s}_1, \ldots, \mathbf{s}_n]$ (11) is equal to λ^* $(\lambda_l = \lambda^*)$.

7 Concluding Remarks

The article contains a proposal of the iterative procedure of the eigenvalues λ_i (4) identification through collinearity models (23) extraction. This procedure uses comparison rule (30) for the regularizing parameters λ (11) evaluation. The proposed procedure of the eigenvalues λ_i (4) extraction could be used for the purpose of collinear biclustering in exploratory data analysis and for modelling of multiple interactions [9].

Acknowledgments. The presented study was supported by the grant S/WI/2/2017 from Bialystok University of Technology and funded from the resources for research by Polish Ministry of Science and Higher Education.

References

1. Golub, G.H., Van Loan, C.F.: Matrix Computations, 4th edn. Johns Hopkins University Press, Baltimore (2013)
2. Duda, O.R., Hart, P.E., Stork, D.G.: Pattern Classification. Wiley, New York (2001)
3. Jolliffe, I.T.: Principal Component Analysis. Springer, New York (2002)
4. Cullum, J., Willoughby, R.A.: Large Scale Eigenvalue Problems. North-Holland (1986)
5. Saad, Y.: Numerical Methods for Large Eigenvalue Problems. Society for Industrial and Applied Mathematics, Philadelphia (2011)
6. Bobrowski, L.: Eigenvalue problem with the basis exchange algorithm. J. Adv. Math. Comput. Sci. **23**(6), 1–12 (2017). (http://www.sciencedomain.org/issue/2877)

7. Bobrowski, L: Large matrices inversion using the basis exchange algorithm. Br. J. Math. Comput. Sci. **21**(1), 1–11 (2017). (http://www.sciencedomain.org/abstract/18203)
8. Bobrowski, L.: Data mining based on convex and piecewise linear criterion functions, Technical University Białystok (2005). (in Polish)
9. Bobrowski, L.: Biclustering based on collinear patterns. In: Rojas, I., Ortuño, F. (eds.) IWBBIO 2017. LNCS, vol. 10208, pp. 134–144. Springer, Cham (2017). https://doi.org/10.1007/978-3-319-56148-6_11

The Shapley Value in Fuzzy Simple Cooperative Games

Barbara Gładysz[1](✉) and Jacek Mercik[2]

[1] Wrocław University of Science and Technology, Wrocław, Poland
barbara.gladysz@pwr.edu.pl
[2] WSB University, Wrocław, Poland
jacek.mercik@wsb.wroclaw.pl

Abstract. This article presents an analysis of the Shapley value of simple cooperative games in situations where the weights of the players and the majority required to accept a decision (in the form of a winning coalition) are given by fuzzy numbers. We propose a modification of the concept of the additivity of the payoff function which is necessary in this framework.

Keywords: Shapley value · Fuzzy numbers · Additivity

1 Introduction

When analysing a decision making body in which group decisions are made on the basis of voting, one uses an a priori approach which assumes that the weights of individual decision makers (players) are predetermined constant values (e.g. according to the result of an election). Observation of real life indicates that this assumption is unreasonable. For various reasons, the real weights of players are variable and rarely take the values assumed by the a priori analysis. For example, if the players are parliamentary parties (whose weights are determined by the number of members of parliament each party has), it is very rare that all the members of a particular party take part in a parliamentary vote. Hence, the weights of parties in particular votes are not constant integer numbers, but variable integer numbers which in theory can vary from 0 (i.e. none of the members of a given party took part in a vote) to the number of members of parliament belonging to that party.

Let's look at the votes in the Senate of the Republic of Poland at the meetings on December 20–22, 2011. At the Senate there were 4 groups at that time: Platforma Obywatelska PO (63 seats), PiS Law and Justice (29 seats), Circle of Independent Senators (4 seats), Parliamentary Club Solidarna Polska SP (2 seats) and Parliamentary Club of the Polish Peasant Party PSL (2 seats). The majority required in the votes are at least 1/2 or at least 2/3 of votes in the presence of at least half of the senators. Not all members were present at the meetings. And so the number of senators participating in the voting ranged from 84 to 95: PO (57–63), PiS (28–29); The senators of the other groups were all present. Hence, in the case of voting that took place then, the required majority (at least 50%) was at the level of 43 to 48 votes. However, the majority of at least 2/3 would range from 57 to 64 votes. And here, PO would not always have the

N. T. Nguyen et al. (Eds.): ACIIDS 2018, LNAI 10751, pp. 410–418, 2018.
https://doi.org/10.1007/978-3-319-75417-8_39

required majority. This example shows that when determining the player strength index, you must take into account possible fluctuations in the number of senators participating in the voting. Hence, the literature proposes the use of fuzzy logic in the analysis of players' strength.

Our approach to the situation described above proposes that the weights of the players are described by fuzzy numbers. As a result, the required majority (quota) is also a fuzzy number.

The structure of the article is as follows: The next section describes such voting procedures in the form of simple cooperative games. After this, we describe the weights of players as fuzzy numbers and give some illustrative examples. Finally, we make some conclusions from this analysis and give some directions for future research.

2 The Shapley Value in Cooperative Games

Let N be a finite set of players, q be a quota, w_j be the voting weight of player $j \in N$. By $(N, q, w) = (N, q, w_1, w_2, \ldots, w_n)$ we shall denote a committee (weighted voting body) with a set of players N, quota q and weights $w_j, j \in N$. We shall assume that the w_j are nonnegative integers. Let $t = \sum_{j=1}^{n} w_j$ be the total weight of the committee.

The theory of cooperative games can be used to describe a decision making boy in which decisions are made according to a voting procedure. Since N is the set of players (decision makers), there exist 2^N possible coalitions, i.e. subsets of N. We define a non-negative real function v (called the characteristic function of the game) on the set of all coalitions of players N, i.e. $v : T \rightarrow v(T)$, where $T \subseteq N, v(T) \geq 0$. The value $v(\emptyset) = 0$, since the coalition of all players (the so called grand coalition) cannot gain more than what is gained when all the players cooperate together.

We call the pair (N, v) a simple cooperative game with the set of players N when the characteristic function v satisfies the following condition:

$$v(T) = \begin{cases} 1 & if \quad \sum_{i \in T} w_i \geq q \\ 0 & otherwise \end{cases}.$$

When $q > \left[\frac{t}{2}\right] + \varepsilon$, where $\varepsilon \geq 0$ and [.] denotes the integer part, such a game is called a simple strict majority game. Note that for a standard majority game, we may define $\varepsilon = 1$.

A cooperative game (N, v) is called supperadditive if for each pair of coalitions S, T we have $v(S \cup T) \geq v(S) + v(T)$ when $S \cap T = \emptyset$. This condition ensures that it pays to cooperate.

Any vector $(x_1, \ldots, x_N) \in R^N$ such that $x_1 + \ldots + x_N = v(N)$ is a payoff vector of the game (N, v). The set of all possible payoff vectors of the game (N, v) given by $P(v)$.

The set of payoff vectors $r(v)$ is called a solution of the cooperative game defined by the characteristic function v on the set of players N when $r(v)$ is a subset of $P(v)$ and this set of payoff vectors satisfies the concept of a cooperative solution accepted by the players. For example, they might only accept egalitarian solutions (where each player obtains the same payoff) or solutions where each player's payoff is proportional to their contribution (this rule is commonly used to split the profits from a joint investment), etc.

A set $r(v)$ is called a value if for any game (N, v) the set $r(v)$ contains exactly one member, i.e. there only exists one solution of a game.

It is commonly thought (see e.g. Aumann 1978; Hart and Mas-Collel 1989; Bertini 2011) that the Shapley value (Shapley 1953) is precisely such a value, $\varphi(v)$, for the cooperative game (N, v): $\varphi(v) = (\varphi_1(v), \varphi_2(v), \ldots, \varphi_N(v))$, where

$$\varphi_i(v) = \sum\nolimits_{T \subseteq N, \quad i \in T} \frac{(t-1)!(n-t)!}{n!} (v(T) - v(T \backslash \{i\})). \tag{1}$$

This the expected marginal value of a player to a coalition under the assumption that each player enters the coalition in a random order, i.e. all coalitions are equally likely, until the grand coalition is formed.

The Shapley value has the following properties:

- effectiveness: for any game (N, v), $\sum_{i=1}^{N} \varphi_i(v) = v(N)$,
- for any insignificant player i, a player who only increases the value of any coalition when entering it by the value that player i can ensure herself, i.e. $\forall T \subset N (i \notin T \Rightarrow v(T \cup \{i\}) = v(T) + v(\{i\})$ (a null player satisfies the further condition that $v(\{i\}) = 0$), the Shapley value of any such game is given by $\varphi_i(v) = v(\{i\}$, for a null player it is given by $\varphi_i(v) = 0$,
- symmetry (also called anonymity): for each game (N, v) and any permutation π of the set N, $\varphi_i(\pi(v)) = \varphi_{\pi(i)}(v)$,
- local monotonicity: for each pair of players $i, j \notin S \subset N : v(S \cup \{i\}) \geq v(S \cup \{j\}) \Rightarrow \varphi_i(v) \geq \varphi_j(v)$, as well as
- strong monotonicity (Young 1985): when two games (N, v) and (N, w) satisfy the condition that if $v(T \cup \{j\}) - v(T) \geq w(T \cup \{j\}) - w(T)$ for each coalition $T \not\ni j$, then $\varphi_j(v) \geq \varphi_j(w)$.

Shapley's theorem (1953): the only effective, additive[1] and symmetric value of a game satisfying the null player condition is the Shapley value[2].

[1] The property of additivity (also called the "law of aggregation" by Shapley (1953)): for any two games $(N, v), (N, w)$: $\varphi(v + w) = \varphi(v) + \varphi(w)$ (i.e. $\varphi_i(v + w) = \varphi_i(v) + \varphi_i(w)$ for all i in N, where the game $(v + w)$ is defined by $(v + w)(S) = v(S) + w(S)$ for any coalition S). This property states that when two independent games are combined, their values must be added player by player. This is a prime requisite for any scheme designed to be eventually applied to systems of interdependent games.

[2] There exist several theorems related to additional properties of the Shapley value. For example, Young (1985) proved that the Shapley value is the only value satisfying the properties of effectiveness, symmetry and strong monotonicity, van den Brink (2001) showed that it is the only value preserving a fairness condition according to a modification of marginal contributions to a coalition and Myerson (1977) showed that it preserves fairness based on balanced contributions to a coalition.

3 The Shapley Value in Fuzzy Cooperative Simple Games

An interval fuzzy number $\widetilde{X}$ is a family of real closed intervals $[\widetilde{X}]_{\lambda}$, where $\lambda \in [0, 1]$, such that: $\lambda_1 < \lambda_2 \Rightarrow [\widetilde{X}]_{\lambda_1} \subset [\widetilde{X}]_{\lambda_2}$, where $I \subseteq [0, 1] \Rightarrow [\widetilde{X}]_{\sup I} = \bigcap_{\lambda \in I} [\widetilde{X}]_{\lambda}$ (Zadeh 1965). The interval $[\widetilde{X}]_{\lambda}$ for a given $\lambda \in [0, 1]$ is called the λ-level of the fuzzy number $\widetilde{X}$. We denote such an interval by $[\widetilde{X}]_{\lambda} = [\underline{x}(\lambda), \bar{x}(\lambda)]$.

Dubois and Prade (1978) introduced the following useful definition of the L-R class of fuzzy variables. The fuzzy variable $\widetilde{X}$ is called an L-R type fuzzy variable when its membership function takes the following form:

$$\mu_X(x) = \begin{cases} L\left(\frac{m-x}{\alpha}\right) & for \quad x < m \\ 1 & for \quad \underline{m} \le x \le \bar{m} \\ R\left(\frac{x-\bar{m}}{\beta}\right) & for \quad x > \bar{m}, \end{cases} \tag{2}$$

where: $L(x)$, $R(x)$ are continuous non-increasing functions x; $\alpha, \beta > 0$.

The functions $L(x)$, $R(x)$ are called the left and the right spread functions, respectively. The most commonly used spread functions are: $\max\{0, 1 - x^p\}$ and $\exp(-x^p)$, $x \in [0, +\infty)$, $p \ge 1$. An interval fuzzy variable for which $L(x)$, $R(x) = \max\{0, 1 - x^p\}$ and $\underline{m} = \bar{m} = m$ is called a triangular fuzzy variable and will be denoted by (m, α, β).

Consider two fuzzy variables X, Y with possibility distributions $\mu_X(x)$, $\mu_Y(y)$, respectively. The possibility distributions of the fuzzy variables $Z = X + Y$ and $V = XY$ are defined by means of Zadeh's extension principle (Zadeh 1965) as follows:

$$\mu_Z(z) = sup_{z=x+y}(\min(\mu_X(x), \mu_Y(y))) \tag{3}$$

$$\mu_V(v) = sup_{v=xy}(\min(\mu_X(x), \mu_Y(y))) \tag{4}$$

If we wish to compare two fuzzy numbers, i.e. ascribe the level of possibility that a realization of $\widetilde{X}$ (the value taken by an observation of $\widetilde{X}$) will be greater than (not less than) a realization of $\widetilde{Y}$, then we can apply the measure of possibility proposed by Dubois and Prade (1978):

$$Pos(X \ge Y) = sup_{x \ge y}(min(\mu_X(x), \mu_Y(y))) \tag{5}$$

In classical cooperative games the values of the characteristic function v in (1) belong to the set of real numbers. In the case of voting games, it is assumed that the number of votes cast by a certain player (hereafter called the weight of a player) is determined. For various reasons, this assumption is often unrealistic and can lead to incorrect conclusions regarding the power of a given player (e.g. regarding any assessment of the ability of a given parliamentary party to become a member of a winning coalition and the resulting division of power between coalition members). Hence, the concept of fuzzy games was introduced.

A review of Shapley's work on fuzzy cooperative games can be found in the paper (Borkotokey 2014). There are two types of games: cooperative games with fuzzy coalitions, and cooperative games with fuzzy characteristic functions. Borkotokey (2008) and Yu and Zhang (2010) discuss cooperative game with fuzzy coalition and fuzzy characteristic function. In co-operative games with real characteristic function, authors propose different approaches in the design of characteristic functions. Tsurumi et al. (2001), Li and Zhang (2009) construct a real game function using Choquet integral and fuzzy proportions of individual players in the coalition. Pang et al. (2014) propose a blurred game in which the character function is described by the concave integral. Meng and Zhang (2011) construct a characteristic function as a fuzzy number support function. We propose a cooperative fuzzy game in which the weights are fuzzy and the characteristic function is a real function. In the modeling of decision problems, the apparatus of fuzzy and probabilistic logic as well as their connections are often used Gładysz (2016). In fuzzy logic, different measures of the majority relation are proposed. We proposed the adoption of the measure of the possibility introduced by Dubois and Prade (1978), see Eq. (5). The main contribution of the paper is a modification of the additivity concept, which is used in payoff function.

This article proposes the use of non-negative triangular fuzzy numbers to describe the weights of players. We thus assume that the weight of a player is given by a triangular fuzzy number (m, α, β), where m denotes the most likely value of the weight of a particular player.

The pair $\widetilde{(N, v)}$ will be called a fuzzy simple cooperative game based on the set of players N and the real characteristic function v if for each coalition $T \subseteq N$ the characteristic function takes the value

$$v(T) = Pos\left(\sum\nolimits_{i \in T} \widetilde{A}_i \geq q\widetilde{A}_0 + \varepsilon\right), \tag{6}$$

where:

$\widetilde{A}_i = \left(m_{A_i}, \alpha_{A_i}, \beta_{A_i}\right)$ – the fuzzy weight of the i-th player, $i = 1, \ldots, N$,
$\widetilde{A}_0 = \sum_{i=1}^{N} \widetilde{A}_i$ – fuzzy quota,
$q \geq \frac{1}{2} + \varepsilon, \quad \varepsilon > 0.$

The fuzzy contribution of player i to a given coalition is given by the increase in the value of the characteristic function when player i enters that coalition.

Since the characteristic function is in this case a real function, then the increase in the value of coalition $T \cup \{i\}$ resulting from player i entering coalition T is the same as the fall in the value of the coalition $T \cup \{i\}$ resulting from player i leaving that coalition.

It should be noted that such a game $\widetilde{(N, v)}$ is not superadditive.

Lemma. For $S \cap T = \emptyset$, it follows that $v(S \cup T) \geq v(S)$ and $v(S \cup T) \geq v(T)$. This is a modified version of the property of superadditivity (called fuzzy superadditivity) which implies the willingness of players and subcoalitions to cooperate.

Proof. It suffices to show that $v(S \cup T) \geq v(T)$.

$$v(S \cup T) = Pos\left(\sum\nolimits_{i \in T \cup S} \widetilde{A}_i \geq Q_0\right) = \max\left\{0, min\left\{1 + \frac{m_T + m_S - m_Q}{\beta_T + \beta_S + \alpha_Q}\right\}\right\}$$

$$v(T) = Pos\left(\sum\nolimits_{i \in T} \widetilde{A}_i \geq Q_0\right) = \max\left\{0, min\left\{1 + \frac{m_T - m_Q}{\beta_T + \alpha_Q}\right\}\right\}$$

Thus $v(S \cup T) \geq v(T) <=> \frac{m_T + m_S - m_Q}{\beta_T + \beta_S + \alpha_Q} \geq \frac{m_T - m_Q}{\beta_T + \alpha_Q} <=> m_S(\beta_T + \alpha_Q) \geq \beta_S(m_T - m_Q)$.

From the definition of the sum of two fuzzy variables in Eq. (3) it follows that $m_Q = \sum_{i \in N} m_i \geq m_T$. Hence, $m_T - m_Q \leq 0$ and thus $v(S \cup T) \geq v(T)$. □

It can also be shown that the function given by (6) satisfies all the other standard properties of a characteristic function. It should thus be expected that the Shapley value appropriate to such a fuzzy game is the unique value resulting from Shapley's theorem.

Example 1. Consider the game $\widetilde{(N, v)}$ defined in the following way: $N = \{1, 2, 3\}$, and the fuzzy weights of the players are given by the triangular fuzzy numbers: $\widetilde{A}(1) = (47, 5, 2)$, $\widetilde{A}(2) = (47, 4, 1)$ and $\widetilde{A}(3) = (2, 0, 1)$.

This means that the weight of player 1 varies between 42 and 49, such that the most likely value is 47. The weight of player 2 can take values between 43 and 48, such that the most likely value is 47. The weight of player 3 belongs to the interval [2, 3], such that the most likely value is 2. The fuzzy weight of the grand coalition in this game is given by (96, 9, 4), i.e. varies in the interval between 87 and 100, with the most likely value being 96.

Assume that a standard majority is required to form a winning coalition[3], i.e. $q = 1/2$ and $\varepsilon = 1$. Hence, the quota may be estimated by the fuzzy number $q\widetilde{A}_0 + \varepsilon = \frac{1}{2}(96, 9, 4) + 1 = (49, 5, 2)$.

Table 1 presents values of the characteristic function for each of the possible coalitions in the game $\widetilde{(N, v)}$ with $q = 0.5$ and $\varepsilon = 1$.

Table 1. Values of the characteristic function for the game $\widetilde{(N, v)}$.

Coalition T	$\widetilde{A}$(T)	$v(T)$
{1}	(47, 5, 2)	0.71
{2}	(47, 4, 1)	0.67
{3}	(2, 0, 1)	0
{1, 2}	(94, 1, 3)	1
{1, 3}	(49, 5, 3)	1
{2, 3}	(49, 4, 2)	1
{1, 2, 3}	(96, 9, 4)	1

[3] q defines what proportion of the votes describes a winning majority.

Table 2 presents the marginal contributions of the player entering a coalition for each of the possible permutations of the players in this game.

Table 2. Contributions of individual players when forming a grand coalition according to a random permutation of the set {1, 2, 3} for $q = 1/2$ and $\varepsilon = 1$.

Order	Contribution of player 1	Contribution of player 2	Contribution of player 3
1, 2, 3	0.71	0.29	0.00
1, 3, 2	0.71	0.00	0.29
2, 1, 3	0.33	0.67	0.00
2, 3, 1	0.00	0.67	0.33
3, 1, 2	1.00	0.00	0.00
3, 2, 1	0.00	1.00	0.00
Mean	0.46	0.44	0.10

Player 1 has the greatest Shapley value, 0.46. Player 2 has a slightly lower value, 0.44. Player 3 has the lowest value, 0.1. It should be noted that in the case of simple games, these values are equal to the (here, fuzzy) Shapley-Shubik power index (1954) and can thus be used to assess the ability of a given player to form a winning coalition.

Example 2. In the classical (non-fuzzy) version of example 1 it is assumed that the weights of these players take their maximum values, i.e. $A(1) = 49$, $A(2) = 48$ and $A(3) = 3$. The appropriate quota is $q = 51$. The corresponding values of the characteristic function for the individual players are given by $v(\{1\}) = v(\{2\}) = v(\{3\}) = 0$. The values for the coalitions are: $(\{1,2\}) = v(\{1,3\}) = v(\{2,3\}) = v(\{1,2,3\}) = 1$. The contributions of individual players are given in Table 3.

Table 3. The marginal contributions of individual players to the grand coalition in example 2.

Order	Contribution of player 1	Contribution of player 2	Contribution of player 3
1, 2, 3	0	1	0
1, 3, 2	0	0	1
2, 1, 3	1	0	0
2, 3, 1	0	0	1
3, 1, 2	1	0	0
3, 2, 1	0	1	0
Mean	1/3	1/3	1/3

The expected contribution of each player is equal to 1/3. Comparing the Shapley values for these two games (Tables 2 and 3), it can be seen that player 3 has a much lower power index in the fuzzy game. This results from the fact that when the weight of player 3 is equal to 2 (the most likely value of this weight), then it is likely that one of the other two players can individually form a majority (e.g. when player 1 has a weight of 48 and player 2 has a weight of 45). Also, even when the weight of player 3 is equal to 3, it is still possible that one of the other players can individually form a majority (e.g. when player 2 has a weight of 48 and player 1 has a weight of 44). Hence, the power of player 3 in the fuzzy game is much lower than the power of the other two players. In the classical form of this game, two parties are needed to form a winning coalition and any such coalition is winning (hence the equality of the power indices). It thus seems that taking into account the fuzziness of a voting game that the power of player 3 in the game described in example 2 is for practical purposes significantly lower than 1/3.

4 Conclusions

This proposed fuzzy approach to simple games, which takes into account the indeterminacy of players' weights, seems to represent real life more accurately than classical models, where the weights are assumed to be fixed. Future research is required to check whether the properties of the Shapley value are preserved for other forms of fuzzy numbers, although the authors do not expect it to be otherwise. Another task is to consider fuzzy versions of the characteristic function, which is here assumed to be a real function due to a specific form of "defuzzification". As a result, it is necessary to take into account, not only the fuzzy form of the property of superadditivity, but to also formulate a version of Shapley's theorem for such fully fuzzy games.

In addition, it seems that such fuzzy approaches to cooperative games should be particularly important when particular players can (and often) do form coalitions, but in reality are very protective of their own interests and often treat coalitions as something temporary. An example of such coalitions can be found in e.g. Mercik and Ramsey (2017), where the concept of pre-coalitions was used to analyse the effects of the United Kingdom leaving the European Union. Such an analysis would be based on a fuzzy simple game with 28 players.

References

Aumann, R.J.: Recent developments in the theory of the Shapley value. In: Proceedings of the International Congress of Mathematicians, Helsinki, pp. 995–1003 (1978)

Bertini, C.: Shapley value. In: Dowding, K. (ed.) Encyclopedia of Power, pp. 600–603. SAGE Publications, Los Angeles (2011)

Borkotokey, S.: Cooperative games with fuzzy coalition and fuzzy characteristic function. Fuzzy Sets Syst. **159**, 138–151 (2008)

Borkotokey, S., Mesiar, R.: The Shapley value of cooperative games under fuzzy setting: a servey. Int. J. Gen. Syst. **43**(1), 75–95 (2014)

van den Brink, R.: An axiomatization of the Shapley value using a fairness property. Int. J. Game Theory **30**, 309–319 (2001)

Dubois, D., Prade, H.: Operations on fuzzy numbers. Int. J. Syst. Sci. **9**(6), 613–626 (1978)

Gładysz, B.: Fuzzy-probabilistic PERT. Ann. Oper. Res. **245**(1/2), 1–16 (2016)

Hart, S., Mas-Collel, A.: Potential, value, and consistency. Econometrica **57**(3), 589–614 (1989)

Li, S., Zhang, Q.: A simplified expression of the Shapley function for fuzzy game. Eur. J. Oper. Res. **196**, 234–245 (2009)

Meng, F., Zhang, Q.: The Shapley value on a kind of cooperative games. J. Comput. Inf. Syst. **7** (6), 1846–1854 (2011)

Mercik, J., Ramsey, D.M.: The effect of Brexit on the balance of power in the European Union Council: an approach based on pre-coalitions. In: Mercik, J. (ed.) Transactions on Computational Collective Intelligence XXVII. LNCS, vol. 10480, pp. 87–107. Springer, Cham (2017). https://doi.org/10.1007/978-3-319-70647-4_7

Myerson, R.: Graphs and cooperation in games. Math. Oper. Res. **2**(3), 225–229 (1977)

Pang, J., Chen, X., Li, S.: The Shapley values on fuzzy coalition games with concave integral form. J. Appl. Math. **2014** (2014). http://dx.doi.org/10.1155/2014/231508

Shapley, L.S.: A value for n-person games. In: Kuhn, H.W., Tucker, A.W. (eds.) Contributions to the Theory of Games II, Annals of Mathematics Studies, vol. 28, pp. 307–317. Princeton University Press, Princeton (1953)

Shapley, L.S., Shubik, M.: A method for evaluating the distribution of power in a committee system. Am. Polit. Sci. Rev. **48**, 787–792 (1954)

Tsurumi, M., Tanino, T., Inuiguchi, M.: A Shapley function on a class of cooperative fuzzy games. Eur. J. Oper. Res. **129**, 596–618 (2001)

Young, H.: Monotonic solutions of cooperative games. Int. J. Game Theory **14**, 65–72 (1985)

Yu, X., Zhang, Q.: An extension of fuzzy cooperative games. Fuzzy Sets Syst. **161**, 1614–1634 (2010)

Zadeh, L.A.: Fuzzy sets. Inf. Control **8**, 338–353 (1965)

A Constraint-Based Framework for Scheduling Problems

Jarosław Wikarek, Paweł Sitek(✉), and Tadeusz Stefański

Institute of Management and Control Systems, Kielce University of Technology, Al. 1000-lecia PP 7, 25-314 Kielce, Poland
{j.wikarek, sitek, t.stefanski}@tu.kielce.pl

Abstract. Scheduling and resource allocation problems are widespread in many areas of today's technology and management. Their different forms and structures appear in production, logistics, software engineering, computer networks, etc. In practice, however, classical scheduling problems with fixed structures and only standard constraints (precedence, disjoint etc.) are rare. Practical scheduling problems include also logical and non-linear constraints and use non-standard criteria of schedule evaluations. In many cases, decision makers are interested in the feasibility and/or optimality of a given schedule for specified conditions formulated as questions, for example, *Is it possible…?*, *What is the minimum/maximum…?*, *What if..? etc.* Thus there is a need to develop a programming framework that will facilitate the modeling and solving a variety of diverse scheduling problems. This paper proposes such a constraint-based framework for modeling and solving scheduling problems. It was built with the CLP (Constraint Logic Programming) environment and supported with MP (Mathematical Programming).

Keywords: Scheduling · Constraint logic programming
Mathematical programming optimization · Hybrid methods
Decision support systems

1 Introduction

Nowadays scheduling is a quite common process, which appears in many areas such as manufacturing, logistics, software engineering, computer networks, inventory management, supply chain management etc. This is the process of deciding how a set of tasks to be performed on a set of resources and at what time. In this process, there are also some constraints (precedence, disjunctive, cumulative, etc.). An assignment over time of tasks to resources is called a schedule. There are different objective functions which evaluate the quality of a schedule (makespan, the number of resources, the mean flow time etc.). Classification of scheduling problems, formal models and solution methods are widely described in the literature [1]. In practice, standard scheduling problems reported in the literature are rare. Additional constraints typically occur, due to specific technologies, business-related conditions, legal contracts, safety reasons, etc. Moreover, the users are more and more interested in the schedule evaluation criteria (e.g. time varied) other than those found in the literature. Another important aspect is

N. T. Nguyen et al. (Eds.): ACIIDS 2018, LNAI 10751, pp. 419–430, 2018.
https://doi.org/10.1007/978-3-319-75417-8_40

related to a possibility of asking questions, which may concern the feasibility of a given schedule under given conditions. Due to the universality and diversity of scheduling problems and users' requirements, there is a need to develop a programming framework capable of modeling and solving diverse and complex problems. This paper proposes a constraint-based approach implemented by constraint logic programming (CLP), which is highly flexible in the modeling of all constraint types. In practice, scheduling problems are often modeled and solved using mathematical programming (MP) methods [2]. Their effectiveness is dependent on the problem size and may be unsatisfactory. Therefore, the proposed programming framework includes a hybrid method, which uses both environments, CLP and MP, to find the solution [3, 4]. This method is enriched by the transformation technique developed by the authors [5, 6] and used in the proposed framework as a presolving method.

2 Constraint-Based Framework for Scheduling Problems

Figure 1 shows a general scheduling process diagram. Scheduling is the process of executing activities (jobs, tasks, etc.) by assigning them to resources (machines, processors, manpower, tools, etc.) while satisfying a number of constraints resulting from the problem character and type. These constraints are of various types, for example, disjunctive, precedence, cumulative, time, etc. The process ends with the development of the schedule defining the allocation of activities to resources in time. Each schedule can be evaluated according to a selected criterion (makespan, number of resources used, job flow time, etc.). The sets of activities, resources, constraints, objectives and the organizational environment are the major factors in defining scheduling problems. The problems are classified into several categories. The most important and most common classification is based on the $\alpha||\beta||\gamma$ system, where α- machine environment, β- job characteristics, γ- objective to be minimized [7]. JSSP (Job Shop Scheduling Problem), FSSP (Flow Shop Scheduling Problem), FFSSP (Flexible Flow Shop Scheduling Problem), project scheduling, etc., are the most widely known scheduling problems [1, 8].

Figure 2 shows the concept of the constraint-based programming framework for modeling and solving scheduling problems. The implementation of the framework used the CLP declarative environment [9] and MP. The CLP environment enables easy modeling constraints of any type and the separation of the modeling process from the process of searching for the solution. Solving mechanisms in CLP, such as constraint propagation, backtracking, etc. are effective in finding feasible solutions to constrained problems [9]. Unfortunately, they are a lot worse at dealing with optimization [2, 10]. Scheduling problems contain multiple constraints particularly important in practical problems. In addition to formal constraints resulting from the scheduling process (sequence, precedence, etc.), other constraints appear due to business, law, marketing or safety related conditions. The CLP is based mostly on CSP (Constraint Satisfaction Problem) [9]. In order to describe a problem, they use facts and rules, i.e., logical structures known, for example, from the PROLOG language [11]. Both structures are written as predicates.

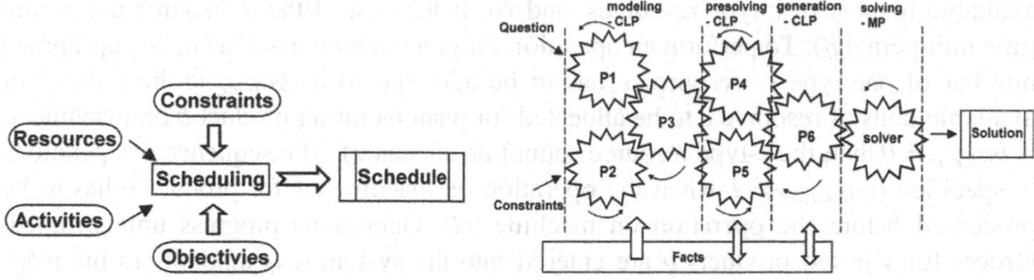

Fig. 1. The schema of general scheduling process

Fig. 2. The schema of programming framework for modeling and solving scheduling problems

A predicate is a function that returns true/false and consists of clauses. The simplest form of predicate is a fact. A fact is a statement about relationship between objects.

In this framework, the CLP environment was used to model the problem, pre-solve it and generate a MP model. The MP environment was used to solve this model. The framework was implemented in ECL^iPS^e [11]. A few sets of predicates (P_1, .., P_6), a set of facts and the MP environment make up the framework. MP may be a built-in environment, e.g., EPLEX, or an external environment, e.g., LINGO, SCIP, and Gurobi. The sets of predicates are described in Table 1.

Table 1. The sets of predicates of the framework

Name	Description
P1	A set of predicates for the modeling of questions (Table 2), which are assessment criterions or conditions to be met if the schedule is to be satisfied
P2	A set of predicates for the modeling of the constraints
P3	A set of predicates, which, based on questions, constraints and facts, generates a CLP model
P4	Built-in predicates responsible for the presolving method of constraint propagation
P5	A set of predicates responsible for the transformation of the CLP model into the transformed CLP^T model (with reduced number of variables, constraints and aggregated parameters)
P6	A set of predicates, which, based on the CLP^T model, generate the final MP

3 Illustrative Example - Statement of the Problem

The illustrative example refers to the scheduling problems with a given set of machines $A = \{a_1, ..., a_{LA}\}$ on which a specified set of products (jobs) $B = \{b_1, ..., b_{LB}\}$ is executed. Each product requires that defined operations ($ex_{b,a}$ amounts to the duration of the operation for product b on machine a, $ex_{b,a} = 0$ means that product b is not performed on machine a) be performed on selected machines. A set of resources (except machines) of different types $C = \{c_1, ..., c_{LC}\}$ is made available in the system (e.g., employees with different skills). The resources are limited (fo_c amounts to the available number of c-type resources, and ro_c is the cost of the c resource use within time unit/period d). To perform an operation for product b on machine a, an appropriate number of one type of resource c has to be allocated to it ($bo_{b,a,c}$ is the number of available units of resource c to be allocated for manufacturing product b on machine a; if $bo_{b,a,c} = 0$ then the c-type resource cannot be allocated). The sequence of operations is specified ($go_{b,a1,a2} = 1$ when the operation on machine $a1$ for product b has to be processed before the operation on machine $a2$). Operations progress uninterrupted. Orders for selected products b are entered into the system (za_b amounts to the magnitude of the demand for product b). Only one operation for product b be can be accomplished on machine a. Time is divided into periods $D = \{d_1, ..., d_{LD}\}$. The problems considered here have constraints (1)–(10). Their formalization is presented in Appendix A. Table 5 summarizes the decision variables, indices and constraints of the modeled problem.

Statement of the problem: A schedule, i.e., the allocation of resources and machines for products/orders in time, has to be found that satisfies all constraints and specified criteria/conditions formulated as questions Qu_1..Qu_12 (Table 2).

Table 2. Criteria of evaluation and feasibility for schedules in the form of questions

Questions	
Qu_1	What is the min C_{max} (makespan)?
Qu_2	What is the minimum makespan (C^1_{max}) if the set of resources is fo_c = N?
Qu_3	What is the minimum set of resources fo_c at C^3_{max}?
Qu_4	Is it possible to schedule orders in C^4_{max} and what are the required sets of resources C^2?
Qu_5	Is it possible to schedule orders in C^5_{max} if the set of resources is C^3?
Qu_6	Is it possible to schedule orders in C^6_{max} if resources c_i and c_j cannot be used simultaneously?
Qu_7	Is it possible to schedule orders in C^7_{max} if machines a_i and a_j cannot be used simultaneously?
Qu_8	What is the min C^8_{max} if resources c_i and c_j cannot be used simultaneously?
Qu_9	What is the min C^9_{max} if machines a_i and a_j cannot be used simultaneously?
Qu_10	What is the min C^{10}_{max} if machines a_i and a_j and resources c_k and c_l cannot be used simultaneously?
Qu_11	Is it possible to schedule orders in C^{11}_{max} if the cost of resources is Wp?
Qu_12	What is the min C^{12}_{max} if the available budget is Wp?

4 Illustrative Example – Transformation Method

The transformation of the model is the authors' original concept [5] implemented in this framework as a presolving method. Indices (dimensions), decision variables, parameters, constraints and facts are all subject to transformation. The process starts with index transformation, which then extends over decision variables, parameters and constraints. Facts are the last transformed item. In general, transformation involves eliminating infeasible points in the multidimensional search space. In practice, infeasible combinations of indices (dimensions) are determined based on the facts and constraints and the corresponding points are removed. The process is similar to the constraint propagation process except that it concerns index (dimension) domains instead of variable domains and is based on constraints and facts instead of constraints alone. This section describes the transformation process for the illustrative example (formulated in Appendix A).

The following indices (scheduling problem dimensions) can be discriminated in the illustrative problem (Appendix A, Sect. 3): *b*-product, *a*-machine, *c*-resource, *d*-period. The facts allow establishing that not every product *b* will be manufactured on machine *a*. And if the product *b* is manufactured on machine *a*, then not all resources *c* are used to do this. In addition, for a given scheduling problem it may be that not all products *b* are manufactured during the given schedule (not all products are considered in a given schedule-eg. not all were ordered).

The information (knowledge) read from the set of facts and constraints is the basis for the transformation of indices in accordance with the $b, a, c \rightarrow f$ regime, where f is the new transformed index of product b execution on machine a with resource c. This new index is developed based on the existing values of indices b, a, c instead of all possible combinations. In the next step, decision variables are transformed as a result of index transformation. Table 6 in Appendix A shows the post-transformation decision variables for the illustrative model. Parameters are transformed in the same way, as are constraints in the next stage. The mode of transformation results from the location of the indices in the constraint (in summation, phrase for, etc.). Appendix A shows the constraints after transformation, (1T)–(10T), used in the illustrative model. The final stage involves the transformation of facts. Figure 3 shows the schematic of fact transformation. Fact-based data representation facilitates its integration with any database system and data warehouse.

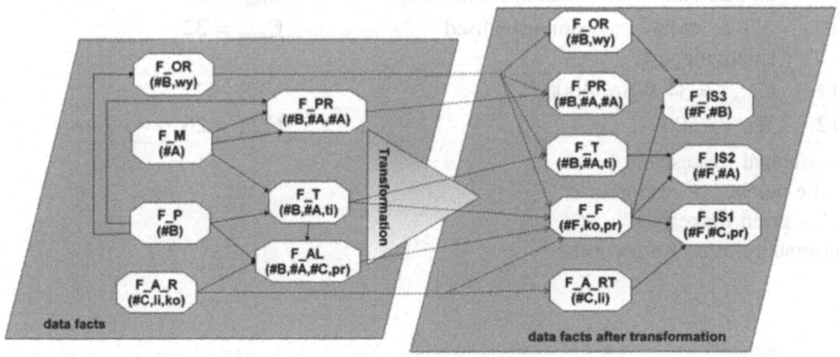

Fig. 3. The schema of transformation of the facts for illustrative example

5 Computational Experiments

Multiple computational experiments were carried out with the use of the framework to evaluate the proposed solution (Sect. 3). During these experiments the illustrative problem was implemented for each question in Table 2. The results are presented in Table 3 and, for questions Qu_1, Qu_10 and Qu_12, in Figs. 4, 5 and 6, in the form of a Gantt charts. The diversity of questions that can be asked and implemented using the framework is indicative of its potential and flexibility. Obtaining answers from one framework is convenient and facilitates decision support within the scheduling area.

Table 3. Obtained results of the numerical experiments for asked questions Qu1..Qu12

Q	Parameters	Result	T
Qu_1	——	$C_{max} = 26$	66
Qu_2$_A$	$fo_1 = 4$, $fo_2 = 5$, $fo_3 = 4$	$C_{max} = 27$	87
Qu_2$_B$	$fo_1 = 40$, $fo_2 = 40$, $fo_3 = 40$	$C_{max} = 26$	54
Qu_3$_A$	C_{max}'' = 26	$fo_1 = 4$	67
Qu_3$_B$	C_{max}'' = 30	$fo_2 = 5$	45
Qu_4	C_{max}'' = 30	Yes $fo_1 = 5$, $fo_2 = 6$, $fo_3 = 6$	34
Qu_4	C_{max}'' = 25	NO	8
Qu_5	C_{max}'' = 25, $fo_1 = 40$, $fo_2 = 40$, $fo_3 = 40$	NO	12
Qu_5	C_{max}'' = 28, $fo_1 = 4$, $fo_2 = 8$, $fo_3 = 4$	YES	24
Qu_6	c_1 i c_3 cannot be used simultaneously and C_{max}' = 30	No	23
Qu_8	c_2 i c_3 cannot be used simultaneously	$C_{max} = 31$	194
Qu_7	a5 i a6 cannot be used simultaneously and C_{max}' = 34	YES	56
Qu_9	a5 i a6 cannot be used simultaneously	$C_{max} = 30$	203
Qu_10	c_2 i c_3 and a_5 i a_6 cannot be used simultaneously	$C_{max} = 32$	210
Qu_11	C_{max}' = 34 Wp = 4000	YES	20
Qu_12	Wp = 3500	$C_{max} = 26$	53

C_{max} – optimal makespan
fo_c – the number of resources c
C_{max}'' – given makespan
T – computing time (in seconds)

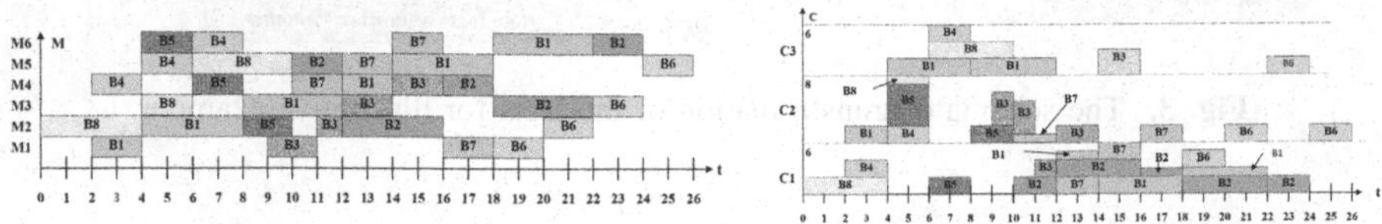

Fig. 4. Example Gantt charts illustrate the answer to the question Qu_1 (the left allocation of machines, the right the use of resources)

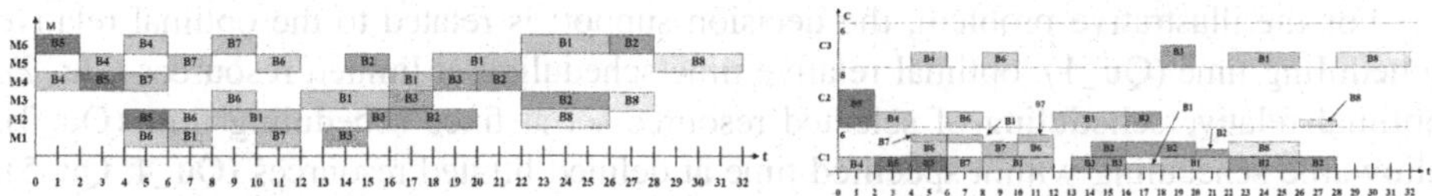

Fig. 5. Gantt charts illustrate the answer to the question Qu_10 (the left allocation of machines, the right the use of resources)

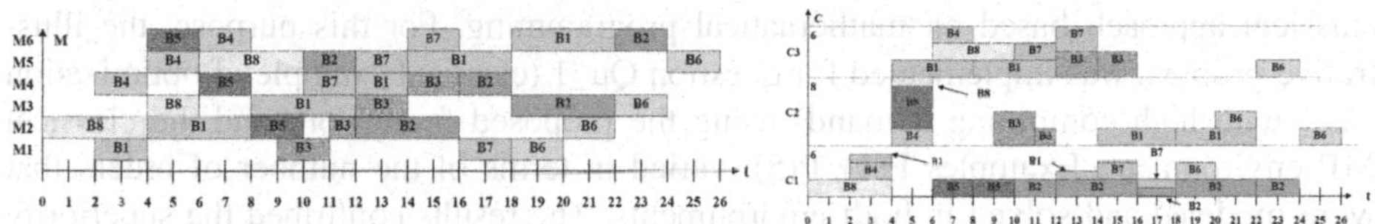

Fig. 6. Gantt charts illustrate the answer to the question Qu_12 (the left allocation of machines, the right the use of resources)

Table 4. Obtained results of the numerical experiments for asked question Qu_1

P(n)	So	V_N	Mathematical programming				Framework			
			V	C	FC	T	V	C	FC	T
1	L	87	8531	9007	22	45	1703	3328	22	34
	S	87	8531	9007	22	10	1703	3328	22	8
2	L	99	8531	9009	22	791	1939	3324	22	40
	S	99	8531	9009	22	345	1939	3324	22	13
3	L	115	8531	9012	24	1136	2234	3319	24	70
	S	115	8531	9012	24	594	2234	3319	24	31
4	L	131	8531	9015	26**	1200*	2706	3311	24	98
	S	131	8531	9015	24	1023	2706	3311	24	48
5	L	147	8531	9017	28**	1200*	2883	3308	26	154
	S	147	8531	9017	26**	1200*	2883	3308	26	85

So – MP solver (L-LINGO, S-SCIP)
V_N – The number of nonzero decision variables
FC – Objective function (makespan) - Qu_1.
* – Interruption of the calculation after 1200 s
V – The number of decision variables
C – The number of constraints.
** – Feasible solution
T – Solution time in seconds

For the illustrative problem, the decision support is related to the optimal relative scheduling time (Qu_1), optimal relative time scheduling at limited resources (Qu_2), optimal relative scheduling of selected resource set at fixed scheduling time (Qu_3), allowable scheduling within specified time at defined limited resources (Qu_4, Qu_5), allowable scheduling at the pre-set time on condition that the selected resources cannot be used simultaneously (Qu_6, Qu_7), optimal scheduling if selected resources cannot be used simultaneously (Qu_8, Qu_9, Qu_10) and allowable and optimal scheduling relative to resource cost, Qu_11 and Qu_12, respectively.

At the further stage of the experiments, the proposed approach was compared to a classical approach based on mathematical programming. For this purpose, the illustrative problem was implemented for question Qu_1 (classical example of optimization C_{max} with high computing demand) using the proposed framework and the classical MP environment. Examples P(1)..P(5), varied in terms of the number of orders that were modeled and solved in both environments. The results confirmed the superiority of the framework, which, depending on the problem size (Table 4), allowed finding the solution from 1.5 to 20 times faster than in the case of the MP environment.

6 Conclusions

The proposed concept in the form of a constraint-based framework has multiple advantages over the MP-based approach in that it is highly flexible, allows modeling a wider spectrum of constraints (not only linear, integer or binary but also non-linear, logical and symbolic constraints) and automatic generation of decision and optimization models based on a set of reference constraints and questions of different type and structure (Tables 2 and 3). This framework is highly efficient and effective in solving the models generated. As shown by computing experiments (Sect. 6, Table 4), finding the solution for the illustrative example (scheduling length optimization) using the framework is many times faster than with the MP environment. Certainly, the NP-completeness of particular cases is not solved, but it allows solving problems of much larger sizes (larger number of machines, resources or products) within acceptable time. In addition to its flexibility and efficiency, the framework offers easier model generation, which is automatic. This considerably reduces the time and effort of a user, who would have to create the model for each question in the MP environment.

Appendix A Models and Facts

(See Tables 5, 6, 7 and 8).

Table 5. The decision variables, indices and constraints of the model for illustrative example

Indices	
a	Machine $a = 1..LA$
b	Product type $b = 1..LB$
c	Resource (in this example employees) $c = 1..LC$
d	Period $d = 1..LD$
Parameters	
$ex_{b,a}$	The time required to make a product b on the machine a
fo_c	The number of resources c
ro_c	The cost of a unit of resource c per period d
$bo_{b,a,c}$	If the resource c can be used to make the product b on the machine a, then $bo_{b,a,c}$ denotes the number of units madc in one run, otherwise $bo_{b,a,c} = 0$
$go_{b,a1,a2}$	If the operation of the product b on the machine $a1$ to be processed before the operation on the machine $a2$ then $go_{b,a1,a2} = 1$, otherwise $go_{b,a1,a2} = 0$
Inputs	
za_b	Demand/order for product b
Auxiliary parameter	
pr_d	Coefficients for conversion numbers of periods d for the variables dimension $pr_d = d$ (necessary for linearizing the model)
Decision variables	
$X_{b,a,c,d}$	If the employee c in period d makes the product b on the machine a then $X_{b,a,c,d} = 1$ otherwise $X_{b,a,c,d} = 0$
$Z_{b,a}$	The number of the period in which machine a finishes its work on product b
$Y_{b,a,c,d}$	If the period d is the latest in which the employee c makes the product b on the machine a then $Y_{b,a,c,d} = 1$ otherwise $Y_{b,a,c,d} = 0$
Wp	The total cost of the resources employed

$$Wp = \sum_{b=1}^{LB}\sum_{a=1}^{LA}\sum_{c=1}^{LC}\sum_{d=1}^{LD} X_{b,a,c,d} \cdot ro_c \tag{1}$$

$$Z_{b,a} \leq Cmax \forall b = 1..LB, a = 1..LA \tag{2}$$

$$\sum_{c=1}^{LC}\sum_{d=1}^{LD} bo_{b,a,c} \cdot X_{b,a,c,d} = ex_{b,a} \cdot za_b \forall b = 1..LB, a = 1..LA : ex_{b,a} > 0 \tag{3}$$

$$\sum_{b=1}^{LB}\sum_{c=1}^{LC} X_{b,a,c,d} \leq 1 \forall a = 1..LA, d = 1..LD \tag{4}$$

$$\sum_{a=1}^{LA}\sum_{b=1}^{LB} bo_{b,a,c} \cdot X_{b,a,c,d} \leq fo_c \forall c = 1..LC, d = 1..LD \tag{5}$$

$$X_{b,a,c,d-1} - X_{b,a,c,d} \leq Y_{b,a,c,d-1} \forall a = 1..LA, b = 1..LB, c = 1..LC, d = 2..LD \tag{6}$$

$$\sum_{d=1}^{LD}\sum_{c=1}^{LC} Y_{b,a,c,d} = 1 \forall a = 1..LA, b = 1..LB : ti_{ba} > 0 \tag{7}$$

$$Y_{b,a,c,1} = 0 \forall a = 1..LA, b = 1..LB, C = 1..LC, Y_{b,a,c,LD} = 0 \forall a = 1..LA, b = 1..LB, C = 1..LC$$

$$Z_{b,a} = \sum_{c=1}^{LC}\sum_{d=1}^{LD} (pr_d \cdot Y_{b,a,c,d}) \forall a = 1..LA, b = 1..LB : ex_{b,a} > 0 \tag{8}$$

$$Z_{b,a2} - za_b \cdot ex_{b,a2} \geq Z_{b,a1} \forall b = 1..LB, a1, a2 = 1..LA : go_{b,a1,a2} = 1 \tag{9}$$

$$X_{b,a,c,d} \in \{0,1\} \forall b = 1..LB, a = 1..LA, c = 1..LC, d = 1..LD; Y_{b,a,c,d} \in \{0,1\} \forall b = 1..LB, a = 1..LA, c = 1..LC, d = 1..LD \tag{10}$$

Table 6. The new decision variables, indices and constraints of the transformed model

Indices	
f	Implementation (new index from the aggregation of *a, b, c*) *f = 1..LF*
Parameters	
ro_f	The cost of implementation *f*
$bo_{f,c}$	If resource c can be used in implementation f, then $bo_{f,c}$ denotes the units of resource c required
Coefficients instance	
$ist1_{f,c}$	If the implementation *f* is done using the resource *c* then $ist_{f,c} = 1$ otherwise $ist_{f,c} = 0$
$ist2_{f,a}$	If the implementation *f* is done using the machine *a* then $ist_{f,a} = 1$ otherwise $ist_{f,a} = 0$
$ist3_{f,b}$	If the implementation *f* is done for product b then $ist_{f,b} = 1$ otherwise $ist_{f,b} = 0$
Decision variables	
$X_{f,d}$	If during the period *d* implementation *f* is performed then $X_{f,d} = 1$ otherwise $X_{f,d} = 0$
Z_f	The number of the period in which the implementation *f* ends
$Y_{f,d}$	If the period *d* is the latest in which implementation *f* is performed then $Y_{f,d} = 1$ otherwise $Y_{fd} = 0$

$$Wp = \sum_{f=1}^{LF} X_{f,d} \cdot ro_f \tag{1T}$$

$$Z_f \leq Cmax \forall f = 1..LF \tag{2T}$$

$$\sum_{f=1}^{LF}\sum_{d=1}^{LD} ist3_{f,b} \cdot ist2_{f,a} \cdot X_{f,d} = ex_{b,a} \cdot za_b \forall b = 1..LB, a = 1..LA : ex_{b,a} > 0 \tag{3T}$$

$$\sum_{f=1}^{LF} ist2_{f,a} \cdot X_{f,d} \leq 1 \forall a = 1..LA, d = 1..LD \tag{4T}$$

$$\sum_{f=1}^{LF} ist1_{f,c} \cdot bo_{f,c} \cdot X_{f,d} \leq fo_c \forall c = 1..LC, d = 1..LD \tag{5T}$$

$$X_{f,d-1} - X_{f,d} \leq Y_{f,d-1} \forall f = 1..F, d = 2..LD \tag{6T}$$

$$\sum_{d=1}^{LD} \sum_{f=1}^{LF} ist2_{f,a} \cdot ist3_{f,b} \cdot Y_{f,d} \leq 1 \forall a = 1..FA, b = 1..FB : ex_{b,a} > 0 Y_{f,1} = 0 \forall f = 1..LF Y_{f.,LD} = 0 \tag{7T}$$

$$Z_f = \sum_{f=1}^{LF} pr_d \cdot X_{f,d} \forall f = 1..LF \tag{8T}$$

$$\sum_{f=1}^{LF} (ist2_{f,a2} \cdot ist3_{f,b} \cdot Z_f) - za_b \cdot ex_{b,a2} \geq \sum_{f=1}^{LF} (ist2_{f,a1} \cdot ist3_{f,b} \cdot Z_f) \forall b = 1..LB, a1, a2 = 1..LA : go_{b,a1,a2} = 1 \tag{9T}$$

$$X_{f,d} \in \{0,1\} \forall f = 1..LF, d = 1..LD; Y_{f,d} \in \{0,1\} \forall f = 1..LF, d = 1..LD \tag{10T}$$

Table 7. Description of the constraints for both models

Constraint	Description
(1) (1T)	Calculation of the total cost of the resources employed
(2) (2T)	Makespan greater than the time of end-use for all machines
(3) (3T)	The schedule satisfies consumer demand
(4) (4T)	A resource cannot be used for two purposes in a single period
(5) (5T)	Only resources from the available set can be used in any period
(6) (6T)	The processing of a product on any machine is uninterrupted
(7) (7T)	The processing of a product on any machine is uninterrupted
(8) (8T)	Determination of the completion time
(9) (9T)	The precedence constraints
(10) (10T)	Binary variables

Table 8. Description of the facts

Fact	Description
Facts – data	
F_M(#A)	Facts about machines/processors
F_P(#B)	Facts about products
F_A_R(#C,fo,ro)	Facts about constrained resources
F_T(#B,#A,ex)	Facts about technology for product
F_AL(#B,#A,#C, bo)	Facts about allocation of product and resources to machines/processors
F_PR(#B,#A,#A)	Facts about precedence
Facts-inputs	
F_OR(#B,za)	Facts about orders
New facts after transformation	
F_F(#F,ko)	Facts about allocation product and resources
F_A_RT(#C,fo)	Facts about constrained resources
F_IS1(#F,#C,bo)	If the implementation *f* is done using the resource *c*
F_IS2(#F,#A)	If the implementation *f* is done using the machine *a*
F_IS3(#F,#B)	If the implementation *f* is done for product *b*

References

1. Błażewicz, J., Ecker, K.H., Pesch, E., Schmidt, G., Weglarz, J.: Handbook on Scheduling From Theory to Applications. Springer, Heidelberg (2007). https://doi.org/10.1007/978-3-540-32220-7. ISBN:978-3-540-28046-0
2. Schrijver, A.: Theory of Linear and Integer Programming. Wiley, New York (1998)
3. Milano, M., Wallace, M.: Integrating operations research. Constraint Program. Ann. Oper. Res. **175**(1), 37–76 (2010)
4. Hooker, J.N.: Logic, optimization and constraint programming. INFORMS J. Comput. **14**, 295–321 (2002)
5. Sitek, P., Wikarek, J.: A hybrid programming framework for modeling and solving constraint satisfaction and optimization problems. Sci. Program. **2016**, 13 (2016). https://doi.org/10.1155/2016/5102616. Article ID 5102616
6. Sitek, P., Wikarek, J., Nielsen, P.: A constraint-driven approach to food supply chain management. Ind. Manag. Data Syst. **117**(9) (2017). https://doi.org/10.1108/imds-10-2016-0465. Article ID 600090
7. Graham, R.L., Lawler, E.L., Lenstra, J.K., Rinnooy Kan, A.H.G.: Optimization and approximation in deterministic sequencing and scheduling: a survey. Ann. Discrete Math. **4**, 287–326 (1979)
8. Coelho, J., Vanhoucke, M.: Multi-mode resource-constrained project scheduling using RCPSP and SAT solvers. Eur. J. Oper. Res. **213**, 73–82 (2011)
9. Rossi, F., Van Beek, P., Walsh, T.: Handbook of Constraint Programming (Foundations of Artificial Intelligence). Elsevier Science Inc., New York (2006)
10. Achterberg, T., Berthold, T., Koch, T., Wolter, K.: Constraint integer programming: a new approach to integrate CP and MIP. In: Perron, L., Trick, M.A. (eds.) CPAIOR 2008. LNCS, vol. 5015, pp. 6–20. Springer, Heidelberg (2008). https://doi.org/10.1007/978-3-540-68155-7_4
11. Eclipse – home. www.eclipse.org. Accessed 5 July 2017

Computer Vision Techniques

A Reliable Image-to-Video Person Re-identification Based on Feature Fusion

Thuy-Binh Nguyen[1,2], Thi-Lan Le[1(✉)], Dinh-Duc Nguyen[1], and Dinh-Tan Pham[3]

[1] International Research Institute MICA, HUST-CNRS/UMI-2954-GRENOBLE INP, Hanoi University of Science and Technology, Hanoi, Vietnam
Thi-Lan.Le@mica.edu.vn

[2] University of Transport and Communications, Hanoi, Vietnam

[3] Hanoi University of Mining and Geology, Hanoi, Vietnam

Abstract. This paper proposes a novel feature-fusion frame work for the image-to-video person re-identification. In this framework, we formulate person re-id problem as a classification-based information retrieval where a person appearance model is learned in the training phase and the identity of an interested person is determined by the probability that his/her probe image belongs to the model. To learn the person appearance model, two features that are Kernel descriptor (KDES) and Convolution Neural Network (CNN) are investigated. Then, three fusion schemes including early fusion, product rule and query-adaptive late fusions are proposed. Extensive experiments have been conducted on two public benchmark datasets: CAVIAR4REID and RAID. The obtained accuracies at rank 1 are 95.00% and 94.29% for CAVIAR4REID and RAID, respectively. Among three proposed fusion schemes, the two late fusion schemes obtain better results than that of the other one. They gain approximately 10% improvement for accuracy at Rank 1.

Keywords: CNN · Kernel descriptor · Person re-identification Feature fusion

1 Introduction

Person re-identification, the task of recognizing people in a non-overlapping camera network, has attracted especial attention of the computer vision and pattern recognition community because of its widely applications in video surveillance. The existing person re-identification methods are classified into two main approaches: image-based and video-based [1]. In the first approach, an individual has only single image on both the probe and the gallery sets while an individual in the second approach contains a set of images. As consequence, the works belonging to the first approach focus mainly on image content analysis and matching while that of the second approach can exploit different information such as temporal and motion. One special case of the video-based approach is the image-to-video person re-identification in which the probe is an image

N. T. Nguyen et al. (Eds.): ACIIDS 2018, LNAI 10751, pp. 433–442, 2018.
https://doi.org/10.1007/978-3-319-75417-8_41

while the gallery has a set of images [2–4]. This reflects the situation in real life such as criminal or suspect search where one sole query image is available. The image-to-video person re-identification shares the same common challenges with the image-based person re-identification (i.e. occlusion, low resolution, large variation in poses and viewpoints, similar appearance) and has its own challenge: matching two different modalities that are an image and a video.

The main contribution of this paper is a novel feature-fusion framework for the image-to-video person re-identification. In this framework, we formulate the image-to-video person re-identification as a classification-based information retrieval problem where the model of person appearance is learned from the gallery images and the identity of interested person is determined by the probability that his/her probe image belonging to the model. In order to learn the person appearance model, two kinds of features that are hand-designed and learning features are investigated. For hand-designed feature, Kernel descriptor (KDES) that has been proved to be robust for representing person appearance is chosen [2]. Concerning learning feature, the features before fully connected layer of Googlenet are extracted [5]. In order to evaluate the influence of classification methods for these features, we employ two classification methods that are Softmax and SVM (Support Vector Machine). Moreover, several data augmentation algorithms are used to enrich training samples. Then, three fusion schemes including early fusion, product rule and query-adaptive late fusions are proposed. The early fusion scheme concatenates both hand-designed and learning features for person representation while two late fusions schemes aim to fuse two lists of retrieved people obtained by applying one kind of feature. Product rule-based late fusion scheme has outperformed other fusion schemes in information retrieval [6], plant identification [7]. In this study, we propose to use this scheme for person re-identification problem. Moreover, inspired by the work of Zheng et al. [8] for query-adaptive late fusion in image-to-image person search, we propose to use it for the image-to-video person re-identification. The main idea is that instead of using fixed weight for each feature, we compute it adaptively according to the query image.

The remainder of this paper is organized as follows. Section 2 describes the related work on the image-to-video and feature fusion approaches. The proposed framework is presented in Sect. 3. And finally, experimental results and conclusion are shown in Sect. 4 and Sect. 5, respectively.

2 Related Work

A large number of works have been dedicated to person re-identification problem [1]. In this section, we focus on the feature fusion and image-to-video person re-identification.

Feature extraction plays an important role in person re-identification. In literature, there are many features that have been proposed for person appearance representation [9]. Taking into account that each feature has its own advantage and drawback, several works aim at fusing different features in order to get a good person re-identification performance. The feature fusion approach is divided into two categories: early fusion (also named feature-level fusion) and late fusion (named score-level fusion). In the first category, features are concatenated to generate a larger dimension vector for representing an image while the methods belonging to the second category compute the weight (score) for each feature in similarity function. Gao et al. [10] proposed an early fusion method for the image-to-image person re-identification. A combined feature is created by concatenating a high-dimensional low-level feature (WHOS formed by combining color histograms and HOG) and a low-dimensional mid-level feature. The authors have proved that the recognition rate is improved when using the combined features with experiments conducted on several datasets. The work of Eisenbach et al. [11] belongs to the late fusion scheme. The authors prove that late fusion methods can achieve better results than early fusion ones for the image-to-image person re-identification. Taking into account that features are not equally important for all queries, in [8], the authors proposed to learn adaptively weight for each query. The obtained results of this technique for the image-to-image person re-identification are very promising.

Concerning the image-to-video person re-identification, Pham et al. [2] have proposed a fully automated person re-identification that also formulates the image-to-video person re-identification as classification-based problem. The authors proposed to employ KDES as a descriptor and SVM as a classifier. This method has outperformed many state of the art works on different benchmark datasets. However, this method uses one sole kind of feature for person representation. In this study, we compare the proposed framework with the method provided by Pham et al. [2] on two benchmark datasets. Zhang et al. [3] introduced a novel temporally memorized similarity learning neural network to solve the challenge in the image-to-video person re-identification. In this work, CNN features are extracted at image level and these feature vectors are fed into LSTM (Long Short-Term Memory) network to generate a unified signature to represent a sequence of images by concatenating all feature vector at each node. Finally, feature vectors of a probe image and a sequence images are forwarded to the similarity sub-network for distance metric learning. Therefore, the image-to-video problem is turned into matching two feature vectors by learning a metric in a new sub-network. By using LSTM, the authors focus on the case that images of the same person in training set have a temporal constraint. In the case that the images of the training set are not temporal-related, LSTM cannot show its advantage.

As analyzed, several works have been dedicated to the feature fusion and image-to-video person re-identification. However, most of feature fusion works focus on the image-to-image person re-identification. While working with the image-to-video person re-identification, fusion feature is not simply a similarity function with appropriate scores/weights.

3 Proposed Approach

3.1 Overall Framework

The proposed framework for the image-to-video person re-identification is shown in Fig. 1. In this framework, we formulate the image-to-video person re-id as a classification-based information retrieval problem where the model of person appearance is learned from the gallery images and the identity of interested person is determined by the probability that his/her probe image belonging to the model. In order to learn the model of person appearance, we compute two kinds of features (KDES and CNN) for each person's image. Then, SVM is employed to learn the model. For each query image, a list of people is ranked by descending order of the similarity between this image and learned model. In the framework, we propose three fusion schemes: early, product rule and query-adaptive. In case of early fusion, KDES and CNN features are concatenated and fed to SVM model while in two late fusion schemes, a new list of people is determined from two different lists obtained by using two kind of features. In the following section, we focus on two main components of the framework that are feature extraction and feature fusion.

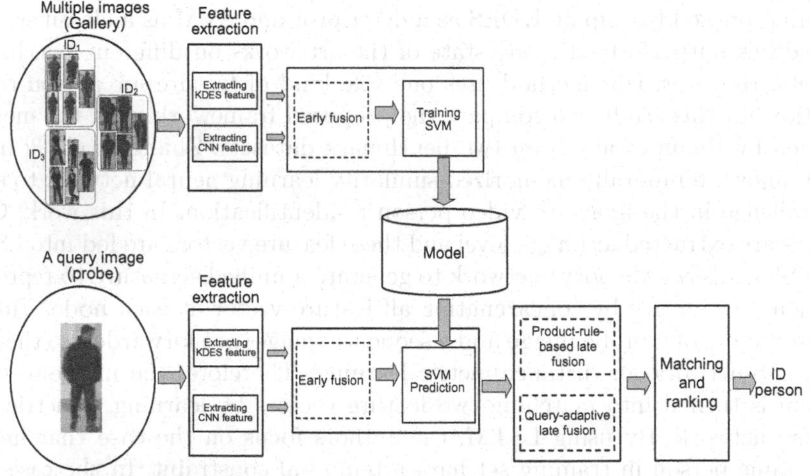

Fig. 1. The proposed framework for the image-to-video person re-identification.

3.2 Feature Extraction

KDES is firstly introduced by Bo et al. [12] which the main idea is to build a compact patch-level features from pixel attributes such as gradient, color and texture. This feature has been proved to be robust for object recognition in general and person re-identification in particular. Pham et al. [2] have improved this feature to make it be invariant to scale change and have shown that this feature outperforms many hand-designed features on a number of benchmark datasets. In this study, we employ the improved version of KDES computed with three kernels provided by Pham et al. [2]. KDES descriptor is computed at three levels: pixel level, patch level and image level. Three features that are gradient, color and texture are calculated at pixel level. Then, patch level feature is determined by matching kernels via kernel approximation. Then, image level is computed by concatenating features from 3 layers-pyramid. By this way, each image is represented by a 63000-dim vector.

Recently, with the impressive results obtained by deep learning, several work try to compute features learned from deep architecture for person appearance representation. Among different architectures, we propose to extracted learned features of GoogleNet [5]. In this paper, we use a pre-trained GoogleNet model on ImageNet dataset. After that, two benchmark datasets (CAVIAR4REID and RAID) are employed to fine-tune the weights of pre-trained network. A vector of 1024-dim is extracted from pool5/7×7 s1 layer for the given image. Some existing works claim that CNN network is not effective on a small dataset. In order to enrich the training set, we apply two data augmentation techniques provided in [13]: rotation and translation. From one image, we generate 3 images by rotating the original image 5 degrees in left and right and translating this image by 1% of image width and 4% of image height.

3.3 Feature Fusion

The image-to-video person re-identification is described as follows. Given a probe (query) image q, its identity is determined by:

$$i^* = \text{argmax}_{i \in 1,2,..,N_g} sim(q, \{g_j\}_{j=1}^{m_g}) \quad (1)$$

where i^* is the identity of probe q, and sim(.,.) is some kinds of similarity functions. The gallery consists of N_g persons. Each person has m_g images, denoted as $\{g_j\}_{j=1}^{m_g}$.

In the early fusion scheme, a larger dimension feature is created by concatenating KDES and CNN features. This complementary combination takes advantages both of above features for representing a person image. After that, extracted feature vectors are forwarded the SVM classifier to match a query image to a set of images in the gallery. In this case, the identity of the query is determined by the identity of the person in gallery having the highest score.

In the late fusion scheme, for each feature, a list of retrieved people and its corresponding score is determined. Then, the final list is determined by the following strategy. Denote $s^{(k)}_{q,\{g_j\}_{j=1}^{m_g}}$ is the similarity score between query q and person represented by $\{g_j\}_{j=1}^{m_g}$ images by using feature k. K is the number of features (K=*2* in our experiments).

- Product rule-based late fusion: In this scheme, the final similarity is determined by product rule [14] as follows:

$$sim(q, \{g_j\}_{j=1}^{m_g})_{product-rule} = \prod_{k=1}^{K} \left(s^{(k)}_{q,\{g_j\}_{j=1}^{m_g}} \right). \quad (2)$$

- Query-adaptive late fusion: Zheng et al. [8] have observed that a feature may be effective for a given query but ineffective for others query. Based on this observation, they try to estimate the effectiveness of a feature based on the score curve. A feature is considered as an effective feature for a given image if its score curve has a L-shape. It means the score corresponding to the first rank is much higher than those of the next ranks. The authors proposed to compute feature weight according to it's shape of score. However, the authors have just applied for the image-to-image approach. Inspired by this work, we propose to apply it for the image-to-video person re-identification.

$$sim(q, \{g_j\}_{j=1}^{m_g})_{query-adaptive} = \prod_{k=1}^{K} \left(s^{(k)}_{q,\{g_j\}_{j=1}^{m_g}} \right)^{\omega_q^{(k)}}, with \sum_{k=1}^{K} \omega_q^{(k)} = 1, \quad (3)$$

  where $\omega_q^{(k)}$ is the feature weighting for the kth feature. This weight is adaptively determined according to the score's shape of feature for each query. The detail of weight computation is presented in [8].

4 Experiments and Results

4.1 Datasets and Evaluation Measure

In order to evaluate the proposed framework, two benchmark datasetes that are CAVIAR4REID [15] and RAID [16] are used. CAVIAR4REID dataset contains multiple images for 72 pedestrians in two non-overlapping camera views. However, only 50 of them have images on both cameras. This dataset is one of the most challenging ones in benchmark datasets because of occlusion, low resolution and strong variation in illumination. The resolution of these images vary from 17×32 to 72×144. With CAVIAR4REID dataset, there exist different ways to create training and testing sets. Hence, in this paper, two scenarios are setup named case A and case B. In case A, a person has 5 images for testing and training sets while in the case B, each individual has 5 images for testing sets and the rest of images are used for training. These setups allow to evaluate the effect of the number of samples in training sets on the person re-identification accuracy. RAID dataset [16] includes 6920 images of 43 people appeared in two indoor and two outdoor cameras. In this paper, the training and testing sets are set the same ones as in [2], in which images from one indoor camera are set for training (210 images) and the other for testing (6710 images). All images in this dataset are the same size as 64×128. The strong variation in illumination of images in this dataset is one of the difficulties for person re-identification problem.

We employ Cumulative Matching Characteristic curve (CMC) as an evaluation measure for person re-identification. The horizontal axis shows ranks while the vertical axis describes the rate of the true matching corresponding to each rank. The value of CMC curve at rank k presents the rate of true matching in the first k images are ranked. Higher CMC curve is, better the person re-identification method is.

4.2 Experimental Results

For CAVIAR4REID dataset, we have performed two experiments: person re-identification without and with data augmentation. The obtained results on CAVIAR4REID for case A (balanced case) and case B (imbalanced case) are shown in Figs. 2 and 3. The results of KDES feature and SDALF are achieved by running the code provided by the authors in [2,17] our own setup. As seen in these figures, the larger number of training sample is, the higher matching rates are. Comparing between two cases, the number of training images in the case B is larger than that of case A. Therefore, the recognition rates of the case B are higher than those of the case A. Data augmentation allows to gain from 4–5% of accuracy at rank 1. It is interesting to see that when applying data augmentation strategy, the matching rate improves for not only learning feature but also hand-designed feature.

We compare the performance of three features: KDES in [2], CNN and SDALF (Symmetry-Driven Accumulation of Local Features) in [17] in the first experiment. The experimental results also show that two chosen features (KDES and CNN) outperform SDALF. KDES obtained better results than CNN feature with Softmax classifier. It is note that CNN with Softmax classifier is the default implementation in GoogleNet. By extracting CNN feature before fully connected layer and putting in SVM classifier, the accuracy at Rank 1 increases by from 5 to 10%.

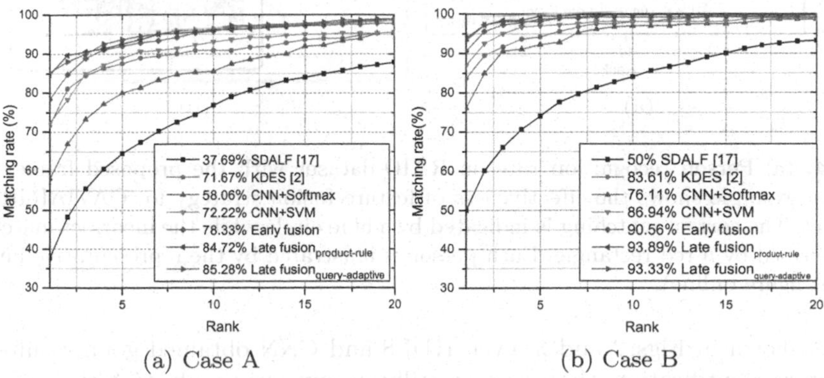

(a) Case A (b) Case B

Fig. 2. CMC curves obtained by the proposed framework for CAVIAR4REID dataset without applying data augmentation.

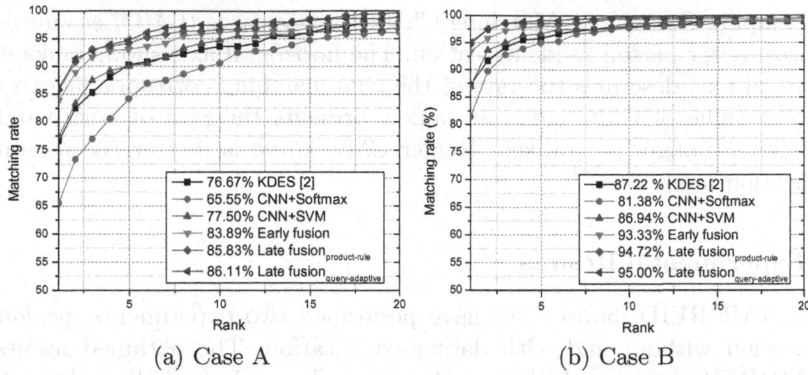

(a) Case A (b) Case B

Fig. 3. CMC curves obtained by the proposed framework for CAVIAR4REID dataset with data augmentation.

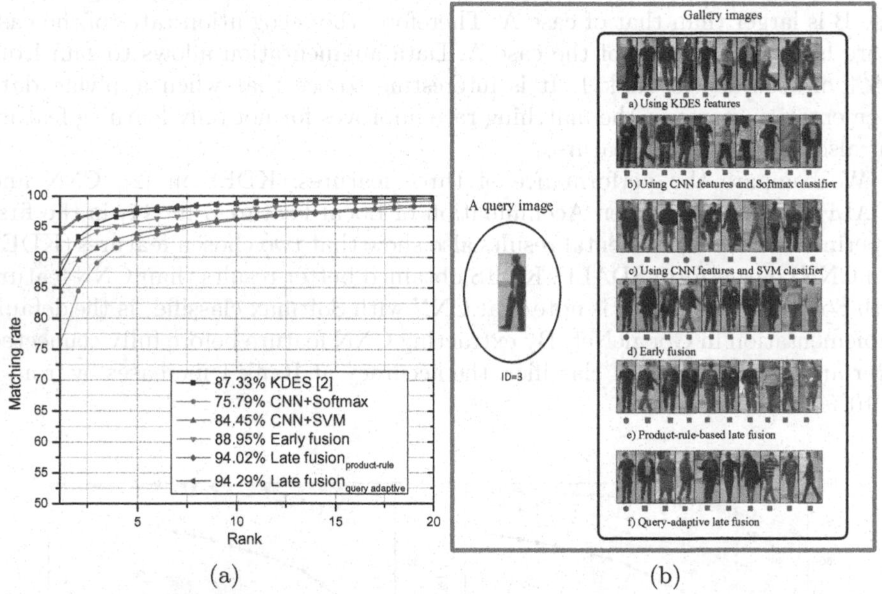

(a) (b)

Fig. 4. (a) Person recognition rate on RAID dataset with the proposed framework; (b) An example shows the effectiveness of feature fusion strategy for CAVIAR4REID dataset. The correct matching is indicated by a blue circle while the incorrect matching is described by a red rectangle. Each person is illustrated by the representative image. (Color figure online)

As shown in Figs. 2 and 3, even KDES and CNN obtained good results for person re-identification, the accuracy still can improve thanks to fusion scheme. The early fusion gots approximately 6% of improvement while the two late fusion schemes obtain 10% of improvements. The query-adaptive late fusion obtains results that are slightly better than those of the product-rule late fusion. This can be explained use of two features that are equally important for query representation. Therefore, the weights for each feature computed by adaptive scheme are similar to the fixed equal weights used in product-rule scheme.

The obtained results for RAID dataset are shown in Fig. 4a. These results are consistent with the results obtained for CAVIAR4REID dataset. This confirms once again the robustness of the proposed framework. The accuracies at rank 1 of two late fusions are greater than 94%. Without using feature fusion, those are 87.33% and 75.79% for KDES and CNN, respectively.

Figure 4b shows an example of person re-identification on CAVIAR4REID dataset. The correct matching is indicated by a blue circle while the incorrect matching is described by a red rectangle. As seen in this figure, fusion schemes allow to get the correct matching at rank 1.

5 Conclusion and Future Work

In this paper, a novel feature fusion framework for the image-to-video person re-identification has been proposed. The results obtained on two benchmark datasets (CAVIAR4REID and RAID) show the robustness of the proposed framework. Three proposed feature fusion schemes have improved significantly the re-identification accuracy (ranging from 6% to 10% at rank 1). Based on this promising results, we will extend our work for the video-to-video person re-identification. Moreover, in this study, we treat the training images for each person as a set of independent images. Temporal and motion information are not exploited yet. In the future, we will investigate these features for the image-to-video person re-identification.

Acknowledgement. This work has been supported by University of Transport and Communications under grant code: T2018-DT-007.

References

1. Zheng, L., Yang, Y., Hauptmann, A.G.: Person re-identification: past, present and future. arXiv preprint arXiv:1610.02984 (2016)
2. Pham, T.T.T., Le, T.L., Vu, H., Dao, T.K., Nguyen, V.T.: Fully-automated person re-identification in multi-camera surveillance system with a robust kernel descriptor and effective shadow removal method. Image Vis. Comput. **59**, 44–62 (2017)
3. Zhang, D., Wu, W., Cheng, H., Zhang, R., Dong, Z., Cai, Z.: Image-to-video person re-identification with temporally memorized similarity learning. IEEE Trans. Circuits Syst. Video Technol. **PP**(99) (2017)
4. Wang, G., Lai, J., Xie, X.: P2SNet: can an image match a video for person re-identification in an end-to-end way? IEEE Trans. Circuits Syst. Video Technol. **PP**(99) (2017)
5. Szegedy, C., Liu, W., Jia, Y., Sermanet, P., Reed, S., Anguelov, D., Erhan, D., Vanhoucke, V., Rabinovich, A.: Going deeper with convolutions. In: Proceedings of the IEEE Conference on Computer Vision and Pattern Recognition, pp. 1–9 (2015)

6. Jović, M., Hatakeyama, Y., Dong, F., Hirota, K.: Image retrieval based on similarity score fusion from feature similarity ranking lists. In: Wang, L., Jiao, L., Shi, G., Li, X., Liu, J. (eds.) FSKD 2006. LNCS (LNAI), vol. 4223, pp. 461–470. Springer, Heidelberg (2006). https://doi.org/10.1007/11881599_54
7. Thanh-Nhan Nguyen, T., Le, T.-L., Vu, H., Nguyen, H.-H., Hoang, V.-S.: A combination of deep learning and hand-designed feature for plant identification based on leaf and flower images. In: Król, D., Nguyen, N.T., Shirai, K. (eds.) ACIIDS 2017. SCI, vol. 710, pp. 223–233. Springer, Cham (2017). https://doi.org/10.1007/978-3-319-56660-3_20
8. Zheng, L., Wang, S., Tian, L., He, F., Liu, Z., Tian, Q.: Query-adaptive late fusion for image search and person re-identification. In: Proceedings of the IEEE conference on Computer Vision and Pattern Recognition, pp. 1741–1750 (2015)
9. Liu, C., Gong, S., Loy, C.C., Lin, X.: Evaluating feature importance for re-identification. In: Gong, S., Cristani, M., Yan, S., Loy, C.C. (eds.) Person Re-identification. ACVPR, pp. 203–228. Springer, London (2014). https://doi.org/10.1007/978-1-4471-6296-4_10
10. Gao, M., Ai, H., Bai, B.: A feature fusion strategy for person re-identification. In: 2016 IEEE International Conference on Image Processing (ICIP), pp. 4274–4278. IEEE (2016)
11. Eisenbach, M., Kolarow, A., Vorndran, A., Niebling, J., Gross, H.M.: Evaluation of multi feature fusion at score-level for appearance-based person re-identification. In: 2015 International Joint Conference on Neural Networks (IJCNN), pp. 1–8. IEEE (2015)
12. Bo, L., Ren, X., Fox, D.: Kernel descriptors for visual recognition. In: Advances in neural information processing systems, pp. 244–252 (2010)
13. McLaughlin, N., Del Rincon, J.M., Miller, P.: Data-augmentation for reducing dataset bias in person re-identification. In: 2015 12th IEEE International Conference on Advanced Video and Signal Based Surveillance (AVSS), pp. 1–6. IEEE (2015)
14. Kittler, J., Hatef, M., Duin, R.P., Matas, J.: On combining classifiers. IEEE Trans. Pattern Anal. Mach. Intell. **20**(3), 226–239 (1998)
15. Cheng, D.S., Cristani, M., Stoppa, M., Bazzani, L., Murino, V.: Custom pictorial structures for re-identification. In: British Machine Vision Conference (BMVC) (2011)
16. Das, A., Chakraborty, A., Roy-Chowdhury, A.K.: Consistent re-identification in a camera network. In: Fleet, D., Pajdla, T., Schiele, B., Tuytelaars, T. (eds.) ECCV 2014. LNCS, vol. 8690, pp. 330–345. Springer, Cham (2014). https://doi.org/10.1007/978-3-319-10605-2_22
17. Bazzani, L., Cristani, M., Murino, V.: Symmetry-driven accumulation of local features for human characterization and re-identification. Comput. Vis. Image Underst. **117**(2), 130–144 (2013)

Video Classification Technology in a Knowledge-Vision-Integration Platform for Personal Protective Equipment Detection: An Evaluation

Caterine Silva de Oliveira[1(✉)], Cesar Sanin[1], and Edward Szczerbicki[2]

[1] The University of Newcastle, Newcastle, NSW, Australia
caterine.silvadeoliveira@uon.edu.au, cesar.maldonadosanin@newcastle.edu.au

[2] Gdansk University of Technology, Gdansk, Poland
edward.szczerbicki@newcastle.edu.au

Abstract. This work is part of an effort for the development of a Knowledge-Vision Integration Platform for Hazard Control (KVIP-HC) in industrial workplaces, adaptable to a wide range of industrial environments. This paper focuses on hazards resulted from the non-use of personal protective equipment (PPE), and examines a few supervised learning techniques to compose the proposed system for the purpose of recognition of three protective equipment: hard hat, gloves and boots. In the KVIP-HC, classifiers, feature images and any context information are represented explicitly using the Set of Experience Knowledge Structure (SOEKS), grouped and stored as Decisional DNA (DDNA). The collected knowledge is used for reasoning and to reinforce the system from time to time, customizing the service according to each scenario and application. Therefore, in choosing the classification methodology that best suits the application, processing time for training (once the system will be eventually reinforced in real time), accuracy, detection time and the predictor sizes (for the purpose of storing data) are analyzed to propose the most reasonable candidates to compose the platform.

Keywords: Decisional DNA (DDNA)
Set of Experience Knowledge (SOEKS) · Industrial hazard control
Machine learning · Deep learning · Classifier performance

1 Introduction

Hazard control is fundamental to ensure occupational health and safety of labors [1]. In industrial environments, workers are exposed to hazards in a variety of situations, such as when accessing controlled zones without authorization, crossing yellow lines, not respecting safe distances from machines and areas, not wearing the required personal protective equipment (PPE), among others. The process of controlling these hazards includes their detection, recognition and evaluation to determine the level of risk they

N. T. Nguyen et al. (Eds.): ACIIDS 2018, LNAI 10751, pp. 443–453, 2018.
https://doi.org/10.1007/978-3-319-75417-8_42

present [1]. Once the risk has been evaluated, it must be minimized in some way so that it does not eventuate.

Among the existing approaches for control of hazards in industries, the real time detection and tracking of workers to indicate when they are exposed to dangerous situations has gained interest. However, the existing sensor data and computer vision technologies are, mostly, not scalable and lack adaptability to broad industrial environments and existing situations. As a result, the existing technologies creates case-based applications that work only for specific circumstances [2].

This work is part of an effort for the development of a Knowledge-Vision Integration Platform for Hazard Control (KVIP-HC) in industrial workplaces, attending a wide range of industrial environments [3]. This paper focuses on hazards resulted from the non-use of personal protective equipment (PPE). For that, a few supervised classification learning techniques are examined to compose the system for the purpose of recognition of three protective equipment: hard hat, gloves and boots. Other market PPEs will be included in the future work, as well as a broader range of risky activities, as mentioned before. In this system, classifiers, feature images and any context information collected through the platform are represented explicitly using the Set of Experience Knowledge Structure (SOEKS) [4], grouped and stored as Decisional DNA (DDNA) [5]. The collected knowledge is used for reasoning and reinforcing the system from time to time, customizing the service according to each scenario and application. In choosing the classification methodology that best suits the application, the processing time for training and detection, accuracy, and predictor sizes are analyzed.

This paper is organized as follow: in section two a background about vision-based classification techniques and the knowledge representation approach in use is introduced; in section three the methodology, including the system design, the process of creating the datasets, and the classification techniques to be investigated are presented; in section four experimental results and performance of each learning technology are discussed; and lastly, in section five, conclusions and future work are presented.

2 Background

The use of sensors data and computer vision techniques support automated detection and tracking of workers indicating potential dangerous situations. However, the existing sensor-based technologies used for this purpose require the devices or markers to be attached on the human body, which can disturb some movements [6]. In contrast, visual sensing facilities, such as video cameras, can monitor workers behavior and environment conditions without any disturbances. In addition, the generated data, such as video sequences or digitized visual data can be processed in powerful computers in real time [7]. For the reasons just described, computer vision approach has been a research focus for a long period of time in surveillance systems, human detection and tracking. The most common vision-based systems use supervised techniques, although some researches have shown interesting results for unsupervised systems such as [8], which presents an efficient and stable system to detect unusual events in a large set of videos based on unsupervised learning approach, using a small set of features.

2.1 Supervised Learning Classification

K-Nearest Neighbors (k-NN) classification technique is conceptually and computationally the simplest technique that provides high classification accuracy [9]. The k-NN algorithm is based on a distance function and a voting function in k-Nearest Neighbors; the metric employed is the Euclidean distance [10]. However, Support Vector Machines (SVM) have high approximation capability and much faster convergence than k-NN for many applications [11].

Another method proven successful on classification problems is the traditional statistical Discriminant Analysis (LDA) [12], and its extension, the Quadratic discriminant analysis (QDA). QDA improves LDA as it works for non-linear problems by separating classes of objects according to quadric surface [13]. However, classification rules based on QDA are known to require generally larger samples than those based on LDA [14].

Recently, there has been a lot of interest in "ensemble learning" methods, with boosting [15] and bagging [16] as well-known classification trees methods. Some examples of algorithms that implement boosting are Adaboost, and Gradient Boosting. AdaBoost constructs a strong classifier as a combination of weak classifiers with proper coefficients as an iterative supervised learning process. Gradient Boosting algorithm determines in each iteration the direction or gradient. There is still randomness, but on average the gradient indicates moving closer to the true regression function when compared to Adaboost [17]. In bagging, each successive tree is independently constructed using a bootstrap sample of the data set and a majority vote is taken for prediction [18]. Some examples of algorithms are Standards Trees, Random Forest and Extremely Randomized trees. In Standard Trees, each node is split using the best split among all variables [19]. In a Random Forests, each node is split using the best among a subset of predictors randomly chosen at that node [18]. In Extreme Randomized trees randomness goes one step further and a reduction of variance is achieved at a cost of slightly increase in bias [20].

Neural networks are also a very powerful Machine Learning classifier. The first computational model for neural networks was based on mathematics and algorithms called threshold logic [21]. In 1965, the first functional networks with many layers (Deep Neural Network DNN) was proposed. However, at that time the computers didn't have enough processing power to effectively handle the work required by large neural networks [22]. A key trigger for the renewed interest in neural networks and learning was the backpropagation algorithm that accelerated the training of multi-layer networks [23]. Another technique called Dropout used to reduce overfitting (a very common problem in deep networks) gave major improvements over other regularization methods [24]. In 2011, Convolutional Neural Network CNN (a class of deep, feed-forward artificial neural networks) was presented [25]. Such supervised deep learning method was the first artificial pattern recognizer to achieve human-competitive performance on certain tasks [26].

2.2 Knowledge Based Approach

Although vision-based approach appears to be a reasonable option for the proposed application, the accuracy of current systems when operating in real life scenarios, subject to change in illumination, variation in backgrounds, noise and different camera resolutions still remains a challenge [27]. Furthermore, the broad industrial environments and situations existing may result in rewriting most of the application code each time the conditions change.

In this context, methods incorporating prior knowledge and context information are gaining interest. With additional contextual information it is possible to enhance the speed and accuracy of the detection algorithm and reduce scalability issues [28]. For instance, an automatic semantic and flexible annotation service able to work in a variety of video analysis with little modification to the code using Set of Experience Knowledge Structure (SOEKS) was proposed in [2]. Though, a lot of information contained in the classifiers was lost by separating the classification of the humans and objects from the event recognition.

Set of Experience Knowledge Structure (SOEKS) and Decisional DNA (DDNA). Set of Experience Knowledge Structure (SOEKS) is a knowledge representation structure designed to obtain and store formal decision events in an explicit way. It is based on four key elements of decision-making actions: variables, functions, constraints and rules. Variables are used to represent knowledge in an attribute-value form; functions define relations between a dependent variable and a set of input variables; constraints are functions that act as a way to limit possibilities, restrict the set of possible solutions and control the performance of the system; finally, rules are relationships that operate in the universe of variables and express the condition-consequence connection as "if-then-else" [4]. A SOEKS from a formal decision event represents a portion of an organization's DDNA (as a gene that guides decision-making). This gene belongs to a decisional chromosome from a certain type or category. A group of chromosomes from different kinds comprise the DDNA, a decisional genetic code of an organization [5].

3 Methodology

In this section, (i) the process of creating the dataset is explained and (ii) the learning algorithms candidates to compose the system are presented. The codes to perform the data-preparation, training, tests and analysis of the methodologies were developed in Python 3 [29] and Jupyter Notebook web application [30] platform.

3.1 Dataset

For the creation of the positive samples 12 videos with different sizes and resolutions were used (containing images of boots, gloves and hard hats). These videos were separated automatically into frames using FFmpeg program [31] and the ones containing the objects of interest were randomly selected. From each selected frame, a Region of Interest (ROI), comprising a PPE was saved as a positive sample. These images were split into *Training* and *Test*. The *Training* positive dataset was composed by 302 samples of boots, 300 of gloves and 310 of hard hats. The *Test* positive dataset by 100 entire or part images of boots, 100 of gloves and 100 of hard hats. For the purpose of training and validation, the positive dataset is considered very small, a couple of thousands images are expected to achieve high accuracy [32]. However, such small quantity of samples is necessary to reduce the amount of data stored in the system and consequently the costs. Thus, a data augmentation was performed on the *Training* dataset and saved temporarily. To augment the data, the samples were rotated, shifted, flipped and saved in a size of 28 × 28 pixels. The images were reduced to 28 × 28 to ensure the recognition time is small enough to be used in real-time applications and also to reduce computation time for training. The final positive dataset was, composed by three subsets containing entire or parts of the object of interest: 3190 of boots, 3206 of gloves, and 3270 of hard hats. To guarantee robustness, the datasets containing samples are subject to low resolution, blur and lens distortion, different lighting conditions and diverse background before being processed. Figure 1 summarizes the process of creating the positive datasets.

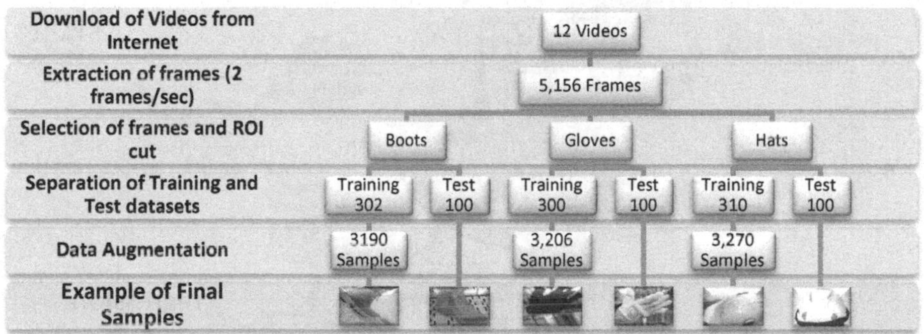

Fig. 1. Process of creation of the positive datasets.

The negative dataset is composed by images of random objects (anything besides gloves, hard hats and boots) and general scenes. The images were standardized, having their sized fixed to 28 × 28 and saved as RGB in .jpg format. The negative dataset is composed by 3000 images for *Training* and 100 images for *Test*.

3.2 Learning Algorithms

Feature-Based Classifiers. Often the hardest part of solving a machine learning problem can be finding the right estimator (or learning algorithm) for the job. Different estimators are better suited for different types of data and different problems. The flowchart shown in Fig. 2 is designed to give users a rough guide on how to approach problems with regard to which learning algorithm to try on their data and the methods presented are available through the Scikit-Learn [33] library. Scikit-Learn is developed by a large community of developers and machine-learning experts and includes tools for a variety of standard machine-learning tasks. For the proposed application, the objective is to predict categories of labeled data. Through Fig. 2, the approach that will suit the application the most is the classification methodology. As previously mentioned, for the training process, a limited number of samples is used (more than 50 per class, but less than 100 K). It is shown in Fig. 2 that the recommended methods are: the Support Vector Machine (SVM) implemented as linear kernel (Linear SVM) multiple kernels (Adjusted SVM), k-Nearest-Neighbors (k-NN) and the Ensemble Classifiers, which includes Decision Tree, Random Forests (RF), Extremely Randomized Trees (ERT), Adaboost, Gradient Boosting, Linear Discriminant Analysis (LDA) and Quadratic Discriminant Analysis (QDA). The implementation of these algorithms in Scikit-Learn library is based on feature vectors, a vector that contains information describing image's important characteristics. In this paper, the raw intensity value of each pixel is used as a feature representation.

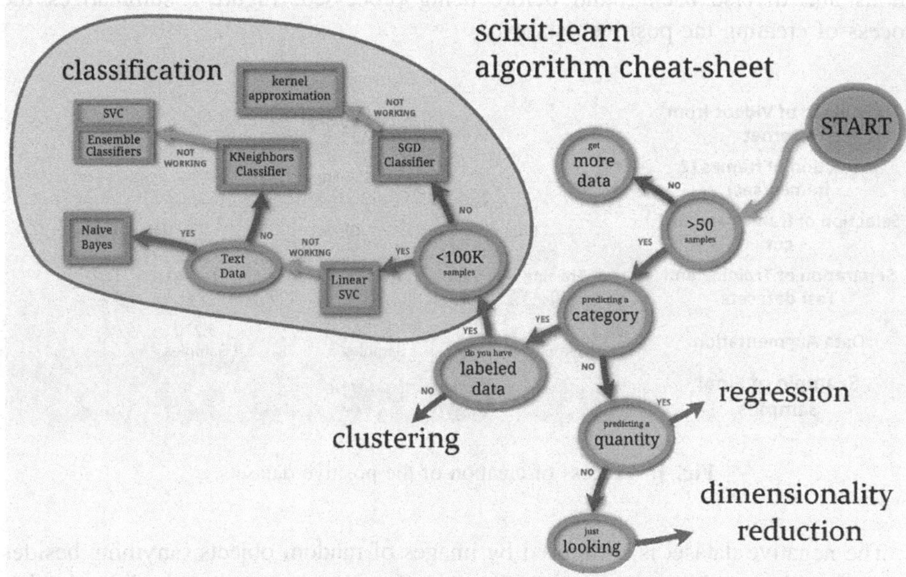

Fig. 2. Flowchart to guide users on choosing estimators [32].

Deep Neural Networks. Neural Networks algorithms are included in Scikit-Learn library but their implementation is not intended for large-scale applications and do not offer support for Graph Processing Units (GPUs) [33]. For that reason, the TensorFlow library [34], an open source library developed to conduct machine learning and deep neural networks research was utilized. Through TensorFlow it was possible to test the performance of a general Deep Neural Network DNN using backpropagation, the Convolutional Neural Network (CNN) and CNN with random regularization to prevent overfitting (CNN Dropout).

4 Experimental Results

Each classifier implemented using Scikit-Learn library has one or more hyper-parameters as input. To optimize the hyper-parameters for the feature-based classifiers random searching was used [35], choosing suitable parameters that maximize the accuracy and minimize the overfitting for all three training datasets.

For the boosting and discriminant analysis algorithms, the default input parameters are chosen. For Linear SVM, the estimation is adjusted by reducing *C* to *0.025*. For the Adjusted SVM, *C* = *1.8* achieve the best performance for kernel *rbf*. For kNN, the number of neighbors *k* chose is *3*. For Standard Decision tree, the optimal maximum depth *max_depth* of each tree is *5*. For Random Forests, the number of jobs to run in parallel *n_jobs* is *10*. Finally, for Extremely Random Forests, the optimal maximum depth *max_depth* is *10*, and minimum number of splits *min_samples_split* of *2*.

For the Deep Neural Networks (DNN, CNN and CNN Dropout), *softmax regression* and *cross-entropy* to compute the loss of the model is used. For the DNN model, a single linear layer, minimizing the loss using *gradient descent algorithm* with a learning rate of *0.01* is utilized. For the CNN model and CNN Dropout, weights and biases are created. The pooling is a max pooling over 2×2 blocks. The convolutional layer has patch size of 5×5 with *3* input channels and *32* output channels. In the fully-connected layer, there are *1024* neurons (to allow processing on the entire image, which is reduced to 14×14), that feed the readout layer together with the number of classes to predict. In addition, the steepest *gradient descent optimizer* is replaced by *ADAM*, a more sophisticated optimizer. For the CNN Dropout, the regularization function is applied before the readout layer.

The results for accuracy, training and test time, and size of classifiers for Boots, Gloves and Hats dataset are presented in Table 1. As can be observed, Adjusted SVM gives better results than Linear SVM, with very good accuracy for all three test datasets. However, in both cases the detection time is high when compared to other methods. As the proposed platform will operate in real time, the detection time must be as small as possible which makes the SVM inappropriate for the proposed application. The k-NN classifier has moderate accuracy and low training time, but the detection time is the highest among all methodologies and the predictors' size are also very large. Due to the slow detection and large predictor size (which must be considered for the purpose storing data for reinforcement) k-NN methodology was also found inappropriate.

Among the ensemble models, Gradient Boosting and RF present the best performances in terms of accuracy. Although, the time required to train a Gradient Boosting classifier is greater than all the other classifiers. It makes RF slightly better for the purpose of eventual re-training of the classifiers. However, RF has the largest predictor size among all the ensemble learning algorithms, which could become an issue as the platform grows and more predictors are incorporated to the system, increasing the volume of data to be stored excessively.

The LDA fitted the datasets better than the QDA. The QDA overfitted all three PPEs for all hyper-parameters verified, resulting in a very low accuracy for the test

Table 1. Classifiers' performance for Boots, Gloves and Hats dataset.

	Classifier	Training acc. (%)	Test acc. (%)	Training time (s)	Test time (s)	Predictor size (KB)
Boots	Linear SVM	*89.22*	*86.50*	*126.22*	*2.28*	*52,831*
	Adjusted SVM	*99.04*	*93.50*	*87.87*	*2.34*	*55,884*
	k-NN	*86.28*	*83.50*	*1.57*	*7.22*	*265,463*
	Decision Tree	*80.98*	*83.00*	*7.45*	*0.00*	*6*
	RF	*99.53*	*89.50*	*1.17*	*0.25*	*658*
	ERT	*95.18*	*87.00*	*1.08*	*0.00*	*479*
	Adaboost	*84.38*	*88.00*	*74.87*	*0.02*	*31*
	Grad. Boosting	*95.23*	*92.50*	*188.60*	*0.00*	*178*
	LDA	*92.35*	*86.00*	*17.15*	*0.00*	*93*
	QDA	*99.84*	*55.50*	*28.94*	*0.25*	*86,511*
	DNN	*99.99*	*89.00*	*3.96*	*0.01*	*1,292*
	CNN	*99.87*	*91.50*	*65.26*	*0.20*	*1,455*
	CNN Dropout	*97.96*	*94.00*	*81.94*	*0.18*	*1,637*
Gloves	Linear SVM	*92.00*	*85.00*	*103.94*	*2.00*	*43,174*
	Adjusted SVM	*99.87*	*95.00*	*69.53*	*1.89*	*46,043*
	k-NN	*75.86*	*67.00*	*1.46*	*7.11*	*263,037*
	Decision Tree	*86.41*	*81.00*	*7.61*	*0.00*	*6*
	RF	*99.77*	*92.00*	*1.46*	*0.23*	*496*
	ERT	*96.74*	*90.00*	*1.17*	*0.01*	*177*
	Adaboost	*90.43*	*86.00*	*77.90*	*0.02*	*31*
	Grad. Boosting	*97.88*	*90.00*	*190.57*	*0.00*	*177*
	LDA	*94.41*	*81.50*	*16.57*	*0.00*	*93*
	QDA	*100.00*	*68.50*	*27.79*	*0.36*	*86,511*
	DNN	*99.75*	*91.00*	*6.15*	*0.01*	*2,551*
	CNN	*99.42*	*96.00*	*124.45*	*0.18*	*2,775*
	CNN Dropout	*99.71*	*95.50*	*82.54*	*0.20*	*3,018*
Hats	Linear SVM	*93.36*	*89.00*	*85.74*	*1.52*	*35,135*
	Adjusted SVM	*97.69*	*95.00*	*62.18*	*1.59*	*40,985*
	k-NN	*95.03*	*86.00*	*1.36*	*7.69*	*266,676*
	Decision Tree	*90.74*	*86.50*	*8.30*	*0.00*	*6*
	RF	*99.49*	*92.50*	*1.00*	*0.23*	*442*
	ERT	*97.67*	*93.50*	*1.14*	*0.00*	*464*
	Adaboost	*91.87*	*88.50*	*76.48*	*0.02*	*31*
	Grad. Boosting	*96.67*	*95.50*	*187.26*	*0.00*	*177*
	LDA	*95.66*	*90.50*	*14.78*	*0.00*	*93*
	QDA	*99.94*	*60.50*	*25.30*	*0.23*	*86,511*
	DNN	*99.11*	*95.00*	*2.01*	*0.01*	*3,232*
	CNN	*98.87*	*97.00*	*50.97*	*0.18*	*3,484*
	CNN Dropout	*99.39*	*97.50*	*62.24*	*0.17*	*3,755*

datasets. Both LDA and QDA classifiers are also large in size, which can lead to the same scalability issues as RF.

Alternatively, the deep neural networks were proven to be universal approximators; all networks were capable of approximating all three measurable functions accurately. The algorithm CNN Dropout presented the best overall accuracy for the test datasets, and detection time for all of the PPEs is less than a sec/100 samples. The training time is moderate but the predictors' size are larger than the ones obtained from the ensemble methods. These results indicates that CNN Dropout is a good option for the system.

5 Conclusion and Future Work

This research is still at its early stages. Classification technologies have been extensively researched and helped on the evaluation of the possibilities of classification technologies to compose the proposed Knowledge-Vision Integration Platform for Hazard Control (KVIP-HC) in industrial workplaces. From the results obtained, it can be concluded that CNN Dropout is a reasonable choice with good average performance for the variables analyzed with low variance among the three PPEs datasets. However, the predictors' size is a concern that will need to be addressed. In addition, ensemble methods such as Random Forest and Gradient Boosting also performed well (in average, slightly lower than CNN Dropout).

In future work, the PPE classification will be expanded. Reinforcement learning will be explored and the capacity of the system to adapt to different set of images with different characteristics will be tested.

References

1. Safetycare Australia: Recognition, evaluation and control of hazards, Malvern East, Victoria (2015)
2. Zambrano, A., Toro, C., Nieto, M., Sotaquirá, R., Sanín, C., Szczerbicki, E.: Video semantic analysis framework based on run-time production rules – towards cognitive vision. J. Univ. Comput. Sci. **21**(6), 856–870 (2015)
3. de Oliveira, C.S., Sanin, C., Szczerbicki, E.: Hazard control in industrial environments: a knowledge-vision-based approach. In: Wilimowska, Z., Borzemski, L., Świątek, J. (eds.) ISAT 2017. AISC, vol. 657, pp. 243–252. Springer, Cham (2018). https://doi.org/10.1007/978-3-319-67223-6_23
4. Sanín, C., Szczerbicki, E.: Set of experience: a knowledge structure for formal decision-events. Found. Control Manag. Sci. **3**, 95–113 (2005)
5. Sanín, C., Szczerbicki, E.: Towards the construction of decisional DNA: a set of experience knowledge structure Java class within an ontology system. Cybern. Syst. **38** (2007)
6. Han, S., Lee, S.: A vision-based motion capture and recognition framework for behavior-based safety management. Autom. Constr. **35**, 131–141 (2013)
7. Chen, L., Hoey, J., Nugent, C.D., Cook, D.J., Yu, Z.: Sensor-based activity recognition. IEEE Trans. Syst. Man Cybern. Part C (Appl. Rev.) **42**(6), 790–808 (2012)

8. Zhong, H., Shi, J., Visontai, M.: Detecting unusual activity in video. In: Proceedings of the 2004 IEEE Computer Society Conference on Computer Vision and Pattern Recognition, CVPR 2004, pp. 819–826. IEEE (2004)
9. Ramteke, R.J., Khachane, M.Y.: Automatic medical image classification and abnormality detection using K-Nearest Neighbour. Int. J. Adv. Comput. Res. **2**(4), 190–196 (2012)
10. Revathy, M.: Image classification with application to MRI brain using 2nd order moment based algorithm. Int. J. Eng. Res. Appl. (IJERA) **2**(3), 1821–1824 (2012)
11. Steinwart, I.: Sparness of support vector machines–some asymptotically sharpbounds. In: Thrun, S., Saul, L., Schölkopf, B. (eds.) Advances in Neural Information Processing Systems, vol. 16, pp. 169–184. MIT Press, Cambridge (2004)
12. Fukunaga, K.: Introduction to Statistical Pattern Recognition, 2nd edn. Academic Press, Inc., Cambridge (1990)
13. Gardner-Lubbe, S., Dube, F.S.: Visualisation of quadratic discriminant analysis and its application in exploration of microbial interactions. BioData Min. **8**, 8 (2015)
14. Wald, P.W., Kronmal, R.A.: Discriminant functions when covariances are unequal and sample sizes are moderate. Biometrics **33**, 479–484 (1977)
15. Shapire, R., Freund, Y., Bartlett, P., Lee, W.: Boosting the margin: a new explanation for the effectiveness of voting methods. Ann. Stat. **26**(5), 1651–1686 (1998)
16. Breiman, L.: Bagging predictors. Mach. Learn. **24**(2), 123–140 (1996)
17. Ridgeway, G.: The state of boosting. Comput. Sci. Stat. **31**, 172–181 (1999)
18. Breiman, L.: Random forests. Mach. Learn. **45**(1), 5–32 (2001)
19. Rokach, L., Maimon, O.: Top-down induction of decision trees classifiers-a survey. IEEE Trans. Syst. Man. Cybern. Part C (Appl. Rev.) **35**(4), 476–487 (2005)
20. Geurts, P., Ernst, D., Wehenkel, L.: Extremely randomized trees. Mach. Learn. **63**(1), 3–42 (2006)
21. Warren, W., Pitts, W.: A logical calculus of ideas immanent in nervous activity. Bull. Math. Biophys. **5**(4), 115–133 (1943)
22. Ivakhnenko, A.G., Lapa, V.G.: Cybernetics and Forecasting Techniques. American Elsevier Publishing Co., New York (1967)
23. Werbos, P.J.: Beyond regression: new tools for prediction and analysis in the behavioral sciences (1975)
24. Srivastava, N.: Dropout: a simple way to prevent neural networks from overfitting. J. Mach. Learn. Res. **15**(1), 1929–1958 (2014)
25. Ciresan, D.C., Meier, U., Masci, J., Gambardella, L.M., Schmidhuber, J.: Flexible, high performance convolutional neural networks for image classification. In: International Joint Conference on Artificial Intelligence (2011)
26. Ciresan, D., Meier, U., Schmidhuber, J.: Multi-column deep neural networks for image classification. In: IEEE Conference on Computer Vision and Pattern Recognition (2012)
27. Mosberger, R., Andreasson, H., Lilienthal, A.J.: Multi-human tracking using high-visibility clothing for industrial safety. In: 2013 IEEE/RSJ International Conference on Intelligent Robots and Systems (IROS 2013), pp. 638–644 (2013)
28. Davis, R., Shrobe, H., Szolovits, P.: What is a knowledge representation? AI Mag. **14**(1), 17–33 (1993)
29. Van Rossum, G., Drake, F.L.: Python 3: Reference Manual. SohoBooks, New York (2009)
30. Kluyver, T., Ragan-Kelley, B., Pérez, F., Granger, B.E., Bussonnier, M., Frederic, J., Ivanov, P.: Jupyter Notebooks-a publishing format for reproducible computational workflows. In: ELPUB, pp. 87–90 (2016)
31. Tomar, S.: Converting video formats with FFmpeg. Linux J. **146**, 10 (2006)
32. Van der Walt, C.M., Barnard, E.: Data characteristics that determine classifier performance (2006)

33. Pedregosa, F., Varoquaux, G., Gramfort, A., Michel, V., Thirion, B., Grisel, O., Vanderplas, J.: Scikit-learn: machine learning in Python. J. Mach. Learn. Res. **12**, 2825–2830 (2011)
34. Abadi, M., Barham, P., Chen, J., Chen, Z., Davis, A., Dean, J., Kudlur, M.: TensorFlow: a system for large-scale machine learning. OSDI **16**, 265–283 (2016)
35. Bergstra, J., Bengio, Y.: Random search for hyper-parameter optimization. J. Mach. Learn. Res. **13**, 281–305 (2012)

Feature-Based Image Compression

Pavel Morozkin[1,2(✉)], Marc Swynghedauw[1], and Maria Trocan[2]

[1] SuriCog, 130 Rue de Lourmel, 75015 Paris, France
{pmor,ms}@suricog.com

[2] Institut Supérieur d'Electronique de Paris, 28 Rue Notre Dame des Champs, 75006 Paris, France
maria.trocan@isep.fr

Abstract. The EyeDee™ embedded eye tracking solution developed by SuriCog is the world's first innovative solution using the eye as a real-time mobile digital cursor, while maintaining full mobility. The system consists in a wearable device capturing images on the human's eye and sending these images over a transmission medium (wire/wireless transmission). One important request of this system is the real-time transmission of the captured images, along with low-power, low-heat, low-MIPS requirements. This work is concentrated around an improvement of the ROI (Region of Interest – region containing image of the human's pupil) image compression performance achieved via extra information removal. The feature based compression lies on ROI image blocks classification, implemented using a neural network.

Keywords: Eye tracking · Image compression · Neural networks

1 Introduction

In [1] the EyeDee™ (Fig. 1) embedded eye tracking solution developed by SuriCog was introduced. The problematic is the deployment of computationally intensive algorithms on a restrained resources embedded platform. Nowadays, most embedded and portable image processing applications require low power consumption and wireless communication and face a common problem of limited CPU resources and limited battery autonomy (measure of the time for which the battery will support the Weetsy™ board operation).

In most cases, these applications have a processing chain as follows:

- Retrieve the image from the sensors (readout);
- Process the image in real time in order to extract relevant features (results of computer vision algorithms: convolution, thresholding, etc.);
- Find parameters of a physical model relevant to the application (for example, SLAM for reconstruction of the 6 degrees of freedom of an object, ellipse fitting for eye tracking, model fitting, etc.) through mathematical optimization with the previously computed features;
- Send the parameters to the end user application.

N. T. Nguyen et al. (Eds.): ACIIDS 2018, LNAI 10751, pp. 454–465, 2018.
https://doi.org/10.1007/978-3-319-75417-8_43

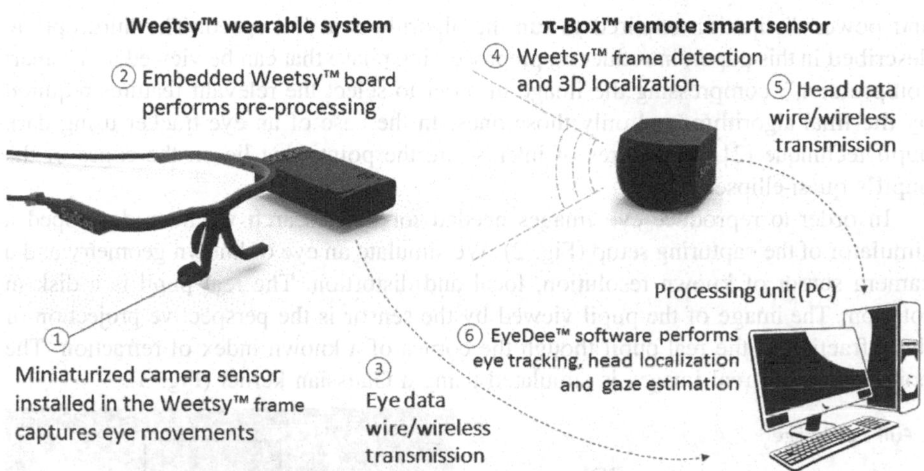

Fig. 1. EyeDee™ eye tracking solution: Weetsy™ frame, Weetsy™ board, π-Box™ remote smart sensor and processing unit (running EyeDee™ software).

Processing algorithms can be highly demanding in CPU resources and can often require different platforms for implementation. For example, FPGA/GPU are optimized for parallel/integer type processing such as image processing, while standard CPU has an easier implementation of floating point mathematical optimizations. Communication/readout stacks can introduce important issues, such as the bottleneck effect due to limited bandwidth/memory, or cost of implementation.

In the case of SuriCog's EyeDee™ solution, the embedded system should be able to capture at high frequency (100 Hz) the image of user's eye (~VGA 8bpp), and broadcast wirelessly in real time to the end application the result of the processing algorithm which consist in the parametrization of a 3D model of the eye. The system should run continuously during more than 3 h, with the lowest latency possible (typically < 10 ms). Three options are possible:

- Locally read the sensors and send the resulting image to the end application to proceed to full algorithm on the client machine;
- Locally read the sensors and locally process the full algorithm, and send the results to the client machine;
- Locally read the sensors, pre-process the image and send these preprocessed images to the client machine for final processing.

SuriCog's EyeDee™ eye tracking algorithm consists in reconstructing the quasi-ellipse of the contour of the pupils in order to fit a 3D model of an eye. The image processing (contour extraction, thresholding, etc.) require to work on a locally preserved region of the image (highest quality image in the vicinity of the pupil), unmodified by any compression algorithm.

The first option is constraint by the limited bandwidth of the wireless channel (Wifi, BT) and the latency/quality loss introduced by standard compression/decompression algorithms. The second option is constraint by the limited resources of CPU, battery and power dissipation required to run the algorithms at full speed. The third option, described in this paper, introduces a preprocessing phase that can be viewed as a "smart compression": compressing the image in order to select the relevant features required by the final algorithm, and only those ones. In the case of an eye tracker using dark pupil technique [2] the features of interest are the points that lie on the edges of the pupil's quasi-ellipse.

In order to reproduce eye images needed for the research we have developed a simulator of the capturing setup (Fig. 2). We simulate an eye of known geometry and a camera sensor of known resolution, focal and distortion. The real pupil is a disk in rotation. The image of the pupil viewed by the sensor is the perspective projection of the refraction of the real pupil though the cornea of a known index of refraction. The noise present in real images is simulated using a Gaussian kernel (Fig. 3).

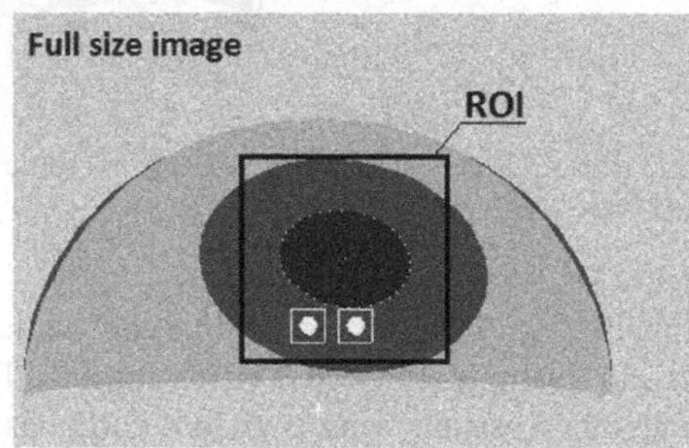

Fig. 2. Ellipse reconstruction on simulator-based eye image.

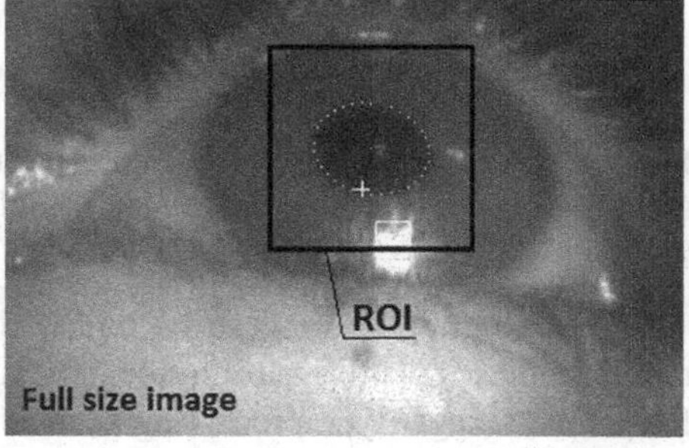

Fig. 3. Ellipse reconstruction based on a real human eye image (videoframe).

In [1] we proposed a deep learning method based on Artificial Neural Network (ANN) used as function regression calculator. It consists in tuning the hyperparameters [3–5] of this network to relate the image of the pupils (inputs) to the 5 parameters of a geometrical ellipse (outputs).

The approach we propose in this paper is to use a deep learning method, based on Artificial Neural Network (ANN), as a classifier. We train the ANN to learn which are the relevant areas of an image (i.e., the edges) and which are not. We then use the ANN output classification as an image compressor to define which area of the image should be broadcasted for further processing.

2 Eye Image Compression Approaches

Herein and after we use the following terminology:

- ROI (Region of Interest) – region containing image of the human's pupil;
- FOI (Features of Interest) – image containing useful (for the eye tracking algorithm) features of ROI.

ROI finding technique is based on Haar-like features for rapid object detection [6–9], but without any machine learning (like Viola Jones object detection framework [10]) applied.

It is necessary to understand the difference of the proposed neural network based approach of ROI eye image compression with respect to the classical one. Classical approach of image compression (Fig. 4), consists in the compression of the ROI image, sending thus the compressed image (bitstream) over a channel, followed by decompression of the bitstream on the remote side to get the original ROI image, which is further used as an input for the eye tracking algorithm. The proposed neural network based approach of image compression (Fig. 5), consisted in the following steps:

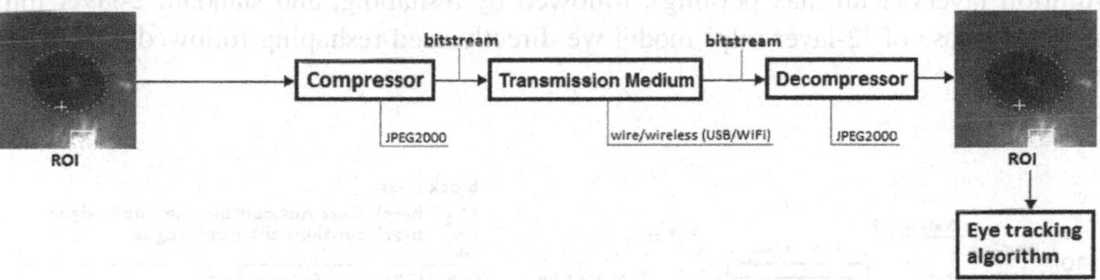

Fig. 4. Image compression: classical approach.

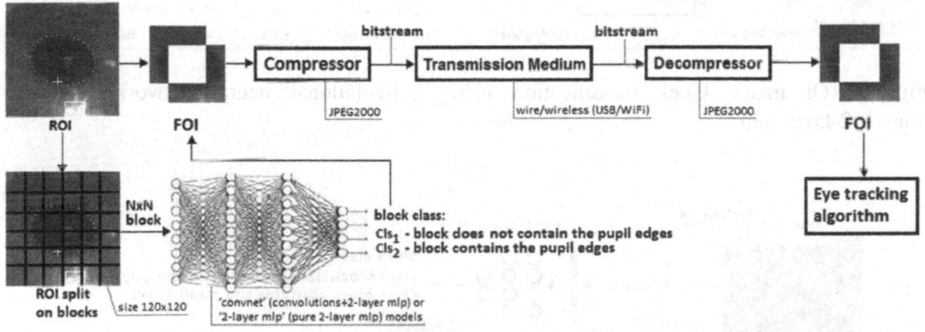

Fig. 5. Image compression: neural network based approach.

– Training a neural network to classify the blocks of a ROI image. The training is based on a set of samples: ROI block itself and a boolean value that indicates if this block contains pupil edges. During the training this value is obtained from the default image processing based eye tracking algorithm.
– Use the trained neural network to perform ROI blocks classification, i.e. to decide if a particular block contains pupil edges.

3 Neural Network Construction and Training

The neural network [11] is aimed on classification of the ROI image blocks into 2 classes (blocks contain/does not contain pupil edges). To this purpose, Torch7 [12] software (neural network 'nn' and optimization 'optim' packages) is used with 'convnet' (convolutions + 2-layer mlp, where 2-layer mlp is *multilayer perceptron*, Fig. 6) and '2-layer mlp' (pure 2-layer mlp, Fig. 7) models. This functionality was further integrated into the EyeDee™ eye tracking software running on Windows platform. During an initial testing we decided to split the ROI image into blocks of size 20 × 20 and 10 × 10 (the block sizes in range 10..20 lead to conditions resulting in maximal benefit of using the presented approach). In case of 'convnet' model we used 2 convolution layers (with max pooling), followed by reshaping, and standard 2-layer mlp model. In case of '2-layer mlp' model we directly used reshaping followed by 2-layer mlp model.

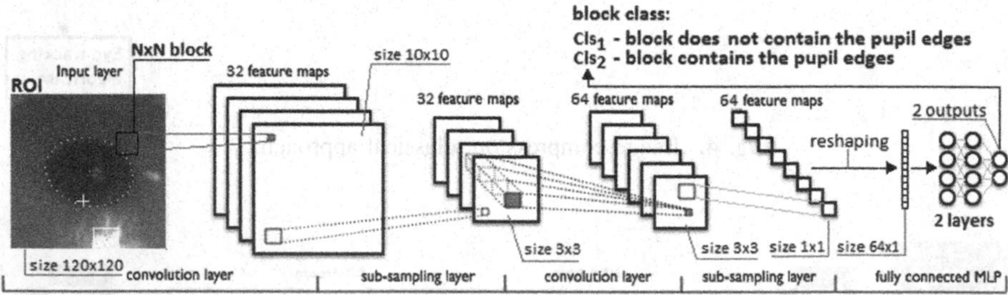

Fig. 6. ROI image block classification using convolutional neural network (convolutions + 2-layer mlp).

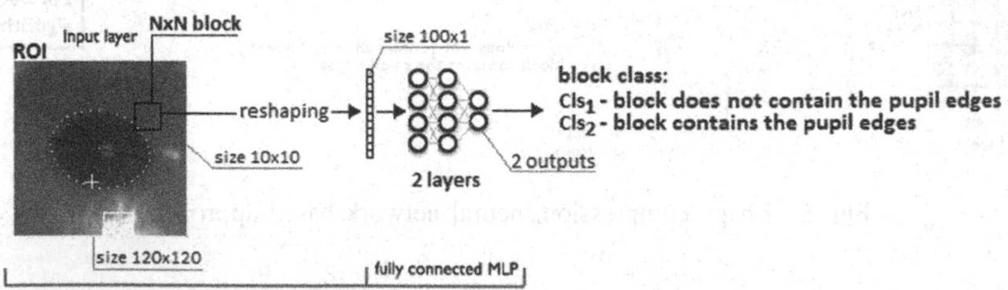

Fig. 7. ROI image block classification using 2-layer mlp neural network.

We used batch learning, i.e. learning on the entire training data set at once. In all tests batch size was set to 10. The training of the neural network is based on the well-known back-propagation approach [13] coupled with a gradient descent optimization method [14]. The output of the network is compared to the desired output using a loss criterion. Therefore, the training can be interpreted as the loss function optimization (error minimization).

4 Testing Framework

After the integration of the Torch7 software into the EyeDee™ eye tracking solution several experiments were completed (Figs. 8 and 9). The first experiment (Fig. 8) is targeted on bitrate (expressed as bpp, bits per pixel) comparison of the eye images in different configurations. This experiment will show the benefit of different compression approaches.

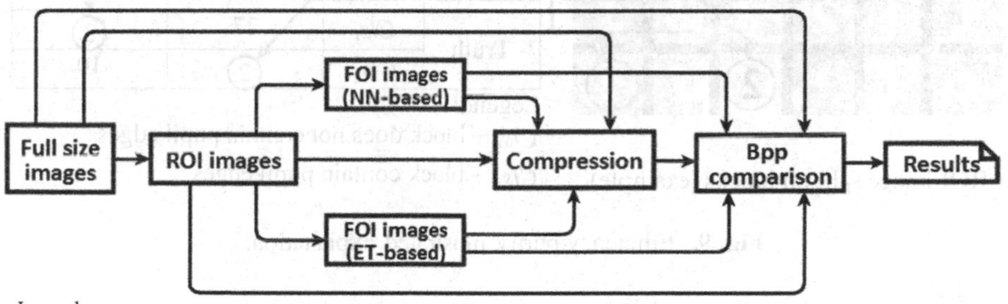

Legend:
FOI images (NN-based) – FOI images obtained with neural network based approach,
FOI images (ET-based) – FOI images obtained with general ET (Eye Tracking) algorithm

Fig. 8. Compressed image comparison scheme.

The second experiment (Fig. 9) is targeted on efficiency (ε) and purity (p) of blocks classification, which are calculated as follows:

$$\varepsilon = \frac{N_{11}}{N_{10} + N_{11}}, \tag{1}$$

$$p = \frac{N_{01}}{N_{01} + N_{11}}, \tag{2}$$

where:

- N_{00} – number of blocks predicted not to contain pupil edges which do not contain edge in reality;
- N_{01} – number of blocks predicted to contain pupil edge which do not contain edge in reality;
- N_{10} – number of blocks predicted not to contain pupil edge which do contain edge in reality;
- N_{11} – number of blocks predicted to contain pupil edge which do contain edge in reality.

These efficiency ε and purity p metrics were selected to estimate the training quality of the neural network, because terminology 'efficiency' and 'purity' is more pertinent (in comparison with commonly used 'precision' and 'recall') in characterizing results of blocks classification.

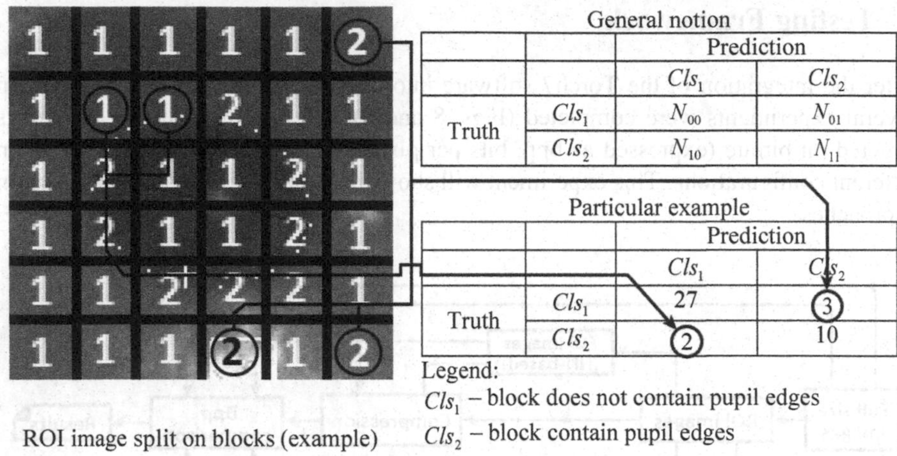

Fig. 9. Efficiency/purity illustrated explanation.

5 Experimental Results

In order to evaluate the learning quality, we have tested the training of the neural network with images taken from a simulator and from previously saved video sequences of eye movements. We finally selected 400 training iterations (epochs), as this amount is reasonable to get satisfying results (i.e., confusion matrixes values). With the same setup we obtained the results for two tested models: 'convnet' (convolutions + 2-layer mlp) and '2-layer mlp' (pure 2-layer mlp).

According to the classification quality results (Figs. 10 and 11), the use of a trained neural network for ROI image blocks classification has a good potential in general. In particular, 'convnet' model shows better results over '2-layer mlp' model. This can be explained by the features extractions followed by the '2-layer mlp' model (instead of pure '2-layer mlp').

		Prediction	
		Cls_1	Cls_2
Truth	Cls_1	67.70%	0.02%
	Cls_2	0.15%	32.05%

Simulator, Block size 20

		Prediction	
		Cls_1	Cls_2
Truth	Cls_1	68.13%	0.14%
	Cls_2	0.08%	31.64%

Video, Block size 20

		Prediction	
		Cls_1	Cls_2
Truth	Cls_1	83.72%	0.53%
	Cls_2	1.13%	14.60%

Simulator, Block size 10

		Prediction	
		Cls_1	Cls_2
Truth	Cls_1	83.62%	0.88%
	Cls_2	1.48%	14.01%

Video, Block size 10

Fig. 10. Confusion matrixes, 400 epochs, '2-layer mlp' model.

According to the application results (Tables 1 and 2), both 'convnet' and '2-layer mlp' models show relatively promising results. In particular, increasing the number of training iterations (from 50 to 400) results in higher efficiency. This is because the number of N_{00} and N_{11} is higher meaning that the model is trained better. For example, using 'convnet' model with ROI blocks of size 10 × 10 at 100 training iterations is enough to reach 100% of both N_{00} and N_{01}.

		Prediction	
		Cls_1	Cls_2
Truth	Cls_1	67.78%	0.00%
	Cls_2	00.00%	32.21%

Simulator, Block size 20

		Prediction	
		Cls_1	Cls_2
Truth	Cls_1	68.27%	0.00%
	Cls_2	0.00%	31.72%

Video, Block size 20

		Prediction	
		Cls_1	Cls_2
Truth	Cls_1	83.87%	0.36%
	Cls_2	0.98%	14.77%

Simulator, Block size 10

		Prediction	
		Cls_1	Cls_2
Truth	Cls_1	83.88%	0.62%
	Cls_2	0.92%	14.56%

Video, Block size 10

Fig. 11. Confusion matrixes, 400 epochs, 'convnet' (convolutions + 2-layer mlp) model.

For the compression of ROI images, we used a JPEG2000 [15] encoder. We configured the encoder to keep a relatively high PSNR (>45 dBs) as the decompressed blocks will be further used in an image processing-based eye tracking algorithm. However, this quality can be lower, as proved in [16]. Results in Fig. 12 show that with the proposed approach it is possible to reach 99.47% of gain in terms of data size reduction.

The visual comparison of ROI image block removal quality (Fig. 13) on the validation set shows that there are more accuracy issues, in comparison with the training set, i.e., the number of N_{01} is increased (Fig. 13b, c). If the neural network is not enough trained the quality is significantly degraded (Fig. 13d) in case of both 'mlp' and 'convnet' models.

Increasing the number of ROI image blocks will result in a more accurate preservation of pupil ellipse edges (less data to transmit). However, for the validation set there are issues of preserved blocks situated in the corners of the ROI images. These issues can be solved by applying some additional logic. For example, if the block is located M-blocks far from the blocks containing pupil edges, this block can be considered as incorrect and should not be used for further image compression.

From the computational point of view (computational complexity), the ROI image compression based on neural network requires just the resources needed to obtain the classification results (i.e., input: N × N ROI image blocks, output: value indicating whether a block contains pupil edges). The neural network has also constant-time response, which can be taken into account for the overall performance estimation. However, the use of neural network has a well known challenge of its training (hyper-parameters tuning, training time reduction, output results quality maximization), as the cost of each training session is time expensive.

Table 1. Average results of ROI image blocks classification quality (expressed as efficiency/purity) and average bit budget (expressed as bits per pixel) needed to transmit the compressed ROI images with extra blocks removed. Use of '2-layer mlp' model.

	Mode	Block size	Epochs	ε%	*p*%	Bits per pixel (bpp)		
						Original ROI	ROI with removed blocks	ROI with removed blocks (reordered)
Training set	Trained on video, tested on video	20	50	97.85	1.24	2.15	1.33 (38.14%)	0.88 (59.07%)
			100	98.89	0.46	2.15	1.33 (38.14%)	0.89 (58.60%)
			200	99.19	0.48	2.15	1.33 (38.14%)	0.89 (58.60%)
			400	99.58	0.49	2.15	1.33 (38.14%)	0.89 (58.60%)
		10	50	70.64	20.12	2.15	1.06 (50.70%)	0.50 (76.74%)
			100	70.14	19.65	2.15	1.04 (51.63%)	0.50 (76.74%)
			200	70.87	17.54	2.15	1.04 (51.63%)	0.49 (77.21%)
			400	76.69	14.06	2.15	1.07 (50.23%)	0.52 (75.81%)
Validation set	Trained on simulator, tested on video	20	50	79.08	20.33	2.15	1.35 (37.20%)	0.90 (58.17%)
			100	81.09	17.74	2.15	1.35 (36.91%)	0.89 (58.49%)
			200	82.05	16.87	2.15	1.35 (36.95%)	0.89 (58.49%)
			400	85.50	20.79	2.15	1.38 (35.55%)	0.94 (56.03%)
		10	50	61.45	28.12	2.15	1.13 (47.41%)	0.51 (76.05%)
			100	63.55	29.64	2.15	1.16 (46.00%)	0.53 (75.12%)
			200	68.42	34.16	2.15	1.23 (42.61%)	0.59 (72.67%)
			400	72.08	37.94	2.15	1.29 (39.80%)	0.63 (70.86%)

Table 2. Average results of ROI image blocks classification quality (expressed as efficiency/purity) and average bit budget (expressed as bits per pixel) needed to transmit compressed ROI images with extra blocks removed. Use of 'convnet' model.

	Mode	Block size	Epochs	ε%	*p*%	Bits per pixel (bpp)		
						Original ROI	ROI with removed blocks	ROI with removed blocks (reordered)
Training set	Trained on video, tested on video	20	50	92.69	2.93	2.15	1.32 (38.66%)	0.86 (60.03%)
			100	93.93	0.27	2.15	1.31 (39.25%)	0.85 (60.48%)
			200	95.44	0.00	2.15	1.31 (38.90%)	0.86 (59.99%)
			400	100.00	0.00	2.15	1.33 (38.10%)	0.89 (58.68%)
		10	50	72.94	28.20	2.15	1.11 (48.50%)	0.55 (74.51%)
			100	70.04	26.66	2.15	1.10 (48.99%)	0.53 (75.45%)
			200	70.48	25.03	2.15	1.09 (49.14%)	0.53 (75.55%)
			400	71.31	23.10	2.15	1.31 (38.90%)	0.86 (59.99%)
Validation set	Trained on simulator, tested on video	20	50	89.90	11.68	2.15	1.37 (36.26%)	0.90 (58.05%)
			100	92.83	16.09	2.15	1.42 (33.67%)	0.96 (55.07%)
			200	93.07	16.51	2.15	1.43 (33.49%)	0.97 (54.81%)
			400	93.15	16.77	2.15	1.43 (33.37%)	0.97 (54.65%)
		10	50	62.06	38.45	2.15	1.18 (44.92%)	0.55 (74.51%)
			100	62.71	44.13	2.15	1.25 (41.94%)	0.59 (72.64%)
			200	62.58	43.91	2.15	1.26 (41.25%)	0.59 (72.57%)
			400	67.72	48.18	2.15	1.32 (38.43%)	0.65 (69.65%)

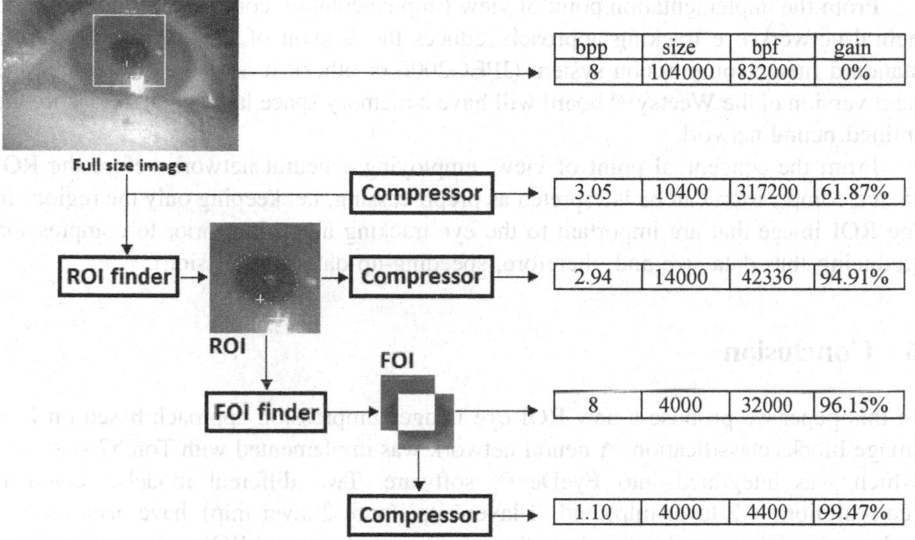

bpp	size	bpf	gain
8	104000	832000	0%
3.05	10400	317200	61.87%
2.94	14000	42336	94.91%
8	4000	32000	96.15%
1.10	4000	4400	99.47%

Fig. 12. Eye image compression configurations. Full image resolution: 400 × 260, ROI image resolution: 120 × 120, compressor: JPEG2000, bpp – bits per pixel, bpf – bits per frame, FOI – Features of Interest.

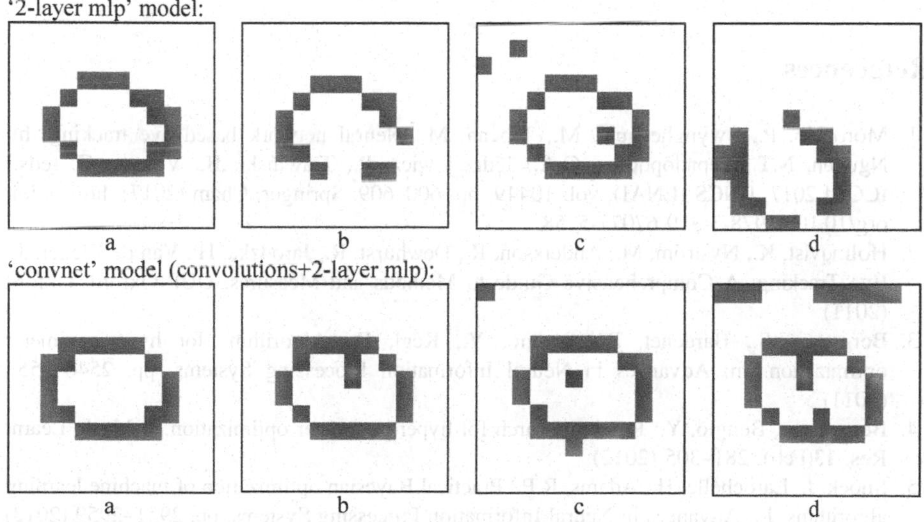

Fig. 13. Visual comparison of blocks removal quality on validation set (increasing degradation order, from best (a) to worst (d)). Block size is 10.

From the implementation point of view (implementation complexity), the proposed neural-network eye tracking approach reduces the amount of data to compress by a standard image compression system (JPEG2000 or other). It is also expected that the next version of the Weetsy™ board will have a memory space large enough to store the trained neural network.

From the conceptual point of view, employing a neural network before the ROI image compression can be interpreted as preprocessing, i.e. keeping only the regions in the ROI image that are important to the eye tracking algorithm, prior to compression (reducing thus data size and, therefore, speeding-up data transmission).

6 Conclusion

In this paper we propose a new ROI eye image compression approach based on ROI image blocks classification. A neural network was implemented with Torch7 software, which was integrated into EyeDee™ software. Two different models, 'convnet' (convolutions + 2-layer mlp) and '2-layer mlp' (pure 2-layer mlp), have been used in order to train the neural network, both with generated and real ROI images. It has been shown that the classification quality is high for these training models. The proposed compression method has an average bitrate gain of $\sim$64%, with respect to direct encoding of ROI images, at the only expense of memory needed to store the trained neural network. Future work is targeted on implementation of the approach in the Weetsy™ board as well as its generalization for usage in wider class of applications.

References

1. Morozkin, P., Swynghedauw, M., Trocan, M.: Neural network based eye tracking. In: Nguyen, N.T., Papadopoulos, G.A., Jędrzejowicz, P., Trawiński, B., Vossen, G. (eds.) ICCCI 2017. LNCS (LNAI), vol. 10449, pp. 600–609. Springer, Cham (2017). https://doi.org/10.1007/978-3-319-67077-5_58
2. Holmqvist, K., Nyström, M., Andersson, R., Dewhurst, R., Jarodzka, H., Van de Weijer, J.: Eye Tracking: A Comprehensive Guide to Methods and Measures. OUP Oxford, Oxford (2011)
3. Bergstra, J.S., Bardenet, R., Bengio, Y., Kégl, B.: Algorithms for hyper-parameter optimization. In: Advances in Neural Information Processing Systems, pp. 2546–2554 (2011)
4. Bergstra, J., Bengio, Y.: Random search for hyper-parameter optimization. J. Mach. Learn. Res. **13**(Feb), 281–305 (2012)
5. Snoek, J., Larochelle, H., Adams, R.P.: Practical Bayesian optimization of machine learning algorithms. In: Advances in Neural Information Processing Systems, pp. 2951–2959 (2012)
6. Lienhart, R., Maydt, J.: An extended set of haar-like features for rapid object detection. In: IEEE ICIP 2002, vol. 1, pp. 900–903 (2002)
7. Messom, C., Barczak, A.: Fast and efficient rotated haar-like features using rotated integral images. In: Australian Conference on Robotics and Automation, pp. 1–6 (2006)
8. Pavani, S.-K., Delgado, D., Frangi, A.: Haar-like features with optimally weighted rectangles for rapid object detection. Pattern Recogn. **43**(1), 160–172 (2010)

9. Barczak, A.L.C., Johnson, M.J., Messom, C.H.: Real-time computation of haar-like features at generic angles for detection algorithms. Res. Lett. Inf. Math. Sci. **9**, 98–111 (2006)
10. Viola, P.A., Jones, M.J.: Rapid object detection using a boosted cascade of simple features. In: Computer Vision and Pattern Recognition, no. 1, pp. 511–518 (2001)
11. Demuth, H.B., Beale, M.H., De Jess, O., Hagan, M.T.: Neural network design. Martin Hagan (2014)
12. Torch framework. www.torch.ch. Accessed 9 Oct 2017
13. Vogl, T.P., Mangis, J.K., Rigler, A.K., Zink, W.T., Alkon, D.L.: Accelerating the convergence of the back-propagation method. Biol. Cybern. **59**(4), 257–263 (1988)
14. Mason, L., Baxter, J., Bartlett, P.L., Frean, M.R.: Boosting algorithms as gradient descent. In: NIPS, pp. 512–518 (1999)
15. ISO/IEC 15444-1:2004 | ITU-T Rec. T.800
16. Morozkin, P., Swynghedauw, M., Trocan, M.: Image quality impact for eye tracking systems accuracy. In: 2016 IEEE International Conference on Electronics, Circuits and Systems (ICECS), pp. 429–431 (2016)

Scoring Photographic Rule of Thirds in a Large MIRFLICKR Dataset: A Showdown Between Machine Perception and Human Perception of Image Aesthetics

Adnan Firoze(✉), Tousif Osman, Shahreen Shahjahan Psyche, and Rashedur M. Rahman

Department of Electrical and Computer Engineering, North South University, Plot-15, Block-B, Bashundhara Residential Area, Dhaka, Bangladesh
{adnan.firoze, tousif.osman, shahreen.psyche, rashedur.rahman}@northsouth.edu

Abstract. In this research we have developed and evaluated a system that uses the image compositional metric called 'Rule of Thirds' used by photographers to grade visual aesthetics of an image. The novel aspect of the work is that it combines quantitative and qualitative aspects of research by taking human psychology into account. The core idea is to identify how similar the perception of a 'good image' and 'bad image' is by machines versus humans (through a user study based on 255 participants on 5000 images from the standard MIRFLICKR database [9]). We have considered the compositional norm, namely 'rule of thirds' used by photographers and inspired by the golden ratio that states that - if an image is segmented on a 3 × 3 grid, then it is appealing to the eye when the most salient object(s) or 'subject(s)' of the image is located precisely on or aligned on the middle grid lines [11]. First, we preprocess the input image by labeling the regions of attraction for human eye using two saliency algorithms namely Graph-Based Visual Saliency (GBVS) [3] and Itti-Koch [4]. Next, we quantify the rule of thirds property in images by mathematically considering the location of salient region(s) adhering to rule of thirds. This is then used to rank or score an input image. To validate, we conducted a user study where 255 human subjects ranked the images and compared our algorithmic results, making it a both a quantitative and qualitative research. We have also analyzed and presented the performance differences between two saliency algorithms and presented ROC plots along with similarity quantification between algorithms and human subjects. Our massive user study and experimental results provides the evidence of modern machine's ability to mimic human-like behavior. Along with it, results computationally prove significance of rule of thirds.

Keywords: Image processing · Visual perception · Image saliency Rule of thirds · Photography · Golden ratio · Computer vision Image score · Image composition · Image · Flickr

N. T. Nguyen et al. (Eds.): ACIIDS 2018, LNAI 10751, pp. 466–475, 2018.
https://doi.org/10.1007/978-3-319-75417-8_44

1 Introduction

One of the most abstract instance of comprehending visual data is 'finding beauty' - only possible by human psychology. In this research, we have devised a synthetic system which can understand visual appeal in a framed image. We have used the basic photographic rules as the foundation and the rules of computer vision and machine learning as the pillars of our system. In this research, first we have identified the salient region of an image using methods derived from research of Itti, Koch, and Harel [3, 4]. Next, we applied photographers' rule of thumb to measure the aesthetic appeal of the image [11]. The metric we have narrowed down to is 'Rule of Thirds' (ROT). Finally, we conducted a survey to evaluate our system's aesthetic measurement of an image with respect to visual aesthetics perceived by human psychology.

2 Related Works

Few have traveled before us in finding automated means to extract compositional properties in images. Here, the novel idea is the human user-study that we conducted. We have incorporated ideas, findings, and bits and pieces of several research in our system to conduct our research. One of these research aims to detect ROT compositions in images [1]. The ROT states that placing important or salient objects along the images' thirds lines (refer to Fig. 1) or around their intersections often produces highly aesthetic photos [2]. In this research, researchers have utilized multiple image saliency algorithms, namely, Fourier Transform (FT) Map, Graph Based Visual Saliency (GBVS) map [3], Global Contrast (GC) map and Objectness (OBJ) map to extract the salient region of an image. Afterwards, several machine learning techniques including the Naïve Bayesian Classifier, Support Vector Machine (SVM) etc. have been used for the rule of thirds detection. This research also shows that GBVS performs the best exhibiting 75% accuracy over a dataset of 2089 images collected from Flickr and Photo.net. We have incorporated the ideas of finding salient regions and ROT compositions of this research into our system.

Fig. 1. Some sample images that follows rule of thirds in photography to enhance appeal or aesthetics

In research [3], authors have proposed a bottom-up saliency model named "Graph Based Visual Saliency" (GBVS). This model mainly works in two steps. While the classical algorithms of Itti and Koch achieved only 84% of the ROC area of human based control, GBVS has achieved 98%. In a prior research [4] by Itti and Koch, authors aimed to solve this same problem of finding salient regions in digital images.

In [5] Mai et al. has used Rule of Thirds composition method to measure the aesthetics of an image. To do that, first they have used a variety of saliency and generic objectness methods to detect the main object of an image and after that they have used the concept of Rule of Thirds to measure the aesthetics of that particular image. For detecting a salient region of an image, in this paper, they have mainly used three algorithms and they are GBVS (Graph-based Visual Saliency), FT (Frequency-tuned salient region) and GC (Global contrast based salient region detection). Furthermore, they have used generic objectness analysis as a complement of saliency analysis. For the detection of ROT they have used various Machine Learning methods, for example, Naïve Bayesian Classifier, Adaboost etc.

Amirshahi et al. [6] have contrasted aesthetically pleasantness of photographs and paintings between computer based scoring and behavioral scoring (scores that are obtained from 30 participants) on the basis of Rule of Thirds compositional method.

Maleš et al. in their paper [7], they have presented a saliency based method to detect the compositional rule – ROT. To detect a salient region they have used two algorithms – Context Aware (CA) salient region detector and Global Contrast (GC) based salient region detector. After that, they have created a training set on which they have applied Principal Component Analysis (PCA). They have used Linear Discriminant Analysis, Mahalanobis Linear Discriminant Analysis, Quadratic Discriminant Analysis and Support Vector Machines to train the classifiers.

In [8], the authors have presented a collection for MIR community which comprises a subset of 25,000 images from the Flickr website under creative commons license that had been a standard dataset for most research in visual data retrieval. These images are also redistributable for research purposes and, also represent a real user community.

3 System Design

Our system takes images as input and produces a score as aesthetic appeal. As mentioned in the abstract and introduction, we have used metric - ROT in our system to produce the beauty measurement. We have accomplished the task of scoring beauty in several stages starting from an input image to final beauty measure. Therefore, we have subdivided the processing of the system into stages presented in Fig. 2. Figure 2 represents simple processing flow of our system. In the initial stage, the system receives image as input from the users via the user interface. Next, that image is fed into the preprocessor which marks the beginning of the second stage of our system. To apply the compositional metric, precondition is to compute the salient regions of the image. Hence the second stage has been developed using GBVS and Itti-Koch saliency algorithms to calculate salient regions.

Based on user selected algorithm out of two (Itti-Koch and GBVS), salient region(s) of the input image are calculated and are passed to the third stage. This stage is also the

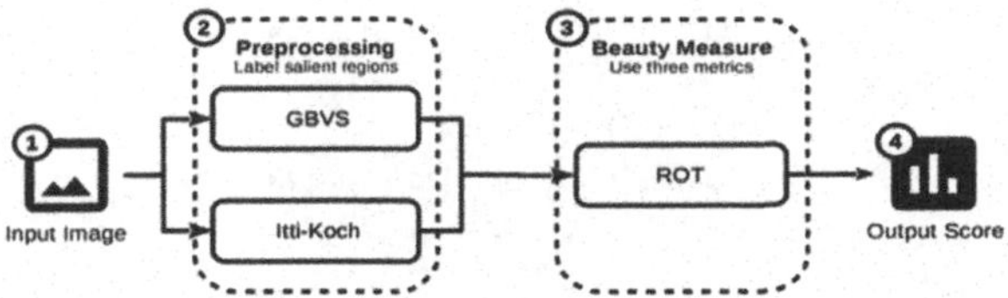

Fig. 2. Simple block diagram of processing stages.

centerpiece of our system. This stage of processing has an engine running as a service in the system, which receives the input image and salient image map from previous stage. Based on the ROT evaluator engine, the system processes it to produce ROT measure. This processing stage is the heart of this research.

Inner workings of this engine will be explained in the following subsections. Finally, in the fourth and final stage the beauty score is transferred to a user interface from the service for representing the output to the users of the system. We have developed a system consisting a web interface powered by Node.JS and an image processing service for ROT metric developed using MATLAB. As a result, users can access this tool just using a web browser without needing access to MATLAB or any other additional software. The web server and the MATLAB engine communicate via JSON. Figure 3 represents a simple overview of our system design.

Fig. 3. Simple overview of system design

For validation, we have conducted a user study to evaluate our system's aesthetic pleasantness measurements against visual aesthetics perceived by humans. We have executed our system on 5000 images from MIR Flickr dataset [11] and analyzed the execution and performance differences between GBVS and Itti-Koch algorithm.

3.1 Workflow of the ROT Engine

Definition of Rule of Thirds or ROT states that in a rectangular frame, human eye tends to perceive objects more appealing - the more they are close to the gridlines having 2/3 area on the greater side and 1/3 area on the shorter side [2]. This visual compositional rule has been established based on the golden ratio [9] proportion guideline by the ancient Greeks [2]. The ROT was first documented and written down in 1797, in the book – 'Remarks on Rural Scenery' by J.T. Smith [10]. In simple terms, if we draw two evenly spaced vertical and two evenly spaced horizontal imaginary gridlines in a rectangle image then each of these lines divides the image according to golden ration. Figure 4 simplifies the understanding of the subdivision.

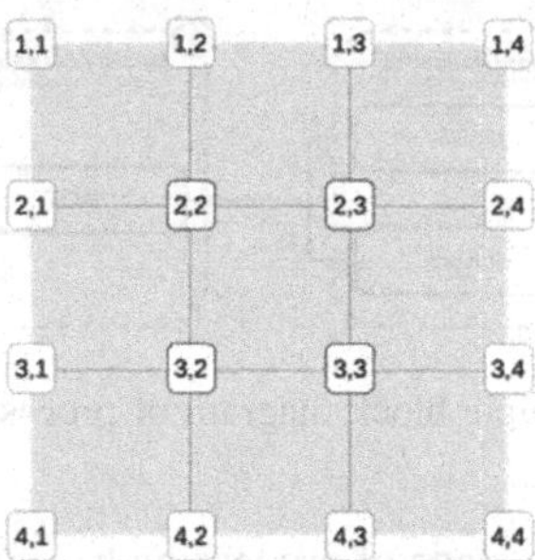

Fig. 4. Corner points of gridlines.

An image having its salient regions closer to any of these lines will have better aesthetic appeal. Furthermore, salient objects closer to the intersecting points of the gridlines will be more appealing to human eye as it means the object is maintaining golden ratio vertically and horizontally both at the same time. To measure distance, we have calculated the center point or centroid ***C (x, y)*** of salient area in the image. Then the distance is calculated with respect to the centroid. Centroid is calculated as follows for a given salient blob:

$$C_x = \frac{X_1 + X_2 + X_3 + \ldots + X_k}{S} \quad (1)$$

$$C_y = \frac{Y_1 + Y_2 + Y_3 + \ldots + Y_k}{S} \quad (2)$$

Where, X_1, X_2, X_3, ... X_k are the x coordinates, Y_1, Y_2, Y_3, ... Y_k are the y coordinates, and S is the sum of all pixels in the salient blob. We have constructed a distance function (Eq. 3) in this research which calculates a distance measurement from each cross sections of gridlines and the centroid of the salient region of the image. This measurement is the proxy of measuring beauty and visual aesthetics.

3.2 Distance that Serves as Scoring Function

To calculate distance, we have measured the distance of centroid from 2/3 cross-sections of each gridline. This gives us a tentative distance measurement from individual gridlines. Pair with the shortest measurement is the candidate axis for further calculation as the salient region would be the closest with that line. An image can have n number of salient regions where n is a positive integer. Therefore, the input image can have n number of distance scores for each subject or salient region. Regarding understandability and writing progression we have used ROI (Region of Interest) as synonym of salient regions in the next few sections. In Fig. 4, (1, 1), (1, 2), (1, 3), ..., (4, 4) are the corner points of gridlines. If g (2, 2), g (2, 3), g (3, 2), g (3, 3) are the cross-sectional points then the distance is measured as Eq. 3.

$$distance_{ROI}(n) = \min(\sqrt{(G_{22x} - C_x)^2 + (G_{22y} - C_y)^2}, \sqrt{(G_{23x} - C_x)^2 + (G_{23y} - C_y)^2}, \sqrt{(G_{33x} - C_x)^2 + (G_{33y} - C_y)^2}, \sqrt{(G_{32x} - C_x)^2 + (G_{32y} - C_y)^2}) \quad (3)$$

3.3 Scoring Function Normalization

As an input image can have more than one ROI, hence it can have more than one distance score. The distance is a measurement in pixels with respect to individual image (where size or dimensions can vary) which is not an absolute measurement that we can compare with scores of other images. Also, the more the Region of Interest (ROI) is closer to prominent gridline, less the distance measurement but aesthetic score is the compliment. Considering these facts, we have constructed the following equation to normalize distance score and produce a score which we can considered as the aesthetic score of an image.

$$score(img) = \left(\sum_{i=1}^{all\,ROIs=N} \left(\frac{Area\ of\ ROI_i \times 0.05}{(img.length \times \mathrm{img}.width)} \times \left(1 - \frac{distance_{ROI}(i)}{\sqrt{img.height^2 + img.width^2}}\right)\right)\right) \times \frac{1}{N} \quad (4)$$

In Eq. 4, distance has been divided by the diagonal of the image to produce a relative measurement and which is then subtracted form one. This measurement is then multiplied with the relative area of ROI to give better score to ROI having bigger area and negate ROIs with smaller areas. Finally scores of all ROIs has been added and divided by the number of ROIs to compute the final aesthetic score.

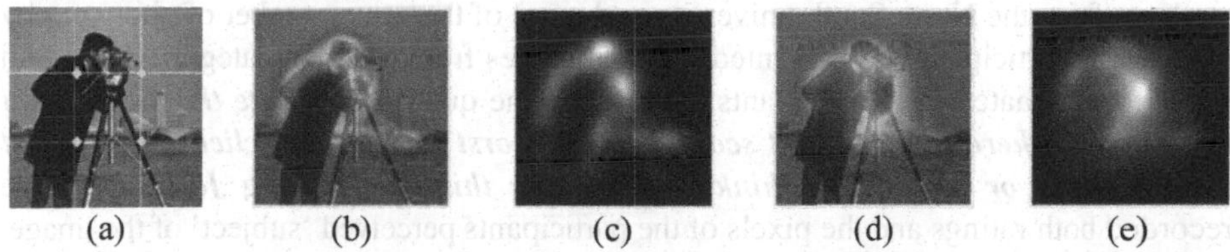

Fig. 5. GBVS vs. Itti-Koch. (a) cameraman.tif image on grid, (b) Itti-Koch Map Overlayed, (c) Itti-Koch Map and (d) GBVS Map Overlayed, and (e) GBVS Map. (Color figure online)

For illustration, we decided to use the standard "cameraman.tif" file (Fig. 5) and produced a running example. Beforehand, if we look closely we can observe this image is not especially good in terms of ROT. In the first step of processing we draw imaginary gridlines over the image. In Fig. 5a, white lines are the gridlines and the yellow dots are the cross-sections of gridlines. Next, we use GBVS (Fig. 5d, e) and Itti-Koch maps (Fig. 5b, c) to produce ROI for the input image maps. Observe that there are one ROI for GBVS and three ROIs for Itti-Koch algorithm. After that, the system uses distance function and next our normalization technique produces final aesthetic score.

From Fig. 6(a) we can see that, we have gotten only one ROI using GBVS algorithm (from 'cameraman.tif'). So our whole procedure is developed surrounding only this ROI. Firstly, we have measured the centroid of the ROI and the axis of the centroid in pixel is ($C_{x,}$ C_y) = (197px, 154px). After that, using Eq. 3 we have calculated the minimum distance between the gridlines and the centroid which is 53.96px. The area of our ROI is roughly 7146px^2 and the height and width of the image are 311px and 376px accordingly. So, finally using Eq. 4 we got the final score for this particular image that is 0.55 (where the scale is as such that 1 is the best and 0 being the least aesthetic).

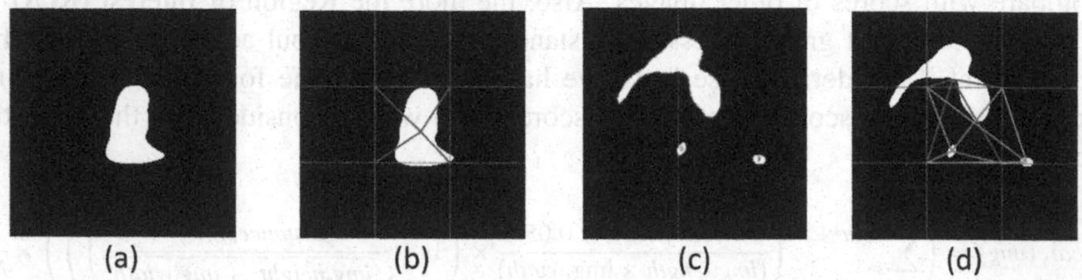

Fig. 6. ROI and distance measurement using GBVS and Itti-Koch. (a) ROI using GBVS, (b) Distance score using GBVS (c) ROI using Itti-Koch, and (d) Distance score using Itti-Koch.

4 User Study

We have conducted a user study to evaluate our system. We used 255 human participants in the study and they were aged between 19 to 29 regardless of gender. They were asked to score randomly presented images. All participants were undergraduate students from the North South university and some of the ware member of photography club. Each participant was presented with 35 images from different category in random order to eliminate bias. Participants were asked the question – ***"Rate the picture you are seeing, where 5 is the best score 1 is the worst score. Also, click on the most salient object or region you think is present in this image using Javascipt."*** We recorded both ratings and the pixels of the participants perceived 'subject' of the image.

We have used 5000 images from this dataset to evaluate our system.

5 Results Analysis

In this research, we have two results and analyzed their accuracy. Firstly, we produced aesthetic score which we measured against aesthetic score response to human psychology, and then, we compared the execution and performance differences of the two saliency algorithms we have used in this research. Our results and analysis have been described in detail in the following subsections.

5.1 Distances from Baseline of Human Users

We have presented the results in Fig. 7. In Fig. 7a, the horizontal axis represents the system given score and the vertical axis represents the user given score. Both of the scores are given in a scale between 1 to 5, and the values in each box represent the number of responses against each score for both system and human. For instance, in position (5,5) the number is 668 means that 668 images were scored 5 by the both system and user. Figure 7b represents score vs. the score difference between the user user and the system. From this figure we can observe, for lower score i.e. 1, the average difference is higher – 2.5 and for higher score i.e. 5, the average difference is lower – 1 (note: lower the better). This gives us the intuition of the fact that in most cases when humans find some image appealing it maintains minimum level of ROT but the opposite is not true.

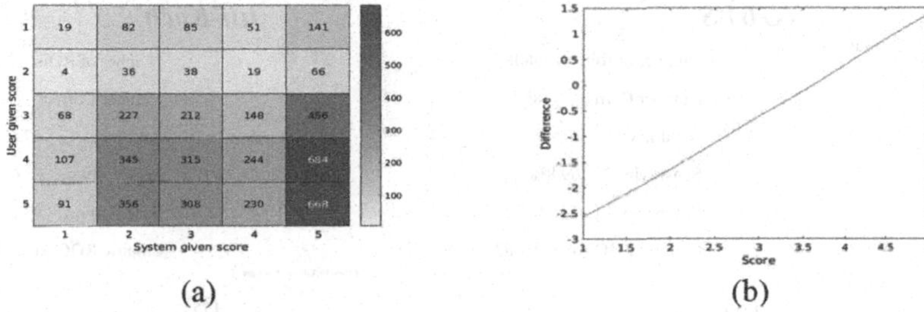

Fig. 7. (a) System generated scores vs. user generated scores and (b) the differences between the score responses.

The average sum of squared difference (SSD) is used to measure distance of score and provide an analytical result as presented in Eq. 5. Here, *i* is image index (1 to 100) and j is category index (1 to 10).

$$SSD_{ij} = \frac{\sqrt{\sum_{i=1}^{for\ all\ in\ j}\left(score_i - score_{human\ user}\right)^2}}{100} \tag{5}$$

5.2 Comparative Analysis of Saliency Algorithms

To do this comparative analysis of saliency algorithms, first we observed that- although the average SSD is low across each category, some results are too far or too close to the baseline. Bounded boxes were drawn around the detected salient regions for the two algorithms- GBVS and Itti-Koch. After that, we went through all of them manually to further validate the system if the saliency algorithms indeed find the salient regions of an image.

5.3 Methodology of the Comparisons

All of the 5000 images were annotated. Parameters that were noted are – salient regions detected by human, GBVS and Itti-Koch Saliency algorithm. We used human perception to detect four scenarios – true positives (properly detected ROI), false positive (any algorithm that detected a salient region which is not ROI), true negatives (non-salient region detected as salient region) and finally false negatives (salient regions missed by the algorithms). If a salient region detected by any of the three algorithms overlapped 50% or more than that of what human perceived as salient region is recognized as true positive. After that, we represented the research findings using ROC plots. They are presented in Fig. 8. The number of regions of interest or ROI's is 2680 which is higher than 5000 (the number of images) is because many single images contained more or less than one salient regions or subjects.

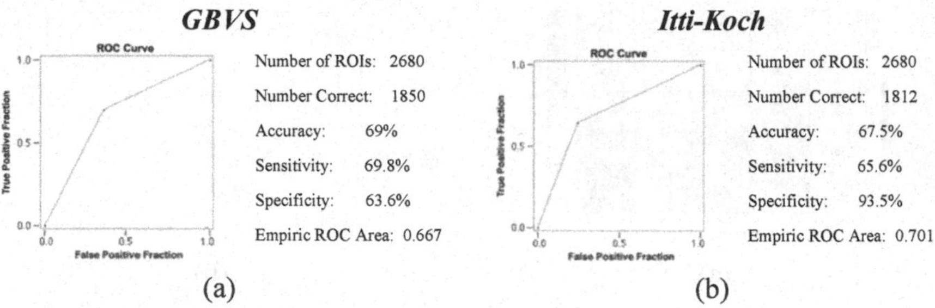

Fig. 8. Comparative ROC plots of between GBVS (a) and Itti-Koch (b) saliency algorithms

6 Discussion and Future Work

In this research, we have successfully devised a system that can perceive beauty with good accuracy. In this regard, we have been successful in most cases. We have analyzed significance of the compositional rule and its combination in different scenarios. We also analyzed differences between GBVS and Itti-Koch saliency algorithm based on real world applications by involving actual human comprehension. Finally, we can conclude that image compositional metrics are a valid and stable mean to understand visual aesthetics and ROT has significance in good magnitude. However, we experimentally validated that GBVS outperforms Itti-Koch to some extent.

Although, we have successfully gone to the closest points with some categories of images, we feel if we use other compositional metrics i.e. Rule of frames, Rule of odds, Rule of space etc. [11] in addition, it would give more insights. A larger user study is coming up as well.

Acknowledgments. This work was funded by North South University's annual research grant for the fiscal year 2017–18. We would like to thank Professor John Kender (http://www.cs.columbia.edu/~jrk/), Professor of Computer Science at Columbia University for his insights.

References

1. Mai, L., Le, H., Niu, Y., Liu, F.: Rule of thirds detection from photograph. In: Proceedings of the IEEE International Symposium on Multimedia (ISM), Dana Point, CA, USA, pp. 91–96 (2011)
2. Peterson, B.: Learning to See Creatively, 1st edn, pp. 92–93. Amphoto Books, New York (2003)
3. Harel, J., Koch, C., Perona, P.: Graph-based visual saliency. In: Advances in Neural Information Processing Systems, pp. 545–552 (2006)
4. Itti, L., Koch, C.: A saliency-based search mechanism for overt and covert shifts of visual attention. Vis. Res. **40**(10–12), 1489–1506 (2000)
5. Mai, L., Le, H., Niu, Y., Liu, F.: Rule of thirds detection from photograph. In: IEEE International Symposium on Multimedia, Portland (2011)
6. Amirshahi, S.A., Hayn-Leichsenring, G.U., Denzler, J., Redies, C.: Evaluating the rule of thirds in photographs and paintings. Art Percept. **2**, 163–182 (2014)
7. Maleš, M., Heđi, A., Grgić, M.: Compositional rule of thirds detection. In: IEEE International Symposium on ELMAR 2011, Dana Point (2011)
8. Murillo, A., Košecká, J., Guerrero, J., Sagüés, C.: Visual door detection integrating appearance and shape cues. Rob. Auton. Syst. **56**(6), 512–521 (2008)
9. Huiskes, M.J., Lew, M.S.: The MIR flickr retrieval evaluation. In: ACM International Conference on Multimedia Information Retrieval (MIR 2008), Vancouver, Canada (2008)
10. Weisstein, E.W.: Golden Ratio. From MathWorld–A Wolfram Web Resource. http://mathworld.wolfram.com/GoldenRatio.html
11. McCurry, S.: 9 Photo Composition Tips (feat. Steve McCurry). Youtube (2015). https://youtu.be/7ZVyNjKSr0M. Accessed 2 March 2016

Towards Robust Evaluation of Super-Resolution Satellite Image Reconstruction

Michał Kawulok[1,2](✉), Paweł Benecki[1,2], Jakub Nalepa[1,2], Daniel Kostrzewa[1,2], and Łukasz Skonieczny[1]

[1] Future Processing, Gliwice, Poland
{mkawulok,pbenecki,jnalepa,dkostrzewa,lskonieczny}@future-processing.com
[2] Institute of Informatics, Silesian University of Technology, Gliwice, Poland
{michal.kawulok,pawel.benecki,jakub.nalepa,daniel.kostrzewa}@polsl.pl

Abstract. Super-resolution reconstruction (SRR) consists in processing an image or a bunch of images to generate a new image of higher spatial resolution. This problem has been intensively studied, but seldom is SRR applied in practice for satellite data. In this paper, we briefly review the state of the art on SRR algorithms and we argue that commonly adopted strategies for their evaluation do not reflect the operational conditions. We report our study on assessing the SRR outcome, relying on new quantitative measures. The obtained results allow us to outline the most important research pathways to improve the performance of SRR.

Keywords: Super-resolution · Image processing · Similarity measures

1 Introduction

In numerous practical applications, images of high resolution are valuable, but are often unavailable due to technological limitations or economic reasons. Obviously, this problem is inherent to satellite imagery, where higher resolution induces higher costs, lower coverage and longer revisit times. In such cases, it is still possible to analyze a low-resolution image or an image sequence to extract the details with the subpixel precision and produce an image of a higher resolution. This process is termed as *super-resolution reconstruction* (SRR) and it is a well-studied problem of computer vision. However, the effectiveness of the state-of-the-art solutions is still insufficient to make the SRR methods capable of enhancing satellite images in real-life scenarios.

Spatial resolution of satellite images is usually expressed in *ground sampling distance* (GSD—the smallest dimension of an object on the ground that can be distinguished from an image[1]). Depending on GSD, the spatial resolution of satellite images is classified as low (LR, GSD > 100 m), medium (MR, GSD of 10–100 m), high (HR, GSD of 1–10 m) and very high (VHR, GSD < 1 m).

[1] It is understood as the distance between the centers of two neighboring pixels, however this simplification may be incorrect in the presence of some distortions.

N. T. Nguyen et al. (Eds.): ACIIDS 2018, LNAI 10751, pp. 476–486, 2018.
https://doi.org/10.1007/978-3-319-75417-8_45

In this paper, we outline the current trends in SRR with particular attention given to the adopted evaluation procedures, including the available benchmarks, which was not discussed in recent surveys on SRR [33]. Our contribution lies in identifying the pivotal obstacles that must be overcome to make the SRR suitable for real-world applications. First, we discuss how the robustness of evaluating the SRR algorithms can be increased with new quantitative measures. Furthermore, we present qualitative examples of several state-of-the-art algorithms in order to confront the numerical measures against the visual impression on the image quality. Finally, we discuss the future research pathways concerned with: (i) enhancing the benchmarks, (ii) introducing new SRR quality measures, better correlated with GSD, and (iii) improving the objective functions, whose optimization is substantial to obtaining a correct reconstruction.

Section 2 is an overview of existing SRR techniques, while the evaluation methodologies and benchmarks are summarized in Sect. 3. Our proposed evaluation framework is presented in Sects. 4 and 5 concludes the paper.

2 Current Trends in Super-Resolution Reconstruction

There have been a number of research works [33] presenting different approaches towards SRR for a variety of image types, modalities and acquisition scenarios. Apart from satellite imagery, SRR was considered for medical imaging [32], analysis of facial images [12], document image processing [3], or microscopy imaging [16]. SRR can be executed given a single image [6], including processing hyper-spectral images [21], or from a sequence of images acquired with some shifts in the spatial domain [13].

2.1 SRR from a Single Image

Single-image SRR is primarily dealt with example-based learning, which consists in establishing the relation between images of low $\mathcal{I}^{(l)}$ and high $\mathcal{I}^{(h)}$ resolution, given some training data. This creates a model that makes it possible to reconstruct a high-resolution image $\mathcal{I}'^{(h)}$ from $\mathcal{I}^{(l)}$. Among many approaches to determine such relation, manifold learning has been extensively exploited [27] to match the patches of high and low resolution. Recently, it has been shown that deep learning can also be highly successful here [7]. For images presenting some repeated patterns at different scales, the patches may also be matched within a single image on the basis of self-similarity [11].

Reconstructing $\mathcal{I}'^{(h)}$ may involve the decomposition of $\mathcal{I}^{(l)}$. This could be aimed at extracting high-frequency components of a single image, which are amplified to increase the spatial resolution, e.g. with the wavelet transform (WT) [6].

2.2 Multiple-Image Fusion

SRR from multiple images is usually addressed by defining a parametrized imaging model (IM) which simulates the process of degrading a hypothetical high-resolution image into the observed $\mathcal{I}^{(l)}$. In general, such models include image warping, blurring, downsampling and finally contamination with the noise [19]. The majority of methods are underpinned with a certain cost function (related to the reconstruction error), whose optimization determines the values of the parameters that control the IM. This, in turn, makes it possible to restore $\mathcal{I}'^{(h)}$. Such approach is applied to hyper-spectral imaging and SRR from multiple images. In the latter case, the images are first registered, and then the subpixel shifts between them are also subject to the optimization. An interesting approach was presented in [5], where the low-resolution images are assessed to select those of good quality and reject the poor ones.

Optimizing the imaging model parameters is an ill-posed problem, which is usually solved by employing the Bayesian framework (BF) or gradient-based techniques with some regularization imposed to provide spatial smoothness of $\mathcal{I}'^{(h)}$. Regularization [20] translates a *maximum likelihood* (ML) problem into *maximum a posteriori* (MAP) estimation. In one of the earliest approaches, the gradient projection algorithm (GPA) [23] was employed. Also, it is often beneficial to incorporate the motion blur into the imaging model [28]. Another fairly popular optimization technique applied here is the projection onto convex sets (POCS) [2], which consists in updating the high-resolution target image iteratively based on the error measured between $\mathcal{I}^{(l)}$ and $\mathcal{I}'^{(l)}$—a downsampled version of reconstructed $\mathcal{I}'^{(h)}$, degraded using the assumed imaging model. Fast and robust super-resolution (FRSR) employs regularization based on total variation combined with bilateral filter [9]—importantly, the error between subsequent reconstructed (and degraded) high-resolution images is minimized, thus avoiding the expensive scaling operation. In [1], a genetic algorithm was applied to determine the parameters that control the subpixel registration. Here, the regularization is ensured by imposing certain constraints on the genetic operators (especially mutation). Furthermore, particle swarm optimization [4], simulated annealing [31] and differential evolution [36] were exploited here.

2.3 SRR for Satellite Imagery

Most of the SRR methods applied to satellite images (see Table 1) either rely on PCA-based decomposition [34] (for hyper-spectral images) or an IM is optimized [2,13,17,18]. Also, the example-based methods [22], including deep learning [14], were found useful for this task. Moreover, an algorithm for SRR can be tightly related with the sensor characteristics—the intrinsic keystone distortion of the IKONOS hyper-spectral sensor was exploited in [21]. Recently, multiple-image SRR for satellite images was proceeded using adaptive detail enhancement (SRR-ADE) [37]—this employs bilateral filter to decompose the input images and amplify the high-frequency detail information.

3 Evaluating Super-Resolution Reconstruction

Applying a proper evaluation strategy is pivotal in achieving progress in numerous domains concerned with advanced data analysis. This encompasses both the

Table 1. SRR methods applied to satellite imagery (the image modalities are abbreviated as SI: single-image, HS: hyper-spectral image, MI: multiple images).

Author	Year	Main concept	Modality	Sensor type
Ducournau and Fablet [8]	2016	Deep learning	SI	AVHRR, AATSR, SEVIRI
Liebel and Körner [14]	2016	Deep learning	HS	Sentinel-2
Zhu et al. [37]	2016	Adaptive enhancement	MI	ZY-3
Qian and Chen [21]	2012	Keystone effect	HS	IKONOS
Wu et al. [31]	2012	Sim. anneal. in RMF	HS	Hyperion, ALOS
Zhang et al. [34]	2012	PCA + IM optimization	HS + MI	SPOT, Landsat-7, IKONOS
Li et al. [13]	2008	IM + Markov trees	MI	Landsat-7
Molina et al. [18]	2008	IM + BF	HS	Landsat-7 ETM+
Miravet and Rodriguez [17]	2007	IM + neural network	MI	QuickBird
Akgun et al. [2]	2005	IM + POCS	HS	AVIRIS
Rubert et al. [22]	2005	Patch-based	SI	Landsat-7
Gonzalez et al. [10]	2004	PCA + WT	HS	SPOT
Zhang [35]	2002	PCA + WT	HS	SPOT, Landsat-7, IKONOS

methodology and benchmarks used for evaluating the emerging solutions. These both aspects in the context of SRR methods are addressed in this section.

3.1 Evaluation Methodology

In most cases, evaluating the developed SRR techniques consists in downscaling $\mathcal{I}^{(h)}$ using different offsets and degradation operators to obtain one or N images of low resolution $\boldsymbol{I}^{(l)} = \{\mathcal{I}_1^{(l)}, \mathcal{I}_2^{(l)}, ..., \mathcal{I}_N^{(l)}\}$. The goal is to reconstruct $\mathcal{I}'^{(h)}$ from $\boldsymbol{I}^{(l)}$. The similarity between $\mathcal{I}'^{(h)}$ and $\mathcal{I}^{(h)}$, commonly measured with peak signal-to-noise ratio—PSNR, structural similarity index—SSIM [30], information fidelity criterion—IFC [25], visual image fidelity—VIF [24], or universal image quality index—UIQI [29], is used to evaluate the reconstruction quality. While such procedure is sufficient to verify certain aspects of the algorithms, the assumptions imposed on the degradation model may actually not hold.

A more realistic scenario was adopted in [36]—the outcome of SRR obtained for Landsat-7 ETM+ images was compared with IKONOS data of higher resolution, but these data were not published as a benchmark. A similar approach was presented in [31], where Hyperion images were upscaled and compared with ALOS data. Unfortunately, such approaches are very uncommon, partially due to the difficulties in comparing the upscaled images with the images obtained using a different sensor. Basically, such metrics as PSNR or SSIM would favor the operations which consist in mapping the image histograms over those that really increase the spatial resolution, but preserve the original histogram. Hence employing some advanced image processing routines, which demand high spatial resolution, is more appropriate here. In the aforementioned works [31,36], $\mathcal{I}'^{(h)}$ is compared with $\mathcal{I}^{(h)}$ based on the subpixel land-cover classification maps. It is also possible to evaluate the SRR outcome without ground truth, based on image entropy [37], but such approach may drift towards amplifying the noise.

3.2 Available Benchmark Data

Virtually all of the SRR benchmarks (e.g., MDSP Super-Resolution And Demosaicing Datasets[2] or Multi-Sensor Super-resolution Datasets[3] contain a few images or image sequences acquired at a single scale, degraded with a certain technique. There are also some widely-used images for testing single-image SRR[4]. The Sun-Hays dataset [26] contains 80 high resolution images and their downsampled counterparts. Very recently, a new DIV2K dataset was published for the NTIRE single-image SRR challenge[5]. With 800 images, each of which is in the original high resolution, as well as after (i) *bicubic* and (ii) *unrevealed* downscaling (2×, 3× and 4×), it is by far the largest benchmark for SRR. Unfortunately, there are no benchmarks encompassing real images acquired at different original resolutions, which could correspond to the real-world conditions, and seldom are the SRR methods validated in such scenarios.

To our best knowledge, there is no well-established benchmark with satellite images at all. On the other hand, there are quite a few free sources with such data, e.g., U.S. Geological Survey[6] (includes images from Sentinel-2, Landsat and Hyperion satellites), Global Land Cover Facility[7] with images acquired by Ikonos, Quickbird, Orbview, Landsat and more, and Copernicus Open Access Hub[8] for all Sentinel data. The aforementioned satellites are equipped with a variety of sensors that have different spatial resolution for specific bands. This makes it possible to match the images covering roughly the same area, acquired at a similar time, using sensors of different resolutions.

---

[2] Available at https://users.soe.ucsc.edu/~milanfar/software/sr-datasets.html (26th Oct 2017).

[3] Available at https://www5.cs.fau.de/research/data/multi-sensor-super-resolution-datasets (26th Oct 2017).

[4] Available at http://www.wisdom.weizmann.ac.il/~vision/SingleImageSR.html (26th Oct 2017).

[5] Available at http://www.vision.ee.ethz.ch/ntire17 (26th Oct 2017).

[6] Available at https://www.usgs.gov (26th Oct 2017).

[7] Available at http://glcf.umd.edu (26th Oct 2017).

[8] Available at https://scihub.copernicus.eu (26th Oct 2017).

4 Proposed SRR Evaluation Framework

One of the main problems with the SRR techniques is that they are usually validated in artificial scenarios (see Sect. 3), which do not reflect the real-life conditions sufficiently. Experimental validation often employs artificially downsampled images, and even if the original images are magnified using SRR, there is usually no ground-truth counterpart available. For many fields of computer vision (facial image analysis being the most prominent), designing proper benchmarks and updating them carefully stimulates the research community and allows for achieving the advancement level required for practical applications. Therefore, creating proper validation procedures and using realistic benchmark data should be an important step towards deploying SRR methods in practice.

4.1 Image Similarity Measures

Using realistic data for validation induces a problem of evaluating the SRR outcome. If two images presenting the same area are captured using different sensors of different resolution, then even high-quality reconstruction of a low-resolution image may be visually very different from the ground-truth image. Relying on standard metrics mentioned in Sect. 3.1 may be misleading, as they may not be sufficiently robust against such variations. Therefore, we investigate two types of image similarity measures based on (i) image filtering and (ii) keypoint detectors. In the first case, we treat an image with a high-pass filter based on the difference of Gaussians to extract the edges, followed by slight blurring to achieve some level of translation invariance. Both $\mathcal{I}^{(h)}$ and $\mathcal{I}'^{(h)}$ are processed in that way and the similarity between the processed maps is measured using PSNR (termed PSNR_{HP}). In a different variant, we apply local standard deviation followed by blurring (PSNR_{σ}).

Furthermore, we employ keypoint detectors (namely, scale invariant feature transform—SIFT, speeded-up robust features—SURF and binary robust invariant scalable keypoints—BRISK) to determine the landmarks in two images that are to be compared. Our general assumption is that if an image is well reconstructed, then the detected landmarks should be similar in $\mathcal{I}^{(h)}$ and $\mathcal{I}'^{(h)}$. We consider both the keypoint locations (we blur the location maps to allow for small displacements and measure PSNR between them)—we term it KP_l, as well as the keypoint features (KP_f). In the latter case, we detect the keypoints in $\mathcal{I}^{(h)}$ only, extract image features in both $\mathcal{I}^{(h)}$ and $\mathcal{I}'^{(h)}$, and the image similarity is obtained based on the distance measured in the feature space.

4.2 Illustrative Examples

In Fig. 1, we present two examples of images reconstructed using different SRR algorithms (selected regions are magnified for better visualization). We implemented all the SRR algorithms in C++ language and here we report the results obtained with FRSR [9], GPA [23] and SRR-ADE [37]. In addition, we initialize FRSR with a high-resolution image obtained with shift-and-add method

(FRSR-SAA)—low-resolution observations are co-registered and mapped into higher-resolution space (the missing values are filled based on median of its neighbors). In the upper row in the figure, we present an artificial image (i.e., $\boldsymbol{I}^{(l)}$ was obtained by degrading $\mathcal{I}^{(h)}$ using different shifts), while in the bottom row, $\mathcal{I}^{(h)}$ and images in $\boldsymbol{I}^{(l)}$ were captured using different satellites (Digital Globe and Sentinel-2, respectively). In both cases, we had $N = 5$ images in the $\boldsymbol{I}^{(l)}$ set. From visual assessment, the artificial image is best enhanced with FRSR-SAA—the digits are clearly visible, though some noise-resembling artifacts can be seen as well. Less artifacts are rendered with GPA, however this is obtained at the cost of the detail level. The results for a real sample of satellite images are quite different—here, FRSR-SAA generates hardly acceptable artifacts and best visual result is obtained using SRR-ADE and GPA.

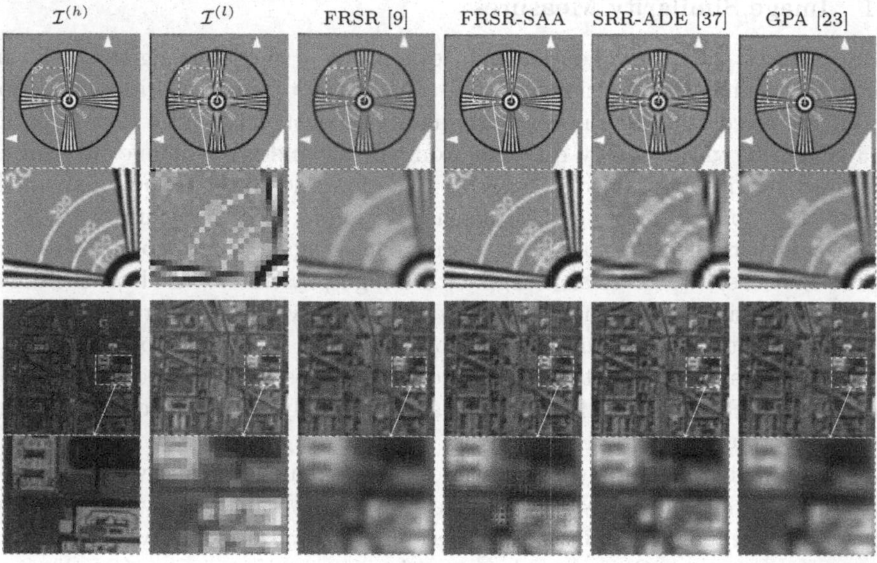

Fig. 1. Examples of reconstruction outcome obtained using different techniques.

Interestingly, the quantitative scores reported in Table 2 confirm only the observation made for the artificial image, while those for the satellite image are somehow surprising. For the artificial image, SSIM, IFC, UIQI are the highest for GPA, while the remaining measures indicate FRSR-SAA as the best method. For the real satellite images, two detail-based measures (KP_f and PSNR_{HP}) indicate the original low-resolution image $\mathcal{I}^{(l)}$ as the most similar to the ground-truth $\mathcal{I}^{(h)}$. KP_l selects FRSR-SAA despite the artifacts (and $\mathcal{I}^{(l)}$ is right behind), and the remaining measures are split between SRR-ADE and GPA. The scores obtained based on the keypoint detectors are important, as they suggest that visual assessment may not correlate with the performance of image understanding algorithms. The latter clearly indicate whether a given SRR method benefits

Table 2. Similarity scores (to $\mathcal{I}^{(h)}$) of the images in Fig. 1 (highest scores are bold).

Image	Measure	Low resol.	FRSR [9]	FRSR-SAA	SRR-ADE [37]	GPA [23]
Artificial	PSNR	19.91	25.90	**30.70**	21.31	26.44
	SSIM	0.72	0.91	0.81	0.81	**0.93**
	IFC	1.32	1.92	1.80	1.61	**1.98**
	UIQI	0.26	0.37	0.37	0.28	**0.38**
	VIF	0.25	0.47	**0.54**	0.30	0.49
	KP_f	4.84	9.69	**14.29**	5.02	9.03
	KP_l	39.40	39.60	**43.64**	38.33	38.00
	$PSNR_\sigma$	30.05	32.24	**37.16**	29.63	32.17
	$PSNR_{HP}$	30.46	33.05	**38.93**	29.87	32.55
Real satellite	PSNR	13.45	16.15	16.05	15.54	**16.22**
	SSIM	0.37	0.46	0.37	0.46	**0.47**
	IFC	0.98	1.21	1.04	**1.28**	1.25
	UIQI	0.25	0.28	0.26	**0.31**	0.30
	VIF	0.12	0.12	0.11	**0.14**	0.13
	KP_f	**3.57**	3.19	3.31	3.35	3.17
	KP_l	40.18	38.84	**40.37**	39.53	38.77
	$PSNR_\sigma$	36.79	36.09	35.27	**36.97**	36.79
	$PSNR_{HP}$	**37.70**	35.52	37.17	36.25	36.43

from the information hidden in multiple low-resolution observations. Naturally, more experiments are necessary, however it appears that making evaluation of SRR robust against real-world sensor variations is very challenging.

Another interesting problem is presented in Fig. 2, where $\mathcal{I}'^{(h)}$ is reconstructed from $N = 5$ low-resolution natural images captured in laboratory conditions (GSD of $\mathcal{I}^{(h)}$ is ca. 50% larger than for $\mathcal{I}^{(l)}$). In this case, the SRR outcome presents more details than $\mathcal{I}^{(h)}$, so during evaluation, poorer algorithms (rendering images of spatial resolution similar to $\mathcal{I}^{(h)}$) may be ranked higher. Overall, such a risk must be considered while preparing a benchmark dataset.

Fig. 2. A reconstruction example of a natural image using the GPA algorithm [23].

5 Conclusions and Outlook

In this paper, we report our study on making SRR evaluation robust against the variations observed in operational conditions. We argue that there are three major research pathways that may be instrumental in developing effective SRR techniques (see Fig. 3). First, we need real-life benchmarks encompassing satellite images of the same region, acquired with sensors of different resolution. This induces the need for robust metrics—we showed that evaluation of SRR applied to real-world images is by no means trivial. We proposed to use keypoint detectors here and our ongoing work is to exploit land-cover classifiers. Also, ensembles of the reported measures may help obtain stable assessment. Finally, having established proper validation framework, we need to revisit the problem of IMs, as they may not reflect the real correspondence between low and high resolution. For tuning the SRR methods (including IMs), we exploit evolutionary algorithms, powerful in hyper-parameter optimization [15]. Out initial results are promising, but as the fitness function is based on the similarity measures, more consideration is necessary to increase their robustness.

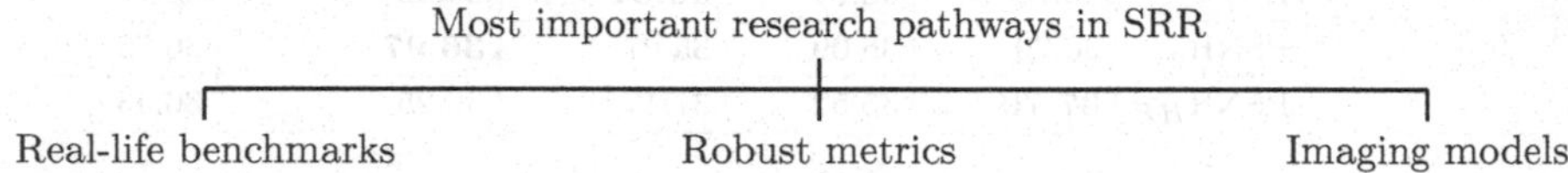

Fig. 3. Most important research pathways in SRR.

Importantly, we plan to publish the code for computing the metrics (as well as our implementation of some SRR techniques) alongside a real-life dataset. We expect that introducing such benchmarks would accelerate the works on bridging the gap between the SRR performance in laboratory and real-world conditions.

Acknowledgments. The reported work is a part of the SISPARE project run by Future Processing and funded by European Space Agency. The authors were partially supported by Institute of Informatics funds no. BK-230/RAu2/2017 (MK) and BKM-509/RAu2/2017 (JN, DK).

References

1. Ahrens, B.: Genetic algorithm optimization of superresolution parameters. In: Proceedings of the GECCO, pp. 2083–2088. ACM (2005)
2. Akgun, T., Altunbasak, Y., Mersereau, R.M.: Super-resolution reconstruction of hyperspectral images. IEEE Trans. Image Process. **14**(11), 1860–1875 (2005)
3. Capel, D., Zisserman, A.: Super-resolution enhancement of text image sequences. In: Proceedings of the IEEE ICPR, vol. 1, pp. 600–605 (2000)
4. Cheng, M.H., Hwang, K.S., Jeng, J.H., Lin, N.W.: PSO-based fusion method for video super-resolution. J. Signal Process. Syst. **73**(1), 25–42 (2013)

5. Del Gallego, N.P., Ilao, J.: Multiple-image super-resolution on mobile devices: an image warping approach. EURASIP J. Image Video Process. **2017**(1), 1–15 (2017)
6. Demirel, H., Anbarjafari, G.: Image resolution enhancement by using discrete and stationary wavelet decomposition. IEEE Trans. Image Process. **20**(5), 1458–1460 (2011)
7. Dong, C., Loy, C.C., He, K., Tang, X.: Image super-resolution using deep convolutional networks. IEEE Trans. Pattern Anal. Mach. Intell. **38**(2), 295–307 (2016)
8. Ducournau, A., Fablet, R.: Deep learning for ocean remote sensing: an application of convolutional neural networks for super-resolution on satellite-derived SST data. In: Proceedings of the IAPR WPRRS, pp. 1–6 (2016)
9. Farsiu, S., Robinson, M.D., Elad, M., Milanfar, P.: Fast and robust multiframe super resolution. IEEE Trans. Image Process. **13**(10), 1327–1344 (2004)
10. González-Audícana, M., Saleta, J.L., Catalán, R.G., García, R.: Fusion of multispectral and panchromatic images using improved IHS and PCA mergers based on wavelet decomposition. IEEE Trans. Geosci. Remote Sens. **42**(6), 1291–1299 (2004)
11. Huang, J.B., Singh, A., Ahuja, N.: Single image super-resolution from transformed self-exemplars. In: Proceedings of the IEEE CVPR, pp. 5197–5206 (2015)
12. Jiang, J., Hu, R., Wang, Z., Han, Z.: Face super-resolution via multilayer locality-constrained iterative neighbor embedding and intermediate dictionary learning. IEEE Trans. Image Process. **23**(10), 4220–4231 (2014)
13. Li, F., Jia, X., Fraser, D.: Universal HMT based super resolution for remote sensing images. In: Proceedings of the IEEE ICIP, pp. 333–336 (2008)
14. Liebel, L., Körner, M.: Single-image super resolution for multispectral remote sensing data using convolutional neural networks. In: Proceedings of the ISPRS Congress, pp. 883–890 (2016)
15. Lorenzo, P.R., Nalepa, J., Kawulok, M., Ramos, L.S., Pastor, J.R.: Particle swarm optimization for hyper-parameter selection in deep neural networks. In: Proceedings of the GECCO, pp. 481–488. ACM, New York (2017)
16. Lukinavičius, G., Umezawa, K., Olivier, N., Honigmann, A., Yang, G., Plass, T., et al.: A near-infrared fluorophore for live-cell super-resolution microscopy of cellular proteins. Nat. Chem. **5**(2), 132–139 (2013)
17. Miravet, C., Rodrıguez, F.B.: A two-step neural-network based algorithm for fast image super-resolution. Image Vis. Comput. **25**(9), 1449–1473 (2007)
18. Molina, R., Vega, M., Mateos, J., Katsaggelos, A.K.: Variational posterior distribution approximation in Bayesian super resolution reconstruction of multispectral images. Appl. Comput. Harmon. Anal. **24**(2), 251–267 (2008)
19. Nasrollahi, K., Moeslund, T.B.: Super-resolution: a comprehensive survey. Mach. Vis. Appl. **25**(6), 1423–1468 (2014)
20. Panagiotopoulou, A., Anastassopoulos, V.: Super-resolution image reconstruction techniques: trade-offs between the data-fidelity and regularization terms. Inf. Fusion **13**(3), 185–195 (2012)
21. Qian, S.E., Chen, G.: Enhancing spatial resolution of hyperspectral imagery using sensor's intrinsic keystone distortion. IEEE Trans. Geosci. Remote Sens. **50**(12), 5033–5048 (2012)
22. Rubert, C., Fonseca, L., Velho, L.: Learning based super-resolution using YUV model for remote sensing images. In: Proceedings of the SIBGRAPI (2005)
23. Schultz, R.R., Stevenson, R.L.: Extraction of high-resolution frames from video sequences. IEEE Trans. Image Process. **5**(6), 996–1011 (1996)
24. Sheikh, H.R., Bovik, A.C.: Image information and visual quality. IEEE Trans. Image Process. **15**(2), 430–444 (2006)

25. Sheikh, H.R., Bovik, A.C., De Veciana, G.: An information fidelity criterion for image quality assessment using natural scene statistics. IEEE Trans. Image Process. **14**(12), 2117–2128 (2005)
26. Sun, L., Hays, J.: Super-resolution from Internet-scale scene matching. In: Proceedings of the IEEE ICCP (2012)
27. Timofte, R., De Smet, V., Van Gool, L.: A+: adjusted anchored neighborhood regression for fast super-resolution. In: Cremers, D., Reid, I., Saito, H., Yang, M.-H. (eds.) ACCV 2014. LNCS, vol. 9006, pp. 111–126. Springer, Cham (2015). https://doi.org/10.1007/978-3-319-16817-3_8
28. Wang, Y., Fevig, R., Schultz, R.R.: Super-resolution mosaicking of UAV surveillance video. In: Proceedings of the IEEE ICIP, pp. 345–348. IEEE (2008)
29. Wang, Z., Bovik, A.: A universal image quality index. IEEE Signal Process. Lett. **9**(3), 81–84 (2002)
30. Wang, Z., Bovik, A.C., Sheikh, H.R., Simoncelli, E.P.: Image quality assessment: from error visibility to structural similarity. IEEE Trans. Image Process. **13**, 600–612 (2004)
31. Wu, B., Li, C., Zhan, X.: Integrating spatial structure in super-resolution mapping of hyper-spectral image. Procedia Eng. **29**, 1957–1962 (2012)
32. Yang, F., Chen, Y., Wang, R., Zhang, Q.: Super-resolution microwave imaging: time-domain tomography using highly accurate evolutionary optimization method. In: Proceedings of the EuCAP, pp. 1–4. IEEE (2015)
33. Yue, L., Shen, H., Li, J., Yuan, Q., Zhang, H., Zhang, L.: Image super-resolution: the techniques, applications, and future. Signal Process. **128**, 389–408 (2016)
34. Zhang, H., Zhang, L., Shen, H.: A super-resolution reconstruction algorithm for hyperspectral images. Signal Process. **92**(9), 2082–2096 (2012)
35. Zhang, Y.: Problems in the fusion of commercial high-resolution satelitte as well as Landsat 7 images and initial solutions. In: Proceedings of the GTPA, pp. 1–6 (2002)
36. Zhong, Y., Zhang, L.: Remote sensing image subpixel mapping based on adaptive differential evolution. IEEE Trans. Syst. Man Cybern. Part B **42**(5), 1306–1329 (2012)
37. Zhu, H., Song, W., Tan, H., Wang, J., Jia, D.: Super resolution reconstruction based on adaptive detail enhancement for ZY-3 satellite images. In: Proceedings of the ISPRS, pp. 213–217 (2016)

Advanced Data Mining Techniques and Applications

An Automated Deployment Scheme with Script-Based Development for Cloud Manufacturing Platforms

Jhang-Jhan Huang[1], Chao-Chun Chen[2], Zhong-Hui Lin[3](✉), Mao-Yuan Pai[4], and Gen-Ming Guo[5]

[1] AndroidVideo, Taipei, Taiwan
jjhuang@androvideo.com
[2] IMIS/CSIE, NCKU, Tainan, Taiwan
chaochun@mail.ncku.edu.tw
[3] IBM, New York, USA
zhlin@us.ibm.com
[4] IAM, CJCU, Tainan, Taiwan
mypai@mail.cjcu.edu.tw
[5] MIS, STUST, Tainan, Taiwan
sambuela@stust.edu.tw

Abstract. In this paper, we propose an automated deployment scheme with script-based development mechanisms aimed at building cloud manufacturing systems by using scripts. The automated deployment scheme cooperates with our previously CMFAS architecture, and consists of three core mechanisms. First, a script-based service deployment mechanism is designed to represent components of a cloud manufacturing service as a set of scripts and standalone packages. Second, an automated workflow generator is designed to compose a workflow-based service request processor in the cloud platform. Thirdly, a service integrated checking mechanism is designed to verify whether manufacturing services are successfully deployed. We deploy an engine diagnosis system to a VMWare-based private cloud for conducting integrated tests. Testing results show that the engine diagnosis cloud service successfully analyzes the engine performance in the CMFAS-based cloud platform.

Keywords: Smart manufacturing · Factory automation
Industry 4.0 · Text-based development

1 Introduction

With the coming of Industry 4.0 era, enterprises are eager to add intelligence to their manufacturing systems, in order to increase their global competitiveness [1,2]. There are many types of manufacturing intelligence for an enterprise to solve various manufacturing challenges, so that rapid cloud prototyping is an eminent solution to enhance the manufacturing automation. For example, some

N. T. Nguyen et al. (Eds.): ACIIDS 2018, LNAI 10751, pp. 489–499, 2018.
https://doi.org/10.1007/978-3-319-75417-8_46

automotive companies are able to diagnose engine performance any time and anywhere through a proper cloud service by just migrating their smart prognosis applications to a cloud platform. Another example is to diagnose the driver behavior from the video data in the cloud. Hence, rapidly cloud service prototyping has become a significant topic in the industry automation, and fits the movement of toward smart manufacturing to earn value-added profits [3–5].

Cloud computing has become a new trend of operating Internet applications in the last decade [6], which allows users to exploit the cloud resources on demand and pay only for how much resource they use. Thus, many small and medium enterprises (SMEs) have high desire to graft their manufacturing intelligence onto cloud computing platforms for bringing commercial values to enterprises [7]. Although the cloud manufacturing concepts attract much attentions from academic and industrial experts, the fundamental issue that how to rapidly and automatically graft existing single-machine applications onto a cloud manufacturing platform, is not thoroughly investigated due to following reasons:

- Re-developing the intelligent manufacturing functions (IMFs), mostly developed in the standalone environment, wastes time and financial cost, and this encounters a more complex environment than the standalone one.
- Grafting IMFs onto a remote and distributed cloud environment increases the deployment barrier.
- Lacking uniform mechanisms to monitor whether the deployed cloud services correctly operate or not.

In this paper, we propose an automated deployment scheme with script-based development mechanisms aimed at building cloud manufacturing systems by using scripts plus existing standalone intelligent manufacturing applications. The automated deployment scheme cooperates with our previously CMFAS architecture, and consists of three core mechanisms. First, a script-based service deployment mechanism is designed to represent components of a cloud manufacturing service as a set of scripts and standalone packages. Second, an automated workflow generator is designed to compose a workflow-based service request processor in the cloud platform. Thirdly, a service integrated checking mechanism is designed to verify whether manufacturing services are successfully deployed. Finally, we deploy an engine diagnosis system to a VMWare-based private cloud for conducting integrated tests. Testing results show that the engine diagnosis cloud service can successfully analyze the engine performance in the CMFAS-based cloud manufacturing platform.

2 Cloud Manufacturing Framework with Auto-Scaling Capability (CMFAS)

Figure 1 shows the design of the auto-scaling cloud manufacturing framework [8]. The main workflow of the cloud manufacturing (CMfg) service is shown in the upper part of the figure. The cloud service GUI is the interface for the user to interact with the CMfg system. Each user will be served by one worker which is matched through the bulletin board-based exchange (BBX) protocol. The advantage of this design is to reduce management efforts for workers. When a user request arrives, the GUI merely posts the request to the task bulletin board (TBB), and a proper worker would take this request to serve through its bulletin-board communication (BBC) module. A worker contains all the manufacturing modules together with the BBC module, shown in the upper middle of the figure. After a task is completely executed, the worker would send the result to the result bulletin board (RBB), and then the GUI can take the results from the RBB. In order to support auto-scaling capability (that is, automatically scaling out/in the cloud resources), the worker controller (WCR) shown in the lower part of Fig. 1, is designed to dynamically adjust the number of workers. The WCR periodically estimates the number of required workers according to the customized scaling algorithm specified in the WKR scaling rule.

3 Overview of Automated Deployment Scheme

Figure 2 shows the reference architecture of the proposed automated deployment scheme with script-based development for CMFAS, which includes three parts (shown in the middle of the figure): the script-based CMS development component, the CMS deployment component and the CMS monitoring component, aiming at rapidly developing and deploying cloud manufacturing services by using scripts plus existing standalone intelligent manufacturing applications. For avoiding re-developing manufacturing functions, the proposed scheme provides developers to create cloud manufacturing services by using scripts, as shown in the bottom of the figure. In addition, the proposed scheme can automatically created cloud manufacturing services based on our previously developed cloud manufacturing paradigm, i.e., CMFAS, shown in the top of the figure.

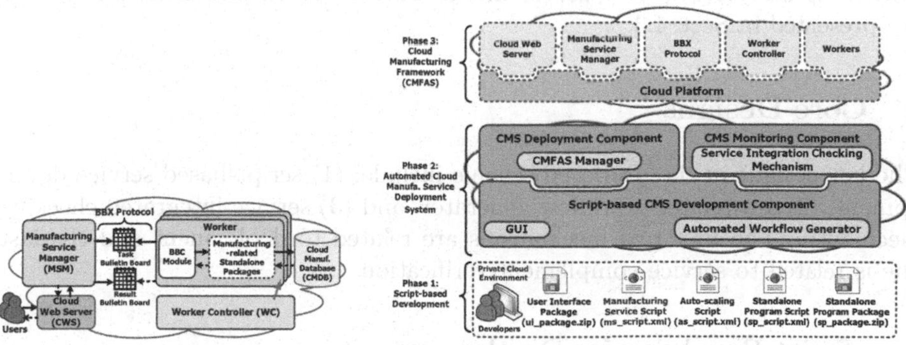

Fig. 1. The architecture of CMFAS, adopted from [8].

Fig. 2. Architecture of the proposed scheme.

The proposed scheme uses three phases to create a cloud manufacturing service. The first phase is to describe the cloud service by using scripts plus related standalone packages. The second phase is to transform these scripts to the required formats of the CMFAS architecture, and deploy them into a cloud platform. Note that development of the cloud systems for achieving the goal of this phase is the main focus of this work. The third phase is the deployed CMFAS-based cloud service, which can served users with using standalone packages in the cloud. The CMFAS-based manufacturing services can be created or removed through our proposed automated deployment scheme.

The script-based CMS development component is designed to automated generate codes associated to the manufacturing service manager and workers from the uploaded scripts, whose details will be presented in Sect. 4.1. This component has a GUI for interacting with developers to receive development scripts. Among these scripts, the auto-scaling capacity and the execution capacity are inherited from CMFAS. In this work, we design a workflow-based manufacturing service manager for improving automation degree of CMFAS. Hence, the automated workflow generator is designed to automatically generate a workflow corresponding to the cloud services that described in script files. The details are presented in Sect. 4.2.

The CMS deployment component is designed to create a CMFAS platform and deploy the cloud service to the CMFAS platform. The core of this component is the CMFAS manager for acquiring virtual machines from the underlying cloud host system, and deploying programs of CMFAS roles to corresponding virtual machines by following [8].

The CMS monitoring component is designed to continuously diagnose aliveness of the deployed cloud manufacturing services. This provides administrators a convenient way to maintain the CMFAS-based service for increase its service-level agreement, meaning that robustness of the cloud manufacturing service is satisfied for more industrial applications. The core of this component is the service integration checking mechanism, which verifies whether a deployed manufacturing service correctly serves in the CMFAS-based platform. The details are presented in Sect. 4.3.

4 Core Designs

The key designs in the proposed scheme include: (1) script-based service development, (2) automated workflow generator, and (3) service integrated checking mechanism. The first two mechanisms are related to deployment and the last one is related to service complement verification.

4.1 Script-Based Service Development

For fitting CMFAS, three scripts and two packages are required for service deployment, including (1) manufacturing service script, (2) auto-scaling rule script, (3) standalone program script, (4) standalone program package, and (5) user interface package. The advantage of the invented script-based service development is to avoid verification and testing costs of re-developing the same manufacturing functions, and to preserve the production confidence given from the existing manufacturing applications. Next, we present the details of scripts, implemented by the XML (eXtensible Markup Language) technique, for service development in our proposed scheme. Note that the standalone program package including standalone-version executable files and the user interface package including the web-based interfaces for manufacturing services are developed by using the well-developed standalone and Web programming techniques, we do not describe them in details due to limit of the paper length.

Figure 3(a) shows an example of a manufacturing service script. There are four components for specifying a manufacturing service, including the name of a manufacturing service, the names of functions in the service, and inputs and outputs of each function. By considering scope covering of these four components, a manufacturing service script is constructed as the hierarchical structure shown in the figure. The 'root' is the root element, indicating that it is the root of the script. It contains a sub-element, 'MS_Class', to enclose a manufacturing service, whose name is specified in the attribute 'name' (e.g., name = "ADAS_Service" in Line 3 of the figure.) Within 'MS_Class', the 'MS_function' is used to enclose a manufacturing function, whose name is specified in the attribute 'name' (e.g., name = "VehicleSensorDataStorage" in Line 5 of the figure.) The inputs and outputs of a manufacturing function are enclosed inside the element 'MS_function'. The 'MS_Function_InputField' encloses the inputs by using the sub-element 'Parameter' with specifying its data type in attribute 'type'. For example, the input parameter 'SensorData' is declared in Line 6 of the figure. Similarly, the 'MS_Function_OutputField' encloses the outputs by using the sub-element 'Parameter' with specifying its data type in attribute 'type'. For example, the input parameter 'StorageResult' is declared in Line 9 of the figure.

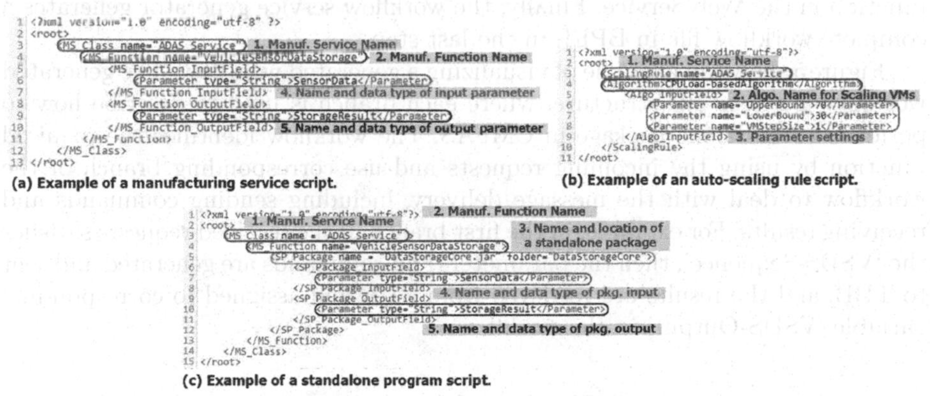

(a) Example of a manufacturing service script.

(b) Example of an auto-scaling rule script.

(c) Example of a standalone program script.

Fig. 3. Illustration of three deployment scripts.

Figure 3(b) shows an example of an auto-scaling rule script. There are four components for specifying an auto-scaling rule, including the name of an applied manufacturing service, and the name and parameters of the auto-scaling algorithm. Figure 3(c) shows an example of a standalone program script. There are five components for specifying a standalone program, including the name of a manufacturing service requiring the standalone program, and the name, the path, the inputs, and the outputs of the standalone program.

4.2 Automated Workflow Generator

Note that each uploaded script is used for building a CMFAS component: the auto-scaling rule script is used for creating the scaling controller, the standalone program script is used for workers, and the manufacturing service script is used for manufacturing service. The first two is inherited from our previous work [8]. In order that the generated CMFAS platform can handle the customized manufacturing services without manual assistance, we invent the automated workflow generator for creating the workflow-based manufacturing service manager in a CMFAS platform. More specifically, the automated workflow generator is designed to compose a workflow-based service request processor with the manufacturing service script in the cloud platform.

Figure 4 shows the flowchart of the automated workflow generator, which consists of five steps and is presented as follows. Firstly, the workflow service generator gets the parameters (such as service name, function name, etc.) for developed manufacturing services from the manufacturing service script. In this step, the manufacturing service script parser will be used to extract parameters from an XML document. Secondly, the workflow service generator generates the workflow template with filling the extracted parameters. In our implementation, the BPEL is adopted as our workflow representation. Thirdly, the workflow service generator generates the WSDL file for describing services and associated function in the Web Service. Finally, the workflow service generator generates a complete workflow file in BPEL in the last step.

Figure 5 shows an example of visualizing a generated workflow. The generated workflow is a tree-like structure, where each branch is used to describe how to perform a standalone package in CMFAS. The workflow identifies the required function by using the incoming requests and use corresponding branch of the workflow to deal with the message delivery, including sending commands and receiving results. For example, in the first branch, if the received requests satisfies the 'VSDS-Sequence', then the parameters and commands are generated and sent to TBB, and the results are retrieved from RBB and assigned to corresponding variable 'VSDS-Output' in the workflow.

4.3 Service Integrated Checking Mechanism

The service integrated checking mechanism is used to verify completeness of message delivery between the web interface and works, so that related commands and corresponding results can be correctly delivered. We proposed two approaches to check completeness of the service integration as follows.

Structure Verification Approach:

This approach is to ensure the generated codes for CMFAS can successfully deliver the request commands and receive the manufacturing results from standalone packages. We first define two key concepts related CMFAS, and describe the condition of achieving the goal in the following theorem.

Definition 1. *A CMFAS-based manufacturing service is consist of five roles:*

$$CMFAS = \{cws, msm, bbx, wkr, wcr\}$$

where cms is the cloud web server, msm is the manufacturing service manager, bbx is the BBX protocol, wkr is the set of workers, wcr is the worker controller.

Definition 2. *A CMFAS-based manufacturing service is* message deliverable *if there always exists a two-way message delivery path between cloud web server (cws) and workers (wkr).*

Theorem 1. *Let (x, y) be the existence of two-way message delivery between components x and y, where $x, y \in \{cws, msm, bbx, wkr, wcr\}$ and $x \neq y$. If a generated CMFAS-based manufacturing service exists (cws, msm), (msm, bbx), and (bbx, wkr), then the generated service is message deliverable.*

Theorem 1 indicates the condition that a generated CMFAS-based manufacturing service can successfully operate. Thus, we can develop a static structure verification module by obeying the theorem to verified the codes (generated or uploaded) whether satisfies completeness of service integration.

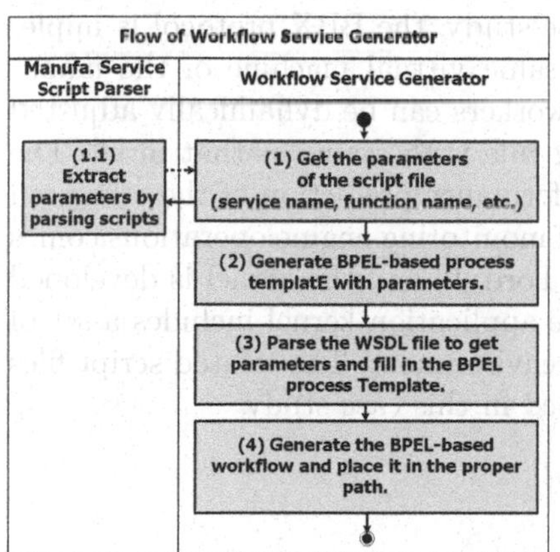

Fig. 4. The flowchart of the automated workflow generator.

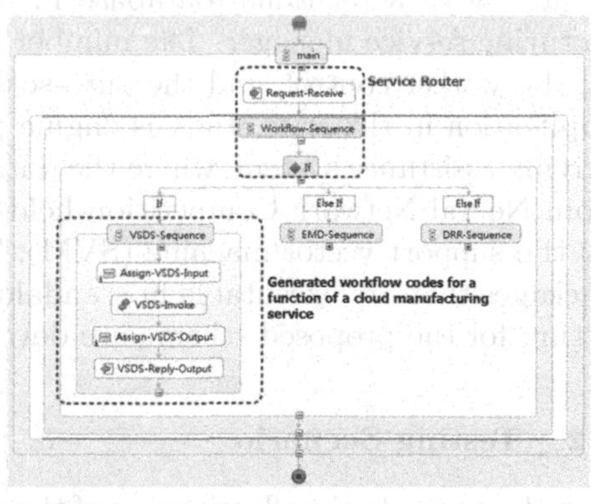

Fig. 5. An example of visualizing a generated workflow.

Checking-by-Testing Approach:
Since the CMFAS provides highly automatic operational mechanisms, we develop a module in this work to physically test a CMFAS-based cloud service through feeding proper requests. In addition, the manufacturing service interface is designed by using the REST (Representational State Transfer) technology, and thus, the approach needs only to send requests and receive results through Web programming techniques.

Figure 6 shows the syntax of a manufacturing function in the URI format. The prefix of the URI is the same to a REST service. The service 'getWorkflowService' is developed to deal with manufacturing requests, and its parameters includes inputs and outputs of a request.

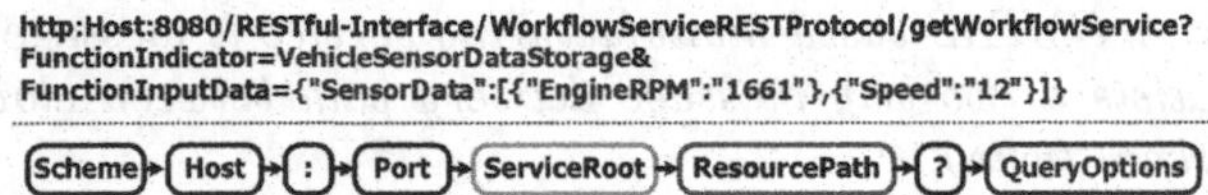

Fig. 6. An example of the REST API format for testing a manufacturing service.

5 Case Study

5.1 System Deployment and Experimental Settings

We deploy the proposed scheme in the VMWare-based private cloud. There are five virtual machines are used to build the components for a CMFAS-based service, including the cloud web server, the manufacturing service manager, a worker, and the worker control. In this case study, the BBX protocol is implemented with a relational database in the same virtual machine of the manufacturing service manager. The number of workers can be dynamically adjusted by the worker control, and the auto-scaling rule is the same as that in [8]. The application in this case study is engine performance prediction of the advanced driving assistance system, where the data of monitoring engine operations comes from Neural Network Competition held by Ford [9] and the kernel is developed by the support vector machine (SVM). The application kernel includes a set of packages that are executable in standalone environment. The related script files fitting for the proposed scheme are developed in this case study.

5.2 Testing Scenario

In order to verify the effectiveness of the proposed automated deployment scheme with service integrated checking mechanism, we design various operational scenarios to test the developed system. Due to space limit, we display only the deployment of the engine diagnosis cloud service. In the scenario, certain engine diagnosis applications are deployed to a CMFAS-based platform, and related scripts are developed, where some snippets are shown previously.

Scenario: Deploying Engine Diagnosis Cloud Service

Step 1: The developer logins the system via the Web GUI.
Step 2: The GUI receives five kinds of scripts from the developer.
Step 3: The cloud development component creates virtual machines to play roles of CMFAS components.
Step 4: The cloud development component creates the workflows from scripts, and the CMS deployment component installs the generated workflows and standalone packages to corresponding virtual machines in the cloud.
Step 5: The CMS monitoring component would verify whether the deployed services correctly work in the generated CMFAS platform.
Step 6: After Step 5, the generated cloud manufacturing services can be provided to users through Web interfaces in the generated CMFAS platform.

5.3 Integrated Testing Results

The integrated testing results of various operational scenarios show that the developed engine diagnosis cloud service works smoothly. Due to space limit, only some GUI snapshots are displayed below for validating efficacy of the scheme.

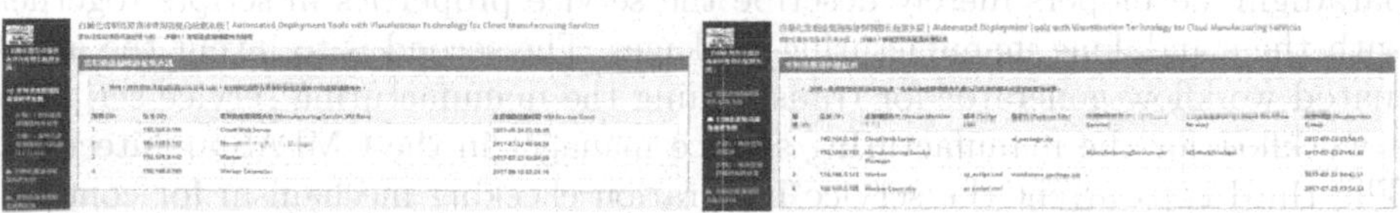

(a) Snapshot of creating a CMFAS platform with 4 virtual machines. (b) Snapshot of deploying scripts to corresponding virtual machines.

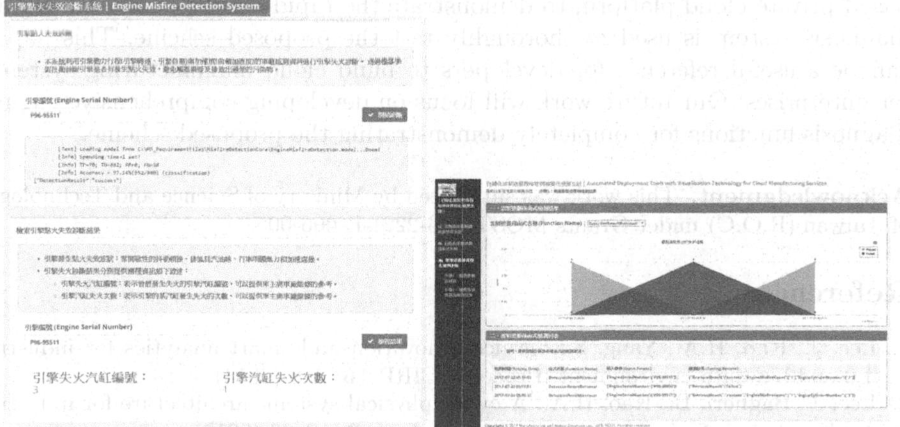

(c) Snapshot of deployed engine diagnosis cloud service. (d) Snapshot of cloud service checking results.

Fig. 7. Screenshots of integrated testing results for the proposed scheme.

Figure 7(a) is a GUI snapshot showing creation of a CMFAS platform with 4 virtual machines, which play roles of *cws*, *msm*, *wkr*, and *wcr*. This indicates the proposed scheme can communicate with the private cloud with creating and deleting CMFAS platforms. Figure 7(b) is a GUI snapshot of deploying scripts to corresponding virtual machines. This indicates the scripts are correctly sent to corresponding virtual machines for setting customized cloud services. Notice that the manufacturing service script has been transformed into a BPEL-formatted worflow and corresponding WAR-formatted executable file by using the automated workflow generator mentioned in Sect. 4.2. Figure 7(c) is a GUI snapshot showing the deployed engine diagnosis cloud service. This indicates that the script-based development paradigm is validated. Figure 7(d) is a GUI snapshot showing cloud service checking results. This indicates that our proposed scheme can continuously monitoring the status of the deployed cloud service with the checking-by-testing approach, mentioned in Sect. 4.3.

6 Conclusions and Future Work

In this paper, an automated deployment scheme with script-based development mechanisms is proposed for rapidly creating cloud manufacturing systems. Three key designs are included in the scheme. The first is to invent the script-based development for a cloud manufacturing service. With such development paradigm, developers merely describe the service properties in scripts together with the standalone manufacturing packages. The second is to invent the automated workflow generator for transforming the manufacturing service script to a workflow for the manufacturing service manager in the CMFAS architecture. The third is to invent the service integration checking mechanism for continuously verifying whether a deployed cloud manufacturing service correctly works. We implement a prototype of the proposed scheme and deploy it on a VMWare-based private cloud platform to demonstrate the rapid development. An engine diagnosis system is used to thoroughly test the proposed scheme. This paper can be a useful reference for developers to build cloud manufacturing systems for enterprises. Our future work will focus on developing comprehensive engine diagnosis functions for completely demonstrating the proposed scheme.

Acknowledgment. This work was supported by Ministry of Science and Technology of Taiwan (R.O.C) under Grants MOST 106-2221-E-006-005.

References

1. Lee, J., Kao, H.A., Yang, S.: Service innovation and smart analytics for industry 4.0 and big data environment. Procedia CIRP **16**, 3–8 (2014)
2. Lee, J., Bagheri, B., Kao, H.A.: A cyber-physical systems architecture for industry 4.0-based manufacturing systems. Manuf. Lett. **3**, 18–23 (2015)
3. Yu, S., Xu, X.: Development of a product configuration system for cloud manufacturing. In: Umeda, S., Nakano, M., Mizuyama, H., Hibino, H., Kiritsis, D., von Cieminski, G. (eds.) APMS 2015. IAICT, vol. 460, pp. 436–443. Springer, Cham (2015). https://doi.org/10.1007/978-3-319-22759-7_51

4. Bekris, K., Shome, R., Krontiris, A., Dobson, A.: Cloud automation: Precomputing roadmaps for flexible manipulation. IEEE Robot. Autom. Mag. **22**(2), 41–50 (2015)
5. Kehoe, B., Abbeel, P.: A survey of research on cloud robotics and automation. IEEE Trans. Autom. Sci. Eng. **12**(2), 398–409 (2015)
6. Xu, X.: From cloud computing to cloud manufacturing. Rob. Comput.-Integr. Manuf. **28**(1), 75–86 (2012)
7. Huang, B., Li, C., Yin, C., Zhao, X.: Cloud manufacturing service platform for small- and medium-sized enterprises. Int. J. Adv. Manuf. Technol. **65**(9), 1261–1272 (2013)
8. Chen, C.C., Lin, Y.C., Hung, M.H., Lin, C.Y., Tsai, Y.J., Cheng, F.T.: A novel cloud manufacturing framework with auto-scaling capability for the machining industry. Int. J. Comput. Integr. Manuf. **29**(7), 786–804 (2016)
9. Feldkamp, T., Feldkamp, L., Marko, K.: IJCNN 2001 Neural Network Competition (2017)

A Parking Occupancy Prediction Approach Based on Spatial and Temporal Analysis

Eric Hsueh-Chan Lu(✉) and Chen-Hao Liao

Department of Geomatics, National Cheng Kung University,
No. 1, University Rd., Tainan City 701, Taiwan (R.O.C.)
luhc@mail.ncku.edu.tw, how880062@gmail.com

Abstract. It's very difficult to find an appropriate parking space in urban area, when drivers are near to their destinations. The literature studies showed that 30% of the traffic congestion, unnecessary fuel consumption and exhaust emissions are caused by searching for the parking spaces. With the ever-changing nature of technology, smart parking systems composed of smart devices and sensor technologies are readily available and provide various information such as locations, real-time available counts, and parking costs. However, the drivers can't know whether there is an available parking space at the arrival time. In this paper, we propose a parking occupancy prediction approach based on spatial and temporal analysis. In this approach, we extract related features and build the parking occupancy prediction model by Naïve Bayes classifier and decision tree. The prediction model can be used to predict the level of parking occupancy rate for each street block in the next hour. To evaluate the performance of proposed approach, we carried out the experiment by the on-street parking data collected by the *SFPark* system in San Francisco, USA. The results show that our proposed smart parking guidance system can significantly improve the prediction accuracy for the level of parking occupancy rate.

Keywords: Smart parking · Occupancy prediction
Spatial and temporal analysis

1 Introduction

Finding an available parking space in urban area is a big problem for the city transportation. Since the resource of parking spaces is limited for a city, the cars will spend lots of time searching for a parking space, and it may lead to congested traffic, fuel consumption and exhaust emissions. The literature showed that about 30% of the traffic congestion is caused by cars which are cruising for parking spaces [5]. With the ever-changing nature of technology, the concept of smart city can be fulfilled by the appearance of *Internet of Things* (*IoT*), which make the current Internet to a network of interconnected objects, such as sensors, parking meters [2], and the service of parking management is included. Melbourne, Australia and Santander, Spain and San Francisco, USA have deployed the sensors to record the parking events and availability of on-street parking spaces [7]. San Francisco even provides the real-time parking information to the public to help them make decisions about parking. Hence, smart parking systems have become a critical issue currently.

N. T. Nguyen et al. (Eds.): ACIIDS 2018, LNAI 10751, pp. 500–509, 2018.
https://doi.org/10.1007/978-3-319-75417-8_47

In the scenario of the city which can provides the real-time information of on-street parking spaces to drivers. Drivers can check the status of the available on-street parking spaces through the smart phones or other smart devices, and each of the drivers can select the parking space that is available at current time. However, the drivers can't know whether there is an available parking space at the arrival time. Therefore, in this paper, we propose a smart parking occupancy prediction approach based on spatial and temporal analysis. The parking occupancy prediction can utilize the historical data of each parking street block. In this approach, we first extract several features including spatial, temporal and other related features and then build a parking occupancy prediction model based on Naïve Bayes and Decision Tree to predict the level of occupancy rate for the next hour. To evaluate the performance of our proposed prediction approach, the real world data of the on-street occupancy of the parking street block data collected from San Francisco is used to conduct a series of experiments, and the experimental results show that the proposed approach can improve the accuracy of parking occupancy level prediction comparing to the previous 7-day method.

The contributions of this paper are listed as follow.

1. We define several features related to parking occupancy level including temporal related features, spatial related features and weather condition.
2. Due to the data contains numeric and categorical attributes, we utilize Naïve Bayes and Decision Tree C5.0 model to learn the occupancy level prediction model.
3. We use the occupancy level prediction model to predict the occupancy level for the next hour.
4. Based on the real dataset, the experimental results show that our approach can improve the prediction accuracy.

The organization of the paper is as follow. Section 2 gives a review of the related work of parking prediction and parking allocation problem. Section 3 gives the detailed explanation on the procedure of the proposed method. In Sect. 4, we evaluate the proposed method by the experiments and analyses the results. In Sect. 5, a conclusion and future work of the paper will be mentioned.

2 Related Work

In this section, we review some important studies related to smart parking issues. Several literatures are related to the issues of parking occupancy prediction which is aim to predict the occupancy status of on-street parking spaces. On-street parking means the parking space which is monitored by the in-ground sensors in urban area. Vlahogianni *et al.* proposed a methodological frame work [6]. They aim to provide two kinds of parking prediction service. It contains the probability of a free space to continue being free in subsequent time intervals, and the short-term parking occupancy prediction in the predefined selected region of the smart city of Santander, Spain. The data of the on-street parking spaces' status was collected by the sensors which are installed on-street. They separate the on-street parking spaces into four regions, and utilizing survival analysis and neural network models for each region. The result shows that the Weibull parametric model can describe the probability of a parking space to continue to be vacant in the approaching time intervals. Richter *et al.* proposed a temporal and spatial clustering method [4] to lower the storage space of the prediction model, in order to save the prediction models in the in-vehicle navigators. To predict the future availability of each on-street road segments, they use the 7-day model to learn one prediction model for each road segment at each day of week and time of day. The spatial and temporal clustering method try to lower the model need to be storage and can still preserve the accuracy of the prediction. They utilized the historical data of the on-street parking spaces of *SFPark* project in San Francisco, USA. Although the storage space can be reduced up to 99%, the accuracy will drop about 10% to around 68%. Zheng *et al.* [7] utilizing the on-street parking occupancy data of two cities, namely San Francisco, USA and Melbourne, Australia, to predict the occupancy rate at certain time of day. They utilized three kinds of machine learning algorithm, Regression Tree, Support Vector Regression and Artificial Neural Network to predict the occupancy rate with three kinds of input features. Liu proposed a methodology [3] to determine the real-time parking availability information for on-street parking operators. The research determines the relationship between over-paid on-street parking time, which means that the owner pays more than the actual parking time needs and how this may make the same parking space available for next vehicle based on actual on-street parking time. Richter *et al.* [4] and Fabusuyi *et al.* [1], they classify the parking occupancy in classes and utilize the classification model to predict the occupancy rate of each on-street parking street block.

3 Proposed Method

The proposed method is to build the prediction model of parking *OCCupancy Level* (*OCCL*). The model can predict the *OCCL* of the parking street blocks at the arrival time. In this paper, the parking occupancy data should be prepared and the features which are correlated to the *OCCL* of the parking street block should be generated. The features will be utilized as the input features of the selected learning algorithm (i.e. Naïve Bayes and Decision Tree). After the evaluation of the learning algorithm, we can use the best prediction model as the *OCCL* prediction model.

Since the goal of the parking occupancy prediction is to use the classification model to predict the parking occupancy levels for each parking street block in the next time period, there are two important factors that may affect the result of the accuracy of the prediction model. First, we have to extract the features which are related to the parking occupancy. We extract three kinds of features including temporal features, spatial features and other features. Second, we utilized classification methods which are Naïve Bayes Classifier and Decision Tree to evaluate the result of the prediction model. The definition of the parking occupancy prediction for each parking street block is defined as follow:

1. t is the time period.
2. b is the parking street block.
3. $F = \{Temporal\ Features,\ Spatial\ Features,\ Other\ Features\}$ is the input feature set consists of Temporal, Spatial and Others features.

4. $O_b(t)$ is the occupancy rate for b at t.
5. $L^n(.)$ is the function that can transform the *OCCR* from continuous values to n discrete levels.
6. $Y_b^n(t)$ is the output of the prediction model which can predict the *OCCL* for b at t and there are n kinds of level can be outputted.

3.1 Temporal Features

For the temporal-related features, they are related to the *OCCL* of the block in the time dimension, and there are 3 kinds of feature were extracted.

1. *Day of Week* (*DOW*): *DOW* consists of Monday to Sunday totally 7 kinds of discrete values, and will be stored as the categorical variable.
2. *Time of Day* (*TOD*): *TOD* consists of the time period from 0_{AM}–1_{AM} (0), 1_{AM}–2_{AM} (1), …, to 23_{PM}–24_{PM} (23), totally 24 discrete values, and will be stored as the categorical variable.
3. K_1 *Occupancy Rate* (K_1*_OCC*): $K_1\_OCC = \{O_b(t-1), O_b(t-2), \ldots, O_b(t-K_1)\}$. The goal is to predict the *OCCL* for b at t ($Y_b^n(t)$), thus we consider K_1 previous observations from t. K_1*_OCC* will be stored in two type of values, one is the original continuous value of occupancy rate from 0 to 1 (i.e., numerical variable), another is to discretized value by $L^n(.)$ (i.e., categorical variable).

3.2 Spatial Features

The meaning of spatial features can be described as finding the parking street block which is similar to the target of the parking street block in the spatial domain. In the spatial-related features part, totally 3 kinds of feature were obtained.

1. K_2 *Nearest parking street Blocks* (K_2*_NB*): $K_2\_NB = \{O_{1NB_b}(t-1), O_{2NB_b}(t-1), \ldots, O_{K_2NB_b}(t-1)\}$. The idea is to extract the current occupancy rate value from the top K_2 nearest parking street block of b, where kNB_b indicates the id of top-k nearest parking street block of b. K_2*_NB* will be stored in two type of values, one is the original continuous value of occupancy rate from 0 to 1 (i.e., numerical variable), another is to discretized value by $L^n(.)$ (i.e., categorical variable).
2. K_3 *Similar parking street blocks based on Root Mean Square Error* (K_3*_SRMSE*): $K_3\_SRMSE = \{O_{1SRMSE_b}(t-1), O_{2SRMSE_b}(t-1), \ldots O_{k_3RMSE_b}(t-1)\}$. The idea is to extract the occupancy rate values from the top K_3 most similar parking street blocks of b based on *RMSE*, where $kSRMSE_b$ indicates the id of top-k most similar parking street block of b. K_3*_SRMSE* will be stored in two type of values, one is the original continuous value of occupancy rate from 0 to 1 (i.e., numerical variable), another is to discretized value by $L^n(.)$ (i.e., categorical variable).
3. K_4 *Similar parking street blocks based on RMSE with different TOD* (K_4*_SRMSETOD*): $K_4\_SRMSETOD = \{O_{1SRMSE_b^{TOD}}(t-1), O_{2SRMSE_b^{TOD}}(t-1), \ldots, O_{k_4SRMSE_b^{TOD}}(t-1)\}$. The idea is just like finding the top K most similar parking street blocks of b based on *RMSE*, but it has to consider the situation of

different *TOD*. Hence, K_4*_SRMSETOD* will return the occupancy rate values from the top K_4 most similar parking street blocks of *b* when the prediction time period *t* = *TOD* based on *RMSE*. K_4*_SRMSETOD* will be stored in two type of values, one is the original continuous value of occupancy rate from 0 to 1 (i.e., numerical variable), another is to discretized value by $L^n(.)$ (i.e., categorical variable).

3.3 Other Related Features

Other related features mean that they are not directly related to the occupancy level of the specific block id, but they may affect the prediction result.

1. *Block_ID* (*BID*): since each block in the urban area may have its own variation of the occupancy level, so the bid is selected to be the static feature. Since they are totally 109 block in the data set, they are totally 109 bid will be stored as the categorical variable.
2. *Hourly Precipitation* (*HP*): it's the weather condition of the city of San Francisco. And the meaning of it is the accumulation of the rainfall in an hour. The unit of it is millimetre (mm), and it will be stored as the numerical variable.

4 Experimental Evaluation

To evaluate the performance, a series of experiments are conducted by utilizing the real on-street parking occupancy data. The experiments are to evaluate the parking occupancy prediction model. We use R 3.4.0 to run the experiment of the model of Naïve Bayes Classifier by the Package 'e1071', and Decision Tree C5.0 by the Package 'C50'. The evaluation for other experiment use Java 1.8 to implement. For the experimental computer, CPU is Intel i5 3.3 GHz and memory is 8 GB.

4.1 Experimental Data

The data is collected from the SFPark Pilot Evaluation. The SFPark pilot project is aim to control the usage of the on-street parking spaces in the City of San Francisco, USA. They installed the in-ground sensors to observe the occupancy status of the on-street parking spaces, and developed a parking fee rate changing strategy to gain the usage of the on-street parking spaces. The original data named "SFpark_ParkingSensorData_HourlyOccupancy_20112013" recorded the total occupied time and the total vacant time for the on-street parking street blocks in the period of an hour by the in-ground sensor which is installed under the on-street parking spaces, and the data recorded the information for totally 410 on-street parking street blocks for 8,228 spaces (about 25% of the on-street parking spaces in San Francisco) from April 1, 2011 00:00 to July 31, 2013 24:00. Although the data records the total occupied time and the total vacant time based on the status of the parking meter, the most important attribute *OCCupancy Level* (*OCCL*) is not contained. According to the literature of Richter *et al.* [4], *OCCL* can be mapped by the *OCCupancy Rate* (*OCCR*) for each parking street block at each time period. Since the data doesn't contain the records of total spaces and the total occupied space of a parking street block at the time period, we have to generate the *OCCR* by the formula (1), where *TOTAL_OCCUPIED_TIME* and *TOTAL_VACANT_TIME* indicate the sum of seconds across all spaces on that parking street block during that hour when a space was occupied and vacant, respectively.

$$OCCR = \frac{TOTAL\_OCCUPIED\_TIME}{TOTAL\_OCCUPIED\_TIME + TOTAL\_VACANT\_TIME} \quad (1)$$

We used two kinds of functions to map the *OCCR* to *OCCL*:

1. *OCCL3:* It consists of 3 discrete levels which are based on the definition of *SFPark* website application, the original definition was based on the availability classes of the parking street block. To explain in a more convenient way, we use the occupancy level to explain. The occupancy level is shown in Table 1.

Table 1. Mapping *OCCR* to *OCCL3*.

OCCL	Low	Medium	High
OCCR range	0–70%	70–85%	85–100%

2. *OCCL4*: Since the *OCCL* defined by *SFPark* has a big range of the level on high availability (i.e., low occupancy), about 60% of the data are in the *OCCL* of low occupancy level with the definition of *OCCL3*, we define the new occupancy level (*OCCL4*) into 4 classes as the Table 2 shows. In Each level of *OCCL4*, it has the equal range of *OCCR* value.

Table 2. Mapping *OCCR* to *OCCL4*.

OCCL	Very Low	Low	Medium	High
OCCR range	0–25%	0–70%	70–85%	85–100%

We found that there are some problems for the data: (1) The sensors were installed in different time, some of them were installed in the December, 2011 and July, 2012, so there are only 211 out of 410 blocks are installed and start record the data from April 1, 2011 00:00 to July 31, 2013 24:00 one record per hour per block (i.e., 20,472 records per block). (2) Due to the early battery failures of the sensors, the data recorded around November 2012 contains lots of *TOTAL_UNKNOWN_TIME* for each block, so we utilized the data from April 1, 2011 to October 8, 2012, and there are 13,362 records per block. (3) In the 211 out of 410 blocks, almost half of them have the problem of the unstable *TOTAL_TIME* due to the malfunction of sensors, thus the *Total number of spaces for a block* will be unstable, so we select only 109 out of 211 blocks which the *TOTAL_TIME* remain the same from April 1, 2011 to October 8, 2012, and there are 13,362 records per block. Finally, we utilized the parking street block data for 109 parking street blocks for totally 1,456,458 records from April 1, 2011 to October 8, 2012, and totally 2,026 on-street parking spaces (about 6% of the on-street parking spaces in San Francisco). For each block, there are totally 13,362 tuples, and we use the front 70% of the data (i.e., 390 days) as the training data, and other 30% of the data (i.e., 167 days) will be the testing data. Since there are 109 blocks, so there are totally 1,456,458 tuples, and 1,020,240 tuples are utilized to train the model.

4.2 Feature Selection

There are 4 features K_1_OCC, K_2_NB, K_3_SRMSE and K_4_SRMSETOD, K_1 to K_4 represent the number of columns of the features. For instance, if K_1 is 4, the feature of K_1_OCC will utilize totally 4 columns of feature as the input attribute for training the model. This experiment will use one feature as the only input feature each time period, and set the value of the column from 1 to 10. The results are shown in Figs. 1 and 2. The number of horizontal axis for every figure means the input values of K_1, K_2, K_3, K_4; the vertical axis means the accuracy of the classification results for (a) *OCCL3* and (b) *OCCL4*. We observe that the temporal feature K_1_OCC always has the best performance than other 3 features. The accuracy will decrease when the number of K_1 increase for Naïve Bayes Classifier and the accuracy won't have the dramatically drop when the K_1 increase for C5.0. Based on the results, K_1 are set as 1 and 3 for Naïve Bayes Classifier and C5.0, respectively. For the spatial features, since K_4_SRMSETOD always has the better results than K_2_NB and K_3_SRMSE, we decide to use K_4_SRMSETOD as the spatial feature. Based on the results, K_4 are set as 5 and 4 for Naïve Bayes Classifier and C5.0, respectively.

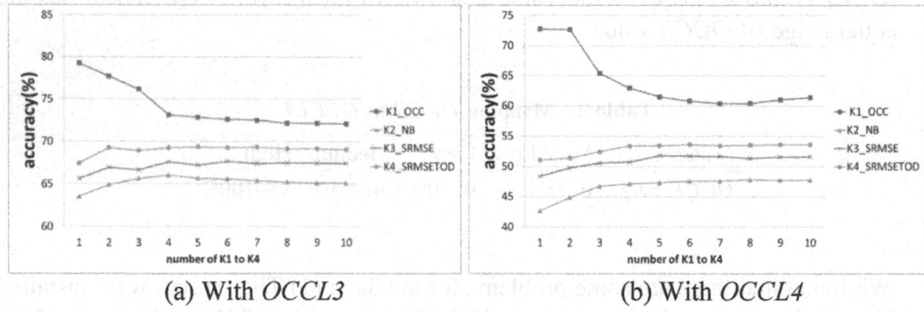

(a) With *OCCL3* (b) With *OCCL4*

Fig. 1. The result of Naïve Bayes using categorical variables.

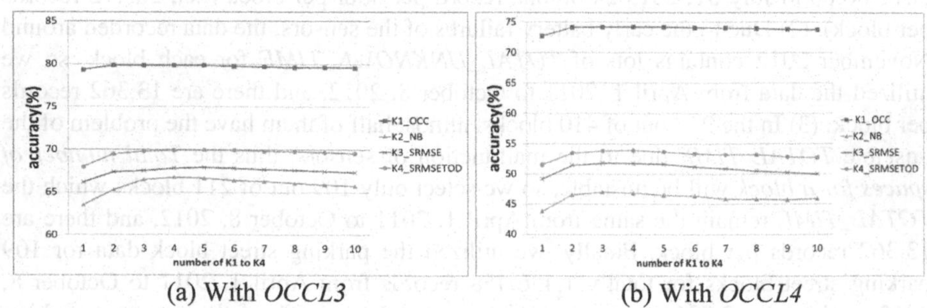

(a) With *OCCL3* (b) With *OCCL4*

Fig. 2. The result of C5.0 using categorical variables.

4.3 Feature Combination

In this part, we will utilize the other related features to combine the features and find the best combination of the feature set. There will be totally 5 kinds of set we utilized for different classification model. The sets with utilizing different combination of input features are shown in Table 3. We will compare all three classification models with utilizing different set with the 7-day model [4], which consider the historical mean of the *OCCR* in different time of day (Monday to Sunday, totally 7 days) and day of week for each block and to map the *OCCR* to *OCCL* by the predefined definition of mapping function for *OCCL3* and *OCCL4*.

Table 3. Attributes utilized in different Set.

Set ID	*DOW*	*TOD*	K_4*_SRMSETOD*	K_1*_OCC*	*HP*	*BID*
Set 1	O	O				
Set 2	O	O	O			
Set 3	O	O	O	O		
Set 4	O	O	O	O	O	
Set 5	O	O	O	O	O	O

In Fig. 3, the result shows that by using the more attributes to train the model, the accuracy of the model will increase. We also compare 3 models (i.e. Naïve Bayes, *C5.0 (categorical)*, and *C5.0 (numerical)*) to 7-day model by utilizing different sets one by one. In Set 1, we utilized only 2 temporal features *DOW* and *TOD* to train the model, and Set 2 we added the temporal feature K_4*_SRMSETOD*, though the accuracy of Naïve Bayes, *C5.0 (categorical)*, and *C5.0 (numerical)* will increase, the accuracy of these 3 models is still worse than the result of 7-day model. When we add K_1*_OCC* (i.e., utilize Set 3), the accuracy of C5.0 (categorical) and C5.0 (numerical) will have a dramatically increase, and have better performance than 7-day model. In Set 4 and Set 5 we add the *HP* and *BID* respectively, and the accuracy of *C5.0 (categorical)* and *C5.0 (numerical)* with the input feature Set 5 compare to that with the input feature Set 3 increase about 1%. In the two cases, *C5.0 (numerical)* is better than *C5.0 (categorical)*. For Naïve Bayes Classifier, the accuracy is only better than the 7-day model, the reason may be related to the assumption of all the input features are independent to each other for Naïve Bayes Classifier, so the performance of Naïve Bayes can't have dramatically improvement like C5.0 models. The *C5.0 (numerical)* can have about 8–15% improvement of accuracy compare to 7-day model. From the experiments, we found that the result of utilizing *C5.0 (numerical)* with input feature Set 5 will have the best performance no matter with *OCCL3* or *OCCL4*.

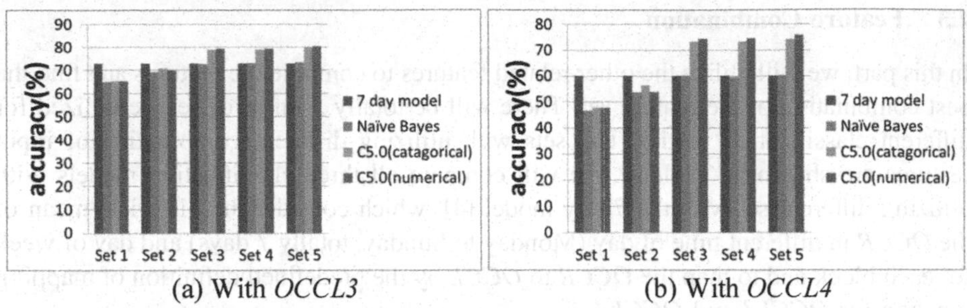

(a) With *OCCL3* (b) With *OCCL4*

Fig. 3. The results of using different models with different Sets.

4.4 Predict the Occupancy Rate Ahead k Hours

Since the sampling rate of the experimental data is an hour, in this experiment, we utilize Set 5 as the input feature set to predict the *OCCL* ahead the prediction time periods from 1 to 5 h. The previous experiments only predict the *OCCL* ahead 1 h. In Fig. 4, we observe that the accuracy of occupancy prediction decreases with the prediction time period increases. It is very reasonable because the unknown information significantly increases when we want to predict the occupancy level after much more hours. We also can see that C5.0 models will have better result than 7-day model when predicting the *OCCL* ahead 1 to 4 h; if we predict the *OCCL* ahead 5 h, the result of using 7-day model will have higher accuracy. As for Naïve Bayes, its performance is worse than 7-day model when predicting the *OCCL* ahead 2 to 5 h.

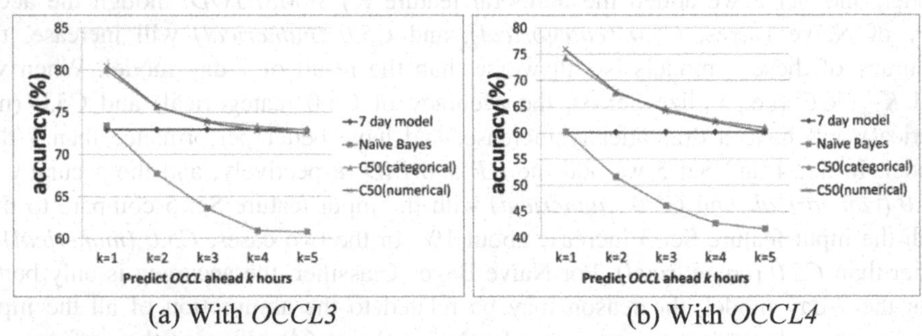

(a) With *OCCL3* (b) With *OCCL4*

Fig. 4. Predict *OCCL* ahead *k* hours by different models.

5 Conclusion and Future Work

In this paper, we have proposed a parking occupancy prediction approach for the on-street parking street block in an urban area. In this approach, we utilize Naïve Bayes and Decision Tree C5.0 model to learn the occupancy level prediction model with several features including temporal related features, spatial related features and weather condition. Based on the real world data of the on-street occupancy of the parking street block data at San Francisco, our proposed approach is better than previous 7-day model in terms of prediction accuracy. In the future, we will try to utilize the regression model to replace the classification model and decrease the error of the prediction of occupancy rate. The data we utilized has a low sampling rate (1 data per hour). We will collect the data with higher sampling rate, so that the occupancy rate of the parking street block can be closer to the real-world condition. We will collect different types of the parking spaces such as parking garages and parking lots, and combine the data with the on-street parking spaces to develop more complete occupancy prediction model.

Acknowledgment. This research was supported by Ministry of Science and Technology, Taiwan, R.O.C. under grant no. MOST 106-2119-M-006-020.

References

1. Fabusuyi, T., Hampshire, R.C., Hill, V.A., Sasanuma, K.: Decision analytics for parking availability in downtown Pittsburgh. Interfaces **44**(3), 286–299 (2014)
2. Jin, J., Gubbi, J., Luo, T., Palaniswami, M.: Network architecture and QoS issues in the internet of things for a smart city. In: Proceedings of the International Symposium on Communications and Information Technologies, pp. 974–979, October 2012
3. Liu, C.: Development of prediction model for real-time parking availability for on-street paid parking. Master's dissertation, University of Pittsburgh (2016)
4. Richter, F., Di Martino, S., Mattfeld, D.C.: Temporal and Spatial clustering for a parking prediction service. In: Proceedings of the 26th IEEE International Conference on Tools with Artificial Intelligence, pp. 278–282, November 2014
5. Shoup, D.: Free parking or free markets. Access Mag. **1**(38), 28–35 (2011)
6. Vlahogianni, E.I., Kepaptsoglou, K., Tsetsos, V., Karlaftis, M.G.: A real-time parking prediction system for smart cities. J. Intell. Transp. Syst. **20**(2), 192–204 (2016)
7. Zheng, Y., Rajasegarar, S., Leckie, C.: Parking availability prediction for sensor-enabled car parks in smart cities. In: Proceedings of the 10th IEEE International Conference on Intelligent Sensors, Sensor Networks and Information Processing, pp. 1–6, April 2015

An Approach for Diverse Group Stock Portfolio Optimization Using the Fuzzy Grouping Genetic Algorithm

Chun-Hao Chen[1], Bing-Yang Chiang[2], Tzung-Pei Hong[2,3](✉), Ding-Chau Wang[4], and Jerry Chun-Wei Lin[5]

[1] Department of Computer Science and Information Engineering, Tamkang University, Taipei, Taiwan
chchen@mail.tku.edu.tw

[2] Department of Computer Science and Engineering, National Sun Yat-sen University, Kaohsiung, Taiwan
m053040005@student.nsysu.edu.tw

[3] Department of Computer Science and Information Engineering, National University of Kaohsiung, Kaohsiung, Taiwan
tphong@nuk.edu.tw

[4] Department of Information Management, Southern Taiwan University of Science and Technology, Tainan, Taiwan
dcwang@stust.edu.tw

[5] School of Computer Science and Technology, Harbin Institute of Technology Shenzhen Graduate School, Shenzhen, China
jerrylin@ieee.org

Abstract. Previously, an algorithm has been proposed for optimizing a diverse group stock portfolio using the grouping genetic algorithm. However, it is not easy to set appropriate parameters for genetic operations when datasets are different in the previous approach. In this paper, we propose an approach that can not only obtain a diverse group stock portfolio but also tune parameters dynamically for the genetic operations using the fuzzy grouping genetic algorithm. Three parts, grouping, stock, and stock portfolio parts, are used as encoding scheme. The fitness function defined in the previous approach is utilized to evaluate quality of a chromosome. To deal with parameter setting for genetic operations, a fuzzy logic controller is designed and employed to dynamically adjust them using the predefined knowledge base for deriving a better solution. Experiments on a real dataset were conducted to show the effectiveness of the proposed approach.

Keywords: Diverse group stock portfolio · Fuzzy logic controller
Fuzzy grouping genetic algorithm · Portfolio optimization

1 Introduction

With economic prosperity, more and more investors utilize various investment strategies to allocate their capital in different assets to earn profit and avoid risk. Since one of the popular assets is stock, how to obtain a stock portfolio that can reach both

N. T. Nguyen et al. (Eds.): ACIIDS 2018, LNAI 10751, pp. 510–518, 2018.
https://doi.org/10.1007/978-3-319-75417-8_48

low risk and high return is an important issue. In other words, to get an actionable portfolio from the given financial dataset is an attractive topic for researchers. Because different expected returns and risks will result in various portfolios, optimization techniques have been designed for obtaining appropriate portfolios [1, 3, 4, 6, 9–12].

To make a portfolio more effective and easy to be used, an approach has been proposed to obtain a diverse group stock portfolio (DGSP) using the grouping genetic algorithm [5]. A stock group contains a set of stocks. A group stock portfolio is consists of a set of stock groups. Thus, various stock portfolios can be generated according to a given DGSP. Since obtaining a DGSP is a grouping problem, the grouping genetic algorithm (GGA) proposed by Falkenauer [7], which has been shown its ability to solve the grouping problems, was utilized in that approach. It first encoded a DGSP into a chromosome using the grouping, stock, and stock portfolio parts. The fitness function that composed of the portfolio satisfaction, group balance, and diversity factor, was designed to evaluate chromosomes. Three genetic operations were applied on the population to generate various offsprings. The evolution process was repeated until the stop conditions were reached.

To ensure metaheuristic algorithms to obtain high quality solutions, how to set appropriate parameters for genetic operations to get various solutions is a critical and important task. Like most of metaheuristic algorithms, the GGA also has difficulty in how to set proper parameters and get a good balance between exploration and exploitation. Because parameters for genetic operations, including crossover, mutation and inversion, were predefined in the previous approach [5], it meant that the searching ability may be limited in certain regions. To conquer this problem, this paper propose a more sophisticated approach that can not only for obtaining a DGSP but also adjusting crossover, mutation, and inversion rates dynamically using the fuzzy grouping genetic algorithm (FGGA) [14]. Hence, the main difference between the proposed and previous approaches is that a fuzzy logic controller is designed and utilized in the proposed approach to set parameters for genetic operations based on fitness values of chromosomes in every iteration. Experimental results on a real dataset were also made to show the effectiveness of the proposed approach.

2 The Used Fuzzy Logic Controller

Based on fuzzy set [19], the fuzzy logic controller, which is famous for its convenient and intuitive, has been designed to solve vague, uncertain problems such as elevator control [8], nuclear reactor control [2] and automatic container crane operation systems [18]. Due to this, in this section, the four components of the fuzzy logic controller used in the proposed approach are stated. Before describing them, the two measurements are described that are the convergence rate M_t and the diversity rate D_t. The convergence rate is the tendency for the optimization search process to approach a particular solution. The convergence rate M at t generation is defined as follows.

$$M_t = \frac{f_b^c}{f_b^p}, \tag{1}$$

where f_b^c is the fitness of current best solution and f_b^p is the fitness of the best solution in the past τ generations. It shows that from generation to generation, when search process can find a solution better than that in pasted generations, the convergence rate will be high. By contrast, when search process cannot find a better one anymore, the convergence rate may be low. When chromosomes are converged, it will close to 1. Besides, the diversity rate is the rich degree of the population. The diversity rate D at t generation is defined as follows.

$$D_t = \frac{f_t^{max} - f_t^{avg}}{f_t^{max}}, \tag{2}$$

where f_t^{max} is the maximum fitness value at generation t and f_t^{avg} is the average fitness value at generation t. When the diversity rate is high, it means that the population is diverse; Otherwise, it indicates the population may converge. Based on these two measurements, the dynamical adjusting process for crossover, mutation, and inversion rates is then handled by a fuzzy logic controller. In the following, the knowledge base contains membership functions, rule base, inference mechanism and defuzzification are described.

2.1 Membership Functions

The membership functions used for the convergence and diversity rates, and for crossover, mutation and inversion rates are shown in Fig. 1(a) and (b), respectively.

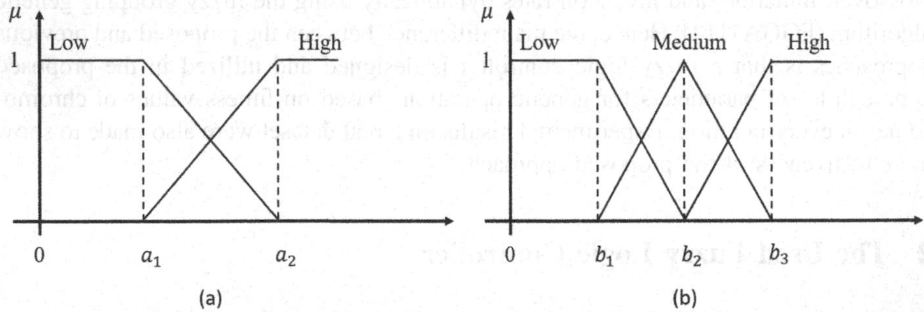

Fig. 1. Membership functions for the proposed approach

Figure 1 shows that two membership functions, *Low* and *High*, are used for the convergence and diversity rates. Three membership functions, *Low*, *Medium* and *High*, are employed for crossover, mutation and inversion rates. According to parameters, a_1, a_2, b_1, b_2 and b_3, the given values of linguistic variables are transformed into fuzzy values. The parameter setting for those membership functions for the two measurements and genetic operations in the proposed approach are shown in Fig. 2(a) and (b).

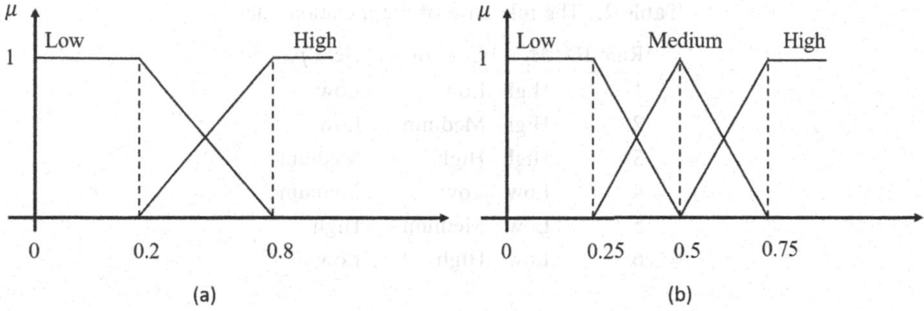

Fig. 2. The parameter setting for membership functions used in the proposed approach.

2.2 Rule Bases

In order to get a good trade-off between exploration and exploitation, the robust rule bases are necessary. The general principles of the associated rules in rule base utilized in the proposed approach are described as follows. In the early iteration when convergence has not yet happened, the fuzzy logic controller will increase crossover rate and decrease mutation and inversion rates. The reason for that is to make the proposed approach could have more powerful local searching ability when solutions are still diverse. In the later iterations, because chromosomes may start to converge, to increase the diversity of chromosomes, the fuzzy logic controller will increase the mutation and inversion rates, and decrease the crossover rate. By designing these two general principles in the fuzzy logic controller, we expect the proposed approach can effectively find much better solutions at the end of search process. To reach the goal, the rule bases of three the crossover rate (p_c), mutation rate (p_m) and inversion rate (p_i) are given in Tables 1, 2 and 3.

Table 1. The rule base of the crossover rate.

Rule ID	M_t	Current p_c	Next p_c
1	High	Low	Medium
2	High	Medium	High
3	High	High	Low
4	Low	Low	Low
5	Low	Medium	Low
6	Low	High	Medium

Based on these rules listed in Tables 1, 2 and 3, the linguistic terms for crossover, mutation and inversion rates at $t + 1$ generation can be determined by their states at t generation.

Table 2. The rule base of the mutation rate.

Rule ID	M_t	Current p_m	Next p_m
1	High	Low	Low
2	High	Medium	Low
3	High	High	Medium
4	Low	Low	Medium
5	Low	Medium	High
6	Low	High	Low

Table 3. The rule base of the inversion rate.

Rule ID	D_t	Current p_i	Next p_i
1	High	Low	Low
2	High	Medium	Low
3	High	High	Medium
4	Low	Low	Medium
5	Low	Medium	High
6	Low	High	Low

2.3 Inference Mechanism

The inference is a process of how to integrate fuzzy outputs derived from rules into a universes of discourse. Three types of inference mechanisms are commonly used: the Mamdani-type inference [13], Sugeno-type inference [14] and Tsukamoto-type inference [15]. The proposed approach uses the Mamdani-type inference [13] as the inference mechanism. Hence, the universes of discourses of crossover, mutation and inversion rates at the *t* generation can be formed.

2.4 Defuzzification

As to the defuzzification operator, the weighted average is adopted in the proposed approach because of its simple and efficiency. The mathematical expression of it is shown as follows.

$$z^* = \frac{\sum \mu_C(\bar{z}) * \bar{z}}{\sum \mu_C(\bar{z})}, \tag{3}$$

where $\bar{z}$ is the centroid of each membership function corresponding to a certain rule and $\mu_C(\bar{z})$ is the membership value of $\bar{z}$ in that membership function. As a result, the crisp values for the crossover, mutation and inversion rates can be derived from their corresponding universes of discourse.

3 The Proposed Mining Algorithm

This section describes the proposed algorithm for obtaining a DGSP by the fuzzy grouping genetic algorithm. The details of the algorithm are described as follows:
Input: A set of stocks $S = \{s_i \mid 1 \leq i \leq n\}$, cash dividends of stocks $Y = \{y_i \mid 1 \leq i \leq n\}$, sets of membership functions for measurements and genetic operations $MF = \{MF_{Mt}, MF_{Dt}, MF_c, MF_m, MF_i\}$, rule bases for crossover rate $Rule_c$, mutation rate $Rule_m$, and inversion rate $Rule_i$.
Parameters: Number of maximum purchased stocks *numCom*, maximum investment capital *maxInves*, number of maximum purchased units *maxUnit*, number of groups K, population size *pSize*, crossover rate p_c, mutation rate p_m, inversion rate p_I, number of iteration *iter*.
Output: A diverse group stock portfolio *DGSP*.
Step 1: Generate initial population based on cash dividends of stocks Y.
Step 2: Evaluate fitness of every chromosome C_q in the population as follows.
Sub-step 2.1: Calculate portfolio satisfaction of each chromosome C_q using the following formula.

$$PS(C_q) = \sum_{p=1}^{NC} subPS(S_p)/NC,$$

where NC is the number of stock portfolios can be generated from the chromosome C_q. The $subPS(SP_p)$ is the portfolio satisfaction of the p-th stock portfolio SP_p, which is evaluated by objective and subjective criteria defined in the previous approach.
Sub-step 2.2: Calculate group balance of each chromosome using the following formula.

$$GB(C_q) = \sum_{i=1}^{K} -\frac{|G_i|}{N}\log\frac{|G_i|}{N},$$

where $|G_i|$ is the number of stocks in the i-th group, N is number of stocks.
Sub-step 2.3: Calculate diversity factor of each chromosome using the following formula.

$$DF(C_q) = \frac{\sum_{i=1}^{K} D_i^q}{K},$$

where D_i^q is the diversity value of group G_i in chromosome C_q.
Sub-step 2.4: Set fitness value of each chromosome C_q using the following formula.

$$f(C_q) = PS(C_q) * GB(C_q)^{\alpha} * DF(C_q)^{\beta}.$$

Step 3: Execute selection operation on the population to form the next population. The elitist selection strategy is used in the proposed approach.

Step 4: Employ fuzzy logic controller to find crossover, mutation and inversion rates.

Sub-step 4.1: Calculate convergence rate M_t using formula (1).

Sub-step 4.2: Calculate diversity rate D_t using formula (2).

Sub-step 4.3: Convert crisp inputs into fuzzy inputs in accordance with the antecedent of rules in rule bases.

Sub-step 4.4: Use the max-min composition of the fuzzy inputs derived from rules to get universe of discourses.

Sub-step 4.5: Perform defuzzification operator, the weighted average, to calculate updated parameters, p_c', p_m' and p_i', by using formula (3).

Step 5: Apply crossover operation on the population with p_c'.

Step 6: Apply mutation operation on the population with p_m'.

Step 7: Apply e inversion operation on the population with p_i'.

Step 8: If the stop criterion is reached, go to the next step. Otherwise, go to Step 2.

Step 9: Output the DGSP with the highest fitness value.

4 Experimental Results

Experiments were made to show the effectiveness of the proposed approach in this section with the following parameter setting: *pSize* was 50, initial values for p_c, p_m and p_i, were 0.8, 0.03 and 0.6. The number of iteration *iter* was 100. The *numCom* was 4, *maxInves* was 1 million, *maxUnit* was 40, and the *K* was set at 6.

The experimental dataset was collected the TSE from 2011/01/01 to 2016/12/31. Thirty stocks used in the experiments were selected from seven categories, namely cement, computer and peripheral equipment, plastic, semiconductor, finance, electron and steel. The dataset contains the stock prices, cash dividends, risk and categories of stocks. The variance of cash dividend of each stock can be calculated by its cash dividends. The risk value of each stock is calculated by history simulation. The dataset from 2011–2013 was used as training period to obtain DGSPs by the previous and proposed approaches. The derived DGSPs were then tested by the datasets from 2014, 2015 and 2016.

Firstly, experiments were conducted to show the convergences of the previous (MDGSP) and proposed (Fuzzy_MDGSP) approaches. The results are shown in Fig. 3.

Figure 3 shows that in training phase, the curves of both the approaches gradually go upward, finally converging to a certain value. And, the proposed approach converges much earlier and obtains result as good as the previous work. It shows that the fuzzy logic controller plays a useful role in adjusting the parameters to enhance the searching ability of the proposed approach.

To verify the derived DGSP by the proposed approach can as good as the previous work, experiments were then made to compare them in terms of return on investment (ROI). The results are shown in Table 4.

Table 4 shows that the proposed approach can have similar returns when the testing periods are 2014 and 2016. However, in 2015, the average ROIs of the proposed and previous approaches are 0.028 and −0.0006, respectively, which indicate the proposed approach is better than the previous approach in terms of the ability to avoid risks.

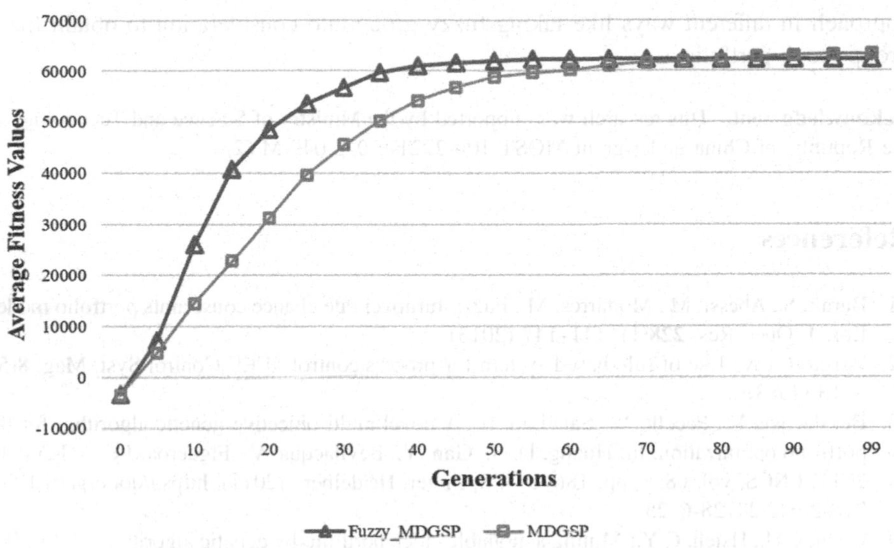

Fig. 3. The convergences of the previous and proposed approaches.

Table 4. Comparing Fuzzy_MDGSP with MDGSP in terms of ROI

Method	Testing period	Avg. ROI	Avg. max ROI	Avg. min ROI
Fuzzy_MDGSP	2014	0.169	0.318	0.033
	2015	**0.028**	**0.144**	**−0.096**
	2016	0.187	0.367	0.068
MDGSP	2014	0.215	0.367	0.084
	2015	−0.006	0.089	−0.142
	2016	0.205	0.355	0.084

According to the experimental results, we can conclude that the proposed approach is efficient and effective to obtain a DGSP.

5 Conclusions and Future Work

In this paper, an enhanced approach for deriving a DGSP is proposed using the fuzzy grouping genetic algorithm for reaching the goal. The proposed approach first generates initial population based on cash dividends of stocks. Each possible DGSP is encoded into a chromosome using the grouping, stock, and stock portfolio parts. The fitness function defined in the previous approach is used to evaluate quality of chromosomes. Before applying genetic operations on population, the fuzzy logic controller is employed to dynamically adjust the crossover, mutation, inversion rates according to the two measurements, the convergence and diversity rates of population. The evolution will be stopped when reaching the stop criteria. Experiments showed that the proposed approach can not only reach similar returns to the previous approach but also have better ability to avoid risk. In the future, we will continue to improve the proposed approach in different ways like taking fuzzy group into consideration to obtain fuzzy group stock portfolio.

Acknowledgment. This research was supported by the Ministry of Science and Technology of the Republic of China under grant MOST 106-2221-E-032-049-MY2.

References

1. Barak, S., Abessi, M., Modarres, M.: Fuzzy turnover rate chance constraints portfolio model. Eur. J. Oper. Res. **228**(1), 141–147 (2013)
2. Bernard, J.A.: Use of rule-based system for process control. IEEE Control Syst. Mag. **8**(5), 3–13 (1988)
3. Bevilacqua, V., Pacelli, V., Saladino, S.: A novel multi objective genetic algorithm for the portfolio optimization. In: Huang, D.-S., Gan, Y., Bevilacqua, V., Figueroa, J.C. (eds.) ICIC 2011. LNCS, vol. 6838, pp. 186–193. Springer, Heidelberg (2011). https://doi.org/10.1007/978-3-642-24728-6_25
4. Chen, C.H., Hsieh, C.Y.: Mining actionable stock portfolio by genetic algorithms. J. Inf. Sci. Eng. **32**, 1657–1678 (2016)
5. Chen, C.H., Lu, C.Y., Hong, T.P., Su, J.H.: Using grouping genetic algorithm to mine diverse group stock portfolio. In: IEEE Congress on Evolutionary Computation, pp. 4734–4738 (2016)
6. Chang, T.J., Yang, S.C., Chang, K.J.: Portfolio optimization problems in different risk measures using genetic algorithm. Expert Syst. Appl. **36**, 10529–10537 (2009)
7. Falkenauer, E.: A genetic algorithm for grouping. In: Proceedings of the Fifth International Symposium on Applied Stochastic Models and Data Analysis, pp. 198–206 (1991)
8. Fujitec, F.: FLEX-8800 Series Elevator Group Control System. Fujitec Co., Ltd., Osaka (1988)
9. Hoklie, L.R.Z.: Resolving multi objective stock portfolio optimization problem using genetic algorithm. In: International Conference on Computer and Automation Engineering, pp. 40–44 (2010)
10. Lin, P.C.: Portfolio optimization and risk measurement based on non-dominated sorting genetic algorithm. J. Ind. Manag. Optim. **8**, 549–564 (2012)
11. Lwin, K., Qu, R., Kendall, G.: A learning-guided multi-objective evolutionary algorithm for constrained portfolio optimization. Appl. Soft Comput. **24**, 757–772 (2014)
12. Liu, Y.J., Zhang, W.G.: Fuzzy portfolio optimization model under real constraints. Insur.: Math. Econ. **53**(3), 704–711 (2013)
13. Mamdani, E.H., Assilian, S.: An experiment in linguistic synthesis with a fuzzy logic controller. Int. J. Man-Mach. Stud. **7**(1), 1–13 (1975)
14. Mutingi, M., Mbohwa, C.: Grouping Genetic Algorithms: Advances and Applications, vol. 666. Springer, Heidelberg (2016). https://doi.org/10.1007/978-3-319-44394-2
15. Roy, A.D.: Safety first and the holding of assets. Econometrica **20**, 431–449 (1952)
16. Takagi, T., Sugeno, M.: Fuzzy identification of systems and its applications to modeling and control. IEEE Trans. Syst. Man Cybern. **15**(1), 116–132 (1985)
17. Tsukamoto, Y.: Fuzzy logic based on Lukasiewicz logic and its applications to diagnosis and control. Doctoral Dissertation, Tokyo Institute of Technology (1979)
18. Yasunobu, S., Hasegawa, T.: Evaluation of an automatic container crane operation system based on predictive fuzzy control. Control Theory Adv. Technol. **2**(3), 419–432 (1986)
19. Zadeh, L.A.: Fuzzy sets. Inf. Control **8**(3), 338–353 (1965)

A Novel Approach for Option Trading Based on Kelly Criterion

Mu-En Wu[1] and Wei-Ho Chung[2(✉)]

[1] Department of Information and Finance Management, National Taipei University of Technology, Taipei, Taiwan
mnwu@ntut.edu.tw

[2] Research Center for Information Technology Innovation, Academia Sinica, Taipei, Taiwan
whc@citi.sinica.edu.tw

Abstract. An option trading model based on Kelly criterion is proposed in this work. Via longing and shorting options at different strike prices, various portfolio strategies which lock the losses and profits in advance can be formed; in other words, we hold a portfolio of options with a fixed profit and loss distribution. We design and use Kelly criterion applied to the options trading, in terms of calculating the optimal bidding fraction. In this paper we provide a model for developing an option trading system for finding the profitable option portfolio with optimal bidding fraction. This is a new approach for option trading with position management, and some future directions are provided.

Keywords: Kelly criterion · Option trading · Profitable gamble
Optimal *f*

1 Introduction

The Kelly criterion [1] can be regarded as an optimization process for wagering ratios in the long term [11–13]. It can be interpreted as follows. Assume a game (e.g., coin-tossing) with win-rate & odds; the game is assumed to be played for an unlimited numbers of rounds. In each round, we have to decide the wagering ratio of our total assets. If the current game is lost, the bid amount is lost. If the game is won, the profit calculated by the odds will be returned. The Kelly criterion is an approach to calculate the optimal bid ratio for the best asset growing rate.

However, the Kelly criterion is usually not applied to real-world markets due to difference between its assumptions and the markets. First, the Kelly criterion is based on the assumption that the games can be continued indefinitely, which is infeasible in reality [12, 15]. Besides, Kelly is not applicable to real-world trading of financial instruments. In real-world trading, the distribution of odds is unavailable; the win rate is also unknown. The win rate and odds are fixed and known in conventional casino games; e.g., in a coin-toss game, the win rate is 0.5 and odds 2. The above arguments show the significant difference in conventional games and financial trading. To solve the issue, Ralph Vince proposed the Holding Period Return in his work [7, 8], which calculated the optimal bid ratio or "Optimal *f*" based on empirical outcomes. The

N. T. Nguyen et al. (Eds.): ACIIDS 2018, LNAI 10751, pp. 519–527, 2018.
https://doi.org/10.1007/978-3-319-75417-8_49

Optimal f can be interpreted as the extension of the Kelly criterion, rendering the Kelly formula as a special scenario with a single set of win rate and odds. The Optimal f by Vince is found to have greater applicability to real-world trading.

In this work, we propose an options trading model. In this model, due to the different strike prices of options, there are many portfolio options, such as Spread Strategy, Bull Spread, and Bear Spread. Once the portfolio is constructed, and as long as the portfolio is held till the settlement date, the maximum loss and profit are fixed. In other words, the odds are known. Therefore, the only unknown parameter remaining under the Kelly criterion is the probability distribution of the underlying index prices. We emphasize that the model proposed in this work operates without any human perspective or intervention needed. The construction of the portfolio and the bid ratio are completely based on historical data, and the quoted prices on the underlying options.

2 Preliminaries

2.1 Kelly Criterion

Consider a game with a win-rate p and odds b, and assume that the gambler has an initial capital of A_0. The t-th step capital is denoted by A_t, and the bidding fraction is f, where $0\% < f < 100\%$. The Kelly formula is derived as follows: If the gambler wins the $(t - 1)$-th round, then $A_t = A_{t-1}(1 + bf)$. If the gambler loses the $(t - 1)$-th round, then $A_t = A_{t-1}(1 - f)$. Since the gambler plays T rounds and has a win-rate of p, we can expect that he/she will win Tp rounds and lose $T \times (1 - p)$ rounds. In theory, the value of A_T should then be:

$$A_T = A_0(1 + bf)^{Tp}(1 - f)^{T(1-p)}.$$

By optimizing A_T based on the above equation, we can find the solution of f. After differentiating the above equation, the optimal f value can be found as follows:

$$f = \frac{p(1 + b) - 1}{b}.$$

Without loss of generality, we start with 1 dollar as the initial capital, and the resulting end capital after the 40 games is shown in Fig. 1.

By Kelly criterion, in a game with win rate of 50% and odds 2, the optimal bid ratio is 25%, which leads to the best capital growth rate. On average, the end capital after 40 games will be 10.5 times the starting capital.

2.2 Holding Period Returns

The work of Ralph Vince extends the Kelly criterion from the scenario involving a single set of win rate and odds to a scenario with multiple sets of empirical outcomes.

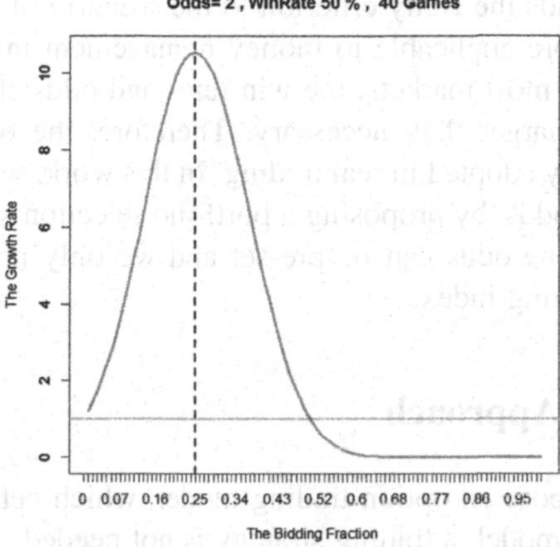

Fig. 1. Coin-Tossing with Win-rate 50%, odds 2.

We consider the game with multiple odds as $(b_1, b_2, \ldots, b_n)$, where $b_i \in \mathbb{Z}$. Vince defines the holding period return as:

$$\mathrm{HPR}_i(f) = \left(1 - f\frac{-b_i}{L}\right),$$

where $L = \min\{b_1, b_2, \ldots, b_n\}$ is the maximum loss and f is the bid ratio. Without loss of generality, we assume $b_i < 0$ for some i. Each b_i corresponds to the probability p_i. Vince defines the Terminal Wealth Relative (TWR) as

$$\mathrm{TWR}(f) = (\mathrm{HPR}_1(f) \times \mathrm{HPR}_2(f)\ldots \times \mathrm{HPR}_n(f))^{1/n}.$$

The f corresponding to the maximum TWR is the optimal f. As an example, under the profit & loss case of (3, −5, −2, 15), based on TWR formula, the different bid ratios will lead to the following TWR distribution (Fig. 2).

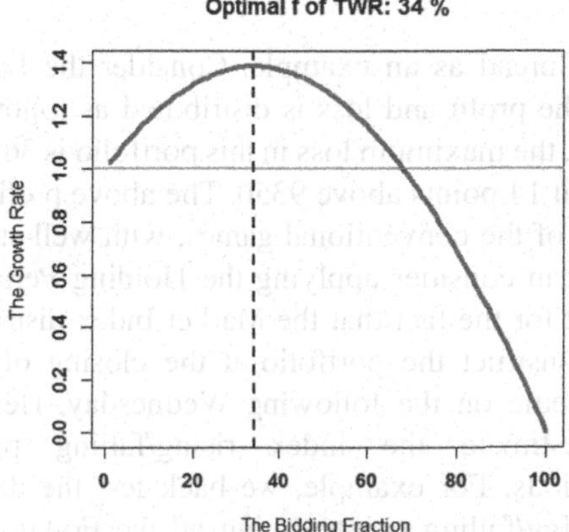

Fig. 2. The return distribution of different bidding fractions on profit vector (3, −5, −2, 15).

Vince's work extends the Kelly criterion to the scenario of multiple profit and loss outcomes, which is more applicable to money management in real trading. However, some issues remain. In most markets, the win rates and odds change with time, which renders the optimal f larger than necessary. Therefore, the optimal f by the above calculation is not widely adopted in real trading. In this work, we overcome the issue of invariant win rate and odds, by proposing a portfolio selection strategy based on option spreads, under which the odds can be pre-set and we only need to estimate the distribution of the underlying index.

3 The Proposed Approach

In this section, we propose an option trading model, which better fits the usage of the Kelly criterion. In this model, a trading strategy is not needed. The trade will based on the most profitable option portfolio.

3.1 Option Spread Trading Based on Kelly Criterion

We use an example to explain the trading model. We consider the Taiwan Stock Exchange TAIEX, at Friday, 01/13/2017. The Weekly Option will reach its settlement date on Wednesday, 01/18/2017. The TAIEX is at 9378 right before closing on 01/13/2017. The prices quoted for the option are as follows (Table 1).

Table 1. Option price on 2017.01.13

Call	Strike price	PUT
90	9300	14.5
54	9350	29.5
29.5	9400	53
13.5	9450	89
5.5	9500	130

We take the Bull Spread as an example. Consider the Long 9300Call@90, and Short 9350Call@54. The profit and loss is distributed as follows (Fig. 3).

By the above figure, the maximum loss in this portfolio is 36 points below 9300, and the maximum profit is at 14 points above 9350. The above profit and loss structure will be more similar to that of the conventional games, with well-structured profit and loss points. Therefore, we can consider applying the Holding Period Return method proposed by Vince, except for the fact that the Market Index distribution is still missing.

Assume that we construct the portfolio at the closing of each Friday, with the weekly option set to settle on the following Wednesday. Here, we need to use the historical data to estimate the index rising/falling points from historical Friday-Wednesday periods. For example, we back test the data from 01/02/2016 to 01/13/2017, for the rising/falling points during all the Friday-Wednesday periods, as shown in the following table.

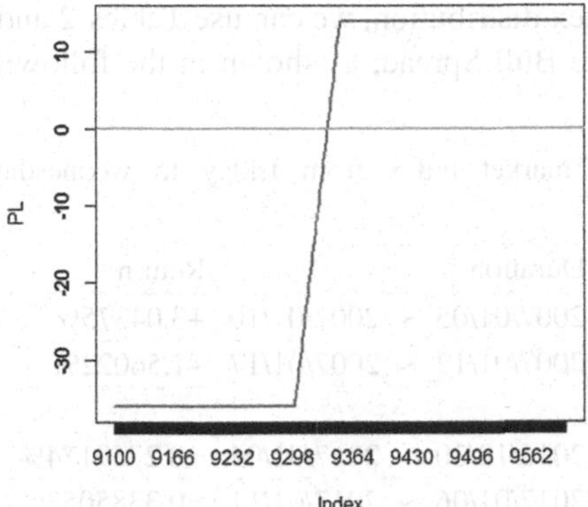

Fig. 3. The profit & loss distribution of a bull spread (Long 9300Call@90; Short 9350 Call@54).

During the historical period considered, we have 490 observations. We consider the historical rising/falling points in the same period as the empirical probability distribution. These 490 observations are shown in the histogram below (Fig. 4).

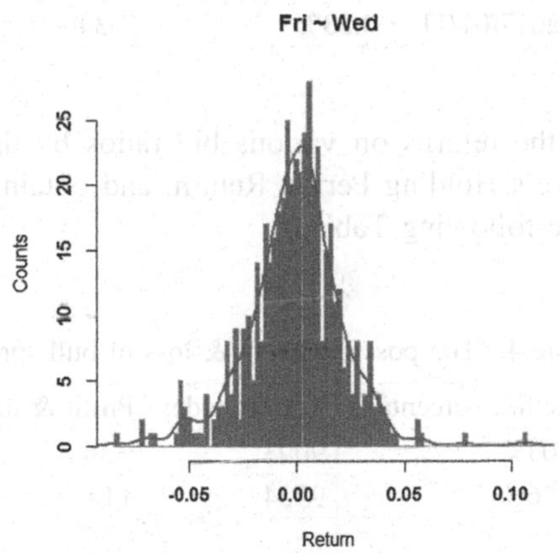

Fig. 4. The distribution of market index from Friday to Wednesday from 2007/01/05 ~ 2017/01/11.

Based on the historical distribution, we then attempt to obtain the distribution for the current market index. We can use current index at 9378, jointly with the historical distribution and this formula

$$9378 \times (1 + \text{Rise/Fall\_Percentage in Table II}).$$

Using the above formula, we can obtain the market index distribution for next Wednesday, shown in the following Table.

With the market index distribution, we can use Tables 2 and 3 to calculate the profit and loss structure of the Bull Spread, as shown in the following Table.

Table 2. The return of market index from friday to wednesday during 2007/01/05 ~ 2017/01/11.

Duration	Return
2007/01/05 ~ 2007/01/10	+3.04375%
2007/01/12 ~ 2007/01/17	+1.56022%
…	
2016/12/30 ~ 2017/01/04	−0.2739174%
2017/01/06 ~ 2017/01/11	−0.3385053%

Table 3. The possible closed prices of market index on 2017/01/18.

Periods	Rise/fall percentage	Market index estimation
2007/01/05 ~ 2007/01/10	+3.04%	9093
2007/01/12 ~ 2007/01/17	+1.56%	9524
…	…	…
2016/12/30 ~ 2017/01/04	−0.27%	9352
2017/01/06 ~ 2017/01/11	−0.33%	9346

Then we calculate the returns on various bid ratios by the profits and losses in Table 4, based on Vince's Holding Period Return, and obtain the best bid ratio. The results are shown in the following Table 5.

Table 4. The possible profit & loss of bull spread.

Rise/fall percentage	Market index	Profit & loss
+3.04%	9093	−36
+1.56%	9524	+14
…	…	…
−0.27%	9352	−14
−0.33%	9346	−10

Table 5. The return of bidding various faction on Bull Spread.

Bidding fraction	1%	2%	…	50%	…	99%	100%
Return	0.998	0.997		0.875		0.269	0

The return distribution on various bid ratios is shown in Fig. 5. We start with the fund of 1 dollar, the final outcome is less than 1 dollar for every bid ratio. In other

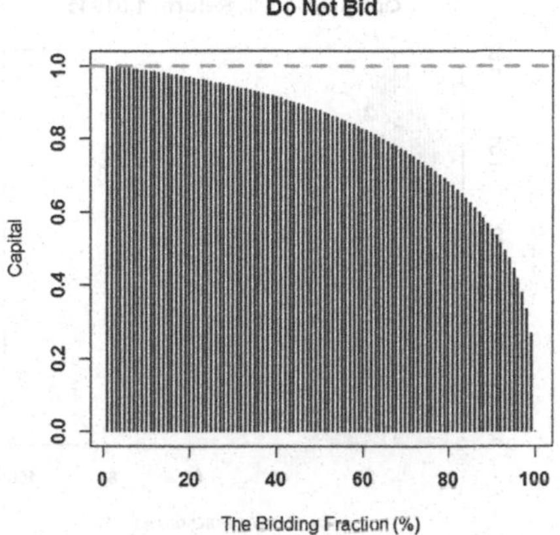

Fig. 5. The return distribution of bull spread for different bidding fractions.

words, no bids will result in a positive outcome; therefore this is a game with negative expected outcome, and is not a favorable portfolio. If a Bull Spread is not favorable, then the Bear Spread has to be profitable. Thus, by reversing the positions of the portfolio, that is, Short 9300Call@90 and Long 9350Call@54, results in the following profit and loss structure, shown in Fig. 6.

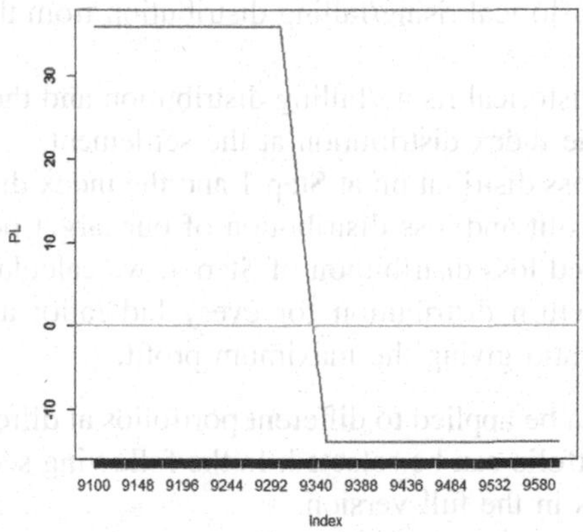

Fig. 6. The profit & loss distribution of a bear spread (Short 9300Call@90; Long 9350Call@54).

Figure 7 shows the various bidding fraction for bear spread in Fig. 6. The optimal bid ratio is 13% and the expected return is 1.01945.

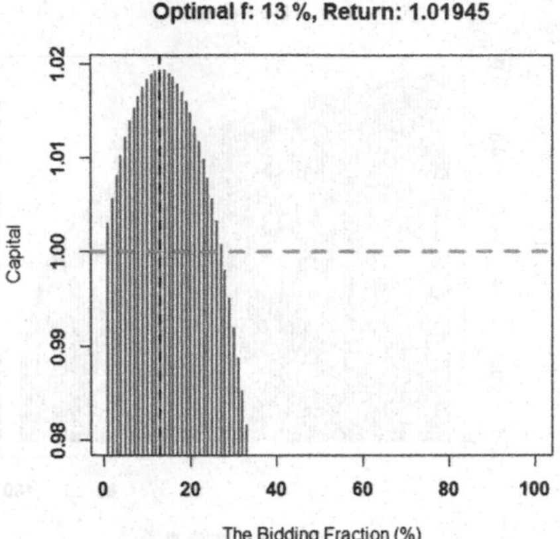

Fig. 7. The return distribution of of bear spread for different bidding fractions.

4 The Proposed Algorithm

In the previous section we propose a trading model based on options, which can be summarized as follows.

Step 1. Calculate the profit and loss distribution of the set of option portfolios, such as the Bull Spread or Bear Spread.
Step 2. Calculate the historical rising/falling distribution from the time of market entry to settlement.
Step 3. Based on the historical rising/falling distribution and the current market index, we calculate the index distribution at the settlement.
Step 4. By profit and loss distribution at Step 1 and the index distribution at Step 3, we calculate the profit and loss distribution of our target portfolio.
Step 5. By the profit and loss distribution of Step 4, we calculate the TWRs, and then calculate the return distribution for every bid ratio; among these ratios, we select the bid ratio giving the maximum profit.

The above steps can be applied to different portfolios at different strike prices; then, the most favorable portfolio can be selected. In the following section, we will show the experimental outcomes in the full version.

5 Conclusions and Future Work

In this work we propose a new portfolio selection and trading approach based on option spreads. Under the proposed method, investors can avoid heuristic construction of trading portfolios. We only need to select the time periods of interest, and back test the historical distributions of rising/falling points during the identical periods. Then, based on the quoted prices, we calculate the profit and loss distributions, and use Kelly criterion to obtain the optimal bid ratio. The advantage of this approach is that we can obtain the return of multiple option portfolios with the optimal bid ratios, and then select the most profitable option portfolio for trading.

Acknowledgments. This work was supported by Ministry of Science and Technology, Taiwan, under Grant MOST 106-2221-E-027-145, 106-3114-E-001-006, 105-2221-E-001-009-MY3, 104-2221-E-001-008-MY3, and Academia Sinica Thematic Project under Grant AS-104-TP-A05.

References

1. Kelly, J.L.: A new interpretation of information rate. Bell Syst. Tech. J. **35**(4), 917–926
2. Thorp, E.O.: The kelly criterion in blackjack, sports betting, and the stock market. In: Zenios, S.A., Ziemba, W. (eds.) Handbook of Asset and Liability Management, vol. 1 (2006)
3. Thorp, E.O.: Understanding the kelly criterion. In: The Kelly Capital Growth Investment Criterion: Theory and Practice. World Scientific Press, Singapore (2010)
4. MacLean, L.C., Thorp, E.O., Ziemba, W.T.: The Kelly Capital Growth Investment Criterion: Theory and Practice, vol. 3. World scientific, Singapore (2011)
5. MacLean, L.C., Thorp, E.O., Ziemba, W.T.: Good and bad properties of the Kelly criterion. Risk **20**(2), 1 (2010)
6. Stutzer, M.: On growth-optimality vs. security against underperformance. The Kelly Capital Growth Investment Criterion: Theory and Practice. 641–653 (2011)
7. Vince, R.: The Mathematics of Money Management: Risk Analysis Techniques for Traders, vol. 18. Wiley, Hoboken (1992)
8. Vince, R.: The New Money Management: a Framework for Asset Allocation, vol. 47. Wiley, Hoboken (1995)
9. Vince, R.: The Leverage Space Trading Model: Reconciling Portfolio Management Strategies and Economic Theory, vol. 425. Wiley, Hoboken (2009)
10. Vince, R.: Portfolio Management Formulas. Willey, Hoboken (1990)
11. Gottlieb, G.: An optimal betting strategy for repeated games. J. Appl. Prob. **22**(4), 787–795 (1985)
12. Wu, M.-E., Tsai, H.-H., Tso, R., Weng, C.-Y.: An adaptive kelly betting strategy for finite repeated games. In: Zin, T.T., Lin, J.C.-W., Pan, J.-S., Tin, P., Yokota, M. (eds.) GEC 2015. AISC, vol. 388, pp. 39–46. Springer, Cham (2016). https://doi.org/10.1007/978-3-319-23207-2_5
13. Chou, J.-H., Lu, C.-J., Wu, M.-E.: Making profit in a prediction market. In: Gudmundsson, J., Mestre, J., Viglas, T. (eds.) COCOON 2012. LNCS, vol. 7434, pp. 556–567. Springer, Heidelberg (2012). https://doi.org/10.1007/978-3-642-32241-9_47
14. Mulvey, J.M., Bilgili, M., Vural, T.: A dynamic portfolio of investment strategies: applying capital growth with drawdown penalties. In: The Kelly Capital Growth Criterion: Theory and Practice (2011)
15. Wu, M.-E., Wang, C.-H., Chung, W.-H.: Using trading mechanisms to investigate large futures data and their implications to market trends. Soft. Comput. **21**(11), 2821–2834 (2017)

Music Recommendation Based on Information of User Profiles, Music Genres and User Ratings

Ja-Hwung Su[1,2(✉)], Chu-Yu Chin[3,5], Hsiao-Chuan Yang[1], Vincent S. Tseng[4], and Sun-Yuan Hsieh[3]

[1] Department of Information Management, Cheng Shiu University, Kaohsiung, Taiwan
bb0820@ms22.hinet.net
[2] Department of Information Management, Kainan University, Taoyuan, Taiwan
[3] Department of Computer Science and Information Engineering, National Cheng Kung University, Tainan, Taiwan
[4] Department of Computer Science, National Chiao Tung University, Hsinchu, Taiwan
[5] Telecommunication Laboratories, Chunghwa Telecom Co., Ltd., Taoyuan, Taiwan

Abstract. Music data has been becoming bigger and bigger in recent years. It makes online music stores hard to provide the users with good personalized services. Therefore, a number of past studies were proposed for effectively retrieving the user preferences on music. However, they countered problems such as new user, new item and rating sparsity. To cope with these problems, in this paper, we propose a creative method that integrates information of user profiles, music genres and user ratings. In terms of solving problem of new user, the user similarities can be calculated by the profiles instead of ratings. By the user similarities, the unknown ratings can be predicted using user-based Collaborative Filtering. In terms of solving problem of rating sparsity, the unknown ratings are initialized by ratings of music genres. Even facing new music items, the rating data will not be sparse due to imputing the initialized ratings. Because the rating data is enriched, the user preference can be retrieved by item-based Collaborative Filtering. The experimental results reveal that, our proposed method performs more promising than the compared methods in terms of Root Mean Squared Error.

Keywords: Collaborative filtering · Music recommendation · New user Rating sparsity · User-based

1 Introduction

In the last few years, music has been more and more popular because it is more available than ever. According to the evolutionary multimedia techniques, the users are allowed to listen to the music through the Web. For this need, a number of online music websites provide the users with online listening services such as Spotify, Pandora,

N. T. Nguyen et al. (Eds.): ACIIDS 2018, LNAI 10751, pp. 528–538, 2018.
https://doi.org/10.1007/978-3-319-75417-8_50

Youtube, Last.fm and so on. Whatever the music website is, it is not easy for a user to find the preferred music from a large amount of music data. Hence, how to effectively retrieve the user's interested music has been a hot topic in recent years. For this topic, lots of recommender systems called Collaborative Filtering (CF) made attempts to discover the user preferences from user behaviors such as ratings. That is, the user preferences in CF can be represented by votes ranged from 1 to 5. Based on the user-to-item rating matrix, the goal of a music recommender system is to provide the user with a music list containing a set of potential music pieces. Figure 1 is the workflow of the traditional Collaborative Filtering which includes two main phases, namely rating prediction and item selection [7]. The first phase concerns with how to derive the unknown ratings, while the second phase concerns with how to generate a good item list. After these two phases, the user can obtain a predicted item list.

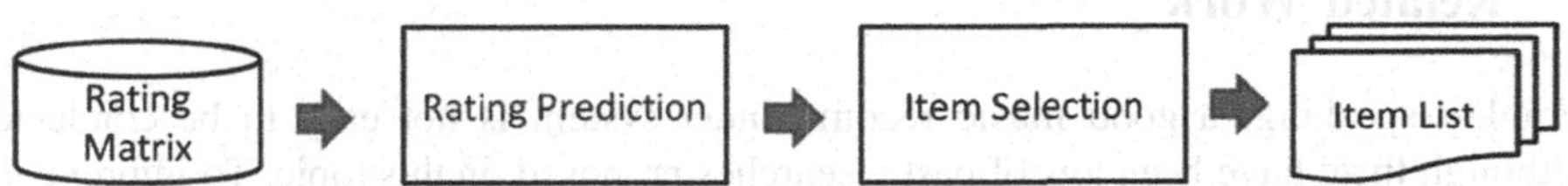

Fig. 1. Workflow of the traditional Collaborative Filtering.

Table 1 is an example of a user-to-item rating matrix. In this matrix, there are 5 users and 6 music items, where zero stands for the unknown item vote. From this example, we can know that, there exist three problems unsolved for traditional music recommender systems, including new user, new item and rating sparsity. For problems of new user and new item, both user 5 and item 3 have no rating to assist the predictions of unknown ratings, respectively. For problem of rating sparsity, it is obvious that, this matrix is too sparse to predict the unknown ratings. Accordingly, these problems motivate us to propose a creative music recommender system that fuses information of user profiles, music genres and user ratings to conduct a better prediction for user preferences on music. Overall, the major contributions can be summarized as follows.

Table 1. Example of a user-to-item rating matrix.

	Item 1	Item 2	Item 3	Item 4	Item 5	Item 6
User 1	0	3	0	0	1	0
User 2	4	0	0	5	4	2
User 3	0	1	0	2	2	3
User 4	4	2	0	2	0	0
User 5	0	0	0	0	0	0

I. To cope with the problem of new user, the user similarity between users is calculated by the user profiles, not the user ratings. Hence, the user-based CF can be performed successfully.

II. To cope with the problems of new item and rating sparsity, the unknown ratings in the matrix are initialized by the virtual genre ratings. With the initialized ratings, the matrix is not sparse anymore and the item-based CF can be performed thereupon.

To know how effective the proposed idea is, a number of experiments were made on real data and the experimental results show that, the proposed recommender system can really estimate the better unknown ratings than the compared methods.

The rest of this paper is structured as follows. In Sect. 2, the related work is introduced briefly. In Sect. 3, the proposed method and examples are presented for predicting the user ratings in great detail. The experimental results are revealed in Sect. 4. Finally, conclusions and future work are made in Sect. 5.

2 Related Work

Frankly speaking, a good music recommender system is not easy to be conducted although there have been lots of past researches proposed on this topic. To improve the performance of recommendation, in this paper, we propose a hybrid music recommender system that combines user-based and item-based CFs with multiple features. To show the relevance between our proposed method and the previous work, we make a brief review in this section.

I. Rating-similarity-based recommender system
Like this work, this type of recommender systems generates the recommendation list based on the item ratings. To this end, the goal of them is to calculate the unknown ratings via similarities of known ratings. Basically, this type of recommender systems can be categorized into two main classes, including item-based [8] and user-based [9] recommender systems. The basic idea of them is that, the unknown ratings for an item can be inferred by similar items or users on ratings. However, whatever for item-based or user-based recommender systems, they always suffer from problems of new item and new user. To address the problems, a hybrid recommender system was proposed by Wang et al. [13], which predicted the unknown ratings by considering similar users and items simultaneously. Although this hybrid recommender system makes advantages of user-based and item-based recommender systems, it cannot still avoid problem of rating sparsity.

II. Rating-learning-based recommender system
Because the performances of the rating-similarity-based recommender systems rely on the rating similarity too much, a different type is the rating-learning-based recommender system. The main concern of this type is how to generate an effective model by learning from the known ratings. For this concern, a set of machine learning methods were applied in this field, including Matrix Factorization [4, 12], SVM (Support Vector Machine) [11], Bayesian [1] and DecisionTree [5]. Without considering rating similarities, the unknown ratings can be estimated. Yet, the improvement of predictions is not significant if facing highly sparse data.

III. Content-filtering-based recommender system
In addition to above methods, the other type is the content-filtering-based recommender system. The major intent of this type is to attack the limitation of ratings by using useful contents in addition to ratings such as social media tags [7], acoustic features [2], playcounts [3], context features [10] and so on. By unifying contents and ratings, the recommendation quality can really be enhanced. Rahman et al. [6] performed a fuzzy inference system by fusing audio and playlist information. However, extra features cannot deal with traditional problems mentioned in above.

3 Proposed Recommender System

3.1 Basic Idea

The goal of this paper is to propose an improved music recommender system that can associate the user's interests with music more precisely than traditional ones. To reach this goal, three traditional problems need to be coped with, including new user, new item and rating sparsity. Clearly, all the problems are caused by insufficient ratings. Therefore, the major concept for dealing with new users is to find the similar users for the active user based on user profiles, while the concept for dealing with new items and rating sparsity is to roughly impute the unknown ratings before performing the item-based prediction. Because the user similarity has been calculated by profiles, new users' preferences can be predicted by considering similar users on profiles. Also, because the unknown ratings are imputed first, the recommender system can learn more from the enriched rating matrix. On the basis of the basic concepts, we have to overcome three challenges:

I. How to calculate the profile similarity between two users?
II. How to approximate the near-real ratings as the imputed ratings?
III. How to integrate above ideas to improve the prediction accuracy?

To aim at the challenges, in this paper, we propose two new techniques, namely "user similarity calculation by profiles" and "unknown rating imputation by genre ratings". In the following sub-sections, we will present the details of how to deal with the challenges.

3.2 Overview of the Proposed Method

To improve the quality of music recommendation, we conduct a hybrid music recommender system that attacks the problems mentioned in above. As shown in Fig. 2, the recommendation process includes two main stages, namely offline preprocessing and online recommendation.

I. Offline preprocessing
In this stage, three matrixes are generated finally. The first matrix is the user similarity matrix and the second matrix is the imputed rating matrix. Then, based on the imputed rating matrix, the item similarity matrix is generated. With these matrixes, the online recommendation can be speeded up.

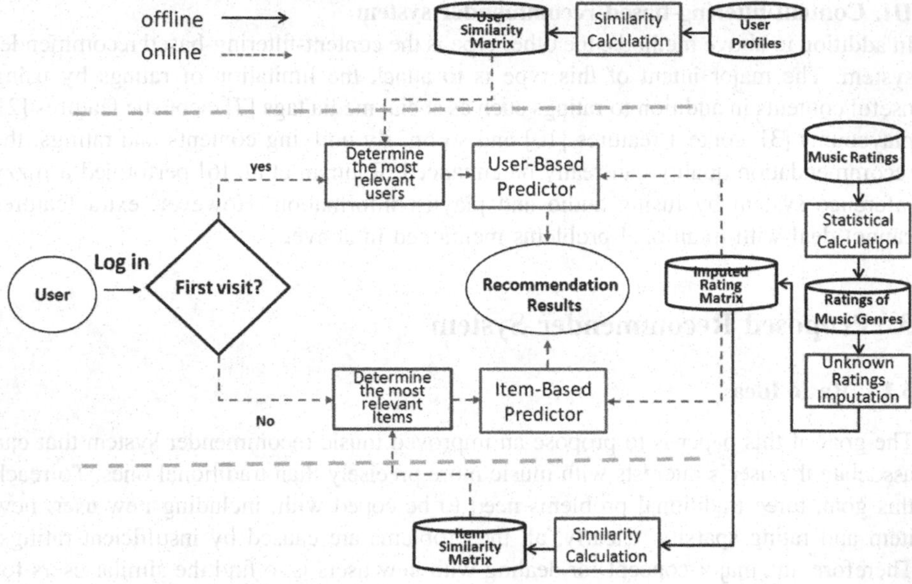

Fig. 2. Framework of the proposed music recommender system.

II. Online recommendation

According to the similarity matrixes and the imputed rating matrix, the hybrid recommender system predicts the unknown ratings in this stage. It starts with an active user visit. If the user is a new user, the user-based predictor is performed to predict the unknown ratings. Otherwise, the item-based predictor is executed. The recommendation quality can be lifted clearly, whether the user is a new user or not. This is because the rating matrix has been imputed.

3.3 Offline Preprocessing

To meet the requirement of the online recommendation, three matrixes need to be generated by three operations, including constructions of user similarity matrix, imputed rating matrix and item similarity matrix.

A. *Construction of user similarity matrix*

The goal of the user similarity matrix is to solve the problem of new user. For an active user, the most relevant users can be determined by this matrix. Figure 3 reveals the workflow of this operation. For each user in the database, the referred semantic profile is converted into a bit vector. Thereupon we can obtain a bit map including all users' vectors. By this map, the user similarities can be calculated. Finally, the similarity matrix is constructed. The similarity is defined in Definition 1.

Fig. 3. Operation of construction of the user similarity matrix.

Definition 1. Assume there are *n* unique attributes in the profile. Given two users {*x*, *y*} with the bit vectors $\{a_{x,1}\, a_{x,2}, \ldots, a_{x,n}\}$ and $\{a_{y,1}\, a_{y,2}, \ldots, a_{y,n}\}$, where $a \in \{0, 1\}$, the user similarity is defined as:

$$sim_{x,y} = \frac{\sum_{0<i\leq n} a_{x,i} * a_{y,i}}{\sqrt{\sum_{0<i\leq n} \left(a_{x,i}\right)^2 * \sum_{0<i\leq n} \left(a_{y,i}\right)^2}}. \tag{1}$$

Example 1. Assume there are 6 attributes in the profile. Table 2 is an example of bit map of profiles. Based on Tables 2 and 3 is the similarity matrix calculated by Eq. (1).

B. *Construction of imputed rating matrix*

As shown in Fig. 2, the main processes in this operation include "statistical calculation" and "unknown rating imputation". In the first process, the genre ratings for each user are calculated. The genre rating is defined as in Definition 2.

Table 2. Example of a bit map of profiles.

	Gender		Bloodtype			
	Male	Female	A	B	O	AB
User 1	0	1	0	0	1	0
User 2	1	0	0	0	0	1
User 3	0	1	0	1	0	0
User 4	1	0	0	0	0	1
User 5	1	0	0	0	1	0

Table 3. Example of the user similarity matrix.

	User 1	User 2	User 3	User 4	User 5
User 1	1	0	0.5	0	0.5
User 2	0	1	0	1	0.5
User 3	0.5	0	1	0	0
User 4	0	1	0	1	0.5
User 5	0.5	0.5	0	0.5	1

Definition 2. Assume there are *p* rated music items (known items) in the qth genre. For the zth user, the qth genre rating is defined as:

$$\overline{g_q^z} = \frac{\sum_{i=1,\, v_q^i \neq 0}^{p} v_q^i}{p}, \tag{2}$$

where v_q^i stands for the ith item rating of the qth genre.

Table 4. Example of a user-to-item-to-genre matrix.

	Genre 1		Genre 2		Genre 3	
	Item 1	Item 2	Item 3	Item 4	Item 5	Item 6
User 1	0	3	0	0	1	0
User 2	4	0	0	5	4	2
User 3	0	1	0	2	2	3
User 4	4	2	0	2	0	0
User 5	0	0	0	0	0	0

Table 5. Example of genre ratings.

	Genre 1	Genre 2	Genre 3
User 1	3	**0**	1
User 2	4	5	3
User 3	1	2	2.5
User 4	3	2	**0**

After the first process, the genre ratings for each user can be generated. However, there probably exist zero ratings in the genre rating results. Hence, for each user, the zero ratings of each genre have further to be predicted by user-based CF [9]. That is, zero rating is not allowed. Note that, user-based CF will be described in Sub-Sect. 3.4. Afterwards, the genre ratings are inserted into the referred unknown (zero) elements of the original user-to-item rating matrix. That is, there is no zero element in the user-to-item rating matrix, which is called "imputed rating matrix".

Example 2. Assume there are three genres in the database. Table 4 is an example of user-to-item-to-genre rating matrix. According to Eq. (2), the resulting genre ratings are shown in Table 5. Because there are two zero elements in the results, we have to calculate the potential values by user-based CF [9], as shown in Table 6. Finally, an example of the imputed rating matrix is shown in Table 7.

Table 6. Example of predicted genre ratings.

	Genre 1	Genre 2	Genre 3
User 1	3	**2**	1
User 2	4	5	3
User 3	1	2	2.5
User 4	3	2	**1**

Table 7. Example of the imputed rating matrix.

	Item 1	Item 2	Item 3	Item 4	Item 5	Item 6
User 1	**3**	3	**2**	**2**	1	**1**
User 2	4	**4**	**5**	5	4	2
User 3	**1**	1	**2**	2	2	3
User 4	4	2	**2**	2	**1**	**1**
User 5	0	0	0	0	0	0

C. *Construction of item similarity matrix*

Because the user-to-item rating matrix has been enriched as the imputed rating matrix, the item similarities are more robust. In this operation, the robust item similarities are inserted into the item similarity matrix. Note that, the item similarity is the same as Eq. (1) and the example is shown in Table 8.

Table 8. Example of an item similarity matrix.

	Item 1	Item 2	Item 3	Item 4	Item 5	Item 6
Item 1	1	0.9578	0.9132	0.9132	0.8224	0.7171
Item 2	0.9578	1	0.9605	0.9605	0.8953	0.7542
Item 3	0.9132	0.9605	1	1	0.9814	0.8490
Item 4	0.9132	0.9605	1	1	0.9814	0.8490
Item 5	0.8224	0.8953	0.9814	0.9814	1	0.8808
Item 6	0.7171	0.7542	0.8490	0.8490	0.8808	1

3.4 Online Recommendation

As mentioned in Sect. 1, this stage can be decomposed into two parts, namely rating prediction and item selection. In this paper, the focus is put on the rating prediction. By referring to Fig. 2, the system needs to know if the active user is a new user. For a new user, the most-relevant users are determined by the user similarity matrix first. Next, the user-based prediction is performed to calculate the unknown ratings by Definition 3.

Definition 3. Assume the most-relevant user set is U, the item set is I and the imputed rating matrix is $R_{U \to I}[r_{b,c}]$, where $r_{b,c}$ indicates the cth item rating of the bth user. Then, the dth item rating of the bth user, which is predicted by user-based CF, is defined as:

$$\widehat{r_{b,d}} = \frac{\sum_{a \in (U-b), d \in I} sim_{b,a} * r_{a,d}}{\sum_{a \in U} sim_{b,a}}, \tag{3}$$

where $sim_{b,a}$ is defined in Eq. (1).

On the other hand, if the active user is not a new user, the related unknown ratings are predicted by the item-based CF, which is defined in Definition 4.

Definition 4. Assume the user set is S, the most-relevant item set is T and the imputed rating matrix is $R_{U\rightarrow I}[r_{b,c}]$, where $r_{b,c}$ indicates the cth item rating of the bth user. Then, the dth item rating of the bth user, which is predicted by item-based CF, is defined as:

$$\widehat{r_{b,d}} = \frac{\sum_{u\in(S-b),c\neq d,c,d\in T} similarity_{d,c} * r_{u,c}}{\sum_{d,c\in T} similarity_{d,c}}, \tag{4}$$

where $similarity_{d,c}$ is the similarity between items d and c and derived from the item similarity matrix.

4 Experiments

4.1 Experimental Settings

To carry out the experiments, we implemented a real system to collect the experimental data. In this experiment, 52 volunteers were invited to provide the ratings and profiles. The profile includes city, gender, age level, blood type, constellation, interest, education level and job. In the experimental data, there are 15 genres and each genre contains 40 music pieces. For each volunteer, 15 music pieces of each genre were randomly selected to be rated. That is, totally there are 3900 ratings in this experimental data. To evaluate our proposed method from viewpoint of new user, 2 users were defined as new users. Additionally, three different percentages of ratings were selected as the unknown ratings to test the sensitivity of data sparsity. Table 9 shows the related data settings. Note that, the evaluation is the Root Mean Squared Error (RMSE). *Moreover, if the unknown rating cannot be predicted, the RMSE is 5 here.*

Table 9. Data settings.

Type	Data #	Testing ratings	Density
Data without new users	Data 1	20% per user	(75 * 80% * 50)/(50 * 600) = 10%
	Data 2	40% per user	(75 * 60% * 50)/(50 * 600) = 7.5%
	Data 3	60% per user	(75 * 40% * 50)/(50 * 600) = 5%
Data with new users	Data 4	20% per user	(75 * 80% * 50)/(52 * 600) = 9.6%
	Data 5	40% per user	(75 * 60% * 50)/(52 * 600) = 7.2%
	Data 6	60% per user	(75 * 40% * 50)/(52 * 600) = 4.8%

4.2 Evaluations of the Proposed Method

Basically, the major intent of the experiment is to evaluate the effectiveness on facing problems of new user, new item and rating sparsity. Figures 4 and 5 showing the experimental results deliver some aspects. First, whether the data is with new users or not, our proposed method performs more stably than compared methods [8, 9, 13]. Second, if the sparsity increases, our proposed method is still much better than compared methods [8, 9, 13]. Third, overall, our proposed method brings out the much

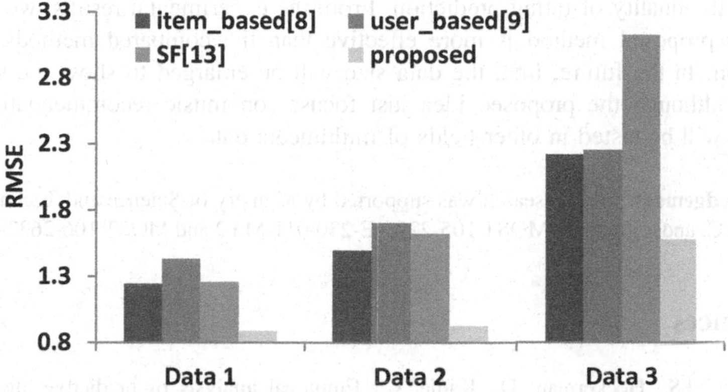

Fig. 4. The comparisons between the proposed method and compared methods without new users in terms of RMSE.

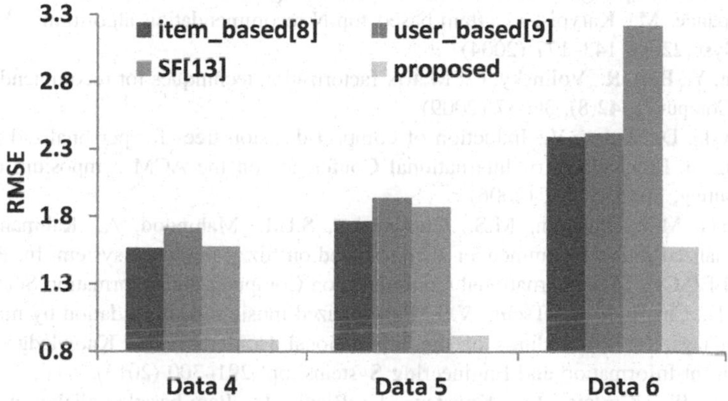

Fig. 5. The comparisons between the proposed method and compared methods with new users in terms of RMSE.

better results whatever the problem faced is. That is, our proposed method can really deal with the traditional problems.

5 Conclusion and Future Work

To tell the truth, issues on music recommendation have been studied for a set of years. Also it is applied to many recent online music websites. However, some problems such as new user, new item and rating sparsity are not effectively settled. To address these problems, a music recommendation algorithm is proposed in this paper. In terms of new user, we provide a bit-profile similarity as a solution to decrease the impact of this problem. In terms of new item and rating sparsity, rating imputation is adopted to increase the quality of rating prediction. From the experimental results, we can know that, our proposed method is more effective than the compared methods on rating prediction. In the future, first, the data size will be enlarged to show the scalability. Second, although the proposed idea just focuses on music recommendation in this paper, it will be tested in other fields of multimedia data.

Acknowledgement. This research was supported by Ministry of Science and Technology, Taiwan, R.O.C. under grant no. MOST 105-2221-E-230-011-MY2 and MOST 106-2632-S-424-001.

References

1. Breese, J.S., Heckerman, D., Kadie, C.: Empirical analysis of predictive algorithms for collaborative filtering. In: Proeedings of International Conference on Uncertainty in Artificial Intelligence, pp. 43–52 (1998)
2. Cheng, R., Tang, B.: A music recommendation system based on acoustic features and user personalities. In: Cao, H., Li, J., Wang, R. (eds.) PAKDD 2016. LNCS (LNAI), vol. 9794, pp. 203–213. Springer, Cham (2016). https://doi.org/10.1007/978-3-319-42996-0_17
3. Deshpande, M., Karypis, G.: Item-based top-N recommendation algorithms. ACM Trans. Inf. Syst. **22**(1), 143–177 (2004)
4. Koren, Y., Bell, R., Volinsky, C.: Matrix factorization techniques for recommender systems. IEE Comput. J. **42**(8), 30–37 (2009)
5. Nikovski, D., Kulev, V.: Induction of compact decision trees for personalized recommendation. In: Proceedings of International Conference on the ACM symposium on Applied computing, pp. 575–581 (2006)
6. Rahman, M.S., Rahman, M.S., Chowdhury, S.U.I., Mahmood, A., Rahman, R.M.: A personalized music recommender service based on fuzzy inference system. In: Proceedings of IEEE/ACIS 15th International Conference on Computer and Information Science (2016)
7. Su, J.H., Chang, W.Y., Tseng, V.S.: Personalized music recommendation by mining social media tags. In: Proceedings of the International Conference on Knowledge-Based and Intelligent Information and Engineering Systems, pp. 291–300 (2013)
8. Sarwar, B., Karypis, G., Konstan, J., Riedl, J.: Item-based collaborative filtering recommendation algorithms. In: Proceedings of International Conference on World Wide Web, pp. 285–295 (2001)
9. Shi, Y., Larson, M., Hanjalic, A.: Collaborative filtering beyond the user-item matrix: a survey of the state of the art and future challenges. ACM Comput. Surv. **47**(1), 143–177 (2014)
10. Su, J.H., Yeh, H.H., Philip, S.Y., Tseng, V.S.: Music recommendation using content and context information mining. IEEE Intell. Syst. **25**(1), 16–26 (2010)
11. Xu, J.A., Araki, K.: A SVM-based personal recommendation system for TV programs. In: Proceedings of International Conference on Multi-Media Modeling Conference, pp. 401–404 (2006)
12. Xue, H.-J., Dai, X.-Y., Zhang, J., Huang, S., Chen, J.: Deep matrix factorization models for recommender systems. In: Proceedings of the Twenty-Sixth International Joint Conference on Artificial Intelligence (IJCAI-17) (2017)
13. Wang, J., De Vries, A.P., Reinders, M.J.: Unifying user-based and item-based collaborative filtering approaches by similarity fusion. In: Proceedings of International Conference on Special Interest Group on Information Retrieval, pp. 501–508 (2006)

W-PathSim: Novel Approach of Weighted Similarity Measure in Content-Based Heterogeneous Information Networks by Applying LDA Topic Modeling

Phu Pham[1], Phuc Do[1(✉)], and Chien D. C. Ta[2]

[1] University of Information Technology (UIT), VNU-HCM, Ho Chi Minh City, Vietnam
phamtheanhphu@gmail.com, phuc.do@uit.edu.vn

[2] Industrial University of Ho Chi Minh City (IUH), Ho Chi Minh City, Vietnam
tdcchien@gmail.com

Abstract. In information retrieval, similarity measure or top-k similarity searching had been extensively researched. Similarity search supports to find the most relevant information in a large-scale collection of datasets, especially, with large-scale **h**eterogeneous **i**nformation **n**etworks (HINs) which is composed by multiple types of object and relation. There are studies related to similarity search applied in HINs, the "***PathSim***" is one of a remarkable work of *Sun et al.* which is based on meta-path for calculating the similarity between objects in multi-typed linking information networks. However, there is also a shortcoming of *PathSim* in weighting the "*path instance(s)*" of defined *meta-paths* in similarity scoring between two objects. The shortage of evaluating the weight of linked connections between objects might influence the output quality. In this paper, we present ***W-PathSim*** model, which applies the ***Latent Dirichlet Allocation*** (***LDA***) topic modeling for generating the weighting attribute for the object's links. We conduct experiments on real DBLP and Aminer datasets in order to demonstrate the effectiveness of our proposed model.

Keywords: Heterogeneous information networks (HINs) · Similarity measurement · Topic modeling · Meta-path based similarity measure

1 Introduction

Discovering the relationship and similarity measure between objects are the fundamental problems of knowledge mining and information retrieval tasks, especially in the heterogeneous information networks (***HINs***) [1, 2]. The measure of similarity between objects plays an important role in multiple fields includes: clustering, classification [3] and community detection [4]. Evaluating the similarity between objects in the information networks also support for the involvements in building the knowledge-based search engine, recommendation system, etc. The similarity between two objects is computed mostly based on evaluating how they are linked to each other in the

N. T. Nguyen et al. (Eds.): ACIIDS 2018, LNAI 10751, pp. 539–549, 2018.
https://doi.org/10.1007/978-3-319-75417-8_51

information networks (*INs*). For example, the users on the social networks such as: *Facebook, Twitter, etc.* are connected via relationship such as "*friend*", "*closest friend*", etc. Or in the bibliographic networks, there are objects such as: "*authors*", "*papers*", "*venue*", etc., the links between objects such as: author (A), venue (V), paper (P) are "*author_of*", "*refer_to*", etc. There are two main types of information networks, includes: homogeneous information networks (***HoINs***) and heterogeneous information networks (***HINs***). In *HoINs*, it consists only single-type of object and link, as shown in Fig. 1(A). But with *HINs* the amount of object and relation type is more than one, as shown in Fig. 1(B).

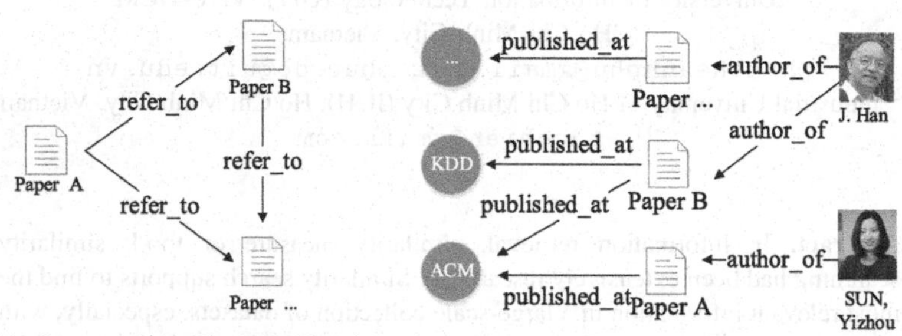

Fig. 1. Illustration of homogeneous information networks (HoINs) and heterogeneous information networks (HINs)

There are some studies related to similarity measure for objects in INs, but most of them are focuses on *HoINs* such as: the proposed models of *Jeh and Widom SimRank* [5], *Personalized PageRank* (also called *P-PageRank*) [6] algorithm, and the *SCAN* [7] model for data clustering in information networks of *Xu et al.*

Due to the complicated structure of HINs (*multi-typed object and relation*), the related mining techniques are necessary to follow the "*schema*" or "*meta*" level in connected components' descriptions [2, 8], the "*network schema*" (as illustrated in Fig. 2A) is defined as the meta-structure of the information networks. Different types of

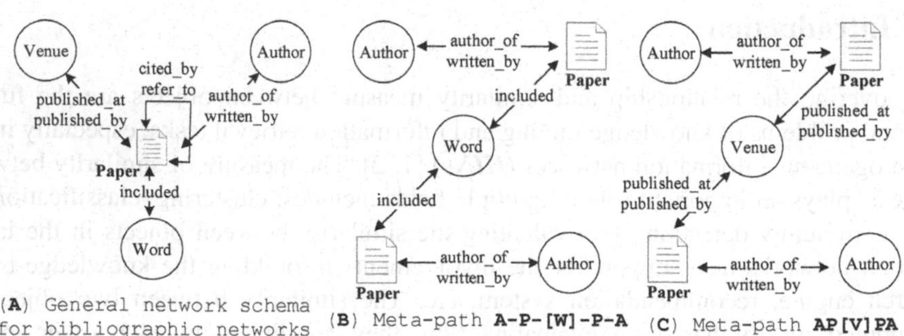

Fig. 2. Illustration of bibliographic network schema and related common meta-paths

HINs are described in different network structural schema (*BINs, social network, encyclopedia, etc.*). However, there are also existed problems related to thoroughly weighting the links between objects rather than merely counting the number of link. For example, in a ***bibliographic information networks*** (*BINs*), two authors are connected via semantic paths, such as: *A-P-[W]-P-A* (as shown in Fig. 2B), or *A-P-[V]-P-A* (as shown in Fig. 2C), etc. In the ***PathSim***, it is impossible to identify how papers which are submitted to the same conferences are similar to each other. There is no doubt that multiple papers which published at specific conference might be different in the content or topic due to the differences in committed tracks. In this paper, we propose the ***W-PathSim*** model which apply the *LDA* topic modeling to score the linked path which connect the objects in HINs. The *LDA* topic modeling proposed by *Blei et al.* [9, 10]. The main contributions of our works in this paper are:

- We propose the approach of extracting latent topics via LDA topic model over the text content-based objects of HINs, in the DBLP we have only the "*paper*".
- The extracted latent topics which distribute over the HINs are used for evaluating the similarity between incorporated objects (related to "*paper*" object) such as: "*author*", "*conference*", etc. The weight similarity score $\left(\boldsymbol{sim_{weight}}\right)$ which calculated via distributed latent topics proportions over documents by applying the cosine similarity.
- Finally, we combine the weight similarity scoring of $\left(\boldsymbol{sim_{weight}}\right)$ with the traditional PathSim $\left(\boldsymbol{sim_{PathSim}}\right)$ in order to produce the final similarity score $\left(\boldsymbol{sim_{W-PathSim}}\right)$.

The rest of our paper are organized in 4 sections. Section 2 represents related works and our work's motivations. In Sect. 3, we define related proposed approaches and concepts of our works. Section 4 illustrates our experimental methods, testing dataset usage and results of our proposed algorithm. In this Sect. 4, we also show our experimental analysis and discussions base on the outputs as well as discussions about our methodology. Finally, Sect. 5 shows our conclusion and future works.

2 Literature Reviews and Motivations

2.1 The PathSim Algorithm and Similarity Measurement in HINs

The principle of ***PathSim*** algorithm is the application of the semantic path to evaluate the similarity of same-type objects in HINs. The *PathSim* [8] based on the number of "path instance(s)" which incorporate with defined meta-path (P) [*] to calculate the similarity between same-type objects.

[*]***Meta-path:*** *in short, a meta-path (P) is a graph-based structure which is defined from (a part of) given HINs "network schema", denoted as: $NS_G = (A, R)$. A meta-path (P) with (n) in length, is represented by: $P = A_1 \xrightarrow{R_1} A_2 \ldots \xrightarrow{R_{n-1}} A_n$ or $P = (A_1, A_2 \ldots A_n)$, in short* [2].

Given a symmetric meta-path (P), the similarity $\boldsymbol{PathSim}(a, b)$ between two same-typed object (a) and (b) is calculated by (as shown in Eq. 1):

$$\boldsymbol{PathSim}(a, b) = 2.\frac{\left|\left\{p_{(a \rightsquigarrow b)} : p_{(a \rightsquigarrow b)} \in P\right\}\right|}{\left|\left\{p_{(a \rightsquigarrow a)} : p_{(a \rightsquigarrow a)} \in P\right\}\right| + \left|\left\{p_{(b \rightsquigarrow b)} : p_{(b \rightsquigarrow b)} \in P\right\}\right|} \quad (1)$$

Where,

- $p_{(a \rightsquigarrow b)}$, stands for the path instance between (a) and (b).
- $p_{(a \rightsquigarrow a)}$ and $p_{(b \rightsquigarrow b)}$, represent for the path instance between (a), (a), and (b), (b), respectively.

2.2 Shortcoming in Path Instance Weighting and Related Works

In fact, the traditional *PathSim* only considers the number of symmetric connected links between objects but does not evaluating how much (weighting) these objects are linked to each other. The link weighting plays an important role evaluate the strength of the connection between objects. Recently, there are remarkable extended versions of *PathSim* which are proposed for resolving similarity searching in specific cases, includes: ***PathSimExt*** of *Yao et al.* [11]. The *PathSimeExt* uses the external knowledge resources to support for enriching or ranking the objects. The proposed model of ***R-PathSim*** by *Chodpathumwan et al.* [12] applying the "*relationship reorganizing*" paradigm for extending the *PathSim* application.

3 Methodology

In this paper, we propose the ***W-PathSim*** model which applies the ***Latent Dirichlet Allocation*** (***LDA***) topic modeling [9, 10] in extracting the probabilistic distributions between the set of keywords $(w : w \in W)$ over latent topics $(t : t \in T)$, called $\left(\phi^{(t)}\right)$, and latent topics (t) over papers/documents $(d : d \in D)$, called $\theta^{(d_i)}$ on the given bibliographic HINs.

3.1 Applying LDA for Latent Topic Extraction in HINs

Latent topic extraction in HINs via LDA topic modeling
Following the LDA theoretical statements of *Blei et al.* [9, 10, 13], there are three main assumptions cover the algorithm, providing a collection of document $(d : d \in D)$ in HINs:

- The documents in corpus (HINs) are mixture of topics so from a collection of documents $(d : d \in D)$ we there is a set of topics can be extracted $(t : t \in T)$.
- Every single extracted topic is a probability distribution over set of terms or vocabulary/words $(w : w \in W)$ – the V is composed as the term-document

co-occurrence matrix - $P(w|z) = \phi^{(t)}$ stand for the distribution of word (w) over the given topic (t).

- The set of extracted topic also are distributed over every single document in the HINs - which means every single document in the collection contained multiple topics which distributed probabilistically - $P(z) = \theta^{(d)}$ for the distribution of topic (z) in specific document $(d : d \in D)$.

Document representation by extracted latent topics. In general, the LDA support to produce the generative models of extracted latent topic probabilities, donated as: $P(z_j|d_i)$ is the probabilistic distribution of *(j)-th* topic over document (d_i), with $(z : z \in T)$. Hence, each the document in HINs is now represented as weighted vector $\overrightarrow{d_i}$ (as shown in Eq. 2):

$$\overrightarrow{d_i} = \begin{bmatrix} P(z_1|d_i) \\ \ldots \\ P(z_{n,n\in|T|}|d_i) \end{bmatrix} \tag{2}$$

Content-based HINs enrichment and link weighting via LDA topic modeling

HINs enrichment via LDA. After completing to generate the probabilistic distributions of [word]-[topic] $\left(\phi^{(t)}\right)$ and [topic]-[document] $\left(\theta^{(d)}\right)$. The given HINs is enriched to more type of objects, the set of keywords $(w : w \in W)$ and extracted latent topics $(t : t \in T)$, these new type of objects are used for evaluating the weight of relationship of other same-type objects through semantic paths.

The link weighting via LDA. For the traditional *PathSim*, the links are unweighted and the similarity score mostly based on the "*number of links*" between two objects. In content-based HINs, such as bibliographic networks (*DBLP, CiteSeerX*), the LDA topic modeling help to extract the topic distributions over documents $\left(\theta^{(d)}\right)$, which used for assigning the weight of links related to text content-based object (in this case is "*paper*"). For example, with the semantic path: *A-P-[V]-P-A*, represents for the papers of two author (a) and (b) submit to the same venue. After applying LDA, every paper has a portion of latent topic distributions, represented as vector: $\overrightarrow{d_i}$ (Eq. 2), this weighted vector is used as the weighting attribute for the relations of "*paper*" and "*venue*". The meta-path is become, $A_a - P \xrightarrow{\overrightarrow{d_a}} [V] \xleftarrow{\overrightarrow{d_b}} A_b$.

3.2 W-PathSim Bases on Latent Topics Distributed Portions

Weighted similarity measure based on cosine similarity

Finally, we use these latent topic distributed proportions as the weights between document/paper in HINs. The similarity between authors' documents are calculated by applying "*cosine similarity*" (CS). To begin with, we consider the distributed proportions of latent topics (T) over the given *(a)-th* document $\theta_{|T|}^{(d_a)}$ and *(b)-th* document $\theta_{|T|}^{(d_b)}$ as vector ($\overrightarrow{z_a}$) and ($\overrightarrow{z_b}$), the similarity is computed via equation (Eq. 3):

$$\boldsymbol{sim_{CS}(d_a, d_b)} = \frac{\overrightarrow{z_i} \cdot \overrightarrow{z_j}}{\|\overrightarrow{z_i}\| \cdot \|\overrightarrow{z_j}\|} = \frac{\sum_{i=1}^{|T|} \theta_{t_i}^{(d_a)} \cdot \theta_{t_i}^{(d_b)}}{\sqrt{\sum_{i=1}^{|T|} \theta_{t_i}^{(d_a)})^2} \cdot \sqrt{\sum_{i=1}^{|T|} \theta_{t_i}^{(d_b)})^2}} \tag{3}$$

Where,

- $\boldsymbol{sim_{CS}(d_a, d_b)}$, is the cosine similarity between *(a)-th* document and *(b)-th* document.
- $\theta_{t_i}^{(d_a)}$ **and** $\theta_{t_i}^{(d_b)}$, is the probabilistic distributed values of *(a)-th* document and *(b)-th* document at *(i)-th* topic, respectively.

The weighted similarity scoring. Assuming that, $D_{A_a}^{V_i}$ and $D_{A_b}^{V_i}$ are the set of documents which *(a)-th* author and *(b)-th* author submitted to the *(i)-th* conference, denoted as: (V_i), respectively. The $\boldsymbol{sim_{weight}}$ between papers of both two given authors which submit to specific conference or journal form a weighted matrix with the size as, denoted as: $|D_{A_a}^{V_i}| \times |D_{A_b}^{V_i}|$. Then, the similarity between shared documents of authors in the same conference is calculated as follow (shown in Eq. 4):

$$\boldsymbol{sim_{weight}}(A_a, A_b)_{V_i} = \boldsymbol{avg}\left(\sum_{|D_{A_a}^{V_i}|}^{k=1} \sum_{|D_{A_b}^{V_i}|}^{j=1} \boldsymbol{sim_{CS}}\left(d_{(A_b)k}^{V_i}, d_{(A_b)j}^{V_i}\right)\right) \tag{4}$$

Where,

- $\boldsymbol{sim_{doc}}(A_a, A_b)_{V_i}$, is the similarity value of author (A_a) and (A_b) based on the distributed portion of topics over submitted papers in a conference (V_i).
- $\boldsymbol{sim_{CS}}\left(d_{(A_b)k}^{V_i}, d_{(A_b)j}^{V_i}\right)$, is the pairwise similarity score of two documents, $d_{(A_b)k}^{V_i}$ is document of author (A_a) at (k), and $d_{(A_b)j}^{V_i}$ is document of author (A_b) at (j).

Combination of PathSim and weighted cosine similarity scoring

Finally, we combine the proposed similarity metrics with the *PathSim*, the finally similarity score is calculated as following (as shown in Eq. 5):

$$\boldsymbol{sim_{W\text{-}PathSim}}(A_a, A_b) = \boldsymbol{avg}\left(\left(\sum_{i=1}^{|\boldsymbol{V}|} \boldsymbol{sim_{weight}}(A_a, A_b)_{V_i}\right), \boldsymbol{PathSim}(A_a, A_b)\right) \tag{5}$$

Where,

- **|V|**, is the total number of conferences which authors (A_b) and (A_b) submitted their paper to.

- $sim(A_a, A_b)_{V_i}$, represents for the final similarity scores of two authors (A_b) and (A_b) in specific conference (i). Implementation and experiments.

The Fig. 3 illustrates the method of weighted similarity measure (based on the Eq. 5), via the extracted latent topics distribution over submitted documents of "*Mike*" and "*Jim*" authors (θ), in "*SIGMOD*" conference.

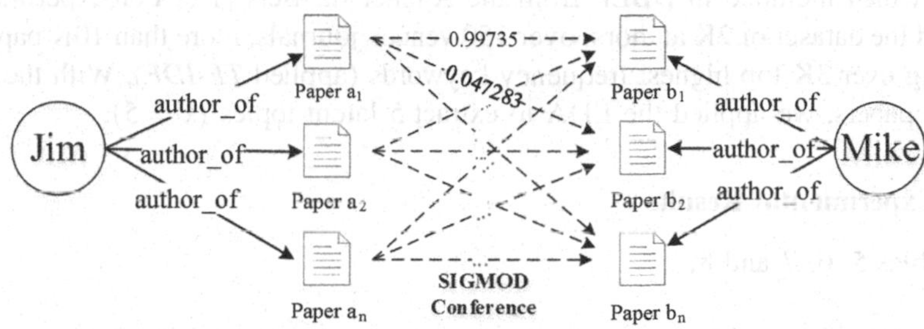

Fig. 3. Illustration of similarity measuring based on latent topics distribution over submitted papers in a specific conference between two authors

For example, Table 1 shows the number of documents which "*Mike*" and "*Jim*" submit to "*VLDB*" and "*SIGMOD*". These 6 documents are applied LDA to extract the document-topic (the number of extracted latent topic is set to 3) probabilistic distribution is shown as Table 2. Based on the probabilistic distributions in Table 2, we can compute the cosine similarity value of pairwise documents in the specific venues (as shown in Table 3). The Table 4 shows similarity scores between "*Mike*" and "*Jim*" which are computed via traditional *PathSim* and *W-PathSim*.

Table 1. Adjacency matrix for meta-path APVPA between two authors "Jim" and "Mike"

	SIGMOD	VLDB
Jim	2 (d1, d2)	1 (d4)
Mike	1 (d3)	2 (d5, d6)

Table 2. The LDA document-topic distribution of *"Mike"* and *"Jim"*

	t1	t2	t3
d1	*0.0923386*	*0.0375083*	*0.870153117*
d2	*0.0126545*	*0.9858852*	*0.001466031*
d3	*0.0582242*	*0.0381457*	*0.903630078*
d4	*0.0144836*	*0.9763511*	*0.009165313*
d5	*0.1858516*	*0.8107631*	*0.003385308*
d6	*0.1807888*	*0.0018485*	*0.817362717*

Table 3. Cosine similarity values between documents of "*Jim*" and "*Mike*" in "*VLDB*" and "*SIGMOD*" venues

	d3	d5	d6
d1	*0.999145*	*0.069342*	*0.99292*
d2	*0.044393*	*0.977503*	*0.006432*
d4	*0.052394*	*0.977913*	*0.014575*

Table 4. Similarity score of "*Jim*" and "*Mike*" between PathSim and W-PathSim

Algorithm	Similarity score	
PathSim	0.4	
W-PathSim	$sim_{weight\ VLDB}$	0.496244
	$sim_{weight\ SIGMOD}$	0.521769
	$sim_{W-PathSim}$	0.454503

4 Experiments and Evaluations

4.1 Experimental Dataset Usage

For experiments, we use the DBLP dataset [14] which downloaded from the in September, 2017. It contains over 1.7M authors, 3.4M papers and 5k conferences/venues. For the document content, we collected the abstract content of over 600K papers which included in DBLP from the Aminer datasets [15]. For experiment, we selected the dataset of 2K authors, over 150 venues/journals, more than 10K papers and selecting over 3K top highest frequency keywords (applied *TF-IDF*). With the dataset of 10K papers, we applied the LDA to extract 5 latent topics ($k = 5$).

4.2 Experimental Results

See Tables 5, 6, 7 and 8.

Table 5. Top-10 most similarity authors to "*Christos Faloutsos*" via PathSim and W-PathSim for A-P-[V]-P-A meta-path in DBLP dataset

PathSim		W-PathSim	
Rakesh Agrawal	*0.766017*	Rakesh Agrawal	*0.779292*
Jiawei Han	*0.740148*	Jiawei Han	*0.764830*
Philip S. Yu	*0.737430*	Wei Wang	*0.763302*
Hans-Peter Kriegel	*0.729180*	Philip S. Yu	*0.763261*
Haixun Wang	*0.728675*	Hans-Peter Kriegel	*0.761627*
Wei Wang	*0.727918*	Haixun Wang	*0.760221*
Jian Pei	*0.695855*	Jian Pei	*0.743087*
H.V. Jagadish	*0.683294*	H.V. Jagadish	*0.734668*
Ming-Syan Chen	*0.682938*	Gerhard Weikum	*0.734358*
Gerhard Weikum	*0.673686*	Raghu Ramakrishnan	*0.733277*

Table 6. Top-10 most similarity authors to "*AnHai Doan*" computed via PathSim and W-PathSim for A-P-[V]-P-A meta-path in DBLP dataset

PathSim		W-PathSim	
Kevin Chen-Chuan Chang	*0.669786*	Sharad Mehrotra	*0.718709*
Alon Y. Halevy	*0.642336*	Kevin Chen-Chuan Chang	*0.716155*
Sharad Mehrotra	*0.640135*	Michalis Vazirgiannis	*0.713262*
Michalis Vazirgiannis	*0.630915*	Alon Y. Halevy	*0.704795*
Krithi Ramamritham	*0.623249*	Krithi Ramamritham	*0.703244*
Stefano Ceri	*0.609137*	Stefano Ceri	*0.698616*
Ji-Rong Wen	*0.603960*	Renee Miller	*0.695583*
Gio Wiederhold	*0.603397*	Oren Etzioni	*0.693257*
RenÃ©e J. Miller	*0.600334*	Ji-Rong Wen	*0.690484*
Ling Liu	*0.599889*	Tie-Yan Liu	*0.689559*

Table 7. Top-10 most similarity authors to "*Christos Faloutsos*" computed via PathSim and W-PathSim for A-P-[W]-P-A meta-path in DBLP dataset

PathSim		W-PathSim	
Jiawei Han	*0.907171*	Rakesh Agrawal	*0.843333*
Rakesh Agrawal	*0.900862*	Jiawei Han	*0.840918*
Hans-Peter Kriegel	*0.839144*	Hans-Peter Kriegel	*0.810152*
Jian Pei	*0.831634*	Wei Wang	*0.807872*
Raghu Ramakrishnan	*0.808722*	H.V. Jagadish	*0.795332*
H.V. Jagadish	*0.804915*	Raghu Ramakrishnan	*0.792045*
Nick Koudas	*0.788012*	Gerhard Weikum	*0.786141*
Gerhard Weikum	*0.787663*	Jennifer Widom	*0.776998*
Hector Garcia-Molina	*0.780158*	Nick Koudas	*0.776615*
Divesh Srivastava	*0.775447*	Divesh Srivastava	*0.772035*

Table 8. Top-10 most similarity authors to "*AnHai Doan*" computed via PathSim and W-PathSim for A-P-[W]-P-A meta-path in DBLP dataset

PathSim		W-PathSim	
Charu C. Aggarwal	*0.982278*	Sharad Mehrotra	*0.878353*
Xuemin Lin	*0.981912*	Kevin Chen-Chuan Chang	*0.875149*
Balakrishna R. Iyer	*0.977654*	Charu C. Aggarwal	*0.874236*
Walid G. Aref	*0.974249*	Carlo Zaniolo	*0.870825*
Jignesh M. Patel	*0.970588*	Jun Yang	*0.870316*
M. Tamer Ã–zsu	*0.970443*	Elisa Bertino	*0.870297*
Ahmed K. Elmagarmid	*0.967568*	Mohamed F. Mokbel	*0.870012*
Jayant R. Haritsa	*0.966469*	Walid G. Aref	*0.869533*
Carlo Zaniolo	*0.962441*	Richard T. Snodgrass	*0.869111*
Jun Yang	*0.961616*	Ahmed K. Elmagarmid	*0.865730*

4.3 Experimental Concluding Remarks

For the experiment with author "*Christos Faloutsos"* with semantic path *A-P-[V]-P-A*, with the traditional *PathSim*, the similarity measurements of author "*Christos Faloutsos*" shows that three authors "*Rakesh Agrawal*", "*Jiawei Han*", "*Philip S. Yu*" are top similar authors to "*Christos Faloutsos*". With the *W-PathSim*, over the extracted latent topic distributions over these authors related papers, it shows that "*Wei Wang*" is much more similar to "*Christos Faloutsos*" than "*Philip S. Yu*", they share much more common keyword and topics in their paper, specifically in "*graph theory*", "*large-graph*", "*graph mining*", etc. (as shown in the Google's scholar profile: *Christos Faloutsos, Wei Wang, Philip S. Yu*). The experiment with meta-path *A-P-[W]-P-A*, The W-PathSim also shows that the author "*Wei Wang*" gains the higher similar score to author "*Christos Faloutsos"* than the traditional *PathSim*. For the experiment with author "*AnHai Doan*", with the meta-path *A-P-[V]-P-A*, the results also shows that papers' distributed latent topics definitely can influence the similarity result output. Comparing two top output similar authors, includes: "*Kevin Chen-Chuan Chang*" (*PathSim*) and "*Sharad Mehrotra*" (*W-PathSim*). If we look at the profile as well as the content of "*Sharad Mehrotra*" published articles, it seem that "*Sharad Mehrotra*" much more similar to "*AnHai Doan*" than "*Kevin Chen-Chuan Chang*", in common topics related to "*cloud computing*", "*big data*", "*distributed systems*", etc. The experimental results with meta-path *A-P-[V]-P-A* also return the same results, with "*Sharad Mehrotra*" as the top similar author to "*AnHai Doan*".

In fact, the current experimental results is not remarkably show a clear improvement between the traditional *PathSim* algorithm and our proposed approach of *W-PathSim* due to the small number as well as the shortage of full-length document's content resource (only the abstract content). However, over experiments, the proposed *W-PathSim* had been proved that the output accuracy can be improved by taking the advantages of content-based evaluation via LDA topic modeling.

5 Conclusion and Future Works

Our works in this paper are the revisions of approaches in object similarity measurement applying in HINs. We propose novel approach of *W-PathSim* for improving the previous traditional *PathSim* algorithm via applying LDA topic modeling. The LDA topic modeling is applied for extracting latent topic distribution over content-based objects in HINs which will be used as the weighting attribute for links between objects. The weighting similarity score is computed via cosine similarity. Over implementation and experiments, it has been shown that the proposed approach of W-Path is effective in object similarity measure in bibliographic information networks (BINs) which the objects majorly incorporates to text documents.

Acknowledgement. This research is funded by Vietnam National University Ho Chi Minh City (VNU-HCMC) under the grant number B2017-26-02.

References

1. Han, J., et al.: Mining knowledge from databases: an information network analysis approach. In: Proceedings of the 2010 ACM SIGMOD International Conference on Management of Data, pp. 1251–1252, ACM (2010)
2. Shi, C., et al.: A survey of heterogeneous information network analysis. IEEE Trans. Knowl. Data Eng. 17–37 (2017)
3. Ji, M., Han, J., Danilevsky, M.: Ranking-based classification of heterogeneous information networks. In: Proceedings of the 17th ACM SIGKDD International Conference on Knowledge Discovery and Data Mining, pp. 1298–1306, ACM (2011)
4. Sun, Y., et al.: Community evolution detection in dynamic heterogeneous information networks. In: Proceedings of the Eighth Workshop on Mining and Learning with Graphs, pp. 137–146, ACM (2010)

5. Jeh, G., Widom, J.: SimRank: a measure of structural-context similarity. In: Proceedings of the Eighth ACM SIGKDD International Conference on Knowledge Discovery and Data Mining, pp. 538–543, ACM (2002)
6. Jeh, G., Widom, J.: Scaling personalized web search. In: Proceedings of the 12th International Conference on World Wide Web, pp. 271–279, ACM (2003)
7. Xu, X., et al.: SCAN: a structural clustering algorithm for networks. In: Proceedings of the 13th ACM SIGKDD International Conference on Knowledge Discovery and Data Mining, pp. 824–833, ACM (2007)
8. Sun, Y., et al.: Pathsim: meta path-based top-k similarity search in heterogeneous information networks. In: Proceedings of the VLDB Endowment, pp. 992–1003 (2011)
9. Blei, D.M., Ng, A.Y., Jordan, M.I.: Latent dirichlet allocation. J. Mach. Learn. Res. 993–1022 (2003)
10. Blei, D.M.: Probabilistic topic models. Commun. ACM 77–84 (2012)
11. Hou, U.L., Yao, K., Mak, H.F.: PathSimExt: revisiting PathSim in heterogeneous information networks. In: Li, F., Li, G., Hwang, S.-w., Yao, B., Zhang, Z. (eds.) WAIM 2014. LNCS, vol. 8485, pp. 38–42. Springer, Cham (2014). https://doi.org/10.1007/978-3-319-08010-9_6
12. Chodpathumwan, Y., et al.: Towards representation independent similarity search over graph databases. In: Proceedings of the 25th ACM International on Conference on Information and Knowledge Management, pp. 2233–2238, ACM (2016)
13. Steyvers, M., Griffiths, T.: Probabilistic topic models. In: Handbook of Latent Semantic Analysis, pp. 424–440 (2007)
14. DBLP - Computer Science Bibliography. http://dblp.uni-trier.de/. Accessed 24 09 2017
15. AMiner Dataset. https://aminer.org/data. Accessed 24 09 2017

Representative Rule Templates for Association Rules Satisfying Multiple Canonical Evaluation Criteria

Marzena Kryszkiewicz(✉)

Institute of Computer Science, Warsaw University of Technology, Nowowiejska 15/19, 00-665 Warsaw, Poland
mkr@ii.pw.edu.pl

Abstract. Originally, strong association rules were defined as those that have sufficiently high values of two parameters: support and confidence. However, it has been shown in the literature that in general neither of these two measures is capable of determining (in)dependence between rules' constituents correctly. Thus, usage of other measures for rule evaluation is also under research. In this paper, we formulate a generic notion of a canonical measure and show important examples of canonical measures. For association rules satisfying any set of criteria based on canonical measures, we offer their concise lossless representation in the form of so called representative rule templates. Also, we derive a number of properties of this representation.

1 Introduction

Originally, strong association rules were defined in [1] as those that have sufficiently high values of two parameters: (relative) support (probability that an association rule is satisfied) and confidence (conditional probability of the rule consequent given its antecedent). However, it has been shown in the literature that in general neither of these two measures is capable of determining (in)dependence between rules' constituents correctly (see e.g. [12]). Thus, usage of other measures for rule evaluation is also under research [3, 5, 9, 10, 13–18]. In this paper, we formulate a generic notion of a *canonical measure*. For a number of example evaluation measures of association rules, we show that they are canonical. For association rules satisfying any set of criteria based on canonical measures, we offer their concise lossless representation in the form of so called *representative rule templates*. Also, we derive a number of properties of this representation.

Our paper has the following layout. In Sect. 2, we briefly recall and comment basic notions of itemsets, (association) rules, support, confidence, dependency of events, ACBC-measures, and a good interestingness measure. Our main contribution is presented in Sects. 3–5. In Sect. 3, we introduce and justify usefulness of the notion of a canonical measure and show that a number of example ACBC-measures are canonical. In Sect. 4, we introduce the notion of rule templates and show that representative rule templates are sufficient to represent all association rules that are strong with respect to any criteria expressible in terms of canonical measures. Also, we derive a number of

N. T. Nguyen et al. (Eds.): ACIIDS 2018, LNAI 10751, pp. 550–561, 2018.
https://doi.org/10.1007/978-3-319-75417-8_52

properties of (representative) rule templates. In particular, we derive a formula determining how many association rules are covered by a rule template. In Sect. 5, we briefly address related work on lossless representations of association rules and make additional claims about our new representation. Section 6 concludes our contribution.

2 Basic Notions and Properties

In this section, we provide definitions and properties related to *rules* and their specific type called *association rules* [1]. Let $\mathcal{I}$ be a set of *items* (e.g. products). Any set $X \subseteq \mathcal{I}$ is called an *itemset*. A *transaction dataset* is denoted by $\mathcal{D}$ and is defined as a set of itemsets. Each itemset in $\mathcal{D}$ is called a *transaction*.

Let X and Y be any itemsets. An expression $X \to Y$ is called a *rule*. If additionally $X \cap Y = \emptyset$, then $X \to Y$ is called an *association rule*. The set of all association rules is denoted by *AR*. Itemsets X and Y, which occur in $X \to Y$, are called its *antecedent* and *consequent*, respectively. The itemset $Z = X \cup Y$ is called the *base* of $X \to Y$.

Itemsets and rules are typically characterized by *support* (or *relative support*) and *confidence* as follows. *Support* of an itemset X is denoted by $sup(X)$ and is defined as the number of transactions in $\mathcal{D}$ that contain X. Alternatively, instead of a support, a notion of a relative support is used: *Relative support* of an itemset X is defined as $sup(X)/|\mathcal{D}|$. The relative support of itemset X can be regarded as the probability of the occurrence of X in a transaction. The relative support of X will be denoted by $P(X)$. Clearly, if $X \subseteq Y$, then $sup(X) \geq sup(Y)$ and $P(X) \geq P(Y)$.

Support of a rule $X \to Y$ is denoted by $sup(X \to Y)$ and is defined as the support of its base $X \cup Y$; that is, $sup(X \to Y) = sup(X \cup Y)$.

The *confidence* of a rule $X \to Y$ is denoted by $conf(X \to Y)$ and is defined as the conditional probability that Y occurs in a transaction provided X occurs in the transaction; that is: $conf(X \to Y) = sup(X \to Y)/sup(X) = P(XY)/P(X)$.

An association rule is defined in [1] as *strong* if its (relative) support and confidence are greater than or equal to user defined *minimum* (*relative*) *support* and *minimum confidence thresholds*, respectively. Nevertheless, antecedents and consequents of rules that are strong with respect to support and confidence are not guaranteed to be dependent. Statistically, *X and Y are dependent* if $P(XY) \neq P(X) \times P(Y)$. Otherwise, X and Y are *independent*. The *dependence* between X and Y is considered as *positive* if $P(XY) > P(X) \times P(Y)$, and *negative* if $P(XY) < P(X) \times P(Y)$. The example beneath shows the case when an antecedent and consequent of a rule are dependent negatively despite high values of (relative) support and confidence.

Example 1. Let $\mathcal{D}$ be a transaction dataset from Table 1 and $\{z\} \to \{v\}$ be a rule of interest. The probabilities of the antecedent, consequent and base of rule $\{z\} \to \{v\}$ are as follows: $P(\{z\}) = 0.5, P(\{v\}) = 0.9, P(\{zv\}) = 0.4$ (see Table 2). Since $P(\{z\}) \times P(\{v\}) = 0.45$, which is greater than $P(\{zv\})$, z and v are dependent negatively. On the other hand, $conf(\{z\} \to \{v\}) = 0.4/0.5 = 0.8$, which is a high value. □

Table 1. Example dataset $\mathcal{D}$

Transaction Id	Transaction
#1	$\{v\}$
#2	$\{v\}$
#3	$\{x\ v\}$
#4	$\{x\ v\}$
#5	$\{x\ y\ v\}$
#6	$\{x\ y\ z\ v\}$
#7	$\{x\ y\ z\ v\}$
#8	$\{y\ z\ v\}$
#9	$\{y\ z\ v\}$
#10	$\{y\ z\}$

Table 2. Example rules found in $\mathcal{D}$ from Table 1

Rule	$\{x\} \to \{y\}$	$\{z\} \to \{y\}$	$\{z\} \to \{v\}$
P(antecedent)	0.5	0.5	0.5
P(consequent)	0.6	0.6	0.9
P(base)	**0.3**	**0.5**	**0.4**
conf(rule)	1/5	1	4/5
P(antecedent) × *P*(consequent)	**0.3**	**0.3**	**0.45**
Are consequent and antecedent dependent?	no	yes - positively	yes – negatively

Please note that (in)dependence of antecedent *X* and consequent *Y* of rule $X \to Y$ is determined based on three probabilities: *P*(*X*), *P*(*Y*) and *P*(*XY*). In fact, large number of rule evaluation measures can be defined in terms of at most these three probabilities. Such measures were called *ACBC-measures* (Antecedent-Consequent-Base-Constants-measures) in [11]. Some of their popular examples are provided in Table 3.

Table 3. Example ACBC-measures of (association) rules

Measure	Definition
relativeSup(*X*→*Y*)	$P(XY)$
conf(*X*→*Y*)	$P(XY) / P(X)$
novelty(*X*→*Y*)	$P(XY) - P(X) \times P(Y)$
lift(*X*→*Y*)	$P(XY) / (P(X) \times P(Y))$
cosine(*X*→*Y*)	$P(XY)/\sqrt{P(X) \times P(Y)}$
Jaccard(*X*→*Y*)	$P(XY)/(P(X) + P(Y) - P(XY))$
accuracy(*X*→*Y*)	$P(XY) + P(\bar{X}\bar{Y}) = 1 + 2P(XY) - P(X) - P(Y)$
F-score(*X*→*Y*)	$\frac{(P(XY)/P(X)) \times (P(XY)/P(Y))}{((P(XY)/P(X)) + (P(XY)/P(Y)))/2} = \frac{2P(XY)}{P(X)+P(Y)}$

In [15], a rule evaluation measure μ was defined as *a good interestingness measure* if the following conditions were fulfilled for any rules $X \to Y$, $X' \to Y'$:

1. If $P(XY) = P(X) \times P(Y)$, then $\mu(X \to Y) = 0$.
2. If $P(X') = P(X)$, $P(Y') = P(Y)$ and $P(X'Y') > P(XY)$, then $\mu(X' \to Y') > \mu(X \to Y)$.
3. If $P(X') = P(X)$, $P(Y') < P(Y)$ and $P(X'Y') = P(XY)$, then $\mu(X' \to Y') > \mu(X \to Y)$.
4. If $P(X') < P(X)$, $P(Y') = P(Y)$ and $P(X'Y') = P(XY)$, then $\mu(X' \to Y') > \mu(X \to Y)$.

Nevertheless, a number of rule evaluation measures, including all measures listed in Table 3, do not satisfy at least one of the postulates. In particular, relative support does not satisfy postulates 1, 3, 4; confidence does not satisfy postulates 1, 3; novelty does not satisfy postulate 3 (if $P(X) = 0$) and postulate 4 (if $P(Y) = 0$); lift, cosine, Jaccard, accuracy and F-score do not fulfill postulate 1.

In the next section, we offer a generic notion of a *canonical measure*, which embraces a large number of ACBC-measures (including all measures from Table 3).

3 Canonical Measures for Evaluating Rules

Let us consider two rules $X \to Y$ and $X' \to Y'$ such that X' occurs in $\mathcal{D}$ no more frequently than X $(P(X') \le P(X))$, Y' occurs in $\mathcal{D}$ no more frequently than Y $(P(Y') \le P(Y))$, but $X' \cup Y'$ occurs in $\mathcal{D}$ no less frequently than $X \cup Y$ $(P(X'Y') \ge P(XY))$. In such a case, if the dependence between Y' and X' is positive, then it is natural to expect that Y and X are either dependent positively, but to degree that is not higher than for Y' and X', or are independent, or are dependent negatively. A rule measure reflecting this expectation would evaluate rule $X' \to Y'$ not lower than rule $X \to Y$. A notion of a *canonical measure*, which we offer in this paper, is based on this observation.

Example 2. Let $\mathcal{D}$ be a transaction dataset from Table 1 and $\{x\} \to \{y\}$, $\{z\} \to \{y\}$ and $\{z\} \to \{v\}$ be rules under investigation. In Table 2, we provide the values of probabilities of their antecedents, consequents and bases. We note that the probabilities of antecedents of $\{x\} \to \{y\}$ and $\{z\} \to \{y\}$ are the same and that the probabilities of their consequents are the same, but the probabilities of their bases are different; namely, $P(\{zy\}) > P(\{xy\})$, and the dependence between y and z is positive, while y and x are independent.

Let us now compare rule $\{z\} \to \{y\}$ with rule $\{z\} \to \{v\}$. The probabilities of their antecedents are the same, but the probability of the consequent of $\{z\} \to \{y\}$ is lower than the probability of the consequent of $\{z\} \to \{v\}$, while the probability of the co-occurrence of $\{zy\}$ is greater than the probability of the co-occurrence of $\{zv\}$. In this case, the dependence between v and z is negative in contrast to the positive dependence between y and z. □

We define *a measure* $\mu : [0..1] \times [0..1] \times [0..1] \to R$ as *canonical* if for any values $p_a, p_c, p_b \in [0,1]$ and $p_{a'}, p_{c'}, p_{b'} \in [0,1]$ such that $(p_a, p_c \ge p_b)$, $(p_{a'}, p_{c'} \ge p_{b'})$, $(p_{a'} \le p_a)$, $(p_{c'} \le p_c)$ and $(p_{b'} \ge p_b)$ the following holds:

$$\mu(p_{a'}, p_{c'}, p_{b'}) \ge \mu(p_a, p_c, p_b).$$

Proposition 1. Let μ be a canonical measure, $X \to Y$ and $X' \to Y'$ be rules such that $P(X') \le P(X)$, $P(Y') \le P(Y)$ and $P(X'Y') \ge P(XY)$. Then

$$\mu(P(X'), P(Y'), P(X'Y')) \ge \mu(P(X), P(Y), P(XY)).$$

Proof. Let μ be a canonical measure, $X \to Y$ and $X' \to Y'$ be rules such that $P(X') \le P(X)$, $P(Y') \le P(Y)$ and $P(X'Y') \ge P(XY)$. Let $p_a = P(X)$, $p_c = P(Y)$, $p_b = P(XY)$, $p_{a'} = P(X')$, $p_{c'} = P(Y')$, $p_{b'} = P(X'Y')$. Then, $p_a, p_c, p_b \in [0,1]$, $p_{a'}, p_{c'}, p_{b'} \in [0,1]$, $(p_a, p_c \ge p_b)$, $(p_{a'}, p_{c'} \ge p_{b'})$, $(p_{a'} \le p_a)$, $(p_{c'} \le p_c)$ and $(p_{b'} \ge p_b)$.

Hence and by the fact that μ is a canonical measure, $\mu(p_{a'}, p_{c'}, p_{b'}) \geq \mu(p_a, p_c, p_b)$. Therefore, $\mu(P(X'), P(Y'), P(X'Y')) \geq \mu(P(X), P(Y), P(XY))$. □

In the remainder of the paper, when using a canonical measure to evaluate a rule $X \rightarrow Y$, we will write interchangeably $\mu(P(X), P(Y), P(XY))$ and $\mu(X \rightarrow Y)$. Thus, Proposition 1 can be rewritten equivalently as Proposition 2:

Proposition 2. Let μ be a canonical measure, $X \rightarrow Y$ and $X' \rightarrow Y'$ be rules such that $P(X') \leq P(X)$, $P(Y') \leq P(Y)$ and $P(X'Y') \geq P(XY)$. Then, $\mu(X' \rightarrow Y') \geq \mu(X \rightarrow Y)$.

It can be easily shown for many ACBC-measures that they are canonical. In Proposition 3, we show that in fact all ACBC-measures from Table 3 are canonical.

Proposition 3. Let $X \rightarrow Y$ and $X' \rightarrow Y'$ be rules such that $P(X') \leq P(X)$, $P(Y') \leq P(Y)$ and $P(X'Y') \geq P(XY)$. Then:

(a) $relativeSup(X' \rightarrow Y') \geq relativeSup(X \rightarrow Y)$
(b) $conf(X' \rightarrow Y') \geq conf(X \rightarrow Y)$
(c) $novelty(X' \rightarrow Y') \geq novelty(X \rightarrow Y)$
(d) $lift(X' \rightarrow Y') \geq lift(X \rightarrow Y)$
(e) $cosine(X' \rightarrow Y') \geq cosine(X \rightarrow Y)$
(f) $Jaccard(X' \rightarrow Y') \geq Jaccard(X \rightarrow Y)$
(g) $accuracy(X' \rightarrow Y') \geq accuracy(X \rightarrow Y)$
(h) $F\text{ - }score(X' \rightarrow Y') \geq F\text{ - }score(X \rightarrow Y)$

Proof. Let $X \rightarrow Y$ and $X' \rightarrow Y'$ are rules such that $P(X') \leq P(X)$, $P(Y') \leq P(Y)$ and $P(X'Y') \geq P(XY)$. Then:

Ad (a) $relativeSup(X' \rightarrow Y') = P(X'Y') \geq P(XY) = relativeSup(X \rightarrow Y)$.

Ad (b) $conf(X' \rightarrow Y') = \frac{P(X'Y')}{P(X')} \geq \frac{P(XY)}{P(X)} = conf(X \rightarrow Y)$.

Ad (c) $novelty(X' \rightarrow Y') = P(X'Y') - P(X') \times P(Y') \geq P(XY) - P(X) \times P(Y)$
$= novelty(X \rightarrow Y)$.

Ad (d–h) Proof in these cases is analogous to the proof of case (c). □

In the remainder of the paper, we assume that a set $\mathcal{M} = \{\mu_1, \ldots, \mu_n\}$ of canonical measures is used to evaluate (association) rules with respect to their corresponding threshold values $E = \{\varepsilon_1, \ldots, \varepsilon_n\}$. An expression $\mu_i(X \rightarrow Y) \geq \varepsilon_i$, where μ_i is a canonical measure and ε_i is a threshold value, will be called a *canonical evaluation criterion*.

An *association rule* $X \rightarrow Y$ will be called *$(\mathcal{M}, E)$-strong* if $\forall \mu_i \in \mathcal{M}$ $(\mu_i(X \rightarrow Y) \geq \varepsilon_i)$. The set of all $(\mathcal{M}, E)$-strong association rules will be denoted by $AR_{(\mathcal{M},E)}$.

4 Representative Rule Templates

In this section, we will first introduce a new notion of a *rule template* and examine its properties and relationship with association rules. Then, we will offer a lossless representation of $(\mathcal{M}, E)$-strong association rules based on so called *$(\mathcal{M}, E)$-representative rule templates.*

4.1 Rule Templates and Association Rules

A construct $X \xrightarrow{Z} Y$ will be called a *rule template* if $X \cap Y = \emptyset$ and $(X \cup Y) \subseteq Z$. Itemsets *X*, *Y* and *Z*, which occur in the rule template $X \xrightarrow{Z} Y$, are called its *antecedent*, *consequent* and *base*, respectively. The set of all rule templates is denoted by *RT*.

Please note that for $Z = X \cup Y$, rule template $X \xrightarrow{Z} Y$ coincides with association rule $X \to Y$. However, in general case, the base *Z* of rule template $X \xrightarrow{Z} Y$ is a (proper or improper) superset of the base $X \cup Y$ of association rule $X \to Y$. In spite of this difference, both the probability of base $X \cup Y$ of rule $X \to Y$ as well as the probability of base *Z* of rule template $X \xrightarrow{Z} Y$ exceeds neither *P*(*X*) nor *P*(*Y*).

Let μ be a canonical measure. We redefine μ for a rule template $X \xrightarrow{Z} Y$ as follows:

$$\mu(X \xrightarrow{Z} Y) = \mu(P(X), P(Y), P(Z)).$$

For example, $novelty(X \to Y) = P(XY) - P(X) \times P(Y)$, while $novelty(X \xrightarrow{Z} Y) = P(Z) - P(X) \times P(Y)$.

A rule template $X \xrightarrow{Z} Y$ *is called* (*M*, *E*)*-strong* if $\forall \mu_i \in \mathcal{M}\ (\mu_i(X \xrightarrow{Z} Y) \geq \varepsilon_i)$. The set of all ($\mathcal{M}$, *E*)-strong rule templates will be denoted by $RT_{(\mathcal{M},E)}$.

Proposition 4. Let $X \xrightarrow{Z} Y$ be a rule template and μ be a canonical measure. Then for each rule $X' \to Y'$ such that $P(X') \leq P(X)$, $P(Y') \leq P(Y)$ and $P(X'Y') \geq P(Z)$, the following holds: $\mu(X' \to Y') \geq \mu(X \xrightarrow{Z} Y)$.

Proof. Let $X \xrightarrow{Z} Y$ be a rule template, μ be a canonical measure and $X' \to Y'$ be a rule such that $P(X') \leq P(X)$, $P(Y') \leq P(Y)$ and $P(X'Y') \geq P(Z)$. Let $p_a = P(X), p_c = P(Y)$, $p_b = P(XY)$, $p_{a'} = P(X')$, $p_{c'} = P(Y')$, $p_{b'} = P(Z)$. Then, $p_a, p_c, p_b \in [0, 1]$, $p_{a'}, p_{c'}, p_{b'} \in [0, 1]$, $(p_a, p_c \geq p_b)$, $(p_{a'}, p_{c'} \geq p_{b'})$, $(p_{a'} \leq p_a)$, $(p_{c'} \leq p_c)$ and $(p_{b'} \geq p_b)$. Hence and by the fact that μ is a canonical measure, $\mu(p_{a'}, p_{c'}, p_{b'}) \geq \mu(p_a, p_c, p_b)$. So, $\mu(X' \to Y') = \mu(P(X'), P(Y'), P(X'Y')) \geq \mu(P(X), P(Y), P(Z)) = \mu(X \xrightarrow{Z} Y)$. □

Proposition 5. Let $X \xrightarrow{Z} Y$ be a rule template and μ be a canonical measure. Then for each rule $X' \to Y'$ such that $X' \supseteq X$, $Y' \supseteq Y$, $(X' \cup Y') \subseteq Z$, the following holds: $\mu(X' \to Y') \geq \mu(X \xrightarrow{Z} Y)$.

Proof. Let $X \xrightarrow{Z} Y$ be a rule template and μ be a canonical measure (*). Let $X' \to Y'$ be a rule such that $X' \supseteq X$, $Y' \supseteq Y$, $(X' \cup Y') \subseteq Z$. Then, $P(X') \leq P(X)$, $P(Y') \leq P(Y)$ and $P(X'Y') \geq P(Z)$. Hence, by (*) and Proposition 4, $\mu(X' \to Y') \geq \mu(X \xrightarrow{Z} Y)$. □

Proposition 5 tells us that whenever a rule template $X \xrightarrow{Z} Y$ reaches some value for a canonical measure, then a number of association rules $X' \to Y'$ associated in a certain way with this rule template $(X' \supseteq X, Y' \supseteq Y, (X' \cup Y') \subseteq Z)$ will also reach at least the same value of this measure. We will use this observation when defining a cover of a rule template.

The *cover of rule template* $X \xrightarrow{Z} Y$ is denoted by $TC(X \xrightarrow{Z} Y)$ and defined as follows:

$$TC(X \xrightarrow{Z} Y) = \{(X' \to Y') \in AR \mid X' \supseteq X, Y' \supseteq Y, (X' \cup Y') \subseteq Z\}.$$

Each association rule in $TC(X \xrightarrow{Z} Y)$ will be called *covered* by $X \xrightarrow{Z} Y$.

Proposition 6. Let $X \xrightarrow{Z} Y$ be an $(\mathcal{M}, E)$-strong rule template. Then, each association rule $(X' \to Y')$ in $TC(X \xrightarrow{Z} Y)$ is $(\mathcal{M}, E)$-strong.

Proof. Follows from definition of the cover of a rule template and Proposition 5. □

Thus, all association rules covered by an ($\mathcal{M}$, E)-strong rule template are ($\mathcal{M}$, E)-strong, too. We will focus now on determining a set of association rules that are covered by a given rule template and on determining its cardinality.

Example 3. Let $\{ab\} \xrightarrow{\{abcde\}} \{e\}$ be a rule template. Then, $TC(\{ab\} \xrightarrow{\{abcde\}} \{e\})$ = $\{\{ab\} \to \{e\}, \{abc\} \to \{e\}, \{abd\} \to \{e\}, \{ab\} \to \{ce\}, \{ab\} \to \{de\},$
$\{abc\} \to \{de\}, \{abd\} \to \{ce\}, \{abcd\} \to \{e\}, \{ab\} \to \{cde\}\}$. □

We find that the number of association rules covered by a single rule template depends on the number of items in its base that occur neither in the antecedent of the rule template nor in its consequent (see Theorem 1).

Theorem 1. Let $X \xrightarrow{Z} Y$ be a rule template. Then:

$$\left|TC(X \xrightarrow{Z} Y)\right| = 3^m, \text{where } m = |Z \backslash (X \cup Y)|.$$

Proof. By definition of TC, $X \to Y$ belongs to $TC(X \xrightarrow{Z} Y)$. In addition, $TC(X \xrightarrow{Z} Y)$ contains association rules having supersets of X as antecedents and supersets of Y as consequents provided additional items in antecedents and consequents come from $Z \backslash (X \cup Y)$. In fact, association rules belonging to $TC(X \xrightarrow{Z} Y)$ could be created from $X \to Y$ by extending its antecedent and consequent in the following way. Let V be a set of items from Z that occur neither in antecedent of $X \xrightarrow{Z} Y$ nor in its consequent; that is $V = Z \backslash (X \cup Y)$. A new rule to be included in $TC(X \xrightarrow{Z} Y)$ can be obtained from rule $X \to Y$ by performing one of the following three operations on each item v from V: (1) add v to the antecedent of $X \to Y$, (2) add v to the consequent of $X \to Y$, (3) do not use v when building a new rule. Thus, there are $3^{|V|} = 3^{|Z \backslash (X \cup Y)|}$ possible ways of building distinct association rules based on given rule template $X \xrightarrow{Z} Y$. □

Example 4. Let $\{ab\} \xrightarrow{\{abcdefghijk\}} \{e\}$ be a rule template and $m = |\{abcdefghijk\} \backslash (\{ab\} \cup \{e\})| = 11\text{-}3 = 8$. So,

$|\ TC(\{ab\} \xrightarrow{\{abcdefghijk\}} \{e\})\ | = 3^8 = 6\,591$. In consequence, if $\{ab\} \xrightarrow{\{abcdefghijk\}} \{e\}$ is ($\mathcal{M}$, E)-strong, then 6 591 association rules are ($\mathcal{M}$, E)-strong, too. □

Theorem 2. Let $X \xrightarrow{Z} Y$ and $V \xrightarrow{U} W$ be distinct rule templates such that $V \supseteq X$, $W \supseteq Y$, and $U \subseteq Z$. Then, $TC(X \xrightarrow{Z} Y) \supset TC(V \xrightarrow{U} W)$.

Proof. Let $X \xrightarrow{Z} Y$ and $V \xrightarrow{U} W$ be distinct rule templates such that $V \supseteq X$, $W \supseteq Y$ and $U \subseteq Z$. $TC(X \xrightarrow{Z} Y) = \{(X' \rightarrow Y') \in AR |\ X' \supseteq X, Y' \supseteq Y, (X' \cup Y') \subseteq Z\} \supseteq \{(X' \rightarrow Y') \in AR |\ X' \supseteq V \supseteq X, Y' \supseteq W \supseteq Y, (X' \cup Y') \subseteq U \subseteq Z\} = TC(V \xrightarrow{U} W)$. Hence, we proved that $TC(X \xrightarrow{Z} Y) \supseteq TC(V \xrightarrow{U} W)$ (*). However, $X \xrightarrow{Z} Y$ and $V \xrightarrow{U} W$ are distinct rule templates, so $V \supset X$ or $W \supset Y$ or $U \supset Z$.

Case $V \supset X$: Then, association rule $X \rightarrow Y$ will belong to $TC(X \xrightarrow{Z} Y)$, but will not belong to $TC(V \xrightarrow{U} W)$ (since no rule in $TC(V \xrightarrow{U} W)$ will have antecedent X). Hence and by (*), $TC(X \xrightarrow{Z} Y) \supset TC(V \xrightarrow{U} W)$.

Case $W \supset Y$: Then, association rule $X \rightarrow Y$ will belong to $TC(X \xrightarrow{Z} Y)$, but will not belong to $TC(V \xrightarrow{U} W)$ (since no rule in $TC(V \xrightarrow{U} W)$ will have consequent Y). Hence and by (*), $TC(X \xrightarrow{Z} Y) \supset TC(V \xrightarrow{U} W)$.

Case $U \subset Z$: Then, association rule $X \rightarrow Z \backslash X$ will belong to $TC(X \xrightarrow{Z} Y)$, but will not belong to $TC(V \xrightarrow{U} W)$ (since no rule in $TC(V \xrightarrow{U} W)$ will have base Z). Hence and by (*), $TC(X \xrightarrow{Z} Y) \supset TC(V \xrightarrow{U} W)$. □

4.2 Representative Rule Templates as a Lossless Representation of Association Rules

We denote ($\mathcal{M}$, E)-*representative rule templates* by $RRT_{(\mathcal{M},E)}$ and define as follows:

$$RRT_{(\mathcal{M},E)} = \{(X \xrightarrow{Z} Y) \in RT_{(\mathcal{M},E)} | \neg\exists (V \xrightarrow{U} W) \in RT_{(\mathcal{M},E)} \text{ such that } (X \xrightarrow{Z} Y) \neq (V \xrightarrow{U} W), V \subseteq X, W \subseteq Y, U \supseteq Z\}.$$

In Theorem 3, we claim that ($\mathcal{M}$, E)-representative rule templates cover all ($\mathcal{M}$, E)-strong association rules.

Theorem 3. $\bigcup_{(X \xrightarrow{Z} Y) \in RRT(\mathcal{M},E)} TC(X \xrightarrow{Z} Y) = AR_{(\mathcal{M},E)}$.

Proof. We will prove Theorem 3 in two steps by showing that: (1) the union of covers of all ($\mathcal{M}$, E)-strong rule templates equals $AR_{(\mathcal{M},E)}$; (2) the union of covers of all ($\mathcal{M}$, E)-representative rule templates equals the union of covers of all ($\mathcal{M}$, E)-strong rule templates, and by this, equals $AR_{(\mathcal{M},E)}$.

(1) We note that for each ($\mathcal{M}$, E)-strong association rule, say $X' \rightarrow Y'$, there is an ($\mathcal{M}$, E)-strong rule template; namely, $X' \overset{X' \cup Y'}{\rightarrow} Y'$ that covers $X' \rightarrow Y'$. On the other hand, for each ($\mathcal{M}$, E)-strong rule template $X \overset{Z}{\rightarrow} Y$, its cover $TC(X \overset{Z}{\rightarrow} Y) \subseteq AR_{(\mathcal{M},E)}$ (by Proposition 6). Thus, $\bigcup_{(X \overset{Z}{\rightarrow} Y) \in RT(\mathcal{M},E)} TC(X \overset{Z}{\rightarrow} Y) = AR_{(\mathcal{M},E)}$.

(2) Let $RRT_{(\mathcal{M},E)}$ do not contain an ($\mathcal{M}$, E)-strong rule template $X \overset{Z}{\rightarrow} Y$. By definition of $RRT_{(\mathcal{M},E)}$, this implies that there is a non-empty set of rule templates $S = \{(V \overset{U}{\rightarrow} W) \in RT_{(\mathcal{M},E)} | (V \overset{U}{\rightarrow} W) \neq (X \overset{Z}{\rightarrow} Y), V \subseteq X, W \subseteq Y, U \supseteq Z\}$. By Theorem 2, the cover of each rule template in S is a proper superset of $TC(X \overset{Z}{\rightarrow} Y)$. Let $U_{\max}$ be the base of a rule template in S with the maximum number of items. Let S' contain all rule templates with base $U_{\max}$. Let $V_{\min}$ be the antecedent of a rule template in S' with the minimum number of items. Let S'' contain all rule templates in S' with antecedent $V_{\min}$. Let $W_{\min}$ be the consequent of a rule template in S'' with the minimum number of items. Let S''' contain all rule templates in S'' with consequent $W_{\min}$. By construction of S''', each rule template in S''' is representative and its cover contains $TC(X \overset{Z}{\rightarrow} Y)$. In this way, for each non-representative rule template $(X \overset{Z}{\rightarrow} Y) \in RT_{(\mathcal{M},E)}$, we will be able to identify at least one representative rule template $V \overset{U}{\rightarrow} W$ such that $TC(V \overset{U}{\rightarrow} W) \supset TC(X \overset{Z}{\rightarrow} Y)$. So, $\bigcup_{(X \overset{Z}{\rightarrow} Y) \in RRT(\mathcal{M},E)} TC(X \overset{Z}{\rightarrow} Y) = \bigcup_{(X \overset{Z}{\rightarrow} Y) \in RT(\mathcal{M},E)} TC(X \overset{Z}{\rightarrow} Y) = AR_{(\mathcal{M},E)}$. □

Theorem 4. Let $X' \rightarrow Y'$ be an association rule. There is an ($\mathcal{M}$, E)-representative rule template covering $X' \rightarrow Y'$ if an only if $X' \rightarrow Y'$ is an ($\mathcal{M}$, E)-strong association rule.

Proof. Let $X' \rightarrow Y'$ be an association rule.

($\Rightarrow$) Follows from Proposition 6.

($\Leftarrow$) By Theorem 3, each ($\mathcal{M}$, E)-strong association rule is covered by at least one ($\mathcal{M}$, E)-representative rule template. □

5 Related Work on Lossless Representations of Association Rules

A number of concise lossless representations of strong association rules with respect to support and confidence have been proposed in the literature (see e.g. [7, 8] for an overview). Such representations typically consist of rules built from particular itemsets called *generators*; that is, itemsets the supports of which are less than the supports of all their proper subsets, and/or *closed itemsets*; that is, itemsets the supports of which are greater than the supports of all their proper supersets [2, 4, 6–8, 19, 20]. For example, in the case of representative association rules [6–8] and minimal non-redundant association rules [2], antecedents of rules are generators, whereas their consequents are set theoretical differences between closed itemsets and rules' antecedents. The set of representative rules is a subset of the set of minimal non-redundant association rules

[7, 8] and typically is by at least on order of magnitude less numerous. On the other hand, representative rules allow determination of pessimistic estimations of supports and confidences of strong association rules, while minimal non-redundant association rules allow exact determination of values of these measures. Yet, these and other developed representations of strong association rules with respect to support and confidence are not flexible enough to represent rules that are strong with respect to arbitrary ACBC/canonical measures.

In [11], we proposed the first representation of strong association rules with respect to any set of ACBC-measures. It consists of rule templates, where each rule template is a pair composed from a *lower rule* built from generators and an *upper rule* built from closed itemsets. For a rule template $(X \rightarrow Y, Z \rightarrow V)$, the following holds: $P(X) = P(Z)$, $P(Y) = P(V)$, $P(XY) = P(ZV)$. For each strong association rule $U \rightarrow W$, there is a rule template $(X \rightarrow Y, Z \rightarrow V)$ such that $X \subseteq U \subseteq Z$, $Y \subseteq W \subseteq V$, which implies that: $P(X) = P(U) = P(Z)$, $P(Y) = P(W) = P(V)$, $P(XY) = P(UW) = P(ZV)$ and that for any ACBC-measure μ: $\mu(X \rightarrow Y) = \mu(U \rightarrow W) = \mu(Z \rightarrow V)$.

In the current paper, we proposed a rule representation which does not require rule templates in the form of pairs of two rules. In addition, a representative rule templates we offered here, may cover strong association rules whose values of canonical measures may be the same or greater than values of respective canonical measures of covering rule templates. This should result in higher conciseness of the new representation. Beneath we formulate additional properties of representative rule templates (lack of space does not allow us to provide their proves).

Proposition 7.

(a) Antecedents and consequents of representative rule templates are generators.
(b) Bases of representative rule templates are closed itemsets.
(c) The set of representative rule templates is never more numerous than the set of rule templates from [11].

6 Summary

In this paper, we offered the notion of a canonical measure and proved that a number of example measures definable (at most) in terms of the probabilities of rule antecedents, consequents and bases are, in fact, canonical. Then we proposed representative rule templates as a representation of association rules satisfying multiple evaluation criteria expressible in terms of canonical measures. We proved that this representation allows deriving all and only association rules strong with respect to a given set of criteria based on canonical measures. We proved a number of properties of rule templates. In particular, we derived the formula determining the number of association rules covered by a rule template. The found formula suggests that representative rule templates can be a very concise representation of strong association rules. By means of an example, we showed that a single rule template may represent thousands of not less strong association rules.

References

1. Agrawal, R., Imielinski, T., Swami, A.N.: Mining association rules between sets of items in large databases. In: ACM SIGMOD International Conference on Management of Data, pp. 207–216 (1993)
2. Bastide, Y., Pasquier, N., Taouil, R., Stumme, G., Lakhal, L.: Mining minimal non-redundant association rules using frequent closed itemsets. In: Lloyd, J., et al. (eds.) CL 2000. LNCS (LNAI), vol. 1861, pp. 972–986. Springer, Heidelberg (2000). https://doi.org/10.1007/3-540-44957-4_65
3. Brin, S., Motwani, R., Ullman, J.D., Tsur, S.: Dynamic itemset counting and implication rules for market basket data. In: ACM SIGMOD 1997 International Conference on Management of Data, pp. 255–264 (1997)
4. Hamrouni, T., Yahia, S.B., Nguifo, E.M.: Succinct minimal generators: theoretical foundations and applications. Int. J. Found. Comput. Sci. **19**(2), 271–296 (2008)
5. Hilderman, R.J., Hamilton, H.J.: Evaluation of interestingness measures for ranking discovered knowledge. In: Cheung, D., Williams, G.J., Li, Q. (eds.) PAKDD 2001. LNCS (LNAI), vol. 2035, pp. 247–259. Springer, Heidelberg (2001). https://doi.org/10.1007/3-540-45357-1_28
6. Kryszkiewicz, M.: Closed set based discovery of representative association rules. In: Hoffmann, F., Hand, D.J., Adams, N., Fisher, D., Guimaraes, G. (eds.) IDA 2001. LNCS, vol. 2189, pp. 350–359. Springer, Heidelberg (2001). https://doi.org/10.1007/3-540-44816-0_35
7. Kryszkiewicz, M.: Concise representations of frequent patterns and association rules. Prace Naukowe Politechniki Warszawskiej. Elektronika **142**, 5–207 (2002)
8. Kryszkiewicz, M.: Concise representations of association rules. In: Hand, D.J., Adams, N.M., Bolton, R.J. (eds.) Pattern Detection and Discovery. LNCS (LNAI), vol. 2447, pp. 92–109. Springer, Heidelberg (2002). https://doi.org/10.1007/3-540-45728-3_8
9. Kryszkiewicz, M.: Dependence factor for association rules. In: Nguyen, N.T., Trawiński, B., Kosala, R. (eds.) ACIIDS 2015. LNCS (LNAI), vol. 9012, pp. 135–145. Springer, Cham (2015). https://doi.org/10.1007/978-3-319-15705-4_14
10. Kryszkiewicz, M.: Dependence factor as a rule evaluation measure. In: Matwin, S., Mielniczuk, J. (eds.) Challenges in Computational Statistics and Data Mining. SCI, vol. 605, pp. 205–223. Springer, Cham (2016). https://doi.org/10.1007/978-3-319-18781-5_12
11. Kryszkiewicz, M.: A lossless representation for association rules satisfying multiple evaluation criteria. In: Nguyen, N.T., Trawiński, B., Fujita, H., Hong, T.-P. (eds.) ACIIDS 2016. LNCS (LNAI), vol. 9622, pp. 147–158. Springer, Heidelberg (2016). https://doi.org/10.1007/978-3-662-49390-8_14
12. Kryszkiewicz, M.: ACBC-adequate association and decision rules versus key generators and rough sets approximations. Fundam. Inform. **148**(1–2), 65–85 (2016)
13. Lavrač, N., Flach, P., Zupan, B.: Rule evaluation measures: a unifying view. In: Džeroski, S., Flach, P. (eds.) ILP 1999. LNCS (LNAI), vol. 1634, pp. 174–185. Springer, Heidelberg (1999). https://doi.org/10.1007/3-540-48751-4_17
14. Lenca, P., Meyer, P., Vaillant, B., Lallich, S.: On selecting interestingness measures for association rules: user oriented description and multiple criteria decision aid. Eur. J. Oper. Res. **184**, 610–626 (2008). Elsevier
15. Piatetsky-Shapiro, G.: Discovery, analysis, and presentation of strong rules. In: Knowledge Discovery in Databases, pp. 229–248, AAAI/MIT Press (1991)
16. Sheikh, L.M., Tanveer, B., Hamdani, S.M.A.: Interesting measures for mining association rules. In: Proceedings of INMIC 2004. IEEE (2004)

17. Shortliffe, E., Buchanan, B.: A model of inexact reasoning in medicine. Math. Biosci. **23**, 351–379 (1975)
18. Suzuki, E.: Interestingness measures - limits, desiderata, and recent results. In: Lenca, P., Lallich, S. (eds.) QIMIE/PAKDD 2009 (2009)
19. Stumme, G., Taouil, R., Bastide, Y., Pasquier, N., Lakhal, L.: Intelligent structuring and reducing of association rules with formal concept analysis. In: Baader, F., Brewka, G., Eiter, T. (eds.) KI 2001. LNCS (LNAI), vol. 2174, pp. 335–350. Springer, Heidelberg (2001). https://doi.org/10.1007/3-540-45422-5_24
20. Zaki, M.J.: Generating non-redundant association rules. In: 6th ACM SIGKDD (2000)

Multiple Model Approach to Machine Learning

Depth Learning with Convolutional Neural Network for Leaves Classifier Based on Shape of Leaf Vein

Tan Kiet Nguyen Thanh[1], Quoc Bao Truong[2(✉)], Quoc Dinh Truong[1], and Hiep Huynh Xuan[1]

[1] College of Information and Communication Technology, Can Tho University, Can Tho, Vietnam

[2] College of Engineering Technology, Can Tho University, Can Tho, Vietnam

tqbao@ctu.edu.vn

Abstract. Nowadays, Depth Learning with Convolution Neural Networks (CNN) has become a well known method in object recognition task to generate descriptions and classify learned features and gradually took over. There are few CNN applied studies on leaves recognition and classifier task that mainly use the existing CNN architecture and pre-trained models. This paper proposes a CNN model for leaves classifier based on thresholding leaf pre-processing extract vein shape data and augmenting training data with reflection and rotation of the image. Preprocessing to reduce the storage capacity of data and increases the computational efficiency of the model. This model was experimented on collector leaves data set in the Mekong Delta of Vietnam, the Flavia leaf data set and the Swedish leaf data set. The classification results indicate that the proposed CNN model is effective for leaf recognition with an accuracy greater than 95%.

Keywords: Depth learning · Convolution neural network
Image segmentation · Leaf classification

1 Introduction

Image classification task is usually based on features engineering such as SIFT, HOG, SURF,… combined with a learning algorithm in these features engineering spaces such as SVM, Neuron, KNN… This leads to the efficiency of all approaches that depend heavily on predefined features. Image features engineering itself is a complex field, needed to be changed and revisited at hand for each problem or data set involved.

Today, with the development of neural networks, neural network architecture has been used as an effective solution to extract high level features from data. Deep Convolutional Neural Networks architectures can accurately portray highly abstract properties with condensed data, while preserving the most up-to-date characteristics of raw data. This is beneficial for classification or prediction. In recent times, CNN has emerged as an effective framework for describing features and identities in image processing. CNN can learn basic filters automatically and combine them hierarchically to describe underlying concepts to identify patterns. CNN does not need computation

N. T. Nguyen et al. (Eds.): ACIIDS 2018, LNAI 10751, pp. 565–575, 2018.
https://doi.org/10.1007/978-3-319-75417-8_53

features engineering. It takes time and effort. The generalization of the method makes it a practical and scalable approach to the various application problems of classification and recognition.

In 2000, Oide and Ninomiya [1] used neural networks to classify soybean leaves using a Hopfield network and a simple perceptron. In 2001, Soderkvist [2] used leaf morphology to train back propagation neurons to classify 15 plants in Sweden. This data set used in the experiment and then become a standard data set - Swish data sets. Many later experiments used this data set. There are studies applying deep architecture to image recognition in advance. Krizhevsky et al. [3] has used Deep Convolutional Neural Networks for ImageNet and their research results have created a new rush for depth learning.

Several publications have suggested the use of CNN in leaf classification in recent years. Jassmann et al. [4] develops an application for classifying plants, based on leaf images. The system uses a CNN in a mobile application for mobile phones to categorize the nature of the leaf, trained with the ImageCLEF data set. The proposed architecture consists of a convoluted layer, followed by a composite layer and two fully connected layers applied to the 60×80 pixel input image. Wu et al. proposed a simplified version of AlexNet for leaf recognition [5]. They have used parametric linear units (PReLU) instead of ReLU. In [6] He et al. proposed a one-to-one connection, named Single Connector (SCL), added to the proposed CNN architecture to create some improvements. This method has been tested on the ICL leaf database and the results reflect an increase in accuracy.

Plant disease identification includes the processing of leaf recognition. Sladojevic et al. [7], have been interested in a new method for developing a disease-identification model based on leaf classification of images, using CNN. The developmental model was able to recognize 13 healthy plant leaf diseases, with the ability to discriminate leaves from the surrounding environment.

In this study, we approached leaf-based visual recognition based on the vein's morphology using the depth learning model. Specifically, Convolutional Neural Network (CNN) is one of the advanced depth learning models that helps us build intelligent systems with high precision as present. Our model implemented is depicted in Fig. 1, the initial image is well-adapted to reduce the amount of unnecessary color information and clarify the vein characteristics then using depth learning neural network for classification.

The rest of the paper is organized as follows. In Sect. 2: neural networks and CNN. In Sect. 3: CNN models for leaves recognition. In Sect. 4: experiments. In Sect. 5: conclusions.

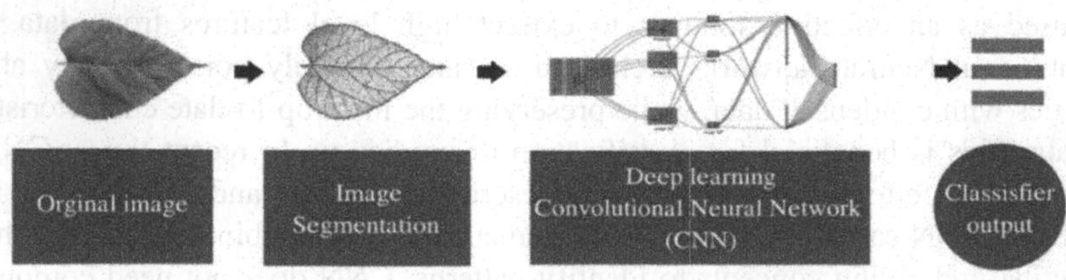

Fig. 1. Our scheme implementation

2 Convolution Neural Network (CNN)

2.1 Artificial Neural Networks

Neural networks are inspired by biological neural systems. The basic computational unit of the brain is a neuron and they are connected with synapses.

In the neural network computational model, the signals that travel along the axons (e.g., ×0) interact multiplicatively (e.g., w0 × 0) with the dendrites of the other neuron based on the synaptic strength at that synapse (e.g., w0). Synaptic weights are learnable and control the influence of one neuron or another. The dendrites carry the signal to the cell body, where they all are summed. If the final sum is above a specified threshold, the neuron fires, sending a spike along its axon. In the computational model, it is assumed that the precise timings of the firing do not matter and only the frequency of the firing communicates information. Based on the rate code interpretation, the firing rate of the neuron is modeled with an activation function f that represents the frequency of the spikes along the axon. A common choice of activation function is sigmoid. In summary, each neuron calculates the dot product of inputs and weights, adds the bias, and applies non-linearity as a trigger function (for example, following a sigmoid response function).

A CNN is a special case of the neural network described above. A CNN consists of one or more convolutional layers, often with a subsampling layer, which are followed by one or more fully connected layers as in a standard neural network.

A neural network is a system of interconnected artificial "neurons" that exchange messages between each other. The connections have numeric weights that are tuned during the training process, so that a properly trained network will respond correctly when presented with an image or pattern to recognize. The network consists of multiple layers of feature-detecting "neurons". Each layer has many neurons that respond to different combinations of inputs from the previous layers. As shown in Fig. 2, the layers are built up so that the first layer detects a set of primitive patterns in the input, the second layer detects patterns of patterns, the third layer detects patterns of those patterns, and so on. Typical CNNs use 5 to 25 distinct layers of pattern recognition.

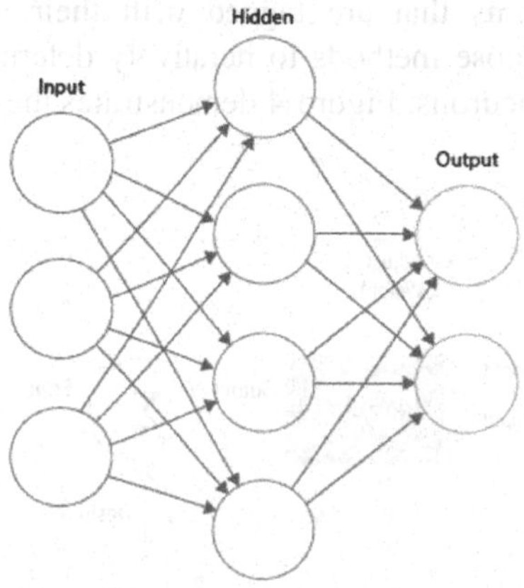

Fig. 2. An artificial neural network [8]

2.2 Convolution Neural Network Structure

The convolution neural network has these components:

- *Convolution layer,* the convolution operation extracts different features of the input. The first convolution layer extracts low-level features like edges, lines, and corners. Higher-level layers extract higher-level features.
- *Non-linear layers,* Neural networks in general and CNNs in particular rely on a non-linear "trigger" function to signal distinct recognition of likely features on each hidden layer. CNNs may use a variety of specific functions such as rectified linear units (ReLUs) and continuous trigger (non-linear) functions to efficiently implement this non-linear triggering. A ReLU implements the function y = max(x,0), so the input and output sizes of this layer are the same. ReLU functionality is illustrated in Fig. 3.

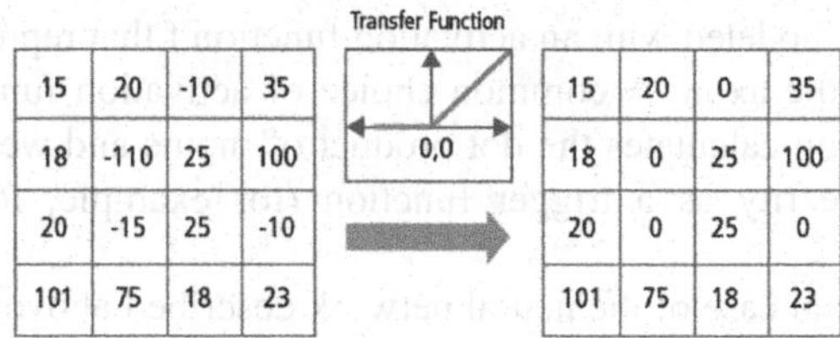

Fig. 3. Pictorial representation of ReLU functionality

- The pooling/subsampling layer reduces the resolution of the features. It makes the features robust against noise and distortion.
- Fully connected layers are often used as the final layers of a CNN. These layers mathematically sum a weighting of the previous layer of features, indicating the precise mix of "ingredients" to determine a specific target output result. In case of a fully connected layer, all the elements of all the features of the previous layer get used in the calculation of each element of each output feature.

Training is performed using a "labeled" data set of inputs in a wide assortment of representative input patterns that are tagged with their intended output response. Training uses general-purpose methods to iteratively determine the weights for intermediate and final feature neurons. Figure 4 demonstrates the training process at a block level.

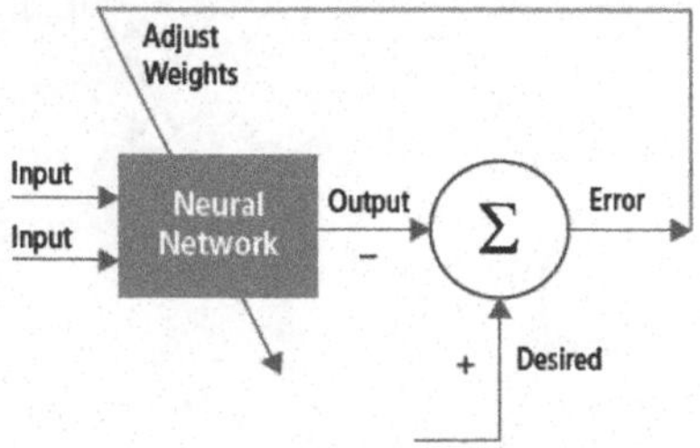

Fig. 4. Training of neural networks [9]

3 Develop CNN Models for Leaves Recognition

3.1 Extract Leaf Vein Shape

The input data of the model is the image, which normally has three RGB channels. However, in order to reduce the storage capacity, extract vein morphology for network training, we propose to extract vein morphology to the remaining one channel input, which will reduce the number of the weight value (associated with the two remaining channels) that the network must learn.

To perform extract leaf vein shape, the image segmentation process involves converting the image to grayscale, and then using adaptive thresholding techniques to segment the image and extract the vein leaf image.

There are many image processing techniques used to segment the image, many researched extracted vein morphology from images obtained by the camera and using Gabor filters, Colony filters, thresholds, independent component analysis,…

We use adaptive local threshold algorithm that decouples object from background with heterogeneous illumination produces a binary image with adjacent thresholds is the *mean (ws) - C*, *ws* is the neighborhood size, *C* is the constant, in this study we use *ws = 10* and *C = 0.2*, these two results in low noise picture most from images in our experiments. Figure 5, presentation of a result illustrating adaptive local threshold.

Fig. 5. Illustrated image with adaptive local threshold to mean (10), C = 0.2

3.2 The CNN Model Classifies Leaves

In the Fig. 6, L1, L2 model, each phase include three transformations. First, the convolution between the input image and n filters (we set n = 100 at L1, n = 250 at L2) is 5 × 5 size. Each filter has a limit size related to a receiving field (5 × 5) in the input image. For each convolution filter generates a feature map. The second transformation is a non-linear function used for all feature maps. We use the ReLU function. Finally, there is a subsampling transformation. In this transformation, each map is divided into a

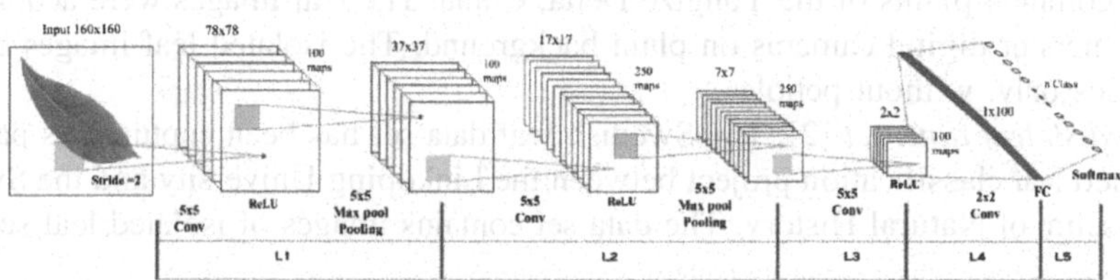

Fig. 6. CNN model for leaves recognition

set of non-overlapping 5×5 square fields, from each point in that field, which retains only the maximum value (or Max Pool).

The CNN model shown in Fig. 6 have layers that include:

[Conv1 - ReLu - Max pool] $\rightarrow$ [Conv2 - ReLu - Max pool] $\rightarrow$ [Conv3 - ReLu] $\rightarrow$ [Conv4 $\rightarrow$ FC] $\rightarrow$ Softmax.

L3, L4 are two convolutional layers to create a fully-enclosed class of filter sizes of 1×1 according to Matconvnet's convention [11]. Finally in the network is a softmax function. It returns the estimated probability of each class, for a particular sample. This layer is fully connected to all the output feature maps of the final convolution layer. A summary of the network parameters is detailed in Table 1.

Table 1. Summary of network parameters

	Input							Output				Parameters		
Name	Width	Height	Depth	nfilter	Filtersize	Stride	Pad	Width	Height	Depth	Size	Weights	Biases	Total
Input								160	160	1	25.600			
Conv1	160	160	1	100	5	2	0					2.5		2.6
pool1	78	78	100		5	2	0	37	37	100	136.900			
Conv2	37	37	100	250	5	2	0	17	17	250	72.250	625	250	625.25
pool2	17	17	250		5	2	0	7	7	250	12.250			
conv3	7	7	250	100	5	2	0	2	2	100	400	625	100	625.1
conv4	2	2	100	52	2	1	0	1	1	52	52	20.8	52	20.852
Softmax	1	1	52					1	1	52	52			
													Total 1.273.802	

The training network updates the network of parameters model. The parameters were optimized using Stochastic gradient descent (SGD) that in contrast performs a parameter update for each training.

4 Experiment and Results

4.1 Experiment Data Sets

In order to test the performance of the vein morphology classification system, we selected three standard sets:

- *Flavia data set* [12]: This data set contains 1907 leaf images of 32 different species and 50–77 images per species. Those leaves were sampled on the campus of the Nanjing University and the Sun Yat-Sen arboretum, Nanking, China. Most of them are common plants of the Yangtze Delta, China. The leaf images were acquired by scanners or digital cameras on plain background. The isolated leaf images contain blades only, without petioles.
- *Swedish leaf data set* [2]: The Swedish leaf data set has been captured as part of a joined leaf classification project between the Linkoping University and the Swedish Museum of Natural History. The data set contains images of isolated leaf scans on

plain background of 15 Swedish tree species, with 75 leaves per species (1125 images in total). This data set is considered very challenging due to its high inter-species similarity.

- *Mekong leaf data set*: this data set was collected by us during the study, in the provinces of the Mekong Delta in Vietnam. Images were taken with a leaf camera in the field on a white background with natural light, and include 52 common trees such as fruit trees, wood trees, medicinal plants, ornamental plants… each with 47–110 images, total 3,921 images.

4.2 Augmentation Data

To limit over-model phenomena of the model due to insufficient data. Augmentation data is an effective solution. We increase number of training images by creating three copies of each image after reflection and rotation.

Thus, each original image creates three image augmentation. Data partitioned for experiment is shown in Table 2.

Table 2. Partition data

Data set	Number layers	Number samples	Train		Val	Test
			60%	Augmentation	20%	20%
Flavia	32	1.907	1.145	4.580	381	381
Swedish	15	1.125	900	3.600	225	225
Mekong	52	3.921	2.353	9.412	784	784

In this study, the input data of model is scanned or taken from leaves of the trees, data is classified into three categories: RGB, graycale and threshold images extract veins morphology, then model the training on these volumes to compare the results. The experimental process is as follows:

- *Data preprocessing:* each data set is divided into three categories: RGB, Graycale, Subdivision.
- *Standardized image size:* After preprocessing, the image resizes to 160 × 160px to match the input of the network.
- *Image Partition:* Each categories is partitioned as shown in Table 2 for experimental purposes.
- *Initialization parameter:*
  - Learning rate: set to 0.0001 (greater than fast convergence network but not very good error rate, smaller than slow convergence network).
  - Weight Decay constant (anti-overfitting) = 0.0005;
  - Momentum constant = 0.9.

The above constants are chosen based on experimental results for proposed model by trial and error Method. Training time depends on computer resources with GPU or CPU, matlab software and Matconvnet tool.

4.3 Experiment Results

The experiment results are aggregated into each test data set and detail in Tables 3, 4 and 5.

Table 3. Experiment results of models on the Swedish data set

Swedish (15 layers)		Model error/Accuracy test set		
Rate Train/Val/Test	Augmentation	RGB	Gray	Subdivision threshold
60-20-20	No	4.89/94.22	5.33/94.66	4.00/96.00
60-20-20	Yes	1.33/95.11	8.00/92.44	2.67/96.44
5 fold	No	3.89/94.93	6.67/93.2	4.51/94.93
5 fold	Yes	2.26/95.33	7.79/93.46	3.89/95.6

Table 4. Experiment results of models on the Flavia data set

Flavia (32 layers)		Model error/Accuracy test set		
Rate Train/Val/Test	Augmentation	RGB	Gray	Subdivision threshold
60-20-20	No	**10.57/91.05**	10.79/88.13	**11.05/86.36**
60-20-20	Yes	**6.44/90.76**	**7.63/90.59**	**7.63/95.11**
5 fold	No	10.78/71.54	12.98/68.78	9.91/72.1
5 fold	Yes	7.84/75.59	10.22/72.30	8.53/77.85

Table 5. Experiment results of models on the Mekong data set

Mekong (52 layers)		Model error/Accuracy test set		
Rate Train/Val/Test	Augmentation	RGB	Gray	Subdivision threshold
60-20-20	No	14.08/87.26	20.54/75.75	10.45/82.47
60-20-20	Yes	**7.75/92.56**	11.93/82.95	**7.27/92.07**
5 fold	No	16.88/77.73	23.56/71.05	13.05/82.49
5 fold	Yes	10.16/86.21	16.17/79.91	8.73/88.42

From the experiment results in Tables 3, 4 and 5, we have following comments:

- The model error rate and test test on RGB data gives better results. However, the storage and training time is longer. On data thresholds that store little data and fast training networks, results equivalent (or less negligible). Augmentation data improves network efficiency, both in terms of model error rate and recognition rate.
- Random partition of data set as 60% train, 20% val, 20% test of the network give results equivalent compare to 5 fold partition on the Swedish data set. For the Flavia and Mekong data set, 5 fold partition has lower performance network. This happens probably due to data sets have unequal data in each layer.

Comparisons of the results were compiled in [13] with the method of leaves recognition on the Swedish and Flavia data sets in the following Tables 6 and 7. Here, the accuracy is **(Number of correct identities/Total leaf of test set) × 100%**.

Table 6. Comparison of results of leaves recognition methods on the Swedish data set

Descriptor	Feature	Classifier	Accuracy
Our method	Shape + vein	CNN	**96.44**
IDSC	Shape	SVM	93.73
IDSC	Shape	k-NN	94.13
TOA	Shape	k-NN	**95.20**
TSL	Shape	k-NN	**95.73**
TSLA	Shape	k-NN	**96.53**
LBP	Shape	SVM	**96.67**
I-IDSC	Shape	1-NN	**97.07**
MARCH	Shape	1-NN	**97.33**
DS-LBP	Shape + texture	Fuzzy k-NN	**99.25**

Table 7. Statistical results of leaves recognition methods on the Flavia data set

Descriptor	Feature	Classifier	Accuracy
Our method	Shape + vein		**95.11**
HOG	Shape		84.70
SIFT	Shape		87.50
SMSD, $Avein/Aleaf$	Shape + vein	PNN	**90.31**
SMSD, FD, CM	Color + shape	k-NN, DT	**91.30**
SMSD	Shape	PNN	91.40
SMSD, CM	Shape + color	RF (k-NN, NB, SVM)	**93.95**
SMSD, $Avein/Aleaf$	Shape + vein	SVM (k-NN)	**94.50**
SIFT	Shape	SVM	**95.47**
SURF	Shape	SVM	**95.94**
SMSD, FD	Shape	BPNN	**96.00**
SMSD, CM, GLCM, A_{vein}/A_{leaf}	Shape + color + texture + vein	SVM	**96.25**

From experiment results with Swedish and Flavia data sets, we can confirm that the CNN-based neural network depth model, which we propose, works very well on classification problem of leaves based on the shape of veins (vein morphology). This result once again confirms the effectiveness and simplicity of the CNN depth geometry model for real-world problems with large data. The recognition process is done by simply building the model and determining the appropriate parameters. The effectiveness of the classification process, recognition is no longer too dependent on finding and identifying image features, a process that takes a lot of time and effort.

5 Conclusion

In this study a novel CNN architecture was proposed for leaf classification task. The model is based on 160 × 160 adaptive threshold images input that obtained from Swedish, Flavia, Mekong leaf dataset. The effect of horizontal reflection and rotation augmentation of data sets is also to further improve the results. The results showed that the proposed architecture for CNN-based leaf classification closely competes with the latest extensive approaches on devising leaf features and classifiers.

In future research, we improve the back propagation algorithm by combining other methods such as genetic algorithms, fuzzy logic, extensive study of network architectures and training algorithms, …

Change the ReLU transfer function with a more flexible function for each class such as ELU (Exponential Linear Unit)… for better training results. The problem of normalizing the size of the subject needs further research to increase its effectiveness on the large and small object images.

References

1. Oide, M., Ninomiya, S.: Discrimination of soybean leaflet shape by neural networks with image input. Comput. Electorn. Agric. **29**(1–2), 59–72 (2000)
2. Soderkvist, O.J.O.: Computer Vision Classification of Leaves from Swedish Trees. Linkoping University, Linkoping (2001)
3. Krizhevsky, A., Sutskever, I., Hinton, G.E.: Imagenet classification with depth convolutional neural networks. In: Advances in Neural Information Processing Systems (2012)
4. Jassmann, T.J., Tashakkori, R., Parry, R.M.: Leaf classification utilizing a convolutional neural network. In: SoutheastCon 2015, pp. 1–3. IEEE (2015)
5. Wu, Y.-H., Shang, L., Huang, Z.-K., Wang, G., Zhang, X.-P.: Convolutional neural network application on leaf classification. In: Huang, D.-S., Bevilacqua, V., Premaratne, P. (eds.) ICIC 2016. LNCS, vol. 9771, pp. 12–17. Springer, Cham (2016). https://doi.org/10.1007/978-3-319-42291-6_2
6. He, X., Wang, G., Zhang, X.-P., Shang, L., Huang, Z.-K.: Leaf classification utilizing a convolutional neural network with a structure of single connected layer. In: Huang, D.-S., Jo, K.-H. (eds.) ICIC 2016. LNCS, vol. 9772, pp. 332–340. Springer, Cham (2016). https://doi.org/10.1007/978-3-319-42294-7_29
7. Sladojevic, S., Arsenovic, M., Anderla, A., Culibrk, D., Stefanovic, D.: Deep neural networks based recognition of plant diseases by leaf image classification. Comput. Intell. Neurosci. 11 (2016)
8. Artificial Neural Network. Wikipedia. https://en.wikipedia.org/wiki/Artificial_neural_network
9. Hijazi, S., Kumar, R., Rowen, C.: Using convolutional neural networks for image recognition. IP Group, Cadence (2015). https://ip.cadence.com/uploads/901/cnn_wp-pdf
10. http://homepages.inf.ed.ac.uk/rbf/HIPR2/adpthrsh.htm

11. Vedaldi, A., Lenc, K.: Matconvnet: convolutional neural networks for matlab. In: Proceedings of the 23rd ACM International Conference on Multimedia, pp. 689–692. ACM (2015)
12. Wu, S.G., Bao, F.S., Xu, E.Y., Wang, Y.X., Chang, Y.F., Xiang, Q.L.: A leaf recognition algorithm for plant classification using probabilistic neural network. In: 2007 IEEE International Symposium on Signal Processing and Information Technology, pp. 11–16. IEEE (2007)
13. Wäldchen, J., Mäder, P.: Plant species identification using computer vision techniques: a systematic literature review. Arch. Comput. Methods Eng. (2017). https://www.researchgate.net/publication/312147459. ISSN: 1134–3060 (Print) 1886-1784 (Online)

An Ensemble System with Random Projection and Dynamic Ensemble Selection

Manh Truong Dang[1], Anh Vu Luong[2], Tuyet-Trinh Vu[1], Quoc Viet Hung Nguyen[3], Tien Thanh Nguyen[2,3](✉), and Bela Stantic[3]

[1] School of Information and Communication Technology, Hanoi University of Science and Technology, Hanoi, Vietnam
[2] School of Applied Mathematics and Informatics, Hanoi University of Science and Technology, Hanoi, Vietnam
[3] School of Information and Communication Technology, Griffith University, Gold Coast, Australia
thanh.nguyen3@griffithuni.edu.au

Abstract. In this paper, we propose using dynamic ensemble selection (DES) method on ensemble generated based on random projection. We first construct the homogeneous ensemble in which a set of base classifier is obtained by a learning algorithm on different training schemes generated by projecting the original training set to lower dimensional down spaces. We then develop a DES method on those base classifiers so that a subset of base classifiers is selected to predict label for each test sample. Here competence of a classifier is evaluated based on its prediction results on the test sample's $k-$ nearest neighbors obtaining from the projected data of validation set. Our proposed method, therefore, gains the benefits not only from the random projection in dimensionality reduction and diverse training schemes generation but also from DES method in choosing an appropriate subset of base classifiers for each test sample. The experiments conducted on some datasets selected from four different sources indicate that our framework is better than many state-of-the-art DES methods concerning to classification accuracy.

Keywords: Ensemble method · Random projection
Multiple classifiers system · Dynamic ensemble selection

1 Introduction

In designing of an ensemble, there are three phases to be considered namely generation, selection, and combination. In the first phase, the learning algorithm(s) learn on the training set(s) to obtain base classifiers. In the second phase, a single classifier of a subset of the best classifier is selected. In the last phase, the decisions made by classifiers of the ensemble are combined to obtain the final one [1].

Homogeneous ensemble methods like Bagging [2] and Random Subspace [3] focus on the generation phase in which these methods concentrate on generating new training schemes from the original training set. In 1984, Johnson and Lindenstrauss (JL)

N. T. Nguyen et al. (Eds.): ACIIDS 2018, LNAI 10751, pp. 576–586, 2018.
https://doi.org/10.1007/978-3-319-75417-8_54

introduced an extending of Lipschitz continuous maps from metric spaces to Euclidean spaces as well as the JL Lemma [4]. The lemma begins with a linear transformation (known as a random projection) from a p-dimensional space $\mathbb{R}^p$ (called up space) to a q-dimensional space $\mathbb{R}^q$ (called down space). Due to the unstable property, random projections have used to construct the homogeneous ensemble [5].

In this study, we first employ random projection to generate the homogeneous ensemble system to solve the classification tasks. In detail, the original training set is projected to many down spaces to generate new training schemes. Due to the unstable property of random projection in which the generated training scheme is different to original training set as well as the other schemes, a learning algorithm can learn on these schemes to obtain the diverse base classifiers. We then consider the selection phase by selecting a subset of classifiers (also called ensemble of classifies or EoC) associated with some random projections to predict class label. Here we propose using a DES method [1, 6] to the random projection-based ensemble in which instead of using all base classifiers for the prediction, only a subset of them is selected to predict the class label for a specific test sample. The selection is based on the neighborhood of the test sample belonging to the validation set in the local region of the projected feature space. The merits of our work lie in the following: to the best of our knowledge, this is the first approach to dynamically select EoC associated with random projections to predict class label for each sample.

The paper is organized as follows. In Sect. 2, random projections and dynamic classifier/ensemble selection are introduced. In Sect. 3, the proposed method based on the combination of random projection and DES is proposed. Experimental results are presented in Sect. 4 in which the results of the proposed method are compared with those produced by some benchmark algorithms on 15 selected datasets. Finally, the conclusions are presented in Sect. 5.

2 Related Methods

2.1 Random Projection

Given a finite set of p-dimension data $\mathcal{D} = \{\mathbf{x}_1, \mathbf{x}_2, \ldots, \mathbf{x}_n\} \subset \mathbb{R}^p$, we consider a linear transformation $\mathrm{T} : \mathbb{R}^p \to \mathbb{R}^q : \mathbf{Z} = \mathrm{T}[\mathcal{D}] = \{\mathbf{z}_1, \mathbf{z}_2, \ldots, \mathbf{z}_n\} \subset \mathbb{R}^q$ and $\mathbf{z}_i = \mathrm{T}(\mathbf{x}_i)$. If the linear transformation T can be represented in the form of matrix $\mathbf{R}$ ($\mathbf{z}_i = \mathrm{T}(\mathbf{x}_i) = \mathbf{R}\mathbf{x}_i$) so that if each element of the matrix is generated according to a specified random distribution, T is known as a random projection. In practice, the random projection is simply obtained by using a random matrix $\mathbf{R} = 1/\sqrt{q}\{r_{ij}\}$ of size $(p \times q)$, where r_{ij} are random variables such that $\mathrm{E}(r_{ij}) = 0$ and $\mathrm{Var}(r_{ij}) = 1$. Several forms of $\mathbf{R}$ are summarized in [7] in which Plus-minus-one and Gaussian are the most popular random projections.

Random projections are useful in dimension reduction since the dimension of the down space can be chosen to be lower than that of up space, i.e., $q < p$. Comparing to Principle Component Analysis (PCA), the directions of random projection are independent of the data while those of PCA are data-dependent and generating the principle components is computationally expensive compare to generating the random matrix in random projection [8]. Furthermore, Fern and Brodley [5] indicated that random projections are very unstable since the dataset schemes generated from an original data source based on random matrices are quite different. This property is important since other sampling methods like bootstrapping only generate slightly different dataset schemes. Thus an ensemble system based on a set of random projections offers a potential for increased diversity. Until now, random projection has been extensively studied and applied to many areas, for example dimensionality reduction in analyzing noisy and noiseless images, and information retrieval in text documents [8], sparse random projection to approximate the $\mathcal{X}^2$ kernel [9], in supervised online machine learning [7, 10], and in analyzing clusters [11].

2.2 Dynamic Ensemble/Classifier Selection

In the selection phase of multiple classifier systems, a single classifier or an EoC can be obtained via static or dynamic approach. While in static approach, the selection is conducted during the training process and then the selected classifier or EoC is used to predict the label of all test sample, the dynamic approach works on the classification process by selecting a different classifier or different EoC for each test sample. We distinguish dynamic classifier selection (DCS) and DES term in which DCS techniques select only one classifier while DES techniques select an EoC for each test sample. Recent research on dynamic selection approaches shows its advantages for classification problems [12].

In dynamic selection approach, we first need to evaluate the competence of each base classifier from the pool of classifier and then select only the most competent or ensemble containing the most competent classifiers to classify each specific test sample. Here the competence is computed according to some criteria on the samples in the local region of feature space which can be defined by k-nearest neighbor techniques (in MCB [13], MLA [14], KNOP [15], META-DES [6], KNORA-Union [16], DES-FA [17]), clustering techniques [18], and potential functions (in DES-RRC [19], DES-KL [20], DES-P [20]). The selection criteria includes the accuracy of base classifiers in the local region [16], or meta-learning [6], or probabilistic-based models by considering posterior probability of the classifier on the neighbors of each test sample [19, 20].

3 Proposed Method

In this paper, we propose an ensemble system for label prediction using KNORA Union [1, 16] method and random projections. The survey in [1] shows that simple DES method like KNORA Union is competitive to many more complex methods. Meanwhile, the random project is advantageous in the homogeneous ensemble generation. In detail, in the training process, K random matrices of size $(p \times q)$ denoted by $\mathbf{R}_j$ $(j = 1,\ldots,K)$ are generated. The new K training schemes $\mathbf{Z}_j$ of size $(N \times q)$ (N is the number of training observations) and then are generated from the original training set $\mathcal{D}$ of size $(N \times p)$ though the projection $\mathcal{D} \xrightarrow{\mathbf{R}_j} \mathbf{Z}_j$ given by:

$$\mathbf{Z}_j = (\mathcal{D}\mathbf{R}_j)/\sqrt{q} \tag{1}$$

The ensemble of classifiers BC_j $(j = 1, \ldots, K)$ is constructed by a learning algorithm $\mathcal{K}$ on training schemes $\mathbf{Z}_j$. As random projection often generates significantly different training schemes from original training set [5, 7, 10], the system diversity is ensure. In DES, each test sample is predicted by selected EoC; and the EoCs for two different test sample may be different. In general, we define the credit of a base classifier on a sample.

Definition 1: The credit of a base classifier BC_j on a sample $\mathbf{x}$ denoted by $w_j(\mathbf{x})$ is the number of times the prediction of BC_j on $\mathbf{x}$ used in the combining algorithm.

We propose using KNORA Union to our system to find the credit of base classifiers on each test sample. The idea of the method is based on the prediction results of base classifiers on the neighbors of each test sample. In this study, instead of getting the neighbors from the validation set, we consider the neighbors in the projected schemes of validation set. Specifically, the validation set $\mathcal{V}$ is projected to the down spaces as:

$$\mathbf{V}_j = (\mathcal{V}\mathbf{R}_j)/\sqrt{q} \tag{2}$$

Denote $kNN_j(\mathbf{x}^u)$ as the k-nearest neighbors of an test sample $\mathbf{x}^u$ in $\mathbf{V}_j$. We select base classifiers for $\mathbf{x}^u$ based on their prediction results on $kNN_j(\mathbf{x}_u)$ as if BC_j gives a corrected prediction on each observation in $kNN_j(\mathbf{x}^u)$, its credit $w_j(\mathbf{x}^u)$ will increase by 1. Based on prediction results on all observations belonging to $kNN_j(\mathbf{x}^u)$, we obtain all $w_j(\mathbf{x}^u)\, j = 1, \ldots, K$. It is noted that there is an exception in which all base classifiers misclassify all observations in $kNN_j(\mathbf{x}^u)$ so that base classifiers contribute nothing to the combination. In this case, we simply set $w_j(\mathbf{x}^u) = 1 \forall j = 1, \ldots, K$ which means that all base classifiers contribute equally to the prediction for $\mathbf{x}^u$.

The output of the base classifiers on $\mathbf{x}^u$ are combined to obtain the predicted class label. Let $\{y_m\}_{m=1,\ldots,M}$ denotes the set of M labels, $\mathrm{P}_j(y_m|\mathbf{x}^u)$ is the probability that $\mathbf{x}^u$ belongs to the class with label y_m given by the BC_j. There are two output types for $\mathbf{x}^u$ namely *Crisp Label* (returns only class label, i.e. $\mathrm{P}_j(y_m|\mathbf{x}^u) \in \{0, 1\}$ and $\sum_m \mathrm{P}_j(y_m|\mathbf{x}^u) = 1$) and *Soft Label* (returns posterior probabilities that $\mathbf{x}^u$ belongs to a class, i.e. $\mathrm{P}_j(y_m|\mathbf{x}^u) \in [0, 1]$ and $\sum_m \mathrm{P}_j(y_m|\mathbf{x}^u) = 1$) [21–24]. In this paper, we propose using fixed combining rules [25–27] to combine the output of base classifiers. As the fixed combining rules apply directly to the output of base classifiers to give the prediction, they are simpler and fast to build and run. Several popular fixed combining methods are Sum, Product, Majority Vote, Max, Min, and Median [25, 26]. In this study because base classifiers set different credits on each test sample, the forms of fixed combining rules applied to the outputs of base classifiers are given by:

$$\text{Sum Rule: } \mathbf{x}^u \in y_t \text{ if } t = \arg\max_{m=1,\ldots,M} \sum_{j=1}^{K} w_j(\mathbf{x}^u)\mathrm{P}_j(y_m|\mathbf{x}^u) \tag{3}$$

$$\text{Product Rule: } \mathbf{x}^u \in y_t \text{ if } t = \arg\max_{m=1,\ldots,M} \prod_{k=1}^{K} w_j(\mathbf{x}^u)\mathrm{P}_j(y_m|\mathbf{x}^u) \tag{4}$$

$$\text{Max Rule: } \mathbf{x}^u \in y_t \text{ if } t = \arg\max_{m=1,\ldots,M} \max_{j=1,\ldots,K} w_j(\mathbf{x}^u)\mathrm{P}_j(y_m|\mathbf{x}^u) \tag{5}$$

$$\text{Min Rule: } \mathbf{x}^u \in y_t \text{ if } t = \arg\max_{m=1,\ldots,M} \min_{j=1,\ldots,K} w_j(\mathbf{x}^u)\mathrm{P}_j(y_m|\mathbf{x}^u) \tag{6}$$

$$\text{Median Rule: } \mathbf{x}^u \in y_t \text{ if } t = \arg\max_{m=1,\ldots,M} \text{median}_{j=1,\ldots,K} w_j(\mathbf{x}^u)\mathrm{P}_j(y_m|\mathbf{x}^u) \tag{7}$$

$$\text{Majority Vote Rule: } \mathbf{x}^u \in y_t \text{ if } t = \arg\max_{m=1,\ldots,M} \sum_{j=1}^{K} w_j(\mathbf{x}^u)\Delta_{jm}$$

$$\Delta_{jt} = \begin{cases} 1 & \text{if } t = \arg\max_{m=1,\ldots,M} P_j(y_m|\mathbf{x}^u) \\ 0 & \text{otherwise} \end{cases} \tag{8}$$

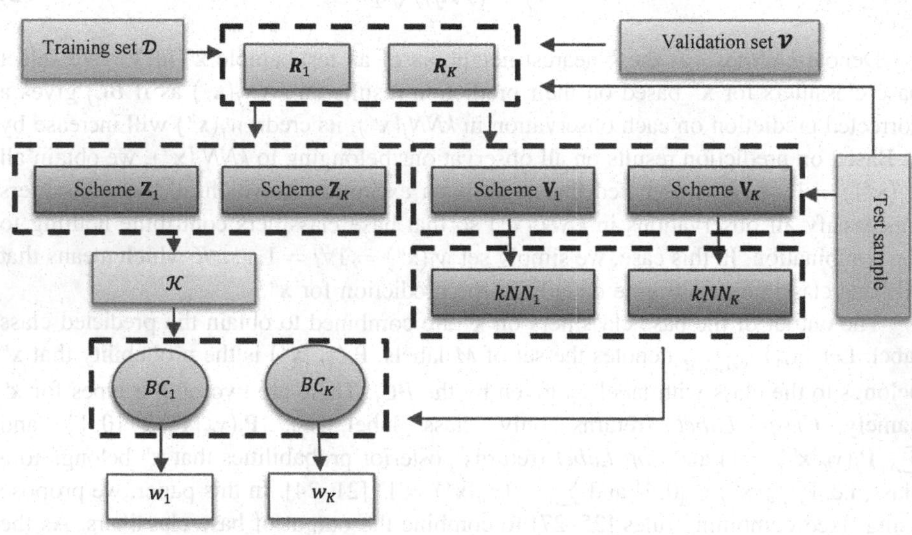

Fig. 1. The process to find the credits of base classifiers

Algorithm: DES with Random Projections-based ensemble

Training Process

Input: Training set: $\mathcal{D}$, Validation set: $\mathcal{V}$, Dimension of down space: q, number of projections: K, learning algorithm: $\mathcal{K}$

Output: Base classifiers: BC_j, Validation schemes $\mathbf{V}_j$, and random matrices: $\mathbf{R}_j$, $(j = 1, \ldots, K)$

```
For j=1 to K
    Generate random matrix R_j;
    Get schemes Z_j by (1)
    BC_j = Learn(K, Z_j);
    Get validation scheme V_j by (2)
End
Return {BC_j}, {V_j}, and {R_j} (j = 1,...,K)
```

Classification process

Input: Test sample $\mathbf{x}^u$, $\{BC_j\}$, $\{\mathbf{R}_j\}$, and $\{\mathbf{V}_j\}$ $(j = 1, \ldots, K)$

Output: Predicted class label for $\mathbf{x}^u$

```
For j=1 to K
    z_j^u = x^u R_j / sqrt(q);  w_j(x) = 0;
End For
```

```
For j=1 to K
    Find kNN_j(x^u) in V_j
    For each v in kNN_j
        For j=1 to K
            Predict label ŷ for v by BC_j;
            If (ŷ==y_v)
                w_j(x^u) = w_j(x^u) +1;
            End
        End
    End
End
If (w_j(x)==0  ∀j = 1,...,K)
    w_j(x)=1  j = 1,...,K;
End
Predict using a fixed combining rule (3)-(8);
```

4 Experimental Studies

4.1 Datasets and Settings

We evaluated the proposed method on 15 datasets from UCI [28], STATLOG project [29], Knowledge Extraction based on Evolutionary Learning (KEEL) [30] and Ludmila Kuncheva Collection of real medical data (denoted by LKC) [31]. Information about the datasets is summarized in Table 1.

Table 1. Information of datasets in evaluation

Datasets	# of features	# of observations	# of classes	Source
Pima	8	768	2	UCI
Bupa	6	345	2	UCI
Wdbc	30	568	2	UCI
Blood	4	748	2	UCI
Sonar	60	208	2	UCI
Ionosphere	34	315	2	UCI
Vertebral	6	310	2	UCI
Waveform without noise (WwtN)	21	5000	3	UCI
Ecoli	7	336	8	UCI
Glass	9	214	6	UCI
Thyroid	3	215	3	LKC
Satimage	19	6435	7	STATLOG
Phoneme	6	5404	2	KEEL
Monk2	6	4322	2	KEEL
Mammographic	5	961	2	KEEL

We performed extensive comparison study with several well-known algorithms to validate our approach. In this study, we compared with several well-known DES methods namely MCB [13], MLA [14], KNOP [15], META-DES [6], DES-FA [17], DES-RRC [19], DES-KL [20], and DES-P [20]. The experiments concerning to those methods and the proposed method are conducted the same as experiments in [6, 12] (the value of k is set to 7). For the proposed method, we used C4.5 learning algorithm as the learning algorithm on 200 new training schemes to construct 200 base classifiers [7, 10, 21]. The new training sets were generated by using Gaussian-based random projections [7, 10] in which q was set as $q = 2 \times log2(p)$. We used Sum Rule to combine the results of EoC on each test sample.

We used Friedman test [32] to assess the statistical significance of the classification results of multiple methods on multiple datasets. Here we test the null hypothesis that "all methods perform equally" on the test datasets. If the null hypothesis is rejected, a post-hoc test is then conducted. In this paper, we used Shaffer's procedure for all pairwise comparisons [32]. The difference in the performance of two methods is treated as statistically significant if the p-value computed from the post-hoc test statistic is smaller than an adjusted value of confident level computed from Shaffer's procedure [32]. We set the confident level α to 0.05.

4.2 Comparing to Benchmark Algorithms

The experimental results of the benchmark algorithms and the proposed method are shown in Table 2. The proposed method obtains the best classification result in 10 datasets. On some datasets, the accuracy of the proposed method is significantly better

Table 2. Mean of accuracy of benchmark algorithms and the proposed method

Database	DES-FA	MLA	MCB	KNOP	DES-RRC	META-DES	DES-KL	DES-P	Proposed method
Pima	73.95	77.08	76.56	73.42	77.64	**79.03**	77.97	76.87	77.87
Bupa	61.62	58	58	65.23	68.01	**70.08**	67.11	67.46	68.53
Wdbc	**97.88**	95.77	97.18	95.42	96.94	97.4	97.13	96.78	97.61
Blood	73.4	76.06	73.4	77.54	78.02	79.14	78.83	77.72	**79.82**
Sonar	78.52	76.91	76.56	75.72	80.77	80.55	78.15	79.49	**88.50**
Ionosphere	88.63	81.81	87.5	85.71	88.8	89.94	88.42	88.42	**96.30**
Vertebral	82.05	77.94	84.61	**86.98**	86.76	86.89	84.19	86.76	86.93
WwtN	84.01	79.95	78.75	84.21	84.63	84.56	84.61	84.59	**86.88**
Ecoli	75.29	76.47	76.47	80	80.66	77.25	79.95	79.83	**82.12**
Glass	55.32	57.6	67.92	62.45	66.04	66.87	63.32	63.13	**67.93**
Thyroid	95.37	94.79	95.95	95.95	97.61	96.78	97.04	96.98	**97.89**
Satimage	93	93.28	95.86	95.86	96.38	96.21	96.2	96.22	**96.52**
Phoneme	79.06	64.94	73.37	78.92	74.65	80.35	77.13	81.64	**96.22**
Monk2	75.92	75.92	74.07	80.55	80.98	83.24	80.85	79.93	**87.44**
Mammographic	80.28	75.55	81.25	82.21	**85**	84.82	84.12	84.98	84.89

**The best results for each dataset are highlighted in bold*

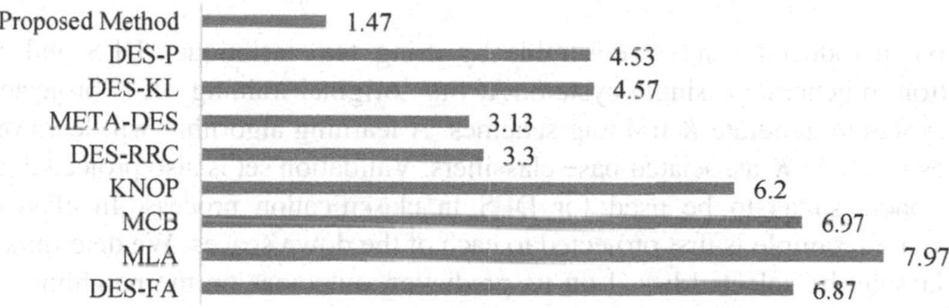

Fig. 2. Average ranking of all methods

than the best result of all benchmark algorithms, for example on Sonar (88.50 vs. 80.77 of DES-RRC), Ionosphere (96.30 vs. 89.94 of META-DES), and Phoneme (96.22 vs. 81.64). For the remaining five datasets, the difference between the accuracy of the proposed method and the best results are not significant except for two datasets, namely Pima (77.87 vs. 79.03 of META-DES) and Bupa (68.53 vs. 70.08 of META-DES).

Figure 2 shows the average rankings of the benchmark algorithms and the proposed method. It can be seen that the proposed method is ranked first (1.47), followed by META-DES (3.13) and DES-RRC (3.3). We conducted the Friedman test base on the rankings of the top five performing algorithms, i.e., DES-RRC, META-DES, DES-KL, DES-P, and the proposed method. In this case, the p-value computed by Friedman test is 1.44E-5. We rejected the null hypothesis of Friedman test and conducted the post-hoc test for all pairwise comparisons among those methods. From the Shaffer's test results shown in Table 3, the proposed method is better than all four benchmark

Table 3. Statistical test result with Shaffer's procedure

Methods	P-value	Reject hypothesis	Methods	P-value	Reject hypothesis
DES-P vs. Proposed method	3.86E-6	Y	DES-RRC vs. DES-P	0.05675	N
DES-KL vs. Proposed method	8.76E-6	Y	META-DES vs. DES-KL	0.07349	N
DES-RRC vs. Proposed method	0.00666	Y	DES-RRC vs. DES-KL	0.08326	N
META-DES vs. Proposed method	0.00791	Y	DES-KL vs. DES-P	0.86249	N
META-DES vs. DES-P	0.04964	N	DES-RRC vs. META-DES	0.95396	N

**Y or N means that the performance of two methods is different or no different respectively*

algorithms. It shows the advantages of combining random projection and DES in a building a high-performance ensemble method.

5 Conclusion

We have introduced a novel ensemble by using two techniques DES and random projection to generate a single system. At first, original training set is projected to K down spaces to generate K training schemes. A learning algorithm will learn on these schemes to obtain K associated base classifiers. Validation set is also projected to the K down spaces so as to be used for DES in classification process. In classification process, a test sample is first projected to each of the down spaces. We determine how a base classifier be selected based on its prediction outcomes on the neighbors of each projected schemes of validation set. The experiments conducted on 15 datasets show that our framework is better than many of the state-of-the-art dynamic classifier/ensemble selection methods concerning to classification accuracy. In the future, the model can be extended to incrementally deal with stream data [33].

References

1. Britto, A.S., Sabourin, R., Oliveira, L.E.S.: Dynamic selection of classifiers—a comprehensive review. Pattern Recog. **47**(11), 3665–3680 (2014)
2. Breiman, L.: Bagging predictors. Mach. Learn. **24**, 123–140 (1996)
3. Ho, T.K.: The random subspace method for constructing decision forests. IEEE Trans. Pattern Anal. Mach. Intell. **20**(8), 832–844 (1998)
4. Johnson, W., Lindenstrauss, J.: Extensions of Lipschitz mapping into Hilbert space. In: Conference in Modern Analysis and Probability, vol. 26, pp. 189–206 (1984). Contemporary Mathematics, American Mathematical Society
5. Fern, X.Z., Brodley, C.E.: Random projection for high dimensional data clustering: a cluster ensemble approach. In: ICML, pp. 186–193 (2003)

6. Cruz, R.M.O., Sabourin, R., Cavalcanti, G.D.C., Ren, T.I.: META-DES: a dynamic ensemble selection framework using meta-learning. Pattern Recog. **48**(5), 1925–1935 (2015)
7. Nguyen, T.T., Nguyen, T.T.T., Pham, X.C., Liew, A.W.-C., Bezdek, J.C.: An ensemble-based online learning algorithm for streaming data. CoRR abs/1704.07938 (2017)
8. Bingham, E., Mannila, H.: Random projection in dimensionality reduction: applications to image and text data. In: ACM SIGKDD, pp. 245–250 (2001)
9. Wang, Z., Yuan, X.-T., Liu, Q.: Sparse random projection for χ^2 kernel linearization: algorithm and applications to image classification. Neurocomputing **151**(1, 3), 327–332 (2015)
10. Pham, X.C., Dang, M.T., Dinh, V.S., Hoang, S., Nguyen, T.T., Liew, A.W.-C.: Learning from data stream based on random projection and Hoeffding tree classifier. In: DICTA 2017 (in press)
11. Rathore, P., Bezdek, J.C., Erfani, S.M., Rajasegarar, S., Palaniswami, M.: Ensemble fuzzy clustering using cumulative aggregation on random projections. IEEE Trans. Fuzzy Syst. (2017, in press). https://doi.org/10.1109/tfuzz.2017.2729501
12. Cruz, R.M.O., Sabourin, R., Cavalcanti, G.D.C.: Dynamic classifier selection: recent advances and perspectives. Inf. Fusion. **41**, 195–216 (2018)
13. Giacinto, G., Roli, F.: Dynamic classifier selection based on multiple classifier behaviour. Pattern Recogn. **34**, 1879–1881 (2001)
14. Smits, P.C.: Multiple classifier systems for supervised remote sensing image classification based on dynamic classifier selection. IEEE Trans. Geosci. Remote Sens. **40**(4), 801–813 (2002)
15. Cavalin, P.R., Sabourin, R., Suen, C.Y.: Dynamic selection approaches for multiple classifier systems. Neural Comput. Appl. **22**(3–4), 673–688 (2013)
16. Ko, A.H.R., Sabourin, R., Britto, A.S.: From dynamic classifier selection to dynamic ensemble selection. Pattern Recogn. **41**, 1735–1748 (2008)
17. Cruz, R.M.O., Cavalcanti, G.D.C., Ren, T.I.: A method for dynamic ensemble selection based on a filter and an adaptive distance to improve the quality of the regions of competence. In: IJCNN, pp. 1126–1133 (2011)
18. Soares, R.G.F., Santana, A., Canuto, A.M.P., de Souto, M.C.P.: Using accuracy and diversity to select classifiers to build ensembles. In: IJCNN, pp. 1310–1316 (2006)
19. Woloszynski, T., Kurzynski, M.: A probabilistic model of classifier competence for dynamic ensemble selection. Pattern Recogn. **44**, 2656–2668 (2011)
20. Woloszynski, T., Kurzynski, M., Podsiadlo, P., Stachowiak, G.W.: A measure of competence based on random classification for dynamic ensemble selection. Inf. Fusion **13**(3), 207–213 (2012)
21. Nguyen, T.T., Nguyen, T.T.T., Pham, X.C., Liew, A.W.-C.: A novel combining classifier method based on variational inference. Pattern Recogn. **49**, 198–212 (2016)
22. Nguyen, T.T., Nguyen, M.P., Pham, X.C., Liew, A.W.-C.: A hybrid classification system with fuzzy rule and classifier ensemble. Inf. Sci. **422**, 144–160 (2018)
23. Nguyen, T.T., Liew, A.W.-C., Pham, X.C., Nguyen, M.P.: A novel 2-stage combining classifier model with stacking and genetic algorithm based feature selection. In: Huang, D.-S., Jo, K.-H., Wang, L. (eds.) ICIC 2014. LNCS (LNAI), vol. 8589, pp. 33–43. Springer, Cham (2014). https://doi.org/10.1007/978-3-319-09339-0_4
24. Nguyen, T.T., Liew, A.W.-C., Tran, M.T., Nguyen, M.P.: Combining multi classifiers based on a genetic algorithm – a Gaussian mixture model framework. In: Huang, D.-S., Jo, K.-H., Wang, L. (eds.) ICIC 2014. LNCS (LNAI), vol. 8589, pp. 56–67. Springer, Cham (2014). https://doi.org/10.1007/978-3-319-09339-0_6

25. Nguyen, T.T., Liew, A.W.-C., Tran, M.T., Pham, X.C., Nguyen, M.P.: A novel genetic algorithm approach for simultaneous feature and classifier selection in multi classifier system. In: CEC, pp. 1698–1705 (2014)
26. Nguyen, T.T., Pham, X.C., Liew, A.W.-C., Pedrycz, W.: Aggregation of classifiers: a justifiable information granularity approach. CoRR abs/1703.05411 (2017)
27. Nguyen, T.T., Pham, X.C., Liew, A.W.-C., Nguyen, M.P.: Optimization of ensemble classifier system based on multiple objectives genetic algorithm. In: ICMLC, vol. 1, pp. 46–51 (2014)
28. Bache, K., Lichman, M.: UCI Machine Learning Repository (2013)
29. King, R.D., Feng, C., Sutherland, A.: STATLOG: comparison of classification algorithms on large real-world problems. Appl. Artif. Intell. Int. J. **9**(3), 289–333 (1995)
30. Alcalá-Fdez, J., Fernández, A., Luengo, J., Derrac, J., García, S., Sánchez, L., Herrera, F.: KEEL data-mining software tool: data set repository, integration of algorithms and experimental analysis framework. Mult. Val. Log. Soft Comput. **17**(2–3), 255–287 (2011)
31. Kuncheva, L.: Ludmila Kuncheva collection LKC (2004)
32. Garcia, S., Herrera, F.: An extension on statistical comparisons of classifiers over multiple data sets for all pairwise comparisons. J. Mach. Learn. Res. **9**, 2579–2596 (2008)
33. Nguyen, T.T., Weidlich, M., Duong, C.T., Yin, H., Nguyen, Q.V.H.: Retaining data from streams of social platforms with minimal regret. In: IJCAI, pp. 2850–2856 (2017)

A Weighted Object-Cluster Association-Based Ensemble Method for Clustering Undergraduate Students

Chau Thi Ngoc Vo(✉) and Phung Hua Nguyen(✉)

Ho Chi Minh City University of Technology, Vietnam National University, Ho Chi Minh City, Vietnam
{chauvtn,nhphung}@hcmut.edu.vn

Abstract. Clustering ensemble methods have received much attention for better clustering quality and robustness as they can exploit the knowledge discovered in their base clusterings. In order to obtain better clusterings on educational data, in this work, we propose a novel clustering ensemble method as the first ensemble-based solution to an educational data clustering task. Different from the existing ensemble methods, our method is based on a weighted object-cluster association matrix. We define this association matrix as a synthesis of the base clusterings. It can capture not only the inherent structure of the data via base clusterings but also the discrimination between the objects via their cluster representatives. As a result, our method effectively groups the students into the clusters each of which have the most similar students based on study performance. This is confirmed by better Normalized Mutual Information values from the experiments on the real educational data sets and the popular Iris data set.

Keywords: Educational data clustering · Clustering ensemble
Co-association matrix · Weighted object-cluster association
K-means · Spectral clustering

1 Introduction

Educational data mining has been of interest due to its capability to discover knowledge hidden in educational data worldwide. Separating clusters of the most similar objects is a kind of knowledge expressed in inherent structures in the data. In the educational domain, such knowledge is widely examined with different clustering tasks for similar student behaviors [2], similar profiles [3], similar performance [8, 9], etc. Although a lot of works have been proposed for such knowledge as a few listed, how good the resulting clusters are questioned because of the use of many fundamental clustering algorithms and few reports on their quality exist. At this moment, *k*-means is one of the most widely-used algorithms in the educational data mining area. It was employed in [2–4, 8] either directly or indirectly. However, we are aware of the Partitional Segmentation algorithm in [7] and the FANNY and AGNES algorithms in [9]. In addition, a comprehensive study on an educational clustering task is needed upon the support required for the students at each institution. Therefore, in this paper,

N. T. Nguyen et al. (Eds.): ACIIDS 2018, LNAI 10751, pp. 587–598, 2018.
https://doi.org/10.1007/978-3-319-75417-8_55

our work is dedicated to this performance-based student clustering task in order to bring a new solution to the task for the discovered clusters of higher quality.

Nowadays, it is not hard to obtain a clustering on a data set with many popular existing algorithms such as: k-means, fuzzy c-means, spectral clustering, Self-Organizing Map, etc. Each single clustering algorithm has its own strengths and weaknesses. Although efficient and effective in some cases, k-means is influenced by the existence of noises and only its intra-cluster compactness is taken into account. Consequently, it can achieve the quality of the clusters to some certain extent. On the other hand, ensemble methods have been considered as a new recent trend in clustering for better clusters, more robustness, and knowledge reuse [15, 18]. Nevertheless, none of the existing works in the educational data mining area has approached the clustering task with ensemble clustering. In contrast, we propose a clustering ensemble method on educational data for the clusters of higher quality in a more efficient and effective manner.

For ensemble clustering, we reviewed some methods in [5, 6, 10, 12–14, 16, 17]. First of all, [5] provided an embedding-based approach to obtain a median consensus clustering. This approach is interesting with no predefined k desired clusters but did not examine the details of each base clustering when embedding the base clusterings in a vector space and treating each of them as a point in that space.

Based on a co-association matrix, i.e. an object-object association matrix, showing the co-occurrence of the objects in the same cluster, [6, 14, 17] have proposed several ensemble methods different from each other in processing each base clustering to add more information into each cell value of the co-association matrix or add weights into the objects. [14] calculated a weight of each object to reflect how confusingly the object is distinguished from the others all over the base clusterings. Their approach can consider the relationship between the base clusterings from the perspective of their agreements on clustering each object; but didn't examine the details of each cluster. With more details, [17] defined a two-level refined co-association matrix. Each cell value was computed by integrating the information at the data point level based on pair-wise Euclidean distances and at the cluster level based on cluster stability. This value had some behind assumptions about the data distribution and decreased the contribution of the varying quality of each cluster when combining them all into a single cell of the object-object matrix. As a more recent work, [6] had a co-association matrix refined by locally weighting each cell value. This value is computed by estimating the uncertainty of each cluster in a base clustering with respect to the entire ensemble via a so-called ensemble-driven cluster index (ECI). The idea behind the local weighting strategy is nice when considering the relationship between the clusters and then integrating it into the association between the objects of the clusters. Nonetheless, their approach didn't detail the relationship between the objects within a cluster. Indeed, like [14, 17], [6] integrated the ECI values over all the clusters into each cell value of the matrix. It is wondered if such integration can reflect the varying quality of each cluster in each base clustering using a co-association matrix. So, our work is based on an object-cluster association matrix so that the contribution of each cluster in a base clustering to consensus clustering can become visible and clearer.

An object-cluster association matrix, i.e. a cluster association matrix, has been used in [10, 13, 16]. In [16], k-means-based consensus clustering was studied. k-means was executed on this matrix taking into consideration the differences between the base clusterings using the user-defined weights. In [10], the authors provided an equivalent version using weighted *k*-means on this matrix to reduce the complexity of spectral ensemble clustering using spectral clustering [12] on a co-association matrix. Similarly, [13] used *k*-means on the matrix instead of weighted *k*-means. Although these works achieved promising results on some real-world data sets, only a binary object-cluster association matrix was defined to show the membership/non-membership of an object to a cluster. It is believed that final clustering with the clusters of higher quality can be attained if more details of each cluster can be explored with this matrix.

Differently, our work defines a weighted object-cluster association matrix as a synthesis of many base clusterings. In this matrix, discrimination between the objects can be captured in connection with their cluster representatives while the inherent structure of the data discovered in each base clustering can be exploited in consensus clustering. Based on this matrix, our weighted object-cluster association-based ensemble method, named WOCA, is proposed with the *k*-means algorithm [11]. As compared to *k*-means, WOCA can derive the final clustering composed of the clusters of higher quality. In addition, WOCA outperforms the existing methods based on either a co-association matrix or a binary object-cluster association one. Its effectiveness has been illustrated with better Normalized Mutual Information values in the various experiments on two real educational data sets as well as on the popular Iris data set.

2 A Clustering Task for Grouping Similar Students

In the educational domain, we can get insights into our students' study by means of the clustering task by exploring each cluster of the similar students. Based on the knowledge discovered, more study predictions and supports can be made for them.

In particular, each student is observed and characterized by means of the results of all the courses he/she has studied already and also those to be studied. This manner allows us to have every group of the students whose studies are similar to each other. Thus, it implies every group of the students who might face the same difficulties in their studies. Such a group helps us to know how successfully the students in the group have studied so far towards the accomplishment of the program for graduation. In a computational form, each student is represented by a vector in a *p*-dimensional data space. Each element of a vector at a dimension is a grade of a subject in the program. Its value is a positive real number in the range [0, 10]. It is supposed that we would like to generate several clusters of *n* students in such a data space.

Our data clustering task on an educational data set is defined as follows.

The input is a data set $\boldsymbol{D}$ of n object vectors in a p-dimensional data space:

$$\boldsymbol{D} = \{X_1, X_2, \ldots, X_n\} \text{ where } X_i = (V_{i,1}, V_{i,2}, \ldots V_{i,p}) \text{ for } i = 1..n.$$

The output is a clustering that captures the inherent structure of the data set $\boldsymbol{D}$ in *k* clusters where *k* is given as the number of clusters of interest. Each cluster has some similar objects who are the students with the most similar study performance.

As previously introduced, this clustering task is resolved by our novel clustering ensemble method, which is proposed in Sect. 3. The ensemble method will attain the task goal with the clusters of higher quality compared to the base clustering ones.

3 The Proposed Ensemble Method

As reviewed in [6], many various ensemble methods have been defined in three main different approaches such as pair-wise co-occurrence-based approach, graph partitioning-based approach, and median partition-based approach. [6] has also analyzed the strengths and weaknesses of some related works with respect to equally treating the given base clusterings and their clusters as well as the assumptions about data distributions and access to the original data features. As for [6], a locally weighted co-association matrix was defined with cluster uncertainty estimation based on entropy. The consensus functions, locally weighted evidence accumulation (LWEA) and locally weighted graph partitioning (LWGP), were then proposed for the final clusterings.

Different from [6] and the other works such as [5, 10, 13, 14, 16, 17], our work follows the pair-wise co-occurrence-based approach. In particular, a weighted object-cluster association matrix is constructed and a clustering ensemble method based on this matrix, named WOCA, is defined for our educational data clustering task.

3.1 The WOCA Clustering Ensemble Algorithm: From Base Clusterings to a Weighted Object-Cluster Association Matrix

Let $\boldsymbol{D} = \{X_1, X_2, \ldots, X_n\}$ be an input data set where $X_i = \{v_{i,1}, v_{i,2}, \ldots, v_{i,p}\}$, $\forall i = 1..n$, n is the number of objects, p is the number of attributes. In addition, let k be the number of desired clusters and e be the ensemble size, i.e. the number of base clusterings. The WOCA method enables us to obtain k clusters in the final clustering after exploiting e base clusterings. It is designed with three main phases as follows.

Phase 1. Obtain the base clusterings

Each base clustering is obtained by running k-means on the original data set with k desired clusters. Initialization is random. The convergence is based on being unchanged of the resulting clusters. This guarantees the minimization of the well-known objective function of k-means:

$$F_r = \sum_{j=1..k} \sum_{i=1..n} \gamma_{i,j} d^2(X_i, C_{r,j}), \forall r = 1..e \quad (1)$$

where $\gamma_{i,j}$ is the membership of X_i to the cluster whose center (representative) is $C_{r,j}$: 1 if a member; otherwise, 0; and $d(X_i, C_{r,j})$ is a distance between X_i and the center $C_{r,j}$.

Phase 2. Construct a weighted object-cluster association matrix

Our weighted object-cluster association (WOCA) matrix, named WOCA_matrix, has n rows corresponding to the number of objects and ek columns corresponding to the union of all the k clusters from each of e base clusterings. Each row represents an object in the ensemble space. Each column is connected to a cluster in a base clustering. A cell value at position (i,l) shows the closeness of the object X_i to the representative of the corresponding cluster C_l, reflecting the association between an object and its cluster. This association is weighted by the value calculated as follows:

$$\text{WOCA\_matrix}[i][l] = d(X_i, C_l)/\text{Radius}_l \quad (2)$$

$$\text{Radius}_l = \max_{i'} = 1..n\, \gamma_{i',l} d(X_{i'}, C_l) \quad (3)$$

Where $d(X_i, C_l)$ is the distance between X_i and the cluster center C_l, Radius_l is the maximum distance between $X_{i'}$ belonging to C_l and the cluster center C_l, and $\gamma_{i',l}$ is the membership of $X_{i'}$ to the cluster center C_l. If X_i also belongs to C_l, this cell value is less than or equal to 1. Otherwise, it is greater than 1.

By means of WOCA_matrix, not only the membership of an object to a cluster but also its degree is recorded. Such a value can be obtained from soft clustering. For demonstration, *k*-means is used for its simplicity and efficiency.

Phase 3. Derive the final clustering

The final clustering is finally derived by executing *k*-means on WOCA_matrix for *k* desired clusters. The convergence is also based on the stability of the *k* resulting clusters, i.e. the minimization of the objective function of consensus clustering:

$$F = \sum_{j=1..k} \sum_{i=1..n} \gamma_{i,j} d^2(X_i, C_j) \quad (4)$$

where $\gamma_{i,j}$ is the membership of X_i to the cluster whose center is C_j: 1 if a member; otherwise, 0; and $d(X_i, C_j)$ is a distance between X_i and the center C_j.

The difference between the minimization of each F_r and F is the computation of the distance: $d(X_i, C_{r,j})$ vs. $d(X_i, C_j)$. Each distance in F_r is computed in the original space while F's distance in the ensemble space. As we integrate the closeness of each object to its base cluster in WOCA_matrix, the smaller cell value implies the more closeness, i.e. the higher membership degree and vice versa, the lower degree for the object with respect to the clusters to which it doesn't belong. Therefore, if two objects have the similar closeness values with respect to all the clusters in the base clusterings, they will get the similar distances to a cluster center in the final clustering. For the clustering process, they are clustered into the same group. As a result, compared to *k*-means on the original data set, *k*-means on WOCA_matrix can achieve higher intra-cluster similarity and higher inter-cluster dissimilarity upon its convergence.

For example, given a data set in a 2-dimensional space: $\boldsymbol{D} = \{X_i\}$, i = 1..8 where $X_1 = (6, 4)$, $X_2 = (3, 2)$, $X_3 = (4, 3)$, $X_4 = (10, 4)$, $X_5 = (9, 6)$, $X_6 = (8, 8)$, $X_7 = (5, 7)$, and $X_8 = (6, 9)$, $n = 8$, $k = 3$, and $e = 2$. Two base clusterings π_1 and π_2 are given in Fig. 1.

The co-association matrix, binary object-cluster association (BOCA) matrix, and our weighted object-cluster association (WOCA) matrix are presented in Tables 1, 2 and 3, respectively. It is shown that the information of our matrix is richer. Hence, it can provide better discriminations.

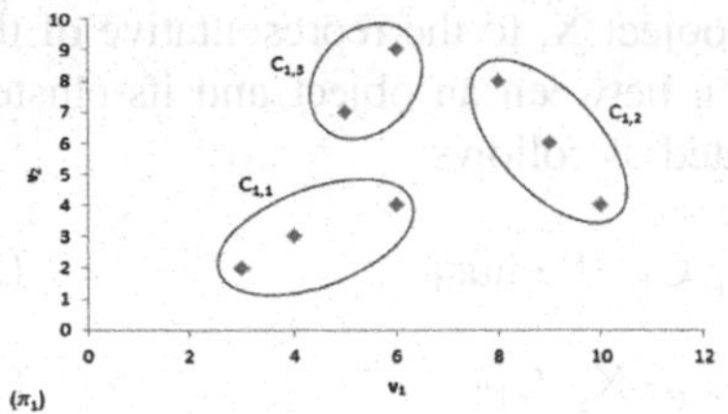
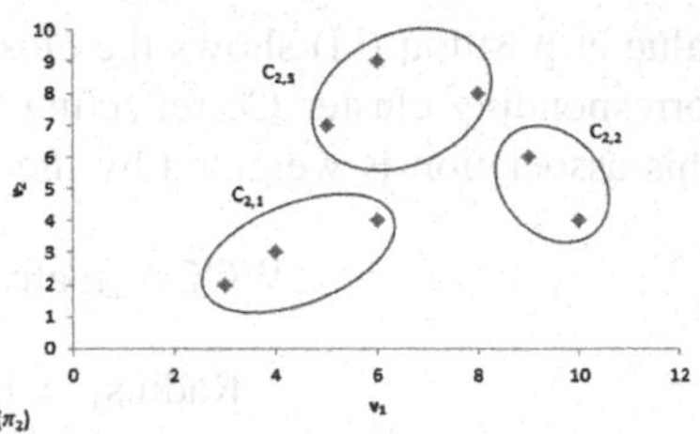

Fig. 1. Two base clusterings π_1 and π_2 for illustration

Table 1. A co-association matrix

	X_1	X_2	X_3	X_4	X_5	X_6	X_7	X_8
X_1	2	2	2	0	0	0	0	0
X_2	2	2	2	0	0	0	0	0
X_3	2	2	2	0	0	0	0	0
X_4	0	0	0	2	2	1	0	0
X_5	0	0	0	2	2	1	0	0
X_6	0	0	0	1	1	2	1	1
X_7	0	0	0	0	0	1	2	2
X_8	0	0	0	0	0	1	2	2

From the co-association matrix in Table 1, no difference can be recognized for the Euclidean distance from X_5 to X_6 and the one from X_6 to X_7:

$$d(X_5, X_6) = 2.24 \text{ while } d(X_6, X_7) = 2.24.$$

This fact might influence the final clustering with random choices in cluster assignment.

Table 2. A BOCA matrix

	$C_{1,1}$	$C_{1,2}$	$C_{1,3}$	$C_{2,1}$	$C_{2,2}$	$C_{2,3}$
X_1	1	0	0	1	0	0
X_2	1	0	0	1	0	0
X_3	1	0	0	1	0	0
X_4	0	1	0	0	1	0
X_5	0	1	0	0	1	0
X_6	0	0	1	0	1	0
X_7	0	0	1	0	0	1
X_8	0	0	1	0	0	1

Table 3. A WOCA matrix

	$C_{1,1}$	$C_{1,2}$	$C_{1,3}$	$C_{2,1}$	$C_{2,2}$	$C_{2,3}$
X_1	1.00	3.26	1.61	1.00	1.61	3.61
X_2	0.86	6.40	2.60	0.86	3.22	5.81
X_3	0.17	5.23	2.09	0.17	2.61	4.67
X_4	2.96	1.00	2.41	2.96	1.00	5.39
X_5	2.85	1.00	1.61	2.85	0.00	3.61
X_6	3.19	3.00	1.00	3.19	1.00	2.24
X_7	2.09	4.40	0.45	2.09	1.84	1.00
X_8	3.20	4.75	0.45	3.20	1.90	1.00

The same fact occurs with the binary object-cluster association matrix in Table 2:

$$d(X_5, X_6) = 1.41 \text{ while } d(X_6, X_7) = 1.41.$$

However, in consensus clustering with our matrix in Table 3, the difference can be described:

$$d(X_5, X_6) = 2.55 \text{ while } d(X_6, X_7) = 2.63.$$

For clustering with higher intra-cluster similarity and higher inter-cluster dissimilarity, it is expected that X_6 should belong to the same cluster as X_5 while belonging to the different cluster from X_7 if the distance between X_6 and X_7 is large enough.

Thus, it is better for a final clustering to be derived from our matrix. Based on this matrix, a base clustering algorithm is executed. In our work, we show a demonstration with *k*-means, which is a popular, efficient, and distance-based algorithm.

3.2 Characteristics of the Proposed Method

Using *k*-means as a base clustering algorithm, our method can inherit the simplicity and efficiency of *k*-means. Analyzed below, our method is more efficient in terms of space and time in comparison with the existing methods based on the co-association matrix and other base clustering algorithms different from *k*-means.

In our method, $O(nek)$ is the size of the weighted object-cluster association matrix while $O(n^2)$ is the size of the co-association matrix. When the ensemble size *e* and the number of clusters *k* are fixed and often smaller than the number of the objects *n*, the cost with our matrix is lower than that with the co-association matrix.

As for time complexity, *k*-means is famous for efficiency to create *k* clusters with $O(nkt)$ time where *t* is the number of iterations if there are *n* objects in the *p*-dimensional space for $p << n$. As the weighted object-cluster association matrix is used, *p* in consensus clustering is now *ek*, where *e* and *k* are given, normally small values as compared to *n*. Thus, the time of ensemble clustering is $O(nekt)$. We also include the time of each base clustering which is $O(nkt')$. The total time is $O(nekt) + eO(nkt') \approx O(nekT)$ where T is the maximum number of iterations. This low complexity is our reason for choosing *k*-means as a base clustering algorithm of our method.

Besides, our method is more practical with only two widely-used parameters: *e* for the number of base clusterings (i.e. the ensemble size) and *k* for the number of desired clusters. These two parameters are ubiquitous with any existing ensemble methods. Meanwhile, in addition to *e* and *k*, some existing works have more parameters, for instance, [6] with ensemble-driven cluster index needs theta, [10] with spectral clustering needs sigma, and [16] requires a user-defined weight for each base clustering. Aware of the availability of hyperparameter tuning methods for choosing appropriate values of a model parameter, we believe that the less number of parameters makes the users more comfortable with the proposed algorithms.

Although our WOCA method is based on *k*-means, other partitioning-based clustering algorithms can be used. In addition, it can be extended with the base clusterings each of which has its own number of clusters. Those changes and extensions are understandable for WOCA as it is among the most recent works considering the difference between the objects in a single cluster of each base clustering via their representatives and thus, better distinguishing between the objects in consensus clustering.

4 An Empirical Evaluation

In this section, we conduct an empirical evaluation on the proposed method in comparison with others using some different approaches.

In particular, we use two real data sets prepared from the study results of the fourth-year regular undergraduate students at Faculty of Computer Science and

Engineering, Ho Chi Minh City University of Technology, Vietnam National University – Ho Chi Minh City [1]. Their programs are Computer Science and Computer Engineering. In addition, we choose different entrance years for data collection so that the data sets can be of different sizes in different periods of time. Each data set is labeled with three predefined classes "Graduating", "Studying", and "Study-Stop" based on students' final study performance. As a result, we get the "Year 4 CS" data set with the Computer Science program and the "Year 4 CE" data set with the Computer Engineering program, respectively. More data descriptions are given in Table 4.

Table 4. Data descriptions

Data set	Program	Entrance year	Number of instances	Number of attributes	Number of classes
Year 4 CE	Computer engineering	2008–2009	186	43	3
Year 4 CS	Computer science	2005–2008	1317	43	3

In addition to educational data sets, we include the Iris[1] data set in this empirical study. This set has 150 objects characterized by 4 attributes, labeled with 3 classes.

For method comparison, we consider three different methods as follows:

– *k*-means: Proposed in [11], this is the base algorithm creating base clusterings.
– OOA ensemble (*k*-means): This is a popular ensemble method which executes the *k*-means algorithm on an object-object association matrix, aka a co-association matrix. This co-association matrix is obtained from the base clusterings where each cell value (i, j) is a cumulative number of co-occurrences in the same cluster of the two corresponding objects X_i and X_j.
– BOCA ensemble (*k*-means): This is an ensemble method adapted from the one proposed in [13]. A binary object-cluster association (BOCA) matrix is formed from the base clusterings. *k*-means is then run on the matrix for the final result.

All the methods are examined with the varying number *e* of base clusterings, i.e. the varying ensemble size *e*, used for consensus clustering. In particular, *e* changes from 10 to 80 with a gap of 10. Each base clustering stems from the execution of *k*-means on an original data set where *k* is 3, the number of classes. Initial clusters of each base clustering are generated from *k* random objects of the data set.

For cluster validation, we use Normalized Mutual Information (NMI). The reason for not using internal measures is that different data spaces have been created in different manners with varying value ranges and number of dimensions. In addition, a predefined cluster of each object is given with our data sets. Besides, NMI has been utilized for evaluation in the existing works on ensemble clustering. Details can be found in [6]. The larger NMI value indicates the better cluster model, i.e. the more

[1] UCI Machine Learning Repository [http://archive.ics.uci.edu/ml].

effective method. Above all, statistical tests with the Paired-Samples T Test method using 95% for the confidence interval of the difference have been applied to check if all the differences between our method and the others are statistically significant.

In the following, Table 5 presents the NMI values from the experiments on the "Year 4 CE" data set, Table 6 on the "Year 4 CS" data set, and Table 7 on the Iris data set. The best values are displayed in bold. For randomness avoidance in an execution of a method, each experiment has been performed 50 times and each value shown in those tables is then an averaged one. Due to space limitation, their corresponding standard deviations are excluded from this section.

With these experimental results, we examine two following questions:

– Is the proposed ensemble method able to generate the cluster models of better quality in comparison to those generated by its base clustering method?
– Is the proposed ensemble method more effective than the others with the traditional object-object association matrix and the binary object-cluster one?

First, it is realized that our ensemble method outperforms its base clustering method, *k*-means, in almost all the cases. Its effectiveness is achieved consistently with the NMI measure on three different data sets with all the various numbers of base clusterings in the experiments. Both the highest and lowest measure values of WOCA are better than those of its base method. Regardless of the number of base clusterings, the difference between WOCA and its base method is stable and clear. On average, it can improve the base method with about 11% and 13% of NMI values for the "Year 4 CE" and "Year 4 CS" data sets, respectively. Such results show the appropriateness of our synthesis of base clusterings. Besides, they confirm the worthiness of an ensemble method on educational data as compared to its base method.

Table 5. NMI values on the "Year 4 CE" data set

e	10	20	30	40	50	60	70	80	Average	Delta
k-means	0.45	0.46	0.47	0.44	0.44	0.46	0.45	0.45	0.45	11%
OOA	0.41	0.40	0.40	0.40	0.41	0.41	0.40	0.40	0.40	25%
BOCA	0.41	0.41	0.38	0.41	0.39	0.41	0.40	0.39	0.40	25%
WOCA	**0.50**	**0.49**	**0.50**	**0.49**	**0.50**	**0.49**	**0.50**	**0.50**	**0.50**	

Table 6. NMI values on the "Year 4 CS" data set

e	10	20	30	40	50	60	70	80	Average	Delta
k-means	0.15	0.15	0.16	0.16	0.15	0.16	0.15	0.16	0.16	13%
OOA	0.13	0.15	0.15	0.15	0.15	0.16	0.15	0.15	0.15	20%
BOCA	0.14	0.14	0.14	0.14	0.14	0.14	0.13	0.15	0.14	29%
WOCA	**0.18**	**0.18**	**0.18**	**0.18**	**0.18**	**0.18**	**0.17**	**0.18**	**0.18**	

Table 7. NMI values on the Iris data set

e	10	20	30	40	50	60	70	80	Average	Delta
k-means	0.72	0.72	0.69	0.71	0.71	0.72	0.71	0.72	0.71	8%
OOA	0.71	0.71	0.71	0.72	0.71	0.70	0.72	0.71	0.71	8%
BOCA	0.70	0.68	0.70	0.69	0.72	0.70	0.68	0.69	0.69	12%
WOCA	**0.76**	**0.77**	**0.76**	**0.76**	**0.76**	**0.77**	**0.77**	**0.77**	**0.77**	

Secondly, our weighted object-cluster association matrix can be concluded to be more effective than the object-object association matrix as well as the binary object-cluster association one for consensus clustering. Indeed, WOCA has very promising results as compared to OOA. It has improved OOA very much. The differences are about 25% and 20% of NMI values on average for the "Year 4 CE" and "Year 4 CS" data sets, respectively. It has also outperformed BOCA with great differences such as about 25% and 29% of NMI values on average for the "Year 4 CE" and "Year 4 CS" data sets, respectively. For a comparison between the object-object association matrix and the binary object-cluster association one, it is found that BOCA is comparable to OOA using the same *k*-means base clustering algorithm. This is because both approaches are based on only the membership of each object with respect to each cluster of every base clustering and treat all the memberships of the base clusterings equally. When we further examine the relationship between each object and its cluster of every base clustering, it is pointed out that the discrimination between objects can be explored. So, WOCA outperforms the OOA and BOCA methods on a consistent basis.

Moreover, we obtain the similar experimental results on the well-known Iris data set in Table 7. In its experiments, WOCA can improve 8% of the NMI values from *k*-means and OOA while 12% of those from BOCA. As the Iris data set is not in the educational domain, the merits of WOCA are shown to some extent. It is also worth noting that the overlapping of the true clusters in a data set has a strong impact on the final clustering. Among the three data sets in our experiments, the Iris data set has the smallest data overlapping percentage; thus, the highest NMI values on average.

In short, our WOCA ensemble method with a weighted object-cluster co-association matrix is more effective than its base clustering method and some existing ones. Better clusters are able to be achieved consistently. Besides, all the differences between our ensemble method and the others have been statistically tested by means of the Paired Samples T Test method using Confidence Interval Percentage = 95%. It is found that they are statistically significant with Sig. (2-tailed) = .000. As a result, we can have the similar students identified in groups together. Support can be then given particularly to each group. More analysis can be made on any group of the students of interest for specific purposes, e.g. a group of the in-trouble students for their ultimate study improvement and graduation.

5 Conclusions

In this paper, we have defined a novel ensemble solution to a student clustering task based on study performance. In our solution, a weighted object-cluster association matrix is proposed as a new consensus clustering scheme. This matrix is based on the current state of each object with respect to its base cluster, discriminated from the objects in the same base cluster and the others not in the same base cluster. Meanwhile, it can examine the inherent structure of the data in each base clustering for a final one of higher quality. With our association matrix, better ensemble clusterings have been achieved with the *k*-means algorithm. As a result, the proposed method can help us group similar students into their more proper clusters. Indeed, our method can produce better clusters on two real educational data sets with different characteristics as compared to its base algorithm and some existing ensemble methods. The clusters of higher quality were also discovered in the well-known Iris data set. Higher NMI values were returned from several corresponding experiments in our empirical study.

As one of the very first works bringing ensemble clustering to the educational domain, our work will further evaluate the generality of our method on more data sets with more methods. Sparse data handling is also considered to further examine the robustness of our method. Besides, making our method parameter-free is of our interest so that it can be more practical for the educational decision support system.

Acknowledgments. This research is funded by Vietnam National University Ho Chi Minh City, Vietnam, under grant number C2017-20-18.

References

1. Academic Affairs Office, Ho Chi Minh City University of Technology, Vietnam. http://www.aao.hcmut.edu.vn. Accessed 29 June 2017
2. Adjei, S., Ostrow, K., Erickson, E., Heffernan, N.: Clustering students in ASSISTments: exploring system and school-level traits to advance personalization. In: Proceedings of the 10th International Conference on Educational Data Mining, pp. 340–341 (2017)
3. Bresfelean, V.P., Bresfelean, M., Ghisoiu, N.: Determining students' academic failure profile founded on data mining methods. In: Proceedings of the ITI 2008 30th International Conference on Information Technology Interfaces, pp. 317–322 (2008)
4. Campagni, R., Merlini, D., Verri, M.C.: Finding regularities in courses evaluation with k-means clustering. In: Proceedings of the 6th International Conference on Computer Supported Education, pp. 26–33 (2014)
5. Franek, L., Jiang, X.: Ensemble clustering by means of clustering embedding in vector spaces. Pattern Recogn. **47**(2), 833–842 (2014)
6. Huang, D., Wang, C-D., Lai, J-H.: Locally weighted ensemble clustering. IEEE Trans. Cybern. **PP**(99), 1–14 (2017)
7. Jayabal, Y., Ramanathan, C.: Clustering students based on student's performance - a partial least squares path modeling (PLS-PM) study. In: Perner, P. (ed.) MLDM 2014. LNCS (LNAI), vol. 8556, pp. 393–407. Springer, Cham (2014). https://doi.org/10.1007/978-3-319-08979-9_29

8. Jovanovic, M., Vukicevic, M., Milovanovic, M., Minovic, M.: Using data mining on student behavior and cognitive style data for improving e-learning systems: a case study. Int. J. Comput. Intell. Syst. **5**, 597–610 (2012)
9. Kerr, D., Chung, G.K.W.K.: Identifying key features of student performance in educational video games and simulations through cluster analysis. J. Educ. Data Mining **4**(1), 144–182 (2012)
10. Liu, H., Wu, J., Liu, T., Tao, D., Fu, Y.: Spectral ensemble clustering via weighted k-means: theoretical and practical evidence. IEEE Trans. Knowl. Data Eng. **29**(5), 1129–1143 (2017)
11. MacQueen, J.: Some methods for classification and analysis of multivariate observations. In: Proceedings of the 5th Berkeley Symposium on Mathematical Statistics and Probability, vol. 1, pp. 281–297 (1967)
12. Ng, A.Y., Jordan, M.I., Weiss, Y.: On spectral clustering: analysis and an algorithm. Adv. Neural. Inf. Process. Syst. **14**, 1–8 (2002)
13. Pattanodom, M., Iam-On, N., Boongoen, T.: Clustering data with the presence of missing values by ensemble approach. In: Proceedings of the 2nd Asian Conference on Defence Technology, pp. 114–119 (2016)
14. Ren, Y., Domeniconi, C., Zhang, G., Yu, G.: Weighted-object ensemble clustering: methods and analysis. Knowl. Inf. Syst. **51**, 1–29 (2016). https://doi.org/10.1007/s10115-016-0988-y
15. Topchy, A., Jain, A.K., Punch, W.: Clustering ensembles: models of consensus and weak partitions. IEEE Trans. Pattern Anal. Mach. Intell. **27**(12), 1866–1881 (2005)
16. Wu, J., Liu, H., Xiong, H., Cao, J., Chen, J.: K-means-based consensus clustering: a unified view. IEEE Trans. Knowl. Data Eng. **27**(1), 155–169 (2015)
17. Zhong, C., Yue, X., Zhang, Z., Lei, J.: A clustering ensemble: two-level-refined co-association matrix with path-based transformation. Pattern Recogn. **48**(8), 2699–2709 (2015)
18. Zhou, Z.H.: Ensemble Methods: Foundations and Algorithms. Chapman and Hall/CRC, Boca Raton (2012)

Automatic Image Region Annotation by Genetic Algorithm-Based Joint Classifier and Feature Selection in Ensemble System

Anh Vu Luong[1], Tien Thanh Nguyen[1,2](✉), Xuan Cuong Pham[3], Thi Thu Thuy Nguyen[2], Alan Wee-Chung Liew[2], and Bela Stantic[2]

[1] School of Applied Mathematics and Informatics, Hanoi University of Science and Technology, Hanoi, Vietnam

[2] School of Information and Communication Technology, Griffith University, Gold Coast, Australia
thanh.nguyen3@griffithuni.edu.au

[3] Department of Computer Science, Water Resource University, Hanoi, Vietnam

Abstract. In this paper, we address the image region tagging procedure in which each image region is annotated by a suitable concept. Specifically, we first extract the feature vector for each segmented region. Then we propose a Genetic Algorithm (GA)-based simultaneous classifier and feature selection method working with ensemble system to learn the relationship between the low-level features and high-level concepts. The extensive experiments conducted on two public datasets namely MSRC v1 and MSRC v2 demonstrate the better performance of our method than several well-known ensemble methods, supervised machine learning methods, and sparse coding-based methods in the regions-in-image classification task.

Keywords: Image regions annotation · Image regions tagging
Genetic Algorithm · Ensemble method · Multi classifiers system

1 Introduction

In this study, we aim to address the automatic image regions annotation (AIRA) problem (it is also called image regions tagging, tagging images at the region-level or assigning tags to image regions problem in literature [1]) in which each region in an image is assigned by a suitable keyword, providing a one-one mapping between the textural word and region. The AIRA framework provides several benefits since it can bridge the semantic-gap problem existing in traditional Content-based Image Retrieval (CBIR) systems [2], gives a better understanding of images content [3], and is more closed to human perception [4]. We distinguish AIRA with approaches namely Automatic Image Annotation [5] in which keywords are assigned to the whole image in general and not related to objects or regions within as well as Region-based Image Annotation [6] where authors concentrate on image-level annotation by using regions as the immediate elements to build the prediction model.

N. T. Nguyen et al. (Eds.): ACIIDS 2018, LNAI 10751, pp. 599–609, 2018.
https://doi.org/10.1007/978-3-319-75417-8_56

Several methods solving the AIRA problem have been proposed recently in the literature. In [4, 7, 8], AIRA is treated as a classification problem where each concept is modeled as an independent class. We address several approaches such as Support Vector Machine (SVM) plus Principle Component Analysis [4], pre-pruning and post-pruning techniques to train a well-behaved Decision Tree (DT) [7], the combination of DT and SVM [8]. Recently, several sparse coding-based approaches solving region tagging have been proposed. The idea behind these approaches is to describe an untagged region by reconstructed sparse coding using knowledge of tagged regions within the training set. The reconstructed coding is obtained by solving objective functions which are proposed in different forms, i.e., putting different penalties on the coding outputs. Several examples of sparse coding-based approaches are Lasso method [9], Group Lasso method [10], Sparse Group Lasso method [11], the graph-guided fusion penalty [12], SGSC algorithm [13], Joint SGSC [13], and graph regularized joint group sparsity [1].

In this study, similar to the other region tagging methods, we assume that the images are already pre-segmented into regions (obtained by some other algorithms), and our work is focused on tagging the concept to the regions. The feature vector is extracted from each image by using the low order moments in HSV color space, texture features in the spectral domain, and the average of pixels position. All the extracted feature vectors and their related region tagged labels are grouped to form the training set. The relationship between low-level features and high-level concepts is then learned by an ensemble method on the training set. We employ the heterogeneous ensemble method [14–20] in which the discriminative hypothesis is generated by combining the outputs of base classifiers which are learned by different learning algorithms on the same training set. Besides, to improve the performance of AIRA, we propose a simultaneous classifier and feature selection method based on GA-based optimization technique applied to ensemble system. The proposed model is shown in Fig. 1.

2 Feature Extraction

We propose a process to obtain the feature vector from each region of each segmented image. In detail, the normalized 256 × 256 image is divided into several meaningful regions by using a segmentation algorithm. For a region $\mathcal{R}$, a represented feature vector is extracted by considering two well-known visual contents namely color and texture. More specifically, we select the color moment [21] on the HSV color space to model the color attributes of $\mathcal{R}$. Here, the three low order moments (mean, variance, and skewness) are extracted for each of the three color planes by the formulas:

$$\mu = \frac{1}{|\mathcal{R}|}\sum\nolimits_{i=1}^{|\mathcal{R}|} c_i \tag{1}$$

$$\sigma = \sqrt{\frac{1}{|\mathcal{R}|}\sum\nolimits_{i=1}^{|\mathcal{R}|} (c_i - \mu)^2} \tag{2}$$

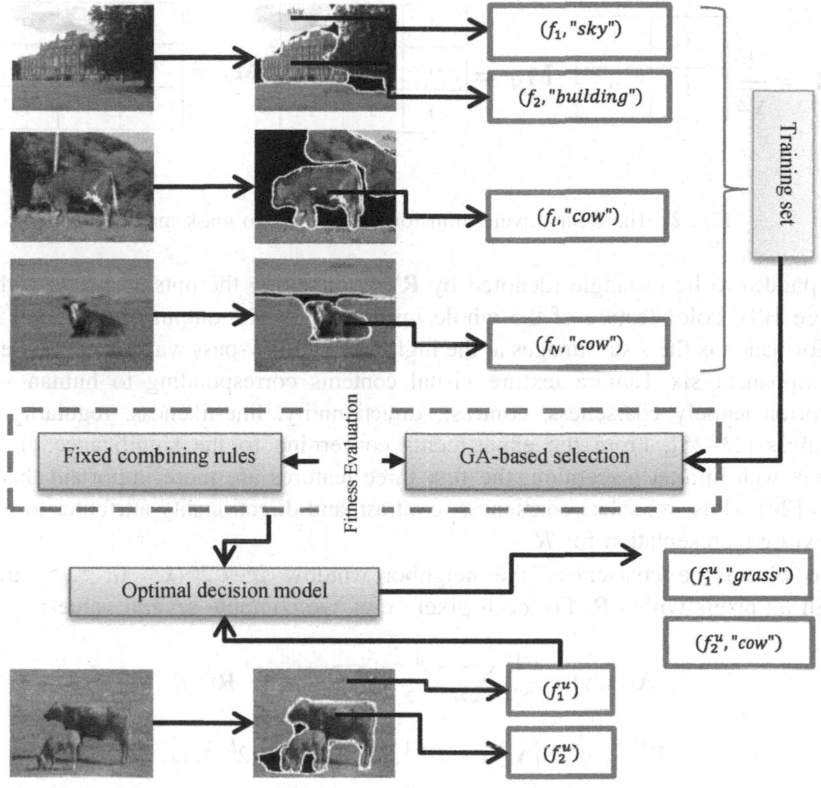

Fig. 1. The proposed AIRA model

$$\theta = \frac{\frac{1}{|\mathcal{R}|}\sum_{i=1}^{|\mathcal{R}|}(c_i - \mu)^3}{\sigma^3} \tag{3}$$

where c_i is the color value in HSV space of ith pixel in $\mathcal{R}$. Totally, the 9-dimension feature vector consisting of color characteristics for $\mathcal{R}$ is obtained after that phase.

In fact, an object might be located based on their pixel coordinates on the image, for instance, sky object has higher averaged coordination that those of sea object. Therefore, we propose adding the average of pixel positions related to two coordinates into the feature vector of the region $\mathcal{R}$:

$$\bar{x} = \frac{1}{|\mathcal{R}|}\sum_{i=1}^{|\mathcal{R}|} x_i \, , \, \bar{y} = \frac{1}{|\mathcal{R}|}\sum_{i=1}^{|\mathcal{R}|} y_i \tag{4}$$

where (x_i, y_i) is the coordination of pixels within $\mathcal{R}$.

We also apply two-level Haar wavelet transformation $\mathbf{H}_2$ (Fig. 2) to $\mathcal{R}$ to analyze its texture on the spectra domain [22]. In detail, the image signal is partitioned into approximate, horizontal, vertical and diagonal components which contain the information of image on the different directions. To support the transformation, input region

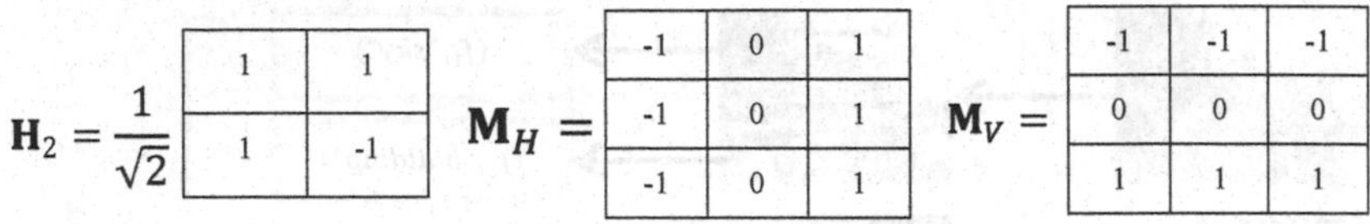

$$\mathbf{H}_2 = \frac{1}{\sqrt{2}}\begin{bmatrix} 1 & 1 \\ 1 & -1 \end{bmatrix} \quad \mathbf{M}_H = \begin{bmatrix} -1 & 0 & 1 \\ -1 & 0 & 1 \\ -1 & 0 & 1 \end{bmatrix} \quad \mathbf{M}_V = \begin{bmatrix} -1 & -1 & -1 \\ 0 & 0 & 0 \\ 1 & 1 & 1 \end{bmatrix}$$

Fig. 2. The Haar wavelet transformation and two mask matrices

$\mathcal{R}$ is padded to be rectangle (denoted by **R**) by offsetting the outside pixels with the average HSV color feature of the whole inside pixels. The output of two-level Haar transformation is the 7 sub-images at the high-pass and low-pass wavelet energy levels. We implement six Tamura texture visual contents corresponding to human visual perception namely coarseness, contrast, directionality, line-likeness, regularity, and roughness [23–25]. From the experiments concerning to the significance of these features with human perception, the first three features are more important than the others [24]. Thus, we select coarseness, contrast, and directionality attributes to model the texture representation for $\mathcal{R}$.

To obtain the coarseness, the neighbor window $2^k \times 2^k$ $(k = 0, \ldots, 5)$ travels though all pixels within **R**. For each pixel (x, y), we compute several values:

$$\mathbf{A}_k(x,y) = \frac{1}{2^{2k}} \sum_{i=x-2^{k-1}}^{x+2^{k-1}-1} \sum_{j=y-2^{k-1}}^{y+2^{k-1}-1} \mathbf{R}(i,j) \tag{5}$$

$$\mathbf{E}_k^H(x,y) = \left|\mathbf{A}_k\left(x+2^{k-1},y\right) - \mathbf{A}_k\left(x-2^{k-1},y\right)\right| \tag{6}$$

$$\mathbf{E}_k^V(x,y) = \left|\mathbf{A}_k\left(x,y+2^{k-1}\right) - \mathbf{A}_k\left(x,y-2^{k-1}\right)\right| \tag{7}$$

$$\mathbf{S}(x,y) = \arg\max_k \max_{i=\{H,V\}} \mathbf{E}_k^i(x,y) \tag{8}$$

Coarseness is defined by:

$$\text{Coarseness} = \frac{1}{|\mathbf{R}|} \sum_{x=1}^{\mathbf{R}_H} \sum_{y=1}^{\mathbf{R}_V} 2^{\mathbf{S}(x,y)} \tag{9}$$

where $\mathbf{R}_H, \mathbf{R}_V$ are the number of pixels on the horizontal and vertical direction of **R**. Contrast attribute is defined by:

$$\text{Contrast} = \frac{\sigma}{\alpha_4^{0.25}} \tag{10}$$

$$\alpha_4 = \frac{\mu_4}{\sigma^4} \tag{11}$$

$$\mu_4 = \frac{1}{|\mathbf{R}|} \sum_{x=1}^{\mathbf{R}_H} \sum_{y=1}^{\mathbf{R}_V} (\mathbf{R}(x,y) - \mu)^4 \tag{12}$$

$$\sigma^2 = \frac{1}{|\mathbf{R}|}\sum_{x=1}^{\mathbf{R}_H}\sum_{y=1}^{\mathbf{R}_V}(\mathbf{R}(x,y)-\mu)^2 \tag{13}$$

$$\mu = \frac{1}{|\mathbf{R}|}\sum_{x=1}^{\mathbf{R}_H}\sum_{y=1}^{\mathbf{R}_V}\mathrm{R}(x,y) \tag{14}$$

The angle direction is defined by:

$$\text{Angle} = \frac{\pi}{2} + \tan^{-1}\frac{\mathbf{L}\otimes\mathbf{M}_V}{\mathbf{L}\otimes\mathbf{M}_H} \tag{15}$$

In this paper, 8-bin histogram on the angle direction, the coarseness, and the contrast are combined to form 10-bin histogram to describe the texture content for the considered region. After performing the above steps, we concatenate the extracted features into a single vector to represent a specific region within the image. That low-level representation includes 81 attributes.

3 Learning System Based on Simultaneous Classifier and Feature Selection Approach

In the heterogeneous ensemble methods [14–20], a fixed set of different learning algorithms is used on the same training set to generate the base classifiers. The discriminative model is then constructed using a combination method on the outputs of these classifiers (the outputs are called Level1 data or meta-data [18]). There are two techniques to combine the outputs of base classifiers, namely fixed combining and trainable combining method [18]. Trainable combining methods [18] work on the meta-data of the training set to form the prediction model. By contrast, fixed combing methods work directly on the meta-data of each observation without using the meta-data of training observations [26]. There are several popular fixed combining methods studied in the literature, namely Sum Rule, Product Rule, Majority Vote Rule, Max Rule, Min Rule, and Median Rule [26]. Of these, Majority Vote and Sum Rules are the most frequently used rules.

In this study, we propose using a joint classifiers and feature selection approach for ensemble learning to improve the accuracy of the classification task. Our approach is based on GA to explore the optimal classifier subset and associated optimal feature set for each classifier to construct the ensemble system.

3.1 Chromosome Design

To solve the simultaneous classifier and feature selection problem, we design the two-part encoding representation for each chromosome as illustrated in (16). Specifically, the first part is the encoding of K base classifiers where each gene has two values '1' and '0' showing which classifiers will be selected or not in the ensemble system (17). The second part contains the feature encoding giving information about which features will be employed by a specific classifier. The genes in feature chromosome encoding also get two values $\{0, 1\}$ (18). Based on this proposed structure, we can select not only classifiers but also their corresponding feature set to construct an optimal ensemble system.

$$E = \begin{bmatrix} e_1 \cdots e_k \\ e_1 = \{e_{11}, e_{12}, \ldots, e_{1D}\} \\ e_2 = \{e_{21}, e_{22}, \ldots, e_{2D}\} \\ \cdots \\ e_k = \{e_{K1}, e_{K2}, \ldots, e_{KD}\} \end{bmatrix} \tag{16}$$

$$e_k = \begin{cases} 1, & \text{if } k^{th} \text{ classifier is selected} \\ 0, & \text{otherwise} \end{cases} \tag{17}$$

$$e_{kd} = \begin{cases} 1, & \text{if } d^{th} \text{ attribute is selected by } k^{th} \text{ classifier} \\ 0, & \text{otherwise} \end{cases} \tag{18}$$

in which D is the dimension of observations.

3.2 Crossover and Mutation Operator

To vary the structure of chromosomes to obtain the new generation, we sequentially use crossover operator in two parts of a pair of chromosomes. In detail, crossover operator is conducted on the classifier encoding first. Here we employ single point splitter where each classifier encoding exchanges its head with the other while retains its tail and their feature encodings are swapped accordingly based on the classifier encoding on the first part. In the second step, single point splitter is applied to feature encoding of kth classifier of two individuals ($k = 1, \ldots, K$).

Mutation operator is conducted on both parts of the chromosome to keep genetic diversity from one generation to the next generation. We defined probability *PMul*1 for the mutation process on the classifier encoding in which mutation occurs by inverting a random gene if the mutation probability is smaller than *PMul*1. Similar mutation operator is conducted on feature encoding associated with each classifier with reference to probability *PMul*2.

3.3 Fitness Computation

We use Sum Rule as the combiner in heterogeneous ensemble system to compute the fitness of each individual:

$$\mathbf{x} \in y_j \text{ if } j = \arg\max_{m=1,\ldots,M} \sum\nolimits_{k=1}^{|\mathcal{K}|} \mathrm{P}_k(y_m|\mathbf{x}) \tag{19}$$

where $\mathrm{P}_k(y_m|\mathbf{x}) \in [0, 1]$ is the prediction of kth base classifier that an observation $\mathbf{x}$ belongs to class y_m ($m = 1, \ldots, M$ and $k = 1, \ldots, |\mathcal{K}|$) and in which M is the number of labels (or concepts or classes). Based on a particular encoding, the selected classifier set $\mathcal{K}$ and their corresponding training set $\mathcal{D}$ with selected features are selected to learn the classifier set $\mathcal{M}$. The fitness of individual is then computed according to the accuracy of the classification task on the sets generated from T-fold Cross Validation : $\frac{1}{T}\sum_{i=1}^{T}\sum_{(\mathbf{x},y)\in\mathcal{D}^i}\frac{[\![h^i(\mathbf{x})=y]\!]}{|\mathcal{D}^i|}$, where h^i is the discriminative model obtained from the combining rule, $\mathcal{D}^i$ is the ith part of $\mathcal{D}$ obtained from Cross Validation procedure, $[\![\cdot]\!]$ returns 1 if the condition is true, otherwise 0, and $|\cdot|$ denotes the cardinality of a set.

The tagging procedure works in a straightforward way which is similar to the fitness computation. For each untagged region $\mathcal{R}$, its feature vector x is extracted. We choose the features corresponding to the optimal feature set and then input directly to $\mathcal{M}$, the concept of $\mathcal{R}$ is chosen by getting the maximal value given by the Sum Rule in (19).

4 Experimental Studies

4.1 Experimental Data and Settings

We chose two imaging datasets namely MSRC v1 and MSRC v2 to evaluate the proposed framework since both are used in region-level approach experiments [27]. The detailed information of the experimental datasets is given in Table 1.

Table 1. Information about the datasets used in the experiment

Dataset	MSRCv1	MSRCv2
# of image	240	591
# of regions	562	1482
# of images in training set	200	471
# of regions in training set	457	1179
# of images in test set	40	120
# of regions in test set	105	303
# of concepts	13	23

We employed five learning algorithms namely Linear Discriminant Analysis (LDA), Naïve Bayes, k Nearest Neighbor (K is set to 5 denoted by 5-NN), DT and Nearest Mean Classifier (NMC) to construct the ensemble system. The fitness of each individual in a generation was computed via 10-fold Cross Validation on the training set. Also, the parameters of the GA-based approach were simply set as: the maximum number of generations is 100, the number of individuals in each generation is 50, $PMul1 = 0.02$, and $PMul2 = 0.03$.

We chose several well-known benchmark algorithms solving the annotation at the region-level problem to compare their performances with those of our method. The four sparse coding-based methods namely Lasso [9], Group Lasso [10], Sparse Group Lasso [11], and SGSC [13] were compared in the experiments. Two state-of-the-art supervisor machine learning methods (DT and Linear SVM [28]) and three well-known ensemble methods (AdaBoost.M2 [29], Random Subspace [30], and Sum Rule [26]) were also evaluated in our experiments.

To set the parameters for these methods, we used the settings in the original papers. For those algorithms in which their parameters got value in a given range, we ran the method with all values and finally reported the best result corresponding with a specific value. The criteria used to compare the performance is the accuracy of region tagging process, given by the number of corrected tagged regions divided by the total number of regions in un-tagged set $\mathcal{T}$: $\sum_{(\mathbf{x},y)} [\![h(\mathbf{x}) = y]\!]/|\mathcal{T}|$ [3].

4.2 Results and Discussions

Tables 2 and 3 show the experimental results of the proposed method, five single classifiers generated by the five learning algorithms, and the selected benchmark algorithms. Our proposed method works on the optimally-selected classifiers and their selected features, so it obtains the best result compared to the base classifiers. For example, comparing to LDA, the best base classifier, the proposed method reaches the error rate that is nearly 3.8% lower on MSRCv1, 0.66% lower on MSRCv2. Consequently, our method meets the target of building an ensemble system where the error rate of the combiner is lower than those of any base classifiers.

Table 2. Classification error rates of five learning algorithms used in the proposed method

Methods	MSRCv1	MSRCv2
5-NN	0.3810	0.5875
DT	0.2191	0.3960
Naïve Bayes	0.4000	0.5446
NMC	0.6952	0.8086
LDA	0.1333	0.2541

Table 3. Classification error rates of the benchmark algorithms and the proposed method

Methods	MSRCv1	MSRCv2
Group Lasso	0.2095	0.4786
Lasso	0.2476	0.4522
Sparse Group Lasso	0.2000	0.4455
SGSC	0.0952	0.3003
Linear SVM	0.2191	0.4455
AdaBoost	0.3905	0.6370
Random Subspace	0.6381	0.7096
Sum Rule	0.1811	0.3300
The proposed method	0.0952	0.2475

Our method also outperforms AdaBoost, Random Subspace, Linear SVM, and Sum Rule. Consequently, our method captures more the relationship between the low-level features and high-level concepts than the other ensemble methods as well as state-of-the-art supervised learning algorithms among the two experimental datasets.

Thus, the proposed method can learn the optimal classifiers and associated features to generate the discriminative model.

It can be seen that the proposed method is better than Lasso, Group Lasso, Sparse Group Lasso on all two datasets while is better than SGSC on MSRCv2 dataset. The significant differences in error rate on MSRCv1 and MSRCv2 are approximately 10% and 21% respectively. It is a remarkable result since we implemented the sparse coding-based methods with a wide range of parameter values and reported the best result but our method is still significantly better (Table 3).

Table 4. Examples of region tagging procedure with the proposed method

	MSRCv1		
Predicted Label			
Human Tagging			
	MSRCv2		
Predicted Label	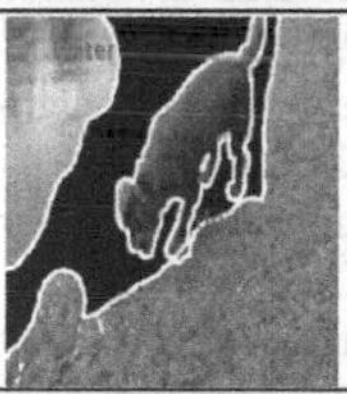		
Human Tagging	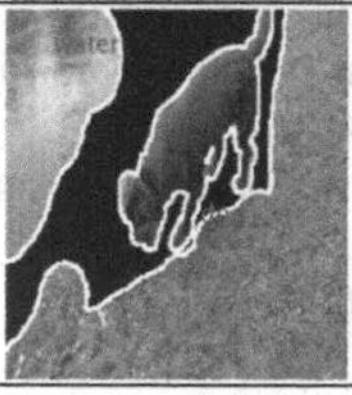		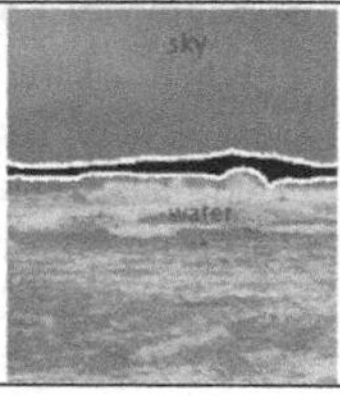

5 Conclusions

Accurate region-level image annotation is important in automatic scene analysis and would facilitate computer vision capability in many applications. In this paper, we have designed a GA based algorithm with novel chromosome structure that can encode both base classifiers and their associated features in a heterogeneous ensemble system. Through the evolution process, the optimal solution is obtained to improve the effectiveness of the region tagging task. The experiments were evaluated on MSRCv1 and MSRCv2 by using a wide range of well-known benchmark algorithms. It was demonstrated that our GA-based method with Sum Rule is better than the benchmark algorithms in the image region tagging task using the two datasets.

References

1. Yang, Y., Huang, Z., Yang, Y., Liu, J., Shen, H.T., Luo, J.: Local image tagging via graph regularized joint group sparsity. Pattern Recognit. **46**, 1358–1368 (2013)
2. Hu, J., Lam, K.-M.: An efficient two-stage framework for image annotation. Pattern Recognit. **46**, 936–947 (2013)
3. Han, Y., Wu, F., Shao, J., Tian, Q., Zhuang, Y.: Graph-guided sparse reconstruction for region tagging. In: CVPR, pp. 2981–2988 (2012)
4. Chang, C.-Y., Wang, H.-J., Li, C.-F.: Semantic analysis of real-world images using support vector machine. Expert Syst. Appl. **36**, 10560–10569 (2009)
5. Zhou, N., Cheung, W.K., Qiu, G., Xue, X.: A hybrid probabilistic model for unified collaborative and content-based image tagging. IEEE Trans. Pattern Anal. Mach. Intell. **33** (7), 1281–1294 (2012)
6. Shi, F., Wang, J., Wang, Z.: Region based supervised annotation for semantic image retrieval. Int. J. Electron. Commun. (AEU) **65**, 929–936 (2011)
7. Liu, Y., Zhang, D., Lu, G.: Region-based image retrieval with high-level semantics using decision tree learning. Pattern Recognit. **41**, 2554–2570 (2008)
8. Chen, Z., Hou, J., Zhang, D., Qiu, X.: An annotation rule extraction algorithm for image retrieval. Pattern Recognit. Lett. **38**, 1257–1268 (2012)
9. Tibshirani, R.: Regression shrinkage and selection via the lasso. J. Roy. Stat. Soc.: Ser. B (Methodol.) **58**(1), 267–288 (1996)
10. Yuan, M., Lin, Y.: Model selection and estimation in regression with grouped variables. J. Roy. Stat. Soc.: Ser. B (Stat. Methodol.) **68**(1), 49–67 (2008)
11. Friedman, J., Hastie, T., Tibshirani, R.: A note on the group lasso and sparse group lasso, arxiv preprint arXiv:1001.0736 (2010)
12. Han, Y., et al.: Graph-guided sparse reconstruction for region tagging. In: CVPR, pp. 2981–2988 (2012)
13. Yang, Y., et al.: Tag localization with spatial correlations and joint group sparsity. In: CVPR, pp. 881–888 (2011)
14. Nguyen, T.T., Liew, A.W.-C., To, C., Pham, X.C., Nguyen, M.P.: Fuzzy If-Then rules classifier on ensemble data. In: Wang, X., Pedrycz, W., Chan, P., He, Q. (eds.) ICMLC 2014. CCIS, vol. 481, pp. 362–370. Springer, Heidelberg (2014). https://doi.org/10.1007/978-3-662-45652-1_36

15. Nguyen, T.T., Pham, X.C., Liew, A.W.-C., Nguyen, M.P.: Optimization of ensemble classifier system based on multiple objectives genetic algorithm. In: ICMLC, vol. 1, pp. 46–51 (2014)
16. Nguyen, T.T., Liew, A.W.-C., Pham, X.C., Nguyen, M.P.: A novel 2-stage combining classifier model with stacking and genetic algorithm based feature selection. In: Huang, D.-S., Jo, K.-H., Wang, L. (eds.) ICIC 2014. LNCS (LNAI), vol. 8589, pp. 33–43. Springer, Cham (2014). https://doi.org/10.1007/978-3-319-09339-0_4
17. Nguyen, T.T., Liew, A.W.-C., Tran, M.T., Nguyen, M.P.: Combining multi classifiers based on a genetic algorithm – a gaussian mixture model framework. In: Huang, D.-S., Jo, K.-H., Wang, L. (eds.) ICIC 2014. LNCS (LNAI), vol. 8589, pp. 56–67. Springer, Cham (2014). https://doi.org/10.1007/978-3-319-09339-0_6
18. Nguyen, T.T., Nguyen, T.T.T., Pham, X.C., Liew, A.W.-C.: A novel combining classifier method based on Variational Inference. Pattern Recognit. **49**, 198–212 (2016)
19. Nguyen, T.T., Liew, A.W.-C., Tran, M.T., Nguyen, T.T.T., Nguyen, M.P.: Fusion of classifiers based on a novel 2-stage model. In: Wang, X., Pedrycz, W., Chan, P., He, Q. (eds.) ICMLC 2014. CCIS, vol. 481, pp. 60–68. Springer, Heidelberg (2014). https://doi.org/10.1007/978-3-662-45652-1_7
20. Nguyen, T.T., Liew, A.W.-C., Tran, M.T., Nguyen, M.P.: Combining classifiers based on gaussian mixture model approach to ensemble data. In: Wang, X., Pedrycz, W., Chan, P., He, Q. (eds.) ICMLC 2014. CCIS, vol. 481, pp. 3–12. Springer, Heidelberg (2014). https://doi.org/10.1007/978-3-662-45652-1_1
21. Yu, H., Li, M., Zhang, H.-J., Feng, J.: Color texture moments for content-based image retrieval. In: IEEE International Conference on Image Processing, pp. 929–932 (2002)
22. Gonzalez, R.C., Woods, R.E.: Digital Image Processing, 3rd edn. Prentice Hall press, Upper Saddle River (2007)
23. Tamura, H., Mori, S., Yamawaki, T.: Texture features corresponding to visual perception. IEEE Trans. Syst. Man Cybern. **8**(6), 460–473 (1978)
24. Deselaers, T., Keysers, D., Ney, H.: Feature for image retrieval: an experimental comparison. Inf. Retr. **11**(2), 77–107 (2008)
25. Thumfart, S., Jacobs, R.H.A.H., Lughofer, E., Eitzinger, C., Cornelissen, F.W., Groissboeck, W., Richter, R.: Modelling human aesthetic perception of visual textures. ACM Trans. Appl. Percept. **8**(4) (2011)
26. Nguyen, T.T., Pham, X.C., Liew, A.W.-C., Pedrycz, W.: Aggregation of Classifiers: A Justifiable Information Granularity Approach. CoRR abs/1703.05411 (2017)
27. Datasets. http://research.microsoft.com/en-us/projects/objectclassrecognition
28. Chang, C.-C., Lin, C.-J.: LIBSVM: a library for support vector machines. ACM Trans. Intell. Syst. Technol. **2**(3), 1–27 (2011)
29. Freund, Y., Schapire, R.E.: Experiments with a new boosting algorithm. In: ICML, pp. 148–156 (1996)
30. Ho, T.K.: The random subspace method for constructing decision forests. IEEE Trans. Pattern Anal. Mach. Intell. **20**(8), 832–844 (1998)

Implementing AI for Non-player Characters in 3D Video Games

Marek Kopel([✉]) and Tomasz Hajas

Faculty of Computer Science and Management, Wroclaw University of Science and Technology, Wybrzeze Wyspiańskiego 27, 50-370 Wroclaw, Poland
marek.kopel@pwr.edu.pl, tomaszhajas@gmail.com
http://ksi.pwr.edu.pl/kopel

Abstract. The purpose of this work was to find a solution for implementing intelligent behavior of independent NPC agents (non-player characters) in video games. NPC is a computer operated character - usually an enemy to the human user player. In modern video games NPCs are programmed to mimic human player behaviour to increase realism. Four approaches to NPC AI implementation were compared: decision tree, genetic algorithm, Q-learning, and a hybrid method. Results were aggregated and discussed along with recommending the best approach.

Keywords: Artificial intelligence · Video game
Non-player character · Neural network · Genetic algorithm
Q-learning · Unreal engine

1 Introduction

The key element of realism in video games is the behavior of computer controlled characters. Gameplay becomes unique, when artificial intelligence (AI) is a part of the immersive, virtual environment. In video games, every snippet of code that targets simulating "intelligent" behavior of virtual players is considered AI. This behavior on its own doesn't have to be sophisticated - its only real role is to fulfill potential player's expectations by providing adequate level of entertainment. Such software is most commonly based on a mere illusion of intelligence produced by a skillful usage of game design techniques, which harness simple controlling algorithms, realistic graphics, convincing character animations and voices borrowed from famous film actors. Often video game characters are controlled by a global algorithm, which has access to all variables inside the program and unjustly uses this knowledge to beat user player.

On average, AI created by video game developers is less sophisticated than techniques used in academic and industrial environments. [1] claims it happens mainly because of the following reasons:

- lack of CPU resources available to AI,
- suspicion in the game development community of using non-deterministic methods,

N. T. Nguyen et al. (Eds.): ACIIDS 2018, LNAI 10751, pp. 610–619, 2018.
https://doi.org/10.1007/978-3-319-75417-8_57

- lack of development time,
- lack of understanding of advanced AI techniques in the game industry,
- fact that efforts to improve the graphics in games usually overshadows all else, including AI.

Today, more and more popularity is gained by goal-driven artificial intelligence. It is a system for computer agents (characters), which allows them to plan an action sequence needed for achieving a given goal. Such sequence not only depends on the goal, but also on agent's actual state and environment. This means, that when two agents with different states share the same goal, they can generate two totally different action sequences. Such approach makes AI more dynamic and realistic.

1.1 Non-player Character

In video games, non-player character (NPC) is a computer operated character - usually an enemy to the human user player. In modern first person shooters (FPS) and other action games they are programmed to mimic human players [10] to increase realism. But their tradition go a long way back to first arcade games, like Space Invaders, where opponents moving pattern is getting more complex with each level. In massively multiplayer online role-playing games (MMORPGs) these characters are usually called mobs (from mobile objects) or bots. This is where "dumb mobs" were commonly used. They had no complex behaviors beyond attacking or moving around. Their implementation was a *static* model with a pre-made waypoints for each map that they followed. More recent implementations in both RPG and FPS also combine *dynamic* approach, which allows for "hunting" state. In this state NPC is actively looking for enemy traces with pathfinding and learning the map in real time. Complex behaviour of NPC based on decision making policy is usually implemented using heuristic, artificial intelligence methods and called AI. Early bot AI method - usually because of performance reasons - was cheating. This means bots were allowed actions and access to information that would be unavailable to the player in the same situation (e.g. the user player position). This technique is considered acceptable as long as the effect is not obvious to the player. But using heuristic AI will always give more human-like behaviour. As Simon Herbert states in [9], AI always had its niche in area, where computer programs use heuristic searches (which he names "the way of human thinking"), without guarantee of complete results, often using only sufficient success criteria.

1.2 Related Works

According to [8], state-of-the-art AI is done mostly using deep learning (DL) and neural networks (NN). But its application to games is still at the 8-bit console stage. Modern 3D games and character behaviours are yet too complex for current AI performance. Google Deepmind implementation of deep reinforcement learning (RL) is showcased in [5] by AI that is learning to play 8-bit Atari 2600 video games. Author of [4] created Arcade Learning Environment (ALE) allowing testing different approaches for achieving human-level performance in Atari 2600 emulator. The same way NN and genetic algorithm (GA) were successfully used in 16-bit platformer Super Mario World. MarI/O is SNES emulator AI plugin implemented by YouTuber SethBling using NEAT method described in [11]. It can outperform human in beating the game.

Earlier works on using RL for both 2D strategy and 3D FPS games, could not create NPC that would outperform human. However the goal there is different. According to [10] in FPS, like Unreal Tournament the goal is to make NPC act similar to human characters. In [2] the AI was applied to NPC in tank battle game. In [13] RL is used to create NPC Team playing Unreal Tournament. As claimed by authors: "the ultimate goal of game AI is to enhance entertainment, not to develop invincible NPCs", because a game you cannot win is no fun. In [6] authors create Genetic Bots, using GA and genetic programming, to outperform default bot AI in Unreal.

One way to deal with AI in modern games is to use a common framework - AI middleware. According to [7] this idea implementations can be found in game engines like Unity, Ureal Engine, CryEngine, and Havok. In popular game mashup platform Garry's Mod ([12]) the default way to create NPC is NextBot - the common model used as AI in games Team Fortress 2 and Left 4 Dead.

2 Method

The method tested as NPC AI starts as a deterministic decision tree (DT) based on finite state machine (FSM) concept. NN is first learned to mimic DT behaviour and then evolves using GA to find better solutions, achievable by DT. GA is a common algorithm to use for control decision making. But using it by itself shows it can stuck at local minima. This is where enhancing it with Q-learning (QL) becomes helpful. Decision making system is modeled as multilayer perceptron (MLP) NN with one layer of hidden neurons. The network uses feed-forward technique to process inputs from agent (NPC) sensors. The sensors feed the network with information like: is NPC in the air/on the ground, can NPC see an enemy, which way and how far is the enemy, what was last action of NPC, etc. NN output is next NPC action selected from predefined set: (turn left/right, move forward, atack, jump).

2.1 Experiment Setup

The model and all tested algorithms are implemented in Unreal Engine (see Fig. 1) using blueprints - a way for programming logic with graph diagrams. An example blueprint is presented in Fig. 2. The algorithms used (DT, GA, QL) are implemented in their basic, most popular versions. For the reinforcement the QL function is iterated as shown in formula 1. It is defined as the reward observed for the action A in current state S plus a fraction $\gamma \in <0, 1>$ of reward for the next state S' achieved by an optimal action A'.

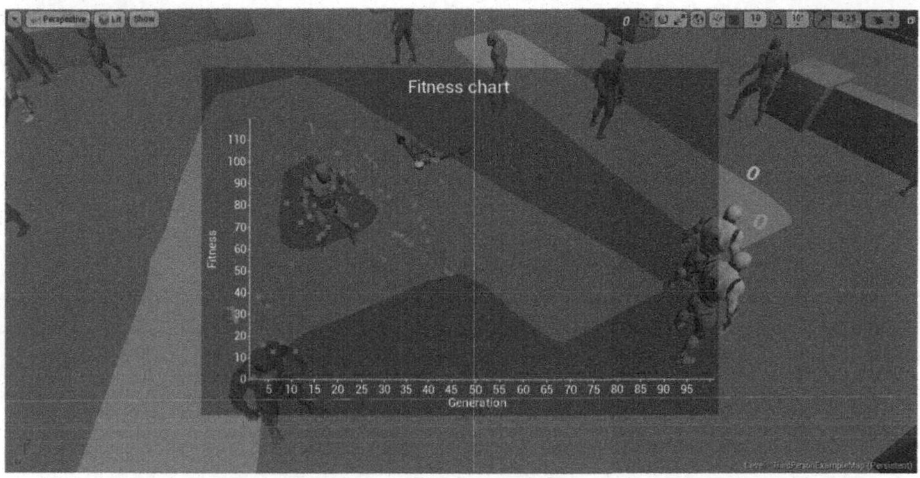

Fig. 1. Graphical interface of the research prototype, showing real time chart with fitness function for all NNs (for each NPC) through progressing generations. Behind the diagram a real time simulation is taking place with NPCs moving and interacting on a level map. Over each NPC head his current fitness function score is shown.

$$Q(S, A) = r + \gamma \cdot \max(Q(S', A')) \tag{1}$$

During the simulation the engine was running on 64-bit system using Intel® CoreTM i7-4810MQ CPU @ 2.80 GHz with 8 GB of RAM. This configuration allowed accelerating the simulation time and make more test possible.

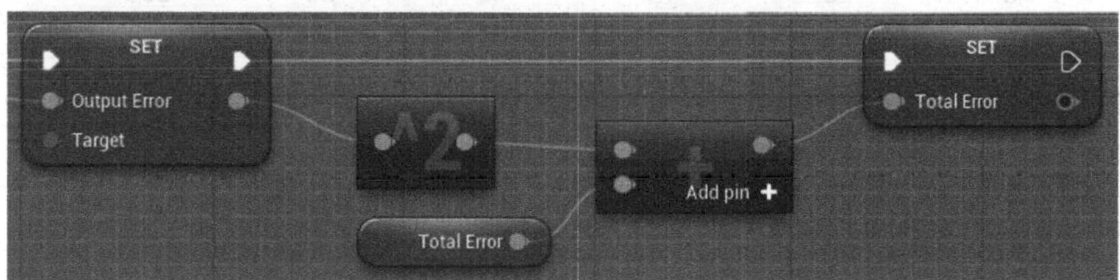

Fig. 2. An example part of blueprints. In Unreal Engine blueprints serve as visual way for representing and developing in-game logic. The algorithms and the NN model researched in this paper were implemented entirely using blueprints.

3 Results

During the experiment four approaches to NPC AI have been implemented. The approaches concern using DT, GA, QL, and a hybrid of GA+QL. Two main measures are used to compare the approaches:

1. learning speed of NPC NN in each generation (presented in Fig. 6)
2. fitness function values for each generation of NPCs (presented in Figs. 3 and 4),
3. total squared error of each NPC in each generation (presented in Fig. 5).

Fitness function values in Figs. 3 and 4 are plotted in 3 series with distinctive colors:

- Best - series for best individual in each generation,
- Average - series for average score of all individuals, in each generation
- Worst - series for worst individual in each generation.

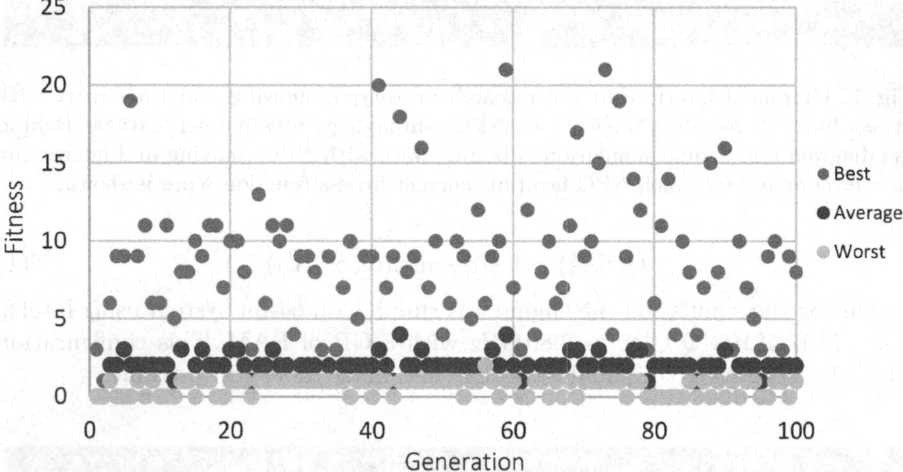

Fig. 3. Fitness function values of individual NPCs using GA approach. (Color figure online)

Individuals trained with QL method – because of processing only their own states – do not have a possibility of predicting other character movements. However, upon considering built-in limitations of recursion tree depth, this flaw does not seem substantial.

NN was able to imitate function Q even after 20 generations of individuals (which lasts 400 s in unaccelerated simulation). It was susceptible to overfitting (or overlearning) phenomenon, resulting in increasing the squared error over time. As stated in [3], overfitting is caused by too accurate adjustment of model to learning data. In this case the learning data are created in real time with QL technique. The more overfitted model is, the more damage causes a single neuron weight mutation. The main bottleneck here are values of weights oscillating near 0 - even a minor change (up or down) can end with rapid changes in general classification-related traits of the NN. Overlearned (overfitted) individuals are

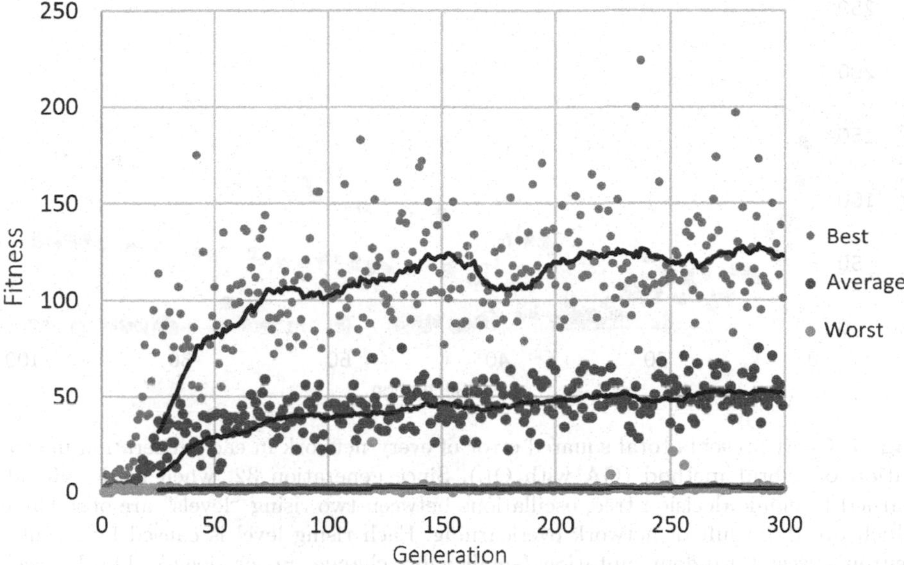

Fig. 4. Fitness function values of individual NPCs using GA+QL hybrid method. QL was stopped in generation 165. Independent GA with success is able to elevate average fitness value, and sometimes reaches even higher levels of evaluation of best individual. Near generation 230 AI instances periodically became similar to each other (GA trait), making fitness values equal. The moment of changing learning policy was empirically unnoticeable. Black lines mark trend lines for corresponding data series (average of 25 neighboring points) (Color figure online)

characterized by lower adaptability and susceptibility to minor model changes, e.g. genetic manipulations or fitness function change.

Early stopping method was used in order to avoid model overfitting. This method was creating backups of NNs just after observing significant increase of total squared error (see Fig. 5). Overfitting not always caused noticeable quality drop in agent's behavior - sometimes GA was creating solutions, which were recognized by strict QL as much worse, although to a human observer (and judging by fitness function values) they were still performing very well.

In order to prevent limitations of evolutionary potential, a decision was made to turn off the weight back-propagation implemented in QL algorithm, just after achieving low and stable total squared error values. GA left alone has partially substituted QL, increasing fitness of individuals even further (by making them diverse and selecting the best networks).

Fitness function charts for networks learning with QL technique were characterized by general increase oscillating around logarithmic curve (see Fig. 4).

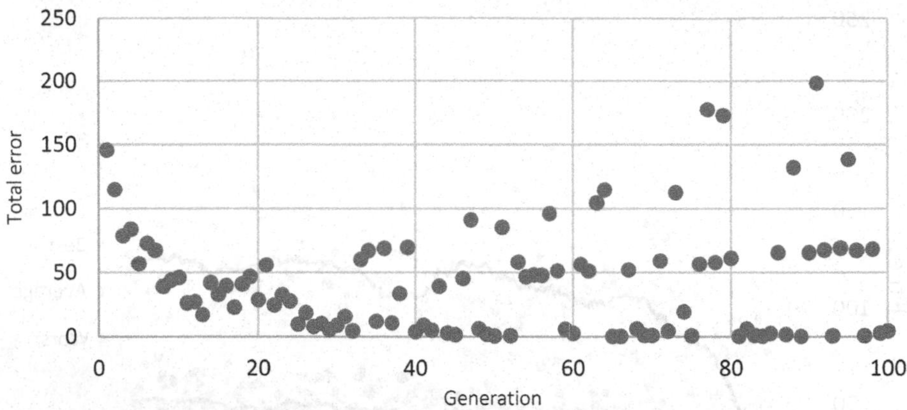

Fig. 5. Chart presents total squared error of every network in each generation in evaluation of hybrid method (GA with QL). Since generation 32, when model already learned to mimic decision tree, oscillations between two rising "levels" are observable, which are the result of network overlearning. Each rising level is caused by essential neuron's weight random mutation (small value change, up or down). Third, lowest level is composed of generations without essential weight mutation inside one of the networks.

4 Discussion

Both methods (GA and QL) were compared with each other and then with a hybrid of the two component methods. As a reference, DT method was used, which is also a key component of the QL method.

The first and the most obvious comparing factor was learning speed of NN models. Then maximal values of fitness function were researched as secondary measures.

Each starting population had weights picked at random from the same seed, and individuals were set to make decisions in a deterministic manner (winning output neuron implies final action). It allowed to avoid excessive randomness and construct comparison conclusions. Fitness values for best individuals are shown in Fig. 7.

Speed of building AI model (learning) with QL was noticeably greater because of a constant flow of data which came from environment (see Fig. 6). Isolated GA approach needs several times more generations to achieve average fitness values comparable with QL results. The results show that simulation delay caused by learning model is acceptable for GA, QL and their hybrid solution.

Maximal fitness values of individuals in both methods were similar, but in GA approach the achieved values should be associated with a great randomness factor – what in Fig. 3 looks like a bunch of oscillating points.

Using GA caused no performance problems. However, for QL - when used with high number of recursive steps - this was not the case. Number of recursive steps in quality function had a great impact on the performance of learning

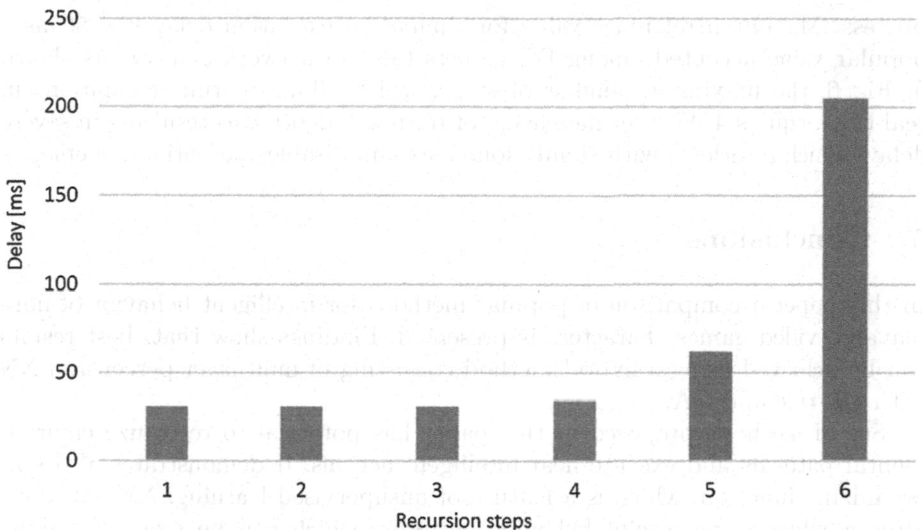

Fig. 6. Average delay (in milliseconds/frame) dependent on recursion depth in Q function. Calculated in the moment of simulation's 10 first seconds with player camera directed towards the center of simulation area. PC specs: Processor: Intel® Core™ i7-4810MQ CPU @ 2.80 GHz, RAM: 8 GB, system architecture: 64-bit

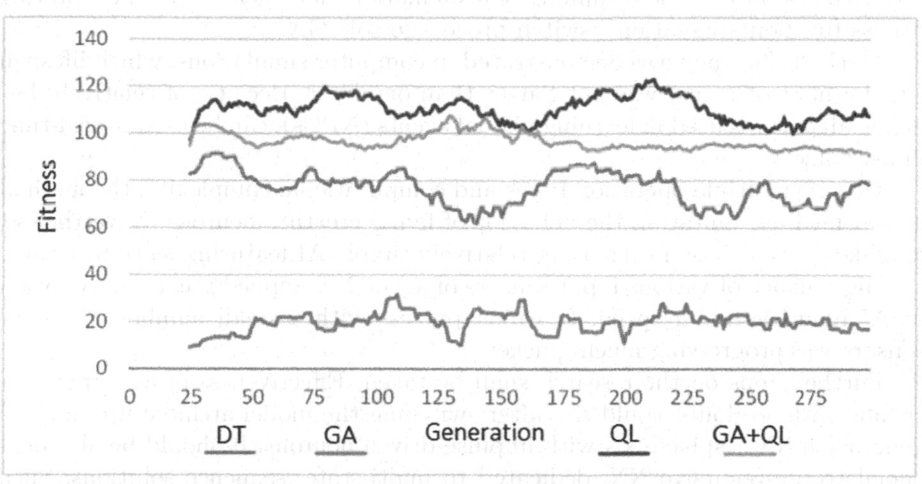

Fig. 7. Fitness function values for best individuals (trend lines of moving average including 25 closest points). Learning algorithms were compared with simple decision tree, which during the learning took a role of Q function. QL algorithm and hybrid solution with GA outperformed decision tree with fitness function values. Sole GA – in opposition to QL – did not expose any ability to develop complex NPC behavior.

process. Maximum reference value for simulation execution delay was 50 ms – popular value accepted among PC gamers (also in network games). As shown in Fig. 6, the maximum number of steps, which allows to run calculations in real time, equals 4. Further increasing of recursion depth was resulting in severe delay, which interferes with simulation clock and disables potential gameplay.

5 Conclusions

In this paper a comparison of popular methods for intelligent behavior of non-playable video games characters is presented. Findings show that, best results can be achieved using a hybrid method consisting of multilayer perceptron NN with QL rule and GA.

Set of methods proposed in this paper has potential to recognize environmental patterns and execute near-intelligent actions. It demonstrates also self-organizing function, which is a feature of unsupervised learning. NN left alone after a while learns useful behavior patterns, which can be seen as making progress in a simulation. A goal achieved this way is analogical to the one achieved by a simple, decision tree-based control algorithm. But this one is enriched with adaptability aspect, common for NNs.

During the experiments, QL properties has shown learning acceleration feature, but its potential is limited to a defined Q function. It makes sense to use QL with GA only at the beginning of a simulation, and then - after encountering fitness function stagnation - switch process to sole GA.

Method efficiency was demonstrated in computer simulations, which lifespan (in the most of cases) was not longer than one hour. Because of relatively low delay, algorithms used in learning virtual agents (NPCs), can be used in real-time video games.

Despite multiple operation types and computational complexity, the method is still far from imitating the behavior of living creature neurons. Nevertheless, it makes a good base for creating relatively simple AI featuring learning traits.

Big number of various input sensors of agent NN implied great deterioration of AI model learning speed. Learning process with a small number of binary sensors was progressing much quicker.

Further steps on the research shall be taken. Effectiveness of learning algorithms with no doubt could rise after switching the model architecture into 3rd generation NN – spiked NN with impulse-driven neurons. It should be also beneficial to use recursive NN, dedicated to multi-state, sequence solutions. Such sequences can be surely found in 3D computer environment. However it should be reminded that the optimal solution is a consensus between accurate representation of human behavior and device-dependent computational performance.

References

1. Fairclough, C., Fagan, M., Mac Namee, B., Cunningham, P.: Research directions for AI in computer games. Technical report, Trinity College Dublin, Department of Computer Science (2001)
2. Fang, Y.P., Ting, I.H.: Applying reinforcement learning for game AI in a tank-battle game. In: 2009 Fourth International Conference on Innovative Computing, Information and Control (ICICIC), pp. 1031–1034. IEEE (2009)
3. Gurney, K.: An Introduction to Neural Networks. CRC Press, Boca Raton (1997)
4. Hosu, I.A., Rebedea, T.: Playing atari games with deep reinforcement learning and human checkpoint replay (2016). arXiv preprint arXiv:1607.05077
5. Mnih, V., Kavukcuoglu, K., Silver, D., Rusu, A.A., Veness, J., Bellemare, M.G., Graves, A., Riedmiller, M., Fidjeland, A.K., Ostrovski, G., et al.: Human-level control through deep reinforcement learning. Nature **518**(7540), 529–533 (2015)
6. Mora-García, A.M., Merelo-Guervós, J.J.: Evolving bot's AI in unreal^{TM}. In: Algorithmic and Architectural Gaming Design: Implementation and Development, pp. 134–157. IGI Global (2012)
7. Safadi, F., Fonteneau, R., Ernst, D.: Artificial intelligence in video games: towards a unified framework. Int. J. Comput. Games Technol. **2015**, 5 (2015)
8. Schmidhuber, J.: Deep learning in neural networks: an overview. Neural Netw. **61**, 85–117 (2015)
9. Simon, H.A.: Artificial intelligence: an empirical science. Artif. Intell. **77**(1), 95–127 (1995)
10. Soni, B., Hingston, P.: Bots trained to play like a human are more fun. In: IEEE International Joint Conference on Neural Networks, IJCNN 2008. (IEEE World Congress on Computational Intelligence), pp. 363–369. IEEE (2008)
11. Stanley, K.O., Miikkulainen, R.: Evolving neural networks through augmenting topologies. Evol. Comput. **10**(2), 99–127 (2002)
12. Studios, F.: Garry's mod (2006)
13. Wang, H., Gao, Y., Chen, X.: Rl-dot: a reinforcement learning NPC team for playing domination games. IEEE Trans. Comput. Intell. AI Games **2**(1), 17–26 (2010)

Sensor Networks and Internet of Things

Comparison of Indoor Positioning System Using Wi-Fi and UWB

Jaemin Hong[1], KyuJin Kim[2], and ChongGun Kim[1(✉)]

[1] Department of Computer Engineering, Yeungnam University, Gyeongsan 38541, Korea
hjm4606@naver.com, cgkim@yu.ac.kr

[2] Department of Nursing, Kyungpook National University, Daegu 41944, Korea
kayjay6t@naver.com

Abstract. Recently, as smart mobile devices become popular, location-based services are wide spreading. In the indoor positioning and guiding system, the accuracy and efficiency of the system are important. Various indoor positioning studies have been studied, and positioning accuracy of various indoor location-based services are also referred. Currently, research using Wi-Fi is the most active, and research using Ultra Wide Band (UWB) is getting attention. In this paper, a design concept of placement devices for indoor positioning based on UWB. The indoor positioning efficiency of Wi-Fi and UWB is also compared.

For calm alarming in the indoor emergency situation, UWB is investigated on acceptability.

Keywords: Indoor positioning · Wi-Fi · UWB

1 Introduction

Recently, location-based services have been attracting attention as smart mobile devices are activated. With the success of location-based services using satellite's Global Positioning System (GPS) outdoors, service providers have begun to pay attention to indoor location-based services. Various indoor positioning studies have been conducted, and indoor location based services are also increasing.

GPS is the most common outdoor positioning system, but it can not be used in-doors and does not work well in areas such as dense buildings. [8] Therefore, various methods for indoor positioning are being studied. Currently, the most popularly known indoor positioning method is a method using a Wi-Fi.

However, indoor positioning using Wi-Fi has a lot of errors. Therefore, indoor positioning using UWB, which is a new indoor positioning method, is getting attention. UWB is suitable for indoor location tracking because of its high multipath resolution and low impact of obstacles.

In this paper, we compare the effectiveness of indoor positioning using Wi-Fi and indoor positioning using UWB, which will be more suitable.

N. T. Nguyen et al. (Eds.): ACIIDS 2018, LNAI 10751, pp. 623–632, 2018.
https://doi.org/10.1007/978-3-319-75417-8_58

2 Related Researches

In this chapter, various indoor positioning methods and their advantages and disadvantages are discussed. In addition, the UWB communication is also discussed.

2.1 Indoor Positioning Methods

Typical positioning methods include triangulation method, and fingerprint method. In order to use the triangulation method, the distance from a plurality of wireless APs is measured, and the accurate position is obtained by summing up the distance information. In indoor environments, RSSI (Received Signal Strength Indication) values are also used.

Fingerprint is a method to measure and store the pattern of RSSI values at various points and calculate the similarity with the measured value in real time to positioning. The data of the RSSI pattern stored in advance is called training data.

Triangulation method. Triangulation is a method of calculating the position from three points that know the distance. If we know the exact distance, we can calculate the exact point. [3] Methods such as AOA, TOA, and TDOA also use triangulation methods [1].

In an indoor environment where Wi-Fi is popularly used, the distance between each AP and the device is calculated by the RSSI value of Wi-Fi. The formula for obtaining the distance between the AP and the device using the RSSI value is as follows.

$$\mathrm{RSSI} = -(10\mathrm{n}\ \log_{10} d + \mathrm{A}) \tag{1}$$

However, even with the same RSSI value, the distance may vary depending on the environment, so it is necessary to calculate the environment variable according to the place or environment. Since the RSSI value is not proportional to the distance and the fluctuation width is very small over a certain distance, the error between the real distance and the measured distance increases.

Fingerprint method. Fingerprint method stores RSSI value patterns of APs collected at various points to create training data, It measures the position by calculating the similarity between the RSSI value measured in real time and the training data (Table 1).

K-Nearest Neighbor (K-NN) algorithm is a method to compare similarity between real time RSSI values and training data. K-NN algorithm is widely used for pattern

Table 1. Example of training data

Point	AP1	AP2	AP3	AP4
0	−61	−63	−55	−47
1	−52	−56	−55	−45
2	−55	−56	−65	−42
…	…	…	…	…

recognition. When the real-time RSSI value is measured in the position measurement step, the data most similar to the real-time RSSI value can be selected from the training data based on the similarity distance. The following formula is used for the similarity distance calculation.

$$D_i = \left(\sum_{j=1}^{n} |S_j - S_{ij}|^q\right)^{1/q} \tag{2}$$

In this formula 'n' is the number of wireless APs used for location measurement, 'j' is the number of the wireless AP, and 'i' is the number of the training data collection point. 'q' has a value of 1 for Manhattan distance method and 2 for Euclidean distance method. S_j is the RSSI value of the j-th AP at the real-time observation point, and S_{ij} is the RSSI value of the j-th AP measured at the i-th collection point. D_i is the similarity distance, and the smaller the value, the closer the point [2].

Since the fingerprint method only requires the RSSI value, it has an advantage that the existing wireless communication environment can be used without any additional setting. And, if accurate training data is collected, positioning accuracy is the highest. However, the RSSI value varies from time to time depending on the environment or the state of the terminal, and many changes occur due to time, temperature, presence of an obstacle, and the like. Therefore, it is difficult to obtain accuracy when collecting training data or measuring real-time RSSI values.

2.2 Ultra Wide Band

It is usually called UWB, and it is one of the recently attracted communication methods. UWB is less impacted by obstacles and radio interference, and is more energy efficient. [6] Because of these advantages, it is suitable for indoor positioning and error is less than 1 m, which is better than other wireless communication methods. The distance is calculated using the Round Trip Time of the signal, not the RSSI value, and the position is measured by triangulation method with the distance. However, since the UWB device is not widely used, it is not possible to use it as a popular device such as a smart phone, and a cost problem arises because a dedicated device is manufactured or purchased.

2.3 Calm Indoor Alarm System

A system that can send emergency event to user by using multiple communication methods when an event occurs and the user can respond to the emergency in real time is proposed. The system is also can record and manage all emergency information.

An indoor calm alarm system in a general hospital, the positioning accuracy is important for guiding the emergency position. In the previous studied system [8], some errors for positioning are shown. In the proposed system, more accurate positioning is very important.

The acceptance of UWB based positioning on the indoor alarm system is a challenge.

3 Design of Positioning System

3.1 Experimental Environment

Samsung Galaxy S3 was used as a Wi-Fi receiving device, and ipTIME N3 was used as a wireless AP for Wi-Fi environment. As the UWB module, DW-1000 of Decawave was used and as a transmitter for transfer from UWB module to Server, Raspberry pi 3 was used.

Each AP or module is arranged as shown in Fig. 1. The training data was collected for each coordinate and the collection interval was 1 m.

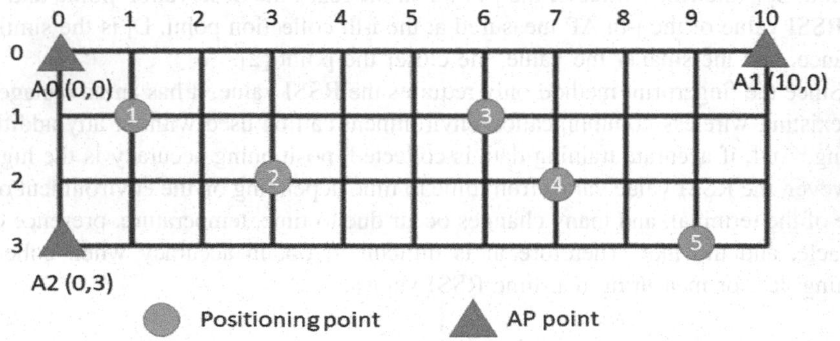

Fig. 1. An experimental environment based on wireless signal

3.2 Design of a System Using Wi-Fi

Figure 2 shows concept of the system using Wi-Fi.

The user's device collects RSSI values of the APs and transmits them through Two-way TCP socket communication using Wi-Fi. The server compares the RSSI

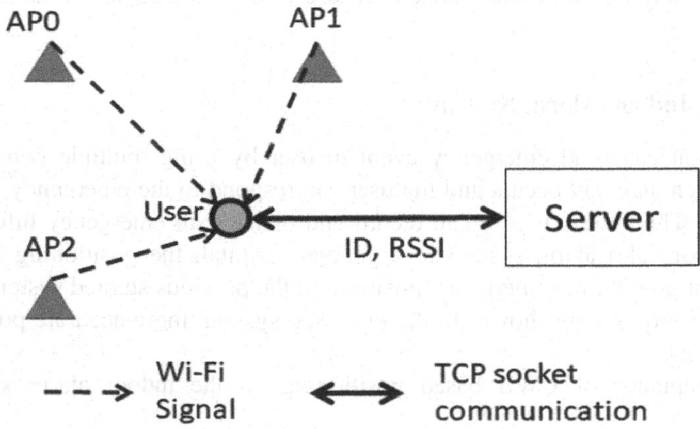

Fig. 2. Concept of the system using Wi-Fi

value with the training data, calculates the similarity distance, searches for coordinates at the nearest point and displays it in the system.

The data exchanged between the server and the user device is ID and RSSI value, and the data transfer format is shown Fig. 3.

ID	/	RSSI0	/	RSSI1	/	RSSI2	/q
	separator	RSSI value of AP0		RSSI value of AP1		RSSI value of AP2	End mark

Example) 0/-51/-48/-37/q --> ID : 0, RSSI0 : -51, RSSI1 : -48, RSSI2 : -37

Fig. 3. Data transfer format between server and user device

3.3 Design of System Using UWB

Concept of the system using UWB is shown in Fig. 4.

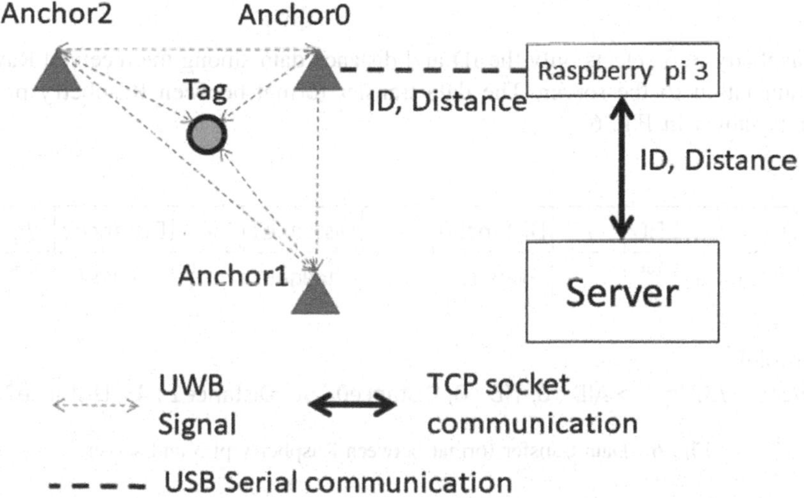

Fig. 4. Concept of the system using UWB

Anchor, which is three fixed modules, and Tag, which is a user device module, share the distance value between each node through UWB communication. Anchor0 sends the ID and distance data between all nodes to Raspberry pi 3 via USB serial communication. Raspberry pi 3 transmits ID and distance data to the server through TCP socket communication. The server calculate the coordinate using the received distance data and displays it in the system.

The data exchanged between Anchor and Raspberry pi 3 is Raw data including ID and distance. Figure 5 shows data transfer format between anchor and Raspberry pi 3.

mc	07	00000000	00000000	00000000	00000000	1d43	6e	00a0ddf7	a0:0
MID	MASK	RANGE				NRANGE	RSEQ	DEBUG	aT:A

MID
- mr : Raw ranges (tag to anchor)
- mc : Bias corrected ranges (tag to anchor)
- ma : Bias corrected ranges (anchor to anchor)

MASK – States which RANGEs are valid
RANGE – Range value of tag to anchor or anchor to anchor
NRANGE – Number of ranges completed by reporting unit
RSEQ – Range sequence number
DEBUG – Time of last range reported
aT:A – T is tag ID, A is the anchor ID

Fig. 5. Data transfer format between anchor and Raspberry pi 3

Raspberry pi 3 selects only the ID and distance data among the received Raw data and transmits it to the server. The data transfer format between Raspberry pi 3 and server is shown in Fig. 6.

AID	/	TID	/	Distance0	/	Distance1	/	Distance2	/q
Anchor ID	separator	Tag ID		RANGE0		RANGE1		RANGE2	End mark

Example)
0/0/50/43/32/q --> AID : 0, TID : 0, Distance0 : 50, Distance1 : 43, Distance2 : 32

Fig. 6. Data transfer format between Raspberry pi 3 and server

4 Efficiency Comparison of Positioning Using Wi-Fi and UWB

4.1 Accuracy Comparison

Training data was collected at each point for positioning using Wi-Fi, Table 2 shows the training data of experimental environment.

The position was measured five times for each positioning point. The measured errors are shown in Table 3.

Table 2. Training data of experimental environment

AP0	−28	−39	−37	−48	−46	−50	−51	−59	−53	−58	−54
	−39	−38	−38	−41	−48	−51	−52	−55	−54	−57	−56
	−37	−38	−40	−48	−48	−50	−52	−56	−57	−56	−57
	−48	−42	−48	−48	−52	−52	−55	−54	−57	−57	−56
AP1	−54	−58	−53	−59	−51	−51	−46	−48	−37	−39	−28
	−56	−57	−54	−55	−52	−51	−48	−41	−38	−38	−39
	−57	−56	−57	−56	−52	−50	−48	−48	−40	−38	−37
	−56	−57	−57	−54	−55	−52	−52	−48	−48	−42	−48
AP2	−48	−42	−48	−48	−52	−52	−55	−54	−57	−57	−56
	−37	−38	−40	−48	−48	−52	−52	−56	−57	−56	−57
	−39	−38	−38	−41	−48	−51	−52	−55	−54	−57	−56
	−28	−39	−37	−48	−46	−50	−51	−59	−53	−58	−54

Table 3. Error of indoor positioning using Wi-Fi

	1	2	3	4	5	Avg.
Point 1	3.6 m	3.2 m	1 m	1 m	1 m	1.95 m
Point 2	2 m	2 m	1.4 m	5.1 m	3 m	2.7 m
Point 3	5.1 m	2.2 m	6.1 m	5.1 m	1 m	3.9 m
Point 4	5.4 m	7.1 m	2.2 m	3.2 m	4.1 m	4.4 m
Point 5	1 m	7.1 m	5.1 m	1.4 m	1.4 m	3.2 m
Avg.	3.42 m	4.32 m	3.16 m	3.16 m	2.1 m	3.23 m

Error of indoor positioning using Wi-Fi is minimum 1 m, maximum 7.1 m and average is 3.23 m. The best result is shown in Fig. 7 as coordinates.

Table 4 shows the error of indoor positioning using UWB in the same experimental environment.

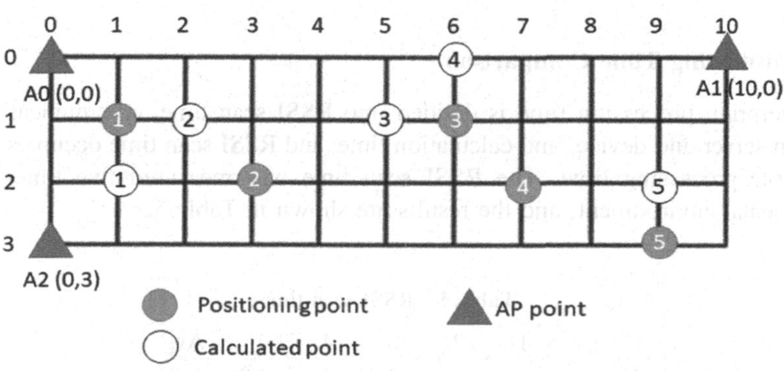

Fig. 7. Results of indoor positioning using Wi-Fi

Table 4. Error of indoor positioning using UWB

	1	2	3	4	5	Avg.
Point 1	0.08 m	0.09 m	0.01 m	0.07 m	0.06 m	0.062 m
Point 2	0.04 m	0.05 m	0.1 m	0.06 m	0.04 m	0.058 m
Point 3	0.03 m	0.02 m	0.04 m	0.04 m	0.06 m	0.038 m
Point 4	0.05 m	0.07 m	0.08 m	0.11 m	0.06 m	0.074 m
Point 5	0.09 m	0.08 m	0.06 m	0.07 m	0.07 m	0.074 m
Avg.	0.058 m	0.062 m	0.058 m	0.07 m	0.058 m	0.061 m

Error of indoor positioning using UWB is minimum 0.01 m, maximum 0.11 m, and average is 0.061 m. The best result is shown in Fig. 8 as coordinates.

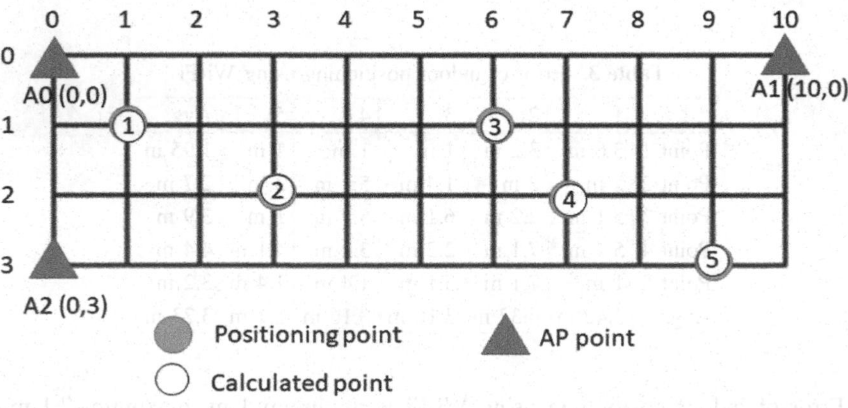

Fig. 8. Result of indoor positioning using UWB

Indoor positioning using UWB is much more accurate than indoor positioning using Wi-Fi.

4.2 Positioning Time Comparison

In Fingerprint, processing time is divided into RSSI scan time, communication time between server and device, and calculation time, and RSSI scan time occupies most of the whole processing time. The RSSI scan time was measured five times in this experimental environment, and the results are shown in Table 5.

Table 5. RSSI scan time

	1	2	3	4	5	Avg.
Scan time	2.6 s	2.5 s	2.6 s	2.4 s	2.8 s	2.58 s

The RSSI scan time, which accounts for the largest portion of the processing time, is about 2.58 s on average, and Fast positioning is difficult due to long scan times. The total processing time is close to about 3 s, and a time error occurs between the actual coordinates and the calculated coordinates. Since a moving person moves about 1 m per second, the time error of 3 s is similar to the distance error of 3 m.

In the indoor positioning using UWB, the processing time is divided into the distance calculation time using the round trip time of the signal, the communication time between modules, and the transmission time to the server, the distance calculation time is negligibly small. The total processing time is about 0.3 s, and the position can be measured up to three times per second. Since the time taken to measure the position is short, it is advantageous in tracking the position of the person on the move.

4.3 Experiment Result

In the indoor positioning using Wi-Fi, there was an average position error of 3.23 m, and the processing time was about 3 s. This is a large error in the indoor, and it is very unsuitable to track people on the move.

However, the average position error in UWB indoor positioning was 0.061 m, and the processing time was about 0.3 s, which is more accurate and faster than Wi-Fi. It is suitable for measuring people on the move because of less error and faster processing time.

5 Conclusions

In this paper, we designed an indoor location tracking system using UWB or Wi-Fi and compared the efficiency of both systems in terms of accuracy and processing time. UWB is more suitable for indoor positioning system because it has better position accuracy and processing time than Wi-Fi.

The positioning effect based on UWB is analyzed to show the likelihood of the indoor location based service. The experiment results show that the proposed system using UWB can be used as a practical one.

For the future research the following have to be studies. We design a system that allows a wider range of indoor positioning. By studying the opinions of doctors and nurses on a general hospital, applicable likelihood has to be studied. In addition, qualitative and quantitative improvement is required.

Acknowledgement. This research was supported by The Leading Human Resource Training Program of Regional Neo industry through the National Research Foundation of Korea (NRF-2016H1D5A1909408) funded by the Ministry of Science, ICT and future Planning, and the BK21 Plus Program funded by the Ministry of Education (MOE, Korea) and National Research Foundation of Korea (NRF).

References

1. Zhou, Y., Law, C.L., Chin, F.: Construction of local anchor map for indoor position measurement system. IEEE Trans. Instrum. Meas. **59**(7), 1986–1988 (2010)
2. Hatami, A., Pahlavan, K.: Comparative statistical analysis of indoor positioning using empirical data and indoor radio channel models. In: IEEE Communications Society Subject Matter Experts for Publication in the IEEE CCNC 2006 Proceedings, pp. 1018–1022 (2006)
3. Dong, Q., Dargie, W.: Evaluation of the reliability of RSSI for indoor localization. In: 2012 International Conference on Wireless Communications in Unusual and Confined Areas (ICWCUCA), pp. 1–6. IEEE (2012)
4. Trevisani, E., Vitaletti, A.: Cell-ID location technique, limits and benefits: an experimental study. In: Proceedings of Mobile Computing Systems and Applications, pp. 51–60 (2004)
5. Hulbert, I.A.R., French, J.: The accuracy of GPS for wildlife telemetry and habitat mapping. J. Appl. Ecol. **38**(4), 869–878 (2001)
6. Di Benedetto, M.-G. (ed.): UWB Communication Systems: A Comprehensive Overview. Vol. 5. Hindawi Publishing Corporation, New York (2006)
7. Park, S.-J., Kim, M.-G.: Implementation of indoor location aware system using 802.11 wireless signal learning algorithm. In: Korean Institute of Information Scientists and Engineers, vol. 34, no. 1C, pp. 361–365. (2007)
8. Hong, J.: Improving trackting performance on indoor positioning system. (MS thesis). Yeungnam University, Korea

A New Approach to Estimate Urban Air Temperature Using Smartphones

Nguyen Hai Chau(✉)

Faculty of Information Technology,
VNU University of Engineering and Technology, Hanoi, Vietnam
chaunh@vnu.edu.vn

Abstract. Measuring urban air temperature at a high spatial resolution is very important for many applications including detection of urban heat islands. However air temperature is currently measured by professional weather stations those are very sparse at 50 km spatial resolution or more. In this paper, we propose a new approach to estimate air temperature from smartphones. Most of the smartphones are not equipped with air temperature sensors. However they are all equipped with battery temperature sensors. When a smartphone is in idle state, its battery temperature is stable and correlated with air temperature around the smartphone. Therefore we have developed a linear regression model to estimate air temperature from the idle smartphones battery temperature. Experiment results show that the new approach is statistically comparable to an existing one using mean error, mean absolute error and coefficient of determination metrics. Advantages of the new approach include simplicity of implementation on smartphones and ability of creating maps of temperature distribution.

Keywords: Smartphone · Battery temperature · Internet of Things
Crowdsourcing

1 Introduction

Temperature is a physical quantity that is very important to human health. In urban areas due to different concentration of roads, buildings and population, there exists urban heat islands (UHI) [1]. An UHI is an urban area that is significantly warmer that its surroundings. The UHI has some disadvantages. It decreases air quality, water quality and directly influences human health. Collection of temperature data for detecting UHI is an important task.

Currently there are approaches for collection of temperature data: direct measurement, using satelite images, estimating temperature from normalized difference vegetable index (NDVI) and crowdsourcing [1]. Direct measurement is often performed by weather stations at very sparse spatial resolution. Satellite data and NDVI approach is limited by temporal resolution that is maximum 2–4 observations per day.

N. T. Nguyen et al. (Eds.): ACIIDS 2018, LNAI 10751, pp. 633–641, 2018.
https://doi.org/10.1007/978-3-319-75417-8_59

In the crowdsourcing approach, Overeem et al. [2], Droste et al. [3] collect battery temperature from a large number of smartphones and use a heat transfer model to estimate air temperature of urban areas. Data is collected when a smartphone is turned on, turned off, plugged in and unplugged [2]. The areas are defined by clusters of smartphones. Experiment results of Overeem et al. are very promising. They published an app on Google Play for data collection since 2014 [4]. However, in an area with a small number of smartphones, the estimated temperature is not very accurate. Furthermore, this approach provides aggregated temperature of an area from a large number of smartphone battery temperature readings rather than the area's temperature distribution.

In this paper, we use the crowdsourcing approach. We use battery smartphone temperature readings to estimate air temperature around the smartphone. In contrast with Overeem et al., we build a statistical model that allows each smartphone to estimate its surrounding air temperature independently. Following this approach we are able to have a temperature distribution rather than aggregated temperature of an area. In the next section, we present our experiments and results.

2 Experiments and Results

2.1 Equipments, Experiment Environment and Data Collection

When a smartphone is in idle state for a while, its battery temperature is stable and is correlated with air temperature around the phone. When the smartphone is not idle, its battery temperature fluctuates and depends highly on many factors such as CPU load, screen brightness level, 3G/WiFi and GPS status. [5]. Thus we consider battery temperature of an idle smartphone is data and that of a non-idle one is noise. We keep the data and omit the noise for model building.

We use three Android smartphones to collect data. Information of the smartphones is in Table 1.

Table 1. List of smartphones used in experiments.

Smartphone model	Manufacturer	OS version	Air temperature sensor
Nexus 4	LG Electronics	Android 5.1	Not available
Galaxy Note 2	Samsung	Android 4.4.2	Not available
Galaxy Note 3	Samsung	Android 4.4.2	Available

Among the three smartphones, Galaxy Note 3 is the only one equipped with an air temperature sensor. We use this smartphone for collection of training and testing data. Nexus 4 and Galaxy Note 2 are used for collection of testing data. Detail of data to collect is in Table 2. Battery information includes battery temperature, charging/discharging status (unplugged/AC plugged/USB plugged/WiFi), voltage and level. Screen status is on or off. For compatibility purpose, we do not collect battery current parameter that is not available for Android 4.4.2 and older versions. The experiments are conducted in Oct–Nov 2015 in Hanoi.

Table 2. Information to collect from each smartphone.

Smartphone	Information to collect
Nexus 4	Battery information, screen status
Galaxy Note 2	Battery information, screen status
Galaxy Note 3	Battery information, screen status, air temperature

When conducting experiments, we position three smartphones close to each others and put them into three environments with different temperature: room (about 20–25° C), fridge (0–4° C) and hot (30–35° C).

2.2 Model Building and Testing

We developed an Android app to collect information listed in Table 2 at 1 Hz frequency. We define idle state when the phone's screen is off, the phone is unplugged, and battery temperature variance in a temporal window is small enough. Thus variables to build our model are air temperature, battery temperature and battery voltage.

We filter data to keep one of idle status and build a linear regression model. The model's outcome is air temperature (variable name is `airtemp`). Its predictors are battery temperature (`battemp`) and battery voltage (`batvoltage`). The model is described in Table 3, its coefficient of determination is $R^2 = 0.7912$. In this table, row names are intercept and variables (predictors) of the model. Values in second, third and fourth columns are estimated regression coefficients, standard errors, t-values of the coefficients, and p-values of the variables, respectively. The model is

$$T_{air} = 2.085 + 0.874 \times T_{battery} - 0.0004 \times V_{battery}, \quad (1)$$

where T_{air} and $T_{battery}$ are air temperature and battery temperature when smartphone is in idle state (Celsius degree), respectively; $V_{battery}$ is battery voltage in mV. We refer this model as the first one.

Table 3. A linear model with battery temperature and voltage as predictors.

	Estimate	Std. error	t value	Pr(>|t|)
(Intercept)	2.0850	0.4284	4.87	0.0000
battemp	0.8740	0.0049	178.25	0.0000
batvoltage	−0.0004	0.0001	−3.99	0.0001

At present, most of the smartphones are equipped with Li-Ion batteries [6]. Since the regression coefficient of `batvoltage` is very small (−0.0004) and voltage of Li-Ion batteries, running from 3700 to 4200 mV [7], changes very slowly when smartphone is idle, we eliminate `batvoltage` from the model and have a new linear regression model in Table 4. This is the second model.

Table 4. A linear model with battery temperature as the only predictor.

	Estimate	Std. error	t value	Pr(>|t|)
(Intercept)	0.4776	0.1469	3.25	0.0012
battemp	0.8765	0.0049	180.13	0.0000

Table 5. Comparison of the first and second models.

Res.Df	RSS	Df	Sum of Sq	F	Pr(>F)
8577	16667.49				
8578	16698.49	−1	−31.00	15.95	0.0001

The second model's R^2 is 0.7909 and can be considered the same as that of the first one. We perform an analysis of variance (ANOVA) [8] test to compare the two models. Test result is in Table 5. The result shows very small difference of the first and second models. We have nearly similar residual sum of squares (16667.49, 16698.49) in **RSS** column and small difference sum of squares (−31.00) in **Sum of Sq** column. We select the second model because it is simpler than the first one but its R^2 is similar to that of the first one. The second model is written as

$$T_{air} = 0.4776 + 0.8765 \times T_{battery}. \quad (2)$$

Testing on Galaxy Note 3 Data. To check the validity of the model in Eq. (2), we randomly divide the data set collected by Galaxy Note 3 into training and testing sets. The training set has 70% number of observations and the testing set is the rest. On the training set, we build a linear model that has `airtemp` as an outcome and `battemp` as the only predictor. We calculate the following metrics for the model's quality assessment: ME (mean error), MAE (mean absolute error) and R^2 (coefficient of determination). This task is repeated 100 times for cross validation. Experiment results show that mean of ME is 0.004 and corresponding 95% confidence interval (CI95) is $[-0.0025, 0.01]$. Means and CI95 of MAE and R^2 are 0.951, $[0.948, 0.955]$ and 0.79, $[0.785, 0.794]$, respectively.

Testing on Nexus 4 and Galaxy Note 2. We use the model in Eq. (2) to predict air temperature from battery temperature readings of Nexus 4 and Galaxy Note 2. Predicted temperature are compared to observed temperature recorded by Galaxy Note 3 using ME, MAE, correlation and R^2 metrics. Graphs showing predicted and observed temperature for Nexus 4 and Galaxy Note 2 are in Figs. 1 and 2. Test results including those of Galaxy Note 3 are described in Table 6.

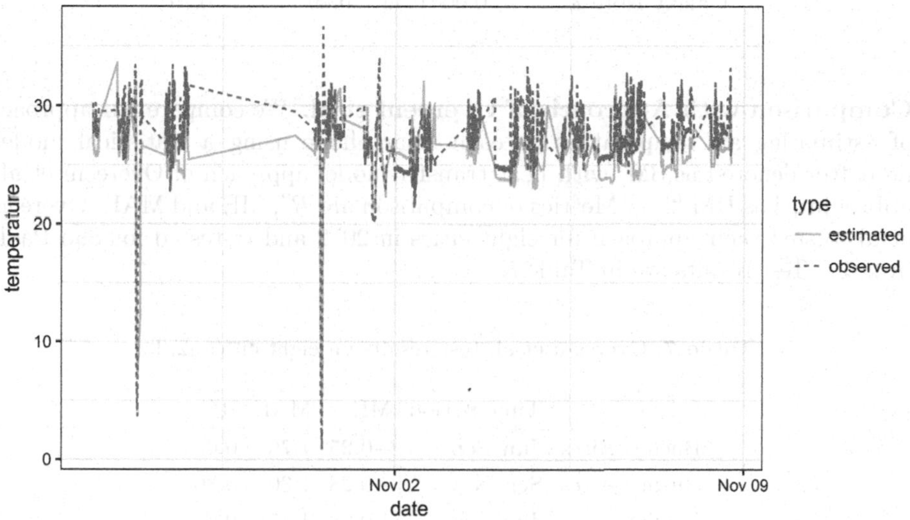

Fig. 1. Predicted temperature on Nexus 4 versus observed temperature; ME = 1.11, MAE = 1.86, correlation = 0.79, R^2 = 0.62.

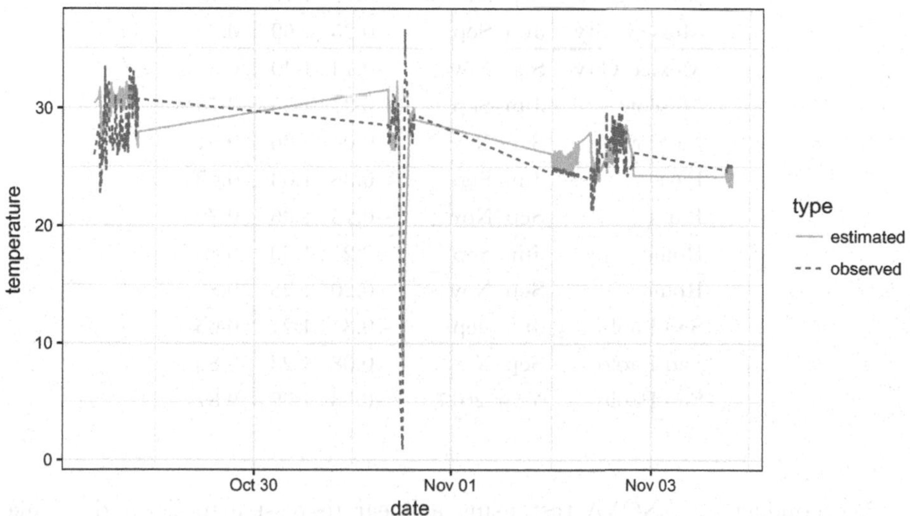

Fig. 2. Predicted temperature on Galaxy Note 2 versus observed temperature; ME = −0.82, MAE = 2.03, correlation = 0.74, R^2 = 0.55.

Table 6. Test results on Nexus 4, Galaxy Note 2 and Galaxy Note 3.

Smartphone model	ME	MAE	Correlation	R^2
Nexus 4	1.11	1.86	0.79	0.62
Galaxy Note 2	−0.82	2.03	0.74	0.55
Galaxy Note 3	0.00	0.95	0.89	0.79

Comparison with Approach of Overeem et al. We compare our approach of estimating air temperature at each smartphone using a statistical model, hereafter denoted as SM, with heat transfer model approach of Overeem et al., abbreviated as HM [2,3]. Metrics of comparison are R^2, ME and MAE. Overeem et al. tested their approach for eight cities in 2012 and re-tested for Sao Paolo in 2017. Test results are in Table 7.

Table 7. Overeem et al. test results on eight cities [2,3].

City	Time period	ME	MAE	R^2
Buenos Aires	Jun–Sep	−0.25	1.76	0.65
Buenos Aires	Sep–Nov	−0.28	1.30	0.86
London	Jun–Sep	−0.28	1.45	0.65
London	Sep–Nov	0.10	1.59	0.72
Los Angeles	Jun–Sep	−0.13	1.57	0.39
Los Angeles	Sep–Nov	0.15	1.16	0.79
Mexico City	Jun–Sep	0.25	1.69	0.36
Mexico City	Sep–Nov	−0.14	1.49	0.33
Moscow	Jun–Sep	−0.27	1.55	0.75
Moscow	Sep–Nov	−0.04	4.00	0.51
Paris	Jun–Sep	−0.38	1.63	0.62
Paris	Sep–Nov	−0.23	1.96	0.70
Rome	Jun–Sep	0.21	1.30	0.86
Rome	Sep–Nov	0.20	1.25	0.83
Sao Paolo	Jun–Sep	−0.31	1.21	0.65
Sao Paolo	Sep–Nov	0.08	1.23	0.85
Sao Paolo	Year 2017	−0.53	1.09	0.87

We conduct an ANOVA test using a linear regression model and a Tukey honest significant difference (Tukey HSD) test to compare means of R^2 of the two approaches.

Comparison results are in Table 8 and Fig. 3. The `approachSM` regression coefficient in second row of Table 8 shows that mean value of SM's R^2 is 0.019

lower than that of HM (CI95: $[-0.25, 0.21]$). However this difference is not statistically significant because the corresponding p-value is $0.87 > 0.05$. Thus we cannot reject a hypothesis that true difference in means R^2 of SM and HM is 0. In other words, mean of R^2 of SM is comparable to mean of R^2 of HM. Figure 3 depicts the difference in means of R^2 of SM and HM approaches and the corresponding CI95. The corresponding CI95 contains 0.

Table 8. A linear model for comparison of SM and HM approaches.

	Estimate	Std. error	t value	Pr(>|t|)
(Intercept)	0.6699	0.0422	15.88	0.0000
approachSM	−0.0187	0.1089	−0.17	0.8658

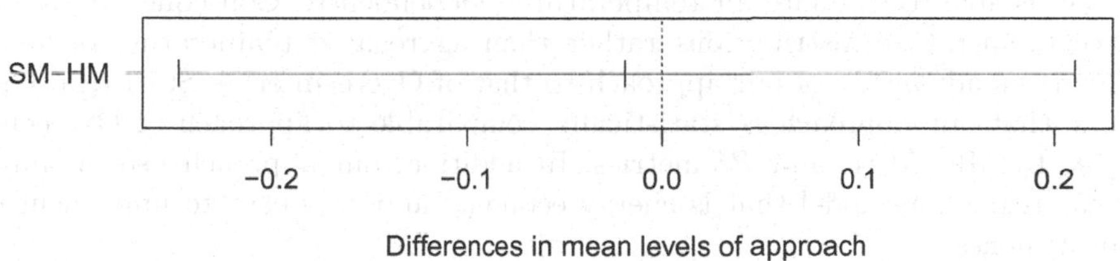

Fig. 3. Difference in means of R^2 of SM and HM approaches.

Following the same test procedure, we compare two approaches by ME and MAE metrics. Comparison results are in Figs. 4 and 5. The results also show that SM approach is comparable to HM approach.

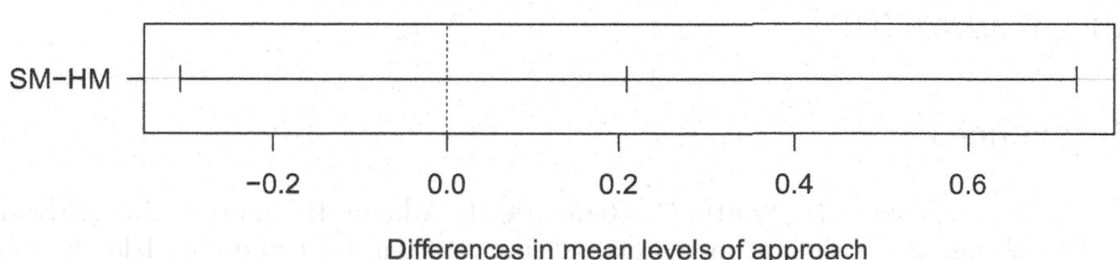

Fig. 4. Difference in means of ME of SM and HM approaches.

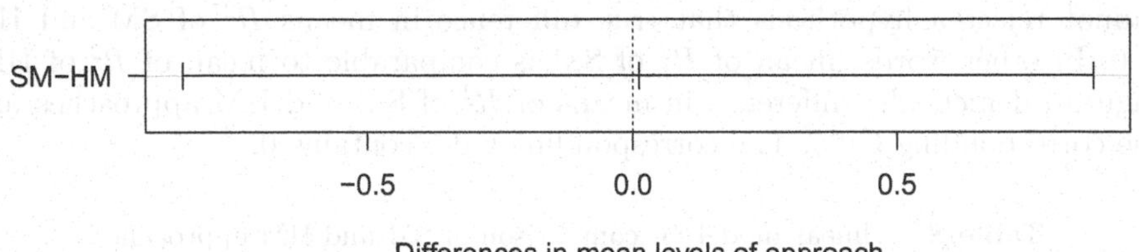

Fig. 5. Difference in means of MAE of SM and HM approaches.

3 Discussion and Future Development

We presented a new approach of using a statistical model to estimate air temperature from smartphone battery temperature. Using our approach, each smartphone is able to estimate air temperature independently. Consequently we can have temperature distributions rather than aggregated temperature of areas. This is an advantage of our approach to that of Overeem et al. Statistical tests show that our approach is statistically comparable to approach of Overeem's el at. by ME, MAE and R^2 metrics. In addition, our approach uses a simple linear regression model that is energy economy and very easy to implement on smartphones.

Our approach is capable for further improvement. Firstly, the temperature sensor inside Galaxy Note 3 is affected by the phone's temperature, it reports unstable air temperature. We will conduct experiments using an independent temperature sensor, for example DHT22 [9]. Secondly, other important factors to temperature of the battery such as battery current will be taken into account to build new statistical models. Finally, we will use a Kalman filter to eliminate noise [10].

We have developed an Android app using the model in Eq. (2). The app's main function is to report estimated air temperature. It does not collect data at this stage. We published the app on Google Play on Mar 2016. To date, the app has more than 60,000 downloads, and approximately 1,200 daily use. Its user rating is 3.2/5.0 [11].

References

1. Rasul, A., Baizter, H., Smith, C., Remedios, J., Adamu, B., Sobrino, J.A., Srivanit, M., Weng, Q.: A review on remote sensing of urban heat and cool Islands. Land **6**(2), 38 (2017)
2. Overeem, A., Robinson, J., Leijinse, H., Steeneveld, G., Horn, B., Ujilenhoet, R.: Crowdsourcing urban air temperatures from smartphone battery temperatures. Geophys. Res. Lett. **40**, 4081–4085 (2013)

3. Droste, A., Pape, J., Overeem, A., Leijinse, H., Steeneveld, G., Van Deldel, A., Uijlenhoet, R.: Crowdsourcing urban air temperatures through smartphone battery temperatures in Sao Paulo, Brazil. J. Atmos. Ocean. Technol. **34**, 1853–1866 (2017)
4. Weather Signal app. https://play.google.com/store/apps/details?id=com.opensignal.weathersignal
5. Milette, G., Stroud, A.: Professional Android Sensor Programming. Wiley, Hoboken (2012)
6. Ferreira, D., Dey, A.K., Kostakos, V.: Understanding human-smartphone concerns: a study of battery life. In: Lyons, K., Hightower, J., Huang, E.M. (eds.) Pervasive 2011. LNCS, vol. 6696, pp. 19–33. Springer, Heidelberg (2011). https://doi.org/10.1007/978-3-642-21726-5_2
7. Liu, G., Ouyang, M., Lu, L., Li, J., Han, X.: Analysis of the heat generation of lithium-ion battery during charging and discharging considering different influencing factors. J. Therm. Anal. Calorim. **116**, 1001–1010 (2014)
8. Faraway, J.: Linear Models with R, 2nd edn. CRC Press, Boca Raton (2015)
9. DHT22 data sheet. https://www.sparkfun.com/datasheets/Sensors/Temperature/DHT22.pdf
10. Sun, J., Wei, G., Pei, L., Lu, R., Song, K., Wu, C., Zhu, C.: Online internal temperature estimation for lithium-ion batteries based on Kalman filter. Energies **8**, 4400–4415 (2015)
11. Smart thermometer app. https://play.google.com/store/apps/details?id=com.naavsystems.smartthermo

Voice Recognition Software on Embedded Devices

Pavel Vojtas[1], Jan Stepan[1], David Sec[2], Richard Cimler[3](✉), and Ondrej Krejcar[1]

[1] Center for Basic and Applied Research (CBAR), Faculty of Informatics and Management, University of Hradec Kralove, Hradec Kralove, Czech Republic
jan.stepan.3@uhk.cz

[2] Department of Information Technologies, Faculty of Informatics and Management, University of Hradec Kralove, Hradec Kralove, Czech Republic

[3] Faculty of Science, University of Hradec Kralove, Hradec Kralove, Czech Republic
richard.cimler@uhk.cz
http://www.uhk.cz

Abstract. This paper deals with an area of voice recognition and its usage for controlling embedded devices and external components. Multiple voice recognition solutions are available right now but many of them are not the best solution for use in performance constrained devices, such as Raspberry Pi. The current state of the art in the field of voice recognition software is covered and appropriate candidates are selected. Tools are tested in terms of quality of evaluation of different commands in areas with various levels of noise. Time requirements for processing are also discussed.

Keywords: Voice recognition · Raspberry Pi · Julius · PocketSphinx

1 Introduction

Nowadays, voice control is an inseparable part of mobile phones, personal computers, laptops and smart devices. Some of them even contain smart algorithms which analyze whole sentences, trying to find contextual information and offer a solution, for example, Siri assistant by Apple, Cortana assistant by Microsoft, and Alexa built into Amazon Echo devices.

Problems emerge when this functionality must be added into custom devices, e.g. voice control capability into the smart home framework [1]. The solutions mentioned above are not opensource and require specific hardware or provide limited API only via cloud services. This might present another limiting requirement: an Internet connection.

Let us define the requirements for voice recognition software which can be run on low-cost hardware such as [2] and can be used in a smart-home concept. It shall be fully scalable, easy to deploy and modify. The requirements for processing power must be as low as possible, with the option to run on popular

N. T. Nguyen et al. (Eds.): ACIIDS 2018, LNAI 10751, pp. 642–650, 2018.
https://doi.org/10.1007/978-3-319-75417-8_60

single board computers, such as the Raspberry Pi or modern routers. No Internet connection must be used, so that the software can run entirely in offline mode. This feature has been selected in order to be able to use voice commands even if an Internet connection is down. Continuous recognition of any words is also not required. Recognizing only a limited set of words is sufficient because there is not a vast number of commands used.

According to [3], which deals with voice processing in the MATLAB environment, the appropriate software for voice recognition should use Hidden Markov Models (HMM) and Gaussian Mixture Models (GMM). The Hidden Markov Model is primarily used for voice detection, and it comprises three phases. The first phase is property extraction. The second phase calculates a vector of probability properties with the use of a Gaussian Mixture Model. The third phase looks for the most probable match with the word in a transcript.

The system for automatic voice recognition described in [4] is also based on HMM and GMM. The role of adaptive tuning, which enables speeding up the decoding process, is mentioned, along with how important a well designed cooperation is between the hardware and software. The implementation and testing on embedded devices can use knowledge from [5] dealing with software design on 32bit ARM microcontrollers.

The combination of HMM and GMM with Weighted Finite Transources usable for the decoding part is discussed in [6]. The whole approach is afterwards tested on the Altera Nios II platform.

All these papers indicate that a solution for voice recognition software runnable on performance constrained devices, such as Raspberry Pi or BeagleBone, should use a Hidden Markov Model and a Gaussian Mixture Model. Another limitation comes from the operating system side, because Linux is the only widely supported system on these platforms. That it is an embedded system also means very limited storage capabilities, so the size of the application is a concern. The performance of ARM processors are also worse than that of a PC, therefore the programming language used for the implementation must be taken into account. The goal is to choose software that fits the requirements and then perform tests which will select the best candidate.

In the following section, the current software technologies are described. The third section deals with the hardware preparation. The testing and its results can be found in the fourth section. The last section presents the conclusions from this research.

2 Current Software Technologies

There are dozens of voice recognition software applications, each of them with a different feature set and in a different state of completeness of its implementation. Those discussed in this paper are all Open-Source solutions. The advantage of this solution is that the source code is freely available, so it is possible to continue its development even if the original author stops working on it. It is also possible to adjust the software for a specific use.

Bavieca: Complete Bavieca software is described in [7]. It contains 25 subprograms, each designed to handle specific tasks. The software itself is written in C++ and uses the LAPACK linear algebra library. A Java API is provided for easy integration into higher level programs. The decoding part is based on the Weighted Finite State Acceptors method.

CMU Sphinx: Another available solution is CMU Sphinx. It has a fork for low performance devices, called Pocket Sphinx. It is discussed in [8,9]. The software consists of three parts. The first is sound transformation and properties vector building. The second part handles the acoustics modeling which transforms the vectors to phonemes using the Gaussian Probability Estimation method. The third part contains a language seeking module, where the phonemes are converted into word sequences. The HMM method is used, see Fig. 1.

Julius: This is covered in [10,11]. A two-pass search is implemented. The first pass sorts the words into a tree structure which is associated with a language model. The language model contains only phrases in word–word form. Two-gram probability is applied and the best approximation is chosen. The second pass uses a three-gram model for recalculating the values from the first path using n-dest (see Fig. 2). Julius uses a rule system based on grammar.

Kaldi: The last solution, described in [12], is Kaldi, written in C++. The architecture is shown in Fig. 3. Its decoding is based on the Weighted Finite State Transducer method. The additional libraries LAPACK, BLAS and Open FST are required.

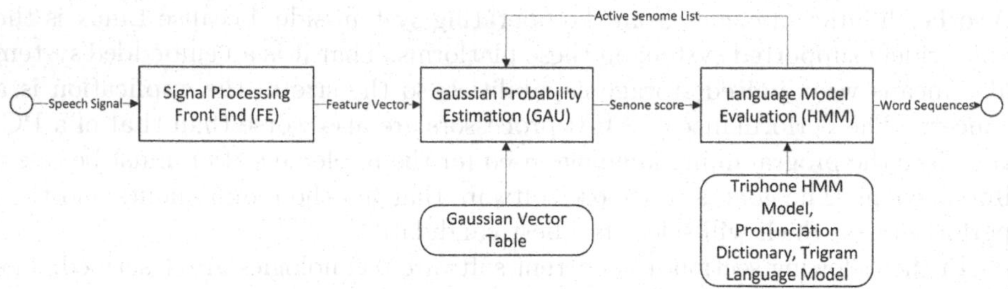

Fig. 1. Sphinx3 structure (redrawn from [9])

The shortcomings of Bavieca and Kaldi lie in the external dependency on LAPACK, which makes these solutions unusable for embedded devices with limited storage memory. The Java Runtime Environment is required for Bavieca to operate, which is another shortcoming if the solution should be used on an embedded device.

The other two lightweight tools (Pocket Sphinx and Julius) were chosen for furthermore implementation and testing. It was necessary to determine which one provides more accurate voice recognition and how much time this task requires.

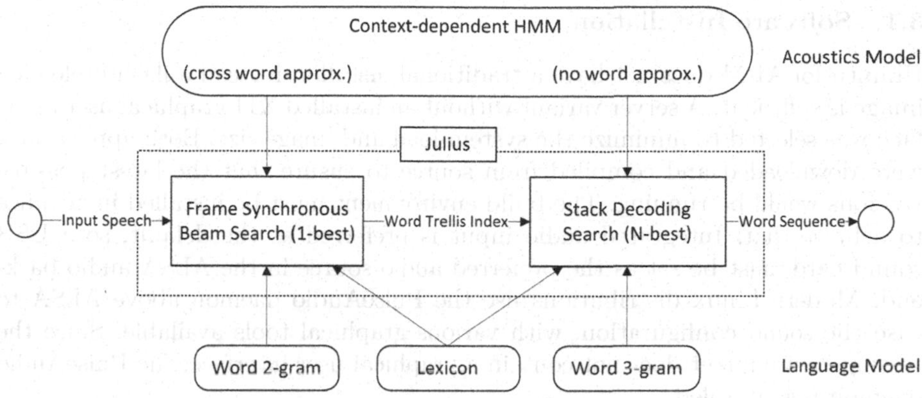

Fig. 2. Julius structure (redrawn from [10])

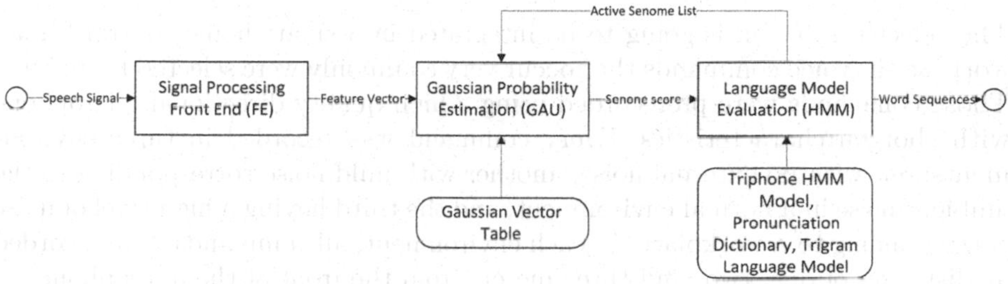

Fig. 3. Kaldi structure (redrawn from [12])

3 Hardware Preparation

The hardware selected corresponds to a low-cost solution which can be used in a smart home concept. Raspberry Pi, in its latest (third) iteration was selected as the platform for implementing the testing environment. It is an affordable singleboard computer based on the Broadcom BCM2837 64bit ARM Cortex-A53 CPU, equipped with 1 GB operating system memory, Ethernet, WiFi and Bluetooth connectivity. Unfortunately, the audio only includes analog output and HDMI digital output. No analog audio input in the form of a microphone or a line is present. This may be solved via a specialized shield with a high fidelity sound card or an external USB one. The Creative Sound Blaster Play2 USB sound card was selected because it is supported out of the box in any recent Linux distribution, which are supported on Raspberry Pi. The choice of the particular distribution is entirely up to the user's preferences and use case scenario. The ARM port of Ubuntu was chosen for testing the voice recognition programs. A high quality power source was used, to filter any undesirable distortion in the audio input.

3.1 Software Installation

Ubuntu for ARM does not have a traditional installer; instead a flash preloaded image is sufficient. A server variant without an installed X11 graphical user interface was selected to minimize the system load and image size. Both applications were downloaded and compiled from source to ensure that the latest possible versions would be running. The build environment must be installed in advance to achieve that. Integrated audio input is preferred as the default, so a USB sound card must be set as the preferred audio source in the ALSA audio backend. Modern Linux distributions use the PulseAudio daemon above ALSA to ease the sound configuration, with various graphical tools available. Since the testing environment does not contain a graphical user interface, the PulseAudio daemon was disabled.

4 Testing

The selected solution is going to be integrated in a smart home control framework, so the voice commands that occur very commonly were selected for testing. Those commands were prerecorded using a high quality directional microphone with shotgun characteristics. Every command was recorded in three environments: one with no external noise, another with mild noise corresponding to the ambient noise in a normal environment, and the third having a high level of noise corresponding to a workplace. In each environment, all commands were recorded at distances of one, two, and three meters from the front of the microphone.

Scripts which initialize each software have been written. Redirecting a microphone input to a wave file is part of these scripts. The time to initialization and to recognize a command was measured. It is important to measure the initialization time, because it is possible to run each software on demand and not as a background service, so as to reduce system load. Julius and Pocket Sphinx return a list of words recognized along with the probability of recognition for each sentence. These probabilities were saved, and the mean was calculated for combination of each software, noise level, and distance.

4.1 Results

The system time scripts have been measured. Each program was launched by a script ten times, and the average values were calculated, because the Linux operating system is not a real-time OS. The results are shown in Table 1. It is obvious that Julius launches and recognizes words more than ten times faster than Pocket Sphinx.

Tables 2 and 3 shows the probability of recognizing a word combination. Again, it can be seen that Julius recognizes words with greater certainty. At three meters distance, Pocket Sphinx recognizes words in a high noise environment with a confidence of 12.5%, but Julius does so with 75%.

Table 1. System times

	Real (ms)	User (ms)	Sys (ms)
Pocket Sphinx	17.82	16.79	0.34
Julius	1.68	1.64	0.02

Table 2. Recognition probability for Pocket Sphinx

Pocket Sphinx	1 m	2 m	3 m
No noise	1.000	0.917	0.625
Mild noise	0.500	0.600	0.063
High noise	0.036	0.208	0.125

Table 3. Recognition probability for Julius

Julius	1 m	2 m	3 m
No noise	1.000	1.000	0.750
Mild noise	0.875	0.875	0.875
High noise	0.500	0.625	0.750

The probability of word recognition is very similar at a distance of one meter up to mild noise (see Fig. 4). It seems that Pocket Sphinx does not handle well a noisy environment.

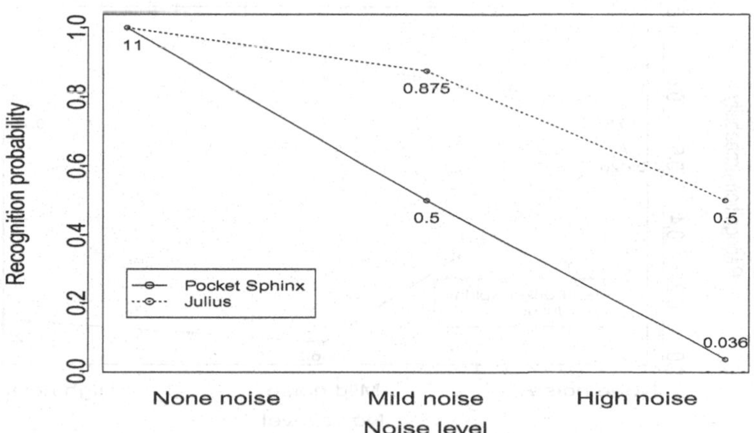

Fig. 4. Results for one meter distance

The difference between the programs starts to be more significant when the speaker is two meters away from the microphone (see Fig. 5).

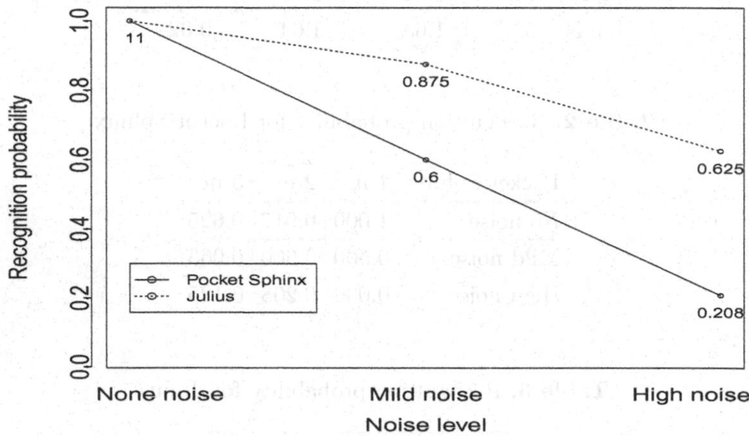

Fig. 5. Results for a distance of two meters

A comparison of results for a distance of three meters is shown in Fig. 6. These results do not mean that Pocket Sphinx may be considered as poorly implemented. The results may be affected by the hardware which has been used. The tests were performed on a specific single board system with Ubuntu 16.04 and external USB sound card. Also, both programs were built from source with default compilation parameters.

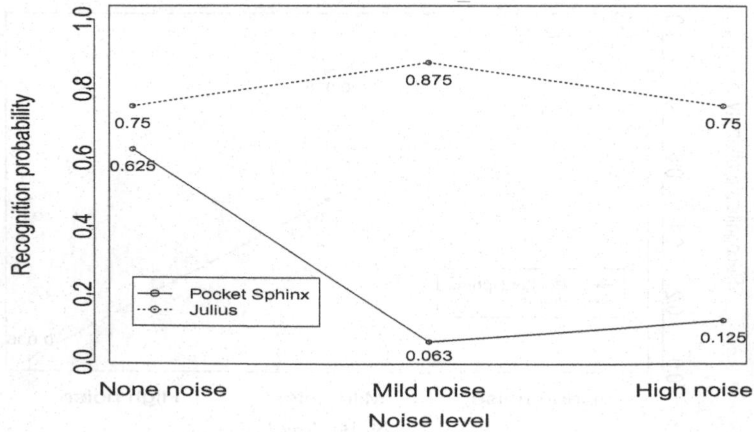

Fig. 6. Results for a distance of three meters

5 Future Work

The main reason of setting constrains for Raspberry Pi minicomputer and low level programming languages is to simplify integration to our smart home automation and Internet of Thing system called HAuSy (Home Automation System). It utilizes three layer architecture instead of two layer which can be found in commercial solution. Whole architecture is described in detail in [13]. The main goal of having central server (top layer), subsystems (middle layer) and nodes (low layer) is to increase stability and robustness. Central server may be located inside the house as ordinary desktop computer or running remotely on any cloud platform. Subsystem is placed in each room and it utilizes Raspberry Pi computer and running custom C++ application. Nodes are placed inside each room and communicate with local subsystem over RS485 wired bus with binary protocol [14], Bluetooth Low Energy or Z-Wave radios. Subsystem acts as a gateway between nodes and server in default online state. Wired nodes are intended to use in time critical application as they outperform wireless technologies [15]. However when any problem occurs on server or with the connection with server, subsystems transition to offline state and perform actions locally. This means that in case of an outage, light switches, door locks, curtain control, smoke detectors and etc. remain functional till outage is fixed.

HAuSy in current state only offers traditional control using switches, dimmers and mobile application. Voice control is achieved with plugin for Amazon Echo devices. However commands are evaluated Amazon cloud service and user must tell every command in form: Alexa, tell HAuSy to..., which is not comfortable and stable Internet connection is required. The results of performed tests (Sect. 4) show that Julius software performance is solid and therefore is going to be integrated into subsystem application to provide voice control in every room equipped with subsystem. Settings and initial calibration of software is going to be performed through server.

6 Conclusion

Apart from commercial cloud solutions, there are many open-source solutions for voice recognition. The most popular have been discussed in this paper and those with the appropriate properties were selected for further testing. Only Julius and Pocket Sphinx met the requirements for a lightweight solution without any external dependencies. A hardware solution for running these libraries has been described. The testing was performed on a Raspberry Pi 3 single board computer. The results showed that Julius has better word recognition probability in the tested scenarios. This solution is ideal for a low-cost platform using affordable hardware where voice recognition is needed.

Acknowledgment. The support of the Specific Research Project at FIM UHK is gratefully acknowledged.

References

1. Dvorak, J., Berger, O., Krejcar, O.: Universal central control of home appliances as an expanding element of the smart home concepts—case study on low cost smart solution. In: Saeed, K., Snášel, V. (eds.) CISIM 2014. LNCS, vol. 8838, pp. 479–488. Springer, Heidelberg (2014). https://doi.org/10.1007/978-3-662-45237-0_44
2. Drabek, A., Krejcar, O., Selamat, A., Kuca, K.: A Smart arduino alarm clock using hypnagogia detection during night. In: Fujita, H., Ali, M., Selamat, A., Sasaki, J., Kurematsu, M. (eds.) IEA/AIE 2016. LNCS (LNAI), vol. 9799, pp. 514–526. Springer, Cham (2016). https://doi.org/10.1007/978-3-319-42007-3_45
3. Preethi, S., Arivu Selvam, B.: Automatic speech recognition system for real time applications. Int. J. Eng. Innov. Res. **2**, 157–160 (2013)
4. Kartheek, K., Srihari Babu, D.: ASR for embedded real time applications. IOSR J. Electr. Electron. Eng. **3**(3), 28–36 (2012)
5. Cheng, O., Abdulla, W., Salcic, Z.: Hardware-software codesign of automatic speech recognition system for embedded real-time applications. IEEE Trans. Industr. Electron. **58**(3), 850–859 (2011)
6. Cheng, O., Abdulla, W., Salcic, Z.: Speech recognition system for embedded real-time applications. In: 2009 IEEE International Symposium on Signal Processing and Information Technology (ISSPIT) (2009)
7. Bolanos, D.: The Bavieca open-source speech recognition toolkit. In: 2012 IEEE Spoken Language Technology Workshop (SLT) (2012)
8. Huggins-Daines, D., Kumar, M., Chan, A., Black, A., Ravishankar, M., Rudnicky, A.: Pocketsphinx: a free, real-time continuous speech recognition system for hand-held devices. In: 2006 IEEE International Conference on Acoustics Speed and Signal Processing Proceedings (2006)
9. Binu, M., Al, D., Zhen, F.: A low-power accelerator for the sphinx 3 speech recognition system. In: Proceedings of the 2003 International Conference on Compilers, Architecture and Synthesis for Embedded Systems, pp. 210–219 (2003)
10. Lee, A., Kawahara, T., Shikano, K.: Julius – an open source real-time large vocabulary recognition engine. In: Proceedings of European Conference on Speech Communication and Technology (EUROSPEECH), pp. 1691–1694 (2001)
11. Lee, A., Kawahara, T.: Recent development of open-source speech recognition engine Julius. In: Asia-Pacific Signal and Information Processing Association Annual Summit and Conference (APSIPA ASC) (2009)
12. Povey, D., Ghoshal, A., Boulianne, G., Burget, L., Glembek, O.: The Kaldi speech recognition toolkit. In: IEEE 2011 Workshop on Automatic Speech Recognition and Understanding (2011)
13. Horálek, J., Matyska, J., Stepan, J., Vancl, M., Cimler, R., Soběslav, V.: Lower layers of a cloud driven smart home system. In: Barbucha, D., Nguyen, N.T., Batubara, J. (eds.) New Trends in Intelligent Information and Database Systems. SCI, vol. 598, pp. 219–228. Springer, Cham (2015). https://doi.org/10.1007/978-3-319-16211-9_23
14. Stepan, J., Matyska, J., Cimler, R., Horalek, J.: Low level communication protocol and hardware for wired sensor networks. J. Telecommun. Electron. Comput. Eng. **9**(2–4) (2017)
15. Stepan, J., Cimler, R., Matyska, J., Sec, D., Krejcar, O.: Lightweight protocol for M2M communication. In: Nguyen, N.T., Papadopoulos, G.A., Jędrzejowicz, P., Trawiński, B., Vossen, G. (eds.) ICCCI 2017. LNCS (LNAI), vol. 10449, pp. 335–344. Springer, Cham (2017). https://doi.org/10.1007/978-3-319-67077-5_32

Intelligent Information Systems

Design and Development of an Online Support System for Elder Care

Sukontip Wongpun(✉) and Sumanta Guha

Computer Science and Information Management Program,
Asian Institute of Technology, Pathum Thani, Thailand
{sukontip.wongpun, guha}@ait.asia

Abstract. An important issue in healthcare nowadays is the tremendous increase in the elder population worldwide. A consequence of this is that a significant amount of care to elders is provided by informal caregivers, such as family members, rather than health professionals. It is vital, therefore, to provide support to informal caregivers in a manner that makes them more effective while, at the same time, easing the burden of constant caregiving. This work presents a prototype of an online support system for informal caregivers to the elderly and its evaluation. The prototype has multiple modules which provide services including patient and caregiver profile management, care recommendation and planning, treatment notification, an information archive, as well as social interaction with other caregivers. Our prototype is web-based and caregivers may access it through a computer or a smartphone. Our evaluation of the prototype indicates more than 70% user satisfaction.

Keywords: Elder care · Informal caregiver · Online support system

1 Introduction

It was reported by the United Nations (UN) that the results of decreasing fertility and increasing life expectancy are the main factors which lead to an aging society [1]. In 2015, the proportion of elders worldwide accounted for 901 million out of a global population of 7.3 billion. Additionally, in 2050 the elder population is projected to more than double from 901 million to be 2.1 billion out of a global population of 9.7 billion [1]. The World Health Organization (WHO) published the active aging policy framework to be used for caring for the older population in 2002 [2]. The WHO has recommended that all countries provide comprehensive policies for their older population. However, many developing countries are facing severe healthcare challenges with the increasing incidence of chronic diseases, disabilities, in addition to the care requirements of an older population. Particularly, challenging when providing care for an older population is balancing support for self-care, informal care and formal care.

"Informal caregiver" refers to a person who does not have professional knowledge about elder care but has a significant personal relationship with an elder, e.g., a daughter, son, relative, or spouse [3]. In addition to medical care, informal caregivers

N. T. Nguyen et al. (Eds.): ACIIDS 2018, LNAI 10751, pp. 653–663, 2018.
https://doi.org/10.1007/978-3-319-75417-8_61

often assist elders with impaired abilities to carry out the activities of daily life. Caring for frail elders is a task which can be complicated, physically demanding and emotionally draining. Elder care burdens affect the quality of life, financial status, mental health and physical wellbeing of informal caregivers [4]. Nevertheless, their personal relationship with an older person often means that the informal caregiver is highly motivated [3]. Consequently, it is to society's benefit to provide them support in the form of information and training, communication with field experts and social support [5].

A huge amount of care information such as a treatment for chronic diseases of elders, as well as nutrition is available on the internet. Awareness has been increased through the health information resources available on websites, which highlight how important the dangers of various diseases are, and provide knowledge to caregivers about caring for patients [6]. For example, the caregiver websites [7, 8] provide online elder care information to support caregivers. These websites have forums featuring topics designed to give caregivers the opportunity to exchange information about health and elder care experiences. Likewise, a communication tool [9] has been created so that frail elders, informal caregivers and primary care professionals can communicate amongst themselves. This tool particularly enhances multi-collaboration between relevant persons involved in elder care, resulting in improved levels of care knowledge for both frail elders and informal caregivers. In addition, many specialized online systems have been developed with the aim of supporting elder caregivers. For instance, the online systems [10, 11] support caregivers of elders with dementia. Education, social support, health care consultation, community resources, self-management program and online coaching are amongst features typically provided by these systems. The results of using these systems can empower the caregiver's abilities, reduce stress from elder care tasks and improve quality of life of caregivers. However, there is no yet an online support system which specific supporting informal caregiver to elder care covers with relevant tasks.

To fill the gap, the main goal of this research is to develop a comprehensive online support system for elder care according to users' requirements, and fulfill the needs of the existing elder care system. The challenge objective of the research is how to design and develop the system, which is to be used to enhance elder care knowledge and abilities of informal caregivers. Therefore, in this work, a prototype of an online support system for elder care has been developed, and has subsequently been evaluated to test the effectiveness and efficiency of the prototype. Our system supports informal caregivers by (a) providing elder care information and knowledge (b) recommending an automatic solution for elder care (c) generating a suitable daily care plan (d) notifying an alert message to each user via mobile phone and (e) connecting them through online social. The rest of the paper is organized as follows. Section 2 proposes the system design. System development illustrates in Sect. 3. In Sect. 4, all detail of a testing system is clearly explained. Finally, in Sect. 5 the conclusion and future work are described.

2 System Design

Our system built on web-based architecture, the system features a responsive user interface design, which can be operated on both computers and mobile devices. The system should to test the system abilities and ease of use before released for real-time use. Thus, the system development model was designed using a prototype model, as shown in Fig. 1.

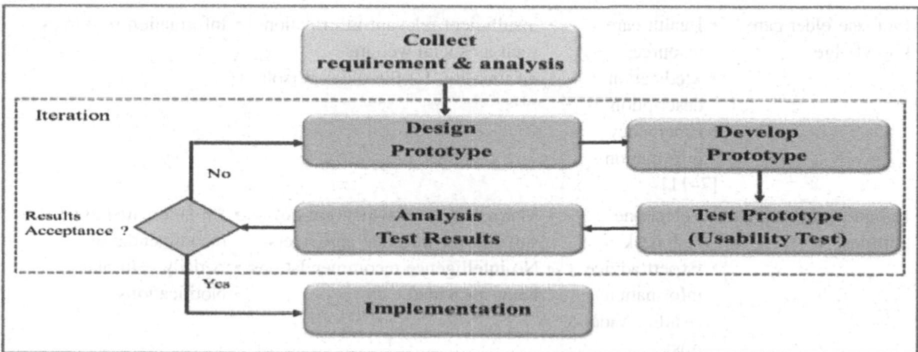

Fig. 1. A system development model based on prototyping model

2.1 Collect Requirements and Analysis

In this step, the main objective was to collect informal caregivers' problems along with the elder's care requirements. As such, it was requisite to understand the elder's health problems and caregivers' responsibilities. In order to carry out the first task of this step an in-depth interview was conducted with relevant health care professionals including doctors, registered nurses, and public health technical officers. After that, an in-depth interview with fifteen informal caregivers was carried out to collect the problems and requirements associated with elder care. The details of the problems and requirements of informal caregivers were published in earlier research [12]. In summary, we analyzed and classified the users' requirements respectively into four main aspects, which are as follows; (1) increase elder care knowledge (2) enhance elder care abilities (3) communicate with their peers and relevant health staff and (4) perceive and understand the elder's health state.

2.2 Design Prototype

In this stage, system modules were designed according to the requirement of users. Informal caregivers' problems and requirements were compared with the existing system information, so as to propose suitable modules and abilities for the system. The comparison results led us to identify comprehensive suitable system modules as shown in Table 1. In summary, this work proposed to implement a comprehensive online support system for elder care consisting of six main modules. Profile management, elder care recommendations, daily care plans, notifications, social interactions and information resources make up the six main modules.

Table 1. Design system modules.

Requirements	Existing information system and features	Weakness of the existing solutions	Proposed system module
Increase elder care knowledge	• Health care resources • Medication description • Emergency information [7–11]	• Insufficient relevant information such as social welfare information for the older person	• Information resources
Enhance elder care abilities	• Telephone Helpdesk • Expert advice information (Audio, Video tips) [9, 11]	• Manual features which are not suitable for informal caregivers • No intelligence recommendations being provided	• An automated elder care recommendation • A daily care plan • Notifications
Communicate with their peers and relevant health staff	• Blogs, Forums, Q&A, • Frequently asked questions [7, 8, 11]	• No solution being provided to encourage caregivers to communicate with a peer group	• Social interaction with feature to recommend new friends for caregivers
Perceive and understand the elder's health state	• Record health of the elder • Shared electronic health record [9]	• Caregivers cannot preliminarily investigate the elder's health status on a continuous basis	• Profile management with geriatric assessment

3 System Development

The prototype was developed with the use of a computer, using the PHP and MySQL database. The system abilities were created by applying different suitable techniques.

3.1 System Modules

The Profile Management Module. It is used to register the user through a mail server, so that they can use the system. Two sub-modules are incorporated into this module, the informal caregiver and elder profile sub-modules. For this module, caregivers are required to fill in their profile, responsibilities, and their elder's profile. Elder profiles include personal data, health, symptoms, and evaluation of an activity of daily living

(ADL). ADL assessment was used to describe the elder health status, and then the ADL assessment was measured by the Barthel Index standard tool [13].

The Elder Care Recommendation Module. The task of this intelligent module is to improve caregiver abilities by providing an automatic elder care recommendation. We classified the elder care recommendations into four aspects, which include; physical therapy, food, exercise and emotional. Each informal caregiver will receive a suitable elder care recommendation that has been generated by applying case-based reasoning (CBR) techniques. CBR is a method that can generate a recommendation using reasoning based on the most similar past cases. CBR plays a major role in healthcare recommendation system because of its similarity with the classical diagnosis and treatment method [14]. The methodology used when creating elder care recommendations was described in earlier research [15].

The Daily Care Plan Module. The daily care plan is a plan for daily elder care which is divided into three periods of time, namely; morning, afternoon and evening. Input, including caregiver responsibilities, elder's health status and the elder care recommendation is processed by this module. After that, the module automatically generates an appropriate daily care plan for a user. Moreover, a user can input medication details, a doctor's appointment and the schedule of an elder. Once that is complete, the daily care plan module processes this information and updates the details of the daily care plan. The Rule-Based Reasoning (RBR) technique is applied to process all of the information and then generate the daily care plan. In this work we generate elder care activity through rules. Rules are programmatic devices containing an if-then-else component which is used to choose each activity in a daily care plan. Included in the components of the daily care plan are various aspects of elder care such as medication, doctor's appointments, activities of daily living (e.g., bathing, dressing, grooming, preparing meals and feeding), an elder care recommendation, and finally, physical therapy or exercises.

The Notification Module. The Notification Module is a module used for alerting a user of significant elder care activities. Data from the daily care plan is processed by this module so that users can be notified via mobile phone. An alert message is also processed and sent out to users by the Line application. The notification has six activities, which are as follows: (1) an alert at time medicine should be taken by an elder (2) an alert about the doctor's appointment of an elder (3) an alert when the status of an elder's health and symptoms should be updated (every 7 days) (4) an alert for evaluating the daily living activities of an elder (every 30 days) (5) an alert for evaluating the satisfaction of the recommendations (7 days after receiving recommendation) and (6) an alert when new elder care information has been input into the system.

The Social Interaction Module. In this module, caregivers are encouraged to participate with their peers when sharing or receiving various experiences and pieces of elder care knowledge. Online spaces are provided for caregivers by the system, by allowing them to create a social network in order to communicate with other caregivers. Also, they can create and join a forum for asking and answering questions.

The Information Resources Module. The module displays a considerable amount of elder care information. Particularly, it has a thorough knowledge base with reliable references. This module provides the search mechanism to allow users to search the knowledge base. The proposed information includes health care information, dietary advice, and social welfare for the elderly.

3.2 System Prototype Interfaces

In this work, we proposed some main screens of the prototype. Figure 2 shows a daily care plan page. This page processes all data from relevant tasks to generate a daily care plan. The care plan will change automatically when a user input or update any relevant data. When a user receives a daily care plan, the notify task processes this data to send an alert message via line application to an individual user. For example, notification of elderly medication time and doctor's appointment as shown in Fig. 3.

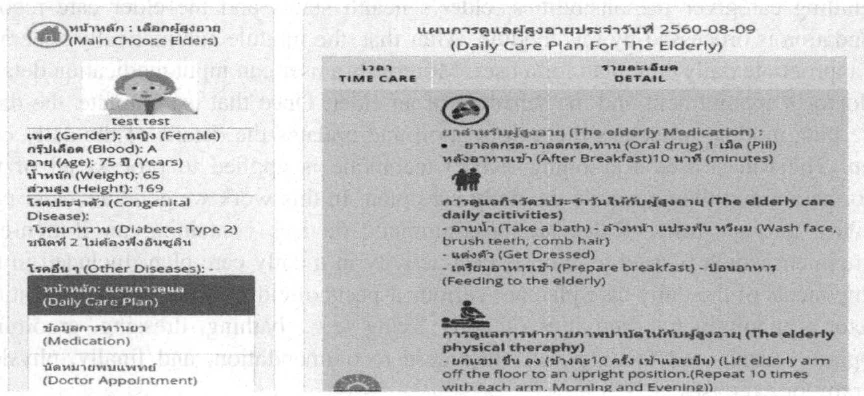

Fig. 2. Example of a daily care plan screen.

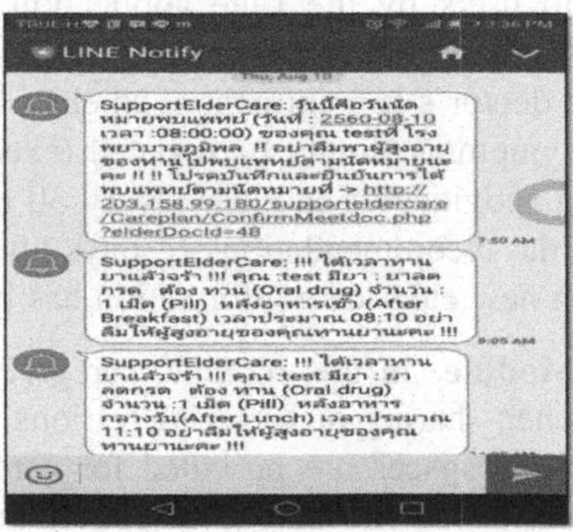

Fig. 3. Example of notification to user

4 System Testing

We invited the targeted users to combine in testing the system in a real environment. During the usability test terms referred to include effectiveness, efficiency, and satisfaction [16]. Effectiveness is defined as how well users can achieve specific goals with accuracy and completion in a specified context of use. The effectiveness calculation is the assessment percentage of users successfully achieving their goals compared to the total number of users [16]. Efficiency is defined as how fast users can accomplish tasks with accuracy and completion. A way to ascertain the efficiency is to calculate the user's effective work time compared to all users' work time [16]. Satisfaction is defined as how comfortable users feel with the system, as well as their perceived acceptability of use. The level of satisfaction can be measured with the use of many formal questionnaires. In this work, we applied the system usability scale (SUS) to 10 questions with a 5-point scale [17]. There are many tools to assess the usability of a system. The SUS is one of the effective and reliable ones that can be used with any type of interface. This tool is cost effective and yields a single score easy to interpret [18].

4.1 Usability Test Procedure

During the test, we focused on user-based evaluation. The user-based test is a method in which users directly participate and are invited to complete typical tasks with a system that can be explored freely [19]. Because of this, we instructed them to "Think aloud by talking aloud". We constructed the usability test procedures, as follows:

1. Identifying a targeted user profile and selecting the user participants. An informal caregiver, who cares for their elders at their home, is the user profile targeted by this system. Once the user profile has been identified, the next step is the selection and invitation of users to participate in testing the system. The number of participants was ten persons. They were divided equally with five users testing the computer-based interface and five the mobile interface. Five participants are the optimal for usability testing and can discover over 80% of the problems with 95% confidence [20]. Below, the details of targeted user profiles, and the demographics of participants are shown in Table 2.
2. Testing the system. During this step, we built seven scenarios which demonstrated all of the tasks that make up the system abilities. In each scenario, a user had to perform a task by his or her self in order to achieve the specific objectives. The results of the usability test were the completion rate of tasks, and relative time-based efficiency, which were collected during observation whilst users operated the system. Moreover, we recorded each instance in which a task was ultimately unsuccessful, and the user failed to achieve the goal in each scenario. These problems were described and classified by each error type. Errors were divided into two brackets; they were categorized as "functionality errors (FE)" and "task errors (TE)". The functionality error is an instance when the system cannot handle the task which user has required performing. The task error is an instance in which the user failed to use any functionality that was provided by the system. Table 3 explains the scenarios and the average of testing results of both devices.

Table 2. Targeted user profile and participants demographic.

Items	Targeted profile	Participants demographic
Gender	N/A	Female 9, Male 1
Age of participants	18 years or over	Between 29 and 40
Education level	N/A	Junior High School 3, Senior High School 3, Bachelor 4
Relationship with an elder	Daughter, son, spouse or relative	Daughter 7, Grandchild 3
Elder age under their care	60 years or over	Between 60 and 93
Elder care time period	At least one time period	All day 4, Morning and Evening 5, Afternoon and Evening 1
Elder health status	Home bound or Bed Bound[a]	Home bound 7, Bed Bound 3

[a]A homebound elder is an elder who has an ADL assessment score between 5 and 11. This elder group can engage in a small amount of self-care. A bed bound elder is an elder who has an ADL assessment score between 0 and 4. This elder group cannot engage in self-care at all [21].

3. Collecting the user satisfaction. Participants were interviewed so that we could receive a recommendation at the end of the testing, by using the SUS questionnaire, [16] along with a set of questions. Some examples of the interview questions are as follows: (1) do you think this system helps to improve your elder care abilities? how does it help? (2) do you think this system increases your elder care knowledge? how does it help? (3) what are the other system abilities that you need? and (4) other suggestions?

4.2 Analysis of Test Results

Overall, 88.57 was the percentage of the system effectiveness, and the relative time-based efficiency was 87.06%. According to the test results (see Table 3) it was indicated that the notification task had the lowest level of effectiveness and efficiency out of all of the tasks, meaning that only 7 users performed without any making errors and were also successful in completing this task. Additionally, the speed of work in this task was the lowest of all the tasks, at 62.6%, to refine the prototype henceforth, the notification task should be modified in the first rank. While an elderly care recommendation task and information resources had the highest levels effectiveness and efficiency. The problems occurred starting in the testing step were analysed to find a suitable methodology, which will be used to improve the system, as shown in Table 4.

Evaluation of user satisfaction by the SUS questionnaire showed that the user satisfaction with the system was a high level at 73.5%. This result was higher than the standard average value of the SUS of 68%. This means user satisfaction was at a high level [18]. Most of the participants thought that they would learn to use this system very quickly. There was a high rate of 90% of users who felt very confident in using the system. However, they needed the support of a technician to be able to use this system.

Table 3. The scenarios and usability test results.

Tasks	Effectiveness	Efficiency	Problems
Task 1: Manage profile Scenario: You would like to record the necessary information to keep and track your elder's health profile. What do you need to do?	90%	87.84%	2 TE
Task 2: Evaluate the ability of activities of daily living (ADL) Scenario: During your day caring for an elder, you need to check and assess your elder's abilities in daily living. What do you need to do?	90%	93.22%	1 TE
Task 3: Receive an automatic elder care recommendation Scenario: You receive a report from the ADL evaluation of your elder and you know about your elder's health status. How can you find out what is the most suitable way to care for your elder?	100%	100%	–
Task 4: Receive a daily Care Plan Scenario: If you would like to receive the details of a suitable way to care for your elder in each time period of each day. What would you do?	90%	89.41%	4 FE
Task 5: Receive a notification via mobile phone Scenario: You need to receive a notification message via a mobile phone such as an alert for when an elder should take their medicine. What do you need to do? How do you check and follow the received messages?	70%	62.26%	1 TE 1 FE
Task 6: Communicate with other informal caregivers Scenario: You would like to share and receive an elder care experience with other informal caregivers who are similarly faced with the same situation of elder care as you. What do you need to do?	80%	80.68%	1 FE
Task 7: Access the information resources Scenario: When you would like to find the necessary information for caring your elders, what would you do?	100%	100%	–

Therefore, to refine the prototype, online help features which support users in using the system should be provided. Included in this are the interview results, which confirmed that if they use the system continuously, it will increase their level of knowledge as well as enhancing their elder care abilities.

Table 4. Example of problems and improvement.

Tasks	Problem	Improvement
Receive a notification via mobile phone	*Task errors* -Some users do not have their line password to confirm for receiving notification services *Functional Errors* -The alert messages sent to users through line application cannot be accessed to the system	-Create an online manual which explains the steps to set up, or reset, line password -Modify an alert message so that it can connect to the system via line application
Communicate with other caregivers	*Functional errors* -The post button on a web board does not work	-Edit the post button on the web board

5 Conclusion and Future Work

This paper demonstrated the system prototype of the online support system for elder care. The system has been developed consistently with the users' requirements and fills the gap of the existing elder care systems. Furthermore, we investigated the system usability and ease of use by means of suitable methods and iterations, until the result was acceptable. The usability test helps find malfunctions and missing system abilities which have failed to meet the users' requirements. Also, the problems of users were discovered while they were using the system. Preliminary test results confirmed that the prototype was acceptable with more than 80% of the effectiveness and efficiency and more than 70% of user satisfactions.

In future work, these test results will be utilized to improve the system, so the user experience and system abilities can be enhanced. We will implement a complete online support system for elder care and it will be provided to the public. Moreover, this work will examine the system's potential and evaluate its impact on caregivers with regards to elder care abilities, elder care knowledge, and the relief of stress.

References

1. United Nations (U.N.). http://www.un.org/en/development/desa/population. Accessed 17 Aug 2017
2. World health organization (WHO). http://www.who.int/ageing/publications/active_ageing/en/. Accessed 21 Feb 2016
3. Knodel, J., Teerawichitchainan, B., Prachuabmoh, V., Pothisiri, W.: The situation of Thailand's older population an update based on the 2014 survey of older Persons in Thailand. Research Collection School of Social Sciences (2015). http://ink.library.smu.edu.sg/soss_research/1948
4. Santini, S., Andersson, G., Lamura, G.: Impact of incontinence on the quality of life of caregivers of older persons with incontinence: a qualitative study in four European countries. Arch. Gerontol. Geriatr. J. **63**, 92–101 (2016)

5. Silva, A.L., Teixeira, H.J., Teixeira, M.J.C., Freitas, S.: The needs of informal caregivers of elderly people living at home: an integrative review. Scandinavian J. Caring Sci. **27**(4), 792–803 (2013)
6. Lim, T.P., Husain, W.: Integrating knowledge-based system in wellness community portal. In: International Conference on Science and Social Research, Malaysia, pp. 350–355 (2010)
7. Family caregiver Alliance. https://www.caregiver.org/. Accessed 20 Aug 2017
8. Caring.com. https://www.caring.com/. Accessed 20 Aug 2017
9. Robben, S.H., Perry, M., Huisjes, M., van Nieuwenhuijzen, L., Schers, H.J., van Weel, C., Melis, R.J.: Implementation of an innovative web-based conference table for community-dwelling frail older people, their informal caregivers and professionals: a process evaluation. BMC Health Serv. Res. J. **12**(1), 251 (2012)
10. Fowler, C., Haney, T., Rutledge, C.M.: An interprofessional virtual healthcare neighborhood for caregivers of elderly with dementia. J. Nurse Practitioners **10**(10), 829–834 (2014)
11. Boots, L.M., de Vugt, M.E., Withagen, H.E., Kempen, G.I., Verhey, F.R.: Development and initial evaluation of the web-based self-management program "partner in balance" for family caregivers of people with early stage dementia: an exploratory mixed-methods study. JMIR Res. Protocols J. **5**(1), e33 (2016)
12. Sukontip, W., Sumanta, G.: Support system architecture for elder care in Thailand. In: 2nd International Conference on Information Technology and Computer Science, Thailand, pp. 60–68 (2016)
13. Wade, D.T., Collin, C.: The Barthel ADL index: a standard measure of physical disability. Int. Disabil. Stud. **10**(2), 64–67 (1988)
14. Begum, S., Ahmed, M.U., Funk, P., Xiong, N., Folke, M.: Case-based reasoning systems in the health sciences: a survey of recent trends and developments. IEEE Trans. Syst. Man Cybern. J. **41**(4), 421–434 (2011)
15. Sukontip, W., Sumanta, G.: Elderly care recommendation system for informal caregiver using case-based reasoning. In: IEEE 2nd Advanced Information Technology, Electronic and Automation Control Conference, China (2017)
16. ISO 9241-11: Ergonomic requirements for office work with visual display terminals (VDTs) – Part 11: Guidance on usability. ISO (1998)
17. John, B.: SUS-A quick and dirty usability scale. In: Jordan, P.W., Thomas, B.A. (eds.) Usability Evaluation in Industry, pp. 189–194. Taylor & Francis (1996)
18. Bangor, A., Kortum, P., Miller, J.: Determining what individual SUS scores mean: adding an adjective rating scale. J. Usability Stud. **4**(3), 114–123 (2009)
19. Bastien, J.C.: Usability testing: a review of some methodological and technical aspects of the method. Int. J. Med. Inform. **79**(4), e18–e23 (2010)
20. Nielsen, J.: Why you only need to test with five users (2000). https://www.nngroup.com/articles/why-you-only-need-to-test-with-5-users/
21. Ministry of Public Health Thailand: Book of senior health record (2016). http://hp.anamai.moph.go.th/ewt_dl_link.php?nid=431

Is Higher Order Mutant Harder to Kill Than First Order Mutant? An Experimental Study

Quang-Vu Nguyen[1(✉)] and Duong-Thu-Hang Pham[2]

[1] Korea-Vietnam Friendship Information Technology College, Da Nang, Vietnam
vunq@viethanit.edu.vn

[2] University of Education – Danang University, Da Nang, Vietnam
hangpdt@ued.udn.vn

Abstract. This paper considers the problem whether higher order mutant is harder to kill than first order mutant or not. Higher order mutation testing has been proposed to overcome the limitations of traditional mutation testing (also called first order mutation testing) such as a large number of generated mutants, limited realism, and equivalent mutants. In this paper, we perform an empirical evaluation to answer the mentioned question with regard to the ratio of number of test cases which can kill a higher order mutant to number of test cases which can kill its constituent first order mutants. Our experimental results indicate that only a half of all generated higher order mutants are harder to kill than its constituent first order mutants.

Keywords: Mutation testing · Higher order mutation testing · Easy to kill
Harder to kill

1 Introduction

Mutation Testing (MT) [1, 2], also called First Order Mutation Testing (FOMT) or Traditional Mutation Testing, has been introduced as a powerful and automated technique to assess the quality of the given set of test cases. One of the MT problems is a large number of generated First Order Mutants (FOMs) [3, 4], which are generated by inserting, via a mutation operator, only one semantic change into the original program. Unfortunately, most of mutants are simple and often easily to detect. In other words, the mentioned FOMs can be killed easily. In mutation testing, the original program and its mutants are executed against the same given set of test cases. The mutant is called "killed" if its output result is different than the output result of original program, with any test case (TC). Higher order mutation testing (HOMT), an approach for generating mutants by applying mutation operators more than once, is an idea presented by Jia and Harman in 2009 [5] and in a manifesto by Harman et al. in 2010 [6]. They believed that the combination of two or more errors to generate mutants can cause limited number of easy to kill mutants. The mutants, which are generated in HOMT, are called Higher Order Mutants (HOMs). In the example given in Table 1, we have the program P, one HOM and two first order mutants FOM1, FOM2 of P. The HOM has two changes compared to original program P, whilst FOM1 or FOM2 has only one change.

N. T. Nguyen et al. (Eds.): ACIIDS 2018, LNAI 10751, pp. 664–673, 2018.
https://doi.org/10.1007/978-3-319-75417-8_62

Table 1. An example of higher order mutant (second order mutant)

Program P	FOM1

...
while (hi<50) && (hi>lo)
{
system.out.print(hi);
hi = lo + hi;
lo = hi - lo;
}
...
	...
	while (**hi>50**) && (hi>lo)
	{
	system.out.print(hi);
	hi = lo + hi;
	lo = hi - lo;
	}
	...

FOM2	**HOM**

...
while (hi<50) && (**hi<lo**)
{
system.out.print(hi);
hi = lo + hi;
lo = hi - lo;
}
...
	...
	while (**hi>50**) && (**hi<lo**)
	{
	system.out.print(hi);
	hi = lo + hi;
	lo = hi - lo;
	}
	...

There are many studies (will be presented in detail in the next Section), in the field of higher order mutation testing (also included second order mutation testing), which are considered the strategies for constructing HOMs in order to overcome the MT's problems [3, 4]. However, most of the studies result only indicates that the mean number of generated HOMs as well as equivalent HOMs is reduced compared to FOMs (see Sect. 2).

The paper aim at answering the following question: Is higher order mutant harder to kill than first order mutant? We will perform an empirical evaluation to answer the mentioned question based on the ratio of number of test cases which can kill a HOM to number of test cases which can kill its constituent FOMs.

The rest of the paper has the following sections. Section 2 summarizes the related works by stressing the answer of the problem "whether HOM is harder to kill than FOM or not". Section 3 explains how our empirical study has been proposed in detail. Section 4 reports and analyzes the results of the experiment. Lastly, Sect. 5 concludes the study and proposes future works.

2 Background and Related Works

The idea of Second Order Mutation Testing (SOMT) was first mentioned by Offutt [7] in 1992. After then, Polo et al. [8] in 2008, Kintis et al. [9] and Papadakis and Malvris [10] in 2010, and Madeyski et al. [11] in 2014 further studied to suggest their algorithms to combine first order mutants (FOMs) for generating second order mutants (SOMs). There are 8 studies in this field of SOMT as follows: Using LastToFirst and RandomMix algorithms [8]; Using RDomF, SDomF, HDom(20%) and HDom(50%) algorithms [9]; Using First2Last, SameNode, SameUnit and SameUnitFirstToLast (SU_F2Last) algorithms [10]; Using RandomMix, Last2First and NeighPair algorithms [11]; Creating SOMs by using 4 difference approaches to insert two faults in Aspect-Oriented programming [12]; Using DifferentOperators algorithm [8]; Using SameUnitDifferentOperators (SU_DiffOp) algorithm [10]; Using JudyDiffOp algorithm [11]. In all of SOMT strategies, the number of generated second order mutants is reduced over 50% (except the result of Kintis et al. [9]), while the mean reduction of number of equivalent second order mutants is about 70%, compared with first order mutants. Unfortunately, there are no empirical study to answer the question "Are SOMs harder to kill than FOMs?".

Higher order mutation testing was first defined by Jia and Harman [5] in 2009 and has been considered as a promising solution for overcoming limitations of first order mutation testing. The studies of single-objective higher order mutation testing have 7 articles, while the studies of multi-objective higher order mutation testing have 8 articles, such as: Using meta-heuristic algorithms to find the Subsuming HOMs [5, 6, 13]; Using Genetic, Local Search and Random Search algorithms to produce subtle HOMs [14, 15]; Using two basic operators (insertion and omission) to apply model-based higher order mutation [16]; Using higher order mutation to reducing equivalent mutant [17]; Using multi-objective optimization algorithm with Genetic Programming to search for HOMs which are harder to kill and more realistic complex faults [6, 18, 19]; Using multi-objective optimization algorithm and HOMT to search high quality and reasonable HOMs [20–24]. According to Jia and Harman [5], who have experimented with 10 benchmark C programs under test (14850 Line of Codes and 35473 test cases in total) and the results indicated that about 67% generated HOMs are harder to kill than their constituent FOMs. This based on the evaluation the number of the subsuming HOMs in general, and the strongly subsuming HOMs in particular [5, 6]. Whilst HOMs which were generated by Akinde's approach [17], are extremely easy to kill because of the MSI of HOMs is close to 0%. MSI (Mutation Score Indicator) is the ratio of killed mutants to all generated mutants [11, 25–28]. The results of study of Omar et al. [14] in 2013 and [15] in 2014 indicated that about 0.0012% generated HOMs are subtle HOMs which are difficult to kill, but they did not mentioned the comparison with FOMs.

For the two given objectives, difficulty to kill and small source changes, Langdon et al. [18] applied NSGA-II, a multi objective optimization algorithm, with genetic programming in the area of higher order mutation testing and the results demonstrated that this approach is able to find higher order mutants that represent more realistic complex faults and are harder to kill than first order mutants. However, the number of live (potentially equivalent) mutants, which cannot be killed by the given test suite (albeit could be killed by new and high quality test cases), in the approach of Langdon et al. is very small (about 1%). It means that about 99% generated mutants are killed by the given set of test cases.

In the works of Nguyen and Madeyski from 2015 to 2017 [20–24], they proposed a new classification of HOMs to cover all of the available cases of generated HOMs, as well as described their objectives and fitness functions. They performed the empirical evaluation of multi objective optimization algorithms in HOMT. The initial achievements demonstrated that their approach leads to generation the harder to kill HOMs (which mimic harder to find defects). The HOMs which were generated in their empirical study, are harder to kill than the generated FOMs in terms of number of test cases that can kill the mutants [24].

3 Experiment Set Up

3.1 Research Problem

In this paper, we focus on answering the main question: Is HOM harder to kill than FOM? We apply the multi objective optimization algorithms and higher order mutation testing to generate and execute mutants (both FOMs and HOMs), and then analysis the results base on the ratio of number of test cases which can kill a HOM to number of test cases which can kill its constituent FOMs. Similar to our previous works [20–24], we also use the same our objectives and fitness functions, which are applied to multi objective optimization algorithms to search for valuable HOMs. In our objective functions, we focus on constructing the HOMs which are killed by as small sets of test cases as possible. It leads to decrease number of generated HOMs.

3.2 Judy Tool and Project Under Test (PUT)

We use and extend Judy tool [28] as the supporting tool to generate HOMs, execute mutation testing and evaluate HOMs with the full set of build-in mutation operators of Judy [28] for 8 selected projects under test to conduct the empirical studies. Judy (http://www.mutationtesting.org/) is a mutation testing tool for Java programs. It supports large set of mutation operators, as well as HOM generation, HOM execution and mutation analysis.

Table 2 shows in short 8 selected real-world, open source projects which were downloaded from the Source Forge website (http://sourceforge.net). The information in this table are the name of projects selected for the experiment along with their number of classes (NOC), lines of code (LOC) and number of given test cases (#TCs). The given set of tests cases were obtained from the selected real-world.

3.3 Multi-objective Optimization Algorithms

We apply 6 different algorithms, 5 multi objective optimization algorithms (including one modified by us) and Random algorithm, into higher order mutation testing for our experimental. The short introductions about these algorithms are as following. NSGA-II is the second version of the Non-dominated Sorting Genetic Algorithm that was proposed by Deb et al. [29, 30] for solving non-convex and non-smooth single and multi-objective optimization problems. Its main features are: it uses an elitist principle;

Table 2. Projects under test

Project under test	NOC	LOC	#TCs
BeanBin	72	5925	68
Barbecue	57	23996	190
JWBF	51	13572	305
CommonsChain 1.2	103	13410	17
CommonsValidator 1.4.1	144	25422	66
CommonsJxPath 1.3	239	41079	28
CommonsFileUploads 1.3.1	69	12321	12
CommonsLang3 3.4	570	122964	126

it emphasizes non-dominated solutions; and it uses an explicit diversity preserving mechanism. The eNSGA-II extends NSGA-II's concepts by adding e-dominance, adaptive population sizing, and self-termination to minimize the need for parameter calibration. e-dominance is a concept where a user is able to specify the precision with which he wants to obtain the Pareto-optimal solutions to a multi objective problem, in essence giving him the ability to assign a relative importance to each objective. NSGA-III is the extension of NSGA-II which is based on the supply of a set of reference points and demonstrated its working in 3 to 15-objective optimization problems [30]. The εMOEA (eMOEA) is a steady state multi-objective evolutionary algorithm that co-evolves both an evolutionary algorithm population and an archive population by randomly mating individuals from the population and the archive to generate new solutions [31, 32]. The **eNSGAII-DiffLOC** algorithm [24] is our proposing one based on modification the eNSGA-II algorithm with the rule *"apply no more than one mutation operator to each line of code"*. We have chosen the *eNSGA-II* algorithm to modify because it is the best algorithm in terms of constructing the "High Quality and Reasonable HOMs" [20–24]. Our experimental results indicated that the proposed eNSGAII-DiffLOC seems to be slightly better than original eNSGA-II algorithm in terms of mutant reduction, generating harder-to-kill mutant and constructing "High Quality and Reasonable HOMs" [24]. Lastly, the Random search is used to generate random solutions, which are evaluated and all non-dominated solutions are retained. The result is the set of all non-dominated solutions.

3.4 Experimental Procedure

For each project under test, we ran the process, which was described in following experimental procedure, 10 times. And then calculate number of test cases which can kill a HOM and number of test cases which can kill its constituent FOMs for each algorithm and PUT.

```
for each PUT do
+ Generate all possible FOMs by applying the set of Judy mutation opera-
tors
+ Execute the FOMs against the given set of test cases
  for each multi-objective optimization and Random algorithm do
    - from the set of all FOMs, generate HOMs guided by objectives and
fitness functions
    - execute the HOMs against the given set of test cases
    - count and save the test case which can kill each HOM belonging to
the set of generated HOMs and the test case which can kill its constitu-
ent FOMs
    - calculate the ratio the number of test cases which can kill HOM
to the number of test cases which can kill its constituent FOMs
  end
end
```

4 Results and Analysis

Experimental results were shown in Tables 3 and 4. Table 3 shows the mean number of generated FOMs (NoF) and Table 4 shows the mean number of generated HOMs (NoH) for each PUT and algorithm.

Table 3. The mean number of generated FOMs

Project under test	NoF
BeanBin	1330
Barbecue	3084
JWBF	1482
CommonsChain 1.2	476
CommonsValidiator 1.4.1	2431
CommonsJxPath 1.3	1453
CommonsFileUploads 1.3.1	1047
CommonsLang3 3.4	9432

As we mentioned before, in this paper we focus on the ratio (named R) of number of test cases which can kill each generated HOM to number of test cases which can kill its constituent FOMs. This ratio will help us to answer the question whether HOM is harder to kill than FOM or not. We divide R into 3 cases: (1) R = 1 when the number of test cases which can kill HOM is equal to the number of test cases which can kill its constituent FOMs; (2) R < 1 when the number of test cases which can kill HOM is smaller than the number of test cases which can kill its constituent FOMs; (3) R > 1 when the number of test cases which can kill HOM is larger than the number of test cases which can kill its constituent FOMs.

Table 4. The mean number of generated HOMs

Project under test	NoH for each algorithm					
	eMOEA	NSGAII	eNSGAII	NSGAIII	eNSGAII DiffLOC	Random
BeanBin	424	402	335	395	339	351
Barbecue	886	905	807	887	433	510
JWBF	216	238	123	230	141	154
CommonsChain 1.2	141	139	127	147	132	144
CommonsValidator 1.4.1	560	557	542	594	564	556
CommonsJxPath 1.3	421	560	490	612	591	652
CommonsFileUploads 1.3.1	116	120	111	116	185	119
CommonsLang3 3.4	2827	2671	2887	2993	2901	2993

In the Fig. 1, sub-figures from (a) to (f) demonstrate the values (%) of 3 cases of R for each PUTs according to the algorithms NSGAII, eNSGAII, NSGAIII, eMOEA,

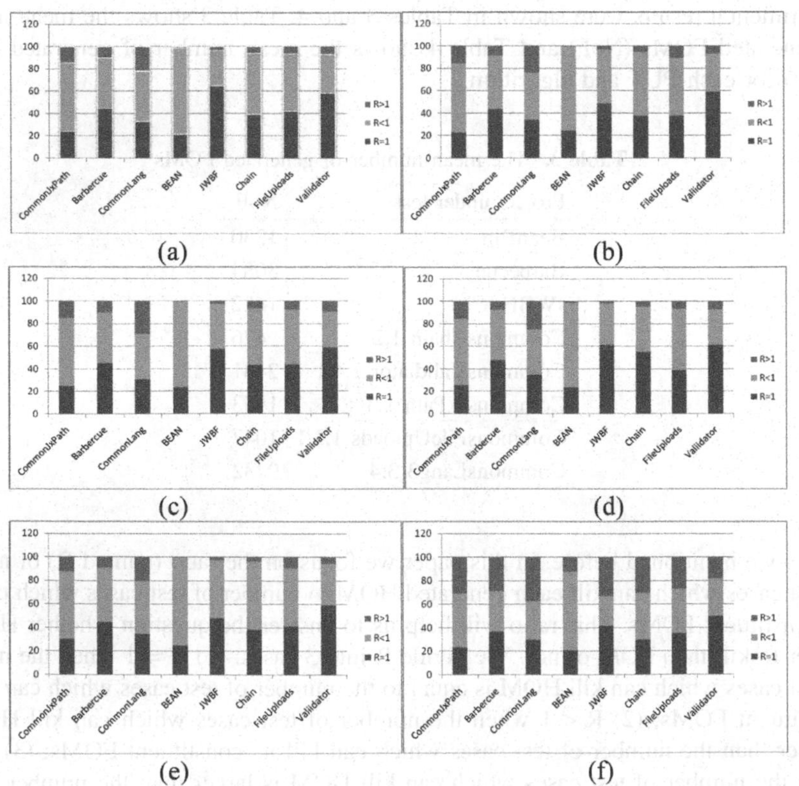

Fig. 1. Values of R for each algorithm.

eNSGA_DiffLOC and Random in turn. In the sub-figures, the column is divided into 3 parts: (1) the part on the top describes the value (%) of the case R > 1; (2) the middle part describes the value (%) of the case R < 1; (3) the part at the bottom describes the value (%) of the case R = 1. The results presented in the above sub-figures show the mean ratio of R = 1, R < 1 and R > 1 cases are about 40%, 50% and 10% in turn for all selected algorithms and PUTs. This means that only a half of generated HOMs are harder to kill than their constituent FOMs in terms of number or test cases that can kill them. Whilst the ratio of HOMs to all generated HOMs are easier to kill than their constituent FOMs is about 10%. And 40% of all generated HOMs are the same their constituent FOMs with regard to number or test cases that can kill them.

5 Conclusions and Future Work

We applied our proposed objectives and fitness functions [20–24] into 6 different algorithms, 5 multi objective optimization algorithms (including one modified by us) and Random algorithm for generating FOMs and then constructing HOMs from the set of generated FOMs. We used the number of test cases which can kill HOMs and their constituent FOMs to find the answering for the question “Is higher order mutant harder to kill than first order mutant?”. Our experimental results indicate that about 50% HOMs (among the set of generated HOMs which were constructed by combining the FOMs) are harder to kill than their constituent FOMs and remaining HOMs are not.

There is always a set of threats to the validity of empirical results. The selected PUTs may not be representative of Java programs in general, and therefore, the results of the study may not be generalizable to all Java programs. In addition, 8 selected PUTs (include the small and big program in term of line of codes) are different with regard to functionality. Another threat to validity stems from the quality and the coverage of test cases. There is a possibility of getting different results if test suites of higher and lower quality or coverage were used.

We hope that the presented research outcome will help us to evaluate the effectiveness of mutation testing in general and higher order mutation testing in particular. However, further research is needed using more and larger projects under test to confirm the preliminary results presented in this paper. Using the more different algorithms is also needed to confirm the correctness of our study.

References

1. DeMillo, R.A., Lipton, R.J., Sayward, F.G.: Hints on test data selection: help for the practicing programmer. IEEE Comput. **11**(4), 34–41 (1978)
2. Hamlet, R.G.: Testing programs with the aid of a compiler. IEEE Trans. Softw. Eng. **SE-3** (4), 279–290 (1977)
3. Nguyen, Q.V., Madeyski, L.: Problems of mutation testing and higher order mutation testing. In: van Do, T., Thi, H.A.L., Nguyen, N.T. (eds.) Advanced Computational Methods for Knowledge Engineering. AISC, vol. 282, pp. 157–172. Springer, Cham (2014). https://doi.org/10.1007/978-3-319-06569-4_12

4. Jia, Y., Harman, M.: An analysis and survey of the development of mutation testing. IEEE Trans. Softw. Eng. **37**(5), 649–678 (2011)
5. Jia, Y., Harman, M.: Higher order mutation testing. Inf. Softw. Technol. **51**, 1379–1393 (2009)
6. Harman, M., Jia, Y., Langdon, W.B.: A manifesto for higher order mutation testing. In: Third International Conference on Software Testing, Verification, and Validation Workshops (2010)
7. Offutt, A.J.: Investigations of the software testing coupling effect. ACM Trans. Softw. Eng. Methodol. **1**, 5–20 (1992)
8. Polo, M., Piattini, M., Garcia-Rodriguez, I.: Decreasing the cost of mutation testing with second-order mutants. Softw. Test. Verif. Reliab. **19**(2), 111–131 (2008)
9. Kintis, M., Papadakis, M., Malevris, N.: Evaluating mutation testing alternatives: a collateral experiment. In: Proceedings of the 17th Asia Pacific Software Engineering Conference (APSEC) (2010)
10. Papadakis, M., Malevris, N.: An empirical evaluation of the first and second order mutation testing strategies. In: Proceedings of the 2010 Third International Conference on Software Testing, Verification, and Validation Workshops, ICSTW 2010, pp. 90–99. IEEE Computer Society (2010)
11. Madeyski, L., Orzeszyna, W., Torkar, R., Józala, M.: Overcoming the equivalent mutant problem: a systematic literature review and a comparative experiment of second order mutation. IEEE Trans. Softw. Eng. **40**(1), 23–42 (2014). https://doi.org/10.1109/tse.2013.44
12. Omar, E., Ghosh, S.: An exploratory study of higher order mutation testing in aspect-oriented programming. In: IEEE 23rd International Symposium on Software Reliability Engineering (2012)
13. Jia, Y., Harman, M.: Constructing subtle faults using higher order mutation testing. In: Proceedings of the Eighth International Working Conference Source Code Analysis and Manipulation (2008)
14. Omar, E., Ghosh, S., Whitley, D.: Constructing subtle higher order mutants for Java and AspectJ programs. In: International Symposium on Software Reliability Engineering, pp. 340–349 (2013)
15. Omar, E., Ghosh, S., Whitley, D.: Comparing search techniques for finding subtle higher order mutants. In: Proceedings of the 2014 Annual Conference on Genetic and Evolutionary Computation, pp. 1271–1278 (2014)
16. Belli, F., Güler, N., Hollmann, A., Suna, G., Yıldız, E.: Model-based higher-order mutation analysis. In: Kim, T.-H., Kim, H.-K., Khan, M.K., Kiumi, A., Fang, W.-C., Ślęzak, D. (eds.) ASEA 2010. CCIS, vol. 117, pp. 164–173. Springer, Heidelberg (2010). https://doi.org/10.1007/978-3-642-17578-7_17
17. Akinde, A.O.: Using higher order mutation for reducing equivalent mutants in mutation testing. Asian J. Comput. Sci. Inf. Technol. **2**(3), 13–18 (2012)
18. Langdon, W.B., Harman, M., Jia, Y.: Multi-objective higher order mutation testing with genetic programming. In: Proceedings of the Fourth Testing: Academic and Industrial Conference Practice and Research (2009)
19. Langdon, W.B., Harman, M., Jia, Y.: Efficient multi-objective higher order mutation testing with genetic programming. J. Syst. Softw. **83**, 2416–2430 (2010)
20. Nguyen, Q.V., Madeyski, L.: Searching for strongly subsuming higher order mutants by applying multi-objective optimization algorithm. In: Le Thi, H.A., Nguyen, N.T., Do, T.V. (eds.) Advanced Computational Methods for Knowledge Engineering. AISC, vol. 358, pp. 391–402. Springer, Cham (2015). https://doi.org/10.1007/978-3-319-17996-4_35
21. Nguyen, Q.V., Madeyski, L.: Empirical evaluation of multi-objective optimization algorithms searching for higher order mutants. Cybern. Syst.: Int. J. (2016)

22. Nguyen, Q.V., Madeyski, L.: Higher order mutation testing to drive development of new test cases: an empirical comparison of three strategies. In: Nguyen, N.T., Trawiński, B., Fujita, H., Hong, T.-P. (eds.) ACIIDS 2016. LNCS (LNAI), vol. 9621, pp. 235–244. Springer, Heidelberg (2016). https://doi.org/10.1007/978-3-662-49381-6_23
23. Nguyen, Q.V., Madeyski, L.: On the relationship between the order of mutation testing and the properties of generated higher order mutants. In: Nguyen, N.T., Trawiński, B., Fujita, H., Hong, T.-P. (eds.) ACIIDS 2016. LNCS (LNAI), vol. 9621, pp. 245–254. Springer, Heidelberg (2016). https://doi.org/10.1007/978-3-662-49381-6_24
24. Nguyen, Q.V., Madeyski, L.: Addressing mutation testing problems by applying multi-objective optimization algorithms and higher order mutation. J. Intell. Fuzzy Syst. **32**, 1173–1182 (2017). https://doi.org/10.3233/jifs-169117
25. Madeyski, L.: On the effects of pair programming on thoroughness and fault-finding effectiveness of unit tests. In: Münch, J., Abrahamsson, P. (eds.) PROFES 2007. LNCS, vol. 4589, pp. 207–221. Springer, Heidelberg (2007). https://doi.org/10.1007/978-3-540-73460-4_20
26. Madeyski, L.: The impact of pair programming on thoroughness and fault detection effectiveness of unit tests suites. Softw. Process: Improv. Pract. **13**(3), 281–295 (2008). https://doi.org/10.1002/spip.382
27. Madeyski, L.: The impact of test-first programming on branch coverage and mutation score indicator of unit tests: an experiment. Inf. Softw. Technol. **52**(2), 169–184 (2010). https://doi.org/10.1016/j.infsof.2009.08.007
28. Madeyski, L., Radyk, N.: Judy - a mutation testing tool for Java. IET Softw. **4**(1), 32–42 (2010). https://doi.org/10.1049/iet-sen.2008.0038
29. Deb, K., Pratap, A., Agarwal, S., Meyarivan, T.: A fast and elitist multi objective genetic algorithm: NSGA-II. IEEE Trans. Evol. Comput. **6**(2), 182–197 (2002)
30. Deb, K., Jain, H.: An evolutionary many-objective optimization algorithm using reference-point-based nondominated sorting approach, part I: solving problems with box constraints. IEEE Trans. Evol. Comput. **18**(4), 577–601 (2014)
31. Kollat, J.B., Reed, P.M.: The value of online adaptive search: a performance comparison of NSGAII, ε-NSGAII and εMOEA. In: Coello Coello, C.A., Hernández Aguirre, A., Zitzler, E. (eds.) EMO 2005. LNCS, vol. 3410, pp. 386–398. Springer, Heidelberg (2005). https://doi.org/10.1007/978-3-540-31880-4_27
32. Deb, K., Mohan, M., Mishra, S.: A fast multi-objective evolutionary algorithm for finding well-spread pareto-optimal solutions. KenGAL, Report No. 2003002. Indian Institute of Technology, Kanpur, India (2003)

Information Systems Development via Model Transformations

Jaroslav Pokorný[1], Karel Richta[2(✉)], and Tomáš Richta[3]

[1] Charles University, Prague, Czech Republic
pokorny@ksi.mff.cuni.cz

[2] Czech Technical University in Prague, Prague, Czech Republic
richta@fel.cvut.cz

[3] Brno Institute of Technology, Brno, Czech Republic
irichta@fit.vutbr.cz

Abstract. Many present systems can be developed by a sequence of transformations from the source specification to the final implementation. An interesting question is whether we can support such a sequence of transformations by some formal apparatus that enables to verify succeeding steps of development, and finally also the whole development process. As an example, we use the transformation of a definition of the set of autonomous agents by classical workflow models and then transform them into a set of Petri nets. Such transformation would support development of software systems, whose specification is based on classical workflow models, but the implementation is based on Petri nets. Each part of the designed system is translated from workflow model into Petri nets, and interpreted by the special Petri Nets Virtual Machines, which are installed on all nodes of the system.

Keywords: Model transformation · Petri net · Formal development

1 Introduction

Many computer driven systems can be developed by a sequence of transformations from the source specification into the final implementation. Software developers call this process as the stepwise refinement. The interesting question is whether we can support such sequence of transformations by some formal apparatus that enables to verify succeeding steps of development, and finally also the whole development process. For this purposes we have to describe formally the semantics of the input specification for each distinguished step, the transformation itself, and also the semantics of transformation result.

Let us suppose the source model as the specification S_{Inp} in some input language L_{Inp}, i.e., $S_{Inp} \in L_{Inp}$. Each individual development step number $k\,(1 \leq k \leq n)$ is the transformation T_k, which transforms the input specification S_{Inp} into the output specification S_{Out} in some output language L_{Out}, i.e., $S_{Out} \in L_{Out}$. The result of this transformation should be correct – so it has to have the same semantics as the input. The transformation T_k could be called a correct transformation with respect to languages

N. T. Nguyen et al. (Eds.): ACIIDS 2018, LNAI 10751, pp. 674–683, 2018.
https://doi.org/10.1007/978-3-319-75417-8_63

L_{Inp} and L_{Out}, iff for all input specifications, the resulting output has the same semantics. The correctness of the transformation should be verified.

The result of transformation is $T_k[\![S_{Inp}]\!] = S_{Out} \in L_{Out}$. We will use special brackets $[\![$and$]\!]$ to distinguish syntactic arguments from semantic ones. Let M_{Inp} be the semantic model of the input language, and $int_{Inp} : L_{Inp} \rightarrow M_{Inp}$ be the mapping that assigns to any input specification $S_{Inp} \in L_{Inp}$ its meaning $int_{Inp}\,[\![S_{Inp}]\!]$ in the input model M_{Inp} – the semantic interpretation of the input specification. We call mapping int_{Inp} the *interpretation* of the input language.

Similarly, let M_{Out} be the semantic model of the output language, and int_{Out} : $L_{Out} \rightarrow M_{Out}$ be the semantic interpretation of the output language. The meaning of S_{Out} in the output model is $int_{Out}[\![S_{Out}]\!] \in M_{Out}$. The transformation T_k is a *correct transformation* with respect to int_{Inp} and int_{Out}, iff for all input specifications S_{Inp}, the resulting output has the same semantics, i.e.

$$\forall S_{Inp} \in L_{Inp} \,.\, \mu(int_{Inp}[\![S_{Inp}]\!]) \; = int_{Out}[\![T_k[\![S_{Inp}]\!]]\!],$$

where μ is the mapping of input model M_{Inp} into the output model M_{Out}. This mapping specifies how to interpret input meaning in the output space.

In our case we have tried (among others) workflow models as the source language, and Petri nets as the output language. This choice was motivated mainly by our interest in Petri nets. So the mapping μ should map all meanings of workflow models into meanings of Petri nets.

Many systems can be considered as a set of autonomous agents that communicate together to solve problems. There are a number of systems that support the provision of such communication on the basis of specifications of the external behavior of agents and a description of their communication. System assembly of a set of agents is relatively well developed and orchestration of such set of agents can be assured and generated from its description by known tools. What is currently not developed and supported by automation is a creation of agents based on their specifications. They are usually implemented manually from the specification without the assistance of an adequate environment.

The purpose of our research is to develop methods and tools that can be used for the creation of autonomous agents from their specification. We investigate the specification of agents based on Petri nets (see [7–9]). We suppose, that all nodes of the distributed system can be equipped by special basic software called Petri Net Operating System (PNOS [10]). Then, all nodes can be supplemented by special software called Petri Net Virtual Machine (PNVM [11]). Such PNVMs can be programmed by uploaded representations of Petri nets in so called Petri Net Byte Code (PNBC), and pose as agents of the whole orchestrated distributed system.

There exist similar approaches based on different versions of Petri nets. Our method is new in several directions - we use version of Petri nets called Reference Petri Nets [11], which is sufficiently complete to describe our expected systems. The second one is the usage of intermediate virtual machines for implementation of partial Petri nets. And the last difference is the usage of formal descriptions on the source and target language, and also for transformations. It enables verification in the future.

The structure of the paper is as follows - in Sect. 2 we define the source specification workflow model. In Sect. 3 we describe the output model - Petri Nets. Section 4 deals with transformations of models in general, and describes the transformation of workflow models into Petri nets. It also introduces the transformation of Petri nets into PNML (see [4]) format for the sake of interpretation by PNVM.

2 The Source Specification Model

Mainstream software processes, like the Unified Process, propose UML diagrams [6] for modelling the integration of conceptual data models and activity diagrams for specifying workflows [3]. Then, these models are used as requirements specification to drive the subsequent development and integration of applications through analysis, design, and implementation activities. We will use workflow specification as the source specification model.

A workflow specification is composed of one or more extended workflow nets (EWF-nets). Therefore, we first formalize the notion of a EWF-net. The definition is an abbreviated version of the classical definition from [1].

Definition 1 (EWF-net).
An *extended workflow net* (EWF-net) N is a tuple $(C, \mathbf{i}, \mathbf{o}, K, F)$ such that:

- C is a set of conditions,
- $\mathbf{i} \in C$ is the input condition,
- $\mathbf{o} \in C$ is the output condition,
- K is a set of tasks,
- $F \subseteq (C \setminus \{\mathbf{o}\} \times K) \cup (K \times C \setminus \{\mathbf{i}\}) \cup (K \times K)$ is the flow relation, and
- every node in the graph $(C \cup K, F)$ is on a directed path from $\mathbf{i}$ to $\mathbf{o}$.

Definition 2 (Workflow specification).
A *workflow specification WS* is a tuple $(Q, top, K^{\diamond}, map)$ such that:

- Q is a set of EWF-nets (components),
- $top \in Q$ is the top level workflow,
- $K^{\diamond} = \cup_{N \in Q} K_N$ is the set of all tasks (data flows),
- $\forall\, n_1, n_2 \in Q.\ n_1 \neq n_2 \Rightarrow (C_{n1} \cup T_{n1}) \cap (C_{n2} \cup T_{n2}) = \emptyset$, i.e., no name clashes,
- *map*: $K \rightarrow Q \setminus \{\text{top}\}$ is a surjective injective function which maps each composite task onto an EWF net, and
- the relation $\{<n_1, n_2> \in Q \times Q \mid \exists t \in \text{dom}(map_{n1}).\ map_{n1}(t) = n2\}$ is a tree.

Q is a non-empty set of EWF-nets with a special EWF-net *top*. Composite tasks are mapped onto EWF-nets such that the set of EWF-nets forms a tree-like structure with *top* as root node (main component). $K^{\diamond}$ is the set of all tasks. Tasks in the domain of *map* are composite tasks which are mapped onto EWF-nets. Throughout this paper we will assume that there are no name clashes, e.g., names of conditions differ from names of tasks and there is no overlap in names of conditions and tasks originating from different EWF-nets. If there are name clashes, tasks/conditions are simply renamed.

Definition 3 (Workflow state).
A *workflow state* s of a specification $WS = (Q, top, K^{\diamond}, map)$ is a multiset (bag) over $Q^{\diamond} \times I$ where $K^{\diamond} = C^{\diamond}$, i.e., $s \in \mathcal{B}(Q^{\diamond} \times I)$.

A workflow state s is a bag of tokens where each token is represented by a pair consisting of a condition from Q and an identifier from I, i.e., $s \in \mathcal{B}(Q^{\diamond} \times I)$. For a token $<x, i> \in s$, x denotes the location of the token and i denotes the identity of the token. Location x is either (1) an implicit or explicit condition ($x \in Q$) or (2) a task state of some task $t \in K^{\diamond}$. When defining the state transitions it will become clear that reachable workflow states will satisfy the input conditions.

Notation: If X is a bag over A and Y is a finite subset of A, then X–Y, X ⊎ Y, Y–X, and Y ⊎ X yield also bags over A with the usual meaning.

Definition 4 (Partial transition relation).
Let $WS = (Q, top, K^{\diamond}, map)$ be a specification and s_1 and s_2 two workflow states of WS. We write $s_1 \rightarrowtail s_2$ if and only if there are $t \in K^{\diamond}$, $i \in I$, c, $p \subset \mathcal{B}(Q^{\diamond} \times I)$ such that binding (t, i, c, p, s_1) and $s_2 = (s_1 - c) \uplus p$.

$\rightarrowtail$ defines a *partial transition relation* on the states of workflow specification.

The reflexive transitive closure of $\rightarrowtail$ is denoted $\rightarrowtail^*$ and $R^{partial}(s) = \{s' \in \mathcal{B}(Q^{\diamond} \times I) \mid s \rightarrowtail^* s'\}$ is the set of states reachable from state s (all in the context of some workflow specification).

Definition 5 (Transition relation).
Let $WS = (Q, top, K^{\diamond}, map)$ be a specification and s_1 and s_2 two workflow states of WS. We write $s_1 \looparrowright s_2$ if and only if $s_1 \rightarrowtail s_2$ or each of the following conditions is satisfied:

- There are $t \in K^{\diamond}$, $i \in I$, $c, p \in \mathcal{B}(Q^{\diamond} \times I)$ such that binding enable (t, i, c, p, s_1), and $s_2 = (s_1 - c) \uplus p$.
- For each $s \in R^{partial}(s_1)$, there is no $c' \in \mathcal{B}(Q^{\diamond} \times I)$ such that binding enable (t, i, c', p, s_1) and $c' > c$.

$\looparrowright$ is the *transition relation* that includes all state transitions in $\rightarrowtail$ and adds transitions of type enable if the number of consumed tokens cannot be increased.

The reflexive transitive closure of $\looparrowright$ is denoted $\looparrowright^*$ and $R(s) = \{s' \in \mathcal{B}(Q^{\diamond} \times I) \mid s \looparrowright^* s'\}$ is the set of states reachable from a state s. If ambiguity is possible, we will add subscripts, i.e., $Q^{\diamond}_{WS}$, $\looparrowright^*WS$, and R_{WS}. The state space $\mathcal{B}(Q^{\diamond}_{WS} \times I)$ and transition relation $\looparrowright$ define a transition system $<\mathcal{B}(Q^{\diamond}_{WS} \times I), \looparrowright_{WS}>$ for WS. Such a transition system can be augmented with different notions of equivalence. We define that the transition system $<\mathcal{B}(Q^{\diamond}_{WS} \times I), \looparrowright_{WS}>$ is the *meaning* of the workflow specification $WS = (Q, top, K^{\diamond}, map)$.

2.1 The Example

As an example of an input model consider some autonomous system, e.g. heating system for a small house, see Fig. 1. This heating system consists of five components

("Hall and stairway", "Kitchen and dining room", "Remote control", "Boiler room", and "Scheduler"). The whole system containing these five components is called "House 1". Components have input and output ports, e.g. components "Hall and stairway" has among other the input port "schedTemp1", and the output port "temp1". The output port "temp1" of "Hall and stairway" is connected to the input port of the "Scheduler", etc. Components contain elements like "thermostat", "knob", whose description and structure is already predefined (Fig. 2).

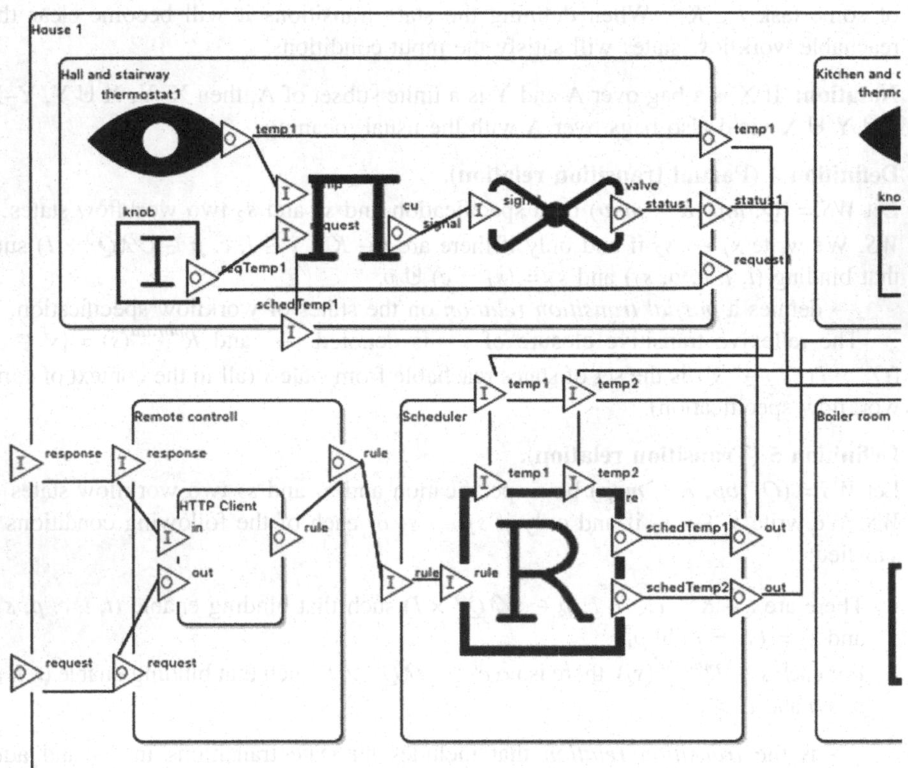

Fig. 1. Autonomous heating system (a part)

Such a description could be created by some CASE tool, and finally exported in the XML format, e.g. as an application of EMF. Look at the Fig. 1, which shows the small part of the representation of this heating system in EMF. This XML document is a valid document according to the XML schema of EMF (see [2]), here **`domotic.ecore`** (see [5]). This schema is designed for so called *domotic systems* – a set of technological components capable of performing functions that can be partially autonomous, programmed by the user, or even completely autonomous [5].

3 The Output Model - Petri Nets

Petri nets are widely used for the specification of problems, mostly in the parallel systems. The following formal definition is loosely based on [9, 10]. Many alternative definitions exist.

3.1 Syntax

A *Petri net graph* (called *Petri net* by some, but see below) is a 3-tuple (*P*, *T*, *W*), where:

- $P = \{P_i\}$ is a finite set of *places*
- $T = \{T_j\}$ is a finite set of *transitions*
- *P* and *T* are disjoint, i.e. no object can be both a place and a transition
- $W: (P \times T) \cup (T \times P) \rightarrow \mathrm{N}$ is a multiset of arcs, i.e. it defines arcs and assigns to each arc a non-negative integer *arc multiplicity*; note that no arc may connect two places or two transitions.

```
<?xml version="1.0" encoding="UTF-8"?>
  <domotic:Component xmi:version="2.0"
xmlns:xmi="http://www.omg.org/XMI"
xmlns:xsi="http://www.w3.org/2001/XMLSchema-instance"
xmlns:domotic="but-hib.domotic" xsi:schemaLocation="but-hib.domotic
../domotic.viewpoint/description/domotic.ecore" name="schedTemp1">
  <contains xsi:type="domotic:Component" name="House 1">
    <inputPorts name="response"/>
    <contains xsi:type="domotic:Component" name="Hall and stairway">
      <inputPorts name="schedTemp1"/>
      <contains xsi:type="domotic:Sensor" name="thermostat1">
        <outputPorts name="temp1"/>
      </contains>
```

Fig. 2. The input model (a part) in EMF format

The *flow relation* is the set of arcs: $F = \{(x, y) \mid W(x, y) > 0\}$. In many textbooks, arcs can only have multiplicity 1, and they often define Petri nets using *F* instead of *W*.

In other words, a Petri net graph is a bipartite multi-digraph $(P \cup T, F)$ with node partitions *P* and *T*.

The *preset* of a transition *t* is the set of its *input places*: ${}^{\bullet}t = \{p \in P \mid W(p, t) > 0\}$; its *postset* is the set of its *output places*: $t^{\bullet} = \{p \in P \mid W(t, p) > 0\}$.

A *marking* of a Petri net (graph) is a multiset of its places, i.e., a mapping M: $P \rightarrow \mathrm{N}$. We say the marking assigns to each place a number of *tokens*.

A *Petri net* (called *marked Petri net* by some, see above) is a 4-tuple (P, T, W, M_0), where:

- (*P*, *T*, *W*) is a Petri net graph;
- M_0 is the *initial marking*, a marking of the Petri net graph.

3.2 Execution Semantics

The behaviour of a Petri net is defined as a relation on its markings, as follows. Note that markings can be added like any multiset:

$$M + M' =^D \{p \to M(p) + M'(p) | p \in P\}$$

The execution of a Petri net graph G = (*P, T, W*), can be defined as the transition relation $\to_G$ on its markings, as follows:

- for any *t* in *T*:

$$M \to_{G,t} M' \Leftrightarrow^D \exists M'' : P \to N : M = M'' + \sum_{p \in P} W(p,t) \wedge M' = M'' + \sum_{p \in P} W(t,p)$$

$$M \to_G M' \Leftrightarrow^D \exists t \in T : M \to_{G,t} M'$$

i.e.

- firing a transition *t* in a marking *M* consumes *W(p, t)* tokens from each of its input places *p*, and produces *W(t, p)* tokens in each of its output places *p*
- a transition is enabled (it may fire) in M if there are enough tokens in its input places for the consumptions to be possible, i.e. iff

$$\forall p : M(p) \geq W(p,t).$$

We are generally interested in what may happen when transitions may continually fire in arbitrary order.

We say that a marking *M' is reachable from* a marking *M in one step* if $M \to_G M'$; we say that it *is reachable from M* if $M \to_G^* M'$, where $\to_G^*$ is the *transitive closure* of $\to_G$; that is, if it is reachable in 0 or more steps.

For a (marked) Petri net *N* = (*P, T, W,* M_0), we are interested in the firings that can be performed starting with the initial marking M_0. Its set of *reachable markings* is the set $R(N) =^D \{M' | M_0 \to^*_{(P,T,W)} M'\}$.

The *reachability graph* of *N* is the transition relation $\to_G$ restricted to its reachable markings *R*(*N*). It is the *state space* of the net.

A *firing sequence* for a Petri net with graph *G* and initial marking M_0 is a sequence of transitions $\vec{\sigma} = \langle t_{i1} \ldots t_{in} \rangle$ such that $M_0 \to_{G,ti1} M_1 \wedge \ldots \wedge M_{n-1} \to_{G,tin} M_n$. The set of firing sequences is denoted as *L*(*N*).

4 Transformation of Petri Nets into PNML

In our case the source language are workflow models, the output language are Petri nets. So the mapping μ should map all meanings of workflow models into meanings of Petri nets. We define the meaning of a workflow specification *WS* = (Q, *top*, $K^\diamond$, *map*) as the transition system $\langle \mathcal{B}(Q_{WS} \times I), \looparrowright_{WS} \rangle$. We also define the meaning of a Petri net *N* = (*P, T, W,* M_0) as the set of reachable markings *R*(*N*). The mapping μ maps sequences of transitions $s_1 \looparrowright s_2 \looparrowright \ldots \looparrowright s_n$ specified by *WS* to firing sequences

$t_1 \ldots t_m$ of Petri net specified by N. For the semantic purposes, the only starting and final states are important, so we can specify mapping $\mu: S \times S \rightarrow M \times M$ by the formula: $\mu(<s_1, s_n>) = <t_1, t_m>$.

To show a high-level implementation of such a transformation, we begin by decom-posing the EMF format as that is our only input. For this purpose let us say that:

- Let $C = \{C_i\}$ be a set of all **<contains>** elements in depth 1, such that they match the xPath expression: c**omponent/contains/contains**.
- Let $D = \{D_i\}$ be a set of all **<dataflow>** elements in depth 1, such that they match the xPath expression: **component/contains/dataflow**.

Now the desired mapping from EMF to PNML can be described by the following algorithm. We suppose that we are able to extract components and data flows from the workflow model in EMF (Fig. 2).

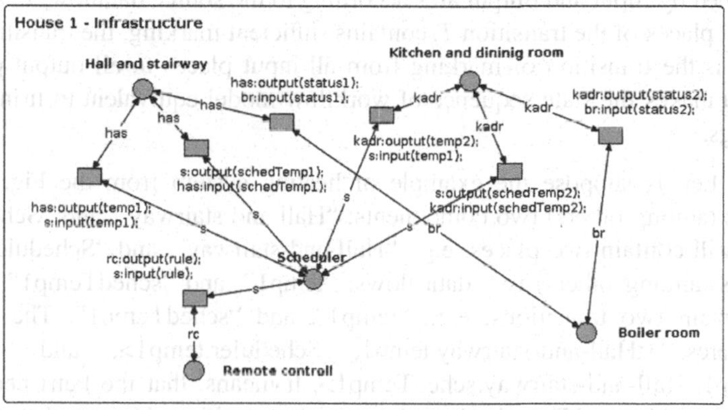

Fig. 3. Resulting Petri net infrastructure of the small house

The Algorithm for transformation of EMF to PNML

Input: The EMF document the workflow description of the agent, and let $C = \{C_i\}$ and $D = \{D_i\}$ be two sets of all components, resp. data flows contained in it.
Output: The PNML document containing the Petri net graph $G = (P, T, W)$, which is behaviorally equivalent to the input.

1. Set $P = \emptyset, T = \emptyset, W = \emptyset$.
2. For every C_i from C create a place P_i such that the attribute id of P_i is unique across the entire document and set $P = P \cup \{P_i\}$.
3. For every D_j from D create a transition T_j such that the attribute id of T_j is unique across the entire document and set $T = T \cup \{T_j\}$.
4. For every D_j from D create a pair of arcs A_1 and A_2, such that the attribute id of A_1 and A_2 is unique across the entire document, set $W = W \cup \{<A_1, A_2>\}$ and:

a. Attribute from of the element A_1 has the value of attribute id of the place P_i which was created from an element C_i referenced by the first expression in attribute ports.
b. Attribute to of the element A_1 has the value of attribute id of the transition T_j which was created from dataflow D_j.
c. Attribute from of element A_2 has the value of attribute id of the place P_i which was created from an element C_i referenced by the second expression in attribute ports of D_j.
d. Attribute to of element A_2 has the value of attribute id of the transition T_j which was created from dataflow D_j.

Sketch of the proof

Due to the step 2, all components of EMF model C will have appropriate place in the Petri net graph G. Due to the step 3, all data flows in D will have appropriate transition in the Petri net graph G. In the step 4, all transitions created in the step 3 will be accomplished by input and output arcs according to the source dataflow. In such a way, when input places of the transition T_j contains sufficient marking, the transition can fire, and result is the transition of marking from all input places to all output places. The mapping μ makes the state sequence of workflow model equivalent to firing sequence of markings.

Example: Let us suppose the example of heating system from the Fig. 1. The set C contains (among others) two components: “Hall and stairway”, and “Scheduler”. So the set P will contain two places, e.g.: “Hall-and-stairway”, and “Scheduler”. The set D contains (among others) two data flows: “temp1”, and “schedTemp1”. So the set T will contain two transitions, e.g.: “temp1”, and “schedTemp1”. The set W will contain arcs <Hall-and-stairway.temp1, Scheduler.temp1>, and <Scheduler. schedTemp1, Hall-and-stairway.schedTemp1>. It means, that the Petri net graph can fire transition $M \rightarrow_G M'$ such that changes the marking M into the marking M' according to the definition of this arc. So the signal of the workflow model is correctly simulated by the constructed Petri net (see Fig. 3).

5 Conclusions

In the foregoing text we described the basics of model transformation and execution-based methodology of distributed embedded control system development. Among the main methods it uses Petri Nets models transformations and target system prototype code generation. So for the description of any system, we can describe particular agents using classical workflow models. These models can be transformed into corresponding Petri nets. The development process starts with the classical workflow model of the system specification. This model of the system describes the functionality from users’ or domain specialist’s point of view. Using our methods, this model is further transformed to the multi-layered architecture based set of Reference Petri Nets. Each layer of the system is translated to the target representation by Petri Nets, which can be interpreted by so called PNVM (Petri Net Virtual Machine), which is installed on all nodes of the system. Targeted system is dynamically reconfigurable by the possibility of instances replacement with its new versions. After the replacement, PNVM interpretation engine starts to perform a new version of partial functionality of the system. The communication between agents will be described using so-called an infrastructure and a platform layer [11]. For the effective usage of such a method we have to create supporting tools. It will be the subject of our further research.

References

1. Adam, N.R., Atluri, V., Huang, W.K.: Modeling and analysis of workflows using Petri nets. J. Intell. Inf. Syst. **10**, 131–158 (1998). https://doi.org/10.1023/A:1008656726700
2. Biermann, E., Ehrig, K., Köhler, C., Kuhns, G., Taentzer, G., Weiss, E.: Graphical definition of in-place transformations in the eclipse modeling framework. In: Nierstrasz, O., Whittle, J., Harel, D., Reggio, G. (eds.) MODELS 2006. LNCS, vol. 4199, pp. 425–439. Springer, Heidelberg (2006). https://doi.org/10.1007/11880240_30
3. van Hee, K.M., Sidorova, N., van der Werf, J.M.: Business process modeling using Petri nets. In: Jensen, K., van der Aalst, W.M.P., Balbo, G., Koutny, M., Wolf, K. (eds.) Transactions on Petri Nets and Other Models of Concurrency VII. LNCS, vol. 7480, pp. 116–161. Springer, Heidelberg (2013). https://doi.org/10.1007/978-3-642-38143-0_4
4. Hillah, L.M., Kindler, E., Kordon, F., Petrucci, L., Trèves, N.: The Petri net markup language and ISO/IEC 15909-2. In: CPN Workshop (2009)
5. Miori, V., Tarrini, L., Manca, M., Tolomei, G.: An open standard solution for domotic interoperability. IEEE Trans. Consum. Electron. **52**(1), 97–103 (2006). https://doi.org/10.1109/tce.2006.1605032
6. OMG: OMG Unified Modeling Language (OMG UML), Superstructure Version 2.2, February 2009. http://www.omg.org/spec/UML/2.2/Superstructure/PDF
7. Peterson, J.: Petri nets. ACM Comput. Surv. **9**(3), 223–252 (1977)
8. Petri, C.A.: Kommunikation mit Automaten. Ph. D. thesis, University of Bonn (1962)
9. Petri, C.A., Reisig, W.: Petri net, vol. 3, no. 4, p. 6477 (2008). http://www.scholarpedia.org/. Accessed 13 July 2017
10. Richta, T., Janoušek, V.: Operating system for Petri nets-specified reconfigurable embedded systems. In: Moreno-Díaz, R., Pichler, F., Quesada-Arencibia, A. (eds.) EUROCAST 2013. LNCS, vol. 8111, pp. 444–451. Springer, Heidelberg (2013). https://doi.org/10.1007/978-3-642-53856-8_56
11. Richta, T., Janoušek, V., Kočí, R.: Code generation for Petri nets-specified reconfigurable distributed control systems. In: Proceedings of 15th International Conference on Mechatronics - Mechatronika 2012, Prague, pp. 263–269 (2012)

A Simply Way for Chronic Disease Prediction and Detection Result Visualization

Dingkun Li, Hyun Woo Park, Erdenebileg Batbaatar, and Keun Ho Ryu(✉)

Database/Bioinformatics Lab, School of Electrical and Computer Engineering, Chungbuk National University, Cheongju, South Korea
{jerryli,hwpark,eegii,khryu}@dblab.chungbuk.ac.kr

Abstract. Disease data provide an abundant source for chronic disease research. Hundreds of applications have been developed to deliver healthcare based on this big data. However, very few applications provide efficient chronic disease data visualization methods to better understand the results. This paper introduces a simple and practical way for visualizing the results of chronic disease detection and prediction. A model called IVIS4BigData has been used to implement the visualization procedure. This model not only demonstrates the historical data but also provides state-of-the-art visualization techniques. An exemplary set of scenarios corresponding to system design as well as visualization evaluation are given at last. Also we consulted several domain experts and common users about our visualization experimental results which satisfied their understanding about our systems. Finally conclusion and overlook of future work complete the paper.

Keywords: Big data analysis · Information visualization · Healthcare IVIS4BigData reference model

1 Introduction

Never before in history data is generated at such high volume, velocity and variety as it is today. This data provides highly valuable information for all users. In healthcare field, data provides vital information for both doctors and patients in all around. Visualization methods play important role for the purpose of better understanding of analyzed result based on big chronic disease data. A standardized user interface can provide inter-operability on the visual level. While visualization is defined as "the process of transforming data, information and knowledge into visual form making use of humans' natural visual capabilities" [6].

Our paper aims to provide methods for chronic disease prediction and detection results visualization. Recently, the clinical and healthcare recommending service is required in medical center for the clinical diagnosis and plan of treatment in connection with chronic disease [16]. Two challenges are confronted with us: 1. for a doctor, how can he view his patients' information efficiently at anytime and anywhere? 2. for a patient, how can he understand his disease information very well even without professional background? Nevertheless, current e-Science research resources and

N. T. Nguyen et al. (Eds.): ACIIDS 2018, LNAI 10751, pp. 684–693, 2018.
https://doi.org/10.1007/978-3-319-75417-8_64

infrastructures (i.e., data, tools, and related Information and Communication Technology (ICT) services) are often confined to computer science expert usage only [13] and turn this hope-filled vision into a reality will take enormous effort from a great amount of designers, analysts, software engineers, usability specialists and medical professionals.

The contributions of our work are: 1. Provide a simple but practical way for chronic disease prediction and detection result visualization (Sect. 4). 2. The visualization results (views) are very easy to understand even detection and prediction procedure are complicated (Sect. 4).

In this paper, Sect. 2 summarizes related works. Section 3 describes our chronic disease healthcare system briefly. Section 4 demonstrates the visualization procedure. Experiment and evaluation are given in Sect. 4. Section 5 completes our work by conclusion and overlook.

2 Related Work

2.1 Big Data

Data scale expanded dramatically in the past few years with the rapid development of emerging information technologies, including cloud computing, social networks, mobile commerce, and the Internet of Things etc. Nevertheless, there is no unambiguous definition about Big Data. One of the widely accepted concept about big data recently applied in the work [1] which depicts it as framework expressing the 3-dimensional increase in data volume, velocity and variety, called 3V's. Perpetually, big data is becoming "bigger" with more volume, velocity and variety since more and more organizations are paying more attention to big data for the sake of the true worth from it, another two properties, veracity and value, have been expanded to Big Data. Hence, Big Data has been evolved to five typical features, volume, velocity, variety, veracity and value, also called 5V's [2]. The information storm resulting from the advent of big data is not only changing people's lives, careers, and ways of thinking but also initiating great transformations [3].

Big data research has become an extremely hot topic in last decade and it will hold this trend, we say at least, for the next decade since the generation of the big data will increase even faster than before. Hundreds of the architectures, frameworks, tools like Lambda architecture [4], Hadoop with it eco-system [5] related to Big Data research have been developed to speed up the utility of these research results.

2.2 Information Visualization

The most precise and common definition of Information Visualization Information System (IV or IVIS) as "the use of computer-supported, interactive, visual representations of abstract data to amplify cognition" stems from Card et al. [7]. A central task in information visualization is to find the appropriate visualization paradigm for both the data and the problem scenario at hand. Many such visual information mappings exist [8]. From Fig. 1 we can see that data visualization is the final step for data

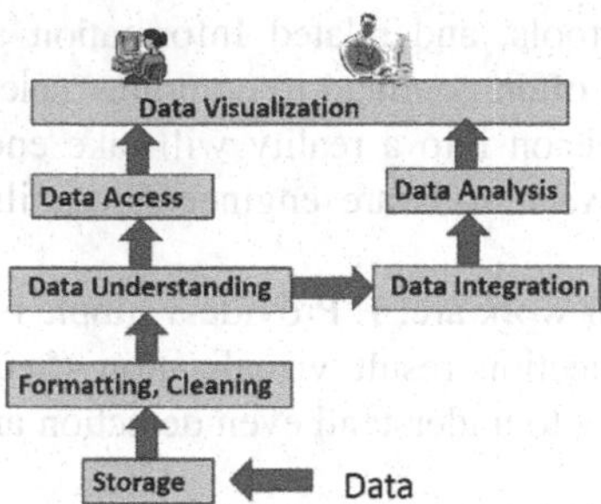

Fig. 1. Data processing procedure

processing. There are two purposes for IV according to authors' understanding, one is to better understand the data with its overlook as well as deep implication, and another one is to provide an evidence for future decision support based on the intuitive data processing result.

In the field of visualization for medical patient records, a number of approaches have been proposed and adopted widely. The usage of these approaches is amplifying the benefits of health informatics databases and networks by dramatically expanding the capacity of patients, clinicians, and public health policy makers to make better decisions [9]. Work [10] provides a useful framework for analyzing health informatics technologies, under the popular term "Health 2.0" which contains three domains: personal health information, clinical health information and public health information. By looking at the ensemble of researches user can easily visualize dynamic patterns of change that may exist in the multivariate data.

2.3 IVIS4BigData Reference Model

A lot of models have been developed for data visualization. For instance, to simplify the discussion about information visualization systems and to compare and contrast them, work [11] defined a reference model for mapping data to visual forms for human perception step by step (total 4 steps), while work [12] defined a multi-dimensional visual medical concept structure which contains the data to implement a presentation layer for archetype based medical data. But most of these models are unable to adapt for covering the recent advancements like cloud technologies or distributed computing technologies. Work [13] developed a hybrid refined and extended IVIS4BigData reference model to cover the new conditions of the present situation with advanced visual interface opportunities for perceiving, managing, and interpreting Big Data analysis results to support insight. The model is depicted in Fig. 2.

Instead of collecting raw data from a single data source, multiple data sources can be connected, integrated by means of mediator architectures, and in this way globally managed in **Data Collections** inside the **Data Collection, Management & Curation** layer. The first transformation, which is located in the **Analytics** layer of the underlying BDM model, maps the data from the connected data sources into **Data Structures**, which represent the first stage in the **Interaction & Perception** layer. The generic term

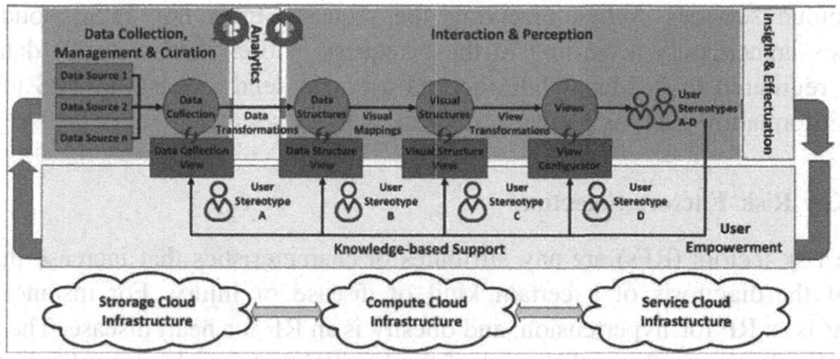

Fig. 2. IVIS4BigData reference model [13]

Data Structures also includes the use of modern **Big Data Storage Technologies** (like, e.g., NoSQL, RDBMS, HDFS), instead of using only data tables with relational schemata. The following steps **Visual Mappings**, which transforms data tables into **Visual Structures**, and **View Transformations**, which creates **Views** of the **Visual Structures**, by specifying graphical parameters such as position, scaling, and clipping, do not differ from the original IVIS reference model. As a consequence, only interacting with analysis results leads not to "added value" for the optimization of, e.g., research results or business objectives. Furthermore, no process steps are currently located within the **Insight & Effectuation** layer because such "added value" is rather generated from knowledge [13]. Our visualization work is based on this reference model.

3 Healthcare System Introduction

This section describes architecture of our healthcare system as well as chronic disease prediction and detection procedure for result visualization in next section.

3.1 System Architecture

We give an overview of the whole system which has been described by our previous work [14]. There are 4 modules included in our system. **Data Collection Module:** The system provides three ways to collect data. First way is using IoT devices, such as phone, watch, ring, etc., to set up a mobile health sensor network (MHSN). Second way is using app to collect data from the user input. Third way is using system interface to import data from public API provided by government, hospital and other organizations. **Data Storage Module:** The data collected is of three types: structured, semi-structured, and unstructured data (Fig. 3). Firstly, all these three kinds of data will be stored in HBase which is quite suitable for mass data preprocessing and storage. Then this data should be converted into structured data for further processing. **Third-party Server (TPS) Module:** TPS response for data statistical analysis, patient emergency detection, disease prediction and detection etc. **Cloud Service Module:** After processing, TPS sends result and processed data to the cloud model, which is used to store data and transfer data. This model is implemented by using GCSql and

GCM cloud services. When receiving the requests from the TPS, cloud model responses immediately according to these requests, stores data or sends data to the devices registered to it. Meanwhile the end user can send requests to TPS to get the related information such as patient statistical analysis result.

3.2 Key Risk Factors Selection

Disease risk factors (RFs) are any attributes or characteristics that increase the possibility of the diagnosis of a certain kind of disease or injury. For instance, heavy drinking is an RF for hypertension, and obesity is an RF for heart disease. The purpose of the RF selection is the seeking out of the key RF that may be more likely than the other factors to develop a certain disease [15]. The Information Gain, GainRatio, and Gini index are the commonly used DM methods for the attribute selection [14].

3.3 Chronic Disease Prediction

In previous sub-section we have found that one key risk factor may be the cause of several diseases, for instance, body fat content is the key risk factor of heart disease, hypertension and obesity, inadequate intake of vitamin B is the key risk fact of asthma, hypertension etc. In order to predict the chronic diseases, long-term observed disease big data including key risk factors will be collected and analyzed. After analyzing big amount of disease data, a basic disease model W.R.T certain kind of disease will be generated. New coming long-term observed patient key risk fact data will be compared with this model to see how similar they are, the higher the similarity, the higher the chance the patient will get this kind of chronic disease. For more information please refer to our previous research [14].

3.4 Chronic Disease Detection

Disease rule has the format like IF THEN rule, for example: IF (age > 46.5, Fat_intake > 42.28 mg/day, married) THEN (hypertension = yes). Our system will mine all these diseases related rules from the training data set. And the rules generated will be stored in the HBase.

Usually, there are more than one rules related to one disease. Compared with these rules, if the matching rate >= β (expert defined threshold, e.g.: 80%), the TPS will treat this observation as the patient. For example, there are 5 rules for heart disease, when there are 4 rules matching above 5 rules, the heart disease will be detected with its expectation as 80%. For more information please refer to our previous research [14].

3.5 System Implementation Based on IVIS4BigData Model

The major difference between our previous work [14] and this work is that it focuses on procedure about system visualization while previous one focuses on algorithm implementation and system development. We describes the procedure for chronic disease prediction and detection result visualization based on IVIS4BigData model in this section.

3.6 Big Data Infrastructure Construction

According to IVIS4BigData model, the first step is to construct the big data analysis infrastructure (platform). It consists of storage, computing and service infrastructure. Figure 3 depicts the structure of our platform.

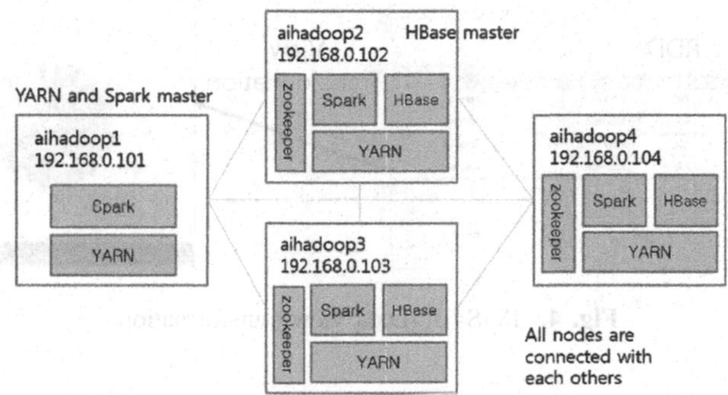

Fig. 3. Big Data infrastructure architecture

HBase cluster is used as the Storage Cloud Infrastructure, all input and out data will be stored in it. Spark on YARN cluster will be used as the compute cloud and service cloud infrastructure.

3.7 Data Collection and Storage

Instead of collecting raw data from a single data source, multiple data sources can be connected, integrated by means of mediator architectures, and in this way globally managed. In our system, three kinds of disease data: unstructured, semi-structured and structured data are collected, preprocessed and stored in HBase. For more information please refer to our previous research [14].

3.8 Interaction and Perception

After data analytics steps, the result will be stored in HBase system, the following steps visual mappings, which transforms data tables into Visual Structures, and View Transformations, which creates Views of the Visual Structures, these views are the app interfaces displayed to end users. Examples will be given in experiment section.

Stereotype users interact with each step of the whole visualization procedure from data collection to view configuration. Their valuable opinions are taken and applied to the system for better results.

3.9 View Transformation

View Transformation creates views from the RDD, these views can be displayed in app interfaces. Figure 4 give one example about this step.

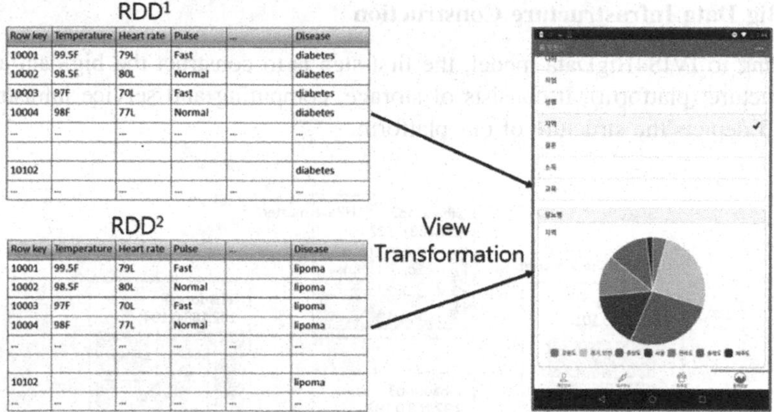

Fig. 4. IVIS4BigData view transformation

4 Experiment Result Visualization and Evaluation

Simulated data and real data downloaded from the Korea National Healthcare Center (KNHC) are used in our experiment. Please refer to work [14] for more detail.

4.1 Chronic Disease Prediction and Detection Result Visualization

The result visualization has several challenges: 1. the result itself is complicated since it contains diseases information, diseases related key risk factors information, risk factor weight (ranking) information, percentage of the risk factors that is above or lower than normal level, etc. 2. Most users are patients who have few medical background to understand complicated result, we should keep the visualization result simple.

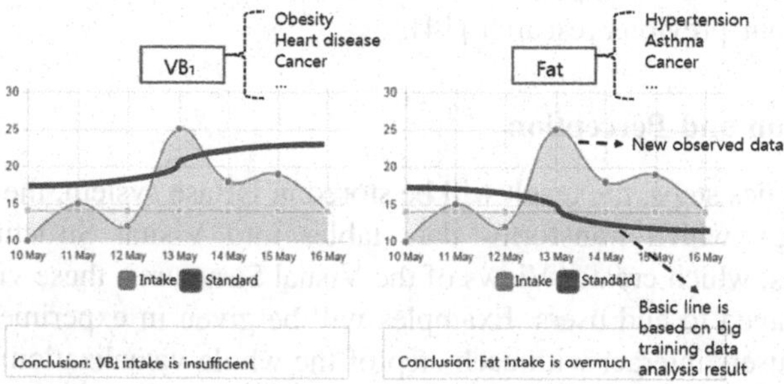

Fig. 5. Chronic disease prediction result interface (Color figure online)

Two examples have been given in Fig. 5 It illustrates the disease prediction result visualization interface of app. VB_1 and fat are key risk factors of certain kinds of diseases which have been shown in the Fig. 5 above. Curve line marked in red is the basic line obtained from big training data analysis result for a specified disease. New observed data marked in blue curve line will be compared with red one to get the similarity percentage. If they are very similar, that means the new observed person may get this factor related diseases in future.

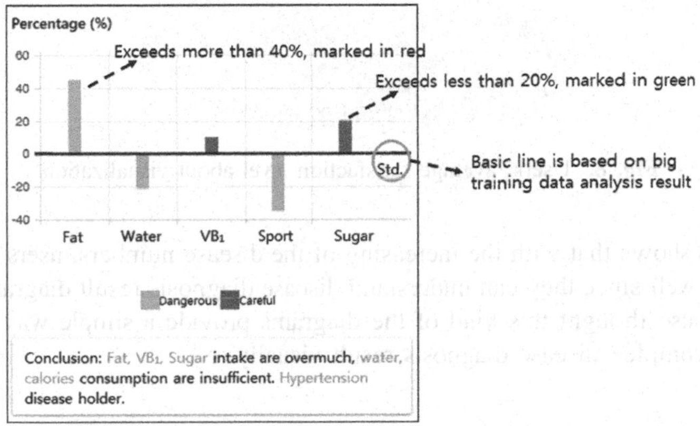

Fig. 6. Chronic disease detection result interface

Figure 6 illustrates the disease detection result visualization interface of app. X-axis of the coordinate lists ranked key risk factors of a certain kind of disease, it is also the basic line based on big training data analysis result, e.g. standard factor (like nutrition) intake for a healthy people, but concrete value will be hidden from figure. Y-axis is the percentage that intake exceeds or inferiors to the standard factor intake. Disease rules consist of these factors. For a certain disease, usually there are more than one rules (consist of risk factors) related to this disease. Compared with these rules, if the matching rate > β (expert defined threshold, e.g.: 80%), the system will treat this people as the disease holder.

4.2 Survey About Visualization Result

A survey about this visualization result has be given to 25 normal users and 5 doctors to see how well they understand these prediction and detection result given in our app.

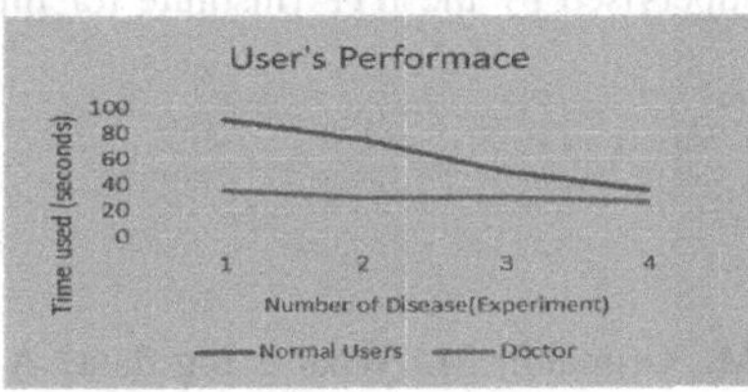

Fig. 7. User performance for understanding visualization output

Figure 7 shows average time used that users can understand the meaning of our result diagrams. Which also shows that both doctors and normal users are getting more and more familiar with our system while increasing disease numbers. Doctors had the concept of disease risk factor so they performs better than normal users.

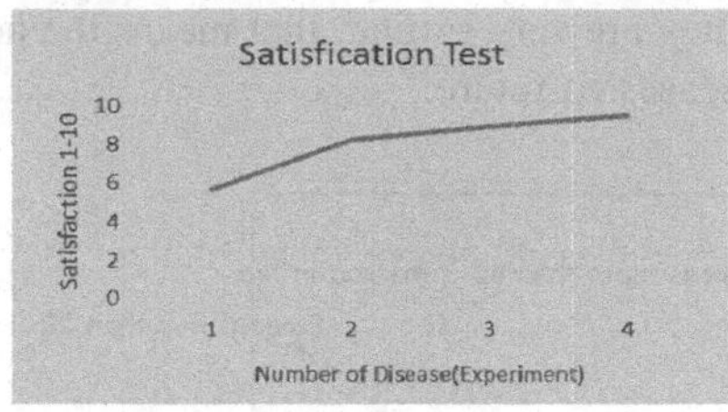

Fig. 8. Users' average satisfaction level about visualization

Figure 8 shows that with the increasing of the disease numbers, users' satisfaction increases as well since they can understand disease diagnosis result diagram faster and faster. And also thought this kind of the diagrams provide a simple way for them to understand complex disease diagnosis result visually.

5 Conclusion and Future Work

Based on IVIS4BigData reference model, we proposed a simple and practical method for chronic disease prediction and detection result visualization. The training data has been used for disease prediction and detection, the result is displayed in our app in the form of diagrams. After asking the opinions from 25 normal users and 5 doctors, we gained a high level of satisfaction about our visualization output.

In next stage of work, we will collect more big chronic disease data, improved algorithms for disease prediction and detection will be used to get precise result for visualization. Also we will rank the importance of the key risk factors according to their effeteness of a specified chronic disease and show them in our app views so users will get a better understanding of our visualization method.

Acknowledgment. This work was supported by the Basic Science Research Program through the National Research Foundation of Korea (NRF) funded by the Ministry of Science, ICT & Future Planning (No. 2017R1A2B4010826) and the MSIP (Ministry of Science, ICT and Future Planning), Korea, under the ITRC (Information Technology Research Center) support program (IITP-2017-2013-0-00881) supervised by the IITP (Institute for Information & communication Technology Promotion).

References

1. De Mauro, A., Greco, M., Grimaldi, M.: What is big data? A consensual definition and a review of key research topics. In: Giannakopoulos, G., Sakas, D.P., Kyriaki-Manessi, D. (eds.) AIP Conference Proceedings, AIP 2015, vol. 1644, no. 1, pp. 97–104 (2012). https://doi.org/10.1063/1.4907823
2. Kuo, M.H., Sahama, T., Kushniruk, A.W., et al.: Health big data analytics: current perspectives, challenges and potential solutions. Int. J. Big Data Intell. **1**(1–2), 114–126 (2014). https://doi.org/10.1504/IJBDI.2014.063835

3. Satram-Hoang, S., Reyes, C., Hoang, K.Q., et al.: Big data analysis of treatment patterns and outcomes among elderly medicare acute Myeloid Leukemia patients. Blood **124**(21), 3698 (2014). https://doi.org/10.1007/s00277-015-2351-x
4. Marz, N., Warren, J.: Big Data: Principles and Best Practices of Scalable Realtime Data Systems. Manning Publications Co., Shelter Island (2015)
5. Sitto, K., Presser, M.: Field Guide to Hadoop: An Introduction to Hadoop, Its Ecosystem, and Aligned Technologies. O'Reilly Media Inc., Newton (2015)
6. Gershon, N., Eick, S.G., Card, S.: Information visualization. Interactions **5**, 9–15 (1998)
7. Card, S.K., Mackinlay, J.D., Shneiderman, B. (eds.): Readings in Information Visualization: Using Vision to Think. Morgan Kaufmann Publishers Inc., San Francisco (1999)
8. Heer, J., Bostock, M., Ogievetsky, V.: A tour through the visualization zoo. Commun. ACM **53**(6), 59–67 (2010)
9. Silverstein, J.C., Foster, I.T.: Computer architectures for health care and biomedicine. In: Shortliffe, E., Cimino, J. (eds.) Biomedical Informatics, pp. 149–184. Springer, London (2014). https://doi.org/10.1007/978-1-4471-4474-8_5
10. Hesse, B.W., Hansen, D., Finholt, T., et al.: Social participation in health 2.0. Computer **43**(11), 45–52 (2010)
11. Barrett, J.C., Fry, B., Maller, J., et al.: Haploview: analysis and visualization of LD and haplotype maps. Bioinformatics **21**(2), 263–265 (2004). https://doi.org/10.1093/bioinformatics/bth457
12. Kopanitsa, G., Veseli, H., Yampolsky, V.: Development, implementation and evaluation of an information model for archetype based user responsive medical data visualization. J. Biomed. Inform. **55**, 196–205 (2015). https://doi.org/10.1016/j.jbi.2015.04.009
13. Bornschlegl, M.X., Berwind, K., Kaufmann, M., et al.: Towards a reference model for advanced visual interfaces supporting big data analysis. In: Proceedings on the International Conference on Internet Computing (ICOMP). The Steering Committee of The World Congress in Computer Science, Computer Engineering and Applied Computing (World-Comp) (2016)
14. Li, D., Park, H.W., Batbaatar, E., Piao, Y., Ryu, K.H.: Design of health care system for disease detection and prediction on Hadoop using DM techniques. In: The 2016 World Congress in Computer Science, Computer Engineering, & Applied Computing (WORLDCOMP 2016), Las Vegas, USA (2016)
15. Piao, Y., Ryu, K.H.: A hybrid feature selection method based on symmetrical uncertainty and support vector machine for high-dimensional data classification. In: Nguyen, N.T., Tojo, S., Nguyen, L.M., Trawiński, B. (eds.) ACIIDS 2017 Part I. LNCS (LNAI), vol. 10191, pp. 721–727. Springer, Cham (2017). https://doi.org/10.1007/978-3-319-54472-4_67
16. Cho, Y.S., Moon, S.C., Ryu, K.S., et al.: A study on clinical and healthcare recommending service based on cardiovascula disease pattern analysis. Int. J. Bio-Sci. Bio-Technol. **8**(2), 287–294 (2016)

An Improved and Tool-Supported Fuzzy Automata Framework to Analyze Heart Data

Iván Calvo, Mercedes G. Merayo, and Manuel Núñez(✉)

Departamento Sistemas Informáticos y Computación,
Universidad Complutense de Madrid, Madrid, Spain
{ivcalvo,mlmgarci,manuelnu}@ucm.es

Abstract. In this paper we present a new formalism that can be used to formally specify complex systems where uncertainty plays an important role. We introduce an improved version of a previous formalism, a *fuzzy* version of finite automata, by defining its syntax and semantics. We successfully applied this formalism to define and analyze information extracted from electrocardiograms (ECGs).

1 Introduction

The use of formal methods in the development of complex systems improves their reliability. The main obstacle to have a widespread use of formal methods is associated with their complexity and the lack of tools to support them. In addition, general purpose formalisms (such as timed automata [1]) are not suitable to be used in specific fields. This is the case of the application considered in this paper: modeling and analyzing the behavior of the heart. There are several approaches to formally model the heart [5,9,10,13] but they fail to take into account common characteristics in biological systems such as uncertainty and imprecision. If we use inaccurate models to analyze a system (whether biological or not), then we will not be able to obtain useful results.

There are many proposals to include fuzzy logic into automata [2,6,16,18]. The last of these proposals, produced in our research group, combined the best features of previous work. However, recent work [4] using this version of fuzzy automata has shown some of its weaknesses, in particular, while modeling and analyzing information about the heart. Actually, we are interested in modeling the behavior of the heart by taking into account data extracted from ECGs (electrocardiograms): heartbeats per minute and RR wave durations. Our model takes normal levels of ECGs from the study of numerous patients [7,17]. In order to assess the usefulness of the model, we will analyze real patients data to check whether our model detects existing illnesses.

We briefly comment on the main improvements of our new fuzzy automata. We have included a variable as a parameter of the actions. This fact strongly

Research partially supported by the projects DArDOS (TIN2015-65845-C3-1-R (MINECO/FEDER)) and SICOMORo-CM (S2013/ICE-3006).

N. T. Nguyen et al. (Eds.): ACIIDS 2018, LNAI 10751, pp. 694–704, 2018.
https://doi.org/10.1007/978-3-319-75417-8_65

simplifies, while keeping the same expressive power, the previous framework based on *fields*. We have simplified the operational semantics by removing an additional clause that offered no substantial benefit. This modification allowed us to fix a potential source of problems in our previous formalism. We have clarified the way in which we obtain and process the data that we feed to the automaton. Moreover, the information returned for each patient after processing their data is structured in a more useful way. Specifically, we have disaggregated the obtained data and we currently provide different alternatives (together with its associated grade of confidence).

Finally, we review the main implementation details of our framework. In order to obtain the data that we feed to our automata, we used the `WFDB Software` [15] to extract *Inter-beat (RR) intervals* from the dataset that we consider in our case study [14]. We included calls to the functionalities `sqrs` and `ann2rr` in our patient data loading script. We obtain several `.cvs` files for each patient's header and record files. These two files are later used by our trace generating script, which produces the data sent to the automaton. Essentially, we format the data in a way that can be easily processed by our automaton. First, the automaton receives the gender and age of the patient. Next, for each minute, the environment sends a sequence of values and the automaton produces a diagnosis (*ok* or *alarm*). These sequences are formed by the actual number of *beats* recorded in the minute (BPM, one value per minute) followed by the length of each *RR interval*. Therefore, for each patient we obtain a sequence of n ok/alarm messages, being n the number of minutes in the record, labeled with the associated minute and the grade of confidence on the validity of the result.

The rest of the paper is structured as follows. Section 2 introduces the syntax of our formalism and its operational semantics. Section 3 defines our model of the heart and evaluates its usefulness with real data. Finally, in Sect. 4 we present our conclusions and some lines for future work.

2 An Extended and Improved Version of Fuzzy Automata

Fuzzy automata [2] have been recently used in our research group and we have detected several deficiencies that we would like to fix in this improved version. In this section we introduce our new formalism. First, we briefly present some concepts related to fuzzy logic. The interested reader is referred to our previous work [3,4] for more details because, due to space limitations, we cannot review all the needed concepts and notations.

Fuzzy relations are similar to *boolean relations* but instead of returning true or false, they return a real value in the interval [0,1]. The idea is that if we are sure that something holds then we have confidence equal to 1; otherwise, we will have a confidence less than 1, in particular, if we are sure that the relation does not hold then we have confidence equal to 0. Usual fuzzy relations can be found in our previous work [4]. In particular, we defined a relation $\overline{\alpha \leq \cdot \leq \beta}^{\delta}$. Intuitively, if a value x is such that $\alpha \leq x \leq \beta$ then we claim that the relation holds with confidence 1. If this is not the case and the distance from x to α or β

is less than δ then we have a positive confidence (the confidence diminishes when the distance increases). Finally, if $x \notin [\alpha, \beta]$ and it is *far* from the interval then we have confidence 0 on x belonging to the interval.

We combine confidence values by using *t-norms*. In this paper we use two of them: the *Gödel* t-norm (computing the minimum of all the values and denoted by $\overline{\wedge}$) and the *Hamacher product* t-norm. This last t-norm (denoted by $\circledast$) is, as usual, associative. Therefore, it is enough to define it for two arguments δ_1 and δ_2 as $\frac{\delta_1 \cdot \delta_2}{\delta_1 + \delta_2 - \delta_1 \cdot \delta_2}$.

After this brief review, we can define our improved version of fuzzy automata. First, we introduce some additional notation.

Definition 1. *Let* Acts *be a finite set of actions (they will be used to model the actions that a system can perform). We will distinguish between* inputs, *preceded by* ?, *and* outputs, *preceded by* !.

A fuzzy constraint *is a formula where fuzzy relations are used instead of boolean relations and t-norms are used to combine relations instead of boolean operators. We denote by $\mathcal{FC}$ the set of fuzzy constraints.*

Let C be a fuzzy constraint with n parameters and $\overline{x} = (x_1, \ldots, x_n) \in \mathbb{R}^n_+$. We have that $\mu_C(\overline{x})$ denotes the satisfaction degree or grade of confidence (GoC) of *C for $\overline{x}$ (a formal definition can be found in our previous work* [4]*).*

In our case study, *inputs* will be used to receive information about the patient (e.g. BPM and RR). We will use *outputs* to send messages to the environment. For example, we can issue an alarm indicating that a potential problem has been found at a certain minute and with a certain grade of confidence. Again, we refer the interested reader to our previous work [4] for longer discussions and examples on fuzzy constraints. A simple example of a fuzzy constraint is $\overline{60 \leq x \leq 69}^{13}$. The idea is that if a patient is in the expected age range, that is [60, 69], then the confidence is equal to 1. Otherwise, if the distance to the interval is more than 13 then the confidence is equal to zero. Finally, if the age is *close* to the interval, then the confidence linearly increases when the distance is reduced. Let us note that if $\delta = 0$ then fuzzy constraints become *usual* constraints.

In order to track some relevant data during the execution of the automaton, we introduce the following notion to deal with *variables* and *variable transformations*.

Definition 2. *Let X be a set of variables taking values in $\mathbb{R}_+$. We define the set of variable transformations $\mathcal{VT}$ as the set of expressions assigning a value to each variable of the set. We will use the following notation*

$$[y_1/x_1, \ldots, y_m/x_m]$$

where each y_i is a real valued expression over the set of variables X and each x_i is a variable in X. The semantics of this transformation is that each x_i takes the value obtained after evaluating y_i (possible taking into account the current values of the variables in X); if a variable x_i does not appear in the expression then we have that the variable does not change its value after the transformation.

Let $\mathfrak{R}$ be equal to $\bigcup_{i \geq 1} \mathbb{R}^i_+$, that is, $\mathfrak{R}$ is a set containing all the tuples, of any arity, with real number values.

Definition 3. *A* fuzzy automaton *is a tuple* $(S, \text{Acts}, X, s_0, T)$ *where:*

- *S is a finite set of states.*
- Acts *is a finite set of actions, partitioned into a set of inputs I and a set of outputs O.*
- *X is a set of variables ranging over $\mathbb{R}_+$. The set includes a variable GoC, which will be used to store the Hamacher grade of confidence associated with sequences of transitions. We assume that the initial value of GoC is 1.*
- *s_0 is the initial state.*
- *$T \subseteq S \times (I \times X \cup O \times \mathfrak{R}) \times \mathcal{FC} \times \mathcal{VT} \times S$ is the set of transitions. We assume that each transition implicitly applies the following variable transformation $[\mu_C \circledast GoC/GoC]$.*

Fuzzy automata are directed graphs where transitions have an associated condition, indicating the grade of confidence with which we can execute the transition, and a transformation of the variables. In addition, transitions can be labeled either by an input or by an output. Intuitively, a transition $(s, (a, \alpha), C, V, s') \in T$ denotes that if the automaton is in state s and receives/sends from/to the environment $a(\alpha)$, where a is an input/output action and α is a variable/tuple of positive real values, then the previous transition can be triggered if $\mu_C(\alpha) > 0$, the new values of the variables will be given by V, and the automaton will move to state s'. Usually, transitions labeled with an output will have a trivial fuzzy constraint (that is, it will be `True`). In order to simplify the graphical representation, if we have two transitions from one state to another one labeled by different fuzzy constraints C_1 and C_2 then we will only draw one transition labeled by $C_1 \| C_2$ (for example, see Fig. 1, transition from q_{38} to q_{39}).

Next, we are going to define the operational behavior of fuzzy automata. This *operational semantics* will be used to obtain their (fuzzy) traces. We start in the initial state of the automaton, produce actions and trigger a transition labelled by the action if the attached value is included in the fuzzy relation induced by the constraint. We decorate transitions with a real number $\epsilon \in [0, 1]$ indicating its certainty. First, we define a single transition and then we concatenate transitions to conform traces.

Definition 4. *Let $A = (S, \text{Acts}, X, s_0, T)$ be a fuzzy automaton and $\triangle$ be a t-norm. Given states $s_1, s_2 \in S$, we have a* transition *from s_1 to s_2, after performing the action $a \in \mathcal{A}ct$ for α with confidence ϵ, denoted by $s_1 \xrightarrow{(a(\alpha),V)}_{\epsilon} s_2$, if the following conditions hold:*

- *There exists $C \in \mathcal{FC}$ such that $(s_1, (a, \alpha), C, V, s_2) \in T$.*
- *$\mu_C(\alpha) = \epsilon$ and $\epsilon > 0$.*
- *The new values of the variables belonging to X are given by $V \in \mathcal{VT}$.*

We say that a sequence $s_0 \xrightarrow{(a_1(\alpha_1),V_1)}_{\epsilon_1} s_1 \xrightarrow{(a_2(\alpha_2),V_2)}_{\epsilon_2} \cdots \xrightarrow{(a_n(\alpha_n),V_n)}_{\epsilon_n} s_n$ of consecutive transitions starting in the initial state of the automaton A is a $\triangle$-trace *of A if $\epsilon = \triangle\{\epsilon_1, \ldots, \epsilon_n\}$ is greater than zero and the values of the variables of X are the result of sequentially applying the variable transformations*

$V_1, \ldots, V_n$ to X. We call this composed variable transformation V. In this case we write $s_0 \xrightarrow{(a_1,\ldots,a_n,\alpha_1,\ldots,\alpha_n,V)}_{\epsilon} s_n$.

Example 1. Consider the component of the automata *Heart* given in Fig. 1 where we assume that the value 0 denotes males and 1 denotes females. For example, we could observe a trace such as

$$(?checkGender(0)), (?checkAge(65)), (?minute(1)), (?readBPM(62)),$$
$$(?readRR(977)), (?readRR(968)), (\cdots), (?noMorePendingRR()),$$
$$(!ok(1, 1.0))$$

as the result of having the automaton working during a minute by analyzing a sample of a 65 years old male patient. As usual, inputs are preceded by ?, outputs are preceded by ! and $(\cdots)$ indicates that some *?readRR* actions have been omitted from the trace due to presentation purposes.

3 Case Study

In this section we present the application of our fuzzy automata in a real scenario: prediction of heart problems. We define the automaton *Heart*, which is able to alert about the level of risk of a patient. In order to produce a diagnosis, we use the available information and physical evidence collected from electrocardiograms (ECGs). The information managed by the automaton is:

- Gender. We have 2 groups: Men and Women.
- Age. We have 8 groups of age.
- Heartbeats. The range of correct heartbeats per minute (BPM) for healthy patients, according to their gender and age.
- RR waves. The range of correct RR waves duration (measured in milliseconds) for healthy patients, according to their gender, age and BPM.

Additionally, we consider that our set of actions consists of the following operations:

- $?checkGender(gen)$. It reads the gender of the patient.
- $?checkAge(age)$. It reads the age of the patient.
- $?minute(m)$. It reads the current minute of the recording.
- $?readBPM(bpm)$. It reads the amount of beats in the current minute.
- $?readRR(rr)$. It reads the next RR interval in the current minute.
- $?noMorePendingRR(\cdot)$. It receives a notification that there are no more RR intervals in the current minute.
- $?endOfRecord(\cdot)$. It receives a notification to denote that there are no more minutes in the analyzed record.
- $!recordAlarm(min, GoC)$. It indicates, with a grade of confidence equal to GoC, that an alarm will be raised in the current minute.
- $!ok(min, GoC)$. It indicates, with a grade of confidence equal to GoC, no alarm will be recorded for the current minute.

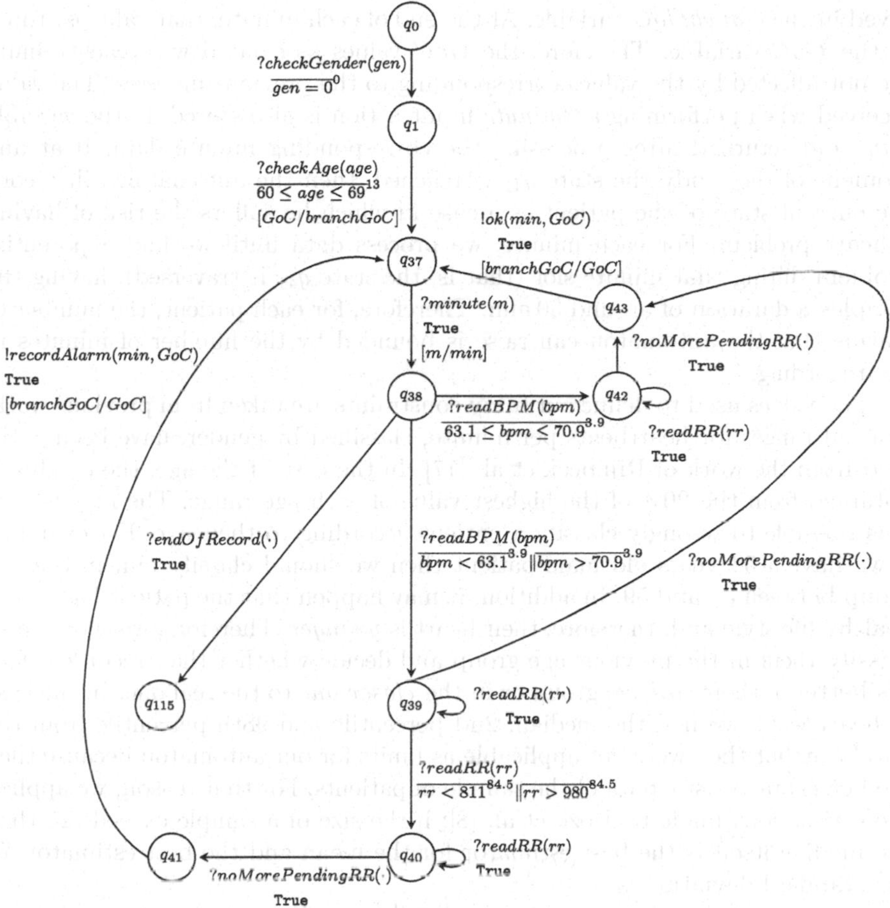

Fig. 1. Sub-automaton corresponding to male patients between 60 and 69 years old.

Our automaton has a total of 116 states and 226 transitions. Initially, our automaton has two transitions: one per gender. After that, each branch has 8 transitions, one per age group. Each of these age groups has an associated sub-automaton with 7 states and a transition to a common *final* state (the state q_{115}) to denote that all the data for this patient have been processed. Each minute, the automaton checks if the number of beats falls within the normal amount of beats per minute in the age range. If it does, then the automaton does not take into account the RR intervals in that minute and signals that minute to be *ok*. Otherwise, the automaton processes each RR interval. If there is at least one interval out of range, an alarm is raised (with a certain grade of confidence). As previously commented, we use variables to track some data. More specifically, in addition to the *built-in* variable *GoC*, we use a variable called *branchGoC*. In every transition entering an *age branch*, the recently computed *GoC* value is saved in the *branchGoC* variable. At the end of each minute, that value is stored in the *GoC* variable. Therefore, the *GoC* values associated with each minute are not affected by the values corresponding to the previous minutes. The value received when performing a *?minute* input action is also stored, in the variable *min*, and returned after processing the corresponding minute data. If at any moment of the study the state q_{41} is reached, then the automaton will record the current state of the patient as a case in which he suffers the risk of having a heart problem. For each minute, we process data until we find a potential problem during that minute slot (that is, the state q_{41} is traversed), having the samples a duration of around 30 min. Therefore, for each patient, the number of alarms that the automaton can raise is bounded by the number of minutes in the recording.

The values used to define our fuzzy constraints are taken from previous work. Normal ranges for heartbeats per minute, classified by gender, have been gathered from the work of Rijnbeek et al. [17]. In the case of the age, the δ value is obtained from the 20% of the highest value of each age range. The idea is that it is possible to wrongly classify a patient according to their age. For example, if we have a 53 years old male patient then we should classify him in the age group between 50 and 59. In addition, it may happen that the patient has a very healthy life style and, therefore, their heart is *younger*. Therefore, we should also classify them in the previous age group and decide whether their recorded data fits better in their *real* age group or in the *closer one* to the real one. In the case of heartbeats, we had the median, 2nd percentile and 98th percentile from the database, but they were not applicable as limits for our automaton because they are not characteristic data of the sample of patients. For that reason, we applied the estimations made by Hozo et al. [8]: if the size of a sample exceeds 25 then the median itself is the best estimator for the mean and the best estimator for the standard deviation is

$$\sigma \approx \frac{b-a}{6}$$

where a is the smallest value of the sample (the 2nd percentile in our case) and b is the largest value of the sample (the 98th percentile in our case). Fortunately, the database that we use [17] has information from 13354 patients and, therefore, we can base our limits on the median of each range while δ is based on the standard deviations of each range.

Concerning RR waves, we have used the data from the work of Haarmark et al. [7]. The problem in this case was that we only had the information of the RR waves duration for the patients of the age range [30–39]. So, if we had used these limits for all the patients, then the prediction would have been erroneous. Therefore, we also considered another related work [11] where the duration of the RR waves is derived from data corresponding to heartbeats. So, our limits are based on the application of the following formula

$$RR_{ms} \approx \frac{60000}{bpm}$$

to the BPM data obtained from the work of Rijnbeek et al. [17].

The python scripts computing the Grades of Confidence given by the automaton and some relevant information used in this study are available at https://github.com/FINDOSKDI/heartdiagnosis.

In order to assess the usefulness of our automaton, we used the MIT/BIH Arrhythmia Database Directory [14] https://physionet.org/physiobank/database/mitdb/. This study includes 48 ECG recordings with a duration of 30 min from the Massachusetts Institute of Technology - Beth Israel Hospital arrhythmia database. All of them present some heart pathology. Specifically, 48% of the samples have been annotated in the database as representative cases of routine clinical recordings while the remaining 52% reflect uncommon cases of

Table 1. Results corresponding to patients #100 and #104. Each cell includes a pair *GoC ok*/*GoC alarm* computed from the data observed during that minute.

	#100				#104			
	(50–59)	(60–69)	(70–79)	(>80)	(50–59)	(60–69)	(70–79)	(>80)
min. 1	0.09/0.15	0.21/1.00	0.28/0.94	0.31/0.00	0.31/0.41	1.00/0.88	0.75/0.47	0.12/0.11
min. 2	0.09/0.15	0.21/1.00	0.28/0.94	0.31/0.00	0.00/0.41	0.54/1.00	0.68/0.75	0.12/0.12
min. 3	0.00/0.15	0.00/1.00	0.00/0.94	0.31/0.22	0.00/0.41	0.00/1.00	0.00/0.75	0.00/0.12
min. 4	0.09/0.15	0.21/1.00	0.28/0.94	0.31/0.00	0.31/0.41	1.00/0.88	0.75/0.47	0.12/0.11
min. 5	0.09/0.15	0.21/1.00	0.28/0.94	0.31/0.00	0.00/0.41	0.00/1.00	0.29/0.75	0.12/0.12
min. 6	0.00/0.15	0.00/1.00	0.04/0.94	0.31/0.14	0.31/0.41	1.00/0.88	0.75/0.47	0.12/0.11
min. 7	0.00/0.15	0.00/1.00	0.00/0.94	0.28/0.31	0.31/0.41	1.00/0.88	0.75/0.47	0.12/0.11
min. 8	0.00/0.15	0.00/1.00	0.00/0.94	0.28/0.31	0.31/0.41	1.00/0.88	0.75/0.47	0.12/0.11
min. 9	0.00/0.15	0.00/1.00	0.00/0.94	0.31/0.22	0.20/0.41	0.83/1.00	0.75/0.65	0.12/0.12
min. 10	0.00/0.15	0.00/1.00	0.00/0.94	0.31/0.26	0.31/0.41	1.00/0.88	0.75/0.47	0.12/0.11
min. 11	0.00/0.15	0.00/1.00	0.00/0.94	0.31/0.26	0.31/0.41	1.00/0.88	0.75/0.47	0.12/0.11
min. 12	0.00/0.15	0.00/1.00	0.00/0.94	0.31/0.31	0.31/0.41	1.00/0.88	0.75/0.47	0.12/0.11
min. 13	0.00/0.15	0.00/1.00	0.00/0.94	0.31/0.22	0.31/0.41	1.00/0.88	0.75/0.47	0.12/0.11
min. 14	0.00/0.15	0.00/1.00	0.04/0.94	0.31/0.14	0.31/0.41	1.00/0.88	0.75/0.47	0.12/0.11
min. 15	0.00/0.15	0.00/1.00	0.04/0.94	0.31/0.14	0.31/0.41	1.00/0.88	0.75/0.47	0.12/0.11
min. 16	0.13/0.15	0.47/1.00	0.51/0.94	0.31/0.14	0.31/0.41	1.00/0.88	0.75/0.47	0.12/0.11
min. 17	0.00/0.15	0.00/1.00	0.04/0.94	0.31/0.14	0.31/0.41	1.00/0.88	0.75/0.47	0.12/0.11
min. 18	0.00/0.15	0.00/1.00	0.04/0.94	0.31/0.14	0.38/0.41	1.00/0.59	0.75/0.25	0.12/0.09
min. 19	0.00/0.15	0.00/1.00	0.04/0.94	0.31/0.14	0.31/0.41	1.00/0.88	0.75/0.47	0.12/0.11
min. 20	0.09/0.15	0.21/1.00	0.28/0.94	0.31/0.00	0.31/0.41	1.00/0.88	0.75/0.47	0.12/0.11
min. 21	0.09/0.15	0.21/1.00	0.28/0.94	0.31/0.00	0.31/0.41	1.00/0.88	0.75/0.47	0.12/0.11
min. 22	0.09/0.15	0.21/1.00	0.28/0.94	0.31/0.00	0.31/0.41	1.00/0.88	0.75/0.47	0.12/0.11
min. 23	0.09/0.15	0.21/1.00	0.28/0.94	0.31/0.00	0.31/0.41	1.00/0.88	0.75/0.47	0.12/0.11
min. 24	0.09/0.15	0.21/1.00	0.28/0.94	0.31/0.00	0.31/0.41	1.00/0.88	0.75/0.47	0.12/0.11
min. 25	0.13/0.15	0.47/1.00	0.51/0.94	0.31/0.14	0.31/0.41	1.00/0.88	0.75/0.47	0.12/0.11
min. 26	0.09/0.15	0.21/1.00	0.28/0.94	0.31/0.00	0.31/0.41	1.00/0.88	0.75/0.47	0.12/0.11
min. 27	0.00/0.15	0.00/1.00	0.04/0.94	0.31/0.14	0.31/0.41	1.00/0.88	0.75/0.47	0.12/0.11
min. 28	0.00/0.15	0.00/1.00	0.00/0.94	0.31/0.31	0.31/0.41	1.00/0.88	0.75/0.47	0.12/0.11
min. 29	0.00/0.15	0.00/1.00	0.00/0.94	0.31/0.22	0.38/0.41	1.00/0.59	0.75/0.25	0.12/0.09
min. 30	0.00/0.15	0.00/1.00	0.00/0.94	0.31/0.30				

arrhythmias. As an example of the obtained results, in Table 1 we show minute data from each applicable age branch of the patients 100 and 104, commented in our previous work [4]. Each cell contains two numbers: the first one is the Hamacher GoC of sending an *ok* signal while the second one is the Hamacher GoC of raising an *alarm*, both referring to that minute and age/gender branch. This table clearly shows why our new approach represents a big step forward with respect to our previous work: we are able to produce and appropriately process several dozens of ok/alarm signals.

We recall that patient 100 is a 69 years old male and patient 104 is a 66 years old female. In both cases, the value of the corresponding *?checkAge* constraint is positive for four branches: $(50 \leq age \leq 59)$, $(60 \leq age \leq 69)$, $(70 \leq age \leq 79)$ and $(age > 80)$. The table is formed by four columns for patient 100 and four columns for patient 104. Each row contains the GoCs obtained in a minute. The record from patient 100 is 30 min long while the record from patient 104 is 29 min long. Let us briefly comment on the results obtained for these two patients. First, we notice that in both cases the maximum confidence is obtained in the 60–69 branch. This means that there are not many *RR* or *BPM* values close to the limit of the normal range. The only case in which there is more confidence outside the 60–69 age branch is in the *!ok* results of the patient 100. We can see that the columns corresponding to the age ranges 70–79 and > 80 present higher confidence in these cases. These values can indicate that the values observed from the heart of patient 100 could be normal for a much older person. This idea is reinforced by the observation that in the > 80 branch the confidence in the *!ok* case is higher that the confidence in the *!alarm* case. The most relevant difference we observe between these two patients is that patient 100 is outside his normal parameters in every minute, while patient 104 is showing a normal behavior most of the time, having some eventual alerts.

4 Conclusions and Future Work

Despite the numerous advances in healthcare, many patients do not receive a correct diagnosis. Some of these problems are provoked by either an incorrect processing of data or by wrong conclusions from relevant observations. Therefore, if data was more accurately processed and analyzed, then it would be possible to obtain improvements in this field. Our proposal goes in this line: automatically process data, extracted from electrocardiograms, to detect potential *malfunctions*. First, we have introduced a variant of finite automata where constraints indicating whether a certain transition can be performed are evaluated under a *fuzzy* point of view. We have modeled the behavior of the heart by taking into account data about the beats per minute and the duration of RR waves. In order to decide whether potential dangers have been observed, we use information about the gender and age of the patients. In the latter case, we also use a *fuzzy* approach because a patient can be classified in several age groups.

We are considering several lines of future work. First, we would like to obtain more data from patients with the aim of applying techniques, such as evolutive algorithms, swarm intelligence and neural networks, to improve the ability of our automata to detect illnesses. In cooperation with researchers in Medicine, we are working on alternative models where the *classification* of patients considers characteristics such as size/weight and medical record. Once we have more complex and complete models, we will test their suitability with alternative data sets [12].

References

1. Alur, R., Dill, D.: A theory of timed automata. Theoret. Comput. Sci. **126**, 183–235 (1994)
2. Andrés, C., Llana, L., Núñez, M.: Self-adaptive fuzzy-timed systems. In: 13th IEEE Congress on Evolutionary Computation, CEC 2011, pp. 115–122. IEEE Computer Society (2011)
3. Boubeta-Puig, J., Camacho, A., Llana, L., Núñez, M.: A formal framework to specify and test systems with fuzzy-time information. In: Rojas, I., Joya, G., Catala, A. (eds.) IWANN 2017. LNCS, vol. 10306, pp. 403–414. Springer, Cham (2017). https://doi.org/10.1007/978-3-319-59147-6_35
4. Camacho, A., Merayo, M.G., Núñez, M.: Using fuzzy automata to diagnose and predict heart problems. In: 19th IEEE Congress on Evolutionary Computation, CEC 2017, pp. 846–853. IEEE Computer Society (2017)
5. Chen, T., Diciolla, M., Kwiatkowska, M., Mereacre, A.: Quantitative verification of implantable cardiac pacemakers over hybrid heart models. Inf. Comput. **236**, 87–101 (2014)
6. Doostfatemeh, M., Kremer, S.C.: New directions in fuzzy automata. Int. J. Approx. Reason. **38**(2), 175–214 (2005)
7. Haarmark, C., Graff, C., Andersen, M., Hardahl, T., Struijk, J., Toft, E., Xue, J., Rowlandson, G., Hansen, P., Kanters, J.: Reference values of electrocardiogram repolarization variables in a healthy population. J. Electrocardiol. **43**(1), 31–39 (2010)
8. Hozo, S., Djulbegovic, B., Hozo, I.: Estimating the mean and variance from the median, range, and the size of a sample. BMC Med. Res. Methodol. **5**(1), 1 (2005)
9. Hunter, P., Pullan, A., Smaill, B.: Modeling total heart function. Annu. Rev. Biomed. Eng. **5**(1), 147–177 (2003)
10. Jiang, Z., Connolly, A., Mangharam, R.: Using the virtual heart model to validate the mode-switch pacemaker operation. In: 32nd Annual International Conference of the IEEE Engineering in Medicine and Biology Society, EMBC 2010, pp. 6690–6693. IEEE Computer Society (2010)
11. Khachaturian, Z., Kerr, J., Kruger, R., Schachter, J.: A methodological note: comparison between period and rate data in studies of cardiac function. Psychophysiology **9**(5), 539–545 (1972)
12. Lichman, M.: UCI machine learning repository (2013). http://archive.ics.uci.edu/ml
13. Méry, D., Singh, N.K.: Formalization of heart models based on the conduction of electrical impulses and cellular automata. In: Liu, Z., Wassyng, A. (eds.) FHIES 2011. LNCS, vol. 7151, pp. 140–159. Springer, Heidelberg (2012). https://doi.org/10.1007/978-3-642-32355-3_9
14. Moody, G.B., Mark, R.G.: The impact of the MIT-BIH arrhythmia database. IEEE Eng. Med. Biol. **20**(3), 45–50 (2001)

15. Moody, G.B.: RR intervals, heart rate and HRV Howto (2016). https://www.physionet.org/tutorials/hrv/
16. Mordeson, J.N., Malik, D.S.: Fuzzy Automata and Languages: Theory and Applications. Chapman & Hall/CRC, London/Boca Raton (2002)
17. Rijnbeek, P., van Herpen, G., Bots, M., Man, S., Verweij, N., Hofman, A., Hillege, H., Numans, M., Swenne, C., Witteman, J., et al.: Normal values of the electrocardiogram for ages 16–90 years. J. Electrocardiol. **47**(6), 914–921 (2014)
18. Wee, W.G., Fu, K.S.: A formulation of fuzzy automata and its application as a model of learning systems. IEEE Trans. Syst. Sci. Cybern. **5**(3), 215–223 (1969)

Usability Testing of a Responsive Web System for a School for Disabled Children

Justyna Krzewińska, Agnieszka Indyka-Piasecka, Marek Kopel, Elżbieta Kukla, Zbigniew Telec, and Bogdan Trawiński(✉)

Faculty of Computer Science and Management, Wrocław University of Science and Technology, Wrocław, Poland
{agnieszka.indyka-piasecka,marek.kopel, elzbieta.kukla,zbigniew.telec, bogdan.trawinski}@pwr.edu.pl

Abstract. The paper presents a responsive website, with the access to a social network, devoted to parents and teachers of a school for disabled children. Usability of the developed website was tested depending on the device, i.e. laptop or smartphone utilized. Two series of usability testing were conducted one week apart. Two groups of potential users of the website took part in each series. One group utilized laptops and the second one used smartphones while completing task scenarios. The groups exchanged devices during the second series. The participants of the study were people aged over 30 who did not have much experience in using mobile devices. Moreover, the expert study was carried out applying the heuristic inspection and checklist methods to detect the main design problems. The results of the study was gathered for usability attributes proposed by Nielsen, namely: efficiency, errors, user's satisfaction, learnability, and memorability. In consequence, elements of the website to be improved were identified.

Keywords: Responsive web design · Usability testing · Web applications Mobile applications

1 Introduction

Responsive Web Design (RWD) is an approach to designing and developing websites which could be accessed and utilized with various devices having different screen sizes, resolutions, proportions, and orientations. RWD employs fluid grids, adjusting screen resolution, automatically resizable images, and media queries [1]. Usability is crucial for success of any software product including both desktop systems and mobile applications. Usable systems enable their users to achieve specific goals effectively and efficiently with a high level of satisfaction. Usability is often presented in literature in terms of models comprising a number of usability attributes. The most popular are ISO 924-11 model which consists of three attributes: effectiveness, efficiency, and satisfaction [2] and Nielsen's model specifying five attributes: efficiency, errors satisfaction, learnability, and memorability [3]. For mobile applications the PACMAD model was devised and it encompasses seven attributes: effectiveness, efficiency, errors,

N. T. Nguyen et al. (Eds.): ACIIDS 2018, LNAI 10751, pp. 705–716, 2018.
https://doi.org/10.1007/978-3-319-75417-8_66

learnability, memorability, satisfaction, and cognitive load [4]. In turn, the consolidated QUIM model decomposes usability into factors, criteria and metrics coming from other models and standards [5]. Numerous works presenting usability testing of mobile and responsive applications with the users as well as experts have been published recently [6–9].

The main goal of the paper is to report the results of usability testing of the website for a school for disabled children which was developed according to the RWD approach. The study was performed employing the Nielsen's model of usability. The usability metrics such as task completion time, number of actions, binary task completion rate, number of errors, and correctness of the first path chosen were considered. User satisfaction was also measured using System Usability Scale (SUS) and Single Ease Question (SEQ) questionnaires.

2 A Website for a School for Disabled Children

A website combined with a social network was developed to serve the teachers and parents of a school for disabled children. The main goal of the website was to promote the school by presenting its mission and objectives. A virtual school chronicle as well as a photo gallery were also available on the website. Moreover, it enabled the users to post and comment announcements, look up an events calendar and class schedules and manage them with ease. A school social network was also worked out where the users could communicate online, create groups, and invite colleagues to join groups. The website was equipped also with an asynchronous communication means such as an email function.

The website was developed also to prepare a tool to conduct usability experiments with different devices including both laptops and smartphones. Therefore, the website was created using the responsive web design. This approach satisfies the contemporary practice when the users utilize various electronic devices with different screen resolutions and sizes ranging from desktop computers with big monitors through laptops and tablets to small smartphones. The responsive design of the website was depicted in Fig. 1.

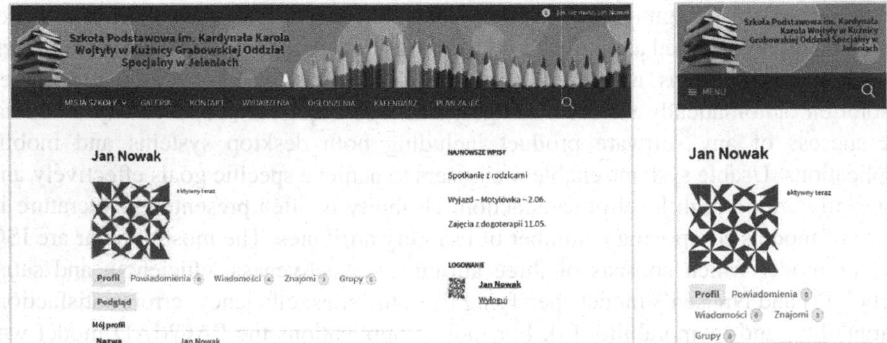

Fig. 1. Illustration of the responsive design of the website.

3 Setup of Usability Tests

The aim of study was to test usability and identify problems of the responsive website for a school for disabled children depending on the device, i.e. laptop or smartphone utilized. The experiments were conducted with two group of users who completed the same tasks using the same application on different devices alternately. Two series of usability testing were carried out one week apart. Two groups of potential users of the website took part in each series. One group utilized laptops and the second one used smartphones while completing task scenarios. The groups exchanged devices during the second series. The participants of the study were people over 30 who did not have much experience in using mobile devices. They were parents and teachers of the school. Moreover, the expert inspection was performed applying the heuristic evaluation and checklist methods to detect the main design problems of the website.

The study was based on the Nielsen's model of usability which consists of five attributes: efficiency, satisfaction, learnability, memorability, and errors [3]. For each attribute at least one metrics was collected during the tests and satisfaction questionnaire was administered after each series of experiment. The metrics collected during usability tests are listed in Table 1.

Table 1. Metrics collected during usability tests by Nielsen's attributes

Attribute	Metrics	Description	Unit
Efficiency	Time	Task completion time by a user	[s]
	Expert time	Ratio of user task completion time to task completion time by an expert	[s/s]
	Clicks	Number of clicks, scrolls, taps, swipes, etc. to complete individual tasks	[n]
Errors	Completion rate	Ratio of successfully completed tasks to all tasks undertaken	[%]
	Errors	Number of errors made by a user when completing individual tasks	[n]
Satisfaction	SUS	System Usability Scale score of a questionnaire administered at the end of the whole test	[n]
	SEQ	Single Ease Question score of a questionnaire administered to individual tasks	1–5
Learnability	First path	Assessment whether the path chosen by a user at the first step was correct or not	{0,1}
Memorability	Comparison	Comparison of results obtained with selected metrics in two series of tests, e.g. time needed to complete the same tasks	

Usability tests were performed with two types of devices: a laptop and smartphone. Each participant used the same laptop Lenovo z510 with Windows 7 and smartphone LG Nexus4 with Android 5.0. The same browser Google Chrome was installed on both devices. Each session was recorded with an application allowing for saving screens, users' actions, comments, and questions. The Morae [10] and Lookback [11] tools were employed for this purpose. The sessions were carried out in natural working environment of the users. The teachers were accomplishing the tasks in the school during the breaks between their classes. In turn, the parents were completing the tasks at their homes. Each session was held with an individual user separately under the supervision of a moderator. The main steps of the study are presented in Table 2. The experimental scenario comprised 10 tasks referring to the most common actions performed by the users of the school website. They are listed in Table 3.

Table 2. Main steps of the study

1.	Preparation of the devices for conducting research including a computer and smartphone
2.	Preparation of the application for conducting research including a user profile, news, etc
3.	Preparation of the materials for users including a personal form, task list, and satisfaction questionnaire
4.	Welcome the participants, informing them about the goals, scope, and course of research
5.	Filling a personal form by a user (before the first series of tests)
6.	Conducting usability tests with a laptop or smartphone
7.	Completing a satisfaction questionnaire by a user after having accomplished all the tasks
8.	Finishing the first series of usability testing
9.	Informing the user about the date of the second series of usability testing

Table 3. 10 tasks to complete by the participants

T1.	Find the names of educational groups the children are assigned to
T2.	Display the first photo in the gallery
T3.	Fill the registration form and send a message "Hello". Give the name "John" and email address "john.smith@gmail.com"
T4.	Find which specialists have classes with the children
T5.	Log in to the portal using User Id.: "John" and Password "johnsmith"
T6.	Display a post on a kynotherapy session accomplished and add a comment "Good fun"
T7.	Look up an events calendar and find the organizers of an event held on May 19
T8.	Look up the latest news and add a comment "I would be pleased to attend"
T9.	Send a message "Call me this afternoon" to the user Joan
T10.	Which lessons do the children have on Thursday?

Teachers and parents of the children going to the school were the participants of the usability study. In total 24 people took part in usability testing. They were split into two separate groups for 12 people. Each participant was asked to complete 10 tasks with different devices during two series of research. Before the study all participants filled a personal form. The basic characteristics of the participants are given in Fig. 2. The majority of the users, i.e. 71%, were between 40 and 60, six were under 40 years of age, and one person was over 60. Thus, all participants were 30-plus. 17 of them were

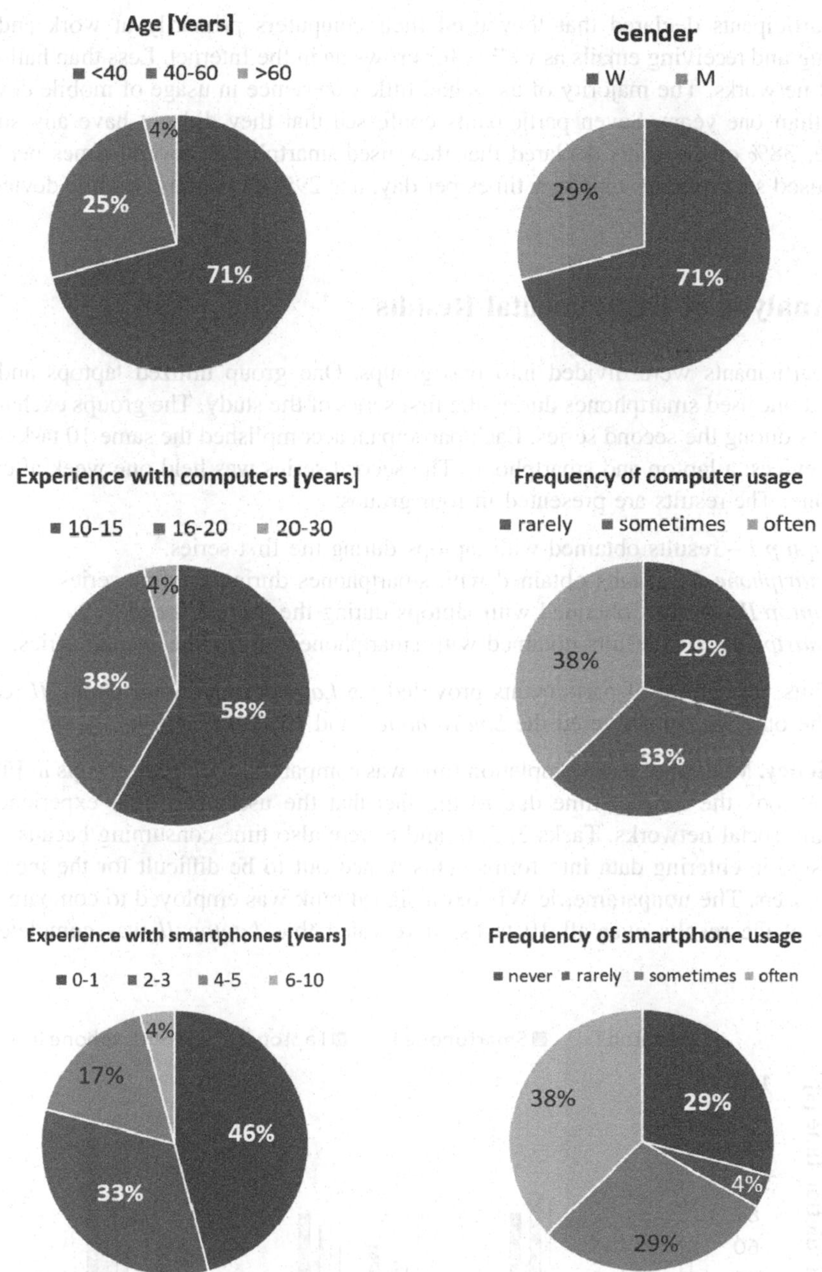

Fig. 2. Characteristics of users taking part in usability testing

women and men constituted 29% of the users. 58% of the participants had between 10 and 15 year experience in using computers. Nine of them had utilized computers for longer than 15 years and one for longer than 20 years. The majority (almost 83%) of the participants declared that they used their computers primarily at work and for sending and receiving emails as well as for browsing in the Internet. Less than half used social networks. The majority of users had little experience in usage of mobile devices (less than one year). Seven participants confessed that they did not have any smartphone. 38% of the users declared that they used smartphones several times per day, 29% used smartphones only few times per day, and 29% did not use mobile devices at all.

4 Analysis of Experimental Results

The participants were divided into two groups. One group utilized laptops and the second one used smartphones during the first series of the study. The groups exchanged devices during the second series. Each participant accomplished the same 10 tasks with two devices: a laptop and smartphone. The second series was held one week after the first one. The results are presented in four groups:

- *Laptop I* – results obtained with laptops during the first series,
- *Smartphone I* – results obtained with smartphones during the first series,
- *Laptop II* – results obtained with laptops during the second series,
- *Smartphone II* – results obtained with smartphones during the second series.

Thus, one group of participants provided the *Laptop I* and *Smartphone II* results and the other group delivered the *Smartphone I* and *Laptop II* results.

Efficiency. Median of task completion time was compared for all four groups in Fig. 3. Task 9 took the longest time due to the fact that the users had little experience in utilizing social networks. Tasks 3, 5, 6, and 8 were also time consuming because they consisted in entering data into forms. This turned out to be difficult for the inexperienced users. The nonparametric Wilcoxon signed-rank was employed to compare each series of the results over all 10 tasks. It revealed that *Laptop II* was completed in

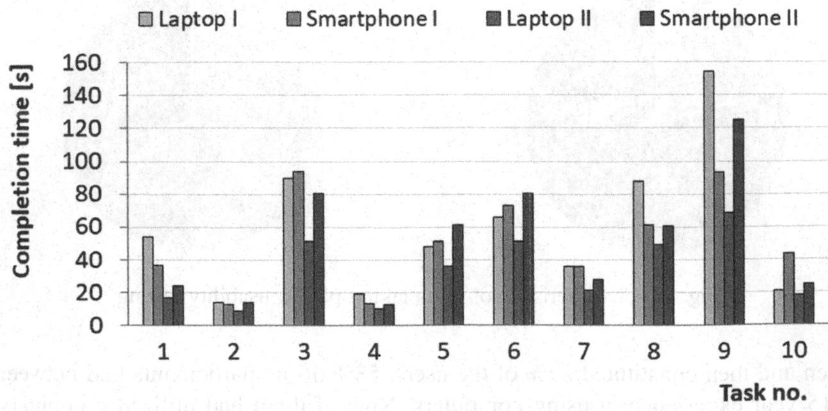

Fig. 3. Median of task completion time

significantly shorter time than both *Laptop I* and *Smartphone II*. In turn, no statistically significant differences in time were observed between *Smartphone I* and both *Laptop I* and *Smartphone II*. Task completion time by the users was compared with task completion time by an expert. The expert was the computer engineer who developed the website. The developer accomplished the tasks on average from 1.4 to 6.8 times faster using a laptop and from 1.4 to 4.6 times faster with a smartphone than the users. The ratio of task completion time by a user to task completion time by an expert is shown in Fig. 4. The median of the number of clicks, scrolls, taps, swipes, etc. made to complete individual tasks is presented in Fig. 5. The number of actions performed was larger with smartphones than with laptops. This is due the fact that the smartphones had smaller size of screens and therefore the completion of tasks required more frequent scrolling the website pages.

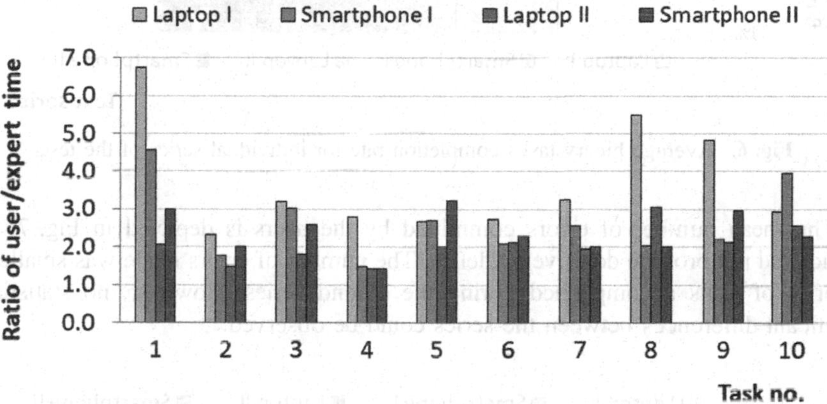

Fig. 4. Ratio of task completion time by a user to task completion time by an expert

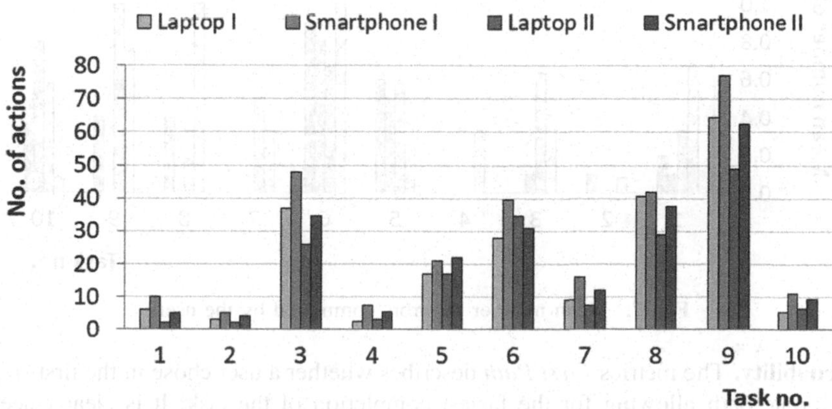

Fig. 5. Median of the number of clicks, scrolls, taps, swipes, etc. made

Errors. Binary tasks completion rate belongs to the most fundamental metrics of errors and effectiveness in usability research. It is expressed by the percentage of tasks completed successfully. As illustrated in Fig. 6 the rate increased in the second series both for laptops and smartphones. However, an increase was larger for laptops than for smartphones and was equal to 21% and 6% respectively.

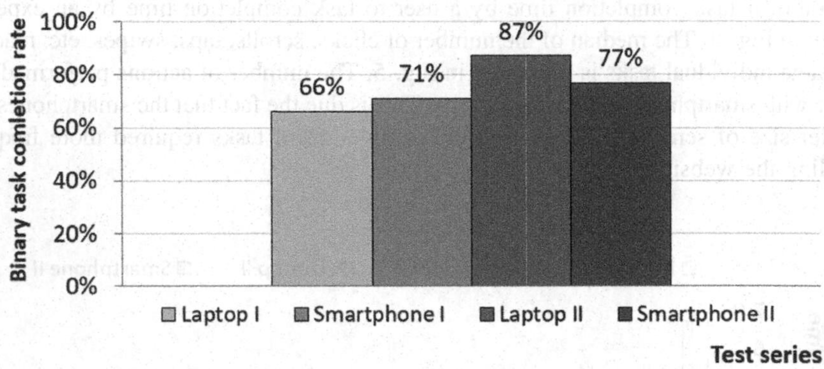

Fig. 6. Average binary tasks completion rate for individual series of the tests

The mean number of errors committed by the users is depicted in Fig. 7. This metrics did not provide decisive evidence. The number of errors made was smaller for majority of tasks accomplished during the second series. However, no statistically significant differences between the series could be observed.

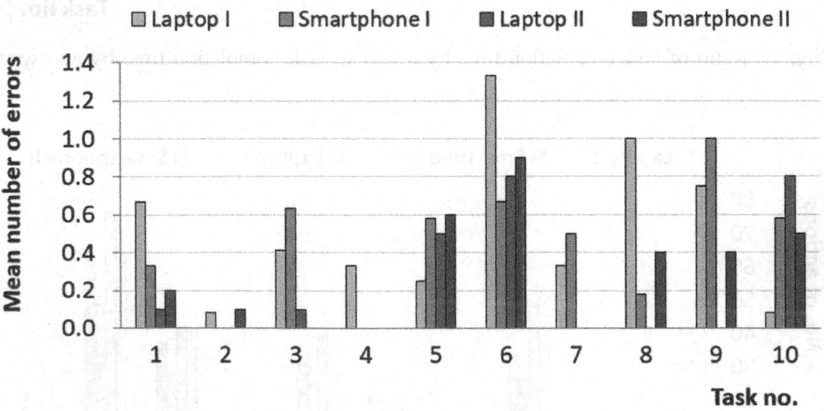

Fig. 7. Mean number of errors committed by the users

Learnability. The metrics *First Path* describes whether a user chose in the first step an appropriate path allowing for the fastest completion of the task. It is clearly seen in Fig. 8 that the percentage of first paths chosen correctly by the users was larger in the second series compared to the first series for both laptops and smartphones.

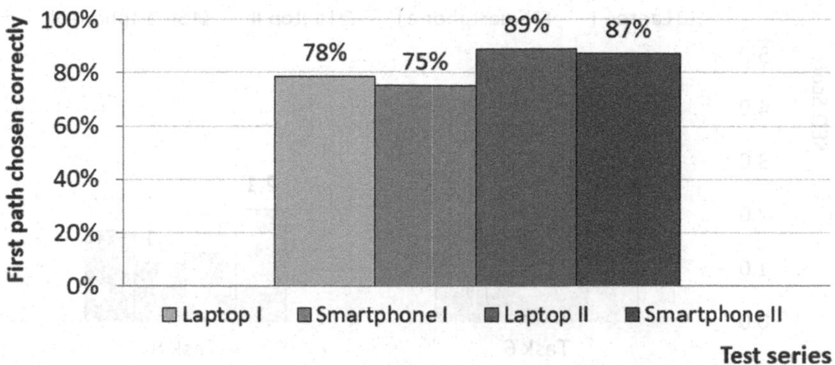

Fig. 8. Percentage of first paths chosen correctly by the users

Satisfaction. The satisfaction was measured using the SUS survey which was conducted after each of four test series. The SUS, i.e. System Usability Scale, is a ten-item questionnaire based on the Likert scale. It was proven that SUS provides also a global view of subjective assessments of usability [12–15]. According to Sauro the average SUS Score is a 68. The SUS score presented in Fig. 9 exceeded the value of 68 by 20.5, 14.3, 24.0, and 19.5, for respective series of the test. This can be regarded as a good result.

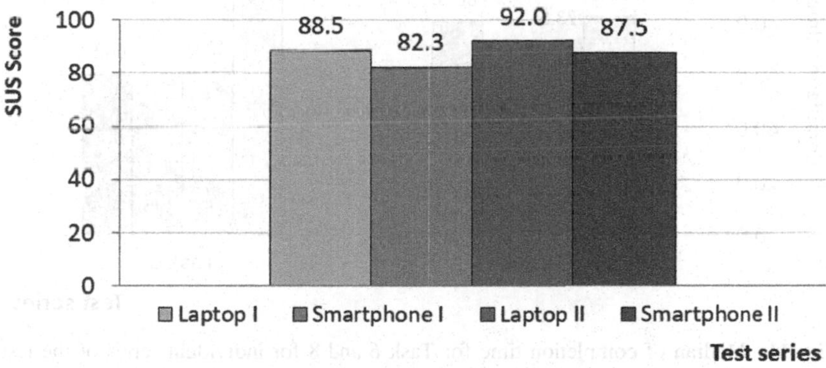

Fig. 9. SUS score for individual series of the tests

The Single Ease Question (SEQ) questionnaire was also used during our study. It contains only one item based on the Likert scale ranging from *very difficult* to *very easy*. It can be employed is to assess the overall ease of the completion of the task [16, 17]. The SEQ questionnaire was administered to the users accomplishing two tasks T6 and T8. The resulting SEQ score based on five point scale is illustrated in Fig. 10. The scores of the second series revealed the higher levels of satisfaction than the scores obtained during the first series.

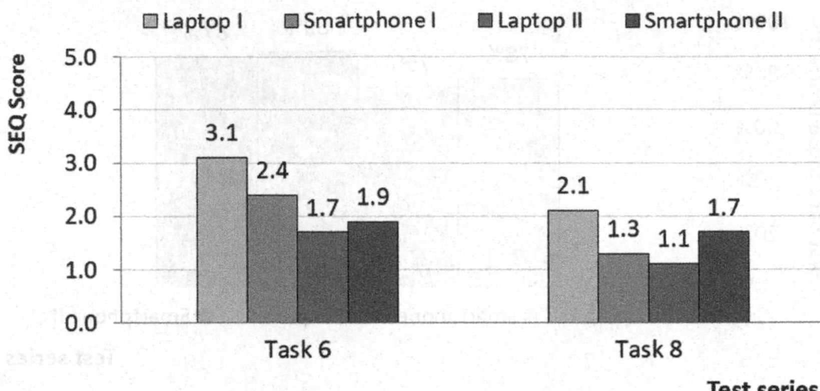

Fig. 10. Mean SEQ score for Task 6 and 8 for individual series of the tests

Memorability. The comparison of time needed to complete the same tasks, i.e. T6 and T8 can be used to assess the *Memorability* factor. The results are depicted in Fig. 11. The users accomplished their tasks faster during the second series.

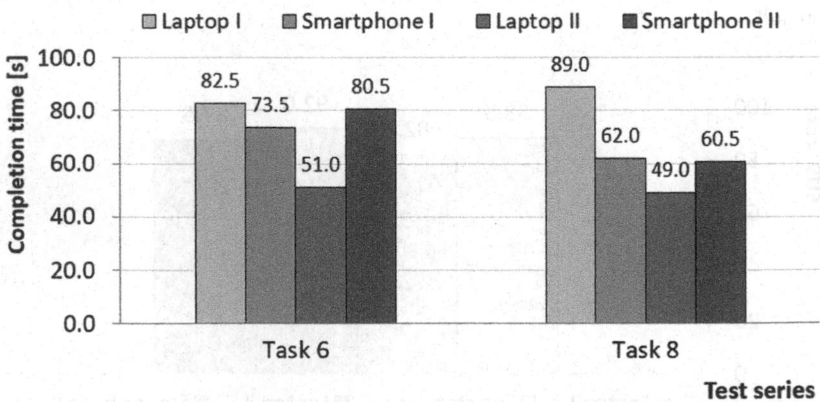

Fig. 11. Median of completion time for Task 6 and 8 for individual series of the tests

Expert inspection. Supplementary study was carried out with experts who performed inspection of the website to discover areas which require to be improved. Moreover, the experts' findings were confronted with difficulties encountered and errors made by the users. Two experts conducted heuristic evaluation [18] and six experts accomplished inspection with a control list worked out on the basis of the well-known 247 web usability guidelines [19]. There is no space to present the details of problems discovered and recommendations formulated by the experts in this paper.

5 Conclusions

The results of usability study of the website for a school for disabled children developed according to responsive web design approach are presented in the paper. During usability testing a series of metrics were collected. They concerned the five Nielsen's attributes of usability, namely efficiency, errors, satisfaction, learnability, and memorability. Moreover the expert examination of the website was conducted using heuristic evaluation and inspection with a control list. The experiments carried out with 24 participants and eight experts provided an extensive list of recommendations on how to improve the website.

The differences between two groups of users could be observed. Better results achieved the group completing the tasks during the *Smartphone I* and *Laptop II* series. The group using laptops in the first series did not improve significantly its results a week later with smartphones. The majority of metrics collected with laptops turned out to have better scores than those obtained with smartphones. The users working with laptops revealed higher satisfaction rate than the ones utilizing smartphones. Generally, the second series of experiments produced more satisfactory outcome. The Nielsen's model turned out to be useful for performing usability tests.

Acknowledgments. This paper was partially supported by the statutory funds of the Wrocław University of Science and Technology, Poland.

References

1. Marcotte, E.: Responsive Web Design, 2nd edn. A Book Apart, New York (2014)
2. ISO 9241-11:1998 - Ergonomic requirements for office work with visual display terminals (VDTs) - part 11: guidance on usability
3. Nielsen, J., Budiu, R.: Mobile Usability. New Riders Press, Berkeley (2012)
4. Harrison, R., Flood, D., Duce, D.: Usability of mobile applications: literature review and rationale for a new usability model. J. Interact. Sci. **1**, 1 (2013). https://doi.org/10.1186/2194-0827-1-1
5. Seffah, A., Donyaee, M., Kline, R.B., Padda, H.K.: Usability measurement and metrics: a consolidated model. Softw. Qual. J. **14**, 159–178 (2006)
6. Saleh, A., Isamil, R.B., Fabil, N.B.: Extension of PACMAD model for usability evaluation metrics using Goal Question Metrics (GQM) approach. J. Theor. Appl. Inf. Technol. **79**(1), 90–100 (2015)
7. Hussain, A., Mkpojiogu, E.: The effect of responsive web design on the user experience with laptop and smartphone devices. Jurnal Teknologi **77**(4), 41–47 (2015)
8. Bernacki, J., Błażejczyk, I., Indyka-Piasecka, A., Kopel, M., Kukla, E., Trawiński, B.: Responsive web design: testing usability of mobile web applications. In: Nguyen, N.T., Trawiński, B., Fujita, H., Hong, T.-P. (eds.) ACIIDS 2016. LNCS (LNAI), vol. 9621, pp. 257–269. Springer, Heidelberg (2016). https://doi.org/10.1007/978-3-662-49381-6_25
9. Błażejczyk, I., Trawiński, B., Indyka-Piasecka, A., Kopel, M., Kukla, E., Bernacki, J.: Usability testing of a mobile friendly web conference service. In: Nguyen, N.-T., Manolopoulos, Y., Iliadis, L., Trawiński, B. (eds.) ICCCI 2016. LNCS (LNAI), vol. 9875, pp. 565–579. Springer, Cham (2016). https://doi.org/10.1007/978-3-319-45243-2_52

10. Morae. https://www.techsmith.com/morae.html. Accessed 31 Oct 2017
11. Lookback. https://lookback.io/. Accessed 31 Oct 2017
12. Brooke, J.: SUS: a quick and dirty usability scale. In: Jordan, P.W., Thomas, B., Weerdmeester, B.A., McClelland, A.L. (eds.) Usability Evaluation in Industry. Taylor and Francis (1996)
13. Bangor, A., Kortum, P.T., Miller, J.T.: An empirical evaluation of the system usability scale. Int. J. Hum.-Comput. Interact. **24**(6), 574–594 (2008)
14. Sauro, J.: A Practical Guide to the System Usability Scale: Background, Benchmarks, and Best Practices. Measuring Usability LLC, Denver (2011)
15. Brooke, J.: SUS: a retrospective. J. Usability Stud. **8**(2), 29–40 (2013)
16. Tedesco, D.P., Tullis, T.S.: A comparison of methods for eliciting post-task subjective ratings in usability testing. In: Usability Professionals Association Conference (UPA 2006), pp. 1–9 (2006)
17. Sauro, J., Lewis, J.R.: Quantifying the User Experience: Practical Statistics for User Research. Morgan Kaufmann, Burlington (2012)
18. Nielsen, J., Molich, R.: Heuristic evaluation of user interfaces. In: Proceedings of the SIGCHI Conference on Human Factors in Computing Systems: Empowering People (CHI 1990), pp. 249–256. ACM, New York, NY, USA (1990)
19. Travis, D.: 247 web usability guidelines. http://www.userfocus.co.uk/resources/guidelines.html. Accessed 31 Oct 2017

Author Index

Printed in the United States
By Bookmasters